Mechanical Drawing

BOARD & CAD TECHNIQUES

Thirteenth Edition

Thomas E. French

Jay D. Helsel

Mc Graw Hill **Glencoe McGraw-Hill**

New York, New York
Columbus, Ohio
Chicago, Illinois
Peoria, Illinois
Woodland Hills, California

Glencoe/McGraw-Hill

A Division of The McGraw-Hill Companies

Copyright © 2003 by Glencoe/McGraw-Hill, a division of the McGraw-Hill Companies. All rights reserved. Except as permitted under the United States Copyright Act, no part of this publication may be reproduced or distributed in any form or by any means, or stored in a database or retrieval system, without prior written permission of the publisher, Glencoe/McGraw-Hill.

Send all inquiries to:
Glencoe/McGraw-Hill
3008 W. Willow Knolls Drive
Peoria, IL 61614-1083

ISBN 0-07-825100-1

Printed in the United States of America

3 4 5 6 7 8 9 10 003/003 07 06 05 04 03

Contents in Brief

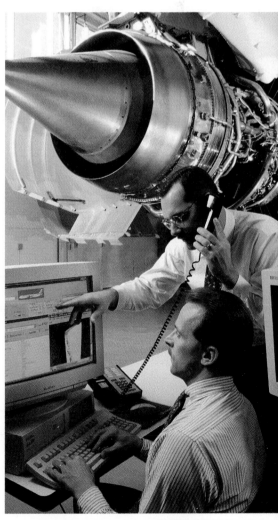

Contents

One Front Street

BOISE, IDAHO
LOMBARD - CONRAD ARCHITECTS
03 NOVEMBER 1999

Chapter 3 Drafting Equipment 85

Chapter 4 Basic Drafting Techniques 111

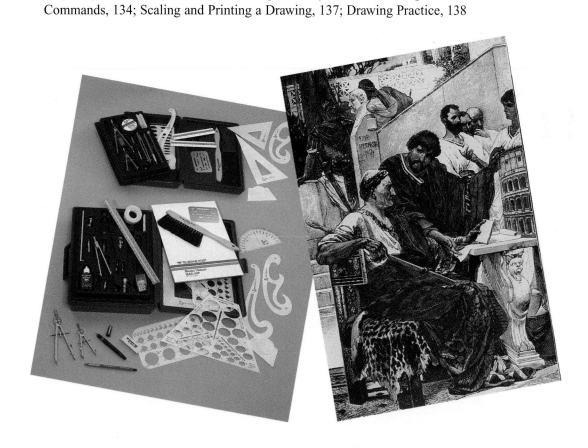

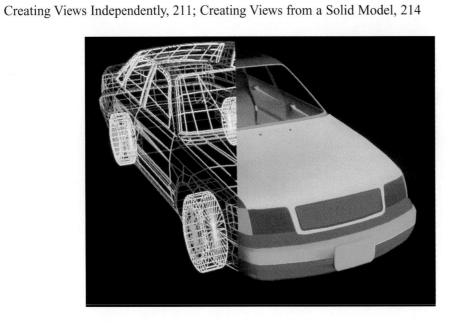

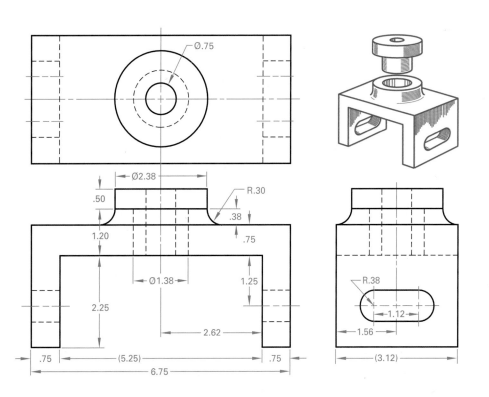

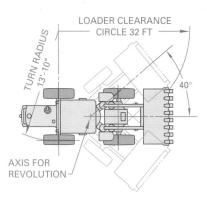

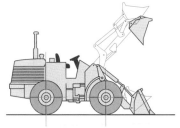

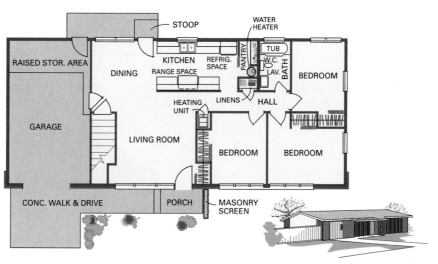

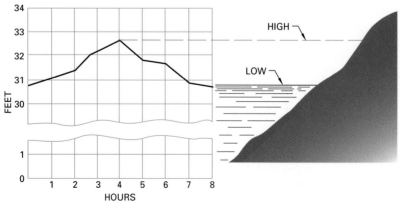

ACTIVITY	APRIL	MAY	JUNE	JULY	AUGUST	SEPT.
Analyze project requirements	▬					
Generate preliminary designs	▬					
Select final design		▬				
Finalize materials/processes		▬				
Create working drawings			▬▬▬			
Submit drawings for approval				▬▬▬		
Send approved drawings to mfg.					▬▬▬	

Appendixes

Drafting Careers

KEY TERMS

career

career plan

computer-aided
   drafting (CAD)

employability

entrepreneur

job

lifelong learning

long-term goals

portfolio

résumé

right-to-know laws

short-term goals

OBJECTIVES

Upon completion of this chapter, you should be able to:

■ Identify current and emerging careers related to drafting.

■ Describe the advantages and disadvantages of being an entrepreneur.

■ Prepare an individual career plan.

■ Explain how to prepare for a drafting career.

■ Demonstrate skills and techniques for applying for a job.

■ Demonstrate workplace skills.

■ Identify the workplace rights and responsibilities of both the employee and the employer.

■ Describe appropriate techniques for finding, adapting to, and resigning from a job.

■ Explain typical uses of board and computer-aided drafting techniques.

We are in the twenty-first century! When we look back at all the changes that occurred in the twentieth century, what can we imagine the future holds for us? In the twentieth century, aviation evolved from the Wright Flyer to the space shuttle, ground transportation evolved from the "horseless carriage" to the Eisenhower Interstate System, and communication evolved from the crystal radio to the cellular telephone. With all these changes, one thing remained constant. Before anything was built, someone designed it, and someone put that design on paper.

In the early days, designers made their own drawings. See Figure 1-1. As technology grew, so did the need for people to turn the design sketches into drawings that could be used to make the product. The drafting career was born.

Career Opportunities

The techniques of drafting have changed radically with the introduction of computers and **computer-aided drafting (CAD)**. But the ties between drafting careers and other careers such as mechanical design, engineering, and architecture are as strong, if not stronger, than ever before.

New technology has expanded the role of drafters in today's workforce. Many drafting and drafting-related careers have developed, and as technology evolves, so will the opportunities for drafters.

Jobs and Careers

It is important to understand that a job is different from a career. A **job** is work that people do for pay. An example of a job is the position of junior drafter in a drafting company. A job is the work you are doing now, did yesterday, or will do tomorrow. According to the U.S. Department of Labor, the average American has at least seven jobs, including part-time jobs as teenagers, before age 30.

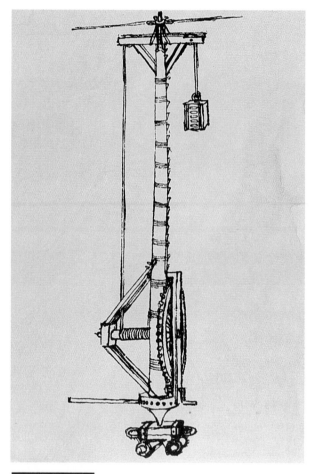

Figure 1-1

A drawing created by Leonardo da Vinci to record an idea for a movable derrick.

A **career** is a series of related jobs built on a foundation of interest, knowledge, training, and experience. During your career, you will probably work at many jobs. If you choose a career in drafting, for example, you may progress from junior drafter to drafting technician to a specialty such as mechanical designer. Table 1-1 describes the basic positions available in a drafting career. However, qualified drafters and designers also have many other exciting career opportunities.

Table 1-1. Basic Drafting Jobs		
Job	**Description**	**Qualifications**
Drafter Trainee	Assists with drawing preparation and performs support tasks such as drawing storage and retrieval.	Some companies require high school graduates who have taken drafting courses. Others allow trainees to learn in apprenticeship settings while they are still enrolled in high school.
Junior Drafter	Prepares drawings under the direction of a drafting technician or senior detailer.	Requires at least one year of high school drafting and an associate degree in drafting technology from a technical or community college.
Drafting Technician (Drafter)	Prepares drawings under the direction of a drafting technician or senior detailer, but requires less supervision than a junior drafter.	Requires an associate degree in drafting technology from a technical or community college and at least one year of drafting experience.
Design Drafting Technician	Capable of combining design skills with drafting skills. This requires the interpretation of the designer's sketches and the engineer's details so that the design drawings can be prepared accurately for the senior detailer.	Requires an associate degree in drafting technology from a technical or community college and at least one year of drafting experience.
Designer	Works with the engineers and drafters to turn a concept design into usable production drawings and specifications.	Requires an associate degree and at least five years industrial experience. Knowledge of the design process and drawing requirements is essential.
Checker	An experienced drafter who checks the drawings created by drafting technicians for accuracy and completeness.	Requires an associate degree and at least five years industrial experience. Detailed knowledge of the design process and drawing requirements is essential.
Senior Detailer	A person who is especially skilled in understanding the details of how things work and go together. Senior detailers are capable of detailing complex parts and making the details understandable.	Requires an associate degree and at least five years industrial experience. Knowledge of drawing requirements is essential.

Extended Career Paths

Over the years, drafters have provided support for engineers, architects, and designers. This is the traditional mainstay of drafting. Companies large and small maintain drafting departments, some with more than 100 drafters, and others with only one drafter. This section discusses career opportunities that are available to drafters who want to specialize.

Engineering

Because of its need for detailed, accurate drawings, engineering is a logical extension for a drafter interested in specialization. An **engineer** is a person who has at least a four-year degree in an engineering specialty. Engineers must be licensed by the states in which they operate. Engineering has evolved into many specialized branches, including aerospace, agricultural, architectural, chemical, civil, electrical, industrial, mechanical, mining and metallurgical, nuclear, petroleum, plastics, and safety. See Figure 1-2. Each type of engineering uses technical drawings to communicate ideas and products for manufacturing or construction.

To help prepare for further studies as an engineer, students should take math courses such as algebra, trigonometry, and calculus. Science and technology courses are also helpful.

Most major corporations employ a complete engineering design team. This team is considered essential in preparing legal documents for contracting services. Large offices may assign more than 100

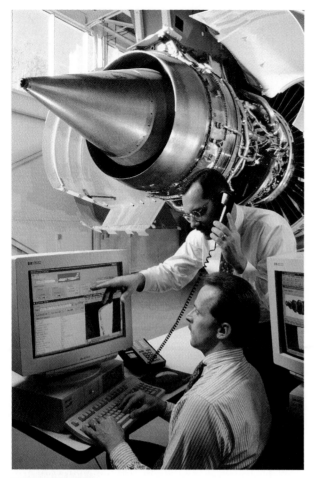

Figure 1-2

Aeronautical engineering is one of many logical extensions of the drafting field.

people to a team; teams in small offices may have five to ten members. A typical engineering design team may include:

- *Research and development personnel*—People who propose concepts for using new technologies through creative sketches and the use of science and mathematics.
- *Development engineer*—An engineer who designs research projects and collects data that can be applied to the development of new products.
- *Project engineer*—A person who coordinates all the specialized areas of engineering and design for production or construction projects.
- *Design engineer*—An engineer who applies mathematics, science, and technology principles to solve problems for production and construction processes.

- *Technical illustrator*—A specialized drafter who can create a pictorial or 3D CAD model from the details of an engineering drawing.

In addition, the design team includes drafters, designers, design drafting technicians, and senior detailers. Refer again to Table 1-1 for a description of these positions.

Architecture

If your interests lie more in building and construction, you might consider a career as an architect or architectural designer. The architect's job is critical in the design and construction of new buildings, both residential and commercial. Architects and their employees create original designs that are pleasing to the eye. However, the buildings they design must also be functional and meet the requirements of the client. See Figure 1-3.

Architectural drawings include floor plans, foundation plans, site plans, elevations, and many specialty plans for electrical, plumbing, and other contractors. If these plans are not extremely accurate, the building cannot be constructed. Therefore, architects must have at least some education and experience in drafting.

Figure 1-3

Architects must design buildings that meet all building codes and requirements, while ensuring that the customer's needs and preferences are met.

Figure 1-4

Mechanical designers turn ideas into plans and sometimes functional prototypes for presentation to supervisors and clients.

In addition to a four-year degree, architects must be licensed by the states in which they operate. Many architects take additional courses of study to specialize in various fields. Examples of specialized architects include landscape architects, city planners, and interior designers. Students who are interested in becoming architects can prepare themselves by taking geometry and other math courses, as well as business courses.

Mechanical Design

The field of mechanical design is similar to engineering, but it places more emphasis on the creative abilities of the drafter. Mechanical designers work from sketches or sometimes even just a company memo describing someone's great new product idea. Some of these ideas may turn out to be much more practical than others. Part of the mechanical designer's job is to determine how, or if, the ideas might work. The designer then provides accurate drawings and specifications to describe the proposed product. See Figure 1-4.

The process of providing drawings and specifications of a new product requires a large amount of creativity, ingenuity, and technical knowledge on the part of the designer. He or she must have the imagination to complete details from a sketchy idea, as well as a solid grasp of materials from which the product might be manufactured.

Technical Illustration

The majority of clients and financial backers for potential products and designs have little or no drafting experience. They find accurate production drawings and specifications difficult to understand. Large organizations often rely on technical illustrators to provide realistic pictorial drawings of proposed new products or construction, as shown in Figure 1-5. These illustrations show how the proposed product will look, how it will work, and so on, in a way that the client can understand easily.

To provide an accurate picture of what the product will look like, the pictorial drawings must be true to the designer's or engineer's specifications. Therefore, the illustrator must have a strong background in drafting principles and understand how to read technical drawings. Illustrators must also have a good imagination, because the products they are drawing usually have not yet been created. How will the product look, given the material specifications of the designer? For example, if a metal has been proposed, the illustrator must make the object look metallic.

Figure 1-5

Technical illustrators help nontechnical people understand a proposed project.

One Front Street

BOISE, IDAHO
LOMBARD - CONRAD ARCHITECTS
03 NOVEMBER 1990

Classes in illustration and graphic arts will help prepare you for a career as a technical illustrator. You may also want to take math, science, and other courses, depending on your area of interest.

Entrepreneurship

An **entrepreneur** is someone who organizes and then runs a business. Large companies have their own drafting departments, but sometimes they subcontract specialized jobs. Small companies may not be able to support a full-time drafting department; therefore, they employ subcontractors. Individuals don't have the skills to do certain jobs, so they look for people who do. These are all opportunities for the entrepreneur. Entrepreneurs are self-employed and take on individual jobs that fit their skills and interests.

Computers have opened up a whole new group of entrepreneurship opportunities for people who prefer self-employment. Many new businesses have been developed with an idea and a computer. These entrepreneurs often work at home, set their own hours, and accept work as it fits into their personal schedules and lifestyles. See Figure 1-6.

There are both positive and negative aspects to being self-employed. On the positive side, you are in charge, job satisfaction is usually high, and entrepreneurship can lead to a good income. On the negative side, there is a financial risk, long hours are required, and you have no guarantee of success. Fluctuations in the need for drafters in a region may also affect your success.

Entrepreneurs play an important role in the American economy. Early in the nation's history, entrepreneurs were spirited individuals who dared to take risks in exchange for a chance to better themselves. Today, entrepreneurship is integrated tightly with market demand and supply. The risks are still present, but as the nation has matured, entrepreneurs have discovered many different avenues for success. The careful entrepreneur can minimize the risks and maximize the profits.

Of course, entrepreneurship is not for everyone. It requires business skills in addition to the technical knowledge needed to provide the product or service. Small-scale entrepreneurs may have to answer the telephone, schedule their own appointments, and do much of their own bookkeeping. Those entrepreneurs who have the capital to employ other workers need personnel management skills. All entrepreneurs need a basic understanding of profit and loss as it applies to their business.

The business pace may vary from frantic activity to almost no activity at all. Either extreme is a cause for concern. As a result, entrepreneurs must also have personal characteristics that allow them to handle a variety of situations without losing control.

Figure 1-6

Computers allow some entrepreneurs, such as this drafter, to work in their homes. This saves the expense of renting office space, which lowers overhead costs and increases the entrepreneur's profit margin.

These characteristics include:

- self-reliance
- enthusiasm
- ability to think calmly and clearly under stressful circumstances
- ability to get along with employees and clients

Entrepreneurs are also responsible for knowing what, if any, occupational licensing is required for their businesses. Also, regardless of the scale of the business, appropriate insurance is absolutely essential. For example, entrepreneurs in every field need to address legal liability issues. The entrepreneur must research the types of professional insurance that are available and choose a policy that fits the company's needs.

In addition, staying current in the field is essential for any entrepreneur. As an entrepreneur, you have a responsibility to know new codes, regulations, and laws that apply to your business, and to take whatever steps are necessary to abide by them.

Trends in Drafting Careers

The role of drafting changes as specialty areas continue to develop. CAD offers a major area of future development, especially in three-dimensional (3D) drawings. Traditional drafting is still done in two dimensions, but the visualization of a 3D walkthrough brings the customer into the project. Using CAD, architects can move customers from room to room. They can change the window treatment, modify sidings, and adjust roofs. Landscapers move plants, shrubs, walks, and gardens, while civil engineers change bridge designs, roadways, and city plans. Mechanical engineers can see immediately the effect of small changes to a part design.

Rapid Prototyping

A technology called **rapid prototyping** now allows engineers in many companies to create a 3D "print" of a proposed part. The engineer converts a carefully constructed 3D CAD model into a format that is acceptable to the rapid prototyping machine. The result is a physical prototype or model of the part that the engineer can inspect and route to others in the department for approval. See Figure 1-7.

Global Economy

As Internet usage becomes more common in industry, the global economy becomes a bigger factor in drafting and many other careers. It is now feasible for companies to have offices in all parts of the world that work together on a single drawing via the Internet. For this reason, the economies of individual countries become more closely linked. As a result, drafters need to be more concerned and informed about international economics.

A company's decision to conduct business internationally should not be made lightly. Factors such as differences in language, culture, and working habits must be taken into consideration. Also, to do business in a given country, the company must understand and abide by the country's codes and legal issues. The company must also take steps to ensure the protection of its intellectual and physical property through international copyrights, etc.

The changes in drafting and engineering graphics technology caused by global use of CAD systems present both advantages and disadvantages for individual drafters. The advantages include extended drawing capabilities, faster drawing completion times, and almost instantaneous communication with

Figure 1-7

Rapid prototyping allows engineers to send a CAD drawing to a 3D printer that creates a physical model of the part. This 3D printer has produced a concept model of an engine block.

coworkers around the world. However, these advantages come at a price. In some situations, drafters may feel pressured to create better drawings faster. On a larger scale, companies are discovering that instant communication and drawing transmission remove the "down time" that was previously caused by mailing or sending drawings by courier services. This may seem minor, but the down time allowed designers and engineers a chance to think about design features and the possible effects of changes. Some companies have actually specified a short waiting period designed to allow workers to think about their designs and discover any potential problems.

Career Plan

An individual **career plan** is your road map to the future. Where do you want to go? How do you want to get there? Remember that a career is a series of related jobs built on a foundation of interest, knowledge, training, and experience. With this in mind, evaluate your choices. Look at all the options, compare them, and determine what will happen if you make various decisions.

Profiling Interests and Aptitudes

Some of the first things you should consider are your personal interests and aptitudes. Are you interested in drawing and illustration techniques? Do you enjoy detailed work such as model building? These are a few of the interests that may be compatible with a career in drafting.

Date:_____

Career Profile

Career:_____

Personal Information	Career Information	Match (1-10)
Values: *List values that are important to you.*	Values: *List values that might be necessary for the career you are profiling.*	Evaluate the "match" between the personal and career information. Write a number from 1 to 10, where the worst match is a 1, and the closest match is a 10.
Interests: *List your interest and hobbies.*	Responsibilities: *List the responsibilities normally assigned to people who work in this field.*	
Personality: *List your personal traits. Be honest, and list both favorable and unfavorable characteristics.*	Personality Needed: *Describe the personality traits needed by a person who works in this field.*	
Personal Relationships: *Do you prefer to work with other people or alone? How important is contact with others?*	Professional Relationships: *Describe the social characteristics needed by someone working in this field. Is a "people person" mentality required?*	
Skills and Aptitudes: *Make a complete list of your skills and aptitudes.*	Skills and Aptitudes: *What skills and aptitudes are needed by people on this career path?*	
Education/Training Preferences: *What kind of education or training experience do you want (or can you afford)?*	Education/Training Required: *List the education and training requirements for people who work in this field.*	

One tool that can help you decide on a career is a career profile. See Figure 1-8. A career profile is a chart you can fill out to help you evaluate various career possibilities. Complete a chart for each career that interests you. Then compare the charts to see which choices might work the best for you.

Evaluating Choices

After you have narrowed the choices, you should research the careers that seem most interesting to you. Pursue detailed information such as:

- potential earnings
- benefits usually offered
- job availability
- specific places of employment
- typical working conditions
- specific educational requirements

Knowing these things will help you choose wisely.

Preparing for a Drafting Career

If you choose a career in drafting or a related area, there are specific things you can do to prepare yourself. Like careers in other areas, drafting careers have minimum requirements for education, experience, and in some cases certification.

Education

As you can see from Table 1-1, the education required varies with different jobs within the drafting field. Regardless of the specific job you want, you should lay a good foundation for it beginning in high school and, if you choose, college. Drafting classes aren't the only requirement. Find out from area businesses what classes are required or encouraged for entry-level employment. You might be surprised at some of them. It is better to know this and take classes when they are offered than to think later, "I wish I had taken that class!"

TECH MATH

Estimating Living Expenses

As you consider your options and explore the career paths that interest you, you should also take your living expenses into consideration. Think about the lifestyle you prefer. Then find out how much it would cost, on a monthly basis, to maintain that lifestyle. To do that, compare your expected monthly income from jobs on each career path with your estimated living expenses. This will help you find out if the career you are considering will support your preferred lifestyle. Follow these steps:

1. Choose a job that would be typical if you followed the career path you are considering, and find out the average yearly salary for that position.
2. Consult tax tables to find out how much tax you would owe on that yearly salary. Subtract the taxes from the total (gross) salary to find your yearly take-home pay, or *net earnings.*
3. Divide your yearly net earnings by 12 to get your monthly net earnings.
4. Prepare a statement of income and expenses, as shown at right. Consider each of the items shown in the illustration, as well as any others, based on your chosen lifestyle.
5. Compare the average monthly income with your total monthly expenses.

Estimated Income and Expenses

Gross salary: Drafting Technician	$32,364.80
Less withheld taxes	−5,737.00
Net yearly earnings	$26,627.80
Expected Monthly Income	
Net monthly earnings: 26,627.80/12=	$2,218.98
Interest on savings	+ 6.50
Total	$2,225.48
Expected Monthly Expenses	
Rent	$725.00
Car loan	254.00
Car insurance	82.00
Food	220.00
Medical and dental care	50.00
Clothing	75.00
Transportation	110.00
Entertainment	100.00
Gifts and contributions	230.00
Miscellaneous	+150.00
Total	$1,996.00

In the example given here, the total monthly income is expected to be $229.48 higher than the monthly expenses. This is a comfortable range, which allows a little extra for emergencies or savings. If your expected monthly expenses exceed the expected monthly income, you may want to reconsider your expenses or consider a different career path.

Experience

While education is necessary, there is no substitute for experience. In many cases, drafters gain experience on the job as junior drafters. However, many companies also cooperate with high schools and community colleges to help students get on-the-job, or work-based, experience while they are still in school. See Figure 1-9. Formal apprenticeships in drafting are rare, although some do exist. Other opportunities include job shadowing, mentoring, and cooperative education programs. Explore the opportunities in your area for work-based experience in drafting.

Getting an early start on experience has two advantages. First, companies that provide students with on-the-job learning experiences are often eager to hire those students after they have achieved the appropriate education. This is partly because the companies become familiar with the students and their individual work ethics. In some cases, companies also like the idea that they have trained the students in the procedures and techniques specific to the company. The new employee's learning curve on the job is therefore not so steep, so the new employee becomes productive in less time.

The other advantage is that experience in the drafting field is a valuable addition to your résumé. Potential employers can see that you are serious about a career in drafting. They know that you are not just blindly pursuing a job. You understand what is involved in drafting and have made an informed choice to pursue a drafting career.

Certification, Licenses, and Permits

The American Design Drafting Association has set up a program for drafting certification. Most employers do not require certification, but if you are looking for a position as a drafter, certification shows potential employers that you have a certain level of drafting skill. Also, the National Skills Standards Board (NSSB) has set up national standards for use by drafters. Because many employers have their own set of company-wide standards, they don't specifically require employees to follow national standards. However, being familiar with these standards is an advantage.

Other organizations have set standards for specific drafting procedures with which every drafter should be familiar. The American National Standards Institute (ANSI) and the International Standards Organization (ISO) are nationally and internationally recognized sources of standards that allow all drafters, regardless of nationality, to read production drawings. As global operations become more common, ISO standards are becoming more popular in the United States. However, ANSI standards (U.S. customary and metric) are in widespread use.

Depending on the specific career path you choose, you may need to acquire a license or permit before you can be employed. As mentioned earlier in this chapter, architects must be licensed by the states in which they work. Construction supervisors and building inspectors may need permits or licenses, depending on the region.

Few drafting careers currently require that you pass a safety test. However, some individual companies do require safety testing. See Figure 1-10. For example, if you were to apply for a job as a designer at a heavy equipment manufacturer, the employer

Figure 1-9

Many companies help students get work-based experience while they are still in school.

might require you to pass a company-sponsored course in heavy equipment manufacturing safety as a condition of employment. As a matter of course, your career preparation should include an investigation of any safety issues.

Setting Your Goals

One of the most important aspects of creating your personal career plan is the setting of goals. By setting goals for yourself, you can plan what you want to accomplish in a certain period of time. Goals also provide a benchmark, or standard, against which you can check your progress.

Short-Term Goals

As you develop your career plan, start with your immediate future. **Short-term goals** are generally considered goals that you can achieve in less than five years. In the next five years, what do you expect to accomplish? Where do you want to be at the end of five years? Short-term goals can often be built around existing knowledge and training. Be careful to keep your goals realistic, though. Unrealistic goals are not only unhelpful; they can damage your self-esteem.

Long-Term Goals

Project short-term goals into the future to form your **long-term goals**. What is your ultimate goal? When you approach retirement, where do you expect to be? There are actually three sides to this question:

■ Where do you want to be professionally?
■ Where do you want to be financially?
■ Where do you want to be emotionally and socially?

All of these need to be considered, but it is the professional aspect that is of immediate concern. In many cases, your professional future will help determine your financial and social future. If you are considering drafting as a career, look at all the plusses and minuses, evaluate your interests and abilities, then go for the training and experience.

Lifelong Learning

We live in a world of dynamic change, and we must keep up with it. Some careers, such as teachers, electricians, and nurses, require continued learning,

Figure 1-10

Some drafting-related jobs require employees to pass safety tests. Why might this company require a quality assurance inspector to pass a course in safety before working in this environment?

or **lifelong learning**, while others assume it. Whatever the case may be, learning does not end with graduation from high school or college. To progress in your career, you must keep abreast of changes.

There are many ways to pursue knowledge throughout your life and career. Some companies provide occasional courses of interest to employees. Others pay part or all of employee costs for continuing education. For busy career people, the Internet now provides a convenient source of courses and even entire degrees.

However, not all knowledge comes from formal schools and coursework. It is up to you to identify opportunities for personal and career growth. In some industries, subscribing to trade journals can help keep you informed of changes in your occupation, as well as sources of further information and education. See Figure 1-11. You might also search on the Internet periodically for news that might affect your career or further education.

Figure 1-11

Trade magazines and journals are available for almost every drafting-related career. You can subscribe to one or more of these, or check with local libraries to see if they carry any that are of interest to you.

Applying for a Job

Competition in the job market varies. Sometimes there are more job openings than there are applicants, but that happens only rarely. You will be better prepared if you assume that competition is high for the job you want. Don't be casual. Approach the process carefully, and plan your strategies.

Job Search

There are many ways to search for a job. On an informal basis, watch the ads in professional journals and in the newspapers. If you want to move out of the area, read newspapers from areas of interest. If you are looking locally, knock on doors. Companies often have unadvertised openings available. Employment agencies can put you in contact with local and distant employers. They are also an excellent resource for specialized, professional, and managerial positions. If you want to move within your company, contact the human resources or personnel department to find out about openings for advancement and transfer.

One area that is often overlooked is a direct reference from relatives and friends. This is called *networking*. Ask people you know about potential job openings in their companies. A recommendation from a respected employee may give you an advantage.

One of the easiest ways to research job openings on a local, national, or even international scale is to use the Internet. Many Web sites are available to help employers find suitable employees, and vice versa. To find job listings in your area of interest, use the following strategies:

- Use a search or metasearch engine to find listings in your area of interest.
- Newspapers from many cities have on-line components that often include job openings. Scan these if you are interested in a particular location.
- If you are working in a specialty area, such as electronics drafting or architectural drafting, go to the Web sites maintained by national or international organizations or trade unions. These sites often list jobs, and sometimes even post résumés for job applicants, as a courtesy to members.

Creating Your Résumé

Employers base their hiring decisions on the information presented by the applicants, so you want to present yourself in the best possible manner. By developing a good résumé, cover letter, and portfolio, you increase the information available to the employer. In addition, well-prepared documentation tells the employer much about your organizational and communication skills.

Your **résumé** is the heart of the application process. Through your résumé, you present yourself as a qualified applicant for employment. Your school guidance department can provide you with a traditional résumé format, but feel free to vary it to suit your skills and needs. Also, many word processing programs provide ready-to-use résumé templates. Keep in mind that the style and format of your résumé must match the requirements of the companies to which you are applying.

The basic parts of a résumé are shown in Figure 1-12. They include:

- *Personal information*—your name, address, and telephone number. Do not include information such as age, gender, or social security number.

- *Job objective*—the position or job for which you wish to be hired.
- *Experience*—a list of relevant jobs you have had in the past. If you are a student, this is where you should list any work-based learning experiences you have had, such as mentoring or apprenticeships.
- *Education*—a list of the schools you have attended or technical certification courses you have taken. Include any honors, diplomas, and certificates you have achieved.
- *References*—People who can recommend you on a personal or professional level. Instructors and former employers are good sources of references. Be sure to ask permission before including anyone's name as a reference.

Figure 1-12

A typical résumé.

Jeremy Williams
555 North Selmington Place
Dubuque, IA 52003
(555) 417-4598

JOB OBJECTIVE
Seeking a position as drafting technician. Desire position with opportunity for career growth.

WORK EXPERIENCE

July 2000–present — Junior Drafter, Benjamin Design & Drafting, Dubuque, Iowa. Responsible for creating original drawings under direct supervision. Used both board drafting and computer-aided drafting systems.

September 1999–June 2000 — Drafting Trainee at Johnson & Co. in association with cooperative education program Fairfield High School.

EDUCATION
ADDA drafter certification March 2001
High School Diploma, Fairfield High School, Dubuque, Iowa

REFERENCES
Available upon request.

E-Mail Résumés

Many companies today request résumés by e-mail. In some cases, companies accept *only* e-mail résumés. This is because having your information in electronic format makes it easier for the company to compare the attributes of all candidates for a specific job. E-mail résumés present a challenge for job applicants, because many e-mail providers do not recognize the boldface, italic, and other styles that make a traditional paper résumé look good. Even if they did, however, many employers would still prefer résumés created in plain ASCII text. This allows them to insert the résumés directly into a database.

Internet Résumés

Another avenue for presenting your résumé is to use the Internet. Many Web sites now allow you to post your résumé. Alternatively, if you know how to create a Web page, you can create one specifically for your résumé.

The advantage of using the Internet to post a résumé is that it gives you much greater flexibility. For example, you can set up the résumé so that interested employers can download it directly from a Web site. You can even create links from your résumé to other sites that offer supporting information. For example, you can include a link to a former employer's Web site to give potential employers a better understanding of your relevant experience. In fact,

Figure 1-13

A typical cover letter.

555 North Selmington Place
Dubuque, IA 52003
(555) 417-4598

Ms. Carol Albiner
Director, Drafting Department
Dubuque Drafting Specialties, Inc.
1667 West Jefferson Street
Dubuque, IA 52012

Dear Ms. Albinger:

Bob Schottland, senior drafter at Hughes Drafting, suggested that I contact you about the position of drafting technician that is open with your company. Please consider me an applicant for this position.

In my current position as junior drafter at Benjamin Drafting & Design, I have had the opportunity to develop skills in both board drafting and computer-aided drafting. Please review my résumé, which is enclosed. It provides more details about my experience and the skills I can bring to your company.

I am especially interested in pursuing a career with Dubuque Drafting Specialties because of your reputation for cutting-edge CAD work. I believe employment with your company would offer me a wonderful opportunity to use my skills and advance my career.

May I have an interview? I shall be glad to call at your convenience.

Sincerely,

Jeremy Williams

you can create links directly to the e-mail addresses or Web pages of the people you list as references. If you do this, however, it is very important to obtain permission from these people *before* you provide a link to them, and *before* you list their addresses or telephone numbers in any on-line document.

In general, maintain a conservative résumé. If you will be using the Internet in your job search, you should research styles for Internet-friendly résumés.

Résumés should be job-specific. That is, they should emphasize the skills and experiences that are important to the job for which you are applying. Don't necessarily omit nonrelevant data, because employers usually like well-rounded employees, but match the emphasis with the position.

Writing the Cover Letter

If you apply for a job by mail or e-mail, you must introduce yourself. The cover letter that accompanies your résumé is often as important as the résumé itself. When a prospective employer receives your application, the first question that comes to mind is, "Why should I read this résumé?" You are important, and you have the skills the employer wants. Say that in one page or less, and be sure the letter contains no errors in spelling or grammar. Your cover letter should give the prospective employer the right impression, but it should not restate your résumé. The cover letter should make the employer want to read your résumé; it should not replace your résumé. Figure 1-13 shows an example of a well-written cover letter.

Assembling Your Portfolio

A **portfolio** is a collection of your best work that you can present to potential employers. If you are seeking a drafting job, a portfolio can showcase your skills much more efficiently than words. It cannot replace a well-written résumé, but it can provide a valuable source of supportive evidence.

What should you put into your portfolio? As you take classes related to drafting, be sure to keep a clean copy of your work. If you have used computer-aided drafting software to create drawings, back up the files that contain your best work. Keep your drawings and files together so that you can find them later. Also keep any work you do for companies through cooperative education arrangements, apprenticeships, and so on. If you have already been employed in a drafting position, keep a copy of your best work for your portfolio.

If you keep the documents as suggested in the previous paragraph, you will have an excellent basis for creating a portfolio. When you create a portfolio for a particular job application, however, be very particular about what you include. A few well-chosen examples of your work, presented neatly, will be much more effective than a large number of drawings chosen at random from your collection. Choose drawings and files that pertain to the job for which you are applying. If you do not have drawings or files that pertain to the job directly, choose those that demonstrate the skills you will need on the job.

Completing the Job Application

One element that is common to almost every job hunt is the job application. Employers use job applications to gather information about applicants in a standardized format. Most applications ask about your skills, work experience, education, and interests. Much of the requested information will probably repeat items that are already in your résumé. However, you should always fill out the job application completely, even if the information overlaps. Here are some other tips for filling out a job application:

- Read and follow the directions exactly.
- Do not leave any unexplained blanks. If an item does not apply to you, write "not applicable" or "will explain in interview."
- Keep the application neat and clean.
- Prepare in advance any lists of information you might need. Examples include schools you have attended and the dates of former employment.

Some job applications ask for information that employers do not have a right to ask. This includes information about your race, religion, gender, whether you have children, and marital status. If you encounter these items on a job application, use your own judgment about how to answer them. One acceptable method is simply to write "will explain in interview." However, if you use this method, you should decide in advance how you will handle the topic during the interview.

Figure 1-14

Find out all you can about the company before the interview takes place.

Interviewing

Companies usually conduct one or more interviews with the job candidates they find most promising. They will call those applicants to request an interview, but you don't have to depend on them to call you. It is good practice to contact the company after your initial application to follow up on your application. Whether you have applied in person or by mail or e-mail, call or write to the company after you apply. Take a positive approach. Express your continuing interest in working for the company, and ask for an interview.

Preparing for the Interview

After you have landed an interview, you should prepare for it thoroughly. Learn about the company. Before you started the job search process, you did a preliminary study of the companies to which you were applying. Now is the time for an in-depth look. You can obtain information by any or all of the following methods:

- Use the public library to find magazine and newspaper articles about the company and the industry.
- Read the company's annual report.
- Visit the company's Web site, if it has one.
- Talk to people who work for the company.

Find out about the company's organizational structure. See Figure 1-14. What is the product line? How is it financed? What is its employee turnover rate? Are you comfortable with its company philosophy?

Why do you need to know so much about the company? The more you know about it, the better you can present yourself as a potentially valuable addition to the company.

Prepare some questions to ask the interviewer. Employee benefits (known as *fringe benefits*), working hours, job security, promotions, and the future of the company form the basis of numerous questions. Be sure you discuss salary. Have a minimum salary in mind, but be flexible. Employee benefits, overtime, and job advancement may have an impact on the salary you decide to accept.

Figure 1-15

Practicing can help you feel more assured during the interview.

After you have prepared the information, you should prepare yourself. Practice answering the "tough" questions typically asked during an interview. Be prepared to answer questions such as:

- What do you consider your greatest strengths?
- What are your most serious weaknesses?
- Where do you plan to be five years from now? Ten years from now?
- What are your career goals?

Plan to answer these and other questions truthfully, but tactfully. You might want to practice in front of a mirror. It may also help you to set up a practice interview, with a friend or family member playing the part of the interviewer. See Figure 1-15. Your body language will also be important during the interview. Ask the friend or family member to critique you. Do you fidget? Do you look directly at the interviewer? Are your answers smooth and coherent? If you find that you need to improve in some areas, this is the time to do so.

Attending the Interview

When the time arrives for the interview, relax! If you have prepared yourself thoroughly, you can face the interviewer confidently. Dress appropriately to look your best, and be sure to arrive with a good night's sleep and a nourishing meal behind you. See Figure 1-16.

Most interviews have several distinct parts. In the first part, the interviewer introduces him- or herself. The interviewer may also tell you a little bit about the company and the job for which the company is hiring. Next, the interviewer asks questions designed to help the company understand your qualifications for the job. If you have not already done so, this is an appropriate time to present your résumé and portfolio. After asking questions about you, most interviewers allow you to ask questions about the company and the job. This is your cue to ask the questions you have prepared. Finally, finish the interview on a positive note. Repeat your interest in the job. Make sure the interviewer has your résumé or business card to facilitate reaching you should the company decide to hire you.

Following Up

After any interview, you should write a cordial letter to the interviewer. Thank the interviewer for the

Figure 1-16

At the interview, your preparation pays off by increasing your self-confidence, which helps make a good impression on the interviewer.

time he or she spent with you. Once again, state your interest in the job, and reinforce how your skills can help the company.

If you do not hear anything further from the company for a few weeks, but you are still interested, you may choose to call the company to inquire about the status of the hiring procedure. Some companies move more quickly than others. In any case, you can determine whether the position has been filled. If it has, you will know to concentrate on other possibilities. If not, then you may decide to write another letter to the interviewer, to keep your name fresh in his or her mind. Again, be careful to be positive but not pushy in your approach.

Workplace Skills

Being a good employee involves more than just technical skill. You also need to be able to get along with other people and work well with them. These workplace skills are often known as **employability skills** or characteristics, or *soft skills*. Your continued employment or promotion sometimes depends on these as much as your technical skill.

In general, employers want employees who follow policies and procedures, including dress code, attendance, and promptness. You can get a head start by making it a habit to comply with your school or vocational center's policies and procedures. By forming good habits while you are still in school, you will find that you don't have to think about these details later. Your good habits will help you comply naturally.

Personal Relationships

Take care to maintain good relationships with your employer and coworkers. Most workplaces can be hectic at times. Abrupt changes in schedules, job assignments, and supervisors can take a toll even on the calmest employee. Even if you become frustrated, try to understand that your coworkers, and possibly your supervisor, are under as much pressure as you are, and act accordingly.

Take a genuine interest in the people who work around you. Consider their work-related problems to be your problems, as far as possible. Help when you can. You may be surprised at how willing they are to return the favor when you need help.

It is also important not to expect everyone to share your views and opinions. In a typical office, people come from a variety of cultures. Maintain your own beliefs, but allow everyone else to maintain theirs, too. If you build a good working relationship with others, then differences involving projects and approaches to assigned tasks can usually be solved to everyone's satisfaction.

Attitude

Your attitude plays an important role in your value to the company. Maintain a positive, enthusiastic attitude, and nurture your self-esteem. If you make an effort and do your work to the best of your ability, you can build your self-confidence as you grow with the company. If the work does not go as expected, keep trying, and be sure to ask for help when you need it.

Be aware that others may judge you on your personal attitude as well as your work. Keep your body language, facial expressions, and gestures appropriate to the workplace. Also, be aware that people have different "personal zones." For example, some people may tend to stand too close to you for your personal comfort, without even realizing that you are uncomfortable. Be aware that this zone varies according to individual preferences as well as cultural background. Respect the personal zones of the people in your workplace.

Try to see yourself as other people see you. Do you take initiatives to help the company provide a better product? On the other hand, do you take responsibility for your choices and the work you do? How do you react to feedback and constructive criticism? Seeing yourself through other people's eyes can reveal ways in which you can improve your general attitude, as well as your self-esteem. See Figure 1-17.

Being willing to learn is another important characteristic of employees with positive attitudes. If you are asked to do something you don't know how to do, say so, but be sure your supervisor understands that you are willing to learn the new skills as necessary. Not only will this help you in your quest for lifelong learning, but it will also help you adapt to change. Few companies go more than a few years without undergoing substantial changes. The changes may be organizational, or they may affect the company's entire product or service line. A willingness to learn new skills will help you remain a valuable employee throughout such changes.

Figure 1-17

Accept constructive criticism from supervisors positively. Ask what you can do to improve your work, and then apply the supervisor's advice.

Communication Skills

Every job requires good communication skills. In a drafting career, much of your communication may be in the form of graphics—drawings and CAD models, for example. However, interpersonal communication skills are also very important. Whether you are interacting with customers or coworkers, you must have the ability to communicate clearly and precisely. Keep in mind that not all customers are external to the company. If you provide drawings or other products that are used by another department, then some of your customers are internal customers. Clear communication is just as important within the company as it is with external clients. See Figure 1-18.

The term *communication* includes many different skills, including:

- understanding spoken and written instructions from other people
- giving clear instructions both verbally and in written form
- explaining potential problems effectively to supervisors, both verbally and in written form
- using body language when appropriate
- demonstrating good telephone etiquette (manners and presentation)
- using good e-mail etiquette

As you can see, communication encompasses a large number of skills. To be able to speak and write effectively, you must understand basic grammar and presentation techniques. You may need to use mathematics or scientific principles to present some concepts clearly.

Self-Management

Employees who can manage their work with minimal supervision are in great demand among employers. You may have heard these employees referred to as "self-starters." In addition to having the skills necessary to perform their job, these employees are problem-solvers. They have the ability to recognize problems related to their work and identify their causes before they become extreme. They then proceed to develop and implement solutions, obtaining permission from supervisors when necessary, and evaluate the results.

Self-management also includes personal issues that affect employee effectiveness. Examples include punctuality, dependability, and reliability. If you know that you are going to be absent or late, call your supervisor and explain. Don't leave people wondering where you are.

Time Management

One important aspect of self-management is time management. In the business world, time really is money in the sense that wasted time decreases output, which decreases profits. With drafting, accuracy

Figure 1-18

This junior designer is using both verbal and written communication techniques to present a new product design to his department.

is of prime importance; however, an accurate drawing must be made within a reasonable time. You may be responsible for more than one project or drawing at a time. In some cases, supervisors may tell you which is most important. In other cases, you may need to make that decision. As a responsible employee, you must manage your time to minimize waste and turn each project around within an acceptable period of time.

One way to help ensure that you meet deadlines is to develop a work schedule. Spend a few minutes with a calendar and a list of projects and deadlines. Determine as realistically as possible how much time each project is likely to take, and pencil it in on your calendar. Then refer to the calendar to help keep yourself on schedule.

Ethical Behavior

The principles of conduct that govern any group or society are known as *ethics*. Ethical behavior includes:

- dealing honestly with employers and coworkers
- respecting company property
- keeping company information confidential
- maintaining your personal integrity, while honoring the values of others

In the workplace, ethical behavior is important both for you and for the company. The company counts on its employees to help maintain the company's good name. Unethical behavior can ruin the company's reputation. On a more personal level, unethical behavior can also ruin your reputation, which can affect your entire career. Would you hire someone who had a reputation for dishonesty or other unethical behavior?

Leadership

For any business to be successful, it needs good leaders. Leaders are not born—they are made; and leadership skills can best be acquired by doing. While you are in school, take responsibility by joining organizations and becoming a worker. Organizations such as SkillsUSA-VICA provide opportunities for students to become concerned citizens and patriots who can assess, predict, control, and adapt to the impact of various technologies on people, society, and the environment. See Figure 1-19. By joining a student organization, you can also learn the basic principles of parliamentary procedure and practice the leadership skills needed to reach an agreement in an orderly manner.

When you feel confident in your ability to work, run for office. You don't have to start by being the president. If you are good with numbers, run for treasurer. Maybe you like writing and transcribing; run for secretary. If you want to see what it's like to be president, run for vice president. Whatever you like to do, take charge.

Good leadership requires people skills. Be sensitive to the wants and needs of others. Work with people, not against them. Don't try to do everything yourself. Those you work with will respond better if they can make their own decisions after you give them the guidelines.

It may surprise you to find that good leaders must know how to follow directions. Most "leaders" are in turn directed by other leaders. For example, your supervisor may report to a vice president or other company officer. Therefore, another characteristic of a successful leader is respect for leadership roles.

Figure 1-19

Participating in a student organization can help you learn leadership and citizenship skills.

Leaders must also be aware of any codes, laws, standards, or regulations that apply to their work and require their followers or employees to follow them. Part of being a leader is accepting responsibility for following all applicable regulations.

Teamwork

The concept of teamwork is closely related to leadership. Every team needs a leader, but the leader must work closely with other team members to achieve positive results. In fact, being a team leader and being a team member require many of the same characteristics.

In today's workforce, the ability to work as part of a team is a critical employability skill. Few people work in isolation. Other people within the company depend on the work of each individual. For example, if you are a computer programmer, you have responsibility for how your part of the program operates. You also have responsibility for how your part works with everyone else's parts. If there is a problem, you may have to make changes or compromises for the good of the product.

Teamwork involves two important concepts: cooperation and communication. You must cooperate with the other members of your team to achieve a common goal, and you must communicate with them so that the work goes smoothly. If members cooperate and share their knowledge and skills within the group, they achieve a higher quality of work, and both the individuals and the company benefit. See Figure 1-20.

For teamwork to be effective, team members must be able to assess the knowledge and skills within the group. This is necessary so that responsibilities can be delegated according to the strengths, skills, and interests of individual group members. Remember that group members do not have to have the same interests and skills. In fact, it's often best if they don't. Individual differences increase the general abilities of the group. This has a positive impact on the way the group performs, whether the group is a student group working on an assignment or a group in the workplace assigned to a specific job. In either case, periodic evaluation of team performance will help the group stay on track and allow reassignment of tasks if necessary.

Figure 1-20

Teamwork skills include being able to determine the right person on the team for each part of the task the team must complete.

Health and Grooming

Good health habits can benefit you in many ways. They can give you the energy you need to do your job well. They can also help you achieve a well-rounded lifestyle, which can be an important factor in your personal happiness.

Although you can't have total control over your health, you can influence it by controlling your diet, exercise, and rest. For example, if you find that you feel tired all the time, take a close look at your daily activities. What do you typically eat during the day? Are you getting enough nutrients?

Some people think they just don't have enough energy to exercise after working or being in school all day. The truth is that, although exercise requires energy, it also energizes you. See Figure 1-21. You may need to experiment with different types of exercise at different times of the day to find an arrangement that meets your needs. You should make the effort—you will feel better, your body will be healthier, and you will have more energy in the long run.

Rest is a factor that many people underestimate. Do you allow enough time to get a good night's sleep? Most people require about 8 hours per night. Too little sleep can keep you from concentrating on your work, which in turn increases the likelihood of

Figure 1-21

Regular exercise can increase your energy level and make you feel better.

accidents and mistakes. Since accuracy is of great importance in drafting, lack of sleep can actually have a negative effect on your work.

Grooming is a general term for your appearance. It includes factors such as the types and styles of clothes you wear, how you wear your hair, and even cleanliness. Most companies have a dress code that determines how employees should dress while they are at work. Some dress codes also address hair length, jewelry, and other aspects of grooming.

Appropriate dress often varies according to the job. Within an architectural firm, for example, the correct attire for a designer who often meets with clients to present new ideas is much more formal than the attire for the building inspector who makes sure the design is being implemented correctly on the building site. See Figure 1-22.

Figure 1-22

Appropriate dress is much different for the building inspector in the photo on the left than for the office worker in the photo on the right, in spite of the fact that both work for the same company.

In some cases, company requirements are actually safety precautions. For example, if you are a quality control member of a design team, you may need to visit areas that contain high-speed machinery. You may be required to wear a minimum of jewelry, keep long hair tied back, and possibly even wear a hard hat when you are in these areas. In any case, you should check with your employer or human resources department about appropriate dress for your job.

Rights and Responsibilities

As an employee of any company, you have both rights and responsibilities. Depending on the industry within which you work, your rights may be governed by federal, state, and local laws; trade unions and organizations; and other associations. You should become familiar with the organizations and institutions that govern your rights as an employee. You should also be careful to fulfill your responsibilities to your employer.

Building skills is the central theme of SkillsUSA-VICA, a national organization of high school students, college students, and working professionals in technical, skilled, and service occupations. More than 245,000 students, 13,000 instructors and school administrators, and 1,000 business, industry, and labor sponsors join together to offer opportunities for leadership, citizenship, character development, skill training, and career awareness and exploration activities.

SkillsUSA-VICA emphasizes respect for the dignity of work, high standards of work ethics, workmanship, scholarship, and safety. In every chapter meeting, contest, leadership conference, or other activity, members develop the skills that enable them to be successful on the job.

An important component of the programs are the local, state, and national competitions to showcase students' occupational and leadership skills. Students compete at the local and state levels, and then each of the states sponsors a team to attend the national championship competition and skills demonstrations. Competitions are specific to the many industries and technical skills supported by SkillsUSA-VICA. Typical competitions for students interested in drafting are listed below.

Technical Drafting Contest—tests student ability to work productively in industrial drafting departments. Students are presented with engineering design problems. They are required to use computer-aided drafting (CAD) skills as well as traditional board drafting to solve the problems. Students are judged on productivity, time management, and the quality of their work.

Architectural Drafting Contest—tests student knowledge of architectural applications of drafting. Students are presented with problems that provide background information, building requirements, and a request for specific types of drawings. They must develop an appropriate plan from the instructions and design notes given in the problem. They are judged on whether they solve the problem correctly, as well as their layout, accuracy, and line work.

You can find out more about opportunities for membership and contest entry through your school. You may also want to visit the SkillsUSA-VICA Web site.

Basic Rights

All employees in this country have certain basic rights, including the right to be free from harassment and discrimination in the workplace. *Harassment* is continued, unwanted attention. We all want attention, but when the attention becomes excessive, you should ask the person to stop. If the unwanted attention is continued, it becomes harassment.

Discrimination involves the isolation of a person or group of people. The usual connotation is negative; that is, we are being discriminated against, rather than for. Discrimination in the workplace isolates employees to the extent that they cannot function effectively on the job. Federal law prohibits companies from discriminating against employees on the basis of race, gender, disability, and many other factors. Employees must be treated with equality, receiving equal pay for equal work and equal opportunities.

Most states also have laws concerning the rights and responsibilities of employees and employers in these areas. As an employee, you need to be aware of the potential problems and the forms of redress if you feel threatened.

Employee Responsibilities

As an employee, you also have certain responsibilities. You are responsible for upholding the good name of the company and for fulfilling the job you have been hired to do. The company depends on you. Your employer pays you an agreed-upon wage, and in return, you should be loyal to the company and do your best work. Your work ethics should reflect well on you and on the company for which you work.

Workplace Safety

Both employers and employees have a responsibility to ensure safety in the workplace. Employers must provide safe working conditions for their employees. In all industries, employees have a right to know about dangers and hazards relative to their place of work. The federal government established the Occupational Safety and Health Administration (OSHA) to monitor standards in the workplace. For details of the OSHA standards relating to employee right-to-know, refer to the OSHA standards for General Industry (29 CFR Part 1910) and Construction (29 CFR Part 1926). Some states, such as Florida, have **right-to-know laws** that exceed those of the federal requirements. As an employee, you need to be aware of the state and federal standards. You should also question any infractions that come to your attention.

Employees have a responsibility to maintain safety standards on the job. The actual rules and regulations vary by industry and by individual company. Drafters, for example, must maintain a clean, orderly work area. It may not sound like it, but this is an important safety issue. Board drafters must handle and store sharp instruments safely to avoid accidents. Drafters using CAD must be careful to keep the work area free of power cords that may cause trips and falls. Wherever you are—in school, at work, or in the community—you should understand the applicable safety rules and work to meet their requirements.

Job and Career Changes

Jobs are not permanent. There are many reasons why you might want to change jobs. For example, you may be unhappy with your current job, your job may be terminated, or you may have simply outgrown it.

There are many different kinds of job changes. For example, you may take the same job you have now, but with a different company. Or, you might take a different job in the same company, or a different job in a different company. Whatever the change you are considering, you should not make the decision overnight. Weigh the advantages and disadvantages carefully. How will the change affect your career prospects? Your benefits? If you will have to move to a new location, how much will it cost? Is the new job worth it?

Career changes are much like job changes, but on a much larger scale. A career change can have a huge impact on your life, so you must be even more careful when considering such a change. This does not mean that you shouldn't change careers. If you are unhappy in your current career path, and you are able to obtain the skills necessary for another career, then changing careers might be the right choice for you.

Whether you are considering a job or career change, it pays to reconsider your short- and long-

Figure 1-23

If others will be affected by your proposed job or career change, discuss the alternatives with them. Doing so helps them feel more involved in (and supportive of) your decision and may also give you needed insight.

term goals. How does your proposed change fit in with your goals? Be flexible as you think about your goals. See Figure 1-23. Do you need to change your goals, or should you retain your current goals but reconsider how to achieve them? Life seldom works out the way we envision it, so flexibility is a key to maintaining and achieving your long-term goals.

Finding a New Job

How do you apply for a new job? Basically, the same way you applied for the first one. (See "Applying for a Job" earlier in this chapter.) Be sure to update your résumé. You now have more (and perhaps different) skills to offer a potential employer. If you are applying for another job in the same career path, you also have more experience.

Most people prefer to have a new job lined up before they resign from their present job. This is a good idea, but be careful not to job-hunt on your current employer's time. If you need to attend an interview during company time, take a vacation day, or make other appropriate arrangements to cover your time away from work.

Resigning from a Job

It is important to resign gracefully from the job you already have. When you decide to make a move, give your current employer sufficient notice to allow the company to find someone to replace you. In most cases, two weeks to a month is a suitable length of time, but this may vary with your individual job. Submit a written letter of resignation stating exactly when you are leaving and why. Be careful not to take a negative tone, even if you are leaving because you are unhappy. Always keep your options open—you may later want a recommendation from this company for a future job.

Adapting to a New Job

No two jobs are exactly alike, so it may take you some time to adjust to your new job and coworkers. Allow yourself time to become comfortable in your new job. Suggestions for easing the transition include:

- Do not expect everything to be done exactly the same way it was done in your previous job. For many tasks, there is more than one way to perform the task correctly, and different companies have different preferences.
- Do not go into a new job expecting to "change the world." It won't happen, and trying to change the way things are done while you are still a new employee tends to irritate both coworkers and employers.
- Do maintain a positive attitude.

Board Drafting

The traditional drafting method, using a drafting board and instruments, is commonly known as **board drafting**. See Figure 1-24. This type of drafting is rapidly being phased out in many industries because computer-based systems allow drafters to complete their work more quickly and with greater accuracy. (See "Computer-Aided Drafting" later in this chapter.)

However, board drafters are still needed for some specific purposes. Small drafting companies and companies who use drafting techniques only occasionally may not feel that it is necessary to spend the money on computer equipment. If they can meet their company's needs using less expensive board drafting equipment, then they increase the company's profit by lowering its expenses.

Figure 1-25

When minor alterations need to be made to a drawing that was originally created using traditional drafting instruments, even CAD operators often make the changes directly on the drawings using board drafting techniques.

Figure 1-24

Board drafting involves drawing by hand, using precision instruments.

Even in large corporations, drafters with excellent board drafting skills are needed in some situations. For example, companies that have been around for more than 10 or 15 years may have hundreds of drawings that were completed using board techniques, and in many cases, those drawings are still in use. See Figure 1-25. When a minor modification is needed quickly, it makes sense to make the modification on the drawing using board techniques.

Board drafting skills are also required in some cases for digitizing drawings that were originally done using board techniques. *Digitizing* is a process by which paper drawings are converted to electronic files that can be manipulated using computer-aided drafting techniques, as shown in Figure 1-26. Although the person who digitizes the drawings may not actually need to use any of the board techniques, a solid understanding of those techniques will help him or her digitize the drawing more quickly and more accurately.

Figure 1-26
Digitizing a drawing doesn't actually involve using board drafting techniques, but the person doing the digitizing must have a solid knowledge of board techniques to digitize a drawing efficiently and accurately.

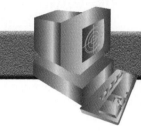

Computer-Aided Drafting

Computers have changed the entire face of drafting in industry. **Computer-aided drafting (CAD)** software has been developed that enables drafters to perform all of the tasks that formerly required board drafting techniques and skills. Over the last several years, as the software has become advanced, more and more companies have converted their drafting rooms to CAD.

Advantages of CAD

CAD can do more than just replace traditional drawing techniques, however. Because of its electronic nature, it is flexible enough to be used interactively in the design process. In fact, many companies use the acronym **CADD** (computer-aided design and drafting) to describe the software they use.

Although CAD software can create traditional two-dimensional drawings, most CAD programs today can also create three-dimensional (3D) models. These models can be created as virtual 3D solids that can have material properties. If such a model is created with extreme care and precision, it can be used to:

- provide the input necessary to drive production machines
- show clients a realistic example of how the finished product will look
- perform various structural and engineering tests without having to build a physical prototype
- extract any two-dimensional drawings needed by engineers, quality control personnel, and other workers

Production Input

Because many production processes are now controlled to some extent by computers, electronic drawings provide an interesting capability. CAD drawings can be interfaced with the computers that control production processes and used to drive the manufacture of a product directly. **CAD/CAM** (computer-aided drafting/computer-aided manufacturing) and **CIM** (computer-integrated manufacturing) are

Figure 1-27

A realistically rendered solid model. This technician is using 3D glasses and a virtual reality setup to design a part. The computer senses movements of the device in the technician's hand and makes appropriate changes on the solid model.

examples of the use of CAD files directly on the production floor. See Figure 1-27.

Older processes, such as NC (numerical control), have also been converted to take advantage of this capability. The NC machines were typically fed cardboard cards punched with holes. The machine interpreted the location of the holes as numerical data and performed its task accordingly. This punched card system has been replaced by **CNC**, or computer-numerical control, systems. These systems still use a numerical basis, but conversion software is now used to convert the electronic CAD files into a numerical format that the CNC machine can read.

Product Visualization

Clients and financial backers for products rarely have the drafting background to be able to read production drawings. To help them understand what a proposed product will look like, drafters and technical illustrators have traditionally created intricate pictorial drawings using airbrushes or watercolors. These techniques often result in beautiful renderings, but they take a large amount of time to create.

CAD solid models provide an alternative that is less time-intensive and therefore less expensive. If a CAD drawing is created as a solid model, it can be rendered to look like the finished product. Most CAD programs have at least primitive rendering capabilities. However, companies that use this method of product visualization usually purchase third-party rendering software that allows for an incredible amount of realism. In many cases, it is hard to tell whether the rendered object is real or CAD-generated. Although the rendering process also takes time, it works with the geometry in the original CAD file, so accuracy is absolutely guaranteed. Because the drafter or renderer doesn't have to worry about accuracy, he or she can concentrate more on the artistic effects. See Figure 1-28.

Another use of visualization involves creating 3D sectional views, such as the half section of a bearing shown in Figure 1-29. These views are often used by engineers and drafters to show pictorially how the pieces of an assembly should work together.

Product Testing

Because it is possible to assign real materials to solid models created in CAD, the models can be used for much of the testing that new products undergo. For example, a civil engineer may create a solid model of a proposed bridge design. The design must be tested for strength and for its behavior under various weather conditions, among other things.

Figure 1-28

Rendering programs allow CAD solid models to be displayed as realistic 3D products.

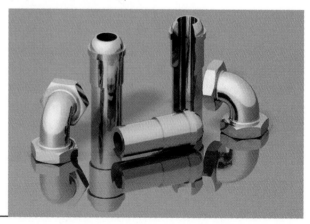

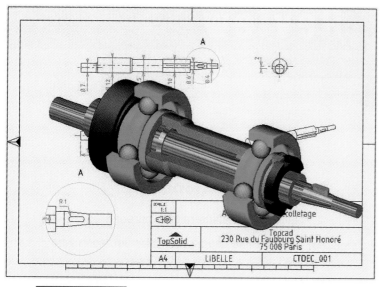

Figure 1-29

Three-dimensional solid sections allow production workers to see at a glance how an assembly should work.

Traditionally, designers built physical scale models and tested them using wind tunnels and other devices. Many tests depended heavily on complicated principles of mathematics and physics.

Fortunately, not only can materials such as steel and concrete be assigned to CAD models; computers can also be programmed to simulate various weather conditions. Much of this type of testing is now done entirely by computer. The tests still involve complex mathematics and physics, but the computer is well equipped to handle such calculations. Designers can determine the strengths and weaknesses of the design at a much lower cost and in a shorter amount of time by allowing the computer to run simulation testing on the CAD models. See Figure 1-30.

Plan Extraction

There are some things a virtual solid model does not do well. For example, it can't tell a quality control inspector at a glance what the diameter of a specific hole should be. Two-dimensional drawings are still widely used for this purpose. Most CAD programs have the ability to extract two-dimensional drawings from a 3D CAD model. These drawings can then be printed as needed.

All of the purposes discussed above can be served with the generation of a single CAD model. The drafter or engineer only has to draw the model once.

All subsequent modifications are reflected automatically in every area of use. It is for these reasons that industry has converted so enthusiastically to CAD or CADD systems for drafting.

Disadvantages of CAD

CAD use does have a few disadvantages. One of the most common problems has nothing to do with the actual CAD software. Instead, it concerns how people *perceive* the CAD system. People who are new to CAD, especially beginning drafters, tend to rely on the software to make drafting decisions that only a drafter can make. However, the software does not enforce good drafting technique. It does whatever the drafter tells it to do. Therefore, CAD software is not a substitute for basic drafting knowledge and skill. It is merely a tool that replaces a multitude of traditional tools. The drafter must still have the knowledge and skill to use that tool correctly.

Another disadvantage of using a CAD system is the initial, or startup, cost. The equipment and software needed for a functional CAD system can run into thousands of dollars. Also, computer equipment may not always be available when and where it is needed. Laptop and palmtop computers are helping to ease this problem, but availability is still a factor.

Figure 1-30

Complex, number-crunching tests such as this computational fluid dynamics (CFD) procedure can be run on CAD solid models.

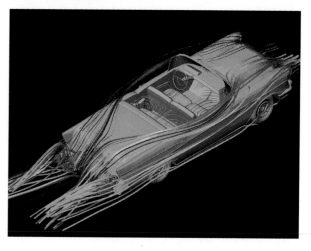

Chapter Summary

- A drafting career may involve many different, and exciting, career paths.

- Being an entrepreneur has both advantages and disadvantages.

- Individual career plans should include both short-term and long-term goals.

- Preparation for a career in drafting includes ensuring that you get the necessary education.

- Applying for a job includes gathering supporting paperwork and presenting yourself successfully as a qualified applicant.

- In addition to drafting skills, a career in drafting requires workplace skills that enhance your employability.

- Leadership, teamwork, and time management are important employability skills.

- Having a job gives you specific rights, but also responsibilities to your employer.

- A change of jobs should be handled systematically to maximize the benefit to you and the companies involved.

- Computer-aided drafting has revolutionized the industry.

Review Questions

1. Name four drafting-related career areas.

2. What term is used to describe someone who organizes and runs a business?

3. Name some of the things you should consider when you develop a career plan.

4. How does a job differ from a career?

5. What qualifications and requirements do you need to consider for a career in drafting?

6. What document should you provide, in addition to a résumé, when you apply for a job by mail or e-mail?

7. Briefly explain what a portfolio should contain if you are seeking employment as a drafter.

8. Name at least three effective strategies for finding a job.

9. What is lifelong learning?

10. List the characteristics and skills, other than technical skill, that make a person employable in a drafting career.

11. What two things does teamwork require?

12. What does "right-to-know" refer to?

13. What federal organization is concerned with dangers and hazards in the workplace?

14. In general, what responsibilities does an employee have toward the employer?

15. Describe the major uses for board drafting in industry today.

16. Name three advantages of using CAD software to create a solid model instead of a traditional two-dimensional drawing.

17. What does CIM stand for?

18. Name at least two disadvantages of using CAD software instead of using traditional board techniques.

CHAPTER 1

Problems

Problems

1. With your instructor's direction, and using the equipment provided, draw a series of rectangles into which you can write your short-term goals.

2. Write your short-term goals in the rectangles from problem 1 and organize them into priorities, with the highest priority on the top or to the left.

3. Modify your work in problems 1 and 2 for long-term goals.

4. Write a career plan consistent with your interests, aptitudes, and abilities. Evaluate the plan, keeping the availability of educational resources and experiences in mind. Revise the plan if necessary.

5. Create a résumé that you could present to a potential employer if you were applying for a drafting job. On the résumé, include education and experience that you would need for the job. Then exchange your résumé with a classmate. Critique each other's work. What strengths do you see? What might be improved?

6. Draft a cover letter to accompany your résumé.

7. Research drafting opportunities in your area. Write a report on a new or emerging technology associated with the field of drafting.

8. Use the Internet to research current trends in drafting technology. Prepare a report and present your findings to the class.

9. Find out which community colleges or technical colleges in your area provide postsecondary drafting education. List their entry requirements.

10. Through your school or on your own, find out about opportunities in your area to gain work-based experience in drafting. Make a list of the opportunities.

11. **Part 1:** You are applying for a job as a design drafter for SnowJet Engineering, a manufacturer of snow machines. The job requires knowledge of both board drafting and CAD skills, as well as basic mechanical design. Prepare your résumé.

    Part 2: SnowJet Engineering is a Canadian company, and you live in northern Vermont. You learned about the job on the Internet. Write a cover letter to include with your résumé.

    Part 3: You have been asked to come to the SnowJet Engineering offices for an interview. Write six questions about the job that you would like to ask your prospective employer.

12. From SkillsUSA-VICA or another student organization, obtain the basic rules of parliamentary procedure. As a group, choose someone to chair a simulated meeting of the organization. Practice making motions and conducting organizational business.

Design and Sketching

OBJECTIVES

Upon completion of this chapter, you should be able to:

- Describe the three basic aspects of design.

- Describe the traditional and concurrent engineering design processes.

- Explain the importance of freehand sketching for communicating technical ideas.

- Develop design ideas using freehand multiview and pictorial sketches.

- Develop techniques for estimating proportions.

- Letter clear, neat freehand notes and dimensions on a technical drawing or sketch.

- Explain the concept of sketching from a CAD operator's point of view.

- Create text appropriate for a mechanical drawing using a CAD system.

Drafting Principles

When you learn to use sketches to bring new ideas to life, you will know that the saying "progress begins on paper" is true. Another saying, "ideas are born in the mind and brought into being at the point of a pencil," also reinforces the importance of learning to sketch. However, freehand sketching is only one way of communicating ideas graphically. The other way is with the use of a computer. Various kinds of graphics software are available that allow you to transform ideas quickly into graphic representations. Practicing the use of computer graphics software will give you the skills necessary to communicate technical ideas quickly and accurately.

When words alone cannot describe new or futuristic forms, sketches are needed to show the thoughts that cannot be said. See Figure 2-1. Also, a quick sketch can be used to simplify a technical discussion. Designers, drafters, technicians, engineers, and architects often explain complicated or unclear thoughts with a freehand sketch. In other words, ideas can be translated into sketches and thus used later for further study.

In this chapter, you will learn about design and sketching. In the first paragraph above, we talked about bringing new ideas to life, which is really the essence of design and creativity. Here, you will learn first about the design process and then about freehand sketching and CAD techniques. When you finish studying this chapter, you should have a good understanding of the importance of sketching (freehand or computer-generated) in the design process, as well as in all aspects of technical communication.

Figure 2-2

The design of the modern automobile began with the invention of the wheel and evolved into an efficient, practical, aesthetically-designed machine.

Aspects of Design

Since the beginning of civilization, men and women have used their imagination, knowledge, and curiosity in creative ways to design and build tools and machines to make their work easier. As time went on, refinements in these creative designs made the tools and machines more efficient and the lives of the people who used them more pleasant.

The design of products need not be entirely original. Rather, ongoing improvements tend to build on one another, and the design evolves. For example, the design, or invention, of the wheel led to the design and development of carts and wagons. Over a very long period of time, with many intermediate steps, it led to the development of modern automobiles like the one shown in Figure 2-2. Each step in the process builds upon the creative efforts of individuals who developed earlier models. Designers do not find it necessary to "reinvent the wheel" continually.

Figure 2-1

Freehand sketches can be used to show new and futuristic ideas quickly.

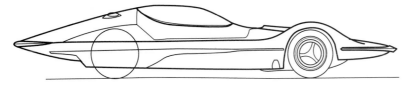

The words *design* and *creativity* are often used in the same sentence. This is because creativity is the key to good design, and the two words are often used interchangeably. **Creativity** is the combination of imagination, knowledge, and curiosity. **Design** is the conception of an idea and its development, through graphic communication, into a practical, producible, and usable product or process.

Designers are concerned with three basic issues:

- the look and feel of the final product
- the successful operation of the final product
- the manufacturing process and choice of materials

The look and feel of the product is the **aesthetic design**. The successful operation of the product is the **functional design**. In a well-designed product, both aesthetic and functional design can be readily seen. The manufacturing process and the choice of materials relate to the feasibility of manufacturing and marketing the product.

Functional Design

The functional aspect of design implies that the product will perform in the manner for which it was intended. In other words, it answers the design question, "Does the product meet the functional design objective?" In the design of any product, however, it is difficult to separate the functional aspects of the design completely from the aesthetics. For example, while the most important aspect of a brake caliper on a bicycle is how it functions, the designer is careful to add an element of beauty to the final design. The finish, color, curvilinear shape, style, and overall appearance may have little to do with the function, but they are evident throughout the design. A famous design architect, Louis Sullivan, coined the phrase "form follows function." This means that while the function of a product is of primary importance, the form (aesthetics) of the product is integrated into the overall design and is not separate.

Aesthetic Design

Aesthetics deals with the form, or overall physical appearance, of the product. Aesthetics involves characteristics such as color, line, style, space, contrast, proportion, and balance. It is relatively easy to apply any of these elements to the design of a sports car. Simply think about any of these elements, and you will quickly visualize some aspect of the design. For example, a sports car painted bright red sends a distinct message that relates to speed. The impressive blend of lines, style, proportion, and balance give the car a pleasing appearance that hints at both speed and excitement.

Engineering Design

In any industry, engineering design is a creative problem-solving process. In the development of any product, designers use knowledge, creativity, experience, and resources to create a new or improved product. The design must be functionally efficient, meet the design objective, and be aesthetically appealing. See Figure 2-3. In addition, the final product must be a financial success.

While the designers and engineers are involved in all aspects of the design process, the drafter also needs to be familiar with each step. Since product design is an evolutionary process, drafters are often called upon to prepare basic or temporary drawings that will undergo refinement at each step of the design process. In many cases, the drafter will play some part in all aspects of the graphic documentation, including sketches, drawings, computer models, and presentation drawings. An understanding of the process will make the drafter more useful as a member of the design team.

Figure 2-3

A successfully designed product.

The Design Process

Many models for the design process are used throughout industry. Designers and engineers tend to work from a basic model and modify it to suit their own particular interests and needs. Two basic approaches are traditional engineering design and concurrent engineering design.

Traditional Engineering Design

Traditional engineering design uses a linear approach. In this method, a design engineer takes the design of a product from the initial design problem or idea stage and carries it through, step by step, until it is turned over to the production division. At this point, the design engineer essentially leaves the project and returns to start another design project. In a linear approach, people from different departments work one after the other on successive phases of development.

The basic steps of the design process are shown in Figure 2-4. Each step is described more fully here.

Step 1 The basic design idea is conceived or simply assigned to the design engineer. This can be done verbally, but it is usually prepared as a document with considerable detail.

Step 2 The design engineer carefully analyzes the need for the product to determine if it is practical and marketable.

Step 3 General and specific objectives are set for the design. Manufacturing and marketing issues may be considered at this point.

Step 4 Serious consideration is given to the actual physical design of the product. Generally, several alternative designs are conceived and sketched or generated on a CAD system. The techniques used are generally a matter of personal preference.

Step 5 The various design concepts are checked for feasibility. The design engineer will want to know which concepts best meet the design objectives, which will be most marketable, etc. Notice in Figure 2-4 that a loop appears to develop between steps 4 and 5. This is the method used to optimize the design solutions.

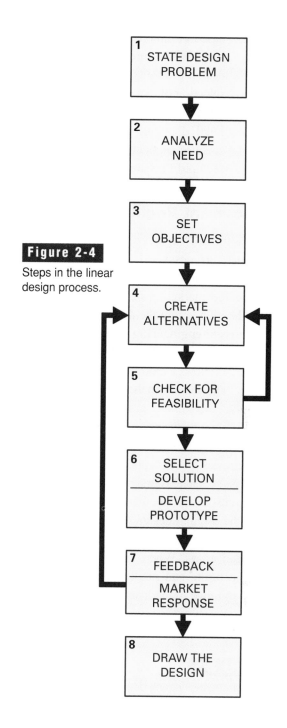

Figure 2-4

Steps in the linear design process.

Step 6 The best and most feasible solution is carefully selected. All aspects of the design are carefully considered at this stage, and a prototype is constructed to test the design. The prototype can be a physical model or a computer-generated model of the design.

Step 7 The prototype is generally submitted to a select group of consumers for market response before final drawings and specifications are prepared. If changes are needed based on this feedback, the design process loops back through step 4. Before the design proceeds to the next step, the final selection of material is generally made, and the manufacturing process is determined.

Step 8 Final drawings are prepared, along with specifications and all other documentation required for manufacture.

Concurrent Engineering Design

Traditional engineering design uses a linear approach, as described in the preceding section. Each step builds upon the previous step. The manufacture and marketing of the product is given little consideration before the design process is complete. **Concurrent engineering design** is just the opposite. It is done in a comprehensive team environment. Figure 2-5 shows a model for concurrent engineering design. The team consists of designers, engineers, drafters, and others associated with the overall design, manufacturing, marketing, and servicing of the product. In addition to the basic functional and aesthetic design concepts, the team considers important issues such as manufacturability, quality, life cycle, costs, and whether the finished product will meet the original design objectives. The entire team meets regularly and considers all of these issues as the need arises. When the design is complete, the manufacturing and marketing departments are prepared to move immediately to their part of the proj-

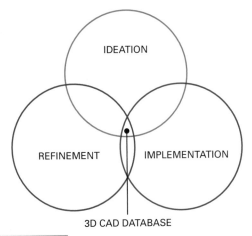

Figure 2-6

The concurrent engineering design process has three overlapping areas with a 3D CAD database as the nucleus.

ect. There are no delays, and there is seldom a need for redesign.

Concurrent engineering design, also called *concurrent engineering*, *concurrent product design*, and *simultaneous engineering*, evolved rather quickly with the development of powerful, computer-based design and manufacturing tools. In the linear design process, an engineer was traditionally responsible for many or all aspects of the design process. The chief engineer would keep several board drafters and countless other documentation technicians busy preparing drawings and developing the associated documents, such as specifications. In concurrent engineering, a comprehensive CAD and computerized engineering database serves as the basis for all aspects of design, manufacture, and marketing of the product. The database can be accessed by anyone on the design team from anywhere in the world through computer networking. There is no longer a need for all team members to be at the same location. The makeup of the team will vary depending on the type of industry and the complexity of the product under consideration. However, the concept and the processes involved in concurrent engineering design remain the same for all applications.

Concurrent engineering is concerned with making better products in less time, so continuous quality improvement techniques are practiced throughout the product's life cycle. In this process, the product's entire life cycle is considered as early as possible. A

Figure 2-5

Concurrent engineering design model.

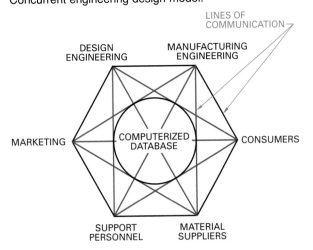

product's **life cycle** is its total life, from the conception of the idea to recycling of the materials from which it is made.

While there are many models that show the process of concurrent engineering, the one shown in Figure 2-6 is one of the more common ones. Ideation, refinement, and implementation overlap throughout the process to guarantee total integration of the three phases. Notice that a 3D CAD database becomes the nucleus of the process. Each of the three phases is further broken down, as shown in Figure 2-7. As you review each of these items, can you imagine how the drafting technician fits into the process?

Since concurrent engineering design is nonlinear, it is difficult to outline the process in a step-by-step manner. However, it is acceptable to view the three areas in the order that they fall into the design process. It all begins with **ideation**. In this phase, the design problem is identified, preliminary solutions are developed, and the preliminary design is agreed upon. Remember that all members of the team are involved in every phase. Following ideation comes **refinement**, which includes:

■ preparation of models and prototypes
■ thorough physical, production, and legal analysis of the design
■ design visualization, or analysis of the aesthetics

Finally, the **implementation** phase takes place. This involves a careful analysis of production, financing, servicing, documenting, final planning, and life cycle issues. Again, while there appear to be three distinct phases in concurrent engineering design, they all overlap, and the entire team is involved in every element of the process. As a result, concurrent engineering design results in a significant improvement in overall quality, as much as a 40% reduction in project time and cost, and as much as an 80% reduction in design changes during production. For these reasons, the concurrent engineering design process is now used in most industrial applications.

Sketching

Throughout the design process, there are many opportunities to use good sketching techniques to capture initial design ideas. You can also use sketches to refine these ideas and generally to communicate technical information. Here you will learn more

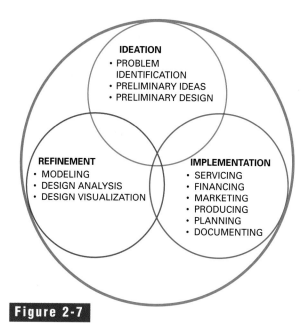

Figure 2-7

Each of the three areas of concurrent engineering can be broken down into specific tasks.

about the various types of sketches, how they are used, and how they are made.

The language of sketching has five basic visual symbols: a point, a line, a surface, a plane, and texture. A **point** is a symbol that describes a location in space. The path between two points, straight or curved, is called a **line**. A flat or nonflat element created from curved lines defines a **surface**. If the surface is flat, such as a circle, triangle, or square, it is referred to as a **plane**. **Texture** refers to the surface quality of an object. Any idea, no matter how simple or complicated or how plain or spectacular, can be sketched using these five visual symbols.

Reasons for Sketching

People create technical sketches for many reasons. The following are the nine most important.

1. To persuade people who make decisions about a product that an idea is good.
2. To develop a refined sketch of a proposed solution to a problem so that a client can respond to it. See Figure 2-8.
3. To clarify a complicated detail of a drawing that has more than one view by enlarging it or by creating a simple pictorial sketch (described later in this chapter).
4. To give design ideas to drafters so that they can do the detail drawings.

Figure 2-8

A refined sketch for the client.

5. To develop a series of ideas for refining a new product or machine part.
6. To develop and analyze the best methods and materials for making a product.
7. To record permanently a design improvement on a project that already exists. The change may result from the need to repair a part that breaks over and over again. It may result from the discovery of an easier or less expensive way to make a part.
8. To show that there are many ways to look at or solve a problem.
9. To spend less time in drawing. It is quicker to make a sketch, which takes only a pencil and a sheet of paper, than to create a mechanical drawing using board drafting techniques.

Multiview Sketches

There are two types of drawings that you can sketch easily: multiview and pictorial. Pictorial sketches are described later in this chapter.

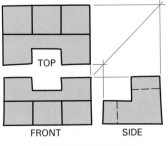

Success on the Job

Sketching for Presentations

As a member of an engineering design team, you may be called upon to present an idea or project to management or a client on a moment's notice. Whether you are a board drafter or CAD operator, skill in sketching can help you make these presentations clearly. Practice sketching and speaking *extemporaneously* (without preparation) so that you can deliver presentations easily and coherently on a moment's notice.

Multiview projection will be discussed in full in Chapter 6. However, you need to know some basic things about views and how they are placed in order to create good multiview sketches.

In multiview drawing, an object is usually shown in more than one view. You do this by drawing sides of the object and relating them to each other, as shown in Figure 2-9. The system by which the views are arranged in relation to each other is known as multiview projection, or **orthographic projection**.

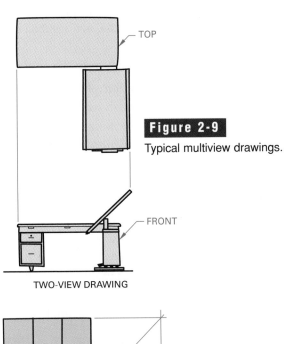

Figure 2-9

Typical multiview drawings.

TOP

FRONT

TWO-VIEW DRAWING

TOP

FRONT   SIDE

THREE-VIEW DRAWING

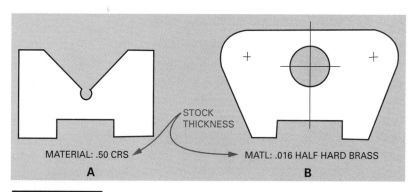

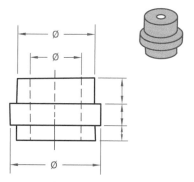

Figure 2-10

Typical one-view drawings. Notice that in each case, the thickness is shown in a note.

Figure 2-11

A cylindrical object may require only one view.

One-View Sketches

If an object can be described in two dimensions (height and width, for example), a one-view drawing is generally sufficient. Objects shown in one-view drawings generally have a depth or thickness that is uniform (the same throughout). In these cases, drafters may give the depth in a note rather than drawing an extra view. Typical one-view drawings are shown in Figure 2-10. The thickness of the stamping is shown by a note on the sketch. Many objects that are shaped like cylinders can also be shown in single views if the diameter of the cylindrical part is noted, as shown in Figure 2-11.

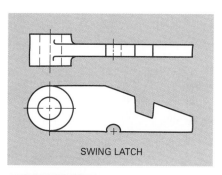

Figure 2-12

A two-view sketch.

Two-View Sketches

Many objects, such as the one shown in Figure 2-12, can be described in two views. If you are careful to select two views that describe the object well, this can help simplify the drawing. Figure 2-13 shows five three-view drawings in which the objects could have been described well in only two views. In drawing A, the top view is not necessary. Which view is not needed in each of the other drawings?

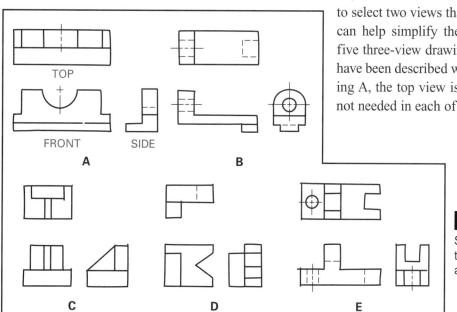

Figure 2-13

Select the two required views for the objects shown in A, B, C, D, and E.

Sketches with Three or More Views

A pictorial drawing shows how the object looks in three-dimensional form. Three directions are suggested for viewing the residence in Figure 2-14. From the front, you can see the width and height. This is the front view. From the side, you can see the depth and height. This is the right-side view. From above, the depth and width show. This is the top view. However, a three-view pictorial does not fully describe this residence. Some of the lines and details are not entirely visible.

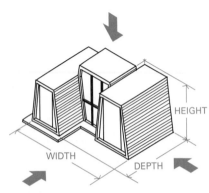

Figure 2-14

A vacation residence in pictorial, showing three dimensions.

The Glass Box

To get an idea of how many views are required to describe an object, many drafters use an imaginary glass box. For example, the residence from Figure 2-14 can be thought of as being inside a transparent glass box, as shown in Figure 2-15. By looking at each side of the building through the glass box, you can see the five views of the building. Figure 2-16 shows the glass box opened up into one plane as the views would be drawn on paper.

Pictorial Sketches

A **pictorial sketch** is a picturelike type of sketch in which the width, height, and depth of an object are shown in one view. The many different kinds of pictorial drawings will be discussed in Chapter 12. For sketching, we will consider only two kinds of pictorial drawings: oblique and isometric. Making oblique and isometric sketches will help you learn how to visualize or "see" objects in your mind. You must be able to do this in order to draw multiview projections. Pictorial drawings also help people who are not trained to read multiview drawings understand basic shapes.

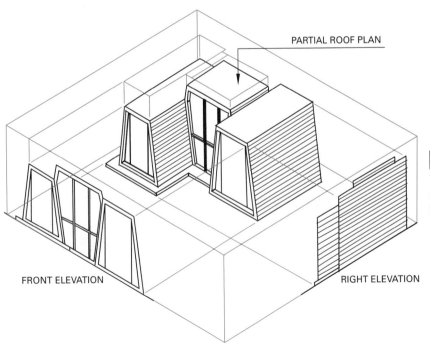

Figure 2-15

The elevations can be projected from the pictorial drawing to the transparent glass box.

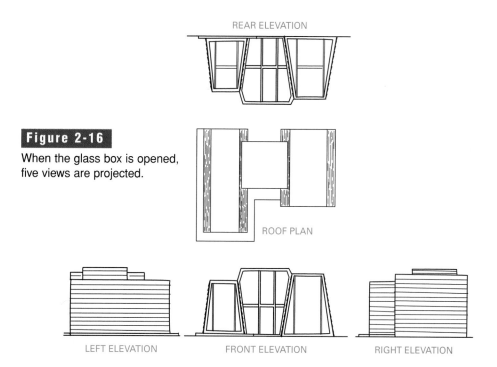

REAR ELEVATION

Figure 2-16

When the glass box is opened, five views are projected.

ROOF PLAN

LEFT ELEVATION   FRONT ELEVATION   RIGHT ELEVATION

Oblique Sketches

Every object has three dimensions: width, height, and depth. Each of these dimensions is called an **axis** (plural, *axes*). In an **oblique sketch**, two of the axes are at right (90°) angles to each other. The third axis can be drawn at any angle to the other two, although an angle of 45° is commonly used. See Figure 2-17.

You may make any side of an object the front view. As you may remember, the front view shows the width and height of the object. You usually make the side with the most detail the front view. The drawing of a digital clock radio in Figure 2-18 is an example of an oblique pictorial sketch. The dial side has been made the front view. This is because it has the most detail and shows the width and height of the radio. The front view is sketched just as the clock radio would appear when you look at it directly from the front.

Figure 2-18

An oblique pictorial sketch.

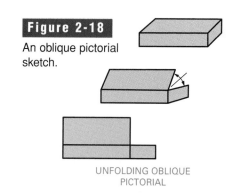

UNFOLDING OBLIQUE PICTORIAL

Figure 2-17

Oblique drawings always have one right-angle corner.

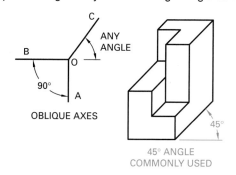

OBLIQUE AXES

45° ANGLE COMMONLY USED

OBLIQUE RENDERING

Isometric Sketches

Like an oblique sketch, an **isometric sketch** is based on three axes to show the width, height, and depth of objects. However, in an isometric sketch, the axes are all spaced equally at 120° apart, as shown in Figure 2-19. You can think of the axes as part of a cube (an object that has six equal square sides). As you can see in Figure 2-20A, an isometric cube has three equal sides. Figure 2-20B shows how to estimate the location of the axes for an isometric sketch. The height OA is laid off on the Y axis. The width OB is laid off to the left on a line 30° above the horizontal. The depth OC is laid off to the right on a line 30° above the horizontal. The 30° lines receding to the left and right can be located by estimating one third of a right angle, as shown in Figure 2-20B.

Lines parallel to the axes are called **isometric lines**. The estimated distances are laid off on the axes, as shown for the cube at Figure 2-20C. Then the rest of the isometric lines are blocked in, as shown in Figure 2-20D.

Figure 2-21

The nonisometric lines form an inclined plane on the isometric drawing.

Lines that are not parallel to the isometric axes are **nonisometric lines**. These lines are usually drawn after the isometric lines are in place, because the isometric lines help define the endpoints of the nonisometric lines. Examples of nonisometric lines are shown in Figure 2-21.

Proportions for Sketching

Sketches are not usually made to scale (exact measure). Nonetheless, it is important to keep sketches in proportion so that each part of the drawing is approximately the right size in relation to other parts of the drawing.

Estimating Proportions

In order to sketch well, you must be able to "eyeball," or estimate by eye, an object's proportions. In preparing the layout, look at the largest overall dimension, usually width, and estimate the size.

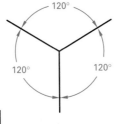

Figure 2-19

The axes for an isometric sketch are located 120° apart.

Figure 2-20

Sketching the isometric axes and an isometric cube.

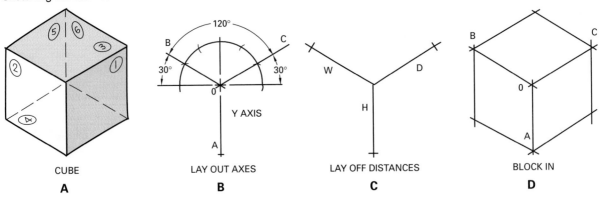

CUBE	LAY OUT AXES	LAY OFF DISTANCES	BLOCK IN
A	B	C	D

Next, determine the proportion of the height to the width. Then, as the front view begins to take shape with the width and height, compare the smaller details with the larger ones and fill them in, too. See Figure 2-22.

It is important that the design drafter, in sketching an object, have a good sense of how various parts of the object relate to each other. This allows the drafter to show the width, height, and depth of an object in the right proportions.

For example, suppose that the design drafter plans a cabinet that is to be 60″ wide and 30″ high. In this case, the proportion is 2 to 1 since it is twice as wide as it is high. If the designer then chooses to make the depth 15″, the proportion of width (60″) to depth (15″) is 4 to 1 (60 ÷ 15 = 4). The proportion of height to depth is 2 to 1 (30 ÷ 15 = 2).

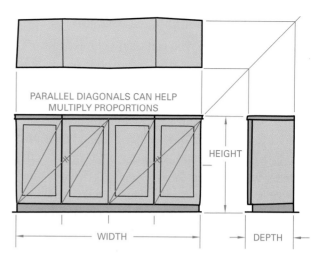

PARALLEL DIAGONALS CAN HELP
MULTIPLY PROPORTIONS

HEIGHT

WIDTH

DEPTH

PROPORTIONS:
WIDTH TO HEIGHT 2:1
HEIGHT TO DEPTH 2:1

Figure 2-22

Sketching a cabinet with proportional units.

TECH●MATH

Ratio and Proportion

Throughout Chapter 2 and at various other places in this textbook, you will find the word *proportion.* However, in order to understand proportion, you must first understand ratio. Theoretically, a **ratio** is the quotient of one quantity divided by another quantity of the same kind. This is written:

$$\frac{a}{b} \quad \text{or} \quad a : b$$

In either case, it is stated, "a is to b." Here is an example:

A playing field is 50 meters long and 25 meters wide. The ratio is 50 : 25. Since ratios are generally reduced to lowest terms, this ratio would be given as 2 : 1.

A **proportion** is the equation obtained when one ratio is set equal to another. For example, if the ratio a : b equals the ratio c : d, we have the proportion:

$$a : b = c : d \quad \text{or} \quad \frac{a}{b} = \frac{c}{d}$$

which reads, "a is to b as c is to d," or "the ratio of a to b equals the ratio of c to d." Here is an example:

Enlarge a rectangle whose sides are 4.50″ and 8.00″. If you want the 4.50″ side to be 6.00″, calculate the size of the other side as follows, using x to represent the unknown quantity:

$$x : 8.00 = 6.00 : 4.50 \quad \text{or}$$

$$\frac{x}{8.00} = \frac{6.00}{4.50}$$

$$x = \frac{(8.00)(6.00)}{4.50}$$

$$x = \frac{48.00}{4.50}$$

$$x = 10.67″$$

The size of the new rectangle is 6.00″ × 10.67″.

Technique in Developing Proportion

Through practice, you can train your eye to work in two directions so that you can both divide and extend lines accurately. For example, you should be able to divide a line in half by estimating. You can divide the halves again to give fourths, and so on. Using a similar technique, you can expand lines one unit at a time. Start by drawing a line of 1 unit. Increase it by one equal unit so that it is twice as long as at first. Practice adding an equal unit and dividing a unit equally in half. Practice developing units on

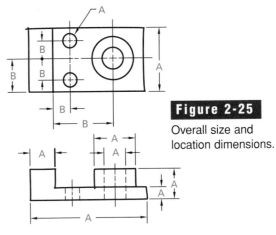

Overall size and location dimensions.

A = OVERALL SIZE DIMENSIONS
B = LOCATION DIMENSIONS

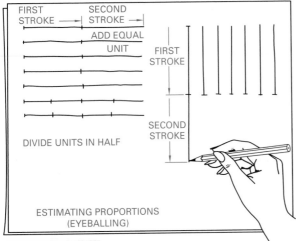

Figure 2-23

Practice estimating proportional units.

Figure 2-24

Using a strip of paper to estimate proportions.

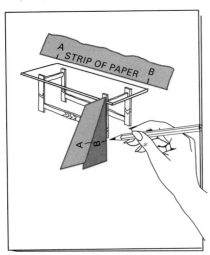

parallel horizontal lines. Then develop them vertically. By learning to compare distances, you can get better and better at estimating. See Figure 2-23.

Design drafters can use scrap paper or a rigid card to measure when they do not have a scale or ruler at hand. For example, the proportion of width to height of the glass-top table in Figure 2-24 is 2 to 1. Mark the width on the scrap of paper, as shown. Then fold the paper in half, aligning the marks that indicate the width, to find the height. In this way, you can draw lines of the same length several times in different directions. This procedure will help you maintain the right proportions.

Dimensioning a Sketch

An effective sketch must fully describe the object. Generally, the sketch is made before the measurements of an object have been decided. After the needed dimensions are determined, they can be recorded on the sketch.

Two types of dimensions are used on sketches. Size dimensions describe the overall geometric elements that give an object form, as shown in Figure 2-25. Location dimensions relate these geometric elements to each other. Together, the two types of dimensions accurately describe the size and shape of the object.

Several types of lines are used to dimension a drawing. Refer to Figure 2-26 as you read the following definitions.

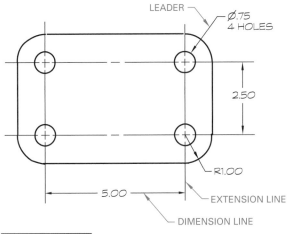

Figure 2-26

Drafters use special kinds of lines to clarify dimensioned drawings.

- A **dimension line** is a thin line used to show the direction of a dimension. Dimension lines generally have an arrowhead at each end to show where the dimension begins and ends. Sometimes the arrowheads are replaced by other symbols such as dots or slash marks.
- **Extension lines** are thin lines used to extend the shape of the object to the dimension line.
- A **leader** is a thin line drawn from a note or dimension to the place where it applies. A leader starts with a horizontal dash and angles off to the part featured, usually at 30°, 45°, or 60°. It ends with an arrowhead.

Board Drafting Techniques

The ability to create good-quality freehand sketches, even on the spur of the moment, is an important skill for drafters. The purpose of this section is to familiarize you with the various types of sketches and sketching techniques. Sketching and lettering are best learned in short, unhurried periods. Take your time, and practice the techniques as you read about them.

Types of Sketches

Any image drawn on paper freehand (with limited use of straightedge or other tools) may be called a sketch. Most drafters use several types of sketches. The type of sketch used depends on the purpose of the sketch and its intended life span.

Rough Sketches

Rough sketches are usually drawn quickly using jagged lines. Their primary purpose is to express thoughts quickly. Figure 2-27 shows rough sketches that were used to develop preliminary designs of a two-position automobile mirror.

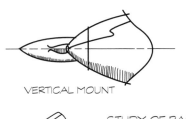

VERTICAL MOUNT

STUDY OF RACING MIRROR

Figure 2-27

Study of two-position mirror for a racing car.

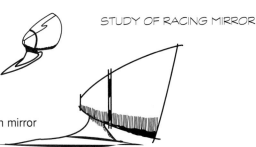

HORIZONTAL MOUNT

Never use instruments or straightedges to prepare a rough sketch. Instruments tend to restrict the creative expressions developed with good pencil techniques. Avoid a mechanical, hard-line look. Concentrate on using good proportions, and add a few choice notes if necessary to clarify the drawing.

Refined Sketches

Refined sketches are drawn more carefully than rough sketches. They show good proportion and excellent line values. They may be more persuasive than an unrefined sketch. Many refined sketches are based on a rough sketch that has captured the general idea.

You may use a straightedge to control long lines on a refined sketch. You may also use a circle template to draw large circles in order to save time.

Presentation Sketches

Pictorial sketches that have been greatly refined are known as *presentation sketches*. These sketches are used to convince a client or management to accept and approve the ideas presented. Pictorial sketches have a three-dimensional view that can be understood easily by nontechnical people. Such sketches are generally drawn so that they look glamorous, artistic, or eye-appealing.

Temporary Sketches

Many technical sketches have short lives. Some are done merely to solve an immediate problem and then are thrown away. Other technical sketches are kept longer. It may take weeks or even months to study some sketches and make mechanical drawings from them. However, these sketches too may eventually be thrown away.

Permanent Sketches

Sometimes the engineering department or the management of a company will include a sketch in a notice or memo to other employees. Such a sketch is an important record and should be kept. Therefore, some sketches are filed as part of a company's permanent records.

The Overlay

A very good way to refine or improve a sketch is to use an overlay. An **overlay** is a piece of translucent tracing paper that is placed on top of a sketch or drawing. Because you can see through the paper, you can quickly trace the best parts of the sketch or drawing underneath. Refining ideas often means sketching over and over again on tracing paper, changing the drawing until the design is final.

Overlays are used in three important ways. The first is reshaping an idea. This might include refining the proportions of the parts of an object or changing its shape entirely. Second, an overlay can be used to refine the drawing itself without really changing the design. Although these two uses may seem very similar, they are both important to sketch development. Finally, an overlay can be used to add various options to a basic drawing. In Figure 2-28, for example, separate overlays could be used to demonstrate various seating and steering wheel designs. This type of overlay is often used to help clients visualize various design possibilities.

Materials for Sketching

Sketching has two major advantages over formal drawings. First, only a few materials are required to create a sketch. Second, you can create a sketch anywhere. You are ready to sketch with a pad of paper, a pencil, and an eraser.

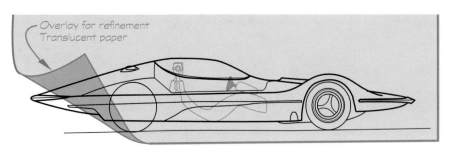

Overlay for refinement
Translucent paper

Figure 2-28

The overlay can speed up the design process by allowing various options to be viewed without redrawing the entire sketch each time.

Paper

You can use plain paper for sketching. If you need to refine the sketch, use tracing paper as an overlay. You may also use graph paper to control proportions while sketching. The most common type of graph paper has heavily ruled 1.00″ squares. The 1.00″ squares are then subdivided into lightly ruled ¹⁄₁₀″, ⅛″, ¼″, or ½″ squares. This paper is called 10 to the inch, 8 to the inch, and so on.

Graph paper ruled in millimeters (mm) is also available, and there are many specially ruled types of graph paper for particular kinds of drawing. For example, you can use special graph paper for isometric or perspective drawings. These kinds of drawings are explained further in Chapter 12.

While you can sketch on any convenient size of paper, standard 8.50″ x 11.00″ letter paper is the best for making small sketches quickly. You can hold the paper on stiff cardboard or on a clipboard while working on it. If you use graph paper, you may want to put it under tracing paper to help guide the lines and spacing.

Pencils and Erasers

Soft lead pencils (grades F, H, or HB), properly sharpened, are best for sketching. Erasers that are good for soft leads, such as a plastic eraser or a kneaded-rubber eraser, are most commonly used.

Use a drafter's pencil sharpener to remove the wood from the *plain* end of a pencil so that you don't remove the grade mark on the other end. Sharpen the lead to a point on a sandpaper block or on a file. If you are using a lead holder, use a lead pointer. Do not forget to adjust the grade mark in the window of the lead holder if it has one. Be careful to remove the needle point that the sharpener leaves by touching it gently on a piece of scrap paper. Refer to Chapter 3 for more detailed information about pencils, leads, lead holders, sharpeners, and erasers.

Four types of points are used for sketching: sharp, near-sharp, near-dull, and dull. The points should make lines of the following kinds:

- Sharp point—thick black lines
- Near-sharp point—thick lines
- Near-dull point—extra-thick lines
- Dull point—light, temporary lines that may later be erased

Sketching Techniques

Sketches drawn freehand have a natural look. The slight unevenness in the direction of lines shows freedom of movement. The guidelines in this section are intended to help you get started using good sketching techniques. As you become a more experienced drafter, you may add techniques and variations of your own.

Straight Lines

To draw a line, hold the pencil far enough from the point that you can move your fingers easily and yet can put enough pressure on the point to make dense, black lines when necessary. Draw light construction lines with very little pressure on the point. They should be light enough that they need not be erased.

You can sketch lines in the following ways, as demonstrated in Figure 2-29:

- Draw one long, continuous line.
- Draw short dashes where the line should start and end. Then place the pencil point on the starting dash and draw toward the ending dash.
- Draw a series of strokes that touch each other or are separated by very small spaces.
- Draw a series of overlapping strokes.

Before you try to draw objects, practice sketching straight lines to improve your line technique. Draw vertical lines from the top down, as shown in Figure 2-30A. Draw horizontal lines from left to right (if you are right-handed). See Figure 2-30B.

Figure 2-29

Four ways to sketch straight lines.

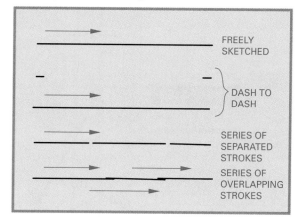

FREELY SKETCHED

DASH TO DASH

SERIES OF SEPARATED STROKES

SERIES OF OVERLAPPING STROKES

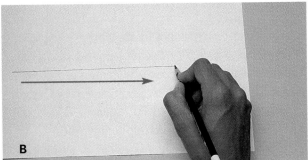

Figure 2-30

(A) Drawing a vertical line. (B) Drawing a horizontal line.

Slanted Lines and Specific Angles

Sketch slanted, or inclined, lines from left to right. It might be easiest to turn the paper and draw an inclined line as if it were a horizontal one. When trying to sketch at a specific angle, first draw a vertical line and a horizontal line to form a right (90°) angle, as shown in Figure 2-31. Divide the right angle in half to form two 45° angles. Or divide it in thirds to form three 30° angles. By starting with these simple angles, you can estimate other angles more exactly. Figure 2-32 shows the direction in which you would sketch to create lines inclined at various angles.

Circles and Arcs

There are several ways to sketch a circle. One way is to start by drawing very light horizontal and vertical lines, as shown in Figure 2-33A. These lines are the vertical and horizontal **centerlines** of the circle. The point at which they cross forms the center of the circle. Then estimate the length of the **radius** (the distance from the center of the circle to its edge; plural *radii*), and mark it off. Using the marks as guides, draw a square in which you can sketch the circle, as shown in Figure 2-33B.

Figure 2-33

Mark off the radii and draw a square in which to sketch a circle.

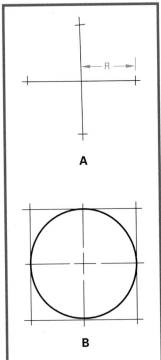

Figure 2-31

Draw horizontal and vertical lines before sketching inclined lines.

Figure 2-32

Sketching inclined lines and angles.

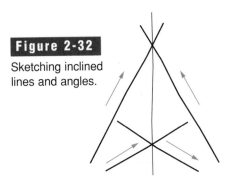

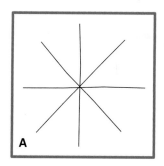

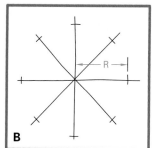

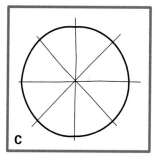

Figure 2-34

(A) Draw centerlines and other lines through the center of the circle. (B) Mark off the estimated radii on all lines. (C) Sketch the circle.

easier to form, so draw it first. Then turn the paper so that the rest of the circle is on the bottom, and finish drawing the circle.

You can use these same methods to sketch variations on circles, as shown in Figure 2-35. These include:

- **arcs**—parts of a circle
- **tangent arcs**—parts of two circles that touch
- **concentric circles**—circles of different sizes that have the same center

Use light, straight construction lines to block in the area of the figure.

For large circles and arcs and for **ellipses** (ovals), use a scrap of paper with the radius marked off along one edge. For circles and arcs, put one end of the marked-off radius on the center point. Draw the circle or arc by placing a pencil at the other end of the radius and turning the scrap paper, as shown in Figure 2-36A.

Two radii are needed for an ellipse, as shown in Figure 2-36B. The shorter radius is the minor radius. The longer radius is the major radius. To draw the ellipse, sketch both centerlines. Mark off both radii on a scrap of paper, as shown in the illustration. Keep both radius points on the centerlines as you draw with the pencil at the other end.

The second way to draw a circle is to draw very light centerlines and extra lines through the center, as shown in Figure 2-34A. Next, estimate the length of the radius and mark off this distance on all the lines, as shown in Figure 2-34B. Then draw a curved line that runs through all the radius marks, as shown in Figure 2-34C. The bottom of the curve is generally

Figure 2-35

Control the shape of arcs and concentric circles by first sketching squares.

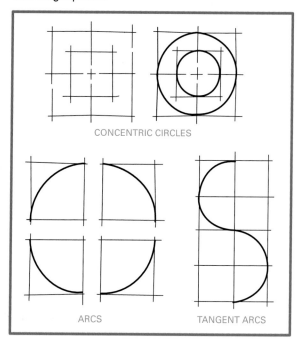

CONCENTRIC CIRCLES

ARCS

TANGENT ARCS

Figure 2-36

Large circles, large arcs, and ellipses can be sketched easily with the aid of a strip of paper.

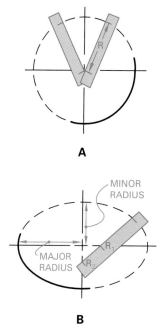

A

B

Figure 2-37

To use your hand as a compass, use your little finger as a pivot point.

Figure 2-38

Two pencils can serve as a compass.

You can also use your hand as a compass. To do this, use your little finger as a pivot (turning point) at the center of the circle. Use your thumb and forefinger to hold the pencil rigidly at the radius you want. Turn the paper carefully under your hand, thereby drawing the circle. See Figure 2-37.

Another way to draw circles is to use two crossed pencils. Hold them rigidly with the two points as far apart as the length of the desired radius. Put one pencil point at the center. Hold it there firmly and turn the paper, drawing the circle with the other pencil point. See Figure 2-38.

Making a Proportional Sketch

To create drawings in the correct proportions, you should follow seven basic steps:

1. Observe the object to be sketched. For this example, observe the chair in Figure 2-39A.

2. Select the views needed to show all shapes. In this case, you can describe the chair fully using three views: top, front, and right-side.

3. Estimate the proportions carefully. On your drawing paper, use light construction lines to mark off major distances for width, height, and depth in all three views. See Figure 2-39B. Notice how the diagonal 45° line is used to project the depth dimensions between the top and right-side views. This technique will be covered in greater detail in a later chapter.

4. Block in the enclosing rectangles, as shown in Figure 2-39C.

5. Locate the details in each of the views. Block them in, as shown in Figure 2-39D.

6. Finish the sketch by darkening the object lines.

7. Add any dimensions and notes required.

Figure 2-39

The development of a multiview sketch.

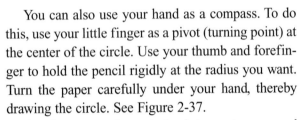

STUDY PROPORTIONS
A

ESTIMATE DIMENSIONS
B

BLOCK IN SHAPE
C

ADD DETAIL
D

Oblique Layout

To make an oblique sketch of an object, always follow a good layout procedure. Follow the steps below to create an oblique sketch of the object shown in Figure 2-40.

1. Estimate the proportions of the object. See Figure 2-40A.
2. Block in lightly the front face of the object, using the estimated distances for the width and height.
3. Decide on the angle of the third (depth) axis depending on which part of the object you want to emphasize or show most clearly. As you can see in Figure 2-41A, if you draw the depth axis at a small angle, such as 30°, the side shows more clearly. If you choose a larger angle, such as 60°, the top shows more clearly. Figure 2-41B shows the effect of using other angles for the depth axis. Lightly sketch the axis at the angle you have chosen.
4. Lightly sketch the object lines along the depth axis. You may draw the depth of the object at the same proportions as the rest of the object, or you can reduce its size by up to one half. When the

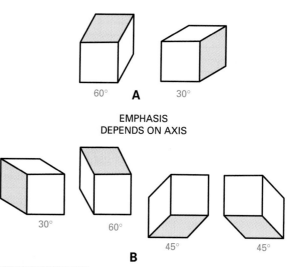

Figure 2-41

(A) The effect of two different angles of projection on oblique sketches. (B) Other examples of angles of projection.

depth dimension of a drawing is exactly one half of the true dimension, it is called a **cabinet oblique sketch**. Using a full-depth dimension produces a **cavalier oblique sketch**. See Figure 2-40B.

5. Darken the final object lines.

Oblique Sketching on Graph Paper

Graph paper is useful for oblique sketching because the front view of the oblique sketch is the same as the front view of a multiview sketch. See Figure 2-42A. If you develop the oblique pictorial drawing on graph paper from a multiview drawing on graph paper, simply transfer the dimensions from one to the other by counting the graph-paper squares:

1. Block in lightly the front face of the object by counting squares.
2. Sketch lightly the depth axis by drawing a line diagonally through the squares. For a cabinet oblique sketch, find the depth by using half as many squares as on the side view. Since the top view of Figure 2-42A shows a depth of four squares, you should use two squares for the depth in the oblique sketch. See Figure 2-42B.
3. Sketch in any arcs and circles, as shown in Figure 2-42C.
4. Darken the final object lines, as shown in Figure 2-42D.

Figure 2-40

Converting a two-view sketch into cabinet and cavalier oblique sketches.

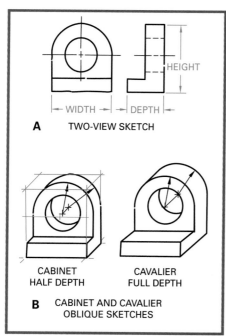

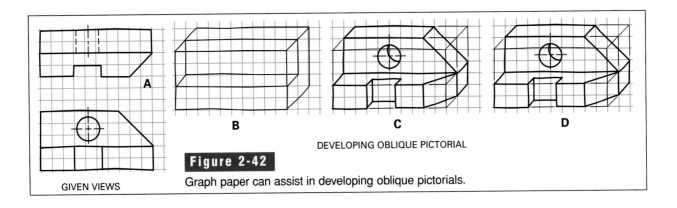

DEVELOPING OBLIQUE PICTORIAL

Figure 2-42

Graph paper can assist in developing oblique pictorials.

GIVEN VIEWS

A   B   C   D

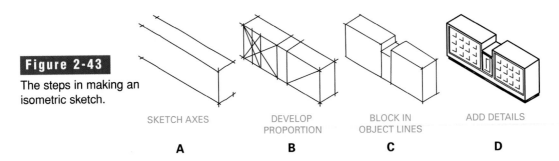

Figure 2-43

The steps in making an isometric sketch.

SKETCH AXES

DEVELOP PROPORTION

BLOCK IN OBJECT LINES

ADD DETAILS

A   B   C   D

Oblique Circles

In oblique sketching, circles in the front view can be drawn in their true shape. However, circles drawn in the top or side views appear distorted. Indeed, you must draw an ellipse to show such circles. Ellipses are more difficult to represent accurately, and in some cases they may confuse the viewer. Therefore, it is better practice to show the circular shapes of important parts in the front view, as shown in Figure 2-40.

Isometric Layout

Figure 2-43 shows the steps in making an isometric sketch.

1. Sketch the isometric axes, as shown in Figure 2-43A.
2. Develop the proportion using the techniques that were described earlier in this chapter. See Figure 2-43B.
3. Block in the major features of the object, as shown in Figure 2-43C.
4. Darken the final object lines and add detail, as shown in Figure 2-43D.

Figure 2-44

Isometric graph paper can be used for quick sketches.

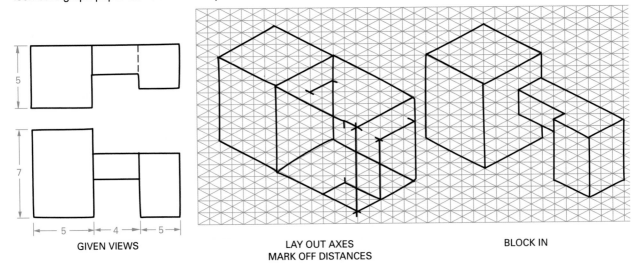

GIVEN VIEWS

LAY OUT AXES
MARK OFF DISTANCES

BLOCK IN

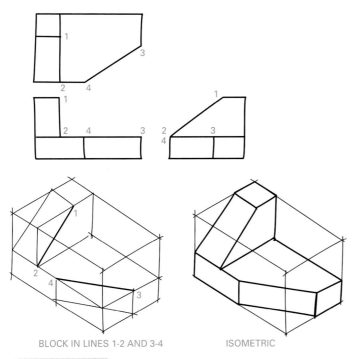

BLOCK IN LINES 1-2 AND 3-4   ISOMETRIC

Figure 2-45

Developing an isometric sketch using an isometric box as reference.

Note that the sketched lines for isometric axes tend to become steeper than 30° if you do not prepare the layout carefully. A better pictorial sketch results when the angle is at 30° or a little less. Using isometric graph paper with 30° ruling, as shown in Figure 2-44, lets you make sketches quickly and easily.

Another way to prepare an isometric drawing is to sketch the object inside a lightly sketched isometric box, as suggested in Figure 2-45. Note, however, that the objects in Figures 2-45 through 2-47 have some nonisometric lines. You can draw these lines by extending their ends to touch the blocked-in box. Locate points at the ends of the lines by estimating measurements parallel to isometric lines. Having located both ends of the nonisometric lines, you can sketch the lines from point to point. Nonisometric lines that are parallel to each other also appear parallel on the sketch, as shown in Figure 2-45. Note how the ends have been located on lines 1-2 and 1-3 in Figure 2-47. Distances *a* and *b* are estimated and transferred from the figure at part A to part B. Any inclined line, plane, or specific angle must be found by locating two points of intersection on isometric lines.

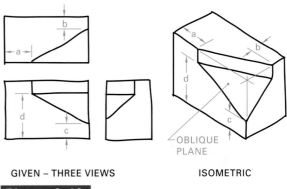

GIVEN – THREE VIEWS   ISOMETRIC

Figure 2-46

Developing an oblique plane in an isometric sketch.

Figure 2-47

Developing three oblique surfaces in an isometric sketch.

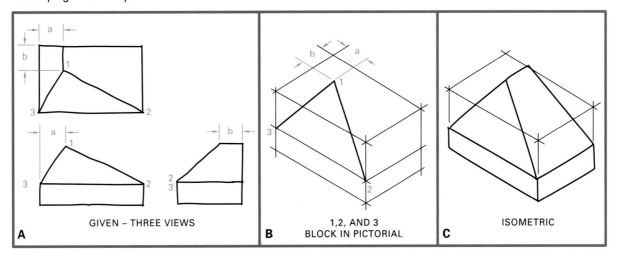

A   GIVEN – THREE VIEWS   B   1,2, AND 3 BLOCK IN PICTORIAL   C   ISOMETRIC

Isometric Circles and Arcs

All circles in an isometric view are drawn as ellipses. To sketch a circle in an isometric view, first sketch an isometric square, as shown in Figure 2-48. Sketch the small-end arcs tangent to (touching) the square. Then sketch the larger arcs tangent at points T to finish the square. Note that the long axis of the ellipse, or isometric circle, is no longer than the true size of the circle. The short axis of the ellipse is shorter. This difference is caused by the isometric angle. Figure 2-48 shows an ellipse for a top view only. Circles on the three faces of an isometric cube are sketched in Figure 2-49.

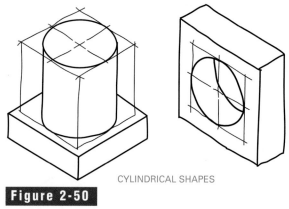

CYLINDRICAL SHAPES

Figure 2-50

Blocking in isometric circles on cylindrical forms.

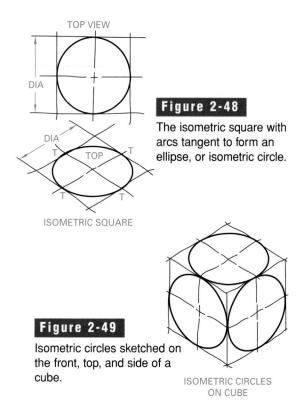

Figure 2-48

The isometric square with arcs tangent to form an ellipse, or isometric circle.

Figure 2-49

Isometric circles sketched on the front, top, and side of a cube.

You will need isometric circles to develop cylindrical and conical shapes in a sketch. Some ways to block in cylindrical shapes are shown in Figure 2-50. Methods of blocking in conical shapes are shown in Figure 2-51.

Arcs developed in an isometric view are shown in Figure 2-52. A semicircular opening and rounded corners appear in the front view. The object is blocked in at Figure 2-52A. The outline of the object is darkened at Figure 2-52B. Note that only partial circles are needed here. The rounded corners take up only a quarter of the full isometric circle that was plotted.

Irregular Curves

Draw irregular (noncircular) curves in pictorial sketches by plotting points along the path of the curve. To locate the curve, you generally transfer the points from a multiview sketch. In Figure 2-53A, a

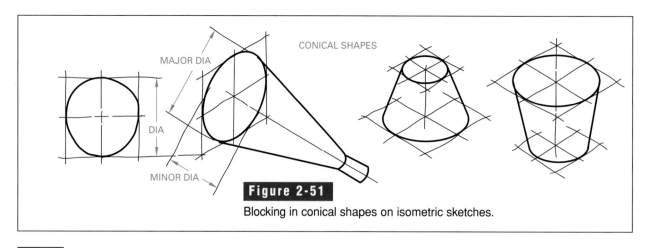

CONICAL SHAPES

Figure 2-51

Blocking in conical shapes on isometric sketches.

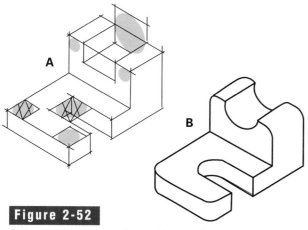

Figure 2-52

To block in arcs on an isometric sketch, use a method similar to the one you use for circles.

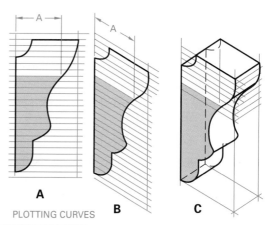

PLOTTING CURVES

Figure 2-53

Plot irregular curves using a coordinated grid.

grid is applied, and points are plotted on the front view where the grid lines intersect the curve on the object. In Figure 2-53B, the grid is rotated to align with the axis in the pictorial view. The points are then plotted at the same distance from the left edge of the object. Similar points are plotted on the pictorial in Figure 2-53C. The intersections serve as points of reference for sketching the pictorial curve.

Success on the Job

Graphic Communication

Developing skills in freehand sketching and lettering will add to your ability to communicate technical ideas. Sketches that are carefully and accurately drawn with lettering that is neat and easy to read will impress others and reduce the chance of communicating errors. Develop these skills early in your career to improve your chance for success on the job.

Lettering

Freehand sketches generally need some freehand lettering to explain features of a new idea or product. **Lettering** is the practice of adding clear, concise words on a drawing to help people understand the drawing. A well-planned sketch may be worth a thousand words, as an old saying goes, but a few choice words, well organized, can explain some important details.

The notes lettered on the rough sketch in Figure 2-54 describe features that are functional and important to operation. On sketches you have prepared or studied in the past, can you recall a need for notes to explain an idea? Simple freehand lettering complements the idea that is captured in a sketch, especially if the lettering is neat and carefully placed on the drawing.

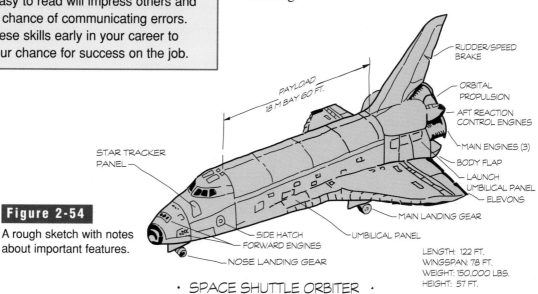

Figure 2-54

A rough sketch with notes about important features.

Composition

In lettering, **composition** means arranging words and lines with letters of the right style and size. Letters in words are not placed at equal distances from each other. They are placed so that the spaces between the letters *look* equal. The distance between words, called *word spacing*, should be about equal to the height of the letters. Figure 2-55 shows examples of proper and improper letter and word spacing. The space between lines of letters is from ½ to 1½ times the height of the letters.

Tools such as lettering triangles and the Ames lettering instrument are available to help board drafters create neat, uniform lettering using the proper spacing. See Figure 2-56. On mechanical drawings, drafters create ruled **guidelines** spaced .12″ apart to help keep their lettering uniform. When you are

sketching, however, you will need to estimate the appropriate distances.

Drafters need good lettering skills throughout their careers. Freehand lettering is one of the first things board drafting students study. This is because you do not learn how to do good lettering all at once. You learn it by practicing it little by little for a long time. Your lettering should get better with every sketch or drawing you do.

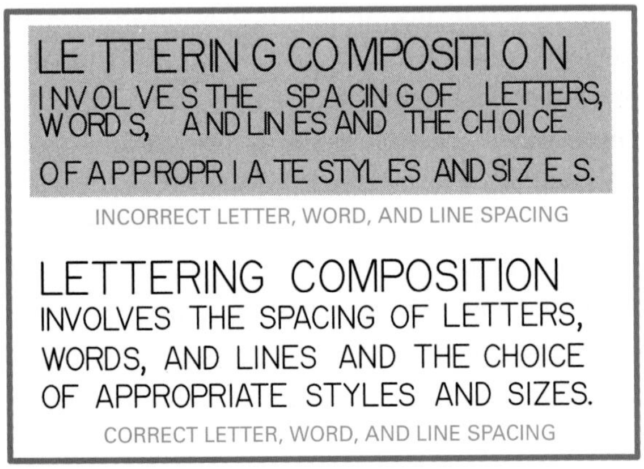

Types of Lettering

The choice of lettering style should be made carefully. For example, it is usually not practical to use fancy roman-style lettering for notes and dimensions on technical drawings. Such lettering requires more pencil strokes, so it is more costly in time and money than the single-stroke styles, yet it would serve the purpose of the drawing no better.

The lettering style most commonly used on working drawings is single-stroke **Gothic lettering**, as shown in Figure 2-57. This style is best because it is easy to read and easy to hand letter. It is made up of

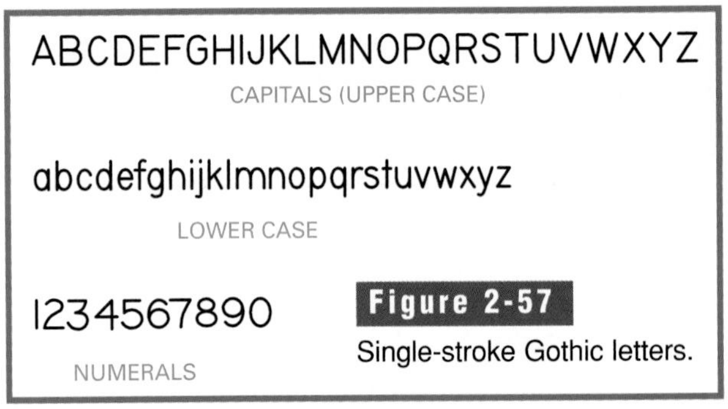

Figure 2-55

Study the word and letter spacing in these examples.

Figure 2-57

Single-stroke Gothic letters.

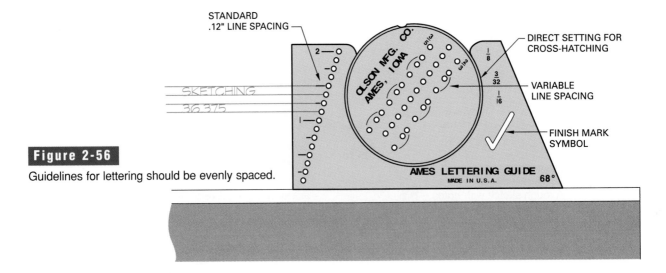

Figure 2-56

Guidelines for lettering should be evenly spaced.

Figure 2-58

Variations in lettering styles for the designer.

ABCDEFGHIJKLMNOPQRSTUVWXYZ 1234567890

ABCDEFGHIJKLMNOPQRSTUVWXYZ 1234567890

A B C D E F G H I J K L M N O P Q R S T U V W X Y Z
1234567890

uppercase (capital) letters, lowercase (small) letters, and numerals. Nearly all companies now use only uppercase lettering. As a result, this book stresses uppercase lettering. Letters and numerals may be either vertical or inclined. However, vertical lettering is most commonly used. You can vary your lettering to make it more individual. Figure 2-58 shows some of the possible variations. They are common styles for designers and architects. Whichever style you

choose, remember that the same style should be followed throughout a set of drawings.

The shapes and proportions of single-stroke vertical letters and numbers are shown in Figure 2-59. The figure also demonstrates the pencil strokes needed to create each letter and the order in which they should be made. Each character is shown in a six-unit grid. By following these, you can easily learn the right shapes, proportions, and strokes.

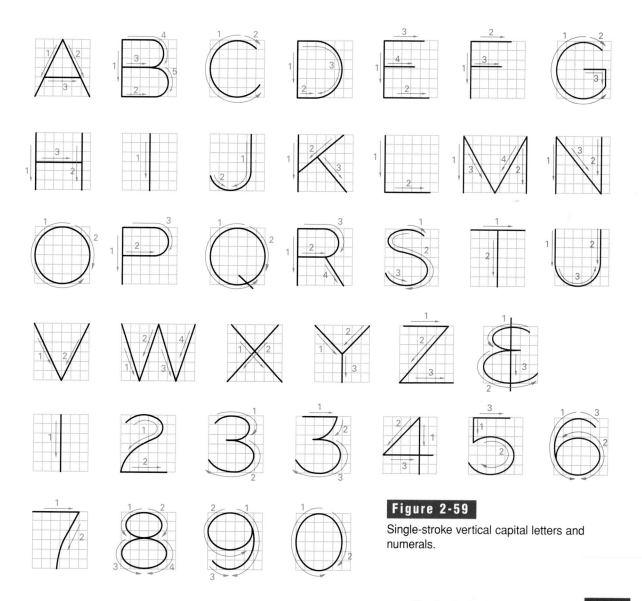

Figure 2-59

Single-stroke vertical capital letters and numerals.

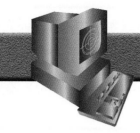
The term *sketching* has a slightly different meaning for drafters using CAD systems. Although AutoCAD and many other CAD programs include a SKETCH command, it is generally not used for the kinds of freehand sketching described in the "Drafting Principles" and "Board Drafting Techniques" portions of this chapter. This is because the concept of freehand sketching is not as practical on a CAD system. It is just as easy, if not easier, to create a perfectly straight line as it is to "sketch" a line. Circles, including isometric circles, can be drawn perfectly using the appropriate CAD commands. Figure 2-60 shows an example of why the SKETCH command is not used for traditional sketching purposes. The sandpaper block in Figure 2-60A, which was drawn using the SKETCH command, took longer to draw and is not as pleasing or accurate as the one in Figure 2-60B.

Therefore, the term *sketching* is generally used by CAD operators to mean rough drawings that will be refined later. These drawings are created with the same commands as finished CAD drawings. In other words, there is no separate procedure for sketching using a CAD program. You will learn the commands and techniques for creating drawings in AutoCAD throughout this textbook.

The same sandpaper block drawn using AutoCAD's SKETCH command (A) and regular CAD commands (B).

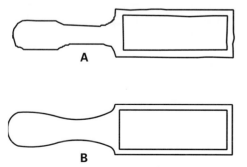

A

B

Sketching for CAD Operators

Creative ideas may strike you at any time. In fact, many good design ideas result from casual conversations among friends or coworkers. The ability to record those ideas immediately and informally using pencil and paper—traditional "sketching"—is just as important for the CAD operator as for the board drafter. Therefore, it is important to work through the "Board Drafting Techniques" section of this chapter. Even if you plan to do nothing but CAD drafting, sooner or later you will need the techniques discussed in "Board Drafting Techniques."

Handheld CAD Systems

One of the drawbacks of using a computer for sketching is that you won't always have one with you when you need it. This is true even though CAD programs are available for new, smaller laptops and handheld computers, or "palmtops." See Figure 2-61. Even palmtops require a battery, which limits the amount of time you can spend "in the field," away from a source of electricity. Also, the amount of memory (file storage space) is limited on these devices. This restricts the number of sketches you can create, which may in turn limit your creative output.

For certain applications, handheld computers are appropriate sketching tools. For example, architects and builders can use a handheld CAD system at a building site to help solve problems that arise during construction. In fact, some architects carry handheld CAD systems instead of bulky blueprints.

The SKETCH Command

As mentioned above, AutoCAD provides a SKETCH command that allows you to draw "freehand" within the CAD system. Although most drafters do not use it for traditional sketching tasks, this command is actually very useful for certain jobs. One of its most important uses is in digitizing irreg-

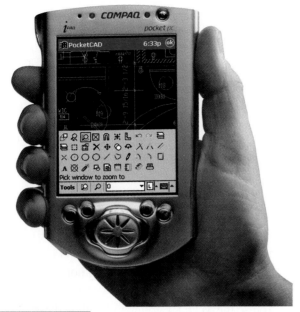

Figure 2-61

CAD systems are now available for handheld computers, increasing the flexibility of CAD systems.

ular surfaces, such as lakes, county or state lines, and other features on maps.

Digitizing is the process of converting a paper copy into electronic format. To digitize a map, drafters use a digitizing tablet (discussed in Chapter 3) and the SKETCH command to trace the boundaries and lines. The advantage of using the SKETCH command for this purpose is that it allows you to draw continuous, irregular lines by holding down the mouse button and tracing the original. Each line created by the SKETCH command is made up of many control points that can later be edited to perfect all of the tiny irregular curves that make up most lines on maps. Map drafting and the SKETCH command are discussed in more detail in Chapter 20, "Map Drafting."

CAD Lettering

Lettering in a CAD drawing is generally referred to as **text**. CAD systems maintain uniform letter and word spacing automatically for notes and dimension text. Text size is also uniform. Most CAD programs, including AutoCAD, also provide several lettering styles. Each style is right for a particular use. It is the drafter's responsibility to set up a text style and size that is appropriate for the drawing.

Lettering styles in CAD and other computer programs are known as **fonts**. AutoCAD provides several types of fonts, including native AutoCAD ("shape compiled") fonts, as well as TrueType® and PostScript® fonts. A few of the more commonly used fonts are shown in Figure 2-62.

Setting the Text Style

The default text style in AutoCAD is the Standard style. This style is boxy and does not look much like hand-lettered text. Therefore, drafters generally use the Roman Simplex font for most mechanical drawings. Follow these steps to set the text of a drawing to Roman Simplex.

1. More information about creating and using files is given in Chapters 3 and 4. For now, start AutoCAD and open a new drawing file by selecting New… from the File menu. What happens next varies depending on your version of AutoCAD. In most versions, you can choose to start a drawing from scratch. If AutoCAD prompts you for a template file, press Enter or OK to accept the default.
2. Enter the STYLE command and press the New… button in the dialog box.
3. Type in a name for the new style, choosing a name (such as "Roman") that will help you remember what font the style uses.
4. Pick the down arrow under Font Name and select romans.shx to activate the Roman Simplex font.
5. Pick the Close button.

The Roman Simplex style is now activated in the current drawing and is set as the current default.

Figure 2-62

Examples of commonly used CAD lettering styles.

ROMAN SIMPLEX IS MOST OFTEN USED FOR MECHANICAL DRAWINGS.

CITYBLUEPRINT IS SOMETIMES USED FOR ARCHITECTURAL DRAWINGS.

COUNTRY BLUEPRINT IS A GOOD ALTERNATIVE FOR ARCHITECTURAL DRAWINGS IF SPACE IS NO OBJECT.

Adding Text to a Drawing

AutoCAD has three commands for adding text to a drawing:

- TEXT
- DTEXT
- MTEXT

TEXT is the oldest command. It has limited capabilities and is rarely used by today's drafters. The only way to activate the TEXT command is to type it at the keyboard.

Using DTEXT

For single-line notes, the DTEXT command is the most efficient command to use. DTEXT stands for "dynamic text." It was called this because, unlike text created with the older TEXT command, DTEXT text appears dynamically on the screen as you type. To use the DTEXT command, follow these steps:

1. Enter the DTEXT command.
2. Pick a point anywhere in the drawing area as a starting point.
3. At the Command line in the lower left part of the screen, AutoCAD prompts for the text height. Use the keyboard to enter .12.
4. The next prompt asks for the angle of rotation for the text. AutoCAD will set text at any angle you specify. Press Enter to accept the default of 0 (normal text).
5. Enter the text for the note. For this example, enter ROMAN SIMPLEX IS AN APPROPRIATE FONT FOR MOST MECHANICAL DRAWINGS. Note that text for mechanical drawings is set in uppercase (capital) letters.

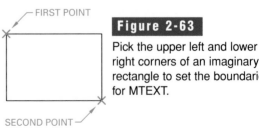

Line Breaks with DTEXT
To start a new line of text when you are using the DTEXT command, simply press Enter. A new line aligns automatically below the previous line. To end the DTEXT command, press Enter twice.

Using MTEXT

The MTEXT command is a more complex command that allows you to insert formatted, multiple-line text into a drawing. It is actually a small text editor that resides in AutoCAD. It allows you to change the font and to use fonts without setting up styles for them. You can set the text in boldface or italic type, underline it, and set up many other characteristics. For most mechanical drawings, these features are unnecessary.

However, MTEXT also wraps text automatically to fit in the space you define. In other words, you don't have to worry about where to press Enter to start a new line of text. This makes the MTEXT command a good choice for adding long notes and specifications to a drawing.

To use the MTEXT command, follow these steps:

1. Enter the MTEXT command.
2. Pick two points on the screen. Pick the upper left and lower right corners of the rectangular space you want the text to fill. See Figure 2-63. For this example, pick points about 2″ apart on the screen.
3. The text style defaults to the current drawing default. In this case, if you followed the steps in "Setting the Text Style" earlier in this chapter, the default is currently Roman Simplex. Enter the text for the note using the default style. Enter THE MTEXT COMMAND ALLOWS GREATER FLEXIBILITY FOR LONG, COMPLEX NOTES AND SPECIFICATIONS.
4. Review all of the options available in the dialog box. Highlight different portions of the text and experiment with the various options. Change the text style, the point size, and other characteristics. You will not often need many of these options, but you should be aware of the software's capabilities.

FIRST POINT

Figure 2-63
Pick the upper left and lower right corners of an imaginary rectangle to set the boundaries for MTEXT.

SECOND POINT

5. When you have finished experimenting, pick the OK button to close the dialog box and display the text on the screen.

Editing Text

The method for editing text is the same for text created with both DTEXT and MTEXT. In both cases, you edit the properties of the text. The actual procedure varies somewhat depending on the version of AutoCAD you are using. Prior to AutoCAD 2000, the most efficient method was to choose Properties… from the Modify menu, and then use the mouse to select the text to be edited. A dialog box appears that allows you to change the various properties of the text, as shown in Figure 2-64.

AutoCAD 2000 and later versions support the use of shortcut menus. First, pick with the mouse to select the text you want to edit. The text becomes dashed to show that you have selected it. Then right-click the mouse to display a shortcut menu. From this menu, select Properties. The Properties dialog box appears, as shown in Figure 2-65.

AutoCAD Versions

This textbook has been developed for use with all versions of AutoCAD beginning with Release 14, including LT versions. Except where noted specifically in the text, all of the commands discussed in the textbook work with all of these versions. Be aware that in some cases, the commands may act slightly differently, or dialog boxes may differ in appearance. The basic function of the commands is the same, however.

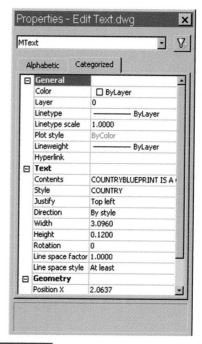

Figure 2-65

The contents of the Properties dialog box vary depending on the object you have selected. The properties shown here are editable for text created with the MTEXT command in AutoCAD 2000 and later.

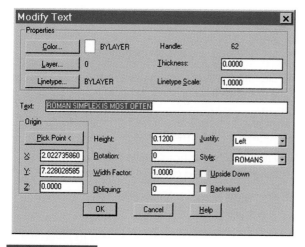

Figure 2-64

The dialog box used to edit DTEXT prior to AutoCAD 2000.

Chapter Summary

- The three aspects of design are functional, aesthetic, and engineering design.

- The traditional design process is basically a linear process, although it allows for feedback in the development process.

- Concurrent engineering is a team approach to making better products in less time.

- Sketches can be made in multiview or pictorial.

- Pictorial sketches are generally done in oblique or isometric.

- The ability to estimate proportions helps a drafter create clear sketches that are easy to understand.

- The type and quality of a sketch depends on the purpose of the sketch and its intended life span.

- Freehand lettering is generally done in single-stroke Gothic style.

- CAD operators need to develop the same freehand sketching skills needed by board drafters.

- AutoCAD's SKETCH command is rarely used for traditional sketching.

- AutoCAD provides several ways to place text on a drawing.

Review Questions

1. What do you call the combination of imagination, knowledge, and curiosity?

2. What word describes the conception of an idea and its development, through graphic communication, into a practical, producible, and usable product or process?

3. What aspect of design deals with the look and feel of a product?

4. What aspect of design deals with the successful operation of a product?

5. Name the two types of engineering design.

6. Why are the five visual symbols in the language of sketching essential in graphic communication?

7. What is another name for multiview drawing?

8. Name two basic types of pictorial sketching.

9. Name two advantages of pictorial sketching.

10. Name five types of freehand sketches.

11. What is the purpose of an overlay on a freehand sketch?

12. Name the materials needed for a freehand sketch.

13. What is the difference between a cavalier oblique and a cabinet oblique sketch?

14. Describe the placement of the three axes for an isometric drawing.

15. In general, how far apart should the guidelines be placed for freehand text?

16. Describe the characteristics of a typical "sketch" created on a CAD system.

17. For what purpose is the SKETCH command most widely used?

18. What three commands allow you to place text on a drawing in AutoCAD?

19. What font is most commonly used for mechanical drawings created on a CAD system?

20. For what kind of text is the DTEXT command more suited? The MTEXT command?

Problems

CHAPTER

Drafting Problems

Problems 1 through 13 in this chapter are designed to be completed using freehand sketching techniques. You should complete them using only paper, pencil, and eraser, regardless of whether you plan to work with traditional board drafting or CAD techniques throughout the rest of the book. Problems 14 through 17 may be completed using freehand or CAD lettering. *Note to CAD students:* After you have studied Chapters 4 and 5, return to the problems in this chapter, complete them using CAD techniques, and compare the advantages and disadvantages of using each method.

For problems 1 through 9, follow the instructions to practice sketching freehand and estimating proportions.

1. Sketch the 2.00″ overlapping squares shown in Figure 2-66 as creative visual studies. Then create two more of your own design.

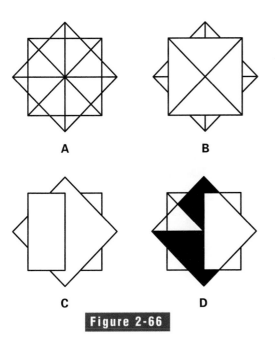

Figure 2-66

2. Sketch the squares shown in Figure 2-67: (A) overlapping; (B and C) diminishing; and (D) as a transparent cube. Sizes are about 38 mm, 28 mm, and 18 mm.

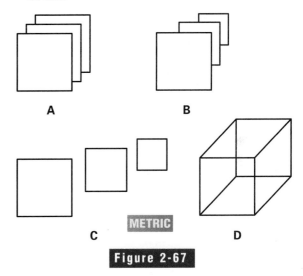

METRIC

Figure 2-67

3. Sketch a cube with the rectangular shape shown in Figure 2-68. Observe the optical illusion.

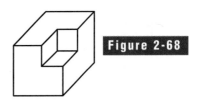

Figure 2-68

4. Sketch the rectangular solid shown in Figure 2-69 with 2-to-1 proportions. Use .50″, 1.00″, and 2.00″.

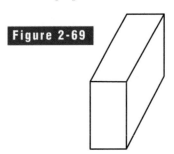

Figure 2-69

Chapter 2 • Design and Sketching **79**

5. Sketch the apparent two-dimensional form shown in Figure 2-70 using six diagonals.

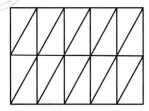

Figure 2-70

6. Sketch a 3.00″ equilateral triangle, as shown in Figure 2-71, with diminishing triangles at midpoints. Note the proportions. How many triangles can you make diminish inside?

Figure 2-71

7. Sketch the 76-mm square shown in Figure 2-72 with diminishing squares at midpoints.

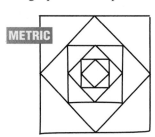

METRIC

Figure 2-72

8. Sketch the pentagon shown in Figure 2-73 using 2.00″ sides. Sketch diminishing five-pointed stars inside the pentagon, as shown.

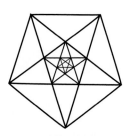

Figure 2-73

9. Sketch the five basic solids shown in Figure 2-74. Use freehand lettering to label each solid.

TETRAHEDRON   OCTAHEDRON

HEXAHEDRON   ICASAHEDRON

DODECAHEDRON

Figure 2-74

For problems 10 through 13, sketch the three views needed for a multiview drawing. Select appropriate proportions and sizes.

10. Sketch three-view drawings for the objects in Figure 2-75A through 2-75P.

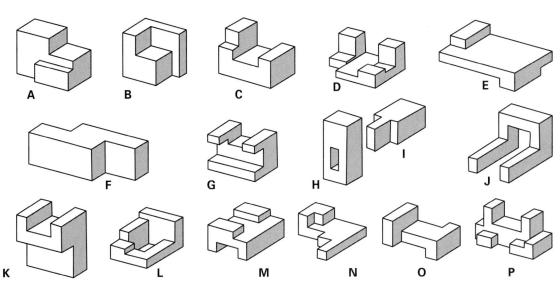
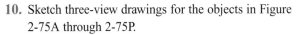

Figure 2-75

11. Sketch three-view drawings for the objects in Figure 2-76A through 2-76T.

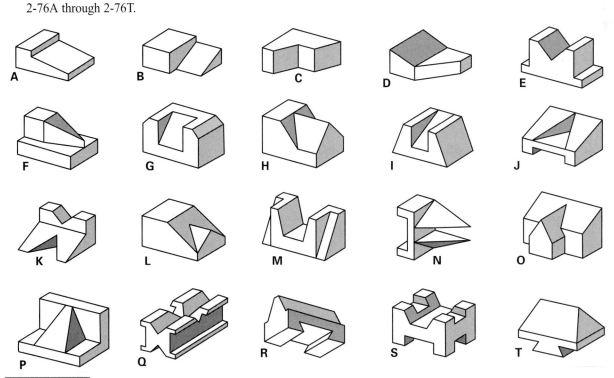

Figure 2-76

12. Sketch three-view drawings of the machine parts
shown in Figure 2-77A through 2-77N.

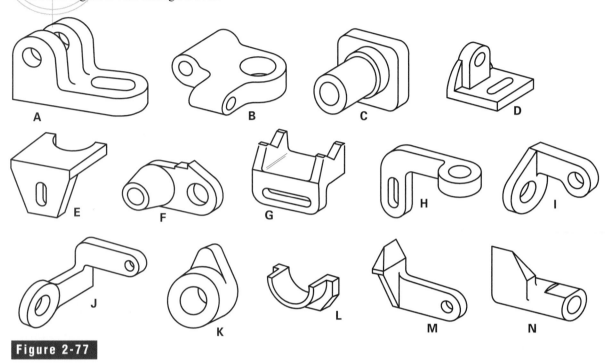

Figure 2-77

13. Sketch three-view drawings of the machine parts
shown in Figure 2-78A through 2-78O.

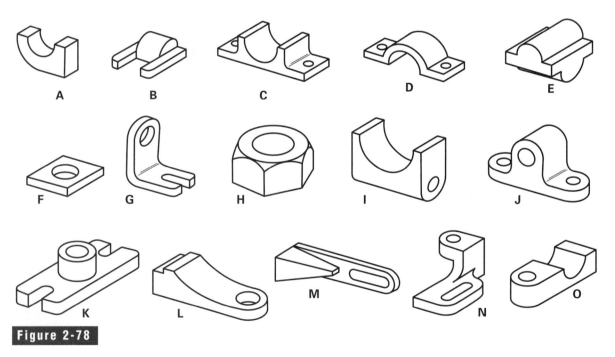

Figure 2-78

14. Draw light guidelines, as shown in Figure 2-79. Reproduce the abbreviations as they appear. Refer to Appendix A and other sources. Next to the abbreviations and symbols, letter the word or words that each represents. *CAD students:* Set up a text style to use the Roman Simplex font. Use DTEXT to complete the exercise. Do not draw guidelines.

.12"
℄ B/M BC CSTG Cl
CONC CBORE EQL SP CSK

.25"
FAO GA HEX
ID LH MST
HT TR MATL

3.5 mm
MAX OD PERP mm PD
PL SQ PC R RH

7 mm
SPHER THD
TOL SFACE
STL USG

Figure 2-79

Design Problems

Design problems have been prepared to challenge individual students or teams of students. Complete all of the problems using freehand sketching techniques. *Note to CAD students:* After you have studied Chapters 4 and 5, return to the problems in this chapter, complete them using CAD techniques, and compare the advantages and disadvantages of using each method. Be creative and have fun!

1. Redesign the digital clock shown in Figure 2-80. Give special attention to the design of the base. Materials optional. Prepare a three-view sketch of your design idea and use overlays to refine the design. Make an oblique or isometric pictorial sketch of your final design. Keep both function and aesthetics in mind as you proceed.

Figure 2-80

2. **TEAMWORK** Work as a team to design a scooter that could be manufactured and sold by your technology club. Apply the eight steps generally used in the traditional (linear) design process. Choose one member of your team to serve as the design engineer. List each step and describe how it would apply to the design of the scooter. Be sure to document each step as you proceed.

   Give every consideration to both function and aesthetics. Design the scooter to fold for easy carrying. Specify lightweight materials. Your final project should include sketches of the product design, general specifications on materials and stock parts, and documentation on the design process.

3. Design a carton (package) for the scooter designed in the previous problem. It should be lightweight, yet durable enough to protect the scooter during shipping. Work along with the design team so that package design is complete when the scooter design is complete.

4. **TEAMWORK** Work as a team to design a three-wheel vehicle. Refer to Figure 2-81 as a starting point. Apply the concept of concurrent engineering design to the process. Choose one team member to serve as the design engineer (project engineer). Discuss and document the procedure you should follow to comply with the principles of concurrent engineering design. How does this process differ from the traditional design process?

Be creative! This vehicle can be pedal-powered, gasoline engine-powered, or simply a freewheeling downgrade coaster. Once the team has developed the basic concept, use overlays to refine the design. Your final project should include sketches of the vehicle design, general specifications on materials and stock parts, and documentation on the design process.

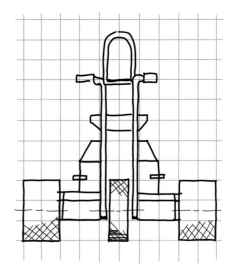

Figure 2-81

Drafting Equipment

CAD software

case instruments

compass

dividers

drawing board

ergonomics

irregular curve

protractor

scale

template

T-square

vellum

OBJECTIVES

Upon completion of this chapter, you should be able to:

- Identify the basic equipment used in board drafting.
- Identify and describe various types of drafting media.
- Identify the purpose of each instrument included in drawing instrument sets.
- Describe the characteristics of drafting pencils, technical pens, and erasers used in drafting.
- Select the appropriate scales for architectural, civil, and mechanical drafting.
- Describe the hardware components of a CAD workstation.
- Identify the three main types of CAD software.
- Identify the characteristics of efficient CAD furniture.
- Describe the ergonomic and personal safety factors to be considered when setting up a CAD workstation.

Drafting Principles

The term "tools of the trade" is commonly used throughout business and industry. Hammers and saws are tools of the trade to the carpenter. Lathes and drill presses are tools of the trade to the machinist. To the drafting technician, drawing instruments and computer-aided drafting (CAD) systems are the tools of the trade. Being able to identify these tools and understanding their purpose will enable you to succeed in the industrial drafting room today and into the future.

Throughout most of the twentieth century, drafting tools changed very little. Many of the board drafting instruments used in the early 1900s are essentially the same as the ones used today. See Figure 3-1. While some instruments have been streamlined and plastics have been added to their design, compasses, dividers, and many other instruments are nearly identical to the ones used a century ago.

The invention of the computer and the development of drafting and design (CADD) software during the late twentieth century have had a major impact on the way technical drawings are made. In most industrial drafting rooms, computer-aided drafting and design systems have replaced board drafting practices, as shown in Figure 3-2. However, board drafting is still alive and well in many industries and will continue to be for some time to come. A clear understanding of the tools used for both board drafting and CAD will help prepare you for a career in drafting or other fields related to technology.

Figure 3-1
Drafting instruments and their uses have changed little since the early 1900s.

Figure 3-2
CAD has replaced board drafting in many industrial drafting rooms.

Board Drafting Equipment

Technical drawings are used in industrial, engineering, and scientific work to communicate technical information for the design and manufacture of products. Designers and scientists often make freehand sketches to help them study new ideas or to show their ideas to other people. However, the final drawings are made with drafting instruments.

Basic Board Drafting Equipment

Figure 3-3 shows a variety of basic board drafting equipment. Many of these items are included in basic student drafting kits. Other common drafting equipment includes:

- drafting board
- T-square, or parallel-ruling straightedge, or drafting machine
- drafting media (paper/vellum or film)
- drafting tape
- drafting pencils
- lead pointer (pencil sharpener)
- erasing shield
- eraser
- triangles, 45° and 30°-60° (not required with drafting machine)
- architect's, engineer's, or metric scale
- irregular curve
- drawing instrument set
- lettering instruments
- black drawing ink
- technical pens
- brush or dust cloth
- protractor
- cleaning powder

Your instructor can tell you exactly what equipment you will need for your course.

Drawing Tables and Desks

Drawing tables and desks come in many different sizes and types, as shown in Figure 3-4. Some drawing tables are made to be used while you are standing up or sitting on a high stool. Other tables are the same height as regular desks. There are also combined tables and desks that you can use either standing up or sitting down. The type that combines a drafting table, desk, and regular office chair is the most comfortable and efficient. The cost of drawing tables and desks varies according to their size, style, and number of components.

Figure 3-3

Basic board drafting equipment.

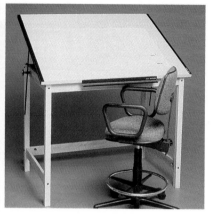

Figure 3-4

Drafting tables are available in a variety of sizes and styles.

Drawing Boards

The drawing sheet is attached to a **drawing board**. Drawing boards used in school or at home usually measure 9″ × 12″ (230 × 300 mm), 16″ × 21″ (400 × 530 mm), or 18″ × 24″ (460 × 600 mm). Boards used to make engineering or architectural drawings are typically larger and may be any size needed. Boards are generally made of soft pine or basswood. They are made so that they will stay flat and so that the guiding edge (or edges) will remain straight. Hardwood or metal (steel or aluminum) strips are used on some boards to provide truer and more durable guiding edges, and the top is often covered with a high-quality vinyl material that is very resilient. The vinyl cover adds considerable life to an ordinary wood drawing surface.

T-Squares

A **T-square** is a drafting instrument that consists of a head that lines up with a true edge of the drafting board and a blade, or straightedge, that provides a true edge. Figure 3-5 shows three common examples of T-squares. Most T-squares have plastic-edged wood or clear plastic blades, and heads of wood or plastic. When extreme accuracy is

needed, stainless steel or hard aluminum blades with metal heads are used. The blade must be very straight. It must be attached securely to the top surface of the T-square head.

You can easily find how accurate your T-square is, as shown in Figure 3-6. First, on a clean sheet of paper, draw a sharp, thin line along the drawing edge of the T-square. Second, turn the drawing sheet upside-down and line up the drawing edge of the T-square with the other side of the line. If the drawing edge and the pencil line do not match, the T-square is not accurate. When you find this condition, you should replace the T-square.

T-squares come in several sizes. The size and type of T-square you choose should be appropriate for the type of drawing you will be doing. It should extend

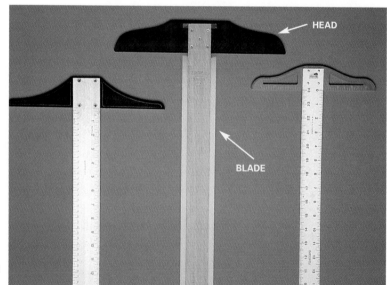

Figure 3-5

T-squares are available in various styles and materials.

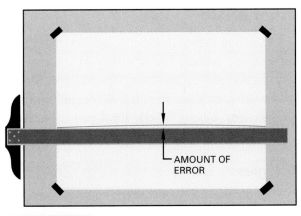

Figure 3-6

Figure 3-6

Check to see that the T-square is accurate.

most of the way across the drafting board you will be using. To find the size of a T-square, measure along the blade from the contact surface of the head to the end of the blade.

Parallel-Ruling Straightedges

Many drafters prefer to use a **parallel-ruling straightedge**, as shown in Figure 3-7, when working on large boards placed vertically or nearly so. A guide cord is clamped to the ends of the straightedge. The cord runs through a series of pulleys attached to the board. This allows the straightedge to slide up and down on the board in parallel positions.

Figure 3-7

A parallel-ruling straightedge is another convenient instrument that can save time.

Figure 3-8

An arm- or elbow-type drafting machine.

Drafting Machines

Drafting machines combine the functions of the T-square, triangles, scales, and protractor. You can draw lines to any desired length, at any location, and at any angle just by moving the scale ruling edge to the desired location. This lets you draw faster and with less work.

Two kinds of drafting machines are currently in wide use. The arm- or elbow-type drafting machine is shown in Figure 3-8. It uses an anchor and two arms to hold a movable protractor head with two scales. The scales are ordinarily at right angles to each other. The arms allow the scales to be moved to any place on the drawing that is parallel to the starting position. The two are parallel when their edges are exactly the same distance apart at all points.

The track-type drafting machine, shown in Figure 3-9, uses a horizontal guide rail at the top of the board and a moving arm rail at right angles to the top rail. An adjustable protractor head and two scales, usually at right angles, move up and down on the arm. The scales may be moved to any place on the drawing that is parallel to the starting position. This type of drafting machine is easy to use on large boards or on boards placed vertically or at a steep angle.

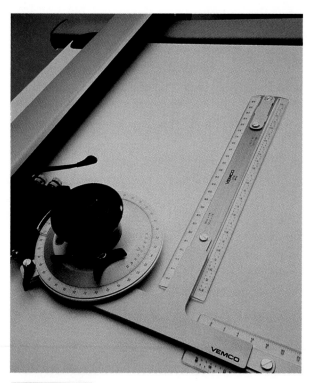

Figure 3-9

The track-type drafting machine is especially adapted for wide drawings. It may also be used for drawings of regular sizes.

Other Basic Tools

In addition to the equipment described thus far in this chapter, drafters use a variety of other tools, instruments, and equipment. Triangles, protractors, erasers and erasing shields, and parallel-ruling straightedges, among others, form part of a board drafter's everyday toolkit.

Triangles

Drafters use two types of **triangles** in combination with the T-square to draw lines at various angles. The 45° triangle has one 90° angle and two 45° angles. The 30°-60° triangle has 30°, 60°, and 90° angles. Techniques for using triangles are presented in Chapter 4.

Protractors

A **protractor** is an instrument that is used to measure or lay out angles. A semicircular protractor is shown in Figure 3-10, where an angle of 43° is measured.

TECH MATH

Measuring Angles

The protractor is designed to measure angles. Angles are measured in a unit called a **degree**. If the **circumference**, or rim, of a circle is divided into 360 parts, each part is an angle of 1°. One fourth of a circle is $^{360}\!/_4 = 90°$, as shown in part A of the illustration on the right. As you may know, a part of a circle is called an **arc**. Two lines drawn to the center of a circle from the ends of a 90° arc on that circle form a **right angle**. An **acute angle**, shown in part B, is less than 90°. An **obtuse angle**, shown in part C, is greater than 90°.

The size of an angle does not depend on the length of its sides. The number of degrees in a given angle is the same no matter what the size of the circle or arc that defines it.

Degrees can be further divided into minutes and seconds. One degree equals 60 minutes (60′). One minute equals 60 seconds (60″). While it is not practical to attempt to measure in

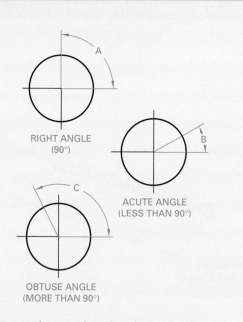

RIGHT ANGLE
(90°)

ACUTE ANGLE
(LESS THAN 90°)

OBTUSE ANGLE
(MORE THAN 90°)

minutes and seconds using the protractor, angular sizes in degrees, minutes, and seconds can be estimated on the drawing and specified accurately in a dimension.

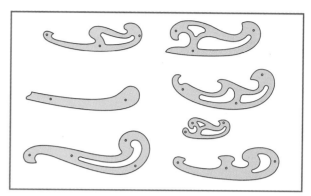

Figure 3-10

A protractor is used to measure or lay out angles.

Figure 3-11

Some examples of irregular, or French, curves.

Irregular Curves

Irregular curves, also called *French curves*, are used to draw noncircular curves such as involutes, spirals, and ellipses. Irregular curves are also used to draw curves on graphs and charts. In addition, they can be used to plot motions and forces and to make some engineering and scientific graphs.

Irregular curves are made of sheet plastic. They come in many different forms, some of which are shown in Figure 3-11. Sets are made for ellipses, parabolas, hyperbolas, and many other special purposes. Many drafters also use flexible curves like those shown in Figure 3-12. Flexible curves can be adjusted to complex curves that may be difficult to draw using other types of irregular curves.

Templates

Templates are an important part of the equipment of engineers and professional drafters. They save a great deal of time in drawing shapes of details. These include bolt heads, nuts, and electrical, architectural, and plumbing symbols. Figure 3-13 shows several examples of templates.

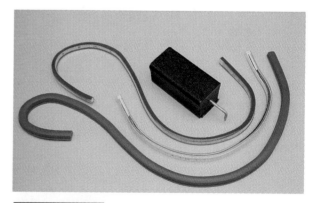

Figure 3-12

Flexible curves for plotting smooth curves. Some drafters use "ducks" like the one shown here (the rectangular object) to position flexible curves accurately.

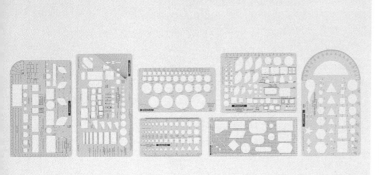

Figure 3-13

Templates are made for many different uses and save a good deal of time.

Tools of the Trade

Drafting instruments and equipment, as well as computer-aided drafting systems, constitute the "tools of the trade" in the engineering, design, and drafting office. Being able to identify them is simply the first step in the process of learning to use them.

Table 3-1. Standard Drawing-Sheet Sizes

ANSI (Inches)		ISO (Millimeters)	
Drawing Size	Overall Paper Size	Drawing Size	Overall Paper Size
A	8.50" × 11.00"	A0	841 × 1189 mm
B	11.00" × 17.00"	A1	594 × 841 mm
C	17.00" × 22.00"	A2	420 × 594 mm
D	22.00" × 34.00"	A3	297 × 420 mm
E	34.00" × 44.00"	A4	210 × 297 mm

Drafting Media

Drawings are made on many different materials. Collectively, these materials are known as *drafting media*, or drawing sheets. Papers may be white, cream, or pale green. They are made in many thicknesses and qualities.

Types of Drafting Media

Most drawings are made in pencil or ink on tracing paper, vellum, or polyester film. When such materials are used, copies can be made by printing or other reproduction methods. Reproduction methods are discussed in Chapter 23.

Vellum is tracing paper that has been treated to make it more transparent. **Polyester drafting films** are widely used in industrial drafting rooms. They are very transparent, strong, and lasting. Drawing films are made with a matte (dull and rough) surface. They are suitable for both pencil and ink work.

Sizes of Drafting Media

Trimmed sizes of drafting media follow standards set by two organizations: the American National Standards Institute (ANSI) and the International Standards Organization (ISO). Table 3-1 lists the specifications by both organizations.

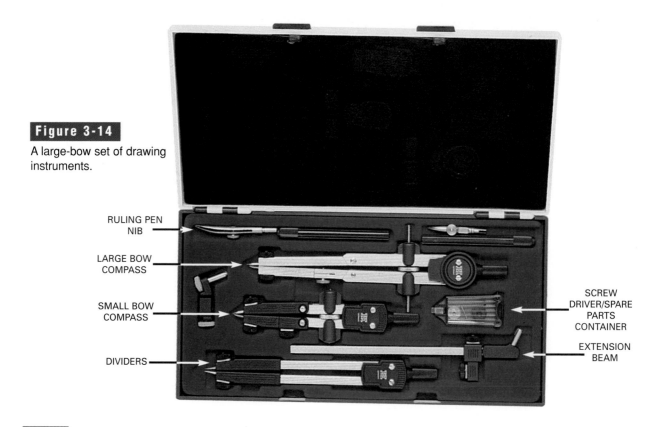

Figure 3-14

A large-bow set of drawing instruments.

RULING PEN NIB

LARGE BOW COMPASS

SMALL BOW COMPASS

DIVIDERS

SCREW DRIVER/SPARE PARTS CONTAINER

EXTENSION BEAM

ANSI standard drawing sheet size A is based on the dimensions of commercial 8.5″ × 11″ letterhead paper. Standard-size sheets can be cut from rolls of film or paper 36″ or 42″ wide with a minimum of waste.

The ISO standard is developed downward in size from a base sheet with an area of about 1 square meter (1 m²). Sheet sizes are based on a length-to-width ratio of one to the square root of two (1:√2). Each smaller size has an area equal to half the preceding size. Multiples of these sizes are used for larger sheets also.

Drawing Instruments

Drawing instruments are necessary to produce accurate technical drawings. These instruments are available as individual items or as a set. Drawing instruments in a set are often called **case instruments**. The basic drawing instruments are shown in Figure 3-14.

A full set of instruments usually includes compasses with pen part, pencil part, lengthening bar, dividers, bow pen, bow pencil, bow dividers, and one or two ruling pens (optional). Large-bow sets like the one shown in Figure 3-14 are favored by most drafters. They are known as master, or giant, bows and are made in several patterns. With large bows, 6″ (152 mm) or longer, circles can be drawn up to 13″ (330 mm) in diameter or, with lengthening bars, up to 40″ (1016 mm) in diameter. Large-bow sets allow you to use a single instrument in place of the regular compasses, dividers, and small-bow instruments. Large-bow instruments let you hold the radius securely at any desired distance, up to their largest possible radius.

The Dividers

Lines can be divided and distances transferred from one place to another with **dividers**. Figure 3-15 shows two types of dividers. Bow dividers are generally more rigid and are therefore more accurate. Once they are set, they remain at that setting until changed. Friction-joint dividers are more quickly adjusted, but are more easily bumped out of adjustment. You can use the dividers to divide a line, arc, or circle into equal parts.

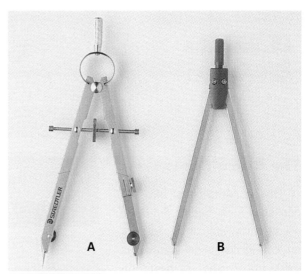

Figure 3-15

Types of dividers: (A) bow dividers; (B) friction-joint dividers.

The Compass

Regular curves are curves that are true circles or arcs. That is, all of their points are exactly the same distance from a center point. These curves can be drawn with a **compass** like the ones shown in Figure 3-16. Bow compasses, like bow dividers, are more rigid than their friction-joint counterparts. Once they are set, they remain at that setting until changed. Friction-joint compasses are easily bumped out of alignment.

Figure 3-16

Compasses: (A) bow compass; (B) friction-joint compass.

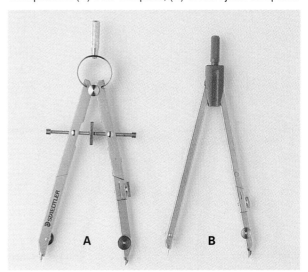

Figure 3-17

Beam compasses are used for large radii.

The Beam Compass

Beam compasses like the one shown in Figure 3-17 are used to draw arcs or circles with large radii. The beam compass is made up of a bar, or beam, on which movable holders for a pencil part and a needle part can be put and fixed as far apart as desired. A pen part may be used in place of the pencil part. By putting a needle point in both holders, you can use a beam compass as a dividers.

The usual bar is about 13″ (330 mm) long. However, by using a coupling to add extra length, you can draw circles of almost any size.

Bow Instruments

The bow instruments include the bow pencil, the bow dividers, the bow pen, and the drop-spring bow compass, as shown in Figure 3-18. They can be adjusted with either a center wheel or a side wheel. They may be of the hook-spring type, as shown in Figure 3-18A, or the fork-spring type, as shown in Figure 3-18B and C. They are usually about 4.00″ (102 mm) long.

The bow instruments are easy to use and are accurate for distances or radii less than 1.25″ or 32 mm. They hold small distances better than the large instruments. Use them for the following purposes:

- Use the bow pencil shown in Figure 3-18A to draw small circles.
- Use the bow dividers shown in Figure 3-18B for transferring small distances or for marking off a series of small distances. You may also use them to divide a line into small spaces.
- Use the bow pen shown in Figure 3-18C to draw small circles. The bow pen is included in a set of bow instruments to offer drafters a choice of marking points in pen or pencil.
- Use the drop-spring bow compass shown in Figure 3-18D to draw very small circles. It is especially useful for drawing many small circles of the same size, such as when drawing rivets.

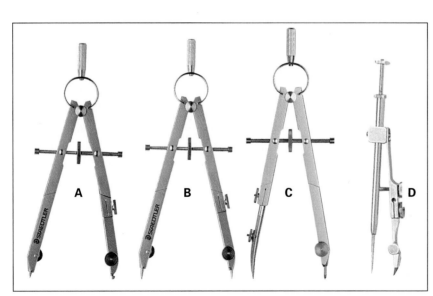

A **B** **C** **D**

Figure 3-18

Bow instruments are used for drawing small circles and arcs and for stepping off short distances: (A) bow pencil; (B) bow dividers; (C) bow pen; (D) drop-spring bow compass.

Drafting Pencils and Pens

Several kinds of pencils and pens have been developed specifically for drafting. The type a drafter selects depends on the type of drawing sheet and the characteristics of the line to be drawn (thick and dark, thin and light, etc.).

Drafting Pencils

Both regular wooden pencils and mechanical pencils are used for technical drawing. Four kinds of lead are now used in drawing pencils. One pencil lead is made from graphite, a form of the element carbon. It also contains clay and resins. Graphite pencils have been used for more than 200 years, and they are still the most important kind.

Grades of Graphite Pencils

Graphite drafting pencils are usually made in 17 degrees of hardness, or grades, as shown in Figure 3-19. The grade of pencil you use depends on the kind of surface on which you are drawing. It also depends on how opaque, or dark, and how thick you want the finished line to be. To lay out views on fairly hard-surfaced drawing paper, use grades 4H and 6H. When you use tracing paper and draw finished views that are to be reproduced by machine, use an H or 2H pencil. Grades HB, F, H, and 2H are sometimes used for sketching and lettering and for drawing arrowheads, symbols, border lines, and so on. Grade selection should be tailored to the drawing and the surface. Very hard and very soft leads are seldom used in ordinary drafting.

Pencils for Film

When film came into use for drawings, new kinds of pencil lead were developed. Three types are described here, based on information furnished by the Joseph Dixon Crucible Company. The first is a plastic pencil. This type is a black crayon. Its lead is extruded, or squeezed out, in a "plasticizing" process. Drawings made with this lead reproduce well on microfilm. The second type of lead is a combination of plastic and graphite and is made by heating. This kind stays sharp, draws a good opaque line, does not smear easily, erases well, and microfilms well. It can be used on paper as well as on film. The third type is used mostly on film. It does not remain as sharp as the other types. However, it draws a fairly opaque line, erases well, does not smear easily, and microfilms well.

These three types of lead are made in only five or six grades. Their grades are not the same as those used for graphite leads. The companies that make these pencils use different systems of letters and numbers to tell what kind of lead is in each pencil and how hard it is. The drafter must experiment with various grades to determine which best suits specific needs.

Technical Pens

Drafting sets like the one shown in Figure 3-20 include technical fountain pens and attachments. These sets allow you to use either ink or lead with the instruments. In addition, entire sets of technical pens are available, as shown in Figure 3-21. Technical pens are precise instruments that provide known line widths. Some technical pens have a refillable cartridge for storing ink. Others have a cartridge that is used once and then replaced. The disposable (stainless steel point) technical pen performs the same tasks as the refillable type and eliminates many maintenance procedures.

Figure 3-19

Standard grades of drawing pencils.

6B   F   9H

6B	softest and blackest
5B	extremely soft
4B	extra soft
3B	very soft
2B	soft, plus
B	soft
HB	medium soft
F	intermediate, between soft and hard
H	medium hard
2H	hard
3H	hard, plus
4H	very hard
5H	extra hard
6H	extra hard, plus
7H	extremely hard
8H	extremely hard, plus
9H	hardest

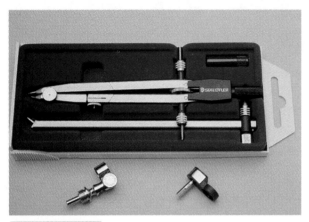

Instruments with technical pen attachments.

Drawing Ink

Ink used for technical drawings is called **drawing ink**. It must be completely opaque in order to produce good, uniform line tone and yet be erasable on all drafting media.

Lettering Guides and Equipment

The lettering set shown in Figure 3-23 has three basic tools: a scriber, lettering templates, and technical pens. Lettering templates come in many styles and lettering sizes, from about .06″ to 2″ (1.5 to 50 mm) high. Other templates are used to draw symbols and other shapes. The width of the pen point used depends on the height of the letters and size of the symbols.

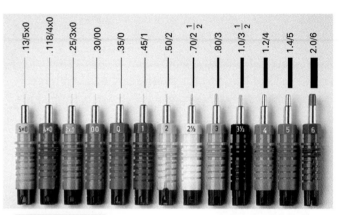

Figure 3-22

The range of lines and point sizes available in technical pens.

Figure 3-21

Technical pen sets include pens with different tips to create various line widths.

Figure 3-23

Three basic parts of a lettering set are the pen, the template, and the scriber.

Technical pens are available with points of various sizes to draw specific line widths. The range of point sizes is shown in Figure 3-22. Points for technical pens are made of different materials for use on different types of media. There are three main types of points:

- Hard-chrome stainless steel for use on paper or vellum
- Tungsten-carbide for long wear on film, vellum, and paper (most commonly used in pen plotters)
- Jewel for long, continuous use on film

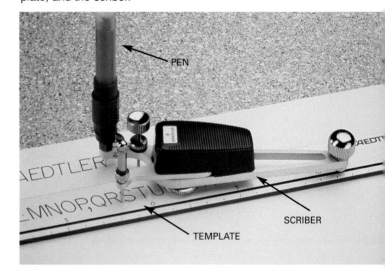

PEN

SCRIBER

TEMPLATE

Erasers and Erasing Shields

Use soft erasers like those shown in Figure 3-24 to clean soiled spots or light pencil marks from drawings. Rubkleen, Ruby, or Emerald erasers are generally good for removing pencil. On film, use a vinyl eraser made especially for erasing on film. Electric erasing machines similar to the one shown in Figure 3-25 can also be used.

Drawing ink is both opaque and waterproof. However, it can easily be removed from polyester drafting film using erasers made especially for that purpose. Plastic erasers, such as the ones shown in Figure 3-26, will either rub away the ink line or absorb it. In addition, drafting ink can be removed from tracing vellum or other drafting media with standard ink erasers or chemically saturated erasers that absorb ink.

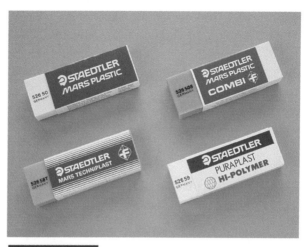

Figure 3-26

Several types of ink erasers are available. Some absorb ink; others remove it from the drawing. The type you choose depends on the medium used for the drawing.

Figure 3-24

Various types of erasers can be used to erase pencil lines on technical drawings.

Figure 3-25

Electric eraser.

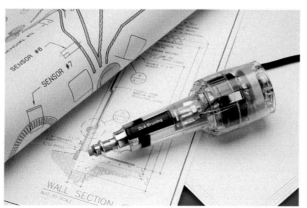

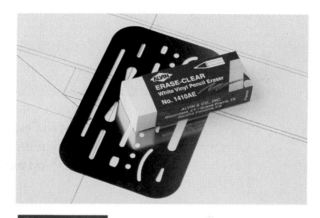

Figure 3-27

An erasing shield protects lines that are not to be erased.

To avoid erasing nearby lines accidentally, most board drafters use an erasing shield. **Erasing shields** are made of metal or plastic and have openings of different sizes and shapes. See Figure 3-27. By positioning the shield so that the part to be erased shows through one of the openings, you can protect lines and areas that you do not want to erase.

Scales

Scales are instruments used to lay off distances and to make measurements. Measurements can be full size or in some exact proportion to full size. When measurements are made in proportion to full size, they are said to be *scaled*.

Scales are made in various shapes, as shown in Figure 3-28. Some scales, like those in Figure 3-29A and B, are open-divided. Only the end units of these scales are subdivided. Other scales are full-divided, as shown in Figure 3-29C and D, with subdivisions over their entire length.

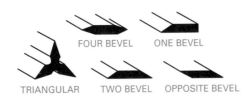

Figure 3-28

Examples of the various scale shapes.

Customary-Inch Scales

Different types of scales are used to make different kinds of drawings. Commonly used customary-inch scales include the architect's scale, mechanical engineer's scale, and civil engineer's scale.

The Architect's Scale

The architect's scale, shown in Figure 3-29A, is divided into proportional feet and inches. The triangular form shown is used in many schools and in some drafting offices because it has many scales on a single stick. However, many drafters prefer flat scales, especially when they do not have to change scales often. The usual proportional scales are listed in Table 3-2.

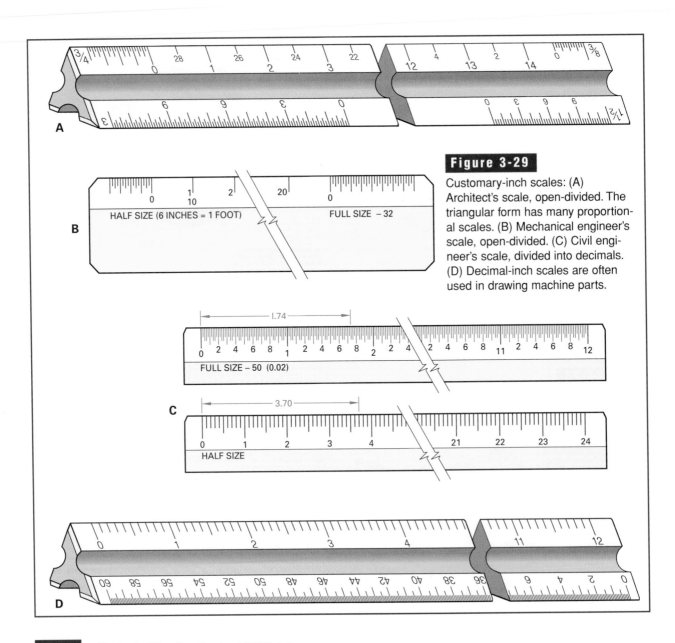

Figure 3-29

Customary-inch scales: (A) Architect's scale, open-divided. The triangular form has many proportional scales. (B) Mechanical engineer's scale, open-divided. (C) Civil engineer's scale, divided into decimals. (D) Decimal-inch scales are often used in drawing machine parts.

Table 3-2. Proportional Scales		
Proportion	**Gradations**	**Ratio**
Full size	12′ = 1′-0″	1:1
1/4 size	3′ = 1′-0″	1:4
1/8 size	1 1/2″ = 1′-0″	1:4
1/12 size	1″ = 1′-0″	1:4
1/16 size	3/4″ = 1′-0″	1:4
1/24 size	1/2″ = 1′-0″	1:4
1/32 size	3/8″ = 1′-0″	1:4
1/48 size	1/4″ = 1′-0″	1:4
1/64 size	3/16″ = 1′-0″	1:4
1/96 size	1/8″ = 1′-0″	1:4
1/128 size	3/32″ = 1′-0″	1:4

The symbol ′ is used for feet and ″ for inches. Thus, three feet four and one-half inches is written 3′-4½″. When all dimensions are in inches, the symbol is usually left out. Also, on architectural and structural drawings, the inch symbol is not shown, and the dimension is given as 3′-4½. See Chapters 18 and 19 for more information about architectural and structural dimensions.

Proportional scales are used in drawing buildings and in making mechanical, electrical, and other engineering drawings. They are also used in drafting in general. The proportional scale to which the views are drawn should be given on the drawing. This is done in the title block if only one scale is used. The title block is the place on a technical drawing where the name of the company, the drafter, the date, and other information about the drawing is located. You will learn more about title blocks in Chapter 4. If different parts of a drawing are in different scales, the scales are given near the views in this way:

- SCALE: 6″ = 1′- 0
- SCALE: 3″ = 1′- 0
- SCALE: 1½″ = 1′- 0

The Mechanical Engineer's Scale

The mechanical engineer's scale, shown in Figure 3-29B, has inches and fractions of an inch divided to represent inches. The usual divisions are:

- Full size—1″ divided into 32nds
- Half size—½″ divided into 16ths
- Quarter size—¼″ divided into 8ths
- Eighth size—⅛″ divided into 4ths

These scales are used for drawing parts of machines or where larger reductions in scale are not needed. The proportional scale to which the views are drawn should be given in the title block of the drawing.

The Civil Engineer's Scale

The civil engineer's scale, shown in Figure 3-29C, has inches divided into decimals. The usual divisions are:

- 10 parts to the inch
- 20 parts to the inch
- 30 parts to the inch
- 40 parts to the inch
- 50 parts to the inch
- 60 parts to the inch

With the civil engineer's scale, 1 inch may stand for feet, rods, miles, and so forth. It may also stand for quantities, time, or other units. The divisions may be single units or multiples of 10, 100, and so on. Thus, the 20-parts-to-an-inch scale may stand for 20, 200, or 2,000 units.

The civil engineer's scale is used for maps and drawings of roads and other public projects. It is also used where decimal-inch divisions are needed. These uses include plotting data and drawing graphic charts.

The scale used should be given on the drawing or work as follows:

- SCALE: 1″ = 500 POUNDS
- SCALE: 1″ = 100 FEET
- SCALE: 1″ = 500 MILES
- SCALE: 1″ = 200 POUNDS

For some uses, a graphic scale is put on a map, drawing, or chart, as shown in Figure 3-30.

Figure 3-30

Civil engineers place graphic scales on maps to show people how to interpret them.

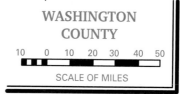

WASHINGTON COUNTY

10 0 10 20 30 40 50

SCALE OF MILES

The Decimal-Inch Scale

Like the civil engineer's scale, the decimal-inch scale is divided into tenths of an inch. Refer to Figure 3-29D. Because many manufacturers now use decimals rather than fractions (4.25″ rather than 4¼″), the decimal-inch scale is used for many machine drawings.

The Metric Scale

Metric scales are divided into millimeters, as shown in Figure 3-31. The usual proportional scales in the metric system are listed as a ratio in Table 3-3. The numbers shown indicate the difference in size between the drawing and the actual part. For exam-

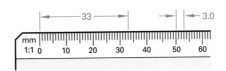

Metric scales are divided into millimeters.

ple, the ratio 10:1 shown on the drawing means that the drawing is 10 times the actual size of the part. A ratio of 1:5 on the drawing means the object is 5 times as large as it is shown on the drawing.

The proportional scale to which the views are drawn is given on the drawing. This is done in the title block if only one scale is used. If different parts on the same drawing are in different scales, the scales are given near the views in this way:

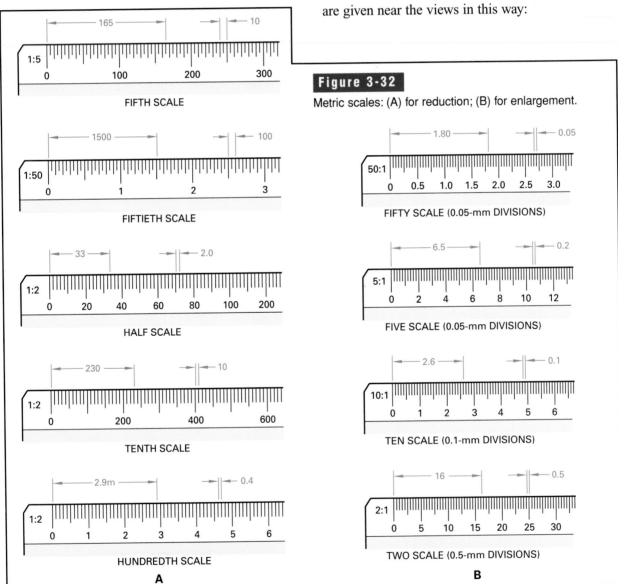

Figure 3-32

Metric scales: (A) for reduction; (B) for enlargement.

Table 3-3. Metric Proportional Scales		
Enlarged	**Same Size**	**Reduced**
1000:1	1:1	1:2
500:1		1:5
200:1		1:10
100:1		1:20
50:1		1:50
20:1		1:100
10:1		1:200
5:1		1:500
2:1		1:1000

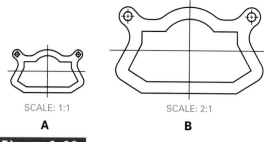

SCALE: 1:1   SCALE: 2:1

A   B

Figure 3-33

A part drawn at two different scales.

- SCALE: 1:2
- SCALE: 1:5
- SCALE: 1:10
- SCALE: 2:1
- SCALE: 20:1

Notice that scales are generally given in multiples of 2 or 5.

To reduce the drawing size of an object, use one of the scales shown in Figure 3-32A. To enlarge the drawing size of an object, use one of the scales shown in Figure 3-32B. An example of an enlarged-scale drawing is shown in Figure 3-33. In Figure 3-33A, the drawing is shown at a scale of 1:1 (same size). In Figure 3-33B, the same part is enlarged to 2:1 (double size).

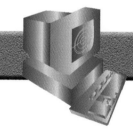

CAD Equipment

In schools and engineering offices around the world, CAD systems are replacing board drafting equipment. Some companies use a mixture of board drafting and CAD equipment. However, other companies have made this transition so completely that they continue to use board equipment only to update drawings that were originally done by hand. All new drawings are created using CAD.

CAD workstations generally require high-end equipment. To purchase and use this equipment effectively, you must understand the components and how they are used in the CAD system.

Hardware

Hardware refers to the computer and all of its accessory equipment, or peripherals. Typical hardware for a CAD workstation includes the CPU, one or more monitors, input devices, and an output device such as a plotter or printer.

The CPU

The central processing unit, or **CPU**, is the "heart" of any computer. This is where the actual computations are done. For CAD, the processing speed of the CPU must be high enough to perform complex mathematical computations quickly. CAD software calculates the location of points, lines, etc., in relation to other objects and points in the drawing. A fast CPU helps the CAD operator avoid lengthy waits while the computer calculates where lines should be displayed on the monitor screen.

Before you buy a computer for a CAD workstation, it is important to be sure that the computer you are considering will run the CAD software you will be using. Different types of CPUs have different operating systems. If you will be using AutoCAD, for example, you should not buy a Macintosh computer, because AutoCAD does not run on the Macintosh operating system.

Monitors

The computer monitor is the display device for a CAD workstation. Some advanced workstations even have two monitors. The first one displays the CAD drawing. The second one displays a three-dimensional, rendered view of the object described by the CAD drawing. See Figure 3-34. A rendered view is one that has been colored and shaded to resemble the "real" finished product.

Figure 3-34

Some high-end CAD workstations have two monitors to help designers visualize the products they are drawing.

Choose the monitor for a CAD workstation very carefully. You should take the following factors into consideration:

- Size
- Resolution
- Refresh rate

▶ **CADTIP**

Monitors and Personal Safety
The size, resolution, and refresh rate of a monitor all affect the image that is displayed. By paying attention to these factors when selecting a monitor, you can reduce or avoid eye fatigue, headaches, and other health problems often associated with computer work.

Size

A CAD monitor must be large enough to display drawings or parts of drawings effectively. Monitor size is measured along the diagonal. For example, a 15″ monitor is one that measures 15″ from the top right to bottom left corner of the screen. The stated size is usually not the actual viewing area. In most cases, the actual viewing area is slightly smaller. Most monitors for CAD workstations are at least 17″. Many CAD operators prefer 21″ monitors.

Resolution

The **resolution** of a monitor is the number of *pixels*, or picture elements, per inch. On a color monitor, a pixel is a set of one red, one blue, and one green dot. The more pixels a monitor displays per inch, the higher the resulting image on the screen. Early monitors had a maximum resolution of 640 × 480. This resolution is good for a 14″ monitor, but it results in jagged edges on a larger monitor.

Most monitors today can be set to more than one resolution. Options range from 640 × 480 to 1600 × 1280 or better, depending on the monitor. Choose a resolution that is appropriate for the size of the monitor. Keep in mind that as resolution increases, total picture size decreases. While you can to buy a 17″ monitor that can display a 1600 × 1280 resolution, this combination is usually not practical because the text becomes too small to read. Table 3-4 shows appropriate resolutions for various monitor sizes.

Table 3-4. Monitor Resolutions	
Monitor Size	Resolution
14″	640 × 480
15″	800 × 600
17″	1024 × 768
20″	1024 × 768 or 1600 × 1280
21″	1024 × 768 or 1600 × 1280
35″	at least 1600 × 1280

Refresh Rates

A monitor's **refresh rate** is the rate at which the image on the screen is redrawn. Early monitors usually had a standard refresh rate of 60 Hz, or 60 times per second. This was sufficient for monitors with a resolution of 640 × 480. However, 60 Hz produces an annoying flickering at higher resolutions, making the images hard to read.

Today's monitors have a range of refresh rates. The same monitor may have a different refresh rate depending on the resolution you select. In general, a monitor for a CAD workstation should have a refresh rate of 70 Hz or higher.

Input Devices

With a few exceptions, input devices for CAD workstations are similar to those used by other types of computers. All systems include a keyboard and mouse. Some also include a digitizer.

Mouse and Keyboard

The CAD software used in schools and many drafting rooms depends heavily on the mouse. Thus, it is important to select a mouse that fits comfortably in your hand (to help avoid cramping) and that can perform the functions required by the software. A standard two-button mouse is acceptable for most software. However, some CAD programs, such as Pro/Engineer, require a three-button mouse.

Although CAD software is evolving to make more use of the mouse, the keyboard is still used extensively. Keyboards are made in many styles and with different "touches" (levels of resistance when you press a key). If possible, you should try several different keyboards before choosing to find one that feels comfortable.

Digitizer

Not every CAD workstation includes a digitizer. Digitizers are specialized equipment used mostly to convert paper drawings to CAD. See Figure 3-35. Many companies keep at least one digitizer so that drawings can be converted on an as-needed basis. These companies may leave older drawings in their paper form until revisions are needed. At that time, the drawings are digitized, the revisions are made, and the drawings are stored in their new digital format.

Printers and Plotters

When CAD systems were first introduced, most CAD operators produced paper drawings using pen plotters. The pen plotter held special technical pens of several different thicknesses and colors. The CAD operator set up the electronic drawing file to plot each line with the appropriate thickness and color.

Figure 3-35

Digitizers allow drafters to convert points on a paper drawing to electronic points in a CAD file.

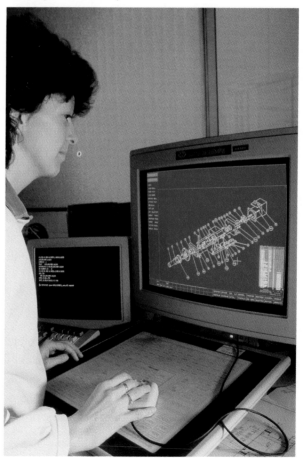

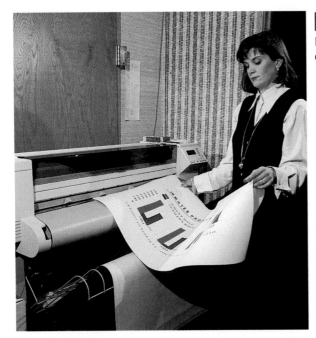

Figure 3-36

Large-format inkjet printers like this one can print E-size drawings.

- Place wires and cords out of the way so that people will not trip over them, causing possible injury to both themselves and the equipment.
- Avoid overloading a circuit by connecting too many electric devices to it.
- Never connect a multiple-outlet extension cord into another multiple-outlet extension cord.
- For equipment such as CPUs and printers that have internal cooling fans, make sure that the vents are not blocked to avoid overheating the equipment.

Hazardous Waste

Some of the equipment included in typical CAD workstations and drafting rooms either use or produce hazardous waste. For example, the toner used by laser printers and copiers may be considered hazardous. Read the instructions that come with toners, inks, and even ordinary batteries to find out how to use and dispose of them properly.

Software

In addition to hardware, CAD systems require **CAD software**—the computer programs used to create technical drawings. The software determines the actual drafting ability of a CAD system, so you should select it carefully. In industry, the software a company uses depends on the products or services the company provides.

While pen plotters are still being used by some companies, inkjet technology has become the standard for printing technical drawings. Inkjet printers are available in a variety of sizes to fit most drafting needs. See Figure 3-36. Prices vary with the size of the printer and the quality of the images it produces.

Before selecting a printer or plotter, consider how you will use it. Will you be producing mostly A- or B-size drawings, or will you need to print E-size drawings, too?

Also consider the level of quality you need. Very accurate, high-resolution printers are available for the most exacting needs, but they are very expensive. Lower-priced printers are suitable for printing proofs and for most school uses.

The terms *printer* and *plotter* are often used interchangeably, although the correct term depends on the actual equipment you use to output drawings. For convenience, the word *printer* will be used in this textbook to refer to both printers and plotters.

Electrical Safety

Most computer CPUs, monitors, external storage devices, digitizers, and printers must be connected to electricity to function. When designing and setting up a CAD workstation, be especially conscious of the hazards involved. Safety should be a primary goal when you design the workstation. To avoid the risk of injury or electrical fire:

 Success on the Job

People vs. Computers
Although selecting the right hardware and software can increase the ability of the CAD workstation, it is not the only requirement for CAD. Remember that the CAD workstation is only a drafting instrument. The skill and technique of the drafter make the difference between a good drawing and a poor one.

Figure 3-37

Furniture for a CAD workstation should provide ample room for the computer and plenty of working space.

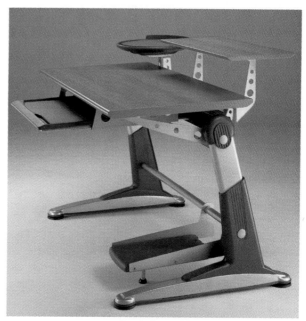

Photo courtesy of www.JustMyOfficeFurniture.com

In general, there are three types of CAD software:

- general-purpose drafting software
- specialty software for architectural, electrical, and other special drafting needs
- third-party software that works with general-purpose drafting software to extend its functionality

AutoCAD, DataCAD, and CADKEY are examples of common general-purpose programs. These programs provide the tools needed to create drawings for many different fields and applications.

Specialized software is often used in fields such as architecture, geographic information systems (GIS), and electronics. Examples include:

- Architectural Desktop, for architectural drafting
- Pro/Engineer, for mechanical drafting
- CATIA, for aeronautical and electronic engineering drafting

Some companies even develop their own software. They use computer languages such as C++ to write CAD programs that will meet their needs.

Third-party software works with a general-use program to provide special features. Symbol libraries for welding, plant processing, electronics, and many other fields are the most common examples of third-party packages.

CAD utility software is another type of third-party software. It is not actually CAD software. Instead, it performs special functions that enable CAD files to be used for specific purposes. For example, utilities are available to:

- prepare CAD files for use with computer-aided manufacturing (CAM) systems
- convert CAD drawings to formats needed by other companies
- convert CAD drawings supplied by other companies to the format needed by your CAD system

CAD Furniture

Standard drafting desks and tables are not required for CAD workstations. Typical computer furniture like the computer desk shown in Figure 3-37 can usually be used. However, you should take your drafting needs into consideration before you select the furniture for your workstation. Ask yourself the following questions:

- **Will the furniture accommodate my monitor?** Make sure the part of the computer desk designed to hold the monitor is large enough and sturdy enough to hold the monitor you will use.
- **Will I be using a digitizer?** Workstations that include a digitizer may require additional table space.
- **Where will I put the printer?** The printer drawers supplied with generic computer desks are not large enough to hold printers used for large drawings. If you plan to use a printer that can print E-size drawings, the printer must stand on the floor. Measure to make sure that the cables connecting the printer to your CPU are long enough to reach through the computer desk to your printer without straining.
- **How much working surface do I need?** Be sure the computer desk provides enough desktop working space for notes, hand-drawn sketches, and other materials you may need.

Ergonomics and Safety

When selecting hardware for a CAD workstation, you should pay close attention to ergonomics. **Ergonomics** is a field of study that investigates ways to design products to promote personal safety and comfort while the products are being used. Ergonomic products "fit the person" instead of requiring that the person adapt to fit the task.

It is important to consider ergonomics when choosing the hardware and furniture for a CAD workstation. Studies have shown that improper use of computer equipment can cause temporary or permanent physical injuries. Table 3-5 lists the causes of computer-related injuries and suggests methods for prevention. Injuries include:

- Musculoskeletal disorders (MSDs)
- Carpal Tunnel Syndrome
- Repetitive stress injuries (RSIs)
- Tennis elbow

Choosing ergonomic input devices makes it easier to avoid stress-related injuries. Many different types of ergonomic keyboards and mice have been developed. Figure 3-38 shows examples of ergonomic keyboards. Some of them can be taken apart so that you can position the pieces where they feel most comfortable for you. Others are designed to keep your hands and forearms straight. Some keyboards even have a built-in rest-period indicator to remind you to take rest breaks periodically.

Figure 3-38

Many types of ergonomic keyboards have been developed to meet different computing needs.

Table 3-5. Computer-Related Injuries		
Type of Injury	**Typical Causes**	**Prevention**
Cumulative injury from stress on the body over a period of time; most commonly affects back and wrists of computer users	• Repetitive motions, such as those involved in using a keyboard or mouse • Sitting or standing in the same position for long periods of time	• Choose hardware and computer furniture that helps reduce stress on the body. * Schedule frequent breaks; during breaks, move away from the workstation and stretch your muscles.
One-time back injury or muscle strain	Improper lifting techniques when moving heavy computer equipment	Keep your back straight while lifting heavy equipment; use your leg muscles to support the weight.
Eye fatigue, headache, and dry eyes	• Staring at a computer monitor for long periods of time • Failure to blink at appropriate intervals. (Studies show that computer users blink less often while they work on a computer.)	• Look away from the monitor frequently to change your focal depth. • Schedule frequent breaks. • Perform the following exercise: Blink rapidly 40 times. Repeat at least once a day.

You should also keep ergonomic factors in mind when you choose CAD furniture. Select a computer desk that has an adjustable keyboard shelf or drawer. Use an office chair with adjustable height. Some chairs have lumbar (lower back) supports. If possible, sit in the chair at the computer desk to see how well they work together before you buy either one.

Figure 3-39 shows the proper positioning of equipment for a CAD workstation. To minimize the risk of computer-related injuries:

■ Place the monitor at or slightly above eye level to help reduce eye fatigue.
■ Adjust the keyboard height so that your forearms (elbow to wrist) are parallel with the floor.
■ If you have an adjustable keyboard, adjust it so that your arms extend straight out when you place your fingers on the keys.

■ Adjust the chair height so that your feet rest comfortably on the floor without cutting off the circulation in your legs.

Another safety practice is to know when your equipment needs maintenance or repair. For example, a decrease in computer performance may indicate the need for hardware or software maintenance. Frayed cords should be replaced immediately. Also, you should read the manuals that come with your equipment carefully, and perform any routine maintenance suggested by the manufacturer. Keeping your equipment in top operating condition can help ensure your comfort and safety.

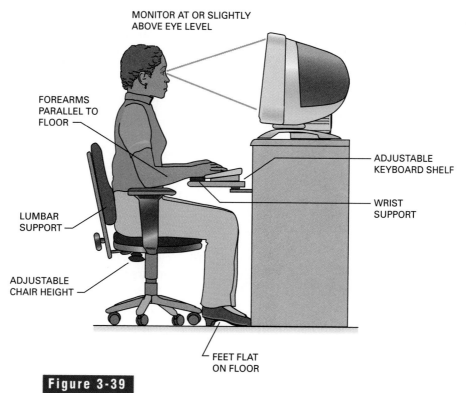

Figure 3-39

Arrange your equipment to minimize stress and potential injury.

Chapter Summary

■ Basic board drafting equipment includes drawing tables or desks, drawing boards, and various instruments that improve drawing accuracy, such as T-squares and triangles.

■ Standard drawing sheet sizes are specified by the American National Standards Institute (ANSI) and the International Standards Organization (ISO).

■ Technical drawings can be made on vellum or polyester film (board drafting) or using computer software.

■ Drawing instruments include the dividers, compass, beam compass, and bow instruments.

■ Board drafters use various drafting pencils and technical pens to create accurate technical drawings.

■ Scales are used to lay off distances and to make measurements.

■ The hardware for a computer workstation includes a CPU, at least one monitor, input devices, and a printer.

■ CAD software is a computer program that allows drafters to create technical drawings using a computer.

■ Before purchasing CAD furniture, you should consider how well the furniture meets your individual drafting needs.

■ Drafters should take ergonomic and personal health concerns into account when considering CAD hardware and furniture.

Review Questions

1. Which instrument is used in measuring, or laying out, angles?

2. Name two types of drafting machines.

3. Explain how you can check the accuracy of a T-square.

4. Name the two most common types of drafting triangles.

5. What is another name for the irregular curve?

6. Name three kinds of drafting media.

7. What term is used to describe the dull, rough surface of polyester drafting film?

8. On what size letterhead is the ANSI standard A-size drawing sheet based?

9. What instruments are commonly included in a drafting instrument set?

10. Describe the range of graphite drafting pencils. Which pencil lead is softest? Which is hardest?

11. List three kinds of points for technical pens.

12. What is an erasing shield, and why is it used?

13. If you create a drawing that shows a part at half its actual size, at what scale have you drawn the part?

14. What type of scale would you use to create a map drawing?

15. Explain why CAD systems require high-end equipment.

16. How does the hardware for a CAD workstation compare with the hardware for computers used for other purposes?

17. What is the primary purpose of a digitizer?

18. Name the three general types of CAD software.

19. Describe the characteristics of good furniture for a CAD workstation.

20. What does the term *ergonomics* mean, and how does it apply to CAD workstations?

Drafting Problems

In Chapter 2 you learned about freehand sketching and lettering. The following problems are designed to give you additional practice in both of these techniques. Additional practice should improve the quality of your work and reduce the time you need to prepare sketches. Each of these sketching problems is designed to fit on an 8.5″ × 11″ drawing sheet. Your instructor may want you to use quarter-inch or half-inch grid paper. Estimate all sizes.

1. Make a freehand sketch of the tic-tac-toe board shown in Figure 3-40. Do not add dimensions (sizes).

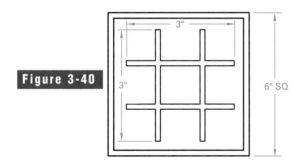

Figure 3-40

2. Make a freehand sketch of the inlay shown in Figure 3-41. Do not add dimensions (sizes).

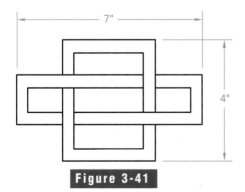

Figure 3-41

3. Make a freehand sketch of the puzzle shown in Figure 3-42. Do not add dimensions (sizes).

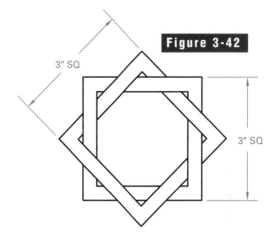

Figure 3-42

4. Make a freehand sketch of the identification plate shown in Figure 3-43. Do not add dimensions. Carefully letter in your own name.

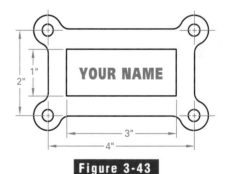

YOUR NAME

Figure 3-43

5. Make a freehand sketch of the bicycle chain link shown in Figure 3-44. If available, use grid paper (quarter-inch or half-inch). The grid lines will help you sketch better circles and arcs.

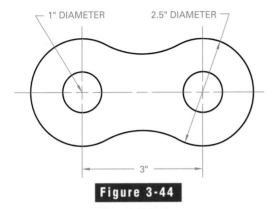

Figure 3-44

Design Problems

1. Design a nameplate for your desk. The base is to be made of walnut wood. Your name is to be engraved on a 1″ × 6″ brass plate attached to the wooden base. Make a freehand sketch of the parts. Estimate all sizes.

2. TEAMWORK Design a sign for your drafting-room door. It should not be larger than 8″ × 22″. Material optional. The lettering, DRAFTING ROOM, can be painted on, or metal letters can be purchased.

3. TEAMWORK Design an ergonomic CAD workstation. Keep in mind all of the ergonomic principles discussed in this chapter. Be ready to present your design to the class for discussion and approval.

Basic Drafting Techniques

KEY TERMS

alphabet of lines

centerlines

coordinate pair

drawing template

hidden lines

inking

line weight

model space

paper space

polar coordinates

sheet layout

viewport

OBJECTIVES

Upon completion of this chapter, you should be able to:

■ Employ safe practices in the drafting room.

■ Prepare a drawing sheet for a technical drawing.

■ Use basic drafting tools and equipment properly and efficiently to produce technical drawings.

■ Identify and use the lines and line symbols recommended by the American National Standards Institute (ANSI).

■ Produce a finished technical drawing using board drafting techniques.

■ Create and set up a drawing file on a CAD system using ANSI or ISO standard layouts.

■ Use CAD commands efficiently to create basic geometry.

■ Produce a technical drawing using CAD.

Since very early in recorded history, technical drawings have been used to communicate and record ideas. Early drawings might simply have been crude markings on stone, done with charcoal sticks. As civilization progressed, parchment, an early form of writing material, provided a medium that was more portable and easier to use. Later, paper as we know it today was developed, along with technical pencils and pens that further improved the communication process.

Improvements in Tools and Techniques

The Greeks and Romans produced and used detailed drawings to guide them in building construction, as shown in the painting in Figure 4-1. They also used technical drawings to guide them in the construction of common items such as carts and wagons.

Since then, technical drawing has become more advanced and more precise. However, the objective has always been to produce technical drawings that are accurate enough to guide workers in the production of a building or product.

In schools and engineering offices around the world, CAD systems are replacing board drafting equipment. In fact, you will now find the latest CAD equipment in all contemporary design studios and corporate design offices. However, it must be clearly understood that the CAD system simply replaces board drafting tools for producing finished technical drawings. The end result of both methods is the same—a detailed drawing used to communicate technical ideas. Whether you use a CAD system or board drafting instruments, you need to know how to use the equipment skillfully, accurately, and quickly.

Drafting Room Safety

Concern for safety and safe work habits is as important in the drafting room as it is in any other type of laboratory or shop situation. Here are some safety tips to remember:

- Keep the area around your workstation clear to avoid tripping.
- Be especially careful when using pencils and other instruments with sharp points.
- Treat paper cuts and puncture wounds from sharp instruments immediately.
- Work only in a well lighted area.
- Adjust the chair at your drafting or CAD station to be comfortable and at the correct height. Take care to avoid physical stress on wrists, arms, neck, and back.

Figure 4-1

The early Romans used detailed drawings, or plans, to guide construction of new buildings.

Board Drafting Techniques

Board drafting involves the use of drafting tools and instruments to produce a technical drawing on carefully chosen drafting media. Developing skill in board drafting techniques involves the following:

- Selecting the size and type of drafting media
- Preparing the sheet with borders and title blocks
- Using tools and instruments proficiently
- Developing skill in the use of standard lines and symbols used on technical drawings throughout the world

Preparing the Drawing Sheet

Proper sheet preparation is an important part of the drafting process. Preparing the drawing sheet includes choosing an appropriate size and type of drawing sheet, fastening it to the drawing board, and laying out the borders and title block.

Choosing the Drawing Sheet

The type of drawing sheet you select depends on how the final drawing will be used. For example, drawing paper is appropriate for short-term use, while polyester film is better for long-term use.

The size of the sheet is determined by the size and complexity of the drawing. It is often useful to make a freehand sketch of the views and notes before proceeding to do the final instrument drawing. Except in rare cases, standard drawing-sheet sizes should be used.

Fastening the Drawing Sheet to the Board

By attaching the drawing sheet to the board, you have the freedom to move the T-square and triangles freely over the whole sheet. The sheet may be held in place on the board in several ways. Some drafters put drafting tape across the corners of the sheet and, if needed, at other places. Others use small, precut,

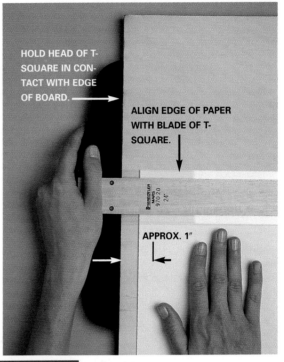

HOLD HEAD OF T-SQUARE IN CONTACT WITH EDGE OF BOARD.

ALIGN EDGE OF PAPER WITH BLADE OF T-SQUARE.

APPROX. 1″

Figure 4-2

To fasten the drawing sheet to the board, first align the sheet with the T-square blade.

circular pieces of tape, called dot tape. Neither of these two methods will damage the corners or the edges of the sheet. They also can be used on composition boards or other boards with hard surfaces. As a result, most drafters prefer to use one of these methods.

To fasten the paper or other drawing sheet, place it on the drawing board with the left edge 1″ (25 mm) or so away from the left edge of the board, as shown in Figure 4-2. (Left-handed students should work from the right edge.) Put the lower edge of the sheet at least 4″ (100 mm) up from the bottom of the board so you can work on it comfortably. Then line up the sheet with the T-square blade, as shown in Figure 4-2. Hold the sheet in position. Move the T-square down, as shown in Figure 4-3, keeping the head of the T-square against the edge of the board. Then fasten each corner of the sheet with drafting tape.

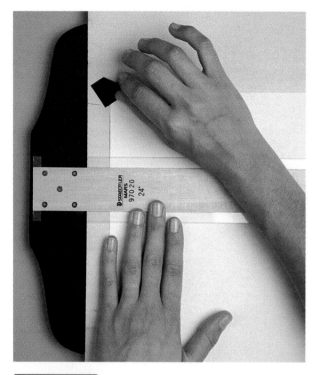

Figure 4-3

Use drafting tape to fasten the drawing sheet to the board.

Sheet Layout

Sheet layout is the process of placing the border and title block on the drawing sheet. Sheet sizes and layout for the U.S. Customary (ANSI) and metric (ISO) drawing sheets are essentially the same. However, there are some differences. For example, U.S. Customary sheets are laid out in decimal inches; metric sheets are laid out in millimeters. Also, the margins for the borders on metric sheets are somewhat uniform in size, while those on U.S. Customary sheets vary.

The sheet sizes and layouts prepared by ANSI and ISO are simply recommendations. They may vary according to the user's requirements. However, all drawing sheets should have a border and title block. Also, it is strongly recommended that the location of various elements of the title block be placed as specified by ANSI or ISO. In many industries, borders and title blocks are printed on the drawing sheets and the drafter simply fills in the blanks in the title block and prepares the drawing within the borderlines.

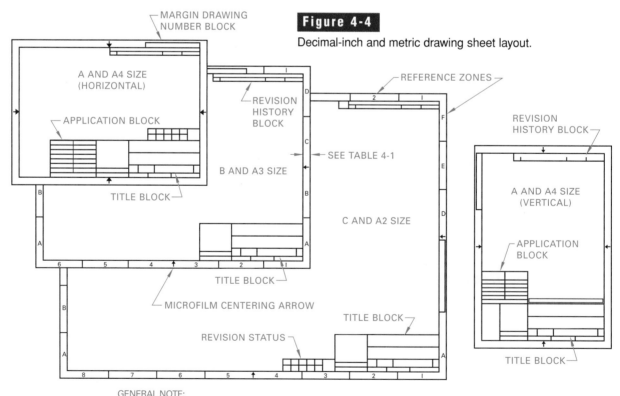

Figure 4-4

Decimal-inch and metric drawing sheet layout.

GENERAL NOTE:
DIMENSIONS SHOWN ARE RECOMMENDED AND MAY BE VARIED
TO ACCOMMODATE THE USER'S REQUIREMENTS.

Table 4-1. Decimal-Inch and Metric Drawing Sheets				
Decimal-Inch Sizes				
Size Designation	Vertical Sheet Size	Horizontal Sheet Size	Margin Sizes	
			Horizontal	Vertical
A (Horizontal)	8.50″	11.00″	.38	.25
A (Vertical)	11.00″	8.50″	.25	.38
B	11.00″	17.00″	.38	.62
C	17.00″	22.00″	.75	.50
Metric Sizes				
A4 (Horizontal)	210 mm	297 mm	10	10
A4 (Vertical)	297 mm	210 mm	10	10
A3	297 mm	420 mm	10	10
A2	420 mm	594 mm	10	10

"who, what, when, and where." The **revision history block** specifies revision dates and related information. The **application blocks** are optional. They provide columns for purposes such as listing specific information used to relate a given drawing to other drawings in a set.

Sheet Layout: U.S. Customary

Table 4-1 gives specific sizes for standard sheets. For example, an A-size sheet placed in the horizontal position is 8.50″ vertically and 11.00″ horizontally (8.50 × 11.00). When placed in the vertical position, it is 11.00″ vertically and 8.50″ horizontally (11.00 × 8.50). B- and C-size sheets are generally not used in the vertical position.

Figure 4-5 shows a recommended layout for the title block, which should be placed in the lower right-hand corner of the drawing. Refer again to Figure 4-4. Since it is only recommended, it can be altered in both size and content. For example, "cage code" is a reference number generally used on drawings prepared for government contracts. It can be eliminated on drawings that are not government-related.

The layout recommendations of both ANSI and ISO are shown in Figure 4-4. ANSI actually lists six standard sizes plus special roll sizes, and ISO lists five standard sizes plus various elongated sizes. The information in Figure 4-4 is limited to the most common sheet sizes used in educational programs.

Reference zones given in the margins are used to locate specific information on the drawing, similar to the way you locate cities and towns on a road map. As you may recall from Chapter 3, the title block provides basic information about the drawing—the

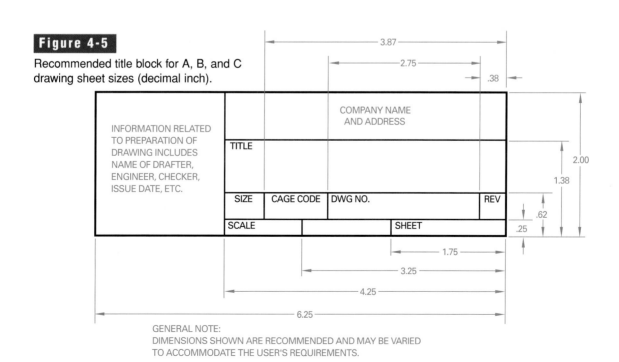

Figure 4-5

Recommended title block for A, B, and C drawing sheet sizes (decimal inch).

GENERAL NOTE:
DIMENSIONS SHOWN ARE RECOMMENDED AND MAY BE VARIED
TO ACCOMMODATE THE USER'S REQUIREMENTS.

Since the trim sizes recommended by ANSI and ISO are in almost universal use in industry, they are also useful sizes to use in drafting courses. Most of the drawing problems throughout this book are planned for A-, B-, A4-, or A3-size sheets. However, to reduce the amount of time and space required to draw the title blocks, you may use the modified version shown in Figure 4-6 for U.S. Customary drawings. Alternate layouts are shown in Figure 4-7. Your instructor may assign one of these or one that he or she has designed.

Sheet Layout: Metric

In Table 4-1, you will see that the smallest metric-size sheet is an A4. Placed in a horizontal position, it is 210 mm vertically and 297 mm horizontally (210 × 297). When placed in a vertical position, it is 297 mm vertically and 210 mm horizontally (297 ×

210). The A4-size sheet may be used in either position. Larger sheets are generally used only in the horizontal position.

Figure 4-8 shows a recommended layout for the metric title block. It is essentially the same as the decimal-inch title block. The difference is in the units of measure used to lay it out. Like the U.S. Customary recommendations, metric recommendations may be altered to accommodate the user's specific requirements.

Figures 4-9 and 4-10 show recommended A4 and A3 drawing-sheet layouts with borders and title blocks. These have been modified from the recommended ISO standard metric sheet layouts. The sheet sizes are standard; the layouts have been modified to save time and space in preparing the sheet. Your instructor may assign one of these or one that he or she has designed.

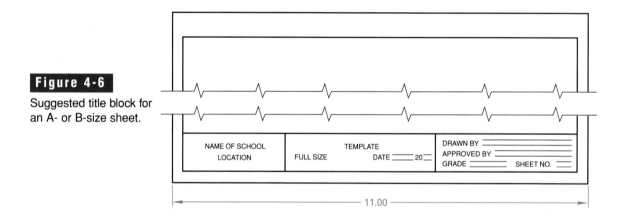

Figure 4-6

Suggested title block for an A- or B-size sheet.

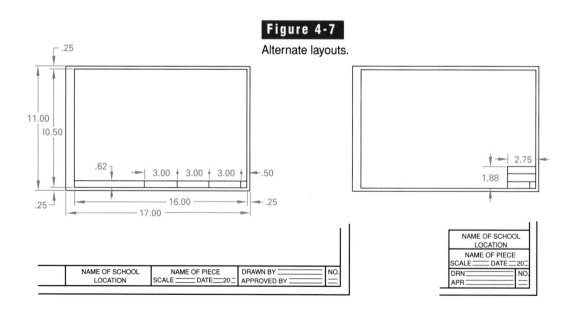

Figure 4-7

Alternate layouts.

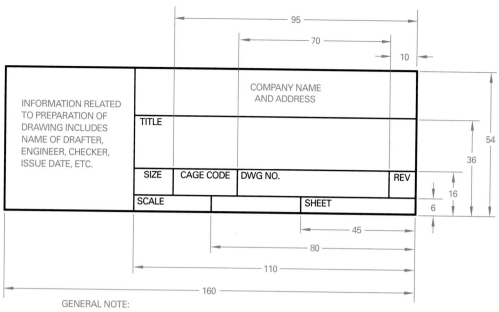

GENERAL NOTE:
DIMENSIONS SHOWN ARE RECOMMENDED AND MAY BE VARIED
TO ACCOMMODATE THE USER'S REQUIREMENTS.

Figure 4-8

Recommended title block for A2, A3, and A4 drawing-sheet sizes (metric).

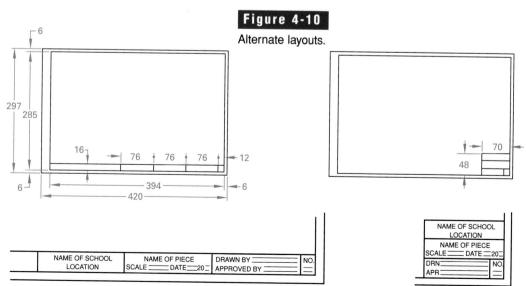

Figure 4-9

Suggested title block for an A4- or A3-size sheet.

Figure 4-10

Alternate layouts.

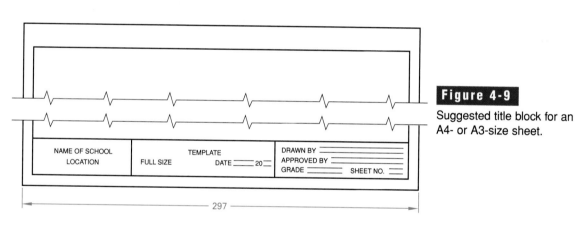

Working with Drafting Pencils and Pens

Techniques for using drafting pencils are somewhat different from those for using technical pens. Unlike pencil marks, ink must be allowed to dry before it can be touched. This affects the way the instrument should be held, as well as the order in which the lines should be drawn.

Sharpening the Pencil

To sharpen a wooden pencil, cut away the wood at a long slope, as shown in Figure 4-11A. Always sharpen the end opposite the grade mark, being careful not to cut the lead. Leave about half an inch (13 mm) exposed. Then shape the lead to a long, conical point. Do this by rubbing the lead back and forth on a sandpaper pad or on a long file, as shown in Figure 4-12, while turning it slowly to form the point, as shown in Figure 4-11B or C. Some drafters prefer the flat point, or chisel point, shown in Figure 4-11D. Keep the sandpaper pad at hand so that you can sharpen the point often. Many drafters like to shape and smooth the point further by "burnishing" it on a piece of rough paper such as drafting paper.

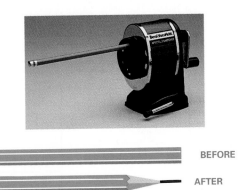

BEFORE

AFTER

Figure 4-13

A drafter's pencil sharpener cuts the wood, not the lead.

BEFORE   AFTER

Figure 4-14

This lead pointer allows a choice of point shapes.

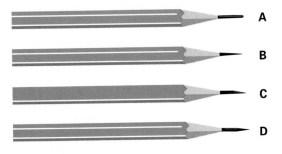

A
B
C
D

Figure 4-11

Sharpening the pencil properly is important.

Figure 4-12

Rub the pencil on a sandpaper pad, turning it slowly, to achieve a conical point.

Mechanical sharpeners have special drafter's cutters that remove the wood, as shown in Figure 4-13. Special pointers are made for shaping the lead, as in Figure 4-14. Such devices may be hand-operated or electrically powered.

Mechanical pencils, also called *lead holders*, are widely used by drafters. They hold plain sticks of lead in a chuck that allows the exposed lead to be extended various lengths. The lead for most lead holders should be shaped in the same way as the lead in wooden pencils. However, some refill pencils have a built-in sharpener that shapes the lead. Still other mechanical pencils use leads made to specific thicknesses for the desired line widths and require no sharpening.

Never sharpen a pencil over the drawing board. After you sharpen a pencil, wipe the lead with a cloth or a Styrofoam-type "stab-it" to remove the dust. Being careful in these ways will help keep the drawing clean and bright. This is important when you plan to use the original pencil drawing to make copies.

Techniques for Using a Drafting Pencil

Pencil lines must be clean and sharp. They must be dark enough for the views to be seen when standard line widths are used. If you use too much pressure, you will groove the drawing surface. You can avoid this by using the correct grade of lead.

Develop the habit of rotating the pencil between your thumb and forefinger as you draw a line. This will help make the line uniform and keep the point from wearing down unevenly.

Inking Techniques

Inking is the process of creating technical drawings using technical pens. Techniques for inking are slightly different from those for drawing in pencil. Hand position and the order in which items are drawn are affected by the fact that ink, unlike pencil, must be allowed to dry to help avoid smudges.

Figure 4-15 shows the correct position for drawing lines with a technical pen. Note the direction of the stroke and the angle of the pen. Hold the technical pen in a nearly vertical position, perpendicular to the media, to get the most uniform line.

Technical pens can be used with compasses, irregular curves, and templates to ink circles, arcs, and other curved lines. Use irregular curves to guide the pen when inking curves that are not true circles or arcs. You can also use either fixed curve templates or flexible curves to create the lines.

Using Erasers

Use soft erasers to clean soiled spots or light pencil marks from drawings. Refer to Chapter 3 for the appropriate erasers to use on different drafting media. Keep in mind that regular ink erasers often contain grit. If you use these erasers at all, use them very carefully to keep from damaging the drawing surface.

The ink used on polyester drafting film is waterproof. However, you can easily remove ink from the film by rubbing it with a moistened plastic eraser or by using an electric erasing machine. Do not use any pressure in rubbing. The polyester film does not absorb ink, so all ink dries on top of its highly fin-

Figure 4-15

The position of the technical pen is important when drawing lines.

ished surface. Remove ink from other surfaces, such as tracing vellum or illustration board, with regular ink erasers or chemically imbibed ink erasers that absorb ink. But be very careful. Press lightly with strokes in the direction of the line to remove ink caked on the surface. Too much pressure damages the surface and makes it hard to revise the drawing.

When working on paper or cloth, erase lines *along* the direction of the work. On film, always erase *across* the direction of the work. Always erase carefully to avoid marring the finish on the drawing sheet. Use an erasing shield to protect nearby lines and areas that you do not want to erase.

Alphabet of Lines

The different lines or line symbols used on drawings form a kind of graphic alphabet commonly known as the **alphabet of lines**. The line symbols recommended by ANSI are shown in Figure 4-16. Two line widths—thick and thin—are generally used. Drawings are easier to read when there is good contrast among different kinds of lines. All lines must be uniformly sharp and black.

Techniques for Drawing Lines

The sections that follow discuss basic drawing techniques. Additional, more complex techniques will be presented in later chapters.

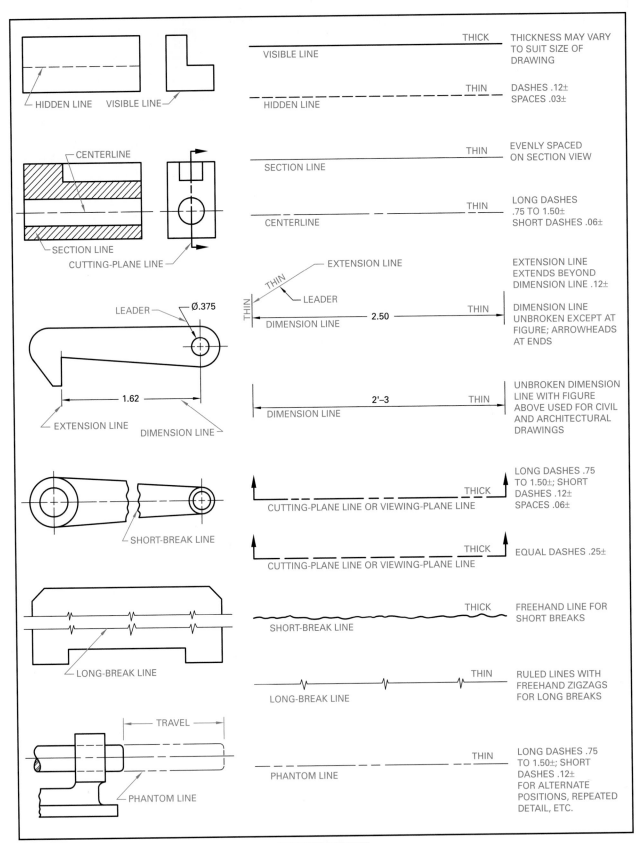

The following are the labels within the figure:

HIDDEN LINE   VISIBLE LINE

CENTERLINE

SECTION LINE

CUTTING-PLANE LINE

LEADER   Ø.375

1.62

EXTENSION LINE   DIMENSION LINE

SHORT-BREAK LINE

LONG-BREAK LINE

TRAVEL

PHANTOM LINE

VISIBLE LINE — THICK — THICKNESS MAY VARY TO SUIT SIZE OF DRAWING

HIDDEN LINE — THIN — DASHES .12± SPACES .03±

SECTION LINE — THIN — EVENLY SPACED ON SECTION VIEW

CENTERLINE — THIN — LONG DASHES .75 TO 1.50± SHORT DASHES .06±

EXTENSION LINE — THIN — LEADER — DIMENSION LINE — EXTENSION LINE EXTENDS BEYOND DIMENSION LINE .12±

2.50 — THIN — DIMENSION LINE UNBROKEN EXCEPT AT FIGURE; ARROWHEADS AT ENDS

2'–3 — THIN — DIMENSION LINE — UNBROKEN DIMENSION LINE WITH FIGURE ABOVE USED FOR CIVIL AND ARCHITECTURAL DRAWINGS

CUTTING-PLANE LINE OR VIEWING-PLANE LINE — THICK — LONG DASHES .75 TO 1.50±; SHORT DASHES .12± SPACES .06±

CUTTING-PLANE LINE OR VIEWING-PLANE LINE — THICK — EQUAL DASHES .25±

SHORT-BREAK LINE — THICK — FREEHAND LINE FOR SHORT BREAKS

LONG-BREAK LINE — THIN — RULED LINES WITH FREEHAND ZIGZAGS FOR LONG BREAKS

PHANTOM LINE — THIN — LONG DASHES .75 TO 1.50±; SHORT DASHES .12± FOR ALTERNATE POSITIONS, REPEATED DETAIL, ETC.

Figure 4-16

Alphabet of lines.

Horizontal Lines

To draw a horizontal line, use the upper edge of the T-square blade as a guide. With your left hand, place the head of the T-square in contact with the left edge of the board. Keeping the head in contact, move the T-square to the place you want to draw the line. Slide your left hand along the blade to hold it firmly against the drawing sheet. Hold the pencil about 1″ (25 mm) from its point. Slant it in the direction in which you are drawing the line. (This direction should be left to right for right-handers and right to left for left-handers.) While drawing the line, rotate the pencil slowly and slide your little finger along the blade of the T-square, as shown in Figure 4-17. This will give you more control over the pencil.

Figure 4-19

Drawing a vertical line.

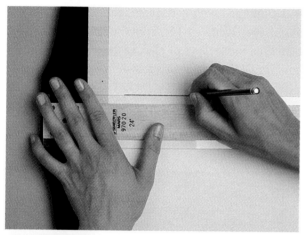

Figure 4-17

Drawing a horizontal line.

Figure 4-18

To ensure accurate drawing, position the pencil as shown here.

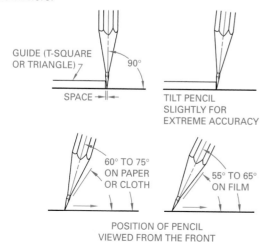

GUIDE (T-SQUARE OR TRIANGLE) 90°

SPACE

TILT PENCIL SLIGHTLY FOR EXTREME ACCURACY

60° TO 75° ON PAPER OR CLOTH

55° TO 65° ON FILM

POSITION OF PENCIL VIEWED FROM THE FRONT

On film, keep the pencil at the same angle (55° to 65°) all along the line. You must also use less pressure on film than on paper or other material. Always keep the point of the lead a little distance away from the corner between the guiding edge and the drawing surface, as shown in Figure 4-18. This will let you see where you are drawing the line. It will also help you avoid a poor or smudged line. Be careful to keep the line parallel to the guiding edge.

Vertical Lines

Use a triangle and a T-square to draw vertical lines, as shown in Figure 4-19. Place the head of the T-square in contact with the left edge of the board. Keeping the T-square in contact, move it to a position below the start of the vertical line. Place a triangle against the T-square blade. Move the triangle to where you want to begin the line. Keeping the vertical edge of the triangle toward the left, draw upward. Slant the pencil in the direction in which you are drawing the line. Be sure to keep this angle the same when you are drawing on film. Keep the point of the lead far enough out from the guiding edge so you can see where you are drawing the line. Be careful to keep the line parallel to the guiding edge.

Inclined Lines

Inclined lines are lines drawn at an angle that is neither horizontal nor vertical. Inclined lines are drawn using triangles, a protractor, or a drafting machine.

30°, 45°, and 60° Lines

Recall from Chapter 3 that angles are measured in degrees, minutes, and seconds. You can draw lines at 30°, 45°, or 60° angles from the horizontal or vertical by using the triangles. Lines inclined at 30° and 60° are drawn with the 30°-60° triangle held against the T-square blade, as shown in Figure 4-20, or against a horizontal straightedge. The 30°-60° triangle can also be used to lay off equal angles, 6 at 60° or 12 at 30°, about a center point.

To draw lines inclined at 45° from horizontal or vertical lines, hold the triangle against the T-square blade, as shown in Figure 4-21, or against a horizontal straightedge. The 45° triangle can also be used to lay off eight equal angles of 45° about a center point.

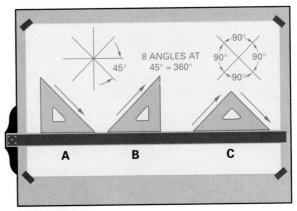

Figure 4-21

The 45° triangle has angles of 45° and 90°.

Lines Inclined at 15° Increments

The 45° and 30-60° triangles, alone or together and combined with a T-square, can be used to draw angles increasing by 15° from the horizontal or vertical line. Some ways of placing the triangles to draw angles of 15° and 75° are shown in Figure 4-22.

Figure 4-22

Drawing lines at 15° and 75° using the two triangles.

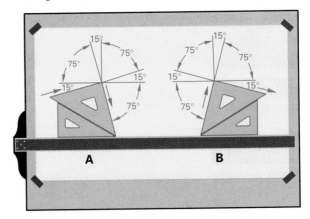

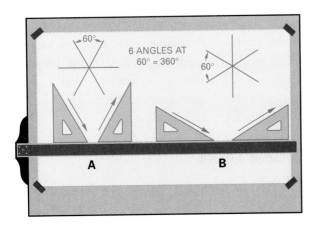

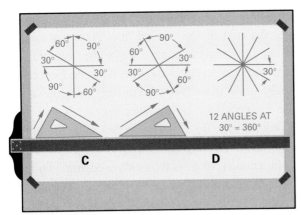

Figure 4-20

The 30°-60° triangle has angles of 30°, 60°, and 90°

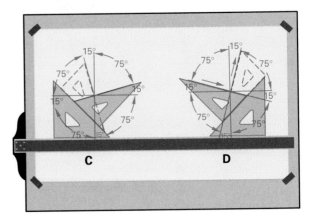

Decimal Degrees

It is becoming a more common practice to specify angles in **decimal degrees** rather than degrees, minutes, and seconds. To convert degrees and minutes to decimal degrees, proceed as follows:

Decimal Degree Equivalents	
Degrees, Minutes and Seconds	Decimal Degrees
0°45′	0.75°
0°0′14″	0.004°
25°30′36″	25.51°
10°±0°30′	10°±0.5°

EXAMPLE: Convert 25°30′ to decimal degrees.

$$25° = 25.00 \text{ degrees}$$
$$+ 30' = .50 \text{ degrees} \quad (30 \text{ minutes}/60 \text{ minutes per degree} = .50)$$
$$25.50 \text{ degrees}$$

EXAMPLE: Convert 25°30′36″ to decimal degrees.

$$25° = 25.00 \text{ degrees}$$
$$30' = .50 \text{ degrees} \quad (30 \text{ minutes}/60 \text{ minutes per degree} = .50)$$
$$+ 36'' = .01 \text{ degrees} \quad (36 \text{ seconds}/3600 \text{ seconds per degree} = .01)$$
$$25.51 \text{ degrees}$$

The table on the right provides several examples of decimal-degree conversions.

Techniques for Special Lines and Surfaces

To describe an object fully, you must show every feature in each view, whether or not it can ordinarily be seen. You must also include other lines that are not actually part of the object to clarify relationships and positions in the drawing. To reduce confusion, special line symbols, or linetypes, are used to differentiate between object lines and lines that have other special meanings.

Hidden Lines

It is necessary to describe every part of an object. Therefore, everything must be represented in each view, whether or not it can be seen. Both interior and exterior features are projected in the same way. Parts that cannot be seen in the views are drawn with **hidden lines** that are made up of short dashes, as shown in Figure 4-23. Notice in Figure 4-23A that the first dash of a hidden line touches the line where it starts. If a hidden line is a continuation of a visible line, space is left between the visible line and the first dash of the hidden line. See Figure 4-23B. If the hidden lines show corners, the dashes touch the corners, as shown in Figure 4-23C.

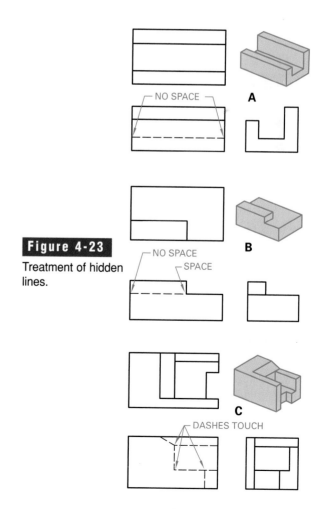

Figure 4-23
Treatment of hidden lines.

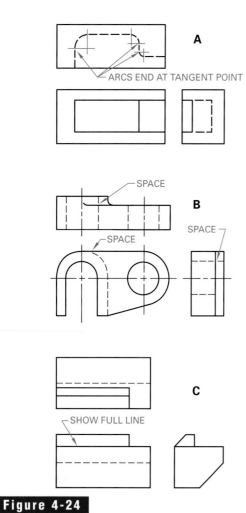

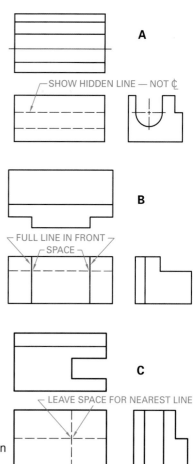

Centerlines

Centerlines are special lines used to locate views and dimensions. (See the alphabet of lines, Figure 4-16.) Primary centerlines, marked P in Figure 4-26, locate the center on symmetrical views in which one part is a mirror image of another. Primary centerlines are used as major locating lines to help in making the views. They are also used as base lines for dimensioning. Secondary centerlines, marked S in Figure 4-26, are used for drawing details of a part.

Primary centerlines are the first lines to be drawn. The views are developed from them. Note that centerlines represent the axes of cylinders in the side view. The centers of circles and arcs are located first so that measurements can be made from them to locate the lines on the various views. As you may recall from the previous section, when a hidden line falls on a centerline, the hidden line is drawn. When a hidden line falls on a visible line, draw the visible line.

Figure 4-24

Treatment of hidden arcs.

Dashes for hidden arcs start and end at the tangent points, as shown in Figure 4-24A. When a hidden arc is tangent to a visible line, leave a space, as shown in Figure 4-24B. When a hidden line and a visible line project at the same place, show the visible line. See Figure 4-24C. When a centerline and a hidden line project at the same place, draw the hidden line, as shown in Figure 4-25A. When a hidden line crosses a visible line as in Figure 4-25B, do not cross the visible line with a dash. When hidden lines cross, the nearest hidden line has the "right of way." Draw the nearer hidden line through a space in the farther hidden line, as in Figure 4-25C.

Figure 4-25

Technique for presenting hidden and visible lines.

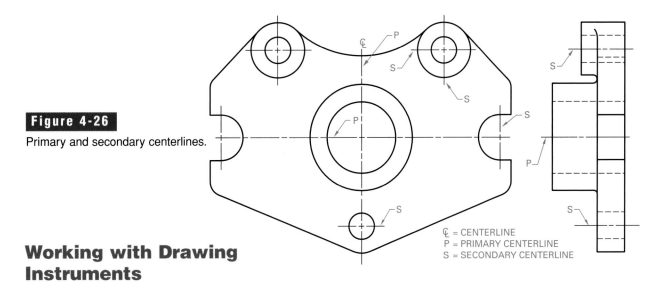

Figure 4-26

Primary and secondary centerlines.

℄ = CENTERLINE
P = PRIMARY CENTERLINE
S = SECONDARY CENTERLINE

Working with Drawing Instruments

The correct use of drawing instruments plays an important role in creating an accurate drawing. The following paragraphs explain how to use the basic drawing instruments.

Using the Dividers

To transfer a distance using the dividers, adjust the points to exactly the length to be transferred, such as the radius of a circle or the length of a line. Transfer the length by positioning the dividers at a new location.

You can also use the dividers to divide a line, arc, or circle into equal parts. For example, to divide a line into three equal parts:

1. Adjust the points of the dividers until they seem to be about one third the length of the line. To adjust the dividers, hold it between your thumb and index finger. Set it to the desired radius using your third and fourth fingers, as shown in Figure 4-27A.
2. Put one point on one end of the line and the other point on the line, as shown in Figure 4-27B.
3. Turn the dividers about the point that rests on the line, as in Figure 4-27C.
4. Then turn it in the alternate direction, as in Figure 4-27D.
5. If the last point falls short of the end of the line, increase the distance between the points of the dividers by an amount about one third the distance *mn*. Then start at the beginning of the line again. You may have to do this several times.
6. If the last point overruns the end of the line, decrease the distance between the points by one third the extra distance.

For four, five, or more spaces, follow the same rules, but correct by one fourth, one fifth, etc., of the overrun or underrun. You can divide an arc or circle in the same way.

Figure 4-27

The dividers are used to divide and transfer distances.

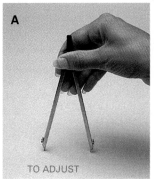

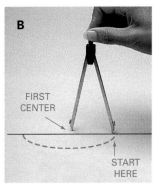

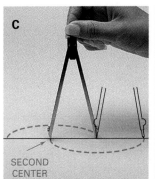

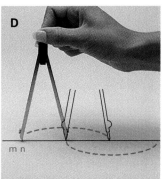

Using the Compass

As you may recall from Chapter 3, the compass is used to draw regular curves, such as circles and circular arcs. Leave the legs of the compass straight for radii under 2″ (50 mm). For larger radii, make the legs perpendicular (at a 90° angle) to the paper, as shown in Figure 4-28. When you need a radius of more than 8″ (200 mm), insert a lengthening bar as shown in Figure 4-29 to increase the length of the pencil leg, or use a beam compass.

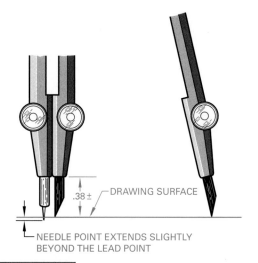

Figure 4-30

Adjusting the point of the compass.

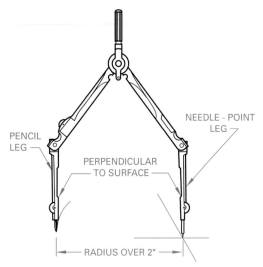

Figure 4-28

Adjusting the compass for large circles.

Figure 4-29

Use the lengthening bar in compasses for circles and arcs of large radii.

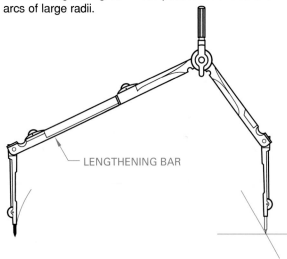

To get the compass ready for use, sharpen the lead as shown in Figure 4-30, allowing it to extend about .38″ (10 mm). Using a long bevel, or slant, on the outside of the lead will keep the edge sharp when you increase the radius. Then adjust the shouldered end of the needle point until it extends slightly beyond the lead point, as shown in Figure 4-30. You cannot use as much pressure on the lead in the compass as you can on a pencil. Therefore, use lead one or two degrees softer in the compass to get the same line weight. **Line weight** refers to the thickness and darkness of a line.

To draw a circle or an arc with the compass, follow these steps:

1. Locate the center of the arc or circle by drawing two intersecting, or crossing, lines.
2. Lay off the radius by a short, light dash, as shown in Figure 4-31A.
3. Adjust the compass setting to the radius, as shown in Figure 4-31B.
4. When the radius is set, raise your fingers to the handle, as shown in Figure 4-31C.
5. Turn the compass by twirling the handle between your thumb and finger. Start the arc near the lower side and turn clockwise, as shown in Figure 4-31D. As you draw the curve, slant the compass a little in the direction of the line. *Do not force the needle point into the paper.* Use only enough pressure to hold the point in place.

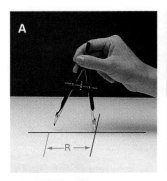

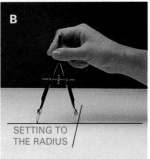

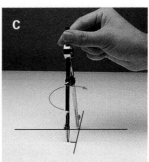

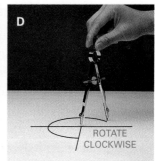

Figure 4-31

Technique for drawing circles and arcs.

Using the Bow Instruments

Bow Pencil

The bow pencil is used to draw small circles. Whether you use instruments with center wheels or with side wheels is up to you. Sharpen and adjust the lead for the bow pencil, as shown in Figure 4-32A. The inside bevel holds an edge for small circles and arcs, as shown in Figure 4-32B. For larger radii, the outside bevel shown in Figure 4-32C is better. Some drafters prefer a conical center point or an off-center point, as shown in Figure 4-32D, E, and F.

Use the bow pencil with one hand. Set the radius as shown in Figure 4-33A. Start the circle near the lower part of the vertical centerline, as shown in Figure 4-33B. Turn clockwise. (Left-handers will need to reverse the above procedure.)

Drop-Spring Bow Compass

Use the drop-spring bow compass to draw very small circles, as shown in Figure 4-34. Attach the marking point to a tube that slides on a pin. Set the radius with the spring screw.

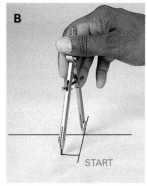

Figure 4-33

Adjusting the radius for the bow pencil compass.

To use the drop-spring bow compass, first set the radius with the adjusting screw. Keep the pin still and turn the lead around it. Hold the marking point up while putting the pin on the center. Then drop the marking point and turn it.

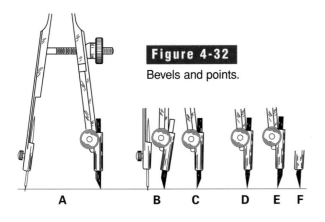

Figure 4-32

Bevels and points.

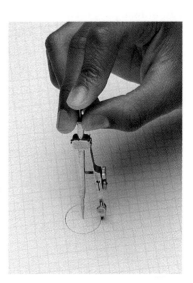

Figure 4-34

The drop-spring bow compass is used for drawing very small circles, especially when there are many to be drawn.

Adjusting Bow Instruments

You can make large adjustments quickly with the side-wheel bows by pressing the fork and spinning the adjusting nut. Some center-wheel bows are also built for making large, rapid adjustments. To do this, hold one leg in each hand and either push to close or pull to open. Make small adjustments with the adjusting nut on both the side-wheel and the center-wheel bows.

Using Irregular Curves

To use an irregular curve, find the points through which a curved line is to pass. Then set the path of the curve by drawing a light line, freehand, through the points. Adjust it as needed to make the curve smooth. Next, match the irregular curve against a part of the curved line, as shown in Figure 4-35A, and draw part of the line. Move the irregular curve to match the next part, and so on, as shown in Figure 4-35B and C. Each new position should fit enough of the part just drawn to make the line smooth. Note whether the radius of the curved line is increasing or decreasing and place the irregular curve in the same way. Do not try to draw too much of the curve with one position. If the curved line is **symmetrical**, or mirrored around an axis, mark the position of the axis or centerline on the irregular curve on one side. Then turn the irregular curve around to match and draw the other side.

Drawing Practice

To achieve skill and accuracy in creating technical drawings, practice is important. The drawing exercise in this section is designed to give you practice using the techniques and instruments discussed in this chapter. Follow the procedure below to complete the template drawing shown in Figure 4-36A.

1. Begin with an 11.00″ × 17.00″ drawing sheet and prepare it as described in the previous section.

2. Measure 3.25″ from the left border line, and from this mark measure 8.75″ toward the right.

3. Lay the scale on the paper vertically near (or on) the left border line, make a mark 2.50″ up, and from this measure 5.50″ more. The sheet will appear as in Figure 4-36B.

4. Draw horizontal lines 1 and 2 with the T-square and triangle, as shown in Figure 4-36C.

5. Lay the scale along the bottom line of the figure, with the measuring edge on the upper side, and make marks 1.75″ apart. Then, with the scale on line 3 and its measuring edge to the left, measure from the bottom line two vertical distances, 2.50″ and 1.50″, as shown in Figure 4-36D.

6. Through the two marks, draw light horizontal lines.

7. Draw the vertical lines with T-square and triangle by setting the pencil on the marks on the bottom line and starting and stopping the lines on the proper horizontal lines, as shown in Figure 4-36E.

8. Erase the lines not wanted (if necessary) and darken the lines of the figure to finish the drawing. Figure 4-36F shows the finished template. Do not add dimensions unless your instructor directs you to do so.

Success on the Job

Skill and Determination
Machinists, mechanics, and cabinetmakers become successful only when they become highly skilled in the use of the tools of their trade. Likewise, developing a high degree of skill in the use of board drafting and CAD equipment will ultimately result in an equally high degree of success on the job. Add to skill the determination to be the best that you can be, and you have achieved the formula for great success as a drafting technician.

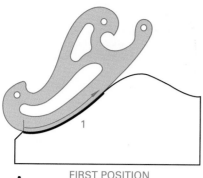

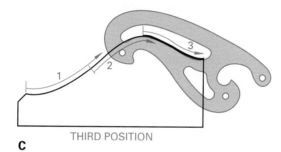

A   FIRST POSITION

B   SECOND POSITION

C   THIRD POSITION

Figure 4-35

Steps in drawing a smooth curve.

Figure 4-36

Template for drawing practice.

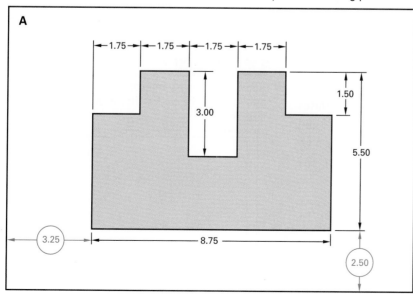

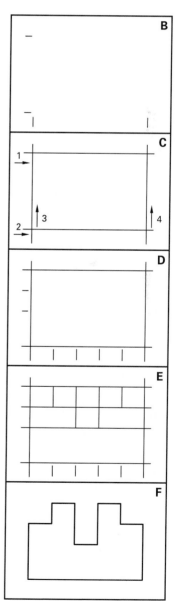

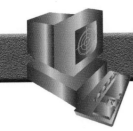

CAD Techniques

Techniques for creating a technical drawing using CAD differ greatly from those used in board drafting. One of the biggest differences is that CAD drawings are drawn at full scale, whether the object being drawn is a microchip or a municipal parking lot.

Obviously, you can't print a drawing of a parking lot at full size. Therefore, you print the drawing at a scale that allows it to fit on the selected drawing sheet. If you set the drawing up properly before you begin to draw, you can both draw at full size and print it at the appropriate scale on the drawing sheet.

In other ways, drafting using a CAD system is similar to board drafting. You must understand drafting concepts before you can create an acceptable CAD drawing. Except where noted, all of the drafting principles discussed in this and other chapters apply equally to CAD and board drafting.

This part of the chapter is designed to be "done," rather than just read. Ideally, you should have access to a CAD workstation so that you can try the techniques as you read about them. If this is not possible, read the contents thoroughly, and then have the book with you the next time you have access to CAD.

Preparing the Drawing File

As in board drafting, you must plan carefully before you begin to draw in CAD. Even though you will be drawing at full size, you must select a sheet size—and a scale for the printed drawing—before you begin. AutoCAD provides **drawing templates** that have already been set up for standard ANSI and ISO sheet sizes. However, it is up to you to make sure your drawing, scaled if necessary, will fit on the selected sheet size.

Creating a New Drawing

The first step in preparing a drawing file is to open a new drawing in AutoCAD. Newer versions of AutoCAD allow you either to base your drawing on a standard drawing template or to use "Quick Setup" or "Advanced Setup" to specify the basic drawing characteristics. We will use a template.

1. Enter the NEW command to create a new drawing.
2. Choose to use a template to begin the drawing.

CADTIP

Entering Commands

AutoCAD allows you to enter commands by picking buttons with a mouse, entering the name of the command at the keyboard, or choosing the command from a pull-down menu. This textbook purposely does not specify the method by which you enter commands. Experiment with all the methods and use a combination that is efficient for you.

3. Select the template based on drawing requirements or your instructor's specification. For this example, select the ANSI B drawing sheet. (Depending on your version of AutoCAD, you may need to select ansi_b.dwt or ANSI B - Named Plot Styles.) The border and title block appear on the screen, as shown in Figure 4-37.

Figure 4-37

AutoCAD provides templates for standard ISO and ANSI borders and title blocks. This is an ANSI layout for a B-size sheet.

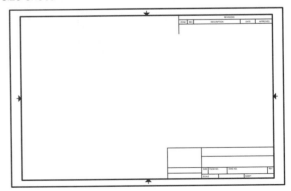

Notice the MODEL button at the bottom of the screen, below the border and title block. AutoCAD has two drawing spaces. **Model space** is a working space, where you will do most of your drawing. Layout, or **paper space**, allows you to position your drawing on the specified drawing sheet.

If you are using AutoCAD 2000 or later, you will also see tabs at the bottom of the drawing area. The current tab displays the name of the template you have selected; in this case, it is "ANSI B Title Block" (or something similar). The other tab is labeled "Model." You should choose the Model tab whenever you are actually drawing or editing geometry. Choose the appropriate layout tab to view or work with the drawing in paper space.

Setting the Drawing Units

Switch to model space by picking the Model tab at the bottom of the drawing area. In model space, you can continue to set up the drawing file. The next task is to specify the drawing units.

In AutoCAD, the term *unit* is purposely vague. AutoCAD ensures that 1 unit = 1 unit. It is up to you to determine whether the unit stands for millimeters, inches, miles, or for some engineering applications, even hours or other nonlinear units. Therefore, before you begin drawing, you must specify what the units will be.

To set the units in a drawing, enter the UNITS command. AutoCAD presents a dialog box that lets you choose from architectural, decimal, engineering, fractional, or scientific units for both length and angle measurements. For the drawings in this textbook, you should choose decimal units unless directed otherwise.

Selecting decimal units in AutoCAD does not limit you to working in decimal inches. By choosing decimal units, you are setting up the drawing correctly for both ANSI and ISO standards. Only the precision differs. For ANSI, select a length precision of two decimal places (0.00). For ISO or metric, select a length precision of one decimal place (0.0). For both standards, set angle precision to no decimal places (0), unless directed otherwise. Be sure you set the precision to match the template you are using. Do not set up ANSI precisions for use on an ISO drawing sheet. In this case, choose ANSI precisions because you are using a B-size sheet.

Setting the Drawing Limits

Next, set the model space **limits**, or physical size, to correspond to the template you selected for paper space. The paper-space layout does not affect model space, so you have to set the limits separately.

Limits for Printing at Full Size

If you have selected a B-size sheet and your drawing will fit on the sheet without scaling, you should set the drawing size to $17'' \times 11''$. To do this, use the LIMITS command. This command works by specifying the lower left and upper right corners of the drawing area.

Enter the LIMITS command, and then look at the Command line at the lower left corner of the screen. It shows that the lower left corner is currently set to 0.00,0.00. This means that the lower left corner of the drawing is set at 0.00 inches horizontally and vertically. In general, you should leave the values at

▶ **CADTIP**

Coordinate Entry

AutoCAD uses the coordinate pair (X,Y) system to specify points on two-dimensional drawings. A **coordinate pair** is a set of two numbers, separated by a comma, that identifies the exact location of a point on a coordinate grid. The grid is made up of a horizontal, or X, axis and a vertical, or Y, axis, as shown in Figure 4-38. The point at which the two axes cross is point (0.00,0.00), which is known as the origin. All points located above the X axis have positive values, and points located below the X axis have negative values. Points to the right of the Y axis have positive Y values, and points to the left of the Y axis have negative values.

The first number in the coordinate pair is the point's location on the X axis, and the second number is its location on the Y axis. If you imagine lines drawn from the X and Y axes at right angles, the point at which the two lines intersect is the location of the point.

Note that there is no space after the comma in a coordinate pair. Also, coordinate pairs are often set in parentheses for clarity: (3.00,4.00). When you see this, do not type the parentheses.

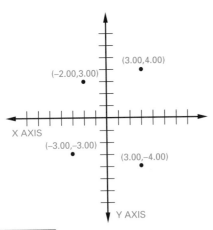

Figure 4-38

A coordinate pair specifies a unique point in two-dimensional space using X and Y coordinates.

0.00,0.00. Press Enter to continue. For a B-size sheet (without scaling), type 17.00,11.00 and press Enter. *Do not* type a space between the comma and the 11.00. Then enter the ZOOM command and the All option to view the entire drawing area.

Limits for Printing a Scaled Drawing

If you plan to scale the drawing to fit on the drawing sheet, you should set the model space limits accordingly. For example, suppose you are working with a B-size sheet, and you plan to print the drawing at a scale of 1:2. In other words, your full-size

Table 4-2. Drawing Limits		
Sheet Size	**Drawing Scale**	**Model Space Limits**
ANSI A	1:1 1:2 1:4	11.00″ × 8.50″ 22.00″ × 17.00″ 44.00″ × 34.00″
ANSI B	1:1 1:2 1:4	17.00″ × 11.00″ 34.00″ × 22.00″ 68.00″ × 44.00″
ANSI C	1:1 1:2 1:4	17.00″ × 22.00″ 34.00″ × 44.00″ 68.00″ × 88.00″
ISO A4	1:1 1:2 1:4	210 mm × 297 mm 420 mm × 594 mm 840 mm × 1188 mm
ISO A3	1:1 1:2 1:4	297 mm × 420 mm 594 mm × 840 mm 1188 mm × 1680 mm
ISO A2	1:1 1:2 1:4	420 mm × 594 mm 840 mm × 1188 mm 1680 mm × 2376 mm

drawing will be twice as big as it will appear on the printed sheet. Therefore, your model-space limits should be twice the size of the B-size sheet, or 34″ × 22″. Table 4-2 shows common drawing limits for drawings of different sizes and scales.

Working with Layers

All CAD programs, including AutoCAD, have a system of layers that gives the CAD operator much greater control over a drawing. A **layer** is similar to a transparent paper overlay. By setting up a layer for dimensions, for example, the CAD operator can control whether dimensions are displayed by turning the layer on and off, or by "freezing" and "thawing" it, as shown in Figure 4-39.

Most companies have rules about what layers to use, what to call them, and what colors should be associated with them. Some companies even use

Figure 4-39

By placing dimensions on a separate layer, you can control whether the dimensions display. In (A), the dimension layer is displayed. In (B), it has been frozen, so it does not show on the screen. If you print the drawing with the dimension layer frozen, the layer will not print.

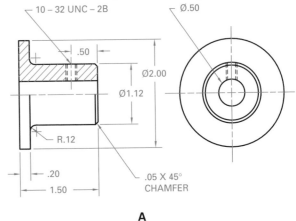

A

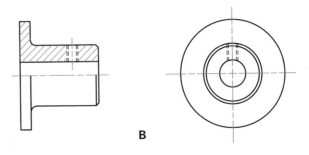

B

x

Table 4-3. Layers for Drafting Problems			
Name	Color	Linetype	Line Weight
Objects	White	Continuous	.30 mm
Dimensions	Red	Continuous	.18 mm
Hidden Lines	White	Hidden	.18 mm
Centerlines	Blue	Center	.18 mm
Notes	Magenta	Continuous	.18 mm

their own drawing templates in which these layers have already been set up. For instructional purposes, this textbook will use a generic set of layers. These layers are shown in Table 4-3.

Creating a New Layer

To set up new layers in a drawing, enter the LAYER command. If you have based the drawing on an ANSI or ISO template, the dialog box that appears should already have layers for the title block and the border. Look closely at the contents of the dialog box. Notice that several properties are listed for each layer, including:

- layer name
- on or off
- frozen or thawed
- layer color
- linetype
- line weight
- plot style
- plot (whether the layer plots when the drawing is printed)

One of the standard layers used in this textbook is the Objects layer. This layer will be used for all of the visible lines of the part or object. Therefore, you already know that it will need to be a solid (or continuous) line that is .12″ thick. To create a new layer named Objects, click the New button in the dialog box. A new layer appears in the window, and the layer name is highlighted. Type the word Objects in the layer name box.

Setting the Layer Color

By default, new layers in AutoCAD are white. To set the color for a layer, pick White or the white color box for that layer. A color palette appears. To choose a different color, just pick a color and pick OK. However, because this is the Objects layer, leave it white.

Colors are used in CAD programs to help the CAD operator distinguish among the layers. See Figure 4-40. Some companies prefer to use white for all of their layers. Others establish company-wide standards. For example, they may declare that all electrical wiring will be on a blue layer named Electr. It is also possible to set up the layers in various colors, but set up a plot style to print them all in black ink. Therefore, the color of a layer may or may not determine the color of the lines on that layer when the drawing is printed. This is up to the individual drafter or company.

Selecting the Linetype

AutoCAD gives new layers a continuous linetype by default, so the Objects layer is already set up for the correct linetype. However, as you can see in Table 4-3, you will need to change it for some of the other layers. To do so, click the word Continuous. A dialog box appears from which you can change the linetype, but notice that you have no other choices. To load

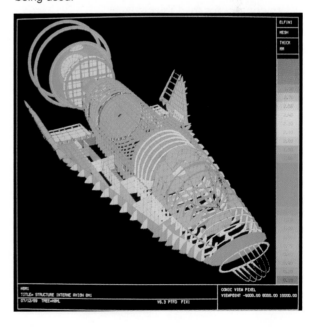

other standard linetypes into the drawing, pick the Load button. Another dialog box appears, allowing you to select from several ISO and ANSI linetypes. To load the ISO standard dashed line, for example, choose ISO02W100 ISO Dash and pick OK. The linetype becomes available for use in the drawing. To choose the ANSI standard dashed line for hidden lines, scroll down to Hidden, pick it, and pick OK.

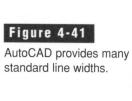

Loading Linetypes

You will usually know in advance which linetypes you will need for a drawing. It is more efficient to load all of the needed lines at one time, before you close the dialog box. After adding each linetype, pick the Load button again to choose another one. Some versions of AutoCAD allow you to load all available linetypes by picking a Load All button. When you have finished, select the linetype you need for the current layer, then pick OK to return to the dialog box for managing layers.

Selecting the Line Width

The default line width in AutoCAD is 0. This does not mean that the line doesn't print. However, the width of the line is not defined. You can and should define the width of the lines on your drawings. To do so, pick the word Default in the Line Width column for the Objects layer.

Notice that AutoCAD specifies all of its line widths in millimeters, as shown in Figure 4-41. Visible lines in CAD are generally made at a width of .12″, or .30 mm. Select .30 mm from the list of line widths, and pick OK to apply it to the Objects layer.

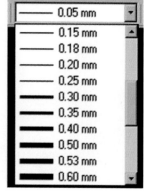

Figure 4-41

AutoCAD provides many standard line widths.

Finishing Layer Setup

Now finish the layer setup for your drawing by creating the other layers listed in Table 4-3. Be sure to give each layer the properties shown in the table. Then enter the Save command to save the drawing file. Because of the settings chosen in this example, you might want to call the file Chapter 4 ANSI B Full Scale, or something similar.

Working with CAD Commands

The commands in a CAD program are the "tools" the CAD operator uses to create drawings. Instead of using a T-square and triangle to create a 45° line, for example, the CAD operator uses the LINE command to create the line and specifies a 45° angle. Commands vary among CAD programs, and sometimes even among different versions of AutoCAD and AutoCAD LT. Therefore, if you see a command in this text that does not appear in your software, use the software's Help feature to find out which command you should use instead. However, within AutoCAD and AutoCAD LT, most of the basic commands are the same.

The sections that follow explain how to use basic drawing commands in AutoCAD. As you read each section, pause and try out the technique in AutoCAD before continuing to read.

Drawing Straight Lines

Draw straight lines in AutoCAD by entering the LINE command. The Command prompt (at the lower left corner of the screen) asks for the first point of the line. For now, use the mouse to pick a point anywhere in the drawing area. The prompt changes to ask for the next point. Pick another point, and another. As you can see, you can continue picking points to create line segments indefinitely. When you are finished, press Enter to leave the LINE command.

Horizontal and Vertical Lines

AutoCAD allows you to create perfectly vertical and horizontal lines with very little effort. The **Ortho** mode forces every line you draw to be either vertical or horizontal, as shown in Figure 4-42. To turn Ortho on, pick the Ortho button at the bottom of the screen, or press the F8 function key on the keyboard.

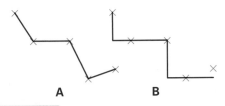

Figure 4-42

The Ortho mode forces every line you draw to be perfectly vertical or horizontal. In this illustration, the points shown in magenta represent the actual points picked by the CAD operator. The black lines show how AutoCAD draws the corresponding lines with Ortho off (A) and with Ortho on (B).

Inclined Lines

The simplest way to create inclined lines accurately is to use **polar coordinates**. Polar coordinates include a specified distance and the angle at which the line should extend. For example, suppose you have already entered the LINE command and specified the first point. Typing @2.50<45 when AutoCAD prompts you for the next point creates a line that extends 2.50 units at 45 degrees.

Notice the required format for polar coordinates. The @ symbol tells AutoCAD that this coordinate will be relative to the last point entered. The @ is followed by the length of the line. The < symbol represents "angle," and the last number is the specified angle.

Polylines

All of the lines discussed so far, even those whose ends join, are actually individual line segments. For many manufacturing uses, the lines must be joined into a single line. AutoCAD makes this possible by providing a polyline. A **polyline** is a line of any length, with any number of defining points, that is considered by the software to be a single line object, as shown in Figure 4-43. Polylines can contain

straight segments, curved segments, or both. To create a polyline, use the PLINE command. Experiment with this useful command until you feel comfortable using it.

The options that appear at the Command line after you enter the first point of a polyline extend its usefulness. The most frequently used options for most applications are the Arc and Close options. Arc allows you to add one or more curved segments to the polyline, and Close joins the last point you entered to the first point of the polyline. This creates a perfectly closed shape, which is very important for use with computer-aided manufacturing (CAM) and computer numerical control (CNC) systems.

Drawing Circles and Arcs

Circles and arcs are easy to create in AutoCAD. In general, to create a circle, use the CIRCLE command. To create an arc, use the ARC command.

The easiest way to create a circle or arc is to specify a center point and a radius. The **radius** of a circle or arc is the distance from its center point to any point on the rim of the circle or arc. See Figure 4-44. However, you can also use other methods, depending on how you need to incorporate the object into the drawing. You can specify a center point and diameter, for instance, or specify two tangent objects and a radius. Experiment with the options of the CIRCLE and ARC commands until you feel comfortable using them.

Using Snap and Grid

To create acceptable technical drawings, you must use techniques that are more accurate than just pointing to a place on the screen to specify endpoints for lines, center points for circles, and so on. One way to produce lines that meet exactly and are exactly the right length is to use coordinate entry. This is a

Figure 4-43

Examples of polylines.

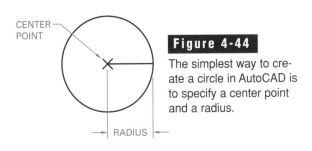

Figure 4-44

The simplest way to create a circle in AutoCAD is to specify a center point and a radius.

fairly slow method, and it has a major disadvantage: you must know or figure out the exact coordinates for every defining point in the drawing.

Fortunately, there are easier ways. AutoCAD provides two tools called *grid* and *snap* to help you select points accurately. **Snap** sets the distance intervals at which the cursor moves when you move the mouse. For example, a snap of .25 allows you to enter points at exactly .25-unit intervals. While the snap is on and set to .25, you can't accidentally enter a point at .23 unit from the previous point. Snap is like a magnet that attracts points to the intervals you specify.

To turn the snap on, pick the Snap button at the bottom of the screen or press the F9 function key. To set the snap interval, enter the SNAP command at the keyboard. You can set the X and Y intervals to the same value or to different values. You can also rotate snap to any angle, and you can specify whether you want a standard or isometric snap.

Grid produces a nonprinting grid of dots on the screen at intervals you specify. These dots provide a visual reference for the CAD operator. To turn grid on or off, pick the GRID button at the bottom of the screen or press the F7 function key on the keyboard. To set the grid spacing, enter the GRID command at the keyboard. You can set the X and Y settings to the same interval or to separate intervals (using the Aspect option). You can also set it to correspond to the snap settings.

By setting up snap and grid intervals that will be useful in your current drawing, you can cut down on drawing time. For example, to create the stencil shown in Figure 4-45, you could set the snap and grid to equal intervals of .50.

Erasing

Erasing in AutoCAD takes two forms. You can either use the ERASE command, or you can simply "undo" one or more of your actions.

When you enter the ERASE command, AutoCAD asks you to select the objects to be erased. You can pick them one by one with the mouse, or you can create a window by picking two diagonal corners of a rectangle. All objects inside the rectangle are selected for erasure. When you have finished selecting the objects to be erased, press Enter to complete the command.

AutoCAD has a formal UNDO command that has several options. However, CAD operators usually just press the "u" key and then Enter. This causes AutoCAD to undo the last action you took. If you press "u" again, AutoCAD continues to undo the previous actions sequentially. In this way, you can undo as many steps as necessary until the drawing reaches the state at which it was last saved. This method can be used even in the middle of many drawing commands, such as the LINE command. This makes it a quick and easy way to correct mistakes without stopping to enter another command or losing your train of thought.

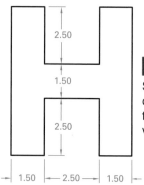

Figure 4-45

Setting snap and grid values of .50 would help you create this stencil for the letter H very quickly and accurately.

Scaling and Printing a Drawing

If you have set up the drawing properly, scaling and printing it is very easy. Before you print, pick the paper space tab to see how the drawing looks on the sheet layout you have selected. You may need to go back to model space and move the entire drawing slightly to accommodate the border and title block.

At this time, fill in the title block with the appropriate information using AutoCAD's TEXT or DTEXT command. Refer to Chapter 2 for more information about using text in AutoCAD.

Printing at Full Size

After you have checked a full-scale drawing in paper space and filled in the title block, enter the PLOT command. Check the printer settings, and make sure the plot scale is set to 1:1. Then pick OK to print the drawing.

Printing a Scaled Drawing

After you have finished a drawing that you want to scale for printing, pick the layout tab at the bottom of the drawing area (ANSI B Title Block, in this case). Notice that the drawing appears at its full size.

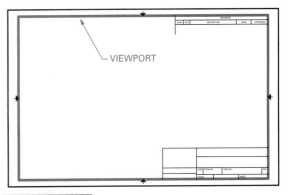

Figure 4-46

Pick the viewport to select it. When it is selected, the line becomes "broken" and small boxes called *grips* appear at key points.

Drawings actually appear in **viewports**, or invisible windows, in paper space. By default, a drawing layout has one viewport. To change the scale of a drawing that appears in a paper-space viewport, you must change the viewport's properties. Pick the viewport to make it active, as shown in Figure 4-46. Then enter the PROPERTIES command. At the top of the Properties dialog box, pick the down arrow to see a list of items whose properties you can change, as shown in Figure 4-47A. Pick Viewport to see a list of the viewport's properties. Scroll down, if necessary, to the item called Standard Scale. Pick Standard Scale to make a down arrow appear next to the current setting. Pick the down arrow to see a list of standard scales, and pick the scale you want to use, as shown in Figure 4-47B. For example, if you are printing at half of full size, the scale is 1:2.

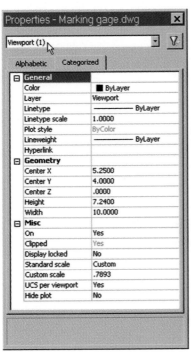

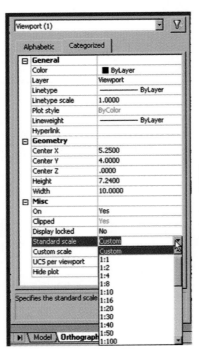

A   **B**

Figure 4-47

To change the scale of a drawing in paper space without changing the size of the border and title block, change the properties of the viewport in which the drawing appears. (A) Select Viewport to see the list of properties for the viewport. (B) Select the scale at which you want the drawing to appear.

Drawing Practice

Practice is required to gain the skill needed to create accurate technical drawings using CAD. To practice drawing and sheet layout in CAD, the following procedure steps you through the process of creating a drawing of the template shown in Figure 4-48. This is the same drawing that is used in the "Board Drafting Techniques" section of this chapter. If you are completing both procedures, notice the differences in the board drafting and CAD techniques.

1. Create a new drawing using AutoCAD's ANSI B template.
2. Switch to model space and set up the units. For this drawing, use decimal units with a length precision of two decimal places and an angle precision of no decimal places.
3. Set the model-space limits. This drawing will be printed at full size, so the limits should equal the sheet size.
4. Create the appropriate layers. This is a fairly simple drawing that contains only visible lines. You will not dimension this drawing, so you really only need an Objects layer. Create the layer and set it up for visible lines .30 mm thick. Leave the layer color at White.
5. Set the snap and grid. Review the template in Figure 4-48 and notice that all of the decimals are in multiples of .25. Therefore, .25 would make a good setting for the snap and grid.

6. Save the drawing before you proceed. Give it a name that is easy to identify, such as Chapter 4 Practice, or name it according to your instructor's directions.

7. Refer again to Figure 4-48 to figure out where to start drawing. In this case, a convenient place to start is the lower left corner of the template. Notice that it is 3.25″ from the left side of the border. This becomes the X coordinate for your starting point. It is 2.50″ from the bottom, so 2.50 becomes the Y coordinate for the starting point. Therefore, the coordinates for the starting point are (3.25,2.50).

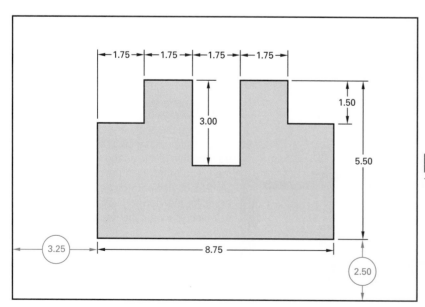

Figure 4-48
Template for drawing practice.

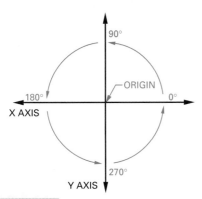

Figure 4-49

Polar coordinates. Notice that 0° lies to the right of the origin along the X axis. The angle value increases counterclockwise through a complete circle.

There are two ways to proceed from here. You can determine the exact coordinates of each of the remaining endpoints on the drawing, or you can use polar coordinates. The easier method, used in this exercise, is the polar coordinates method. Figure 4-49 shows the polar values for the angles you will need to specify. In general:

- A line drawn horizontally to the right has an angle of 0°.
- A line drawn horizontally to the left has an angle of 180°.
- A line drawn vertically bottom-to-top has an angle of 90°.
- A line drawn vertically top-to-bottom has an angle of 270°.

8. Enter the LINE command, and use the mouse to move the cursor until the coordinate display in the lower left corner shows the coordinates to be 3.25,2.50,0.00. (The third number is for three-dimensional drawings only, so you can ignore it for this drawing. Its value will always be 0.00.) Click to set the first point of the line at 3.25,2.50.

9. Work counterclockwise to draw the lines for the template. At the first Next point prompt, for example, enter @8.75<0. This draws an 8.75″ horizontal line to the right from the first point.

10. For the second point, you will need to do a minor calculation. The length of the line is the total length 5.50 less the 1.50 inset. Subtracting 1.50 from 5.50 equals 4.00, so the next line should be 4.00″ drawn vertically bottom-to-top. Therefore, you should enter @4.00<90.

11. Calculate and enter the remaining values on your own. After you have entered the last value, press Enter to end the LINE command. Your finished template should look like the one in Figure 4-48, without the dimensions.

12. Pick the layout view tab to see how the drawing looks on the sheet layout. The drawing should appear to be the correct size for the drawing sheet, but it may appear a little off-center. You can change the position of the drawing on the drawing sheet by selecting the viewport and entering the PROPERTIES command. At the top of the dialog box, select Viewport. Scroll down the list until you see Center X and Center Y, as shown in Figure 4-50. These refer to the horizontal and vertical centers of the drawing. You can change the defaults either by entering a new numerical value or by using the mouse. To use the mouse, pick the icon at the right of the current value. A line appears in the viewport. The origin of the line is the current horizontal or vertical center. Move the mouse and click to select a new center.

13. Save the drawing file. If your instructor requires a printed copy, print the drawing.

Figure 4-50

The properties Center X and Center Y allow you to center the drawing on the drawing sheet for a pleasing display.

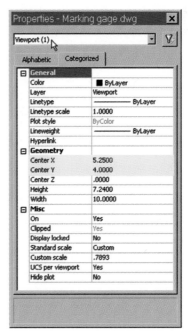

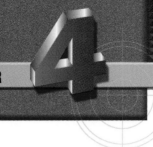
Chapter Summary

- Safety practices apply to the drafting room and should always be observed.

- Preparing a drawing sheet includes choosing an appropriate size and type of drawing sheet, fastening it to the drawing board, and laying out the borders and title block.

- Dividers are used to transfer lengths and to divide geometry into equal parts.

- Regular curves (circles and arcs) are drawn with a compass or circle template.

- A drawing is made "to scale" when the object being drawn is too large for the drawing sheet or too small to be read easily.

- The different lines or line symbols used on drawings form a kind of graphical alphabet commonly known as the alphabet of lines.

- Preparing a drawing file in CAD includes creating a new drawing, selecting a drawing template for the appropriate standard sheet size, and setting up units, limits, and layers.

- Layers in AutoCAD have specific properties such as line width, linetype, and color.

- Commands are the drafting tools used in a CAD program.

- Technical drawings created in CAD can be printed using ANSI or ISO standard layouts in paper space.

Review Questions

1. List at least four safe practices that should be observed in the drafting room.

2. What two organizations publish standard layouts for use with technical drawings?

3. What material is most commonly used to fasten the drawing sheet to the board?

4. Briefly describe how to use dividers to divide a line into four equal parts.

5. What two shapes are commonly used on drafting pencil leads?

6. What is the best position in which to hold a technical pen when inking?

7. How many widths, or thicknesses, of lines are generally used in drafting?

8. Briefly describe the procedure for using an irregular curve.

9. What name is given to the different lines and line symbols used on technical drawings?

10. At what scale are CAD drawings created?

11. In CAD, what is a drawing template?

12. Explain the purpose of using layers in a CAD drawing.

13. What should the drawing limits be for an AutoCAD drawing that will be printed on an A4-size sheet at a scale of 1:2? Assume that the lower left corner is at 0.00,0.00.

14. How can AutoCAD's snap and grid features help when you create a drawing?

15. Explain how to draw a perfectly vertical or horizontal line in AutoCAD.

16. In AutoCAD, what is the appropriate format for specifying a polar coordinate to draw a line 3.50″ long at an angle of 33°?

17. What is the fastest way to erase the last line segment you have drawn in AutoCAD without leaving the LINE command?

18. Briefly describe how to print an AutoCAD drawing at a scale of 1:4.

Problems

Drafting Problems

The problems in this chapter can be performed using board drafting or CAD techniques. The problems are presented in order of difficulty, from least to most difficult.

1. Draw each item shown in Figure 4-51 using the specified sheet size and scale. Do not dimension.

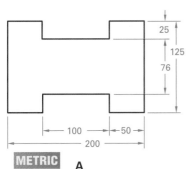

A
SHEET SIZE: A4
SCALE: 1:1

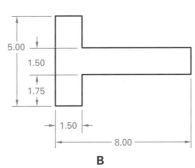

B
SHEET SIZE: A
SCALE: FULL SIZE

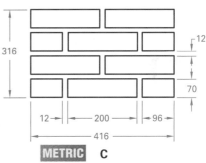

C
SHEET SIZE: A4
SCALE: 1:2

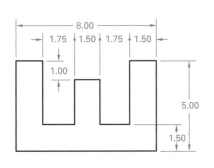

D
SHEET SIZE: A
SCALE: FULL SIZE

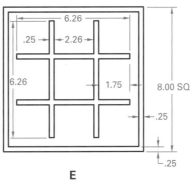

E
SHEET SIZE: A
SCALE: 3/4 SIZE

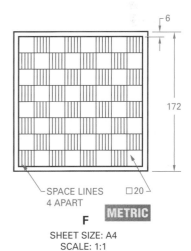

F
SHEET SIZE: A4
SCALE: 1:1

Figure 4-51

2. Draw each item shown in Figure 4-52 using the specified sheet size and scale. Do not dimension.

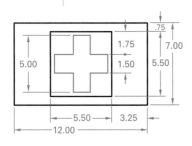

A
SHEET SIZE: B
SCALE: FULL SIZE

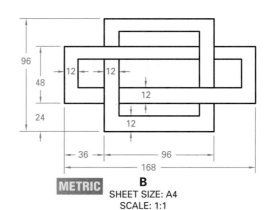

METRIC   **B**
SHEET SIZE: A4
SCALE: 1:1

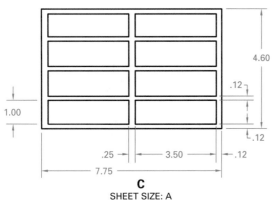

C
SHEET SIZE: A
SCALE: FULL SIZE

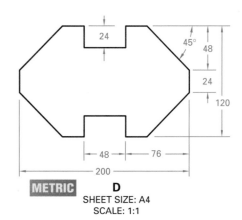

METRIC   **D**
SHEET SIZE: A4
SCALE: 1:1

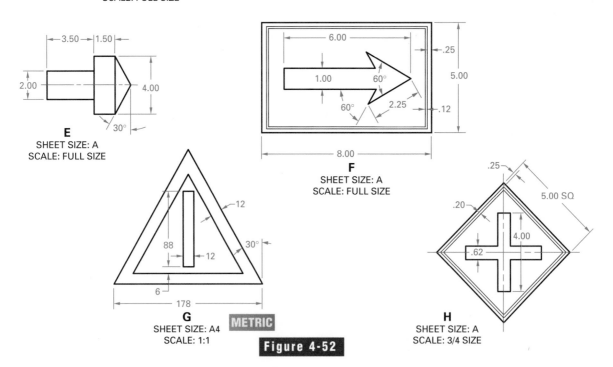

E
SHEET SIZE: A
SCALE: FULL SIZE

F
SHEET SIZE: A
SCALE: FULL SIZE

G   METRIC
SHEET SIZE: A4
SCALE: 1:1

Figure 4-52

H
SHEET SIZE: A
SCALE: 3/4 SIZE

3. Draw the grill plate shown in Figure 4-53. Make all ribs 12 mm wide. The distance AB is 59 mm; BC is 88 mm; AD is 64 mm. The diamond shapes are 38 mm square. Sheet size: A4. Scale: 1:1.

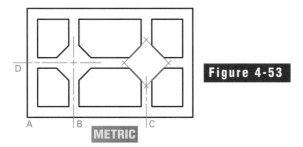

Figure 4-53

METRIC

4. The drawing in Figure 4-54 provides practice in metric measurement. Draw horizontal line AB 180 mm long. Work clockwise around the layout. Remember: Angular dimensions are the same in the U.S. Customary and metric systems. BC = 60 mm; CD = 48 mm; DE = 42 mm; EF = 74 mm; FG = 50 mm; GH = 90 mm. Measure the closing line and measure and label the angle at H. Sheet size: A4. Scale: 1:1.

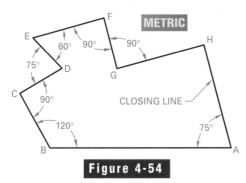

Figure 4-54

5. Practice using a civil engineer's scale by drawing the land parcel shown in Figure 4-55. Measure the length of the closing line to the nearest tenth of a foot and note it on your drawing. Sheet size: B. 1″ = 40′-0.

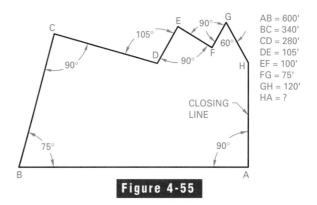

Figure 4-55

6. Draw the head gasket shown in Figure 4-56. The distance between center points A and B is 45.5 mm. The radius of arc C is 30 mm. The radius of arc D is 43 mm. The diameter of circle E is 15 mm. The diameter of circle F is 45.5 mm. Holes labeled G have a diameter of 7 mm. Scale: As assigned.

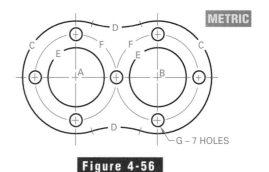

Figure 4-56

7. Draw each item in Figure 4-57. Before you begin each drawing, determine a suitable sheet size and scale. Include all centerlines. Do not dimension.

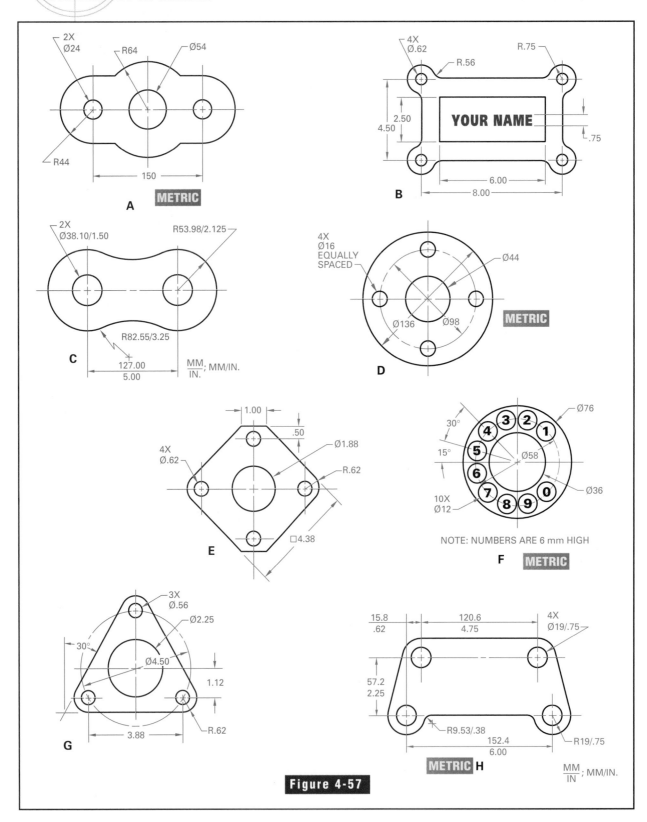

Figure 4-57

8. Draw each item in Figure 4-58. Determine a suitable sheet size and scale. Include all centerlines. Do not dimension.

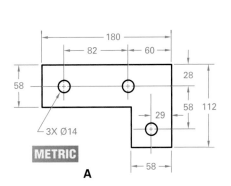

A

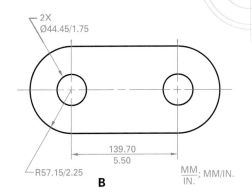

B

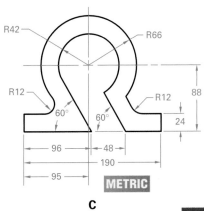

C

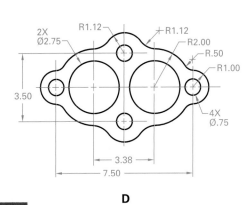

D

Figure 4-58

9. Draw the offset bracket shown in Figure 4-59. Locate all center points before beginning to draw circles and arcs. Do not dimension.

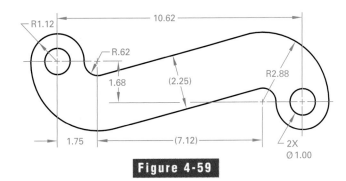

Figure 4-59

10. Draw each item shown in Figure 4-60. Locate points of tangency when required. Determine an appropriate sheet size and scale. Include all centerlines. Do not dimension.

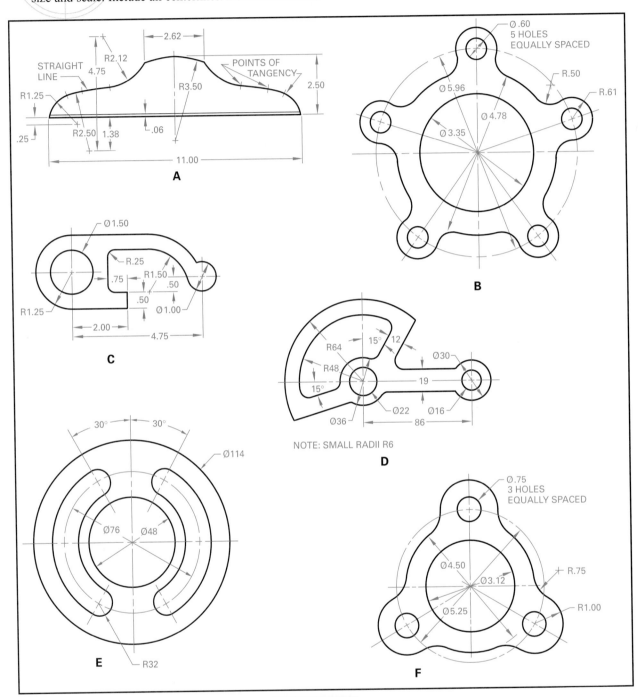

Figure 4-60

11. Draw the frame shown in Figure 4-61. Locate all centerlines before beginning to draw the frame. Lines AO and OB are 134 mm each. Lines OC and OD are 58 mm each. Radii from points A, B, C, and D are 28 mm, 38 mm, and 58 mm. Scale: 1:2.

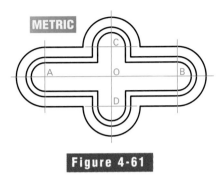

Figure 4-61

12. Draw the multiple dial plate shown in Figure 4-62. Create the centerlines at right angles. Distances FC, FD, FG, FE, EA, and GB are 6.00″ each. The diameter of the inner ring with center F is 4.50″. Diameters of all other inner rings are 4.00″. Scale: 3″ = 1′-0 (1:4).

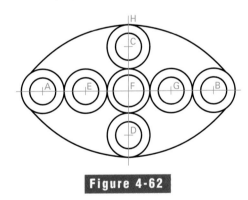

Figure 4-62

13. Draw the double dial plate shown in Figure 4-63. Line AB = 7.00″, and distances AC, CD, and DB are equal. Radii of inner arcs with centers at C and D = 1.50″ and 1.75″, respectively. Radii of outer arcs with centers E and F are 3.75″ and 4.00″ respectively. Scale: Full size.

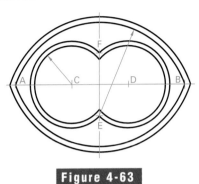

Figure 4-63

14. Draw the paul shown in Figure 4-64. Scale: 1:1.

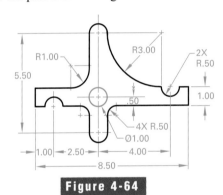

Figure 4-64

15. Draw the tilt scale shown in Figure 4-65. Scale: 1:1.

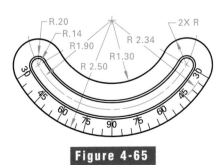

Figure 4-65

16. Construct the irregular polygon shown in Figure 4-66. Begin by drawing line AA centered near the bottom of the sheet. The length of each line is given at the left of the polygon.

Figure 4-66

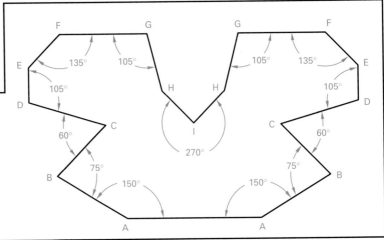

```
AA  =  11.00
AB  =   8.00
BC  =   7.00
CD  =   8.00
DE  =   3.00
EF  =   4.00
FG  =   8.50
GH  =   5.00
HI  =   3.75
```

17. Draw the wire rope hook shown in Figure 4-67 using the dimensions selected by your instructor. Determine the radii necessary for smooth tangencies. If you are using board drafting techniques, prepare this drawing in ink or as specified by the instructor.

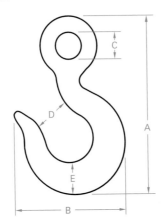

Wire Rope Hook				
A	B	C	D	E
4.94	3.20	.88	1.06	.84
5.44	3.50	1.00	1.12	.90
6.25	4.10	1.12	1.25	1.12
6.88	4.54	1.25	1.38	1.30
7.62	4.88	1.38	1.50	1.38
8.60	5.75	1.50	1.70	1.56
9.50	6.38	1.16	1.88	1.70

Figure 4-67

18. Prepare a two-view drawing of the solid cast-steel thimble shown in Figure 4-68 using the dimensions selected by your instructor. Select suitable radii for all curves. Note the points of tangencies to determine the tapered shape. If you are using board drafting techniques, prepare this drawing in ink or as specified by the instructor.

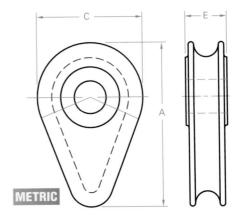

METRIC

Cast-Steel Thimble			
Wire Rope Ø	A	C	E
12	78	54	22
16	128	86	32
20	128	86	32
22	166	114	36
24	166	114	36

Figure 4-68

19. Draw the gasket shown in Figure 4-69. Scale: 1:2.

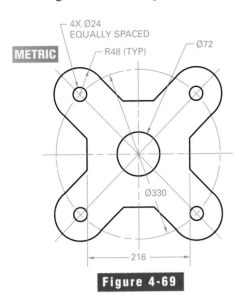

4X Ø24
EQUALLY SPACED

METRIC

R48 (TYP)

Ø72

Ø330

216

Figure 4-69

20. Draw a gasket for the bottom of the guide block shown in Figure 4-70. It should be shaped so that when cut out, it will touch only the metal surface on the bottom. Scale: 1:1.

Figure 4-70

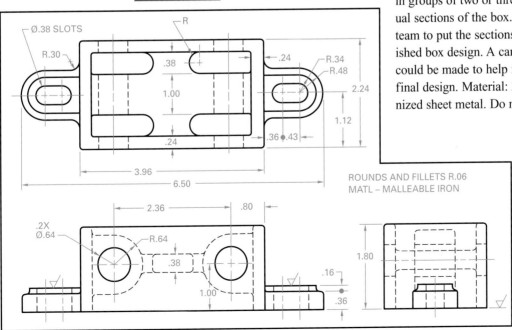

Ø.38 SLOTS

R

R.30

.38

.24

R.34

R.48

1.00

2.24

1.12

Ø330

.24

.36 .43

3.96

6.50

ROUNDS AND FILLETS R.06
MATL – MALLEABLE IRON

2.36

.80

.2X
Ø.64

R.64

.38

1.80

.16

1.00

.36

Design Problems

Design problems have been prepared to challenge individual students or teams of students. In these problems, you are to apply skills learned mainly in this chapter, but also in other chapters throughout the text. The problems are designed to be completed using board drafting, CAD, or a combination of the two. Be creative and have fun!

1. Design a nightstand caddy to hold a watch, wallet, coins, jewelry, and other items. First, sketch your design ideas and then prepare instrument or CAD drawings as assigned. Do not dimension. Material: Optional.

2. **TEAMWORK** Design a drawer divider to hold 3½″ disks, CDs, audiocassette tapes, and DVDs. Inside drawer size: 6″ deep × 12″ wide × 26″ long. Material: Optional. Do not dimension. Each team member should first develop design sketches. The entire team should then select the best design (or combined design). Finally, each team member should prepare finished drawings of the final design.

3. **TEAMWORK** Design a fishing-tackle box. The design team should first make a list of items and quantities of items that the box is to accommodate. Next, work in groups of two or three to design individual sections of the box. Then work as a full team to put the sections together into a finished box design. A cardboard version could be made to help in visualizing the final design. Material: 22-gauge galvanized sheet metal. Do not dimension.

CHAPTER 5

Geometry for Drafting

OBJECTIVES

Upon completion of this chapter, you should be able to:

■ Identify and describe various geometric shapes and constructions used by drafters.

■ Construct various geometric shapes accurately.

■ Solve technical and mathematical problems through geometric constructions using drafting instruments.

■ Solve technical and mathematical problems through geometric constructions using a CAD system.

■ Use geometry to reduce or enlarge a drawing or to change the proportions of a drawing.

Geometry has always been important to people. It was used in ancient times for measuring land and making right-angle corners for buildings and other kinds of construction. Egyptian rope-stretchers used rope with marks or knots at 12 equal spaces. The rope was divided into 3-, 4-, and 5-space sections, as shown in Figure 5-1. A square corner was made by stretching the rope and driving pegs into the ground at the 3-, 4-, and 5-space marks. This was one way an ancient people used geometry.

The theory that the 3-4-5 triangle makes a right angle was proved by the mathematician Pythagoras in the sixth century B.C. This proof is called the **Pythagorean theorem**. The theorem is shown graphically and mathematically in Figure 5-2.

This method also works well for triangles that have the same proportions, such as 6, 8, and 10 units:

$$6^2 + 8^2 = 100$$
$$36 + 64 = 100$$
$$100 = 100$$

The units may be millimeters, meters, inches, fractions of an inch, or any other units of measure.

Geometry and Geometric Constructions

Geometry is the study of the size and shape of things. The relationship of straight and curved lines in drawing shapes is also a part of geometry. Geometric figures used in drafting include circles, squares, triangles, hexagons, and octagons. Many other shapes are shown in Figure 5-3.

Figure 5-1

Egyptian rope-stretchers used 3-4-5 triangles to lay out square corners for buildings.

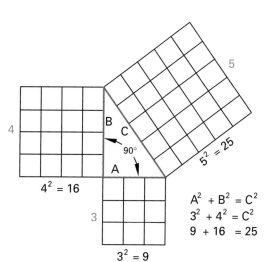

$A^2 + B^2 = C^2$
$3^2 + 4^2 = C^2$
$9 + 16 = 25$

Figure 5-2

Pythagorean theorem shown graphically and mathematically.

Figure 5-3

Dictionary of drafting geometry.

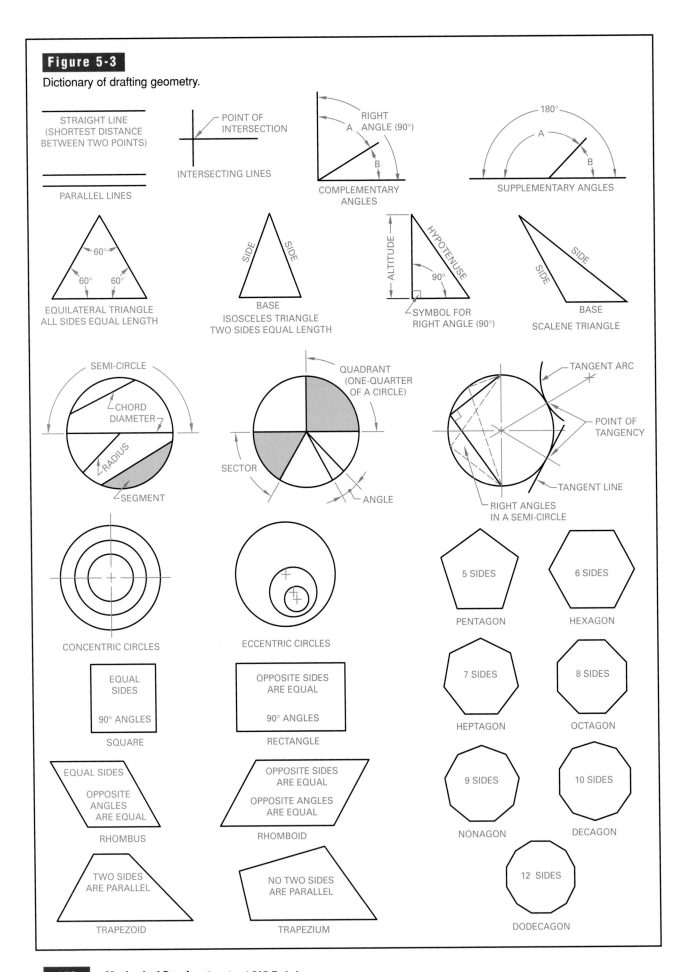

Geometry Formulas

In addition to solving drafting problems using geometric constructions, drafters often need to be able to calculate various aspects of geometric constructions. For example, earlier in this chapter, the Pythagorean theorem ($a^2 + b^2 = c^2$) was used to determine the length of the hypotenuse of a right triangle. While hundreds of these formulas exist, a few are given here as examples. Others can be found in any geometry textbook and should be looked up as needed.

- To find the area of any triangle:
  Multiply the base (b) times the height (h) and divide by two.
  Area = $bh/2$
  Area = $2 \times 6/2$
  Area = 6 square inches

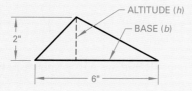

- To find the circumference of a circle:
  Multiply pi (π) times the diameter of the circle. The approximate decimal equivalent of pi is 3.1416.
  Circumference = πd
  Circumference = 3.1416×2.50
  Circumference = 7.85

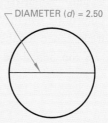

Illustrations made of individual lines and points drawn in proper relationship to one another are known as **geometric constructions**. Geometric constructions are used by drafters, surveyors, engineers, architects, scientists, mathematicians, and designers. Nearly everyone in all technical fields needs to understand the constructions explained in this chapter. Study Figure 5-3 before beginning the geometric constructions in the Board Drafting Techniques and CAD Techniques parts of this chapter.

To understand the geometric constructions, you must understand how the lines, arcs, and other shapes are described. This chapter follows the identification rules used in geometry. Lines and arcs are described using their endpoints. Therefore, line AB is a line segment that extends from point A to point B. Arc AB is an arc whose endpoints are A and B. Angles are described using three points: both endpoints and the **vertex**, or the point at which the two

arms of the angle meet. Angle ABC is an angle whose endpoints are A and C and whose vertex is at point B. Circles are usually specified using their center points, so circle A is a circle whose center is at point A. See Figure 5-4.

Figure 5-4

Identification of lines, angles, arcs, and circles.

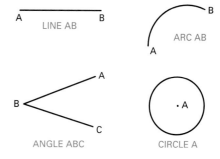

Solving Problems Using Geometry

One of the purposes of geometric construction is to help drafters solve design problems. The constructions you will practice in this chapter may seem at first to have little application to everyday problems. However, drafters frequently use these techniques and others to help solve practical design problems.

For example, suppose a designer is designing a complicated mechanical assembly that has moving parts. The designer must determine whether the current design allows enough clearance for the parts to move freely. Depending on the actual design, the drafter may need to construct accurate circles and various tangent, parallel, and perpendicular lines to measure clearances.

In other cases, designers are given very specific parameters, or guidelines, within which they must work. For example, a designer may be told that the part must fit within a predefined area. Again, the drafter uses principles of geometric construction to meet the specifications.

What other practical design and drafting problems can be solved using geometric constructions? As you complete the constructions in this chapter, think about practical drawing problems that could be solved using each technique.

Board Drafting Techniques

This section consists of a series of example exercises in which you will use board drafting techniques to create various geometric constructions. By working through these constructions, you will begin to understand how to draw the basic geometry used in drafting.

Bisect a Line or Arc

This construction demonstrates how to bisect a line or arc. **Bisect** means to divide into two equal parts. Follow these steps to bisect the line and arc.

1. Draw line AB and arc AB as shown in Figure 5-5A.
2. With points A and B as centers and any radius R greater than one half of AB, draw arcs to **intersect**, or cross, as shown in Figure 5-5B. The **radius** is the distance from the center of an arc or circle to any point on the arc or circle. The two places where the arcs intersect create points C and D.
3. Draw line EF through points C and D, as shown in Figure 5-5C.

Figure 5-5

Bisecting a straight line or arc.

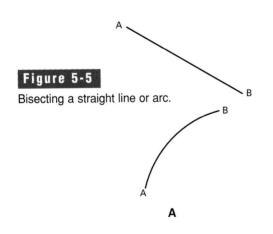

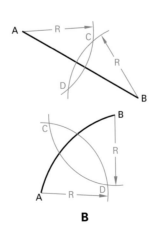

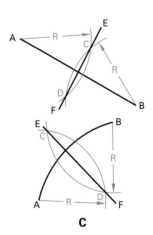

A

B

C

Divide a Line into Equal Parts

Two methods of dividing a line into equal parts are described below. Try both methods. Can you think of situations in which you would need to use one method instead of the other?

Method 1

In this construction, you will divide a straight line into eight equal parts. This method can be applied to create any number of equal divisions. Refer to Figure 5-6 and follow these steps:

1. Draw a line of any length at A perpendicular to line AB, as shown in Figure 5-6A. Lines are **perpendicular** when they cross at 90° angles.

2. Position the scale, placing zero on line AC at such an angle that the scale touches point B, as shown in Figure 5-6B. Keeping zero on line AC, adjust the angle of the scale until any eight equal divisions are included between line AC and point B (in this case, 8″). Mark the divisions.

3. Draw lines parallel to AC through the division marks to intersect line AB, as shown in Figure 5-6C. Two lines are **parallel** when they are always the same distance apart.

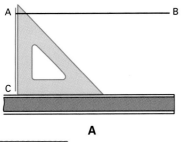
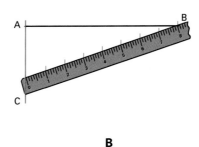
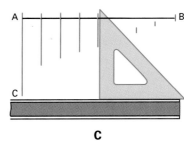

A   **B**   **C**

Figure 5-6
Dividing a straight line into any number of equal parts (first method).

Method 2

Follow the instructions below to divide a line into five equal parts.

1. Draw line BC from point B at any convenient angle and length, as shown in Figure 5-7A.

2. Use dividers or a scale to step off five equal spaces on line BC beginning at point B, as shown in Figure 5-7B.

3. Draw a line connecting point A and the last point on line BC, as shown in Figure 5-7C. Draw lines through each point on BC parallel to this line as shown.

Figure 5-7
Dividing a straight line into any number of equal parts (second method).

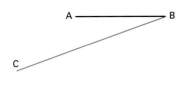
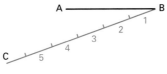
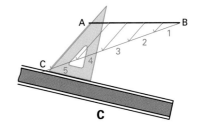

A   **B**   **C**

Draw a Perpendicular Line

Many procedures exist for constructing a line that is perpendicular to another line. Each method is useful in certain drafting situations. Five methods are shown in this section.

Method 1

Figure 5-8A shows given line AB and point O that lies on line AB. Follow the steps below to draw a line at point O on line AB so that the two lines are perpendicular.

1. Draw line AB and point O, as shown in Figure 5-8A.
2. With O as the center and any convenient radius R_1, construct an arc intersecting line AB, locating points C and D. See Figure 5-8B.
3. With C and D as centers and any radius R_2 greater than OC, draw arcs intersecting at point E, as shown in Figure 5-8C.
4. Draw a line connecting points E and O to form the perpendicular line, as shown in Figure 5-8C.

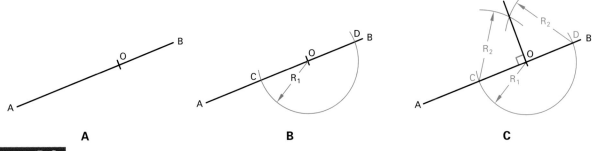

Figure 5-8

Constructing a line perpendicular to a given line through a given point on the line (first method).

Method 2

This construction presents another method of drawing a line perpendicular to a given line at a given point on that line. Use this method when the given point lies near one end of the line.

1. Draw given line AB and point O, as shown in Figure 5-9A.
2. From any point C above line AB, construct an arc using CO as the radius and passing through line AB to locate point D, as shown in Figure 5-9B.
3. Draw a line through points D and C, extending it through the arc to locate point E, as shown in Figure 5-9C.
4. Connect points E and O to form the perpendicular line, as shown in Figure 5-9C.

Figure 5-9

Constructing a line perpendicular to a given line through a given point on the line (second method).

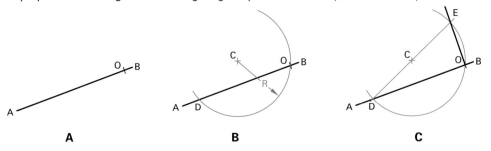

Method 3

This construction demonstrates another way to draw a line perpendicular to a given line through a given point on the line. Follow the steps to create a line at O that is perpendicular to line AB. Refer to Figure 5-10.

1. Draw given line AB and point O, as shown in Figure 5-10A.

2. Place the T-square and triangle, as shown in Figure 5-10B.
3. Slide the triangle along the T-square until the edge aligns with point O on line AB, as shown in Figure 5-10C.
4. Draw a perpendicular line through point O, as shown in Figure 5-10C.

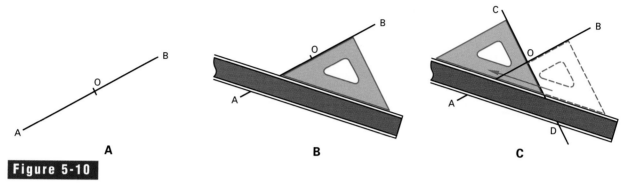

Figure 5-10

Constructing a line perpendicular to a given line through a given point on the line (third method).

Method 4

This exercise demonstrates a method of constructing a line perpendicular to a given line through a point that does *not* lie on the given line. Follow the instructions below to construct a line through point O that is perpendicular to line AB. See Figure 5-11.

1. Draw given line AB and point O, as shown in Figure 5-11A.

2. With O as the center, draw an arc with radius R_1, long enough to intersect line AB to locate points C and D. See Figure 5-11B.
3. With C and D as centers and radius R_2 greater than one half of CD, draw intersecting arcs to locate point E, as shown in Figure 5-11C.
4. Draw a line through points O and E to form the perpendicular line, as shown in Figure 5-11C.

Figure 5-11

Constructing a line perpendicular to a given line through a point that is not on the given line (first method).

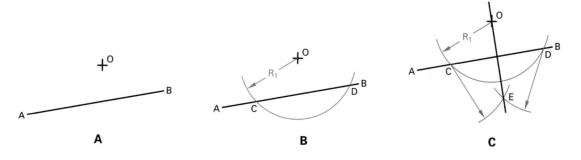

Method 5

Figure 5-12A shows another given line AB and point O that is not on the line. Follow the steps to practice another way to draw a line perpendicular to a given line through a point that is not on the line.

1. Draw given line AB and point O, as shown in Figure 5-12A.
2. Draw lines from point O to any two points on line AB, locating points C and D, as shown in Figure 5-12B.
3. With C and D as centers and CO and DO as radii, draw arcs to intersect, locating point E, as shown in Figure 5-12C.
4. Connect points O and E to form the perpendicular line, as shown in Figure 5-12C.

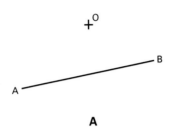
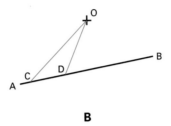
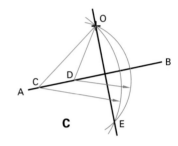

A **B** **C**

Figure 5-12

Constructing a line perpendicular to a given line through a point that is not on the given line (second method).

Draw a Parallel Line

The following construction methods create a line that is parallel to another line. Recall that lines are parallel when they are always the same distance apart.

Method 1

This construction allows you to place a line parallel to a given line. Refer to Figure 5-13.

1. Draw given line AB and point P, as shown in Figure 5-13A.
2. With point P as the center and any convenient radius R_1, draw an arc intersecting line AB to locate point C, as shown in Figure 5-13B.
3. With point C as the center and the same radius R_1, draw an arc through point P and line AB to locate point D. See Figure 5-13B.
4. With C as the center and radius R_2 equal to chord PD, draw an arc to locate point E. A **chord** is a straight line between two points on a circle. See Figure 5-13C.
5. Draw a line through points P and E. Line PE is parallel to line AB, as shown in Figure 5-13C.

Figure 5-13

Using a compass to construct a line parallel to a given line through a given point.

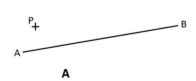
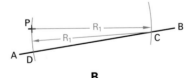
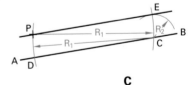

A **B** **C**

Method 2

The following steps demonstrate another way to construct a line parallel to another line through a given point. Refer to Figure 5-14.

1. Draw given line AB and point P, as shown in Figure 5-14A.

2. Place the T-square and triangle, as shown in Figure 5-14B.

3. Slide the triangle until the edge aligns with point P, as shown in Figure 5-14C.

4. Draw a parallel line through point P. See Figure 5-14C.

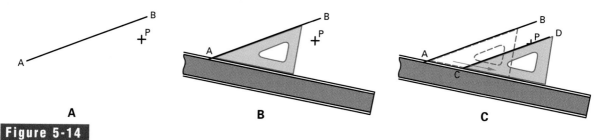

Figure 5-14

Using a triangle and T-square to construct a line parallel to a given line through a given point.

Method 3

Use this method to construct a line parallel to a given line at a specified distance from the given line. Refer to Figure 5-15. *Note:* See "Construct a Tangent Line" later in this chapter for instructions to create a tangent line.

1. Draw given line AB, as shown in Figure 5-15A.

2. Draw two arcs with centers anywhere along line AB. The arcs should have a radius R equal to the specified distance between the two parallel lines. See Figure 5-15B.

3. Draw a parallel line CD tangent to the arcs. Recall that a line is tangent to an arc or circle when it touches the arc or circle at one point only. See Figure 5-15C.

Figure 5-15

Constructing a line parallel to a given line at a given distance from the line.

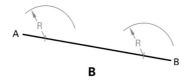

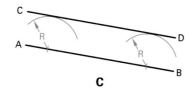

Bisect an Angle

This construction demonstrates a method of bisecting a given angle. Refer to Figure 5-16.

1. Draw given angle AOB, as shown in Figure 5-16A.
2. With point O as the center and any convenient radius R_1, draw an arc to intersect AO and OB to locate points C and D. See Figure 5-16B.
3. With C and D as centers and any radius R_2 greater than one half the radius of arc CD, draw two arcs to intersect, locating point E. See Figure 5-16C.
4. Draw a line through points O and E to bisect angle AOB, as shown in Figure 5-16C.

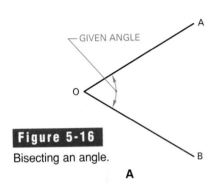

Figure 5-16
Bisecting an angle.

A

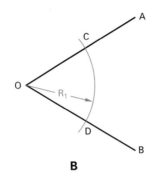

B

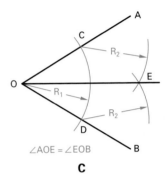

$\angle AOE = \angle EOB$

C

Construct an Angle

This construction demonstrates a method of copying a given angle to a new location and orientation. Refer to Figure 5-17.

1. Draw given angle AOB, as shown in Figure 5-17A.
2. Draw one side O_1A_1 in the new position, as shown in Figure 5-17B.
3. With O and O_1 as centers and any convenient radius R_1, construct arcs to intersect BO and AO at C and D and A_1O_1 at D_1. Refer again to Figure 5-17B.
4. With D_1 as the center and radius R_2 equal to chord DC, draw an arc to locate point C_1 at the intersection of the two arcs, as shown in Figure 5-17C.
5. Draw a line through points O_1 and C_1 to complete the angle. See Figure 5-17C.

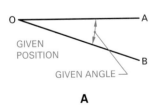

A

Figure 5-17
Constructing an angle.

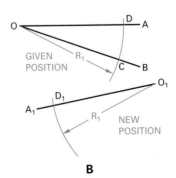

B

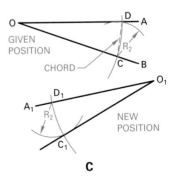

C

Construct a Triangle

A **triangle** is a **polygon**, or closed figure, that contains three sides. The following constructions show methods for drawing various types of triangles.

Method 1

This method constructs an isosceles triangle. An **isosceles** triangle is one in which two sides are of equal length. Refer to Figure 5-18.

1. Draw base line AB, as shown in Figure 5-18A.

2. With points A and B as centers and a radius R equal to the length of the sides you want, draw intersecting arcs to locate the third vertex of the triangle. See Figure 5-18B. The other two vertices (plural of *vertex*) are at the endpoints of the base line.

3. Draw lines through point A and the vertex and through point B and the vertex to complete the triangle. See Figure 5-18C.

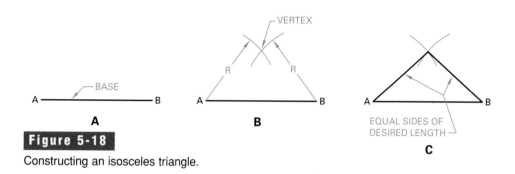

Figure 5-18

Constructing an isosceles triangle.

Method 2

This method constructs an equilateral triangle. An **equilateral** triangle is one in which all three sides are of equal length and all three angles are equal. Refer to Figure 5-19.

1. Draw base line AB as shown in Figure 5-19A.

2. With points A and B as centers and a radius R equal to the length of line AB, draw intersecting arcs to locate the third vertex. See Figure 5-19B.

3. Draw lines through point A and the vertex and through point B and the vertex to complete the triangle, as shown in Figure 5-19C.

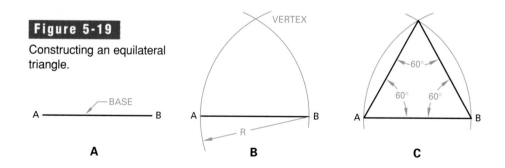

Figure 5-19

Constructing an equilateral triangle.

Method 3

An equilateral triangle may also be constructed by drawing 60° lines through the ends of the base line with the 30°-60° triangle. This method is demonstrated in Figure 5-20.

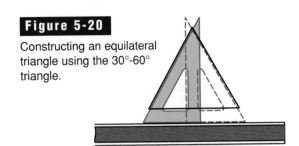

Figure 5-20

Constructing an equilateral triangle using the 30°-60° triangle.

Method 4

Construct a right triangle using this method when you know the length of two sides of the triangle. A **right triangle** is one that has a right (90°) angle at one of its vertices. Given sides AB and BC are shown in Figure 5-21A.

1. Draw side AB in the desired position, as shown in Figure 5-21B.

2. Draw a line perpendicular to AB at B equal to BC. *Note:* Construct the perpendicular line using the method shown in Figure 5-9 or the method shown in Figure 5-12.

3. Draw a line connecting points A and C to complete the right triangle. See Figure 5-21C.

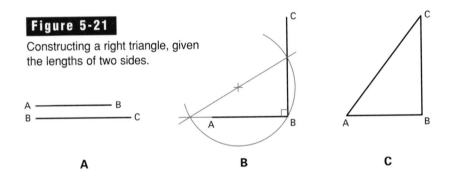

Figure 5-21

Constructing a right triangle, given the lengths of two sides.

A   **B**   **C**

Method 5

Use this method to construct a right triangle when you know the length of one side and the length of the hypotenuse. The **hypotenuse** of a right triangle is the side opposite the 90° angle. Given side AB and hypotenuse AC are shown in Figure 5-22A.

1. Draw the hypotenuse AC in the desired location, as shown in Figure 5-22B.

2. Draw a semicircle on AC using ½AC as the radius. Refer again to Figure 5-22B.

3. With point A as the center and a radius equal to side AB, draw an arc to intersect the semicircle to locate point B, as shown in Figure 5-22C.

4. Draw line AB and then draw a line to connect B and C to complete the triangle, as shown in Figure 5-22C.

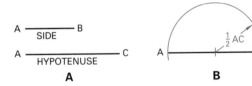

A   **B**   **C**

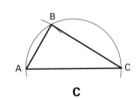

Figure 5-22

Constructing a right triangle, given the length of one side and the length of the hypotenuse.

Method 6

This construction illustrates the 3-4-5 method of drawing a right triangle. Refer to Figure 5-23.

1. Draw a base line AB that is 3 units long, as shown in Figure 5-23A.

2. With A and B as centers and radii 4 and 5 units long, draw intersecting arcs to locate point C, as shown in Figure 5-23B.

3. Draw lines AC and BC to complete the triangle. See Figure 5-23C.

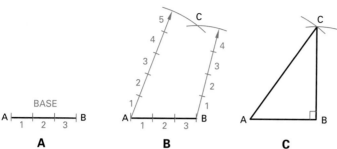

Figure 5-23

Constructing a right triangle using the 3-4-5 method, given a three-unit base.

Method 7

Use this method to construct a triangle when you know the lengths of all three sides. This construction is useful for scalene triangles. **Scalene** triangles are those that include three different angles and sides of three different lengths. Figure 5-24A shows given triangle sides AB, BC, and AC.

1. Draw base line AB in the desired location.

2. Construct arcs from the ends of line AB with radii equal to lines BC and AC to locate point C, as shown in Figure 5-24B.

3. Connect both ends of line AB with point C to complete the triangle, as shown in Figure 5-24C.

Figure 5-24

Constructing a triangle, given the lengths of all three sides.

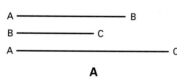

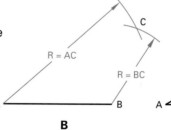

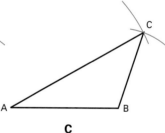

Construct a Circle

This construction describes a method of creating a circle given three points that lie on the circle. Refer to Figure 5-25.

1. Given points A, B, and C, draw lines AB and BC, as shown in Figure 5-25A.

2. Draw perpendicular bisectors of AB and BC to intersect at point O, as shown in Figure 5-25B.

3. Draw the required circle with point O as the center and radius R = OA = OB = OC, as shown in Figure 5-25C.

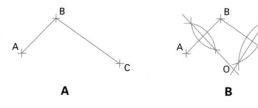

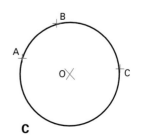

Figure 5-25

Constructing a circle, given three points that lie on the circle.

Construct a Tangent Line

The constructions that follow present methods of creating lines tangent to a circle. As you may recall, a line is tangent to a circle if the line touches the circle at one point only.

Method 1

Use this method to construct a line tangent to a given point on a circle without using a triangle or T-square. Refer to Figure 5-26.

1. Given circle with center point O and tangent point P, as shown in Figure 5-26A, draw line OA from the center of the circle to extend beyond the circle through point P.
2. Draw a line perpendicular to line OA at P, as shown in Figure 5-26B. The perpendicular line is the tangent line.

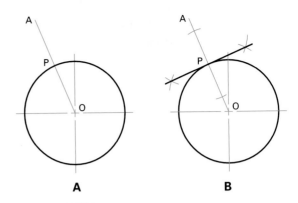

A   **B**

Figure 5-26

Constructing a line tangent to a circle through a given point on the circle (first method).

Method 2

Use this method to construct a line tangent to a given point on a circle using a 30°-60° triangle and a T-square. Refer to Figure 5-27.

1. Given a circle with center point O and tangent point P, place a T-square and triangle so that the hypotenuse of the triangle passes through points P and O.
2. Hold the T-square, turn the triangle to the second position at point P, and draw the tangent line.

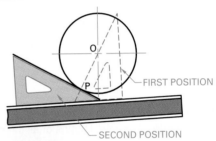

Figure 5-27

Constructing a line tangent to a given point on a circle (second method).

Method 3

This construction creates lines tangent to a circle from a given point outside the circle. Refer to Figure 5-28.

1. Draw a circle with center point O and point P outside the circle, as shown in Figure 5-28A.
2. Draw line OP and bisect it to locate point A, as shown in Figure 5-28B.
3. Draw a circle with center A and radius R = AP = AO to locate tangent points T_1 and T_2. Refer again to Figure 5-28B.
4. Draw lines PT_1 and PT_2 as shown in Figure 5-28C. These lines are tangent to the circle.

A

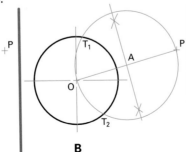

B

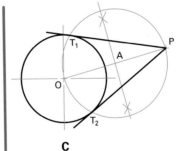

C

Figure 5-28

Constructing a line tangent to a circle from a given point outside the circle.

Method 4

Use this method to construct a line tangent to the exterior of two circles. Refer to Figure 5-29.
1. Draw the two given circles with centers O_1 and O_2 and radii R_1 and R_2. See Figure 5-29A.
2. Draw a circle with center O_1 and a radius R, where $R = R_1 - R_2$. Refer again to Figure 5-29A.

3. From center point O_2, draw a tangent O_2T to the circle of radius R, as shown in Figure 5-29B.
4. Draw radius O_1T as shown in Figure 5-29B, and extend it to locate point T_1.
5. Draw the needed tangent T_1T_2 parallel to TO_2, as shown in Figure 5-29C.

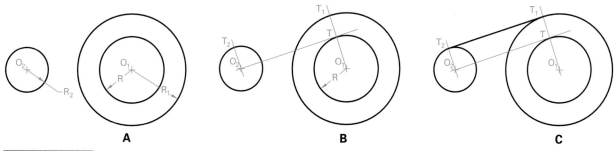

Figure 5-29

Constructing an exterior common tangent to two circles of unequal radii.

Method 5

Use this method to construct a line tangent to the interior of two circles. Refer to Figure 5-30.
1. Draw the two given circles with centers O_1 and O_2 and radii R_1 and R_2, as shown in Figure 5-30A.
2. Draw a circle with center O_1 and a radius R, where $R = R_1 + R_2$. Refer again to Figure 5-30A.

3. From center point O_2, draw a tangent O_2T to the circle of radius R, as shown in Figure 5-30B.
4. Draw radius O_1T to locate point T_1, as shown in Figure 5-30B.
5. Draw O_2T_2 parallel to O_1T.
6. Draw the needed tangent T_1T_2 parallel to TO_2, as shown in Figure 5-30C.

Figure 5-30

Constructing an interior common tangent to two circles of unequal radii.

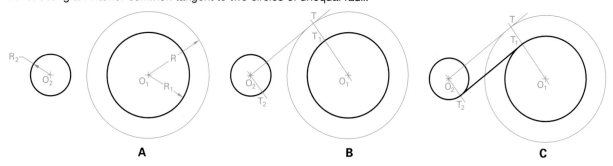

Construct a Tangent Arc

The following constructions demonstrate methods to draw arcs tangent to other geometry, such as straight lines and other arcs.

Method 1

Use this method to construct an arc tangent to two straight lines. The technique is shown for two lines at an acute angle, an obtuse angle, and a right angle. Refer to Figure 5-31.

1. Given lines AB and CD, as shown in Figure 5-31A, draw lines parallel to AB and CD at a distance R from them on the inside of the angle. The intersection O will be the center of the arc you need.
2. Draw perpendicular lines from O to AB and CD to locate the points of tangency T, as shown in Figure 5-31B.
3. With O as the center and radius R, draw the needed arc, as shown in Figure 5-31C.

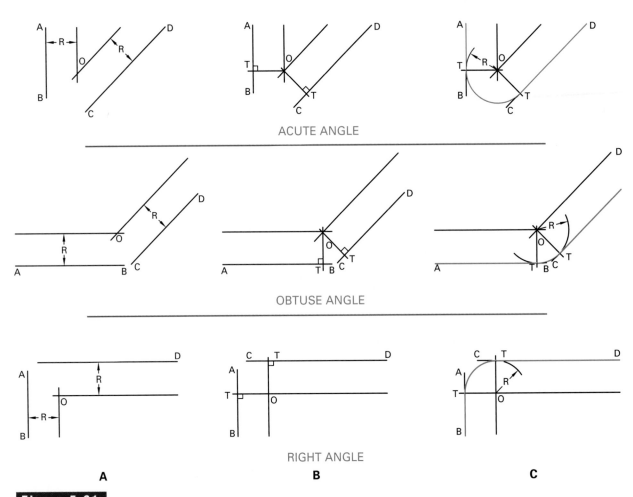

ACUTE ANGLE

OBTUSE ANGLE

RIGHT ANGLE

A   B   C

Figure 5-31

Constructing an arc tangent to two straight lines at an acute angle, an obtuse angle, and a right angle.

Method 2

Use this method to construct an arc tangent to two given arcs. Refer to Figure 5-32.

1. Draw two arcs having radii R_1 and R_2, as shown in Figure 5-32A. The radii R_1 and R_2 may be equal or unequal.
2. Draw an arc with center O_1 and radius $= R + R_1$, where R is the radius of the desired tangent arc.

See Figure 5-32B. The intersection O is the center of the tangent arc.

3. Draw lines O_1O and O_2O to locate tangent points T_1 and T_2, as shown in Figure 5-32C.
4. With point O as the center and radius R, draw the tangent arc needed.

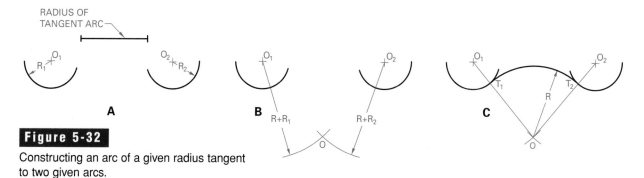

Figure 5-32

Constructing an arc of a given radius tangent to two given arcs.

Method 3

Use this method to construct an arc tangent to a line and an arc, given the line, the arc, and the radius R of the desired tangent arc. Refer to Figure 5-33.

1. Draw given line AB and arc CD as shown in Figure 5-33A.
2. Draw a line parallel to line AB, at distance R, toward arc CD. See Figure 5-33B.

3. Use radius $R_1 + R$ to locate point O_1. Refer again to Figure 5-33B.
4. Draw a line from O_1 perpendicular to AB to locate tangent point T.
5. Draw a line from O to O_1 to locate tangent point T_1 on CD, as shown in Figure 5-33C.
6. With point O_1 as the center and radius R, draw the tangent arc.

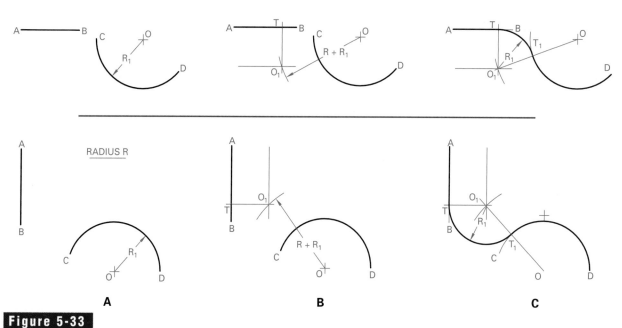

Figure 5-33

Constructing an arc of given radius tangent to an arc and a straight line.

Construct an Ogee Curve

An **ogee curve** is a reverse curve that looks something like an S. To construct an ogee curve, refer to Figure 5-34 and follow the steps below.

1. Draw the given lines AB and CD, as shown in Figure 5-34A, and connect them by drawing line BC.
2. Select a point E on line BC, through which the curve is to pass.
3. Draw perpendicular bisectors of BE and EC, as shown in Figure 5-34B.
4. Draw lines perpendicular to AB at B and to CD at C. Refer again to Figure 5-34B. The lines must cross the bisectors of BE and EC at O_1 and O_2, respectively.
5. To complete the curve, draw one arc with center O_1 and radius O_1E, and the other with center O_2 and radius O_2E. See Figure 5-34C.

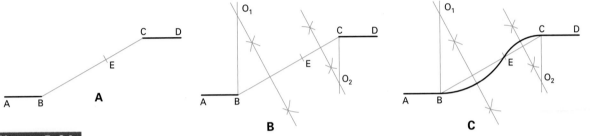

Figure 5-34

Constructing a reverse, or ogee, curve.

Construct a Square

There are several ways to construct a square. The method you choose depends on the rest of the geometry in the drawing.

Method 1

Use this method to construct a square when you know the length of a side. Refer to Figure 5-35.

1. Given the length of the side AB, draw line AB.
2. Construct 45° diagonals from the ends of line AB, as shown in Figure 5-35.
3. Complete the square by drawing perpendicular lines at each end of line AB to intersect the diagonals. Draw the last line from the intersection of the diagonal and the vertical lines. Draw the lines in the order shown by the numbered arrows.

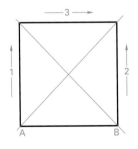

Figure 5-35

Constructing a square, given the length of a side.

Method 2

Use this method to construct a square inscribed in a circle. A square or other polygon is **inscribed** in a circle when its four corners are tangent to the circle. Refer to Figure 5-36.
1. Draw the given circle with center point O.
2. Draw 45° diagonals through the center point O to locate points A, B, C, and D, as shown in Figure 5-36.
3. Connect points A and B, B and C, C and D, and D and A to complete the square.

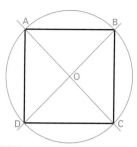

Figure 5-36
Constructing a square inscribed within a circle.

Method 3

Use this method to construct a square circumscribed in a circle. A square or other polygon is **circumscribed** about a circle when the square fully encloses the circle and the circle is tangent to the square on all four sides. Refer to Figure 5-37.
1. Draw the given circle with center point O.
2. Draw 45° diagonals through the center point O.
3. Draw sides tangent to the circle, intersecting at the 45° diagonals, to complete the square.

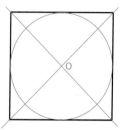

Figure 5-37
Constructing a square circumscribed about a circle.

Construct a Pentagon

A **pentagon** is a five-sided polygon. When all five sides of a polygon are exactly the same length and all its angles are equal, it is called a **regular polygon**. The following constructions demonstrate methods for constructing regular pentagons.

Method 1

Use this method to construct a regular pentagon when you know the length of one side. Refer to Figure 5-38.

1. Given line AB, construct a perpendicular line AC equal to one half the length of AB, as shown in Figure 5-38A.
2. Draw line BC and extend it to make line CD equal to AC. Refer again to Figure 5-38A.
3. With radius AD and points A and B as centers, draw intersecting arcs to locate point O, as shown in Figure 5-38B.
4. With the same radius and O as the center, draw a circle.
5. Step off AB as a chord to locate points E, F, and G. Connect the points to complete the pentagon, as shown in Figure 5-38C.

Figure 5-38
Constructing a regular pentagon, given the length of one side.

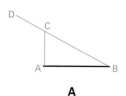

A

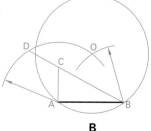

B

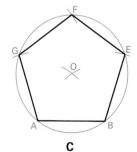

C

Chapter 5 • Geometry for Drafting   **169**

Method 2

This construction demonstrates a method of inscribing a pentagon within a circle. Refer to Figure 5-39.

1. Draw the given circle with diameter AB and radius OC, as shown in Figure 5-39A. The **diameter** of a circle is the distance across the circle through its center point. The symbol for diameter is ⌀.
2. Bisect radius OB to locate point D, as shown in Figure 5-39B.
3. With D as center and radius DC, draw an arc to locate point E.
4. With C as center and radius CE, draw an arc to locate point F.
5. Draw chord CF. This chord is one side of the pentagon.
6. Step off chord CF around the circle to locate points G, H, and J. Draw the chords to complete the pentagon, as shown in Figure 5-39C.

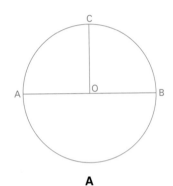
A

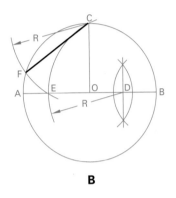
B

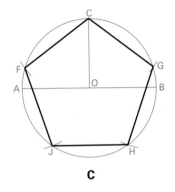
C

Figure 5-39

Inscribing a regular pentagon within a circle, given the circle.

Construct a Hexagon

A **hexagon** is a six-sided polygon. The following constructions demonstrate methods for constructing regular hexagons.

Method 1

Use this method to construct a regular hexagon when you know the distance across the flats, or sides. The distance across the flats is the distance from the midpoint of one side of the polygon through the center point to the midpoint of the opposite side of the polygon. Refer to Figure 5-40.

1. Given the distance across the flats of a regular hexagon, draw centerlines and a circle with a diameter equal to the distance across the flats.
2. With the T-square and 30°-60° triangle, draw the tangents in the order shown in Figure 5-40.

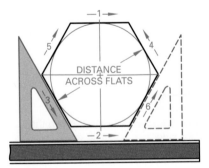

Figure 5-40

Constructing a regular hexagon, given the distance across the flats.

Method 2

Use this method to construct a regular hexagon when you know the distance across the corners. The distance across the corners is the distance from one vertex of a polygon through the center point to the opposite vertex. Refer to Figure 5-41.

1. Given the distance AB across the corners, draw a circle with AB as the diameter.
2. With A and B as centers and the same radius, draw arcs to intersect the circle at points C, D, E, and F.
3. Connect the points to complete the hexagon.

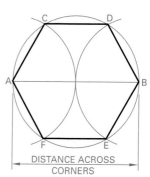

Figure 5-41

Constructing a regular hexagon, given the distance across the corners (first method).

Method 3

This construction demonstrates another method of constructing a regular hexagon, given the distance across the corners. Refer to Figure 5-42.

1. Given the distance AB across the corners, draw lines from points A and B at 30° to line AB. The lines can be any convenient length.
2. With the T-square and 30°-60° triangle, draw the sides of the hexagon in the order shown.

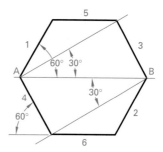

Figure 5-42

Construct a regular hexagon, given the distance across the corners (second method).

Construct an Octagon

An **octagon** is an eight-sided polygon. The following constructions demonstrate methods of drawing regular octagons.

Method 1

Use this method to construct an octagon circumscribed about a circle. Refer to Figure 5-43.

1. Given the distance across the flats, draw centerlines and a circle with a diameter equal to the distance across the flats.
2. With the T-square and 45° triangle, draw lines tangent to the circle in the order shown to complete the octagon.

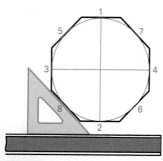

Figure 5-43

Construct a regular octagon circumscribed about a circle, given the distance across the flats.

Method 2

Use this method to construct an octagon inscribed within a circle. Refer to Figure 5-44.

1. Given the distance across the corners, draw centerlines AB and CD and a circle with a diameter equal to the distance across the corners.
2. With the T-square and 45° triangle, draw diagonals EF and GH.
3. Connect the points to complete the octagon.

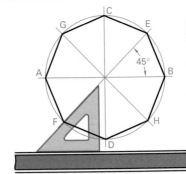

Inscribing a regular octagon within a circle, given the distance across the corners of the octagon.

Method 3

Use this method to construct an octagon inscribed within a square. Refer to Figure 5-45.

1. Given the distance across the flats, construct a square having sides equal to AB.
2. Draw diagonals AD and BC with their intersection at O. With A, B, C, and D as centers and radius R = AO, draw arcs to intersect the sides of the square.
3. Connect the points to complete the octagon.

Figure 5-45

Inscribing a regular octagon within a square, given the distance across the flats.

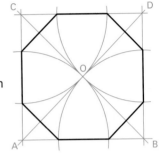

Construct an Ellipse

An **ellipse** is a regular oval. Ellipses are symmetrical around two axes that form a right angle. The shorter axis is the minor axis, and the longer one is the major axis. The constructions in this section demonstrate methods to draw an ellipse.

Method 1

This construction demonstrates the use of the pin-and-string method to draw a large ellipse. Figure 5-46A shows given major axis AB and minor axis CD, intersecting at O.

1. With C as center and radius R = AO, draw an arc to locate points F_1 and F_2, as shown in Figure 5-46A.
2. Place pins at points F_1, C, and F_2, as shown in Figure 5-46B.
3. Tie a string around the three pins and remove pin C.
4. Put the point of a pencil in the loop and draw the ellipse. Keep the string tight when moving the pencil, as shown in Figure 5-46C.

Figure 5-46

Constructing an ellipse by the pin-and-string method.

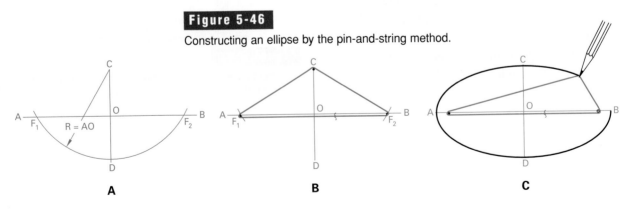

A   B   C

Method 2

This construction demonstrates the use of the trammel method to draw an ellipse. A **trammel** is a piece of paper or plastic on which specific distances have been marked off. Figure 5-47A shows given major axis AB and minor axis CD, intersecting at O.

1. Cut a strip of paper or plastic to use as a trammel. Mark off distances AO and OD on the trammel, as shown in Figure 5-47A.

2. On the trammel, move point O along minor axis CD and point D along major axis AB and mark points at A, as shown in Figure 5-47B.

3. Use a French curve or flexible curve to connect the points to draw the ellipse, as shown in Figure 5-47C.

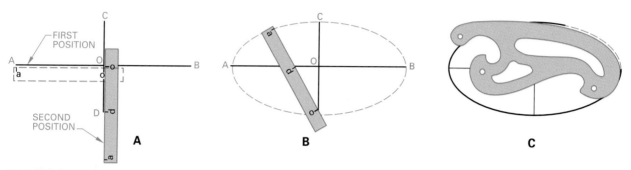

Figure 5-47

Constructing an ellipse by the trammel method.

Method 3

Use this construction method to draw an approximate ellipse using the major and minor axes of the ellipse. This method works when the minor axis is at least two-thirds the size of the major axis. Figure 5-48A shows given major axis AB and minor axis CD, intersecting at O.

1. Lay off OF and OG, each equal to AB - CD, as shown in Figure 5-48A.

2. Lay off OJ and OH, each equal to three fourths of OF.

3. Draw and extend lines GJ, GH, FJ, and FH, as shown in Figure 5-48B.

4. Draw arcs with centers F and G and radii FD and GC to the points of tangency, as shown in Figure 5-48C.

5. Draw arcs with centers J and H and radii JA and HB to complete the ellipse. The points of tangency are marked T in Figure 5-48C.

Figure 5-48

Constructing an approximate ellipse when the minor axis is at least two-thirds the size of the major axis.

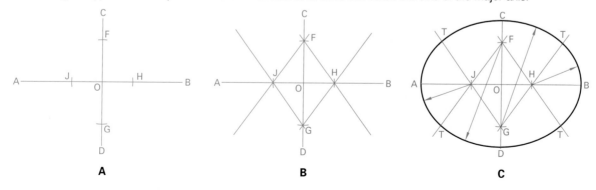

Method 4

Use this construction method to draw an approximate ellipse when the minor axis is less than two-thirds the size of the major axis. Figure 5-49A shows given major axis AB and minor axis CD intersecting at point O.

1. Draw line AC, as shown in Figure 5-49A.
2. Draw an arc with point O as the center and radius OA and extend line CD to locate point E.

3. Draw an arc with point C as the center and radius CE to locate point F, as shown in Figure 5-49B.
4. Draw the perpendicular bisector of AF to locate points J and H.
5. Locate points L and K. OL = OH and OK = OJ.
6. Draw arcs with J and K as centers and radii JA and KB, as shown in Figure 5-49C.
7. Draw arcs with H and L as centers and radii HC and LD to complete the ellipse.

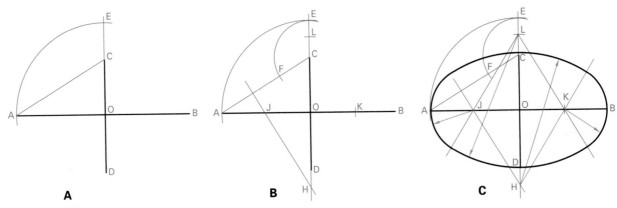

A **B** **C**

Figure 5-49

Constructing an approximate ellipse when the minor axis is less than two-thirds the size of the major axis.

Reduce or Enlarge a Drawing

The following techniques reduce or enlarge an existing drawing.

Method 1

If the drawing is square or rectangular, use a diagonal line to enlarge or reduce the drawing. Refer to Figure 5-50.

1. Draw a diagonal through corners D and B.
2. Measure the width or height you need along DC or DA (example: DG).

3. Draw a perpendicular line from that point (G) to the diagonal.
4. Draw a line perpendicular to DE intersecting at point F.

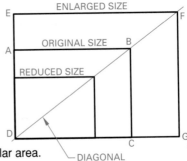

Figure 5-50

Reducing or enlarging a square or rectangular area.

Method 2

Use this method if the drawing is not square or rectangular. Refer to Figure 5-51.

1. Draw a grid larger or smaller than the one shown at C. The size of the grid depends on the amount of enlargement or reduction needed.
2. Use dots to mark key points on the second grid corresponding to points on the original drawing at C.
3. Connect the points and darken the lines to complete the new drawing.

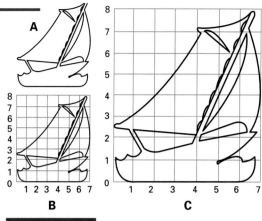

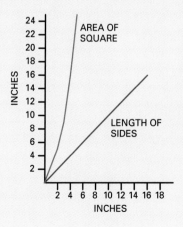

Figure 5-51

Reducing or enlarging a drawing of a sailboat.

Change the Proportion of a Drawing

Occasionally, you may need to change the proportion of a drawing. Use the following technique.

1. Draw a grid over the original drawing, as shown in Figure 5-52A.
2. Draw a grid on a separate sheet of paper in the needed proportion, as shown in Figure 5-52B and C.
3. Use dots to mark key points from the original drawing.
4. Connect the dots to complete the new drawing.

Figure 5-52

Changing the proportion of a drawing.

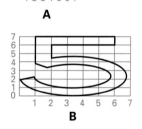

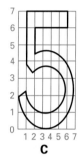

TECH▪MATH

Size and Area Relationships

To find the area of a regular rectangle (square), you simply square the length of one side. For example, a square with 2″ sides has an area of 2^2, or 4″. A square with 3.5″ sides has an area of 3.5^2, or 12.25″.

You might think that doubling the length of the sides of a square would double its area. This is not true. For example, the square on the left in the following illustration has 1.5″ sides. The area of this square is 1.5^2, or 2.25″. The square on the right has 3″ sides—twice the length of the first square. The area of the second square is 3^2, or 9″.

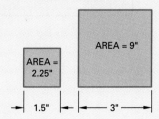

The following graph provides a visual demonstration of the relationship between the length of a side and the area of a square.

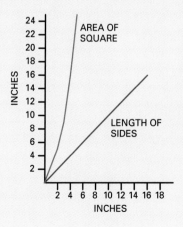

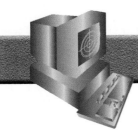

CAD Techniques

The techniques for creating geometry in AutoCAD and other CAD programs differ significantly from board techniques. It may seem at first as though you do not need the detailed knowledge of geometry that is necessary for board drafting. This is not true. Although the software creates the geometry, you must understand the geometric principles before you can direct the software to achieve the correct result.

This section consists of a series of example exercises in which you will use CAD techniques to create the same geometry described in the Board Drafting Techniques section of this chapter. By working through these constructions, you will begin to understand how to draw the basic geometry used in drafting. *Note:* You can also use the same techniques described in the Board Drafting Techniques section of this chapter. However, drafters who use CAD systems usually take advantage of the streamlined methods wherever they are offered by the software.

To work through the constructions, open a new drawing in AutoCAD. Use the template specified by your instructor, or choose to start a new drawing "from scratch." Your instructor will advise you on how many constructions to include in each drawing file. Be sure to save your work frequently.

Object Snaps

AutoCAD has a set of features known as **object snaps** that allow you to "snap" automatically to important points on any AutoCAD object. Object snaps you will use in this chapter include:

- Midpoint
- Nearest
- Endpoint
- Center
- Intersection
- Quadrant
- Perpendicular
- Tangent

Specifying the Intersection object snap, for example, allows you to snap to the intersection of two existing lines or arcs. This can be useful if you have used two arcs to locate the beginning of a new line. Object snaps have many other uses, too, as you will see as you work through the following constructions. To specify an object snap, type the first three letters of its name.

 Success on the Job

Efficiency

Efficiency in the drafting room results from learning basic concepts and skills and then becoming proficient at using them. For CAD users, object snaps greatly increase the efficiency of the drawing process. Using object snaps for geometric constructions reduces the time involved in preparing an accurate, high-quality drawing.

Bisect a Line or Arc

Bisect means to divide into two equal parts. Lines and arcs are usually bisected to find a beginning point for a new line or arc. In AutoCAD, the point that lies at the exact middle of a line or arc is known as the **midpoint**. Because AutoCAD has a Midpoint object snap, bisecting a line or arc—finding its midpoint—is simply part of the construction of the new line or arc.

1. Draw a line and an arc, as shown in Figure 5-53A.
2. Enter the LINE command, but do not enter a first point. Instead, type MID (for Midpoint) and press Enter.
3. At the "of" prompt, select the line you drew in step 1. Depending on the version of AutoCAD you are using, you may see a yellow triangle appear at the midpoint of the line. In any case, the first point of the new line you are creating begins at the exact midpoint of the original line, shown as point C in Figure 5-53B.

4. Pick another point anywhere in the drawing area and press Enter to end the LINE command.
5. Repeat steps 2 through 4, but select the arc in step 3. This results in a line that starts at point C and bisects the arc, as shown in Figure 5-53B.

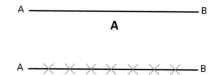

Presetting Object Snaps
If you know that you will be using certain object snaps frequently for a particular drawing, you can set AutoCAD to use them automatically, without having to specify them each time you use them. Object snaps that have been preset in this way are known as *running object snaps*. To set running object snaps, enter the OSNAP command. A dialog box appears. Pick the Object Snap tab of the dialog box to see the available object snaps. Pick the check boxes next to the object snaps you want to run automatically and pick OK to close the dialog box. *Note:* In older versions of AutoCAD, you do not have to select the Object Snap tab. The object snaps appear in the initial dialog box.

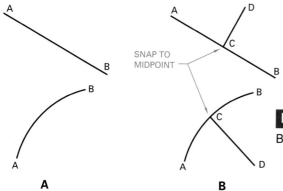

Figure 5-53
Bisecting a line or arc in AutoCAD.

Divide a Line into Equal Parts

AutoCAD includes a DIVIDE command that divides lines, arcs, and other geometry into equal parts. The following procedure divides a line into eight equal parts. Refer to Figure 5-54.

1. Draw a line of any length, as shown in Figure 5-54A.
2. Enter the DIVIDE command.
3. When prompted for the number of segments, type 8 and press Enter.

Markers appear at equal intervals along the line to divide it into eight parts. If you cannot see these markers, you will need to change the point style. To

do so, enter DDPTYPE at the keyboard and select a different point style from the dialog box that appears. See Figure 5-54B.

A ——————————— B

A

A ─×─×─×─×─×─×─×─ B

Figure 5-54   **B**

Dividing a line into equal parts in AutoCAD.

Draw a Perpendicular Line

A line is **perpendicular** to another line when the lines cross at right angles. The following methods are used to create perpendicular lines in AutoCAD.

Method 1

This method creates a line at a given point on another line so that the two lines are perpendicular. Refer to Figure 5-55.

1. Draw a line of any length, as shown in Figure 5-55A.
2. Enter the CIRCLE command. For the center of the circle, use the Nearest object snap to select point O on the given line. The Nearest object snap selects the nearest geometry to the point you pick on the screen. Therefore, using the Nearest object snap ensures that the point you pick is actually on the first line.
3. Move the cursor to control the radius of the circle. Select any circle size that intersects the line at two points and pick (click the left mouse button) to finish the circle. See Figure 5-55B.
4. Reenter the CIRCLE command and enter INT to activate the Intersection object snap. Select point C, where the circle intersects the line. Use the cursor to specify a radius that is larger than the radius of the first circle.
5. Reenter the CIRCLE command and use the Intersection object snap to select point D as the center point. This time, *do not* use the cursor to specify the radius of the circle. Notice at the Command prompt that AutoCAD remembers the radius of your previous circle. Press Enter to accept that radius, because you want this circle to be exactly the same size as the circle from step 4. The drawing should look similar to the one in Figure 5-55C.
6. Erase the original circle from step 2 to lessen confusion on the screen.
7. Draw a line with endpoints at the two intersections of the two circles. Use the Intersection object snap to snap to the exact intersections, as shown in Figure 5-55D. This line is perpendicular to the given line.
8. Erase the two remaining circles.

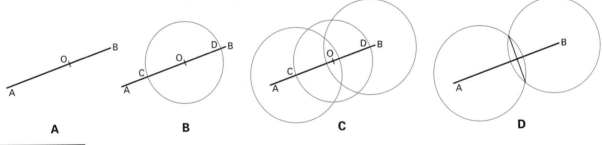

Figure 5-55

Creating a perpendicular line through a point on a given line using AutoCAD.

Method 2

Use this method to create a line perpendicular to a given line through a point that does *not* lie on the given line. Refer to Figure 5-56.

1. Draw given line AB.
2. Reenter the LINE command and pick point O as the first point of the new line.
3. Before specifying the second point of the line, type PER to enter the Perpendicular object snap. Then pick a point on line AB and press Enter. The resulting line is perpendicular to line AB.

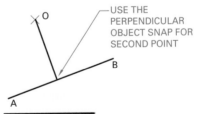

USE THE PERPENDICULAR OBJECT SNAP FOR SECOND POINT

Figure 5-56

Creating a line perpendicular to a given line through a point that does not lie on the given line.

Draw a Parallel Line

Recall that two lines are **parallel** when they are always the same distance apart. To create parallel lines in AutoCAD, use the OFFSET command. Refer to Figure 5-57.

1. Draw given line AB.
2. Enter the OFFSET command and enter an offset distance of 1. This will place the second line 1 unit from line AB.
3. When prompted to select the object to offset, use the mouse to pick line AB.
4. When prompted for the side to offset, pick a point anywhere above line AB. The parallel line CD appears.

Notice that the OFFSET command is still active. You can offset as many lines or arcs as you want without reentering the command. This can save time when you are working on a technical drawing.

5. Press Enter to end the command.

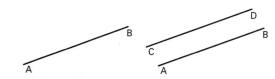

Figure 5-57

Creating a line parallel to a given line using the OFFSET command.

Bisect an Angle

The CAD method for bisecting an angle is very similar to the board drafting method. Refer to Figure 5-58.

1. Use the LINE command to draw two connected line segments to create given angle AOB, as shown in Figure 5-58A.
2. Enter the CIRCLE command and specify point O as its center point. Use the cursor to specify a radius similar to the one shown in Figure 5-58B.
3. Enter the TRIM command and press Enter to select all of the objects on the screen automati-

cally. Then pick any point on the circle *outside* of angle AOB. This procedure trims away all of the circle except for an arc that extends from one arm of angle AOB to the other. See Figure 5-58C.

4. Enter the LINE command. Use the Intersection object snap to place the first point of the line at point O. Then use the Midpoint object snap to place the second point of the line at the exact midpoint of the arc. Refer again to Figure 5-58C. This line bisects angle AOB.

Figure 5-58

Bisecting an angle in AutoCAD.

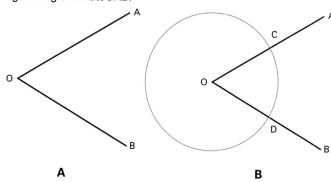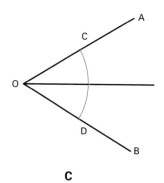

A   B   C

Copy an Angle

This construction demonstrates a method of copying a given angle to a new location and orientation. Refer to Figure 5-59.

1. Draw given angle AOB, as shown in Figure 5-59A.
2. Enter the COPY command and use a window to select both arms of the angle. To do this, pick a point below and to the right of the angle, and then pick another point above and to the left of the angle. The selected lines become dashed to show that they are selected. Press Enter to proceed to the next prompt.
3. For the point of displacement, pick point O.
4. When asked for the second point of displacement, pick another point anywhere on the screen. An exact copy of angle AOB appears, as shown in Figure 5-59B.
5. To change the orientation of the second angle, enter the ROTATE command, select both legs of the second angle, and press Enter.

6. Specify a point anywhere on the angle as the base point. This is the point about which the angle will rotate.
7. Move the cursor to reposition the angle at a new orientation, as shown in Figure 5-59C.

Note that you can control the orientation of the angle by entering a numerical value for the angle of rotation instead of using the cursor.

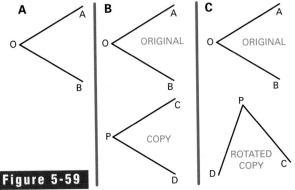

Figure 5-59

Copying and changing the orientation of an angle in AutoCAD.

Construct a Triangle

A **triangle** is a **polygon**, or closed figure, that contains three sides. The following constructions show methods for drawing various types of triangles.

Method 1

This method constructs an isosceles triangle. An **isosceles** triangle is one in which two sides are of equal length. Refer to Figure 5-60.

1. Draw given base line AB, as shown in Figure 5-60A.
2. Create a circle with its center point at point A and a radius equal to the length of the sides you want. See Figure 5-60B.

3. Create a second circle with the same radius, placing its center point at point B, as shown in Figure 5-60B.
4. Enter the LINE command and enter END to use the Endpoint object snap to place the first point of the line at point A. Use the Intersection object snap to place the second point of the line at the upper intersection of the two circles. Then use the Endpoint object snap to snap to point B. See Figure 5-60C.
5. Erase the two circles. The remaining triangle is an isosceles triangle.

Figure 5-60

Constructing an isosceles triangle using AutoCAD.

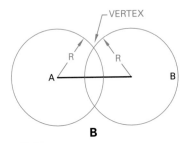

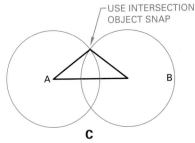

Method 2

This method constructs an equilateral triangle. An equilateral triangle is one in which all three sides are of equal length and all three angles are equal. Refer to Figure 5-61.

1. Enter the LINE command and pick a first point anywhere in the lower part of the drawing area. Enter @2<0 to create a horizontal line that is 2 units long. This is given line AB.
2. Create a circle with its center point at point A and a radius equal to the length of line AB.
3. Create a second circle with the same radius, placing its center point at point B. The drawing should now look like the one in Figure 5-61B.

4. Enter the LINE command and enter END to use the Endpoint object snap to place the first point of the line at point A. Use the Intersection object snap to place the second point of the line at the upper intersection of the two circles. Then use the Endpoint object snap to snap to point B. See Figure 5-61C.
5. Erase the two construction circles. The remaining triangle is an equilateral triangle.

   Note: You can also create an equilateral triangle using the POLYGON command. See "Construct a Polygon" later in this chapter for the procedure.

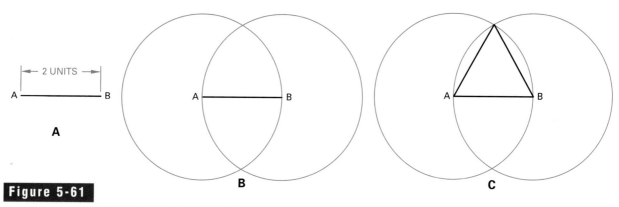

Figure 5-61

Creating an equilateral triangle in AutoCAD using traditional techniques.

Method 3

Construct a right triangle using this method when you know the length of two sides of the triangle. A right triangle is one that has a right angle at one of its vertices. (The **vertex** of a triangle or other polygon is the point at which two sides meet.) In this construction, sides AB and BC are given. Side AB is 2.50 units long, and side BC is 3.25 units long. Refer to Figure 5-62.

1. Draw side AB using the LINE command and polar coordinates: @2.50<0. Leave the LINE command active.

2. Specify the coordinates for side BC: @3.25<90. This creates line BC perpendicular, or at right angles, to side AB. Leave the LINE command active.
3. Use the Endpoint object snap to place the third point at point A, completing the right triangle.

Figure 5-62

Constructing a right triangle in AutoCAD, given the length of two sides.

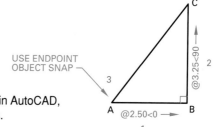

Method 4

Use this method to construct a right triangle when you know the length of one side and the length of the hypotenuse. The **hypotenuse** of a right triangle is the side opposite the 90° angle. In this construction, hypotenuse AC and side AB are given. AC is 4.00 units long, and side AB is 2.25 units long. Refer to Figure 5-63.

1. Draw hypotenuse AC using the LINE command and polar coordinates: @4.00<0. Press Enter to end the LINE command.
2. Enter the CIRCLE command. Instead of selecting a center point, enter 2P to activate the 2 Points option. Choose points A and C as the two points to define the circle, as shown in cyan (blue).
3. Draw another circle with point A as the center and a radius equal to side AB, as shown in magenta.

4. Enter the LINE command with the Intersection and Midpoint object snaps to connect points A and B and points B and C.

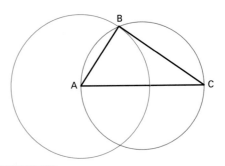

Figure 5-63

Constructing a right angle in AutoCAD, given the length of one side and the length of the hypotenuse.

Method 5

Use this method to construct a triangle when you know the lengths of all three sides. This construction is useful for scalene triangles. **Scalene** triangles are those that include three different angles and sides of three different lengths. In this construction, side AB = 2.35, BC = 1.65, and AC = 3.75. Refer to Figure 5-64.

1. Draw base line AB using polar coordinates. End the LINE command.
2. Draw a circle with a center point at A and a radius equal to AC, as shown in cyan.
3. Draw another circle with a center point at B and a radius equal to BC, as shown in magenta. The upper intersection of these two circles locates point C.
4. Use the Endpoint and Intersection object snaps to draw AC and BC, and erase the construction circles.

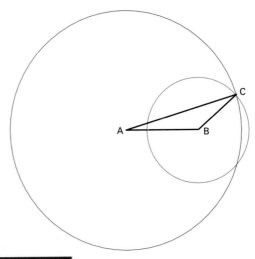

Figure 5-64

Constructing a triangle in AutoCAD, given the length of all three sides.

Construct a Tangent Circle

You already know several methods for creating a circle:

- specify a center point and a radius
- substitute the diameter for the radius by pressing the D key before entering the numerical value
- specify two points on the diameter of the circle
- specify three points on the diameter of the circle

AutoCAD also allows you to create a circle that is tangent to two other objects in AutoCAD by specifying the tangent objects and a value for the radius of the circle. As you may recall, a line is tangent to a circle if the line touches the circle at one point only. Refer to Figure 5-65.

1. Before you can use this option, you must have at least two lines in the drawing to specify as tangents. Use the LINE command to create the two lines. Use coordinate values to place the endpoints of the lines at the coordinates shown in the illustration.

2. Enter the CIRCLE command. Enter T at the keyboard to select the tan tan radius option.

3. At the appropriate prompts, pick anywhere on the two lines as the two tangents. Specify a radius of 1.00. The circle appears, as shown in Figure 5-65.

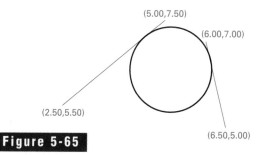

Figure 5-65

Creating a circle tangent to two other objects, given the radius of the circle.

Construct a Tangent Line

Because AutoCAD has a Tangent object snap, creating tangent lines is fairly easy. Refer to Figure 5-66.

1. Draw a circle anywhere in the drawing area, using the cursor to specify any radius.

2. Enter the LINE command. Pick any point outside the circle as the first point of the line.

3. Enter the Tangent object snap and move the cursor near the circle. Select a point on the circle. The line automatically snaps to the tangent point on the circle.

4. To extend the line beyond the tangent point, keep moving the cursor in the same general direction. AutoCAD displays an "Extension" message that shows the length and angle of the extended line.

►CADTIP

Object Tracking

The process for extending a line that is described in step 4 of "Construct a Tangent Line" is known as object tracking. If this doesn't seem to work for you, enter the OSNAP command, go to the Object Snap tab, and make sure the Object Snap Tracking On check box is checked. If this option is not available in your version of AutoCAD, you can achieve the same effect by using the EXTEND command.

Figure 5-66

Using the Tangent object snap to create a line tangent to a circle.

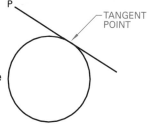

Construct a Tangent Arc

AutoCAD has an ARC command that gives CAD users great flexibility in creating arcs. However, sometimes the best solution is to use the tan tan radius (Ttr) option of the CIRCLE command, trimming away the unneeded parts of the circle. This section illustrates a few of the ways to create arcs in AutoCAD.

Method 1

The procedure for constructing an arc tangent to two lines in AutoCAD is similar to the board drafting procedure. In CAD, the procedure is the same whether the angle is an acute, obtuse, or right angle. Therefore, only an acute angle is shown in Figure 5-67.

1. Draw given lines AB and CD, as shown in Figure 5-67A.

2. Use the OFFSET command to offset both lines 1 unit to the inside.

3. Enter the ARC command. At the prompt, enter C (Center), and use the Intersection object snap to snap to the intersection of the two lines you offset in step 2. Then use the Perpendicular object snap to place the ends of the arc perpendicular to lines AB and CD, as shown in Figure 5-67B.

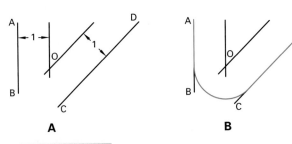

Figure 5-67

Using AutoCAD to create an arc tangent to two lines.

Method 2

This method uses the CIRCLE command to construct an arc tangent to two given arcs. Refer to Figure 5-68.

1. Enter the ARC command and follow the prompts to enter the start point, second point, and endpoint of arcs AB and CD. The radii of the arcs may be equal or unequal.

2. Enter the CIRCLE command. At the prompt, enter T (tan tan radius). Select points on the given arcs near the tangent locations. Note that you have only to pick a point somewhere near the tangent point. AutoCAD calculates the exact tangents for you.

3. Specify a radius of 1.50 to make the tangent circle appear.

4. Enter the BREAK command, and pick two points on the circle to break the arc out of the circle. Use the ERASE command to erase the unwanted portion of the circle. The remaining arc is tangent to the two given arcs.

Figure 5-68

Using the CIRCLE command to construct an arc tangent to two given arcs.

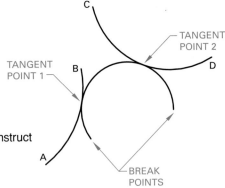

Construct an Ogee Curve

An **ogee curve** is a reverse curve that looks something like an S. The CAD procedure for drawing an ogee curve is similar to the board drafting procedure. Refer to Figure 5-69.

1. With Ortho on, draw given lines AB and CD, as shown in Figure 5-69A. Then turn Ortho off and use the Endpoint object snap with the LINE command to draw line BC.

2. Enter the BREAK command. This command is used to "break" a single line, arc, circle, or other geometry into two distinct objects. At the prompt, enter F (First), and use the Nearest object snap to pick a point E on line BC through which the curve is to pass. Refer again to Figure 5-69A. Line BC becomes two lines: BE and EC.

3. Construct perpendiculars at the midpoints of lines BE and EC. The length of the perpendicular lines does not matter. Erase any circles or arcs used for construction before continuing to step 4. See Figure 5-69B.

4. Turn Ortho on and create a vertical line with its lower endpoint at point B. Create another vertical line with its upper endpoint at point C. The exact length of these lines does not matter, but the lines should be long enough to intersect the perpendiculars you created in step 3 when the perpendiculars are extended. See Figure 5-69B.

5. Enter the EXTEND command and press Enter again to select all the objects. Then pick the perpendiculars as shown in Figure 5-69C to extend them to the vertical lines you created in step 4.

6. Create two circles. For the first, use the intersection of the vertical line from point B and the lower perpendicular as the center point. For the radius, enter the Endpoint object snap and snap to point E. For the second circle, use the intersection of the vertical line from point C and the upper perpendicular as the center point. For the radius, use the Endpoint object snap to snap to point E. It doesn't matter if the circles extend off the screen. Refer again to Figure 5-69C.

7. Notice that the two circles are tangent to each other at point E. One circle is also tangent to line AB, and the other is tangent to line CD. To finish the ogee curve, enter the TRIM command, press Enter to select all the objects, and trim away the unwanted parts of the circle. Erase lines BE, EC, and the vertical and perpendicular lines. Figure 5-69D shows the finished curve.

Figure 5-69

Creating an ogee curve in AutoCAD.

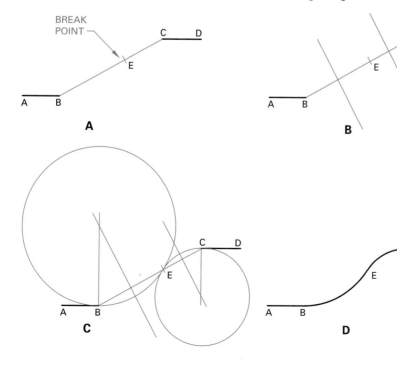

Construct a Polygon

AutoCAD provides a POLYGON command to create regular polygons with 3 to 1,024 sides. A **polygon** is a closed geometric figure with straight sides. A **regular polygon** is one in which the sides are all the same length and all of the angles are equal. Equilateral triangles and squares are examples of regular polygons that have three and four sides, respectively.

The constructions in this section use the following geometry:

- **square**, or four-sided polygon
- **pentagon**, or five-sided polygon
- **hexagon**, or six-sided polygon

Method 1

Use this method to construct a polygon, in this case a square, when you know the length of one of its sides. This method can be very useful when you need to construct a polygon that shares a line with other geometry in the drawing.

1. Enter the POLYGON command, and specify 4 as the number of sides. Press E (Edge) and pick a point on the screen.
2. Either pick another point on the screen for the second endpoint of the edge, or use polar coordinates to specify where the endpoint should be. If you use polar coordinates, the length of the line you specify becomes the length of one side of the square. The square appears on the screen.

Method 2

Use this method to inscribe a pentagon in a circle with a known center point and radius. A polygon is **inscribed** in a circle when all of its corners are tangent to the circle. Refer to Figure 5-70.

1. Create the given circle.
2. Enter the POLYGON command and specify 5 sides.
3. Use the Center object snap to select the center of the circle as the center point of the pentagon.
4. Enter I (Inscribed) to inscribe the polygon in the circle. When prompted for the radius of the circle, use the Nearest object snap to snap to a point on the circle. The pentagon appears inside the circle, with the point you picked using the Nearest object snap as one of the vertices.

You can use this method to "inscribe" a polygon in a circle even if the circle doesn't exist. Follow the steps above, but for the center point, pick a point where you want the center of the polygon to be. Instead of picking a point on the circle to define the radius, enter a numerical value at the keyboard.

Figure 5-70

Using the POLYGON command to inscribe a regular pentagon within a circle.

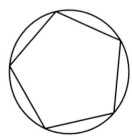

Method 3

Use this method to circumscribe a hexagon about a circle with a known center point and radius. A polygon is **circumscribed** about a circle when the polygon fully encloses the circle and the circle is tangent to the polygon on all of its sides. Refer to Figure 5-71.

1. Create the given circle.
2. Enter the POLYGON command, and specify 6 sides.
3. Use the Center object snap to select the center of the circle as the center point of the hexagon.
4. Enter C (Circumscribed) to circumscribe the polygon about the circle. When prompted for the radius of the circle, use the Nearest object snap to snap to a point on the circle. The hexagon appears inside the circle, with the point you picked on the circle as one of the vertices.

You can use this method to "circumscribe" a polygon about a circle even if the circle doesn't exist. Follow the steps above, but for the center point, pick a point where you want the center of the polygon to be. Instead of picking a point on the circle to define the radius, enter a numerical value at the keyboard.

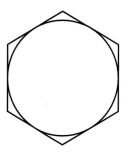

Figure 5-71

Using the POLYGON command to circumscribe a hexagon about a circle.

Construct an Ellipse

An **ellipse** is a regular oval. Ellipses are symmetrical around two axes that form a right angle. The shorter axis is the minor axis, and the longer one is the major axis.

In AutoCAD, the ELLIPSE command allows you to create ellipses of any size by defining the axes. Refer to Figure 5-72.

1. Enter the ELLIPSE command and pick a point anywhere in the drawing area as the first endpoint of the first axis.
2. Pick another point as the second endpoint of the first axis.
3. As the ellipse begins to appear on the screen, select a third point to specify the other axis.

Notice that you don't have to specify two points for the second axis. When you specify the third point, AutoCAD calculates the last point automatically so that the second axis is at right angles to the first.

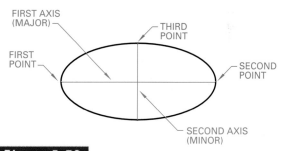

Figure 5-72

Using the ELLIPSE command.

Reduce or Enlarge a Drawing

To change the size of objects in an AutoCAD drawing, you can use the SCALE command. Note that this is different from using the ZOOM command to make objects appear larger or smaller on the screen. It is also different from choosing a standard scale in paper space to scale a drawing for printing. When you use the SCALE command, you change the actual dimensions of the objects. You can scale all of the objects in the drawing at once, or scale only those objects that you select.

This construction demonstrates the effect of scaling objects in AutoCAD. Refer to Figure 5-73.

1. Set the snap and grid to .50. Use the snap, grid, and coordinate display to create two concentric circles (both with the same center point). Make the radius of one circle 2.00 units, and make the radius of the second circle 1.00 unit, as shown in Figure 5-73A.
2. Enter the SCALE command. Pick both circles to scale, and press Enter.
3. The base point is the point around which the scaling will occur. Use the Center object snap to select the center of the circles for the base point.
4. Enter a scale factor of .75 to scale the circles to 75% of their original size, as shown in Figure 5-73B. You can check their size by using the grid, remembering that the dots on the grid are spaced at intervals of .50.

Notice that you must enter a decimal fraction. The number 1 stands for 100%, or full size. If you enter 75, the circles will enlarge to 75 times their original size.

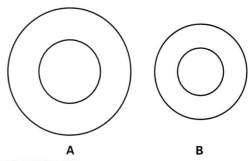

Figure 5-73

Using the SCALE command to reduce or enlarge the physical size of a drawing in AutoCAD.

Success on the Job

Characteristics of Successful Drawings

As you work, remember that one of the most important concepts in drafting is accuracy. Other key concepts include clarity, legibility, and neatness. One reason these concepts are important is that sloppy work may give the designer the wrong answers, which may lead to a part being manufactured with the wrong dimensions or specifications. Lack of clarity may cause people to misunderstand the information, again causing errors in the manufactured product.

Therefore, it is a good idea to practice these concepts even when the work is not critical. Practice working accurately, legibly, and neatly as you perform the following constructions and throughout the rest of the course.

Chapter Summary

- Even the most complex objects can be broken down into geometric shapes such as lines, circles, arcs, and polygons.

- Geometry is the study of the size and shape of things and their relationship to each other.

- Geometric constructions are illustrations made of individual lines and points drawn in proper relationship to one another.

- Geometric constructions are used by drafters, surveyors, engineers, architects, scientists, mathematicians, and designers.

- One of the purposes of geometric construction is to help drafters solve practical design problems.

- In board drafting, the dividers, compass, French curve, scales, and other drafting instruments help drafters create geometric constructions accurately.

- In CAD, many commands are available for drawing basic geometric shapes. Examples include the CIRCLE, POLYGON, ARC, and ELLIPSE commands.

Review Questions

1. How can geometric constructions be helpful to designers and engineers?

2. What kind of triangle has one right (90°) angle?

3. Name three of the most important concepts in drafting. Why are they important?

4. How many sides does a hexagon have?

5. What basic geometric shape has both a major and a minor axis?

6. What is an ogee curve?

7. In what type of triangle are two sides equal?

8. What name is given to a polygon in which all sides are of equal length and all of the angles are 90°?

9. What is the longest side of a right triangle called?

10. What name is given to a triangle in which all three sides are of equal length?

11. Explain how to bisect a line using board drafting techniques.

12. Explain the 3-4-5 method of creating a triangle. What type of triangle does this method create?

13. Name three methods of creating an ellipse using board drafting techniques.

14. Describe the procedure for creating an arc tangent to two straight lines using board drafting techniques.

15. What is the purpose of using object snaps in AutoCAD?

16. How would you construct a line parallel to a given line using AutoCAD?

17. In AutoCAD, what term is used for the point that bisects a line?

18. Describe the procedure for inscribing a hexagon in a circle using CAD techniques.

19. What information do you need to construct an ellipse using AutoCAD?

20. In AutoCAD, what command allows you to reduce or enlarge the physical size (dimensions) of a drawing?

Drafting Problems

The problems in this chapter can be performed using board drafting or CAD techniques. The problems are presented in order of difficulty, from least to most difficult.

Problems 1 through 19: These problems are designed for working four problems on an A-size sheet, laid out as shown in Figure 5-74. Draw each problem three times the size shown. If you are using board drafting, use dividers to pick up the dimensions from the problems, and step off each measurement three times. If you are using a CAD system, use a scale to measure the dimensions, and create the geometry in the CAD system at three times the measured size.

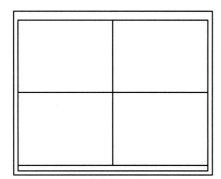

Figure 5-74

1. Draw and bisect line AB, as shown in Figure 5-75A.

2. Draw line AB, as shown in Figure 5-75B. Construct a perpendicular at point P.

3. Draw line AB, as shown in Figure 5-75C. Divide line AB into five equal parts.

4. Draw line AB, as shown in Figure 5-75D. Construct line CD through point P so that CD is parallel to AB and equal in length to line AB.

5. Draw angle ABC, as shown in Figure 5-75E. Bisect angle ABC.

6. Draw angle ABC, as shown in Figure 5-75F. Copy the angle in a new location, beginning with line A_1B_1.

7. Draw base line AB, as shown in Figure 5-75G. Construct an isosceles triangle using base line AB and sides equal to line CD.

8. Draw base line AB, as shown in Figure 5-75H. Construct a triangle on base AB with sides equal to BC and AC.

9. Draw a circle with a 3″ diameter, as shown in Figure 5-75I. Inscribe a square in the circle.

10. Draw a circle with a 3″ diameter, as shown in Figure 5-75I. Inscribe a regular pentagon in the circle.

11. Draw a circle with a 3″ diameter, as shown in Figure 5-75I. Circumscribe a regular hexagon about the circle.

12. Draw a circle with a 3″ diameter, as shown in Figure 5-75I. Circumscribe a regular octagon about the circle.

13. Draw a circle with a 3″ diameter, as shown in Figure 5-75J. Construct a tangent line through point P.

14. Locate points A, B, and C on the drawing sheet, as shown in Figure 5-75K. Construct a circle through these three points.

15. Draw the two lines shown in Figure 5-75L. Construct an arc having a radius R tangent to the two lines.

16. Draw the two arcs shown in Figure 5-75M. Construct an arc having a radius R tangent to the first two arcs.

17. Draw a 3.00″ square, as shown in Figure 5-75N. Construct a regular octagon within the square.

18. Construct an ellipse that has a 4.00″ major axis and a 2.50″ minor axis.

19. Draw a 2.50″ circle, as shown in Figure 5-75O. Construct two lines from point P tangent to the circle.

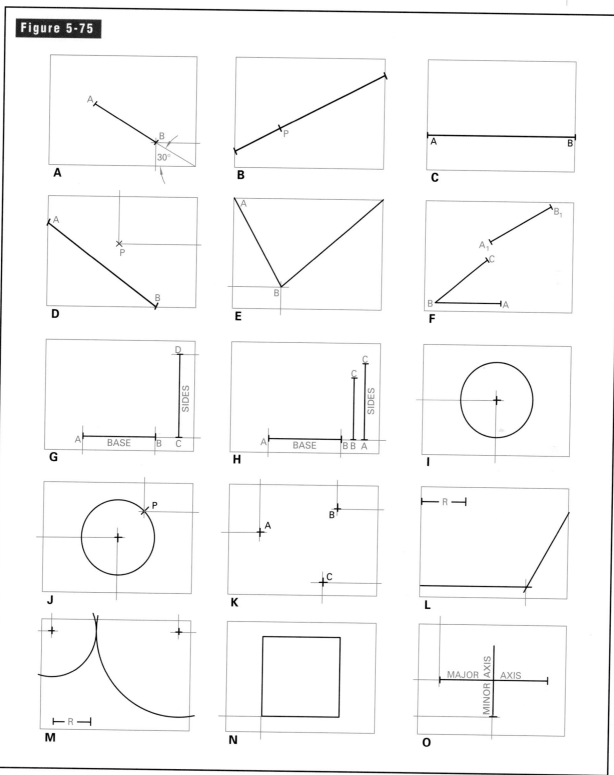

Figure 5-75

Problems 20 through 32: These problems provide additional practice in geometric constructions. They are designed to be drawn one per drawing sheet. Before beginning each drawing, determine an appropriate scale and sheet size. Do not dimension.

20. Draw the gasket shown in Figure 5-76.

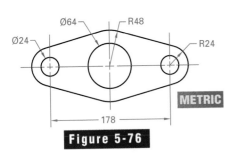

Ø64 R48
Ø24 R24
METRIC
178

Figure 5-76

21. Draw the pipe support shown in Figure 5-77. Locate all centers and points of tangency.

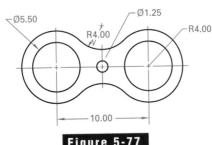

Ø5.50 Ø1.25
R4.00 R4.00
10.00

Figure 5-77

22. Draw the handwheel shown in Figure 5-78. Use the following dimensions: A = Ø7.00″; B = Ø6.12″; C = Ø5.50″; D = R1.25″; E = Ø2.00″; F = Ø1.00″; G (keyway) = .20″ wide x .10″ deep; H = Ø.38″; I = R.38″; J = R.20″; K = 1.00″.

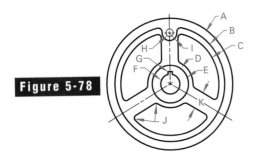

Figure 5-78

23. Draw the combination wrench shown in Figure 5-79. Use the following dimensions: square: 1.00″; octagon: 1.38″ across flats; isosceles triangle: 2.75″ base, 2.00″ sides; pentagon: inscribed within Ø1.38″ circle; hexagon: 1.25″ across flats. If you are using board drafting techniques, do not erase construction lines.

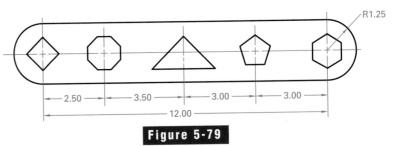

R1.25
2.50 3.50 3.00 3.00
12.00

Figure 5-79

24. Draw the adjustable fork shown in Figure 5-80. Use the following dimensions: A = 220 mm; B = 80 mm; C = 40 mm; D = 27 mm; E = 64 mm; F = 20 mm; G = 8 mm; H = 10 mm.

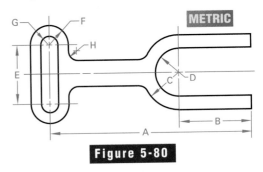

G F
H
E C D
B
A
METRIC

Figure 5-80

25. Draw the rod support shown in Figure 5-81.

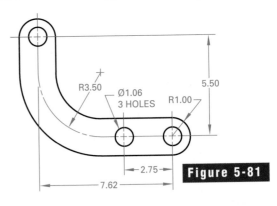

R3.50 Ø1.06 3 HOLES R1.00
5.50
2.75
7.62

Figure 5-81

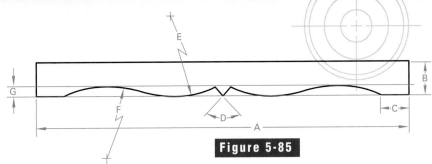

Figure 5-85

26. Draw the rocker arm shown in Figure 5-82.

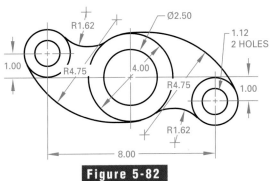

Figure 5-82

29. Draw the valance board shown in Figure 5-85. Use the following dimensions: A = 8′-0; B = 0′-8; C = 0′-7; D = 90°; E = 2′-6; F = 2′-6; G = 0′-2. If you are using board drafting techniques, mark all tangent points, and do not erase construction lines.

30. Draw the kidney-shaped table top shown in Figure 5-86.

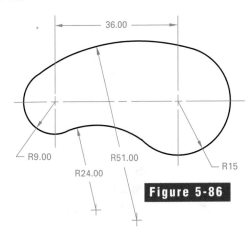

Figure 5-86

27. Draw the hex wrench shown in Figure 5-83. If you are using board drafting techniques, mark all tangent points, and do not erase construction lines.

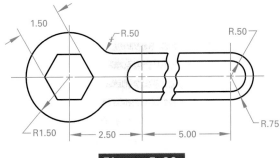

Figure 5-83

28. Draw the offset link shown in Figure 5-84. If you are using board drafting techniques, mark all points of tangency.

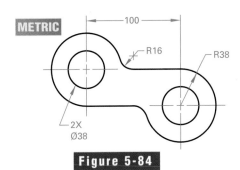

Figure 5-84

31. Draw the adjustable table support shown in Figure 5-87.

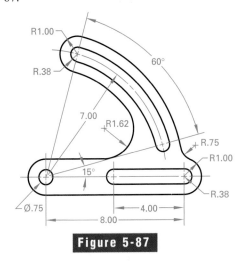

Figure 5-87

32. Draw the tilt scale shown in Figure 5-88. Use the following dimensions: AB = 44 mm; AX = 66 mm; AC = 140 mm; AD = 184 mm; AE = 216 mm; AF = 222 mm; AG = 236 mm; H = R24 mm; I = R16 mm; J = R5 mm; K = ∅12 mm.

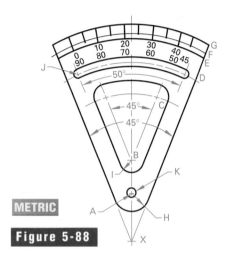

METRIC

Figure 5-88

Design Problems

Design problems have been prepared to challenge individual students or teams of students. In these problems, you will apply skills learned mostly in this chapter but also in other chapters throughout the text. The problems are designed to be completed using board drafting, CAD, or a combination of the two. Be creative and have fun!

1. Design an educational toy used to help toddlers develop manual dexterity, spatial relationships, and color association. The toy should be similar to Figure 5-89, but expanded to include at least six geometric shapes of different colors. Material: 1″ thick pine.

2. Design an octagon-shaped jewelry box with a hinged lid. The overall size should not exceed 160 mm across the corners of the octagon by 90 mm high. Material: optional. Do not dimension.

3. **TEAMWORK** Design a hexagon-shaped picnic table and bench. It should be 32″ high and 60″ across the flats of the hexagon. The base of the table should also be hexagonal.

4. **TEAMWORK** Design and draw a cover for your 8.50″ × 11.00″ or 11.00″ × 17.00″ set of technical drawings. Use various geometric shapes in the design. Geometric shapes, such as circles, squares, hexagons, octagons, ellipses, etc., can be used to enhance the design. Use colors where desired. Use block letters to add information on the cover, such as your name, the school name, the course title, the instructor's name, and the year.

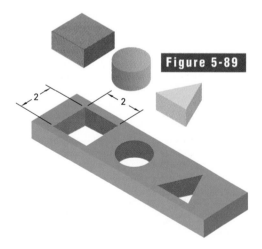

Figure 5-89

CHAPTER 6

Multiview Drawing

KEY TERMS

first-angle projection

implementation

multiview drawing

normal views

orthographic projection

third-angle projection

visualization

OBJECTIVES

Upon completion of this chapter, you should be able to:

■ Explain the relationship of orthographic projection to multiview drawing.

■ Describe the difference between first- and third-angle projection.

■ Determine the number of views needed to describe fully the shape and size of an object.

■ Locate multiple views on a drawing according to accepted principles of drafting.

■ Create the various views of an object.

■ Develop a multiview drawing from the initial idea to a finished drawing using board drafting.

■ Develop a multiview drawing from the initial idea to a finished drawing using CAD techniques.

Drafting Principles

People communicate ideas by verbal and written language and by pictorial, or graphic, means. One of the graphic means is technical drawing. It is a language used and understood in all countries. When accurate visual understanding is necessary, technical drawing is the most exact method that can be used.

Technical drawing involves visualization and implementation. **Visualization** is the ability to see clearly in the mind's eye what a machine, device, or other object looks like. **Implementation** is the process of drawing the object that has been visualized. The designer, engineer, or drafter first visualizes an object and then explains it pictorially using technical drawing.

A properly made technical drawing, such as the one in Figure 6-1, gives a clearer, more accurate description of an object than a photograph or written explanation. Technical drawings made according to standard principles result in views that give an exact visual description of an object.

This type of drawing is different from photographs and pictorial drawings. A pictorial view of a V-block is shown in Figure 6-2. A pictorial drawing is a drawing that shows an object as it appears to the human eye—as it would appear in a photograph. Like a photograph, it shows the object as it appears

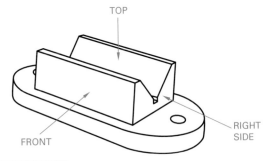

Figure 6-2

This pictorial shows the top, front, and right side in a single view, but it does not supply all the information necessary to build a V-block.

Figure 6-1

This three-view drawing gives an accurate description of the object.

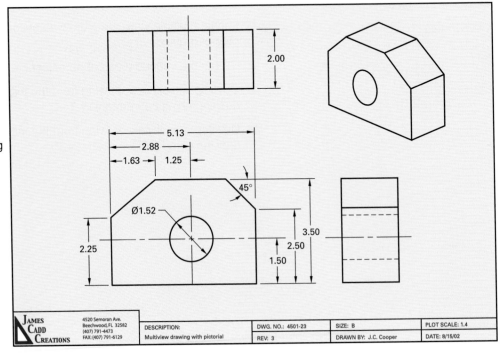

to the eye. Notice that three sides of the V-block are shown in a single view. However, photographs and pictorial drawings do not show the exact forms and relationships of the parts that make up the object. They show the V-block as it appears, not as it really is. For example, the holes in the base appear as ellipses, not as true circles.

Multiview Drawing

The problem, then, is to represent an object on a sheet of paper in a way that will describe its exact shape and proportions. This is done by drawing views of the object as it is seen from different positions. These views are then arranged in a standard order so that anyone familiar with drafting practices can understand them immediately. This type of graphic representation is called **multiview drawing**.

Normal Views

The front, top, and right-side views are the ones most often used to describe an object in a technical drawing. They are therefore known as the **normal views**. The normal views of the V-block from Figure 6-2 are shown in Figure 6-3.

In order to describe accurately the shape of each view, imagine a position directly in front of the object, then above it, and finally at the right side of it. This is where the ability to visualize is important.

■ The **front view** of the V-block shows the exact width and the height of the object. The dashed lines show the outline of details hidden behind the front surface.

■ The **top view** of the V-block shows the exact shape of the top. Therefore, it provides the width

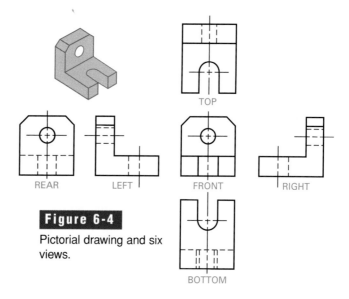

Figure 6-4

Pictorial drawing and six views.

and depth of the V-block. Notice that the holes are true circles in this view, and the rounded ends of the base are true radii. In the pictorial drawing, these appear as elliptical shapes.

■ The **right-side view** of the V-block shows the depth and height accurately. Notice that the shape of the V appears to be symmetrical in the right-side view. It appears distorted in the pictorial drawing.

Relationship of Views

Views must be placed in proper relationship to one another. Only in this way can technical drawings be read and understood properly. Figure 6-3 shows the three normal views of the V-block in their correct places. The top view is directly above the front view. The right-side view is to the right of the front view. Each view is where it logically belongs. Notice that the edges of the views line up exactly. For example, the center of a hole in the front view is directly in line with the center of the hole in the top view.

Other Views

Most objects have six sides, or six views: top, front, bottom, rear, right-side, and left-side, as shown in Figure 6-4. In most cases, two or three views can be used to describe completely the shape and size of all parts of an object. However, in some cases it may be necessary to show views other than the front, top, and right side. Only in very unusual cases are six views necessary.

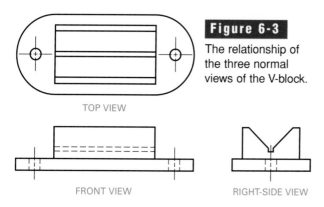

Figure 6-3

The relationship of the three normal views of the V-block.

TOP VIEW

FRONT VIEW

RIGHT-SIDE VIEW

Orthographic Projection

The views that make up multiview drawings are developed through the principles of orthographic projection. **Orthographic projection** is the process of projecting two or more views of an object onto imaginary planes by drawing lines perpendicularly from the object to the planes. As you may recall, a plane is an imaginary flat surface that has no thickness.

Orthographic projection involves the use of three planes. They are the vertical plane, the horizontal plane, and the profile plane, or side view. These planes are shown in Figure 6-5. A view of an object is projected and drawn upon each of the three planes.

Angles of Projection

Notice that the vertical and horizontal planes divide the space into four quadrants, or quarters of a circle. In orthographic projection, quadrants are often referred to as angles. Thus, we get the names *first-angle projection* and *third-angle projection.* First-angle projection is used in European countries. Third-angle projection is used in the United States and Canada. Second- and fourth-angle projections are not used.

First-Angle Projection

In **first-angle projection**, the object is projected onto the planes from the first angle, or quadrant, as shown in Figure 6-6. The front view is projected to the vertical plane. The top view is projected to the horizontal plane. The left-side view is projected to the profile plane. To create a multiview drawing, the horizontal and profile planes are rotated so that all the views lie in a single plane, as shown in Figure 6-7. Therefore, in first-angle projection, the front view is located above the top view. The left-side view is to the right of the front view.

Third-Angle Projection

Third-angle projection uses the same basic principles as first-angle projection. The main difference is that the third quadrant, not the first, is used for projection. Figure 6-8 shows the same object projected onto the planes within the third quadrant. In this case, the front view is projected to the vertical plane. The top view is projected to the horizontal plane. The right-side view is projected to the profile plane. Again, the planes are rotated to lie in a single plane. The result is a drawing in which the top view is above the front view. See Figure 6-9.

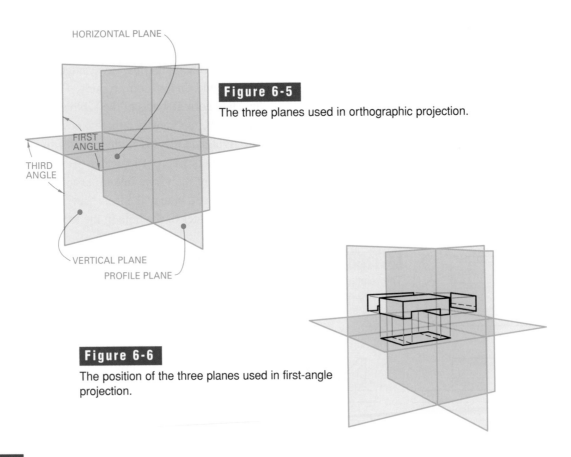

HORIZONTAL PLANE

FIRST ANGLE

THIRD ANGLE

VERTICAL PLANE

PROFILE PLANE

Figure 6-5

The three planes used in orthographic projection.

Figure 6-6

The position of the three planes used in first-angle projection.

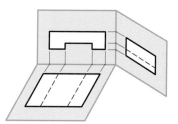

Figure 6-7

Three views in first-angle projection.

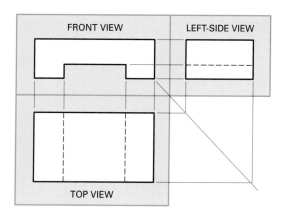

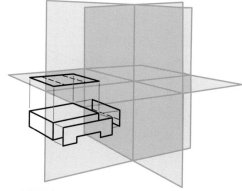

Figure 6-8

The position of the three planes used in third-angle projection.

The Glass Box

In each case, the three views of an object are developed using imaginary transparent, or see-through, planes. The views are projected onto these planes. We mentioned earlier that most objects have six sides. Therefore, six views may result. To explain the theory of projecting all six views, we will use an imaginary glass box.

Imagine a transparent glass box around the nearest bookend shown in Figure 6-10A. In your mind's eye, project the views of the bookend onto the sides of the glass box. When the glass box is opened, or unhinged, the six views begin to unfold. Figure 6-10B shows the glass box partially opened.

When the box is fully opened onto a single plane, the views are in their relative positions as if they had been drawn on paper, as shown in Figure 6-10C. These views are arranged according to proper order for the six views in third-angle projection. Notice that the back view is located to the left of the left-side view.

Figure 6-9

Three views in third-angle projection.

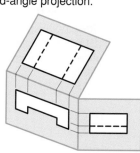

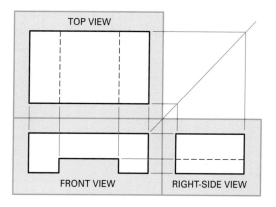

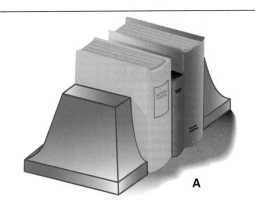

A

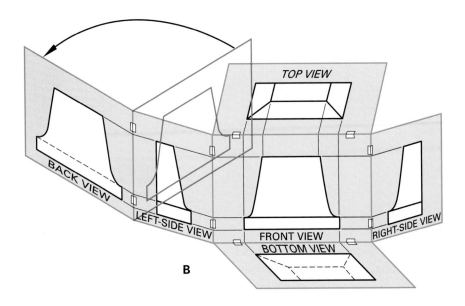

B

Figure 6-10

(A) Imagine the nearest bookend inside a glass box. (B) Then imagine the glass box opening as shown here. (C) When the glass box has been completely opened, you can see all six sides of the bookend easily.

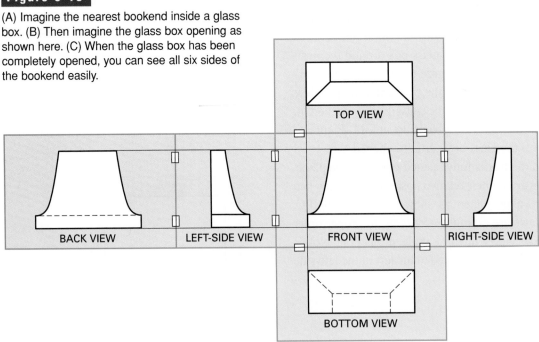

C

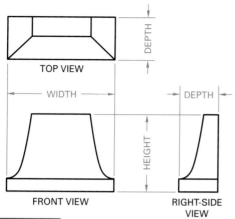

Figure 6-11

The front, top, and right-side views describe the bookend completely.

Look carefully at each of the six views. Notice that some views give the same information found in other views. Some views may also be mirror images of one another. Thus, it is not necessary to show all six views for a complete description of the object. In this case, the object can be described completely using the three normal views, as shown in Figure 6-11.

Deciding Which Views to Draw

We have established that fewer than six views are needed to describe most objects. Usually, three views are sufficient. Which views should you choose?

The general characteristics of an object often suggest the three views required to describe its shape. Sometimes, however, an object has features that can be more clearly described by using more views or parts of extra views.

Most pieces can be recognized because they have a characteristic view. This is the first view to consider. The characteristic view usually becomes the front view, and it is often the first view to draw. Next, consider the normal position of the part when it is in use. It is often desirable to draw the part in its normal position. However, it is not always necessary. For example, tall parts, such as vertical shafts, can be drawn more easily in a horizontal position. Views with the fewest hidden lines are easiest to read. They also take much less time to lay out and draw.

Number of Views

The main purpose of drawing views is to describe the shape of something. Therefore, it is a waste of time to make more views than are necessary to describe an object. In fact, some objects require only one view. For example, parts of uniform thickness, such as the latch and stamping in Figure 6-12, require one view. The thickness of the material is given in a note. Circular objects such as the bushing and sleeve shown in Figure 6-13 can also be shown in one view. Dimensions for diameters are marked with the diameter symbol [$\emptyset$].

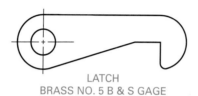

LATCH
BRASS NO. 5 B & S GAGE

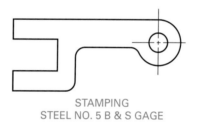

STAMPING
STEEL NO. 5 B & S GAGE

Figure 6-12

Parts of uniform thickness require only one view.

Figure 6-13

Cylindrical objects such as these can be shown in one view. Diameters must be marked clearly.

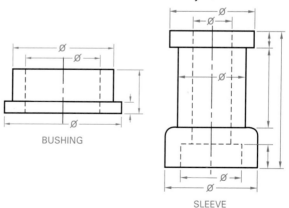

BUSHING

SLEEVE

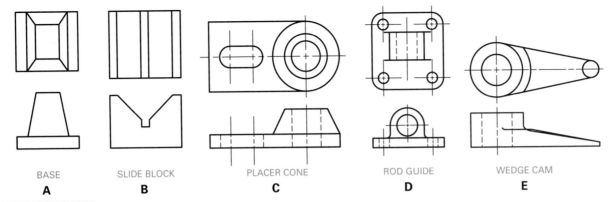

BASE
A

SLIDE BLOCK
B

PLACER CONE
C

ROD GUIDE
D

WEDGE CAM
E

Figure 6-14

Two-view drawings.

Examples of objects that can be described in two views are shown in Figure 6-14. When you use two views, you must select them carefully so that they describe the shape of the object accurately. For the base in Figure 6-14A, there is no question about what views to draw. The top view and either front or side view are enough. The top view and front view describe the slide block shown in Figure 6-14B adequately. A third view would add nothing to the description of the placer cone in Figure 6-14C. There should be no question about the selection of views for the rod guide shown in Figure 6-14D or the wedge cam in Figure 6-14E.

Figure 6-15

Each of these objects can be described adequately in two views. Which views would you choose?

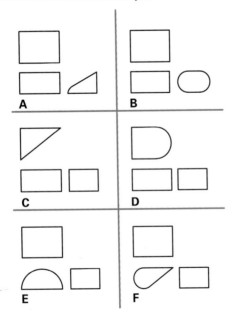

A

B

C

D

E

F

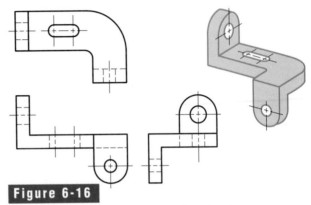

Figure 6-16

This angle plate requires a three-view drawing.

Figure 6-15 shows three views of some objects that can be described adequately in two views. In parts A and B of Figure 6-15, the top and front views are the same. Since the side views are necessary, the front and side views are adequate. In parts C and D of the figure, the top views are necessary. Therefore, the top and front views are sufficient. In parts E and F of the figure, the front views are necessary. Therefore, the front and top views or the front and side views are sufficient.

Many objects, such as the angle plate in Figure 6-16, require three views to describe the shape completely. The sliding base in Figure 6-17 also requires three views. Six views are shown, but as you look at the pictorial view, you will see that the top, front, and right-side views give the best shape description. These views also have the fewest hidden lines. The six views are shown here simply to illustrate the selection of views. In practice, you would only draw the necessary views. Careful thought about your "mind's-eye picture" of an object will help you decide which views best describe its shape.

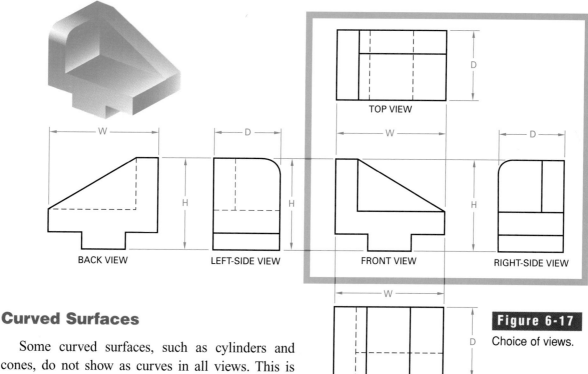

Figure 6-17

Choice of views.

TOP VIEW

BACK VIEW   LEFT-SIDE VIEW   FRONT VIEW   RIGHT-SIDE VIEW

BOTTOM VIEW

Curved Surfaces

Some curved surfaces, such as cylinders and cones, do not show as curves in all views. This is illustrated in Figure 6-18. A cylinder with its axis, or centerline, perpendicular to a plane shows as a circle on that plane, as shown in Figure 6-18A and B. It shows as a rectangle in the other two planes. When choosing the views for an object that contains curved surfaces, be sure to include a view that defines the curved surfaces accurately.

A cone appears as a circle in one view. It appears as a triangle in the other, as shown in Figure 6-18C. The top view of a frustum of a cone appears as two circles. Refer to Figure 6-18D. In this view, the conical surface is represented by the space between the two circles.

Cylinders, cones, and frustums of cones have single curved surfaces. They appear as circles in one view and straight lines in the others. Other objects, such as the handles in Figure 6-19A, have double curved surfaces that appear as curves in both views. The ball handle has spherical ends. Thus, both views of the ends are circles because a sphere appears as a circle when viewed from any direction. The slotted link in Figure 6-19B is an example of tangent plane and curved surfaces. The rounded ends are tangent to the sides of the link, and the ends of the slot are tangent to the sides. Therefore, the surfaces are smooth. There is no line of separation.

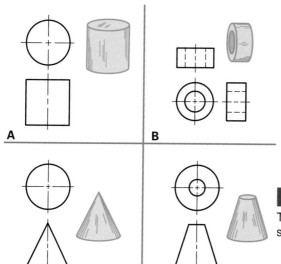

A

B

C

D

Figure 6-18

The curved surfaces of cylinders and cones appear as straight lines in some views.

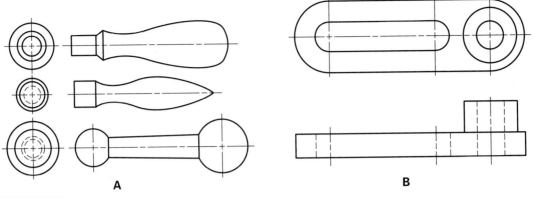

Figure 6-19

Two-view drawings of curved surfaces: (A) double curves; (B) single curves.

Placing Views

The size of the drawing sheet should allow enough space for the number of views needed to give a clear description of the part. Working space for an A-size drawing sheet is shown by the sheet layout in Figure 6-20. Chapter 4 provides a more detailed discussion of working space for different sheet sizes. The method for working out the positions of the views is the same for any space.

Locating the Views

In Figure 6-21A, a pictorial drawing of a slide stop is shown with its overall width, height, and depth dimensions. Some simple arithmetic is needed to place the three normal views properly. It may also be helpful to make a rough layout on scrap paper, as shown in Figure 6-21B. This rough layout need not be made to scale.

A working space of 10.50″ × 7.00″ is used to explain how to place the views of the slide stop. The width, depth, and height dimensions are given in magenta on the sketch. The dimensions in blue indicate the spacing at the top, bottom, side, and between views. Use the following procedure to determine spacing. Figure 6-21C shows the actual calculations.

Vertical Placement

1. Add the height and depth of the object to find the total vertical space needed to draw the front and top views.
2. Subtract the result of step 1 from the total vertical drawing space (7.00″).

The calculations show that a total of 2.25″ is left for the space between views and the space at the top and bottom of the sheet. If you specify a space of .75″ between the front and top views, you will have 1.50″ left for spaces above the top view and below the front view. These could be .75″ each, but a better visual balance will result if .88″ is used below and .62″ above.

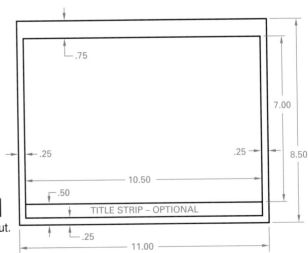

Figure 6-20

A-size sheet layout.

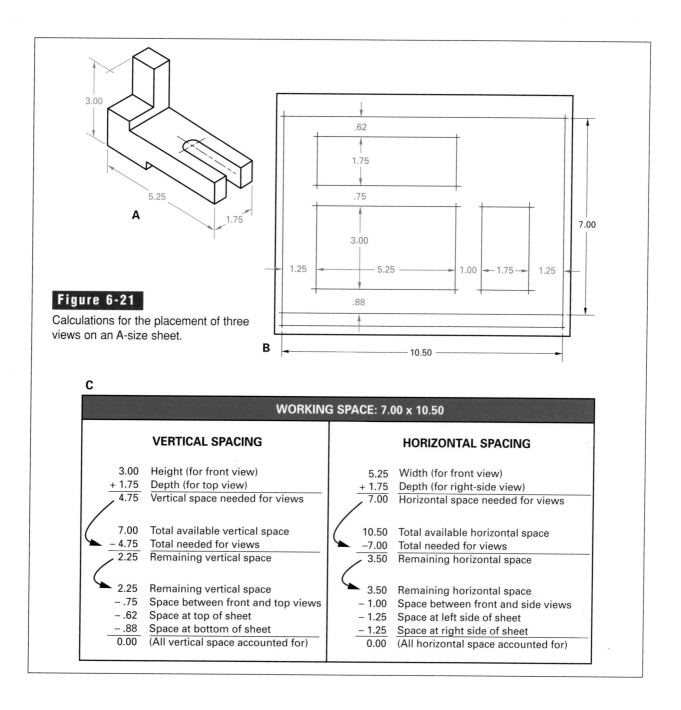

Figure 6-21

Calculations for the placement of three views on an A-size sheet.

Horizontal Placement

1. Add the width and depth to find the total horizontal space needed to draw the front and right-side views.

2. Subtract the result of step 1 from the width of the drawing space (10.50″).

The remaining 3.50″ is the amount left for the space between the front and right-side views and the space at the left and right of the sheet. If you specify 1.00″ between the front and side views, you will have 2.50″ remaining. You can divide this equally for the left and right sides. Keep in mind that you may need to make these spaces larger or smaller for other drawings, depending upon the shapes of the views, the space available, and the space needed for dimensions and notes when added.

Regardless of the number of views, the basic procedure does not change. Figures 6-22 and 6-23 show the same procedure being used for a two-view drawing. When only the front and side views are necessary, you can arrange the views as shown in Figure 6-22D. When the drawing includes only the front and top views, arrange them as shown in Figure 6-23D.

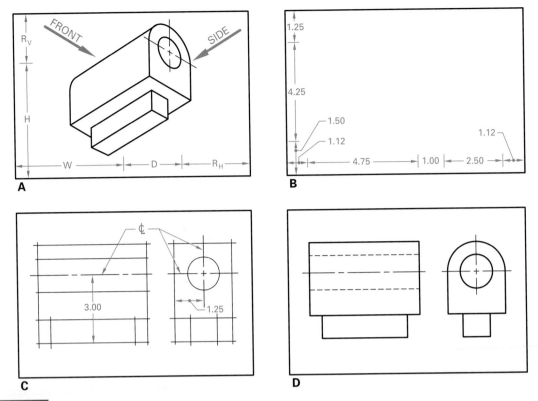

Figure 6-22

Calculations for the placement of views for a front- and right-side-view drawing.

Figure 6-23

Calculations for the placement of views for a front- and top-view drawing.

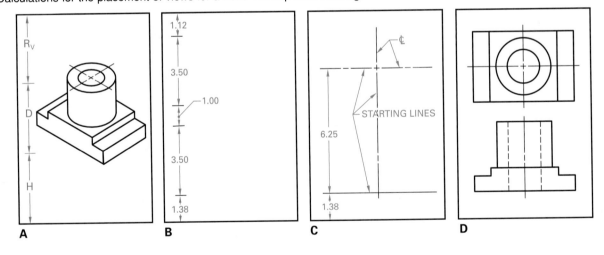

A   B   C   D

Metric View Locations

The procedure for locating views on ISO sheets is exactly the same as the procedure described in the text, except the calculations are performed using the metric system. For example, the dimensions of an A4-size sheet are 210 mm × 297 mm. The effective working area is approximately 172 mm × 270 mm. The metric equivalents of the overall measurements of the slide stop are shown in part A of the figure on the right. Part B shows the corresponding calculations.

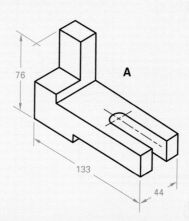

A

76

133

44

B

WORKING SPACE: 172 mm x 270 mm	
VERTICAL SPACING	**HORIZONTAL SPACING**

VERTICAL SPACING		HORIZONTAL SPACING	
76	Height (for front view)	133	Width (for front view)
+ 44	Depth (for top view)	+ 44	Depth (for right-side view)
120	Total vertical space needed for views	177	Total horizontal space needed for views
172	Total available vertical space	270	Total available horizontal space
– 120	Total needed for views	–177	Total needed for views
52	Remaining vertical space	93	Remaining horizontal space
52	Remaining vertical space	93	Remaining horizontal space
– 20	Space between front and top views	– 25	Space between front and side views
– 14	Space at top of sheet	– 34	Space at left side of sheet
– 18	Space at bottom of sheet	– 34	Space at right side of sheet
0	(All vertical space accounted for)	0	(All horizontal space accounted for)

Second Position of the Side View

An alternate position for the side view in third-angle projection is to the right of the top view. See Figure 6-24. This second position may be necessary due to the proportions of the object or the size of the drawing sheet. Going back to the glass box model, you can achieve this position by revolving the side plane around its intersection with the top plane.

Success on the Job

Attention to Detail

Employers, clients, and coworkers depend on drafters to create clear, easy-to-understand drawings that contain all required information. Paying attention to details and accuracy is therefore critical to the drafter's work. Neatly drawn, accurate views placed in standard locations reduce the chance of misunderstandings and errors.

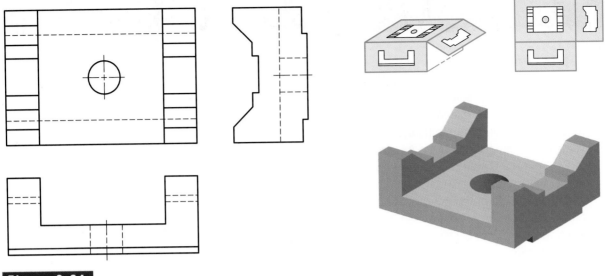

Figure 6-24

Second position of the side view in third-angle projection.

Board Drafting Techniques

The following procedures apply to board drafting. For techniques that apply to computer-aided drafting (CAD) systems, please refer to "CAD Techniques" later in this chapter.

Creating the Drawing

Before you begin to develop a multiview drawing, carefully review the earlier section on placing views. Then read both "Laying Out the Views" and "Adding Details," paying careful attention to the illustrations. Be sure you are familiar with all the steps before you begin.

Laying Out the Views

To lay out a multiview drawing of the rod support shown in Figure 6-25, first perform the calculations to locate the views according to the instructions in the previous section. Then follow these steps:

1. Study Figure 6-25 and determine which view should be used as the front view.

2. Determine which views will be required to describe the part fully. In this case, the front, top, and right-side views will be used.

3. Locate the views as shown in Figure 6-26A, referring to Figure 6-22 or 6-23 as necessary. Use only short, light pencil marks.

4. Block in the views with light, thin layout lines, as shown in Figure 6-26B.

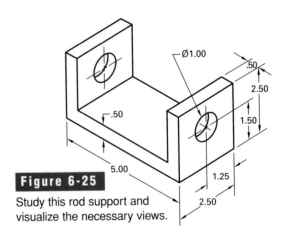

Figure 6-25

Study this rod support and visualize the necessary views.

Figure 6-26

Steps for creating a multiview drawing.

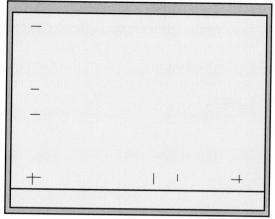

A

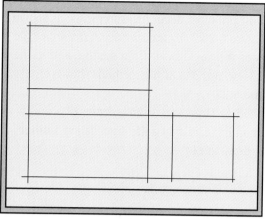

B

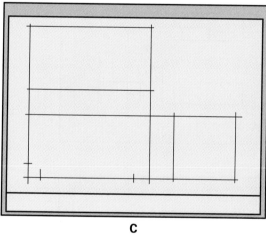

C

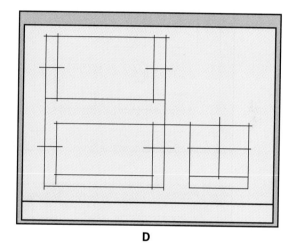

D

E

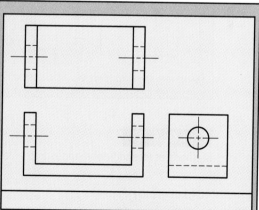

F

Adding Details

After you have located and blocked in the views, you must finish the details. Follow these steps:

1. Lay off principal detail measurements, as shown in Figure 6-26C. To transfer depth from the side view to the top view or from the top view to the side view, use one of the methods shown in Figure 6-27:

   Method A: Draw arcs from a center *O*.

   Method B: Construct a 45° line, called a **miter line**, through *O*.

   Method C: Use dividers to transfer distances.

   Method D: Use a scale to measure distances.

2. Draw the principal detail lines or layout lines, including centerlines, as shown in Figure 6-26D.

3. Draw circles and other details needed to complete the views. See Figure 6-26E.

4. Darken all lines to make them sharp and black and of the proper thickness, as shown in Figure 6-26F.

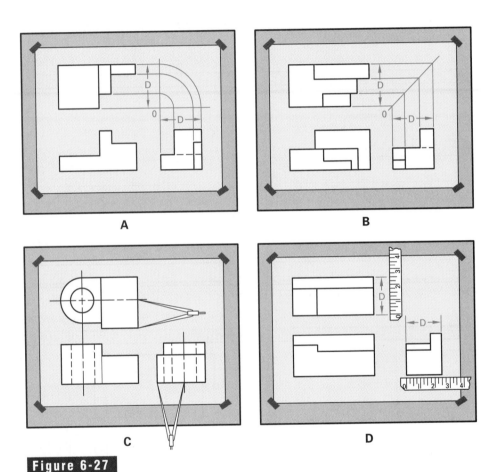

Figure 6-27

Methods of locating depth measurements.

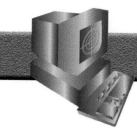

CAD Techniques

The following discussion applies specifically to computer-aided drafting (CAD). Although CAD techniques may differ in their approach from board drafting techniques, it is important to keep the basic drafting principles covered in the first portion of this chapter in mind. Unless specifically stated otherwise, all drafting principles apply equally to board and CAD drawings.

Creating Views Independently

One way to create a multiview drawing using a CAD system is to create the necessary views independently. The process of developing them is much the same as it is on paper. Like their paper counterparts, these drawings are two-dimensional. However, CAD offers a more efficient way to create and align the views.

Laying Out the Views

To lay out a multiview drawing of the rod support shown in Figure 6-25 using CAD techniques, first set up a drawing file according to the instructions in Chapter 4. Be sure to set up Hidden, Centerline, and Object layers. You may also want to set the grid and snap to a convenient spacing. Determine the location of the views according to the instructions earlier in this chapter. Then follow these steps:

1. With Ortho on, use vertical and horizontal construction lines (XLINE) to block in main object lines for the front view, as shown in Figure 6-28. Because construction lines extend infinitely, they automatically form the basis for the top and side views.

▶ **CADTIP**

Using OFFSET
After you have placed the first horizontal construction line, use the OFFSET command to place the other horizontal construction lines quickly and accurately. Repeat the procedure for the vertical construction lines.

2. Add horizontal construction lines at the proper distances to finish blocking in the top view.
3. Add vertical construction lines at the proper distances to finish blocking in the right-side view. Figure 6-29A shows the appearance of the drawing when all construction lines are in place.
4. Trim the construction lines to create the main object lines for all views. Note that this requires careful study of the rod support to determine the appropriate places to trim. Remember to trim *between* the views. Figure 6-29B shows the appropriate trimming operations for the rod support, and Figure 6-29C shows the result after trimming.

Figure 6-28

Use construction lines, rather than lines, to block in the front views. This provides the basis for the top and side views automatically.

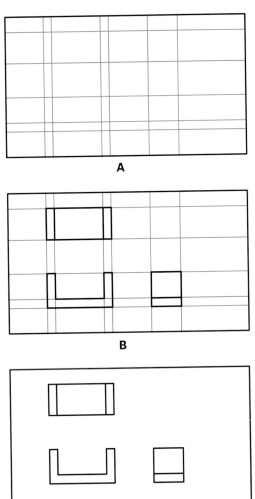

A

B

C

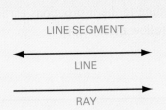

Figure 6-29

(A) Finish placing the construction lines. (B) Visualize the three views before beginning the trimming operation. Trim away all parts of the lines except those shown in black. (C) The result of the trimming operation.

Adding Details

Add the holes and finish the views by following these steps:

1. Pick the interior line in the right-side view to select it. Then pick the Layer Control dropdown box at the top of the drawing area, as shown in Figure 6-30, and pick the Hidden layer. This moves the line to that layer.

2. In the right-side view, use the OFFSET command to offset the top object line down by 1.00. Offset the left object line to the right by 1.25. These temporary lines will intersect at the correct position for the center of the circle that represents the hole. Refer again to Figure 6-25. Add the circle, using the intersection of the temporary lines as its center point. Then delete the temporary lines.

3. Use horizontal construction lines to determine the placement of the hidden lines that represent the circle on the front view. Use the Quadrant object snap to place the construction lines at the exact top and bottom of the circle in the right-

TECH•MATH

Lines, Line Segments, and Rays

Lines created with AutoCAD's LINE command have definite starting and ending points. Because of this, mathematically they are not lines at all. They are line segments.

Construction lines created using the XLINE command are true lines in the mathematical sense. Unlike lines created with the LINE command, they extend infinitely in both directions. This characteristic makes them very useful for laying out multiview drawings. Views are easier to align using construction lines, and the construction lines can be edited to become parts of several different views.

LINE SEGMENT

LINE

RAY

Rays are mathematically similar to lines. However, a ray has a definite starting point and extends infinitely in one direction only. AutoCAD's RAY command allows you to create mathematically correct rays. The illustration above compares line segments, lines, and rays. When might you prefer to use a ray instead of a line or line segment?

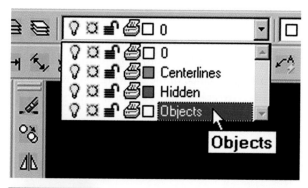

Figure 6-30

The Layer Control dropdown box allows you to select a new current layer or change the layer of selected objects.

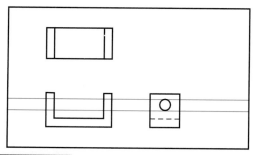

Figure 6-31

Construction lines drawn tangent to the exact top and bottom of the circle show the correct placement of the holes in the front view. Use the Quadrant object snap to snap to these points correctly.

side view. See Figure 6-31. Trim the construction lines to finish the hidden lines on both arms of the rod, and move them to the Hidden layer.

4. In the top view, use the Midpoint object snap and LINE command to create a temporary line from the midpoint of the left side to the midpoint of the right side. This line places the center of the holes in the top view. Offset the line .50 above and below, as shown in Figure 6-32A. Trim the lines and change the linetype to hidden. Figure 6-32B shows the finished hidden lines.

5. Use the centerline you created in step 4 as a basis for the centerlines through the holes. Extend the line .50 on each end and trim the center of the line so that .50 extends past the inside of each arm. Select the centerlines and move them to the Centerlines layer. The finished top view is shown in Figure 6-32C.

6. Use a horizontal construction line through the center of the hole in the right-side view to place the centerlines in the front view. Trim the con-

struction line and move it to the Centerlines layer to finish the centerlines for both holes.

7. Enter the DIMCENTER command and select the hole in the right-side view to create the centerlines for that view, as shown in Figure 6-32D. *Note:* If only a center mark, or cross, appears, enter the DIMSTYLE command and modify the current dimension style to show "lines" for centers of circles.

Figure 6-32

Transferring the circle to the top view.

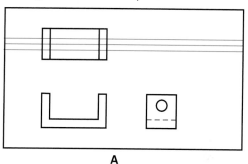

A

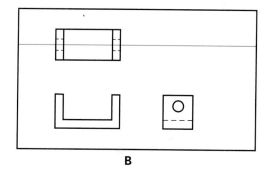

B

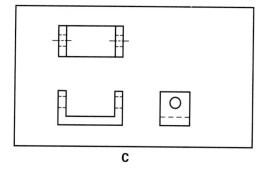

C

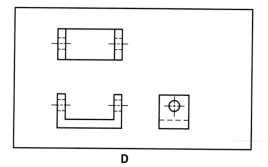

D

Creating Views from a Solid Model

Another way to create a multiview drawing is to extract the necessary views from a **solid model**. A solid model actually exists in three dimensions within the CAD drawing file. It adds a depth dimension (Z axis) to the width (X axis) and height (Y axis) present in a two-dimensional CAD drawing. Figure 6-33 illustrates the relationship of the Z axis to the X and Y axes. Because CAD programs allow you to turn the model around or change your point of view, you can create all the views you need from one solid model.

AutoCAD includes a set of predefined solid primitives, as shown in Figure 6-34. These primitives provide the basis for more complex solid models. By placing two or more primitives in the correct positions, many different complex shapes can be created.

To create the normal views from a solid model, you can use AutoCAD's predefined 3D views. These views, available on the View pulldown menu, include all of the six basic views and four isometric views. AutoCAD also allows you to have more than one viewport, or instance of the drawing, on the screen at one time. To create the two-view drawing of the

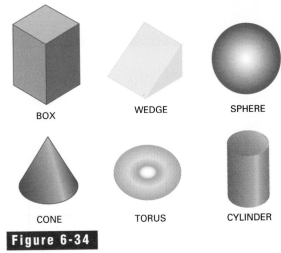

Figure 6-34

AutoCAD's six solid primitives.

Figure 6-33

(A) Two-dimensional drawings use the X axis (horizontal) and the Y axis (vertical). (B) Three-dimensional drawings use three axes. The third, called the Z axis, is at 90° angles to both the X axis and the Y axis and is used for depth.

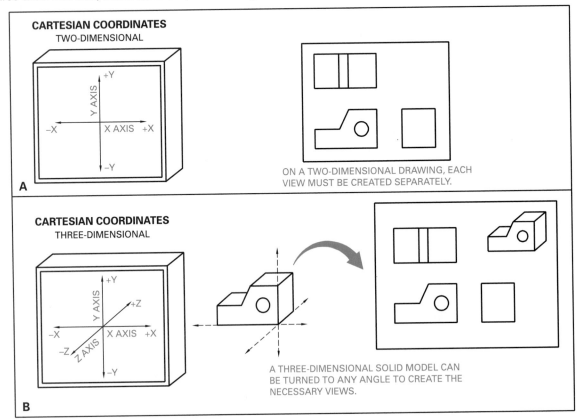

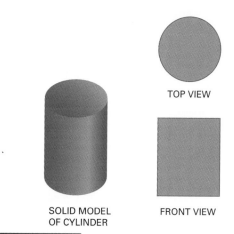

TOP VIEW

SOLID MODEL
OF CYLINDER

FRONT VIEW

Figure 6-35

You can easily create top and front views of the solid cylinder on the left by changing to AutoCAD's predefined Top and Front views.

cylinder shown in Figure 6-35, you would first create the cylinder as a solid primitive. Then select Viewports from the View pulldown menu and select 2 Viewports. This places two copies of the cylinder on the screen. Change the top viewport to the Top predefined view by selecting 3D Views and then Top from the View menu. Then change the bottom viewport to the front view. This provides the multiview drawing. You can move the viewports around to place the views more precisely.

This method of producing a multiview drawing has both advantages and disadvantages when compared to 2D orthographic projections. One advantage is that it can save time because you only have to draw the object once, no matter how many views you need. Also, the model itself becomes the "pictorial view." You can position it at any angle to create the best pictorial representation. Because the views are generated directly from the solid model, there is

no chance that the orthographic views will contain inaccuracies when compared with the pictorial.

In industry, solid models are often preferred because of their flexibility. They can be **rendered** to look almost like photographs. See Figure 6-36. These renderings can help clients visualize how a part, machine, building, or any other object will look when it is finished. They can also be used by the marketing department to help promote the item.

In addition, because they are "solid," solid models have mass properties such as mass, volume, and moments of inertia. Engineers can assign a material (or materials) to a model and ask the CAD software to calculate the mass properties. In some companies, engineers develop automation or animation programs that allow them to test the solid model in ways that could previously be done only using a prototype.

Finally, many companies now use solid models directly to generate prototypes and even the actual products. Links between CAD software and CAM (computer-aided manufacturing) software allow information from the solid model to drive the manufacturing equipment. Changes to the solid model result in changes to the programming instructions that are passed to the CAM software.

The major disadvantage of using solid models is that they are somewhat more complex to create. You have to be able to think in three dimensions. This is actually something drafters do all the time—it just takes practice to work in all three dimensions at once.

Figure 6-36

A solid model created using a CAD program can be rendered to look realistic. In this case, a different color has been used to render each material that will be used in the final product.

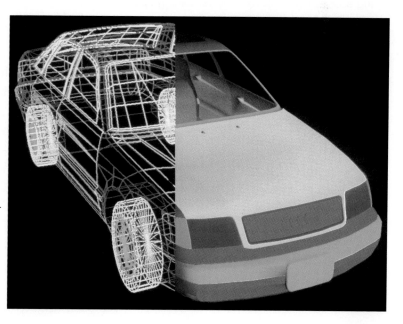

Chapter Summary

- Technical drawings describe three-dimensional objects completely using only two dimensions on a flat surface.

- Multiview drawings are technical drawings that include all the views necessary to create an object, arranged in a standard order.

- The three most often-used views for a multiview drawing are known as the normal views.

- Orthographic projection is the process of projecting the exact form of an object onto vertical, horizontal, and profile planes.

- The number of views needed to describe an object completely depends on its shape and characteristics.

- Views are chosen based on an object's characteristic view, its normal position, and the relative number of hidden lines in the views that are being considered.

- To locate views on a layout, subtract the total vertical and horizontal space needed to draw all views of the object from the total available working space. Then allocate the remaining space among the views.

Review Questions

1. What is the purpose of a multiview drawing?

2. What is visualization? Why is it an important tool for drafters?

3. Most objects have six sides and six possible views. Name them.

4. Name the three normal views in a technical drawing created in the United States.

5. What is the role of orthographic projection in multiview drawing?

6. What is the difference between first- and third-angle projection? Which is used in the United States?

7. Explain why it is important to place each view in a multiview drawing in its proper place.

8. How many views does a sphere usually require?

9. How many views are required to represent a handle that has a double curve?

10. In third-angle projection, where should you place the top view?

11. Briefly describe how to calculate the space between views on a multiview drawing.

12. Briefly explain how to transfer distances from the front view to the top and side views of a drawing using board drafting techniques.

13. Explain the use of construction lines in creating a multiview drawing using CAD techniques.

14. What is a solid model? Why is information from solid models preferred to two-dimensional drawings by some manufacturing companies?

Drafting Problems

The problems in this chapter can be performed using board drafting or CAD techniques. The problems within each part are presented in order of difficulty, from least to most difficult.

Problems 1 through 9 provide practice in visualizing and creating normal views of objects. Do not draw the pictorial views, and do not dimension the drawings.

1. Draw the two views of the sanding block shown in Figure 6-37, and complete the third (top) view. The block is .75″ × 1.75″ × 3.50″. Scale: Full size.

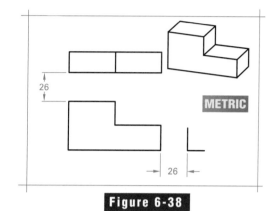

Figure 6-38

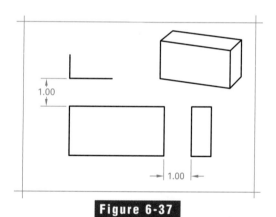

Figure 6-37

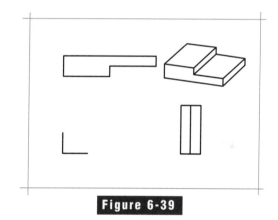

Figure 6-39

2. Draw the front and top views of the step block shown in Figure 6-38. Complete the right-side view in its proper location. The step block is 20 mm × 44 mm × 88 mm. The notch is 22 mm × 44 mm. Scale: Full size.

3. Draw the top and right-side views of the half lap shown in Figure 6-39. Complete the front view in its proper shape and location. The half lap is .75″ × 1.75″ × 3.50″. The notch is .38″ × 1.75″. Allow 1.00″ between views. Scale: Full size.

4. Draw the front and right-side views of the V-block shown in Figure 6-40. Complete the top view in its proper location. The overall size is 1.25″ × 2.00″ × 4.00″. Allow 1.00″ between views. Scale: Full size.

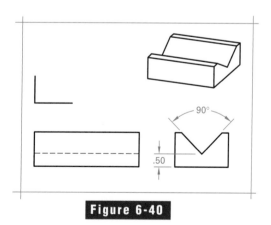

Figure 6-40

5. Draw the front and top views of the slide shown in Figure 6-41. Complete the right-side view in its proper location. The overall size is 2.12″ square × 3.75″. The slots are .38″ deep and .50″ wide. Allow 1.00″ between views. Scale: Full size.

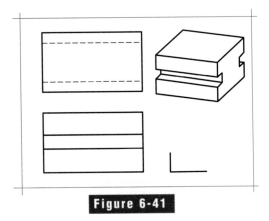

Figure 6-41

6. Draw the top and right-side views of the rod support shown in Figure 6-42. Complete the front view. The overall sizes are 2.00″ square × 3.50″. Bottom and ends are .50″ thick. The holes are 1.00″ square and are centered on the upper portions. Allow 1.00″ between views. Scale: Full size.

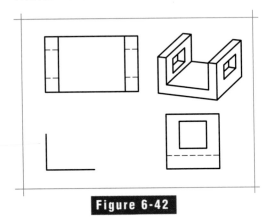

Figure 6-42

7. Draw the front view of the cradle shown in Figure 6-43. Complete the top view in the proper shape and location. Height = 50 mm; width = 150 mm; depth = 58 mm; base = 12 mm thick; A = 76 mm; B = 26 mm. Allow 26 mm between views. Scale: Full size.

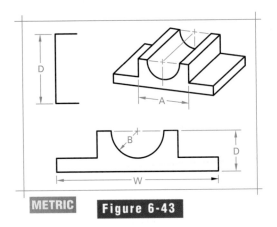

METRIC   **Figure 6-43**

8. Draw the top view of the spacer shown in Figure 6-44. Complete the front view. Base = 2.50″ × 1″; top = Ø1.50″ × .75″; hole = Ø1.00″. A vertical sheet will permit a larger scale. Allow 1.00″ between views. Scale: As assigned.

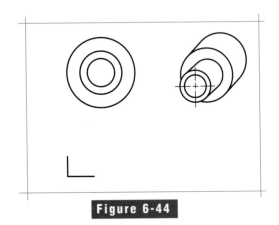

Figure 6-44

9. Draw the front view of the strap shown in Figure 6-45. Complete the top view in the proper shape and location. Allow 1.00″ between views. Overall width is 6.00″. Scale: Full size.

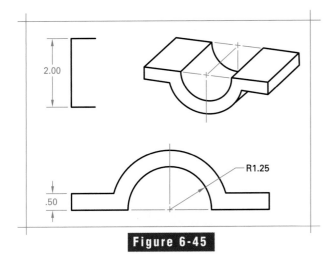

2.00

R1.25

.50

Figure 6-45

Problems 10 through 15 are two- and three-view problems. Each problem has one view missing. Draw the view or views given and complete the remaining view in the proper shape and location. Scale: Full size or as assigned.

10. Stop. See Figure 6-46. W = 5.00″; H = 2.00″; D = 2.00″; base = 1.00″ × 2.00″ × 5.00″; part = .75″ × 1.00″ × 3.00″.

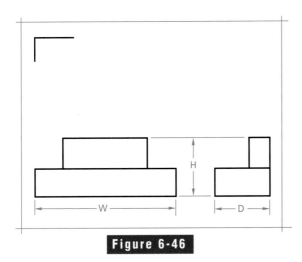

H

W

D

Figure 6-46

11. Link. See Figure 6-47. W = 7.50″; D = 2.50″; H = 1.25″; holes = Ø1.06″.

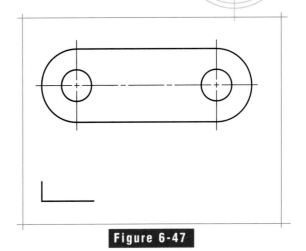

Figure 6-47

12. Angle bracket. See Figure 6-48. W = 5.00″; D = 2.00″; H = 2.25″; material thickness = .50″; holes = Ø1.00″.

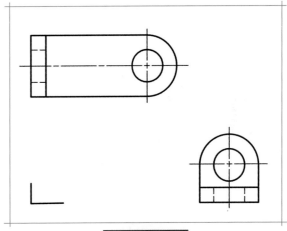

Figure 6-48

13. Saddle. See Figure 6-49. W = 140 mm; D = 50 mm; H = 56 mm; material thickness = 12 mm; hole = Ø26 mm.

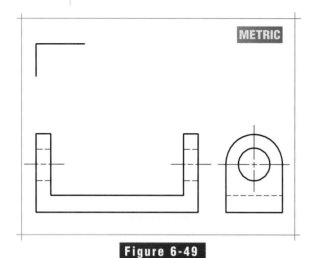

METRIC

Figure 6-49

14. Spacer. See Figure 6-50. W = 6.50″; D = 3.25″; thickness = 1.00″; holes = Ø2.38″ and Ø.75″; A = R.75″.

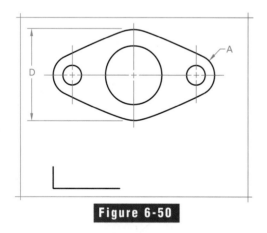

Figure 6-50

15. Dovetail slide. See Figure 6-51. W = 4.25″; D = 2.50″; H = 2.00″; base thickness = .75″; upright thickness = 1.25″; holes = Ø.62″; A = .50″; B = .50″; CD = 1.50″; DE = .75″.

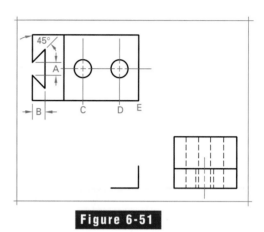

45°

A

B   C   D E

Figure 6-51

For problems 16 through 26, create two- or three-view drawings of the objects shown. Do not draw the pictorial views, and do not dimension the drawings.

16. Stop. See Figure 6-52.

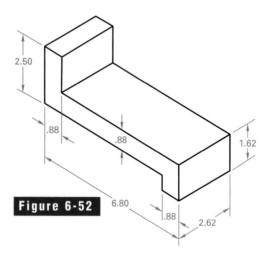

2.50

.88

.88

6.80

.88

2.62

1.62

Figure 6-52

17. Dovetail slide. See Figure 6-53.

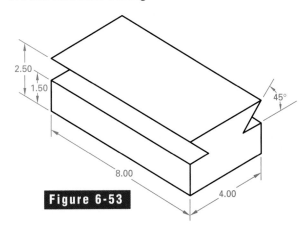

Figure 6-53

18. Slide. See Figure 6-54.

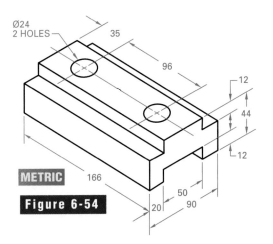

METRIC

Figure 6-54

19. Pivot arm. See Figure 6-55.

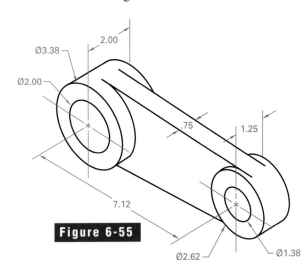

Figure 6-55

20. Shaft support. See Figure 6-56.

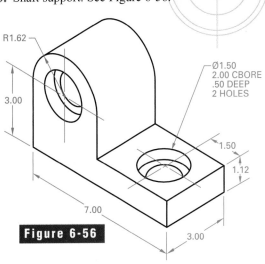

Figure 6-56

21. Edge protector. See Figure 6-57.

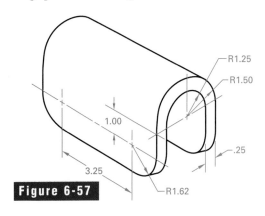

Figure 6-57

22. Socket. See Figure 6-58. A = Ø50.5 mm × 7 mm; B = 38 mm; C = 25.2 × 17 mm long with Ø13 mm hole through; slots = 4.5 mm wide × 8 mm deep; D = Ø6 mm, 4 holes equally spaced.

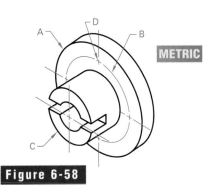

METRIC

Figure 6-58

23. Cam. See Figure 6-59. A = Ø2.62″; B = Ø1.25″, 1.75″ counterbore .25″ deep on both ends; C = 2.12″; D = R.56″; E = 1.75″; F = .88″.

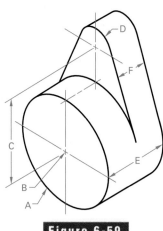

Figure 6-59

24. Camera swivel base. See Figure 6-60. AB = 40 mm; AC = 38 mm; BD = 5.5 mm; BE = 20 mm; EF = 45 mm; H = R12 mm; G = Ø10 mm; J = 6 mm. Boss: K = Ø24 mm × 9 mm long; L = Ø28 mm × 4 mm long; hole = Ø12 mm × 14 mm deep; counterbore = 18 mm × 3 mm deep.

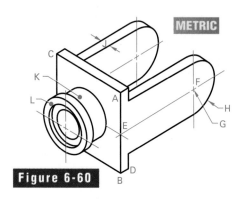

METRIC

Figure 6-60

25. Pipe support. See Figure 6-61. AB = 8.00″; BC = 4.00″; AD = .75″; E = 3.88″ centered; F = Ø2.75″ × 2.25″ long; G = Ø2.23″ hole through centered; slots = 1.00″ wide centered.

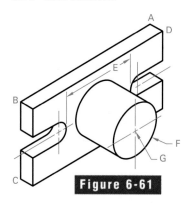

Figure 6-61

26. Angle plate. See Figure 6-62. AB = 6.00″; BC = 6.50″; AD = 9.75″; DE = 1.00″; CF = 1.00″; G = Ø.75″, 2 holes; EH = 2.00″; EJ = 2.50″; FL = 1.12″; LO = 2.00″; FM = 1.50″; MN = 2.50″; P = 1.12″; K = Ø.50″, 8 holes.

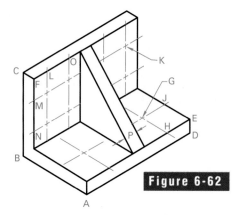

Figure 6-62

For problems 27 through 29, draw all necessary views of the objects shown. Do not draw the pictorial views, and do not dimension the drawings.

27. Knife rack. See Figure 6-63. Back is .50″ × 9.00″ × 18.00″. Front is 1.50″ × 7.00″ × 10.00″ with 30° bevels on each end. Slots for knife blades are .12″ wide × 1.00″ deep. Grooves on front are .12″ wide × .12″ deep. Estimate all sizes not given. Scale: Half size.

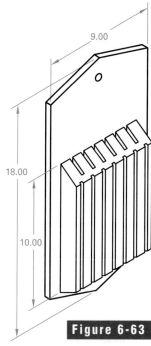

9.00

18.00

10.00

Figure 6-63

28. Mini sawhorse. See Figure 6-64. Top rail is 2.00″ × 4.00″ × 24.00″. Legs are cut from 2.00″ × 12.00″ stock, 14.50″ long. Scale: ¼″ = 1″.

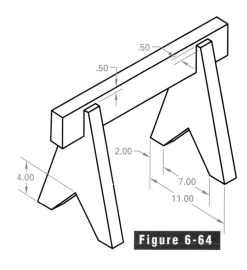

.50

.50

2.00

4.00

7.00

11.00

Figure 6-64

29. Note-paper box. See Figure 6-65. All stock is 6 mm thick. AB = 168 mm, BC = 118 mm, BD = 24 mm, DE = 12 mm, F = 6 mm, GH = 38 mm. Initial inlay is optional. Scale: Full size or as assigned.

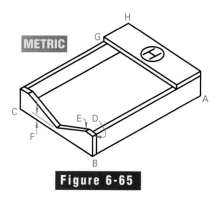

METRIC

H

G

A

C

E D

F

B

Figure 6-65

Problems 30 through 36 show two views of various objects. From the views given, visualize the object. Then draw *three* views of each object.

30. Bracket. See Figure 6-66.

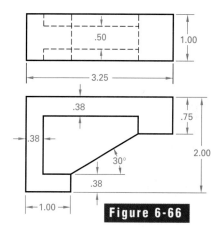

.50

1.00

3.25

.38

.75

.38

30°

2.00

.38

1.00

Figure 6-66

31. Locator. See Figure 6-67.

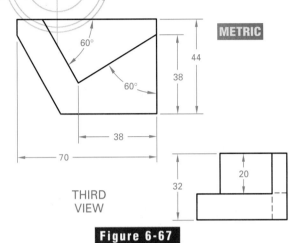

METRIC

THIRD
VIEW

Figure 6-67

32. Locating support. See Figure 6-68.

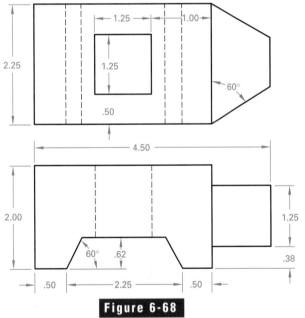

Figure 6-68

33. Pivot. See Figure 6-69.

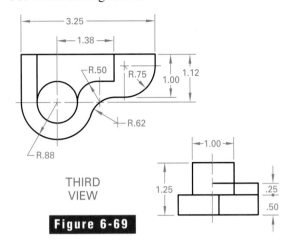

THIRD
VIEW

Figure 6-69

34. Link. See Figure 6-70.

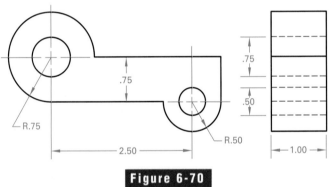

Figure 6-70

35. Locating plate. See Figure 6-71.

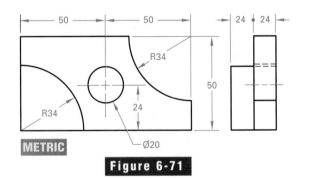

METRIC

Figure 6-71

4

36. Separator. See Figure 6-72.

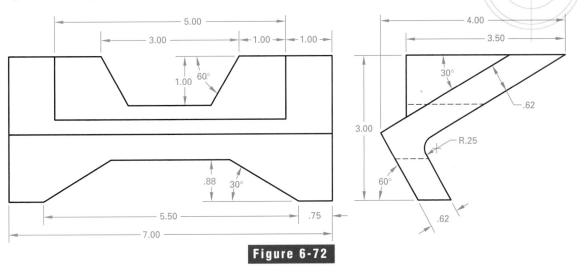

Figure 6-72

Problems 37 through 43 show pictorial views of various objects. Visualize and draw three views of each object.

37. Vertical stop. See Figure 6-73.

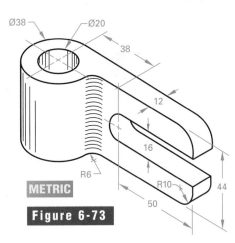

METRIC
Figure 6-73

38. Side cap. See Figure 6-74.

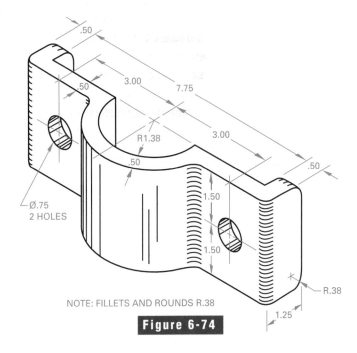

NOTE: FILLETS AND ROUNDS R.38

Figure 6-74

39. V-block base. See Figure 6-75.

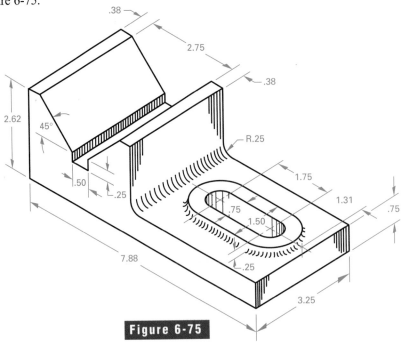

.38

2.75

.38

2.62

45°

.50

.25

R.25

1.75

1.31

.75

.75

1.50

7.88

.25

3.25

Figure 6-75

40. Cross slide. See Figure 6-76.

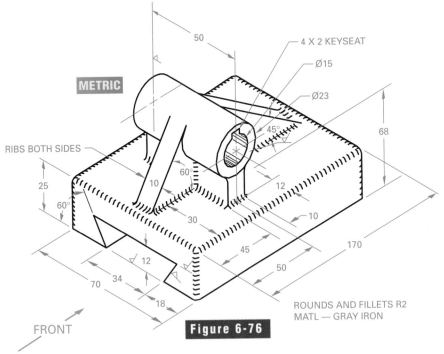

50

4 X 2 KEYSEAT

Ø15

METRIC

Ø23

45°

68

RIBS BOTH SIDES

25

10

60°

12

60°

30

10

12

45

170

70

34

50

18

FRONT

ROUNDS AND FILLETS R2
MATL — GRAY IRON

Figure 6-76

41. Shaft support. See Figure 6-77.

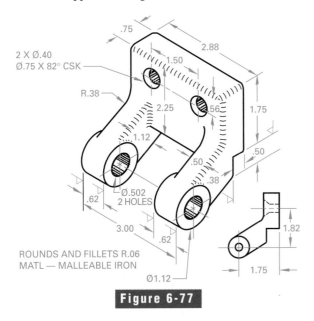

2 X Ø.40
Ø.75 X 82° CSK

R.38

.75
2.88
1.50
2.25
.56
1.12
1.75
.50
.50
.38
.62
Ø.502
2 HOLES
3.00
.62
1.82
1.75
Ø1.12

ROUNDS AND FILLETS R.06
MATL — MALLEABLE IRON

Figure 6-77

42. Flanged coupling. See Figure 6-78.

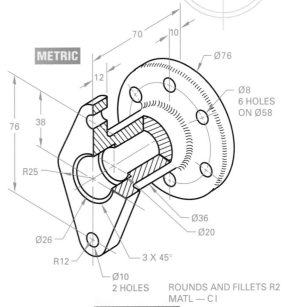

METRIC

70
10
12
Ø76
Ø8
6 HOLES
ON Ø58
76
38
R25
Ø36
Ø20
Ø26
R12
3 X 45°
Ø10
2 HOLES
ROUNDS AND FILLETS R2
MATL — C I

Figure 6-78

43. Adapter. See Figure 6-79.

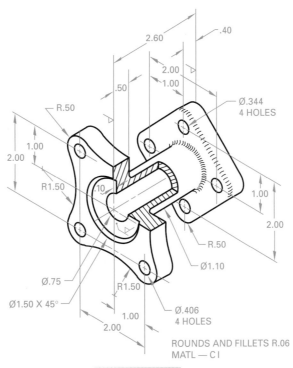

2.60
.40
2.00
1.00
.50
R.50
Ø.344
4 HOLES
1.00
2.00
R1.50
.10
1.00
2.00
R.50
Ø1.10
Ø.75
R1.50
Ø1.50 X 45°
1.00
2.00
Ø.406
4 HOLES

ROUNDS AND FILLETS R.06
MATL — C I

Figure 6-79

Design Problems

Designing new or improved products requires creativity, problem-solving skills, and drafting skills to communicate the new ideas to other members of the team. The following problems provide practice to help you improve these important skills. Use board or CAD techniques to present your designs.

1. Design a tic-tac-toe board similar to the one shown in Figure 6-80 to use golf tees. Devise a convenient way to store the golf tees with the board. Draw all necessary views.

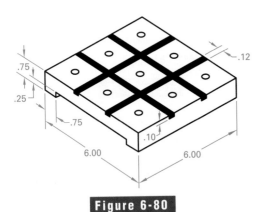

Figure 6-80

2. Design a knife rack similar to the one shown in Figure 6-63. Design it to sit on a counter or table rather than hang on the wall. Draw all necessary views and specify the materials from which it is to be made.

3. **TEAMWORK** Design a shelf to hold sport trophies. The shelf should hang on a wall and should be able to support at least three trophies. Draw all necessary views and specify the materials from which it is to be made. Dimension all views using metric measurements.

4. **TEAMWORK** Design a rack to support a book on a table or counter. The rack should support books at a convenient angle for reading. Draw all necessary views and specify the materials from which it is to be made.

5. **TEAMWORK** Design a storage case for CDs or DVDs. The case should be able to hold up to 50 of the "jewel cases" in which CDs and DVDs are commonly sold. Draw all necessary views and specify the materials from which it is to be made.

7

Dimensioning

KEY TERMS

aligned system

basic hole system

basic shaft system

bilateral tolerances

datums

dimension line

dual dimensioning
system

finish mark

geometric dimensioning
and tolerancing

tolerance

unidirectional system

unilateral tolerances

OBJECTIVES

Upon completion of this chapter, you should be able to:

- Apply measurements, notes, and symbols to a technical drawing.

- Use ANSI and ISO standards for dimensions and notes.

- Differentiate between size dimensions and location dimensions.

- Determine appropriate sizes for precision fits between inter-changeable mating parts.

- Specify geometric tolerances using symbols and notes.

- Designate appropriate surface textures.

- Use board drafting techniques to add dimensions, notes, and geometric tolerances to a technical drawing.

- Use a CAD system to add dimensions, notes, and geometric tolerances to a technical drawing.

Drafting Principles

To describe an object completely, a drafter needs to define both the shape and the size of the object. Most of this book deals with ways of describing shape. This chapter, however, discusses how to show the size of the objects that you draw. It is very important to understand clearly the rules and principles of size description. After all, a machinist cannot make parts correctly unless all the sizes on the drawing are accurate and complete.

Another name for size description is **dimensioning**. Dimensions, or sizes, are measured in either U.S. customary or metric (SI) units. Decimal divisions and metric units are now most commonly used throughout industry and are used exclusively in ANSI Y14.5M (the drafting standard on dimensioning). Refer to Chapters 3 and 4 for more information about U.S. customary and metric units.

Notes and symbols that show the kind of finish, materials, and other information needed to make a part are also part of dimensioning. A complete set of **working drawings** (the drawing or set of drawings from which the part is manufactured) includes shape description, measurements, notes, and symbols. See Figure 7-1. Chapter 13 provides more information about working drawings.

Dimensions on working drawings must be as precise as necessary to allow the manufacturer to create the part or object. When dimensions must be precise, they are given in hundredths, thousandths, or ten-thousandths of an inch. If the metric system is being used, the measurements may be in tenths, hundredths, or even thousandths of a millimeter.

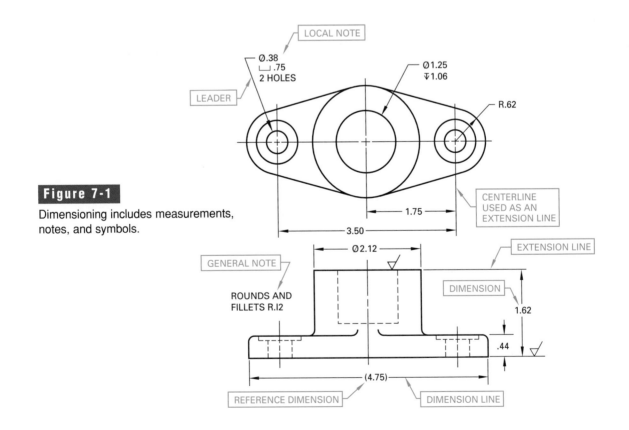

Figure 7-1

Dimensioning includes measurements, notes, and symbols.

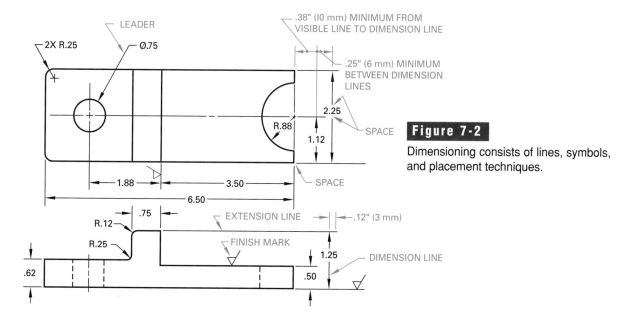

Figure 7-2

Dimensioning consists of lines, symbols, and placement techniques.

Lines and Symbols for Dimensioning

The views on drawings describe the shape of an object. In theory, size could be found by measuring the drawing and applying a scale. In reality, though, this is not practical, even when the views are drawn at full size. The measuring simply takes too much time. More important, it is impossible to measure a drawing accurately enough for many interchangeable parts that must fit closely together. To ensure accuracy and efficiency, the drafter adds size information to the drawing using a system of lines, symbols, and numerical values.

Lines and symbols are used on drawings to show where the dimensions apply, as shown in Figure 7-2. Professional and trade associations, engineering societies, and certain industries have agreed upon the symbols so that people who use the drawings can recognize their meaning. The latest standards information on drawings and symbols can be found in publications from the American National Standards Institute (ANSI), the Society of Automotive Engineers (SAE), the Military Standards, and the International Standards Organization (ISO).

Dimension Lines

A **dimension line** is a thin line that shows where a measurement begins and where it ends. Dimension lines are also used to show the size of angles. The dimension line should have a break in it for the dimension numbers. To keep the numbers from getting crowded, dimension lines should be at least .38″ from the lines of the drawing and at least .25″ from each other. Spacing of dimension lines should be as consistent as possible within a drawing. See Figure 7-3. On metric drawings, dimension lines should be at least 10 mm from the lines of the drawing and 6 mm from each other. In general, dimension lines should be placed outside the view outlines.

Arrowheads

Arrowheads are placed at the ends of dimension lines to show where a dimension begins and ends. They are also used at the end of a leader to show where a note or dimension applies to a drawing. Refer again to Figure 7-1.

Figure 7-3

Dimension lines must be spaced to provide clarity.

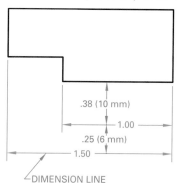

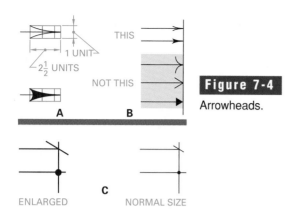

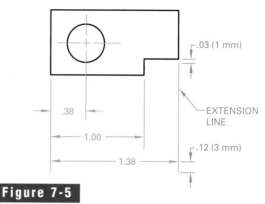

Figure 7-4

Arrowheads.

Figure 7-5

Extension lines. A centerline may be used as an extension line.

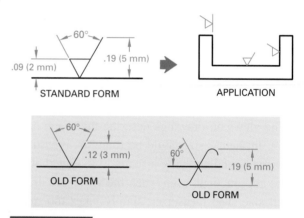

Figure 7-6

The finish mark tells which surfaces are to be machined.

Arrowheads can be open or solid. Their shapes are shown enlarged in Figure 7-4A and reduced to actual size in Figure 7-4B. In any one drawing, they should all be the same size and shape. However, in a small space, you may have to make the arrowheads slightly smaller.

Some industries use other means to point out the endpoint of a dimension line or leader. Figure 7-4C shows some examples of these. These symbols do the same job as arrowheads. For example, slash marks are often used instead of arrowheads in architectural drafting. For most mechanical working drawings, however, the arrowheads shown in Figure 7-4A and B are preferred.

Extension Lines

Extension lines are thin lines that extend the lines or edges of views. They are used to locate center points and to provide space for dimension lines. Since extension lines are not part of the views, they should not touch the outline. Start the extension line about .03″ to .06″ (1 to 1.5 mm) from the part, and extend it about .12″ (3 mm) beyond the last dimension line, as shown in Figure 7-5. Avoid drawing extension lines that cross each other or that cross dimension lines.

Numerals and Notes

Numerals and notes have to be easy to read, so you must create them carefully. Do not make them unnecessarily large, however. Capital letters are preferred on most drawings. In general, make numerals about .12″ (3 mm) high. Refer to Chapter 2 for more information about lettering.

Sometimes drawings are made to be microfilmed or to be reduced photographically and used at a smaller size. When this is the case, make the numerals larger and with heavier strokes so that they will be clear and easy to read when reduced.

The Finish Mark

To dimension a drawing correctly, drafters must know the correct symbols to use as well as the principles of dimensioning. In most cases, drafters must also know the shop processes that are used to build or make the products they draw. Sometimes drafters include symbols on the drawing to show which processes are needed.

The **finish mark**, or surface-texture symbol, shows that a surface is to be machined, or finished. Figure 7-6 shows the standard finish mark now in general use, as well as two older forms of the symbol.

The point of the finish mark symbol should touch the edge view of the surface to be finished or an adjacent extension line. Also, it should be positioned to read from the bottom of the sheet or from the right side of the sheet, as shown in Figure 7-6. Modified forms of this symbol can be used to show that allowance for machining is needed, that a certain surface condition is needed, and other information.

Leaders

A leader is a thin line drawn from a note or dimension to the place where it applies, as shown in Figure 7-7. Always place leaders at an angle to the horizontal. An angle of 60° is preferred, but 45°, 30°, or other angles may be used. A leader starts with a dash, or short horizontal line. This line should be about .12″ (3 mm) long, but it may be longer if needed. CAD systems set the length of the dash automatically. A leader generally ends with an arrowhead. However, a dot is used if the leader is pointing to a surface rather than an edge, as shown in Figure 7-7.

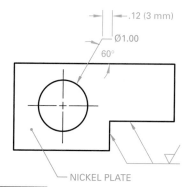

Figure 7-7

Leaders point to the place where a note or dimension applies.

Other rules for drawing leaders include:

- When a number of leaders must be placed close together, draw them parallel to each other.
- Draw a leader to a circle or arc so that the arrowhead points to its center. (CAD systems do this automatically.)
- Keep leaders as short as possible while keeping the note a minimum of .38" (10 mm) away from the view.
- Avoid drawing leaders horizontally, vertically, or at a small angle.
- Avoid drawing leaders parallel to dimension, extension, or section lines.

Units

When you use the U.S. customary system, give the measurements in inches and decimals of an inch. When customary dimensions are in inches, omit the inch symbol (″). Add a note to the drawing: UNLESS OTHERWISE SPECIFIED, ALL DIMENSIONS ARE IN INCHES.

Sometimes, parts must fit together with extreme accuracy. In that case, the machinist must work within specified limits. If the measurements are customary, the decimal inch is used. This is called *decimal dimensioning*. Such dimensions are used between finished surfaces, center distances, and pieces that must be held in a definite, accurate relationship to each other.

With customary measures, you may use decimals to two places where limits of ±.01″ are close enough, as shown in Figure 7-8A, B, and C. Use decimals to three or more places where limits smaller than ±.01″ are required, as shown in Figure 7-9A and B. For two-place decimals, fiftieths, such as .02, .04, or .24 (even numbers) are preferred over decimals such as .03 and .05 (odd numbers).

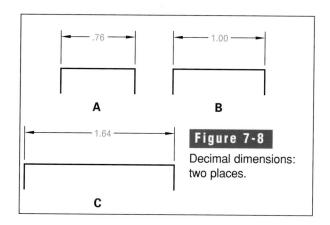

Figure 7-8

Decimal dimensions: two places.

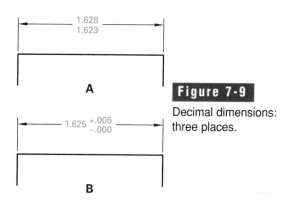

Figure 7-9

Decimal dimensions: three places.

Decimal dimensioning is used in most industries. It is the preferred method of drafting with U.S. Customary measure. A **dual dimensioning system** is sometimes used in industries involved in international trade. This system uses both the decimal inch and the millimeter, as shown in Figure 7-10. However, most international industries now use the metric system alone.

When you use the metric system, give the dimensions in millimeters, meters or, for special applications, micrometers. With metric dimensions, use decimals to one place where limits of ±0.1 mm are close enough. Use decimals to two places or more where limits smaller than ±0.1 mm are required.

Whole numbers do not need a decimal point or zero. A millimeter value less than 1 is shown with a zero to the left of the decimal point; for example, 0.2 mm. A decimal inch value less than 1 does not require a zero to the left of the decimal point. When all dimensions are in millimeters, the mm symbol is omitted. Instead, add a note to the drawing such as UNLESS OTHERWISE SPECIFIED, ALL DIMENSIONS ARE IN MILLIMETERS.

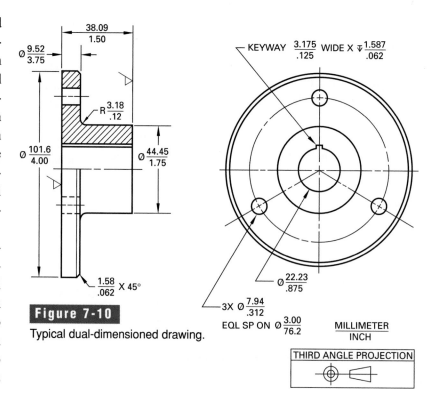

Figure 7-10

Typical dual-dimensioned drawing.

Figure 7-11

Parts can usually be broken down into basic geometric shapes for dimensioning.

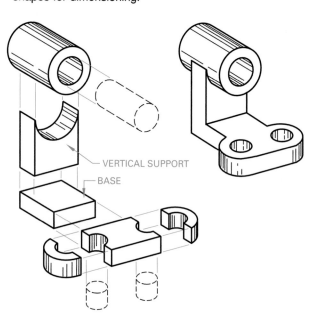

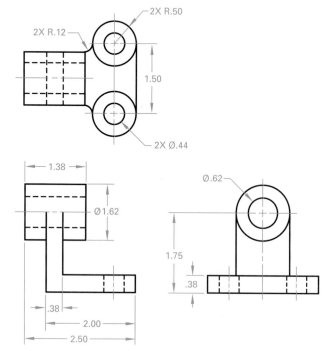

Theory of Dimensioning

There are two basic kinds of dimensions: size dimensions and location dimensions. **Size dimensions** define each piece. Giving size dimensions is really a matter of giving the dimensions of a number of simple shapes. Every object is broken down into its geometric forms, such as prisms, cylinders, pyramids, cones, and so forth, or into parts of such shapes. This is shown in Figure 7-11, where the bearing is separated into simple parts. A hole or hollow part has the same outlines as one of the geometric shapes. Think of such open spaces in an object as *negative* shapes.

The idea of open spaces is especially valuable to certain industries. Drafters in the aircraft industry need to know the weights of parts. These weights are worked out from the volumes of the parts as solids. From these solids, the volumes of the holes and hollow or open spaces (negative shapes) are subtracted to get the total weight per cubic inch or cubic millimeter of the material.

When the object being dimensioned has a number of pieces, the positions of each piece must also be given. These are given by **location dimensions**. Each piece is first considered separately and then in relation to the other pieces. When the size and location dimensions of each piece are given, the size description is complete. Dimensioning a whole machine, a piece of furniture, or a building is just a matter of following the same orderly pattern that is used for a single part.

TECH MATH

Volume and Weight

The weight of a part is sometimes critical in the design process. To determine the weight of the cylinder shown here and in Figure 7-11, first calculate the weight of the Ø1.62″ × 1.38″ cast iron cylinder as a solid.

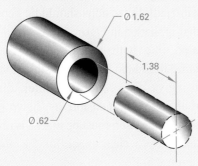

1. Determine the area of the Ø1.62″ circle.
   Area = πr^2
   Area = 3.1416 × .81²
   Area = 3.1416 × .66
   Area = 2.07 square inches
2. Determine the volume of the large cylinder.
   Volume = Area × Length
   Volume = 1.38 × 2.07
   Volume = 2.86 cubic inches

3. Multiply the weight per cubic inch for cast iron (.26) by the volume of the cylinder.*
   Weight = .26 × 2.86
   Weight = .74 pound

Next, calculate the weight of the small cylinder removed to create the hole and subtract that weight from the weight of the large cylinder.
   Area of small cylinder
      = 3.1416 × .31² = .31 square inches
   Volume of small cylinder
      = 1.38 × .31 = .43 cubic inches
   Weight of small cylinder
      = .26 × .43 = .11 pound

Finally, subtract the weight of the small cylinder from the weight of the large cylinder.

    .74 pound (weight of large cylinder)
   − .11 pound (weight of small cylinder)
    .63 pound (net weight of part)

*Note: The weight of various materials can be found in the *Machinery's Handbook* or *The American Machinist's Handbook*.

Placing Dimensions

Two methods of placing dimensions are currently in use: the aligned system and the unidirectional system. In the **aligned system** of dimensioning, the dimensions are placed in line with the dimension lines, as shown in Figure 7-12. Horizontal dimensions always read from the bottom of the sheet. Vertical dimensions read from the right. Inclined dimensions read in line with the inclined dimension line. The aligned system was once the only system in use.

In the **unidirectional system** of dimensioning, all dimensions read from the bottom of the sheet, no matter where they appear, as shown in Figure 7-13. In both systems, notes and dimensions with leaders should read from the bottom of the drawing. The uni-

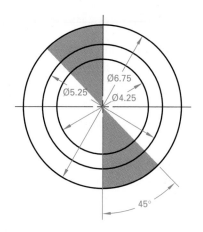

Figure 7-15

Avoid placing dimensions in the shaded area.

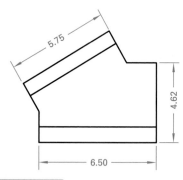

Figure 7-12

The aligned system of placing dimensions.

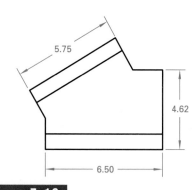

Figure 7-13

The unidirectional system of placing dimensions.

directional system has now replaced the aligned system in most industries. ANSI Y14.5M uses the unidirectional system exclusively.

Place overall dimensions outside the smaller dimensions, as shown in Figure 7-14. When you give the overall dimension, leave out the dimension of one of the smaller distances unless it is needed for reference. If a dimension is needed for reference, put parentheses around it to show that it is for reference only, as shown in Figure 7-14. Reference dimensions are used to help clarify a drawing but are never used in manufacturing a part.

If possible, all dimensions should be kept outside the area that is shaded in Figure 7-15. Avoid crossing a dimension line with another line. Also, avoid dimensioning to hidden lines if possible. On circular end parts, give the center-to-center dimension instead of an overall dimension, as shown in Figure 7-16. When a dimension must be placed within a sectioned area, leave a clear space for the number, as shown in Figure 7-17.

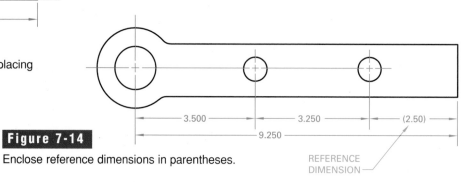

Figure 7-14

Enclose reference dimensions in parentheses.

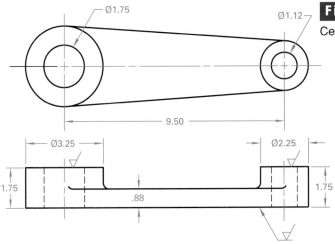

Figure 7-16
Center-to-center dimensions.

Always give the diameter of a circle, not the radius. Use the symbol Ø before the dimension. When dimensioning an arc, use the radius, not the diameter, and place the abbreviation R before the value.

Do not repeat the same dimension on different views. If you need to do so for clarity's sake, make one of them a reference dimension. Also, do not give dimensions that are not needed. This is especially important in interchangeable manufacturing processes in which limits are used. Figure 7-18A shows unnecessary dimensions. They have been omitted in Figure 7-18B.

Identifying the Drawing Scale

The scale used on a drawing should be given in or near the title. If a drawing has views of more than one part and different scales are used, the scale should be given close to the views. Scales are stated as full or full size, 1:1; half size, 1:2; and so forth. If enlarged views are used, the scale is shown as 2 times

full size, 2:1; 4 times full size, 4:1; and so forth. The scales used on metric drawings are based on divisions of 10. Scales such as 10:1, 1:50, and 1:100 are examples.

Size Dimensions

The first shape we will consider is the prism. For a rectangular prism, as shown in Figure 7-19, the width W, the height H, and the depth D are needed. This basic shape may appear in a great many ways. A few of these are shown in Figure 7-20. Flat pieces of irregular shape are dimensioned in a similar way. See Figure 7-21. The rule for dimensioning prisms is: *For any flat piece, give the thickness in the edge view and all other dimensions in the outline view.*

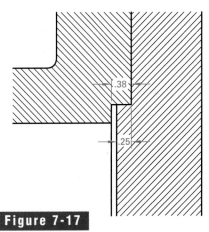

Figure 7-17
Dimensions within a sectioned area.

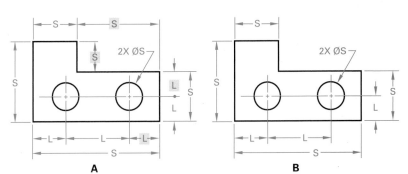

Figure 7-18
The dimensions that are shaded in (A) are not needed and should be omitted (B).

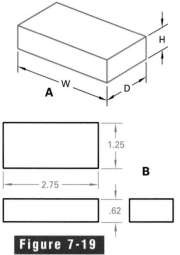

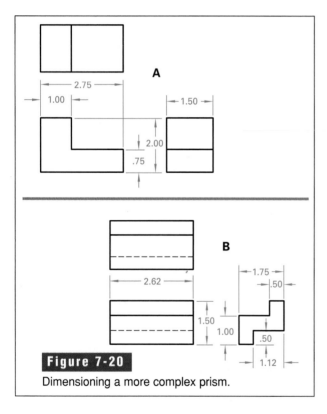

Figure 7-19

Dimensioning a simple rectangular prism.

Figure 7-20

Dimensioning a more complex prism.

The outline view is the one that shows the shape of the flat surface or surfaces. The front views in Figure 7-21 are the outline views. Methods for dimensioning other prisms are shown in Figure 7-22.

The second shape is the cylinder. The cylinder needs two dimensions: diameter and length. See Figure 7-23. Three cylinders are dimensioned in Figure 7-24A. One of these is the hole. Remember that a hollow cylinder can be thought of as two cylinders of the same length, as shown by the washer in Figure 7-24B. The rule for dimensioning cylinders is: *For cylindrical pieces, give the diameter and the length on the same view.* For holes, give the diameter and depth in the end view or section view.

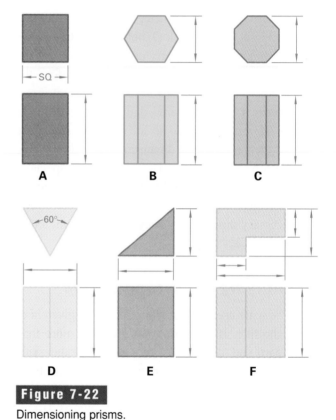

Figure 7-22

Dimensioning prisms.

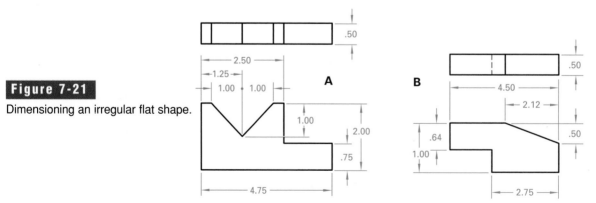

Figure 7-21

Dimensioning an irregular flat shape.

The symbol Ø is always placed with the diameter dimension. On simple drawings, the circular view may be eliminated by using the Ø symbol, as shown in Figure 7-24C. The abbreviation DIA may be found on older drawings instead of the symbol.

Notes are generally used to give the sizes of holes. Such a note is usually placed on the outline view, as shown in Figure 7-25A. These notes are sometimes used to show what operations are needed to form or finish the hole. For example, drilling, punching, reaming, lapping, tapping, countersinking, spotfacing, and so forth may be specified, as shown in Figure 7-26. The symbols used for these operations are defined in Table 7-1.

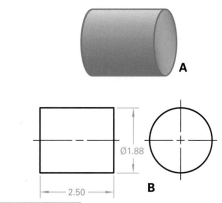

Figure 7-23

Dimensioning a cylinder.

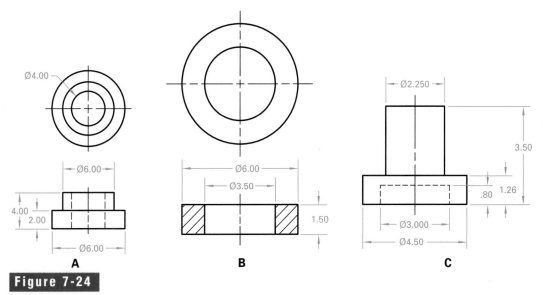

Figure 7-24

Dimensioning holes in cylinders.

Figure 7-25

Dimensioning (A) holes; (B) rounds; and (C) radii.

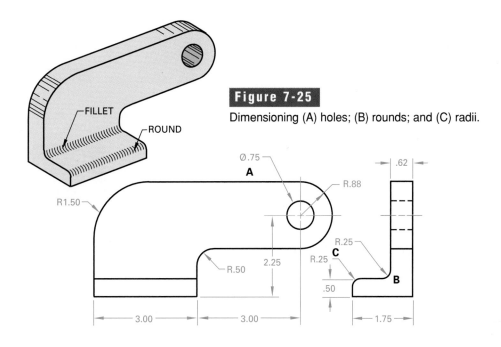

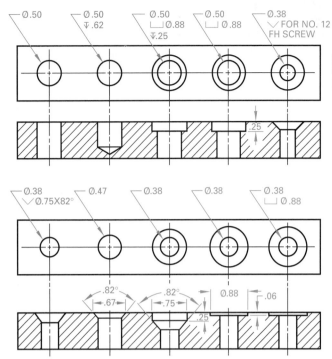

Ø.50 | Ø.50 ⊽.62 | Ø.50 ⊔Ø.88 ⊽.25 | Ø.50 ⊔Ø.88 | Ø.38 ∨FOR NO. 12 FH SCREW

Ø.38 ∨.75X82° | Ø.47 | Ø.38 | Ø.38 | Ø.38 ⊔Ø.88

Figure 7-26

Methods for specifying dimensions and operations.

Table 7-1. Notes and Symbols for Machining Operations	
Traditional Method	**Preferred Method**
½ Drill or .50 Drill	Ø.50
.48 Drill .500 Ream	Ø.500
½ Drill, ⅞ Cbore ¼ Deep	Ø.500 ⊔.875 ⊽ .25
.38 Drill 82° CSK To .75 DIA	Ø.38 Ø.75 x V82°
.38 Drill .88 Spotface .06 Deep	Ø.38 ⊔.88 ⊽ .06

When parts of cylinders occur, such as fillets and rounds, they are dimensioned in the views in which the curves show. Figure 7-25B shows how to dimension a **round**, or internal curve, and Figure 7-25C shows how to dimension a **fillet**, or external curve. The radius dimension is given and is preceded by the abbreviation R.

Some of the other shapes are the cone, the pyramid, and the sphere. The cone, **frustum** (truncated cone), square pyramid, and sphere can be dimensioned in one view, as shown in Figure 7-27. To dimension rectangular or other pyramids and parts of pyramids, two views are needed.

Location Dimensions

Location dimensions are used to show the relative positions of the basic shapes. They are also used to locate holes, surfaces, and other features. In general, location dimensions are needed in three mutually perpendicular directions: up and down, crossways, and forward and backward.

Finished surfaces and centerlines, or axes, are important for fixing the positions of parts by location dimensions. In fact, finished surfaces and axes are used to define positions. There are two general rules for showing location dimensions, as shown in Figure 7-28:

Figure 7-27

Dimensioning some elementary shapes.

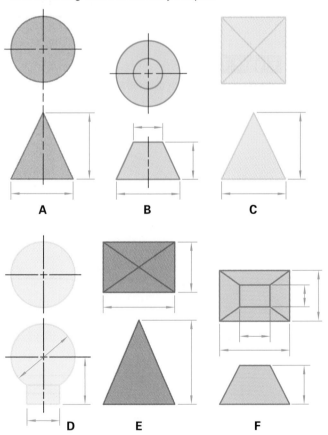

A   B   C

D   E   F

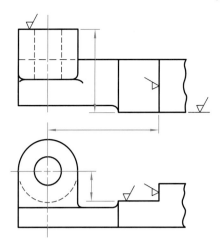

Figure 7-28

Locating dimensions for prisms and cylinders.

- *Prism forms are located by the axes and the surfaces. Three dimensions are needed.*
- *Cylinder forms are located by the axis and the base. Three dimensions are needed.*

Combinations of prisms and cylinders are shown in Figure 7-29. The dimensions marked L in Figure 7-29B are location dimensions.

All the surfaces and axes must be studied together so that the parts will go together as accurately as necessary. That means that in order to include the right location dimensions and notes on a drawing, drafters must be familiar with the engineering practices needed for the manufacture, assembly, and use of a product.

Datum Dimensioning

Datums are points, lines, and surfaces that are assumed to be exact. Examples of datums are shown in Figure 7-30. Such datums are used to compute or locate other dimensions. Location dimensions are given from them. When positions are located from datums, the different features of a part are all located from the same datum.

Two surfaces, two centerlines, or a surface and a centerline are typical datums. In Figure 7-30A, two surface datums are used. In Figure 7-30B, two centerlines are used. In Figure 7-30C, a surface and a centerline are used. A datum must be clear and visible while the part is being made. Mating parts should have the same datums because they fit together.

Standard Details

The shape, methods of manufacture, and use of a part generally tell you which dimensions must be given and how accurate they must be. A knowledge of manufacturing methods, pattern-making, foundry and machine-shop procedures, forging, welding, and so on is very useful when you are choosing and placing dimensions. The quantity of parts to be made must also be considered. If many copies are to be made, quantity-production methods have to be used. In addition, some items may incorporate purchased parts, identified by name or brand, that call for few, if any, dimensions.

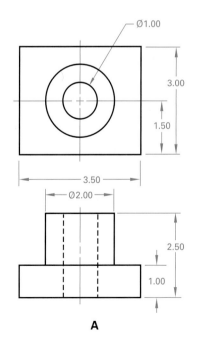

A

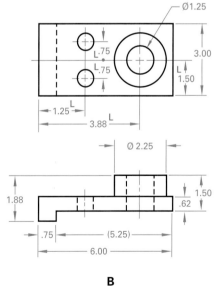

B

Figure 7-29

Examples of dimensioning prisms and cylinders.

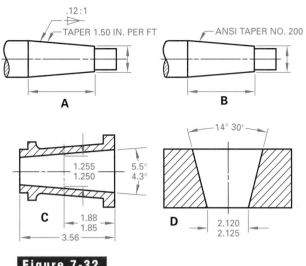

Figure 7-32
Dimensioning tapers.

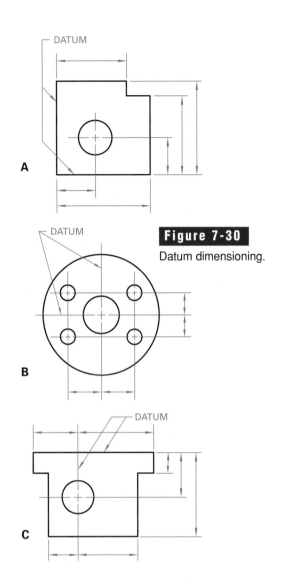

Figure 7-30
Datum dimensioning.

Some companies have their own standard parts for use in different machines or constructions. The dimensioning of these parts depends on how they are used and produced. There are, however, certain more-or-less standard details or conditions. For these, there are suggested ways of dimensioning.

Chamfers

Chamfers are angled corners, or bevels. Figure 7-31 shows two standard methods for dimensioning chamfers.

Tapers

Tapers can be dimensioned by giving the length, one diameter, and the taper as a ratio, as shown in Figure 7-32A. Another method, shown in Figure 7-32B, gives one diameter or width, the length, and either the American National Standard or another standard taper number. For a close fit, the taper is dimensioned as shown in Figure 7-32C. In Figure 7-32D, one diameter and the angle are given. In certain cases, the beginning and ending diameters are given.

Curves

A curve made up of arcs is dimensioned by the radii that have centers located by points of tangency, as shown in Figure 7-33. Noncircular, or irregular, curves can be dimensioned as shown in Figure 7-34A. They can also be dimensioned from datum lines, as shown in Figure 7-34B. A regular curve can be described and dimensioned by showing the construction or naming the curve, as shown in Figure 7-34C. The basic dimensions must also be given.

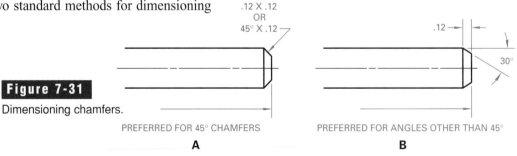

Figure 7-31
Dimensioning chamfers.

PREFERRED FOR 45° CHAMFERS

A

PREFERRED FOR ANGLES OTHER THAN 45°

B

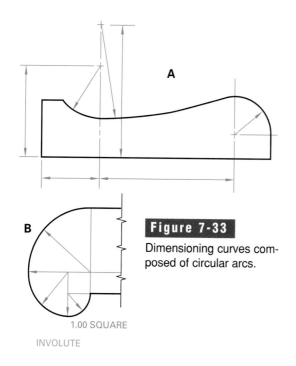

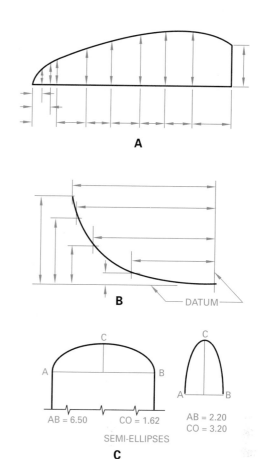

Figure 7-33

Dimensioning curves composed of circular arcs.

Figure 7-34

Dimensioning noncircular curves.

Dimensioning a Detail Drawing

A drawing for a single part that includes all the dimensions, notes, and information needed to make that part is called a **detail drawing**. The dimensioning should be done in the following order:

1. Complete all the views of a drawing before adding any dimensions or notes. Plan ahead so there is plenty of space between views for all necessary dimensions.
2. Think about the actual shape of the part and its characteristic views. With this in mind, draw all of the extension lines and lengthen any centerlines that may be needed.
3. Think about the size dimensions and the related location dimensions. Draw the dimension lines, leaders, and arrowheads.
4. After considering any changes, put in the dimensions and add any notes that may be needed.

Dimensioning an Assembly Drawing

When the parts of a machine are shown together in their relative positions, the drawing is called an **assembly drawing**. If an assembly drawing needs a complete description of size, the rules and methods of dimensioning apply.

Drawings of complete machines, constructions, and so on are made for different uses. The dimensioning must show the information that the drawing is designed to supply.

- If the purpose of the drawing is only to show the appearance or arrangement of parts, the dimensions can be left off.
- If the drawing is needed to tell the space a product requires, give overall dimensions.
- If parts have to be located in relation to each other without giving all the detail dimensions, center-to-center distances are usually given. Dimensions needed for putting the machine together or erecting it in position may also be given.
- In some industries, assembly drawings are completely dimensioned. These composite drawings are used as both detail and assembly drawings. See Chapter 12 for more information about composite drawings.

Photo-drawings are photographs of products with dimensions, notes, and other details drawn on them. Photo-drawings can be substituted for any of the first three uses listed on page 243.

For furniture and cabinet work, sometimes only the major dimensions are given. For example, length, height, and sizes of stock may be given. The details of joints are left to the cabinet maker or to the standard practice of the company. This is especially true if construction details are standard.

Limit Dimensioning

When one part is to be assembled with other parts, it must be made to fit into place without further machining or handwork. These parts are called *mating parts*, or *interchangeable parts*. For mating parts to fit together, specified size allowances are necessary. For example, suppose two mating parts are a rod or shaft and the hole in which it fits or turns. For these parts to fit together, variation in the diameter of the rod and the hole must be limited. If the rod is too large in diameter, it will not turn. If it is too small, the rod will be too loose and will not work properly.

Absolute accuracy cannot be expected. Instead, workers have to keep within a fixed limit of accuracy. They are given a number of tenths, hundredths, thousandths, or ten-thousandths of an inch or millimeter that the part is allowed to vary from the absolute measurements. This permitted variance is called the **tolerance**. The tolerance may be stated in a note on the drawing or written in a space in the title block. An example would be DIMENSION TOLERANCE ±.01 UNLESS OTHERWISE SPECIFIED.

Limit dimensions, or limits that give the maximum and minimum dimensions allowed, are also used to show the needed degree of accuracy. This is illustrated in Figure 7-35A. Note that the maximum limiting dimension is placed above the minimum dimension for both the shaft (external dimension) and the hole in the ring (internal dimension).

In Figure 7-35B and C, the basic sizes are given, and the plus-or-minus tolerance is shown. Consecutive dimensions are shown in Figure 7-35B. In this case, the dimension X could have some variation. This dimension should not be given unless it is needed for reference. If it is given, it should be enclosed in parentheses as a reference dimension.

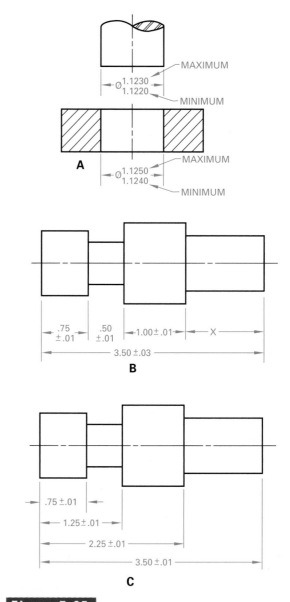

Figure 7-35

Limit dimensions.

Progressive dimensions (each starting at the same place) are shown in Figure 7-35C. Here they are all given from a single surface. This kind of dimensioning is called **baseline dimensioning**.

Very accurate or limiting dimensions should not be called for unless they are truly needed, because they greatly increase the cost of making a part. The detail drawing in Figure 7-36 has limits for only two dimensions. All the others are **nominal dimensions**. The amount of variation in these parts depends on their use. In this case, the general note calls for a tolerance of ±0.1″.

Precision

The latest edition of ANSI Y14.5M gives precise information on accurate measurement and position dimensioning. The following paragraphs are adapted from *American National Standard Drafting Practices* with the permission of the publisher, The American Society of Mechanical Engineers.

Expressing Size

Size is a designation of magnitude. When a value is given in a dimension, it is called the *size* of the dimension. *Note:* The words *dimensions* and *size* are both used to convey the meaning of magnitude. Several different size descriptions can be used to describe a part. Study the following definitions.

- **nominal size** The nominal size is used for general identification. Example: .5″ (13 mm) pipe.
- **basic size** The basic size is the size to which allowances and tolerances are added to get the limits of size.
- **design size** The design size is the size to which tolerances are added to get the limits of size. When there is no allowance, the design size equals the basic size.
- **actual size** An actual size is a measured size.
- **limits of size** The limits of size, usually called *limits*, are maximum and minimum sizes.

Expressing Position

Dimensions that fix position usually call for more analysis than size dimensions. Linear and angular sizes locate features in relation to one another (point-to-point) or from a datum. Point-to-point distances may be enough to describe simple parts. If a part with more than one critical dimension must mate with another part, dimensions from a datum may be needed.

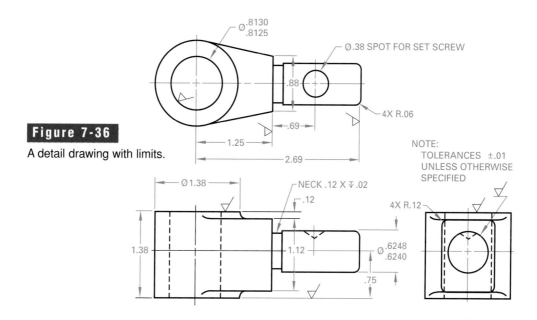

Figure 7-36
A detail drawing with limits.

Locating Round Holes

Figures 7-37 through 7-42 show how to position round holes by giving distances, or distances and directions, to the hole centers. These methods can also be used to locate round pins and other features. Allowable variations for the positioning dimensions are shown by stating limits of dimensions or angles, or by true position expressions.

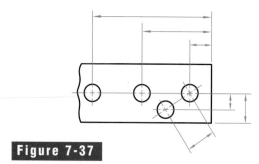

Figure 7-37

Locating holes by linear dimensions.

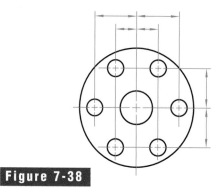

Figure 7-38

Locating holes by rectangular coordinates.

Figure 7-39

Locating holes on a circle by polar coordinates.

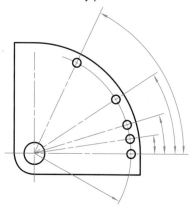

Tolerance

A tolerance is the total amount a given dimension may vary. A tolerance should be expressed in the same form as its dimension. The tolerance of a decimal dimension should be expressed by a decimal to the same number of places.

In a "chain" of dimensions with tolerances, the last dimension may have a tolerance equal to the sum of the tolerances between it and the first dimension. In other words, tolerances accumulate; they are

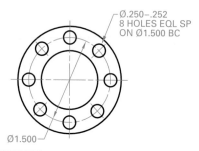

Figure 7-40

Locating holes on a circle by radius or diameter and the words "equally spaced."

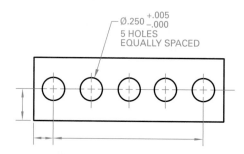

Figure 7-41

"Equally spaced" holes in a line.

Figure 7-42

Dimensions for datum lines.

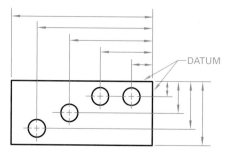

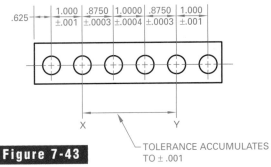

Figure 7-43

Point-to-point, or chain, dimensioning.

TOLERANCE ACCUMULATES TO ± .001

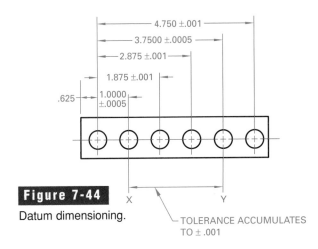

Figure 7-44

Datum dimensioning.

TOLERANCE ACCUMULATES TO ± .001

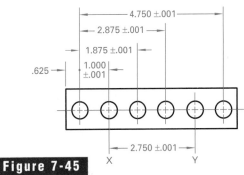

Figure 7-45

Dimensioning to prevent tolerance accumulation between X and Y.

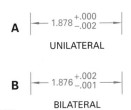

Figure 7-46

(A) A unilateral tolerance allows deviation in one direction only. (B) A bilateral tolerance allows deviation on both sides of the design size.

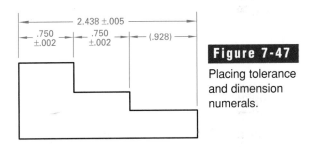

Figure 7-47

Placing tolerance and dimension numerals.

added together. The tolerance on the distance between two features (first and second hole, for example) is equal to the tolerances on the two dimensions from the datum added together. Where the distance between two points must be controlled closely, the distance between the two points should be dimensioned directly, with a tolerance. Figure 7-43 illustrates a series of chain dimensions where tolerances accumulate between points X and Y. Datum dimensions in Figure 7-44 show the same accumulation with larger tolerances. Figure 7-45 shows how to avoid the accumulation without the use of extremely small tolerances. The datum dimensioning method of Figure 7-42 also avoids overall accumulations.

Unilateral Tolerance System

Unilateral tolerances allow variations in only one direction from a design size. This way of stating a tolerance is often helpful where a critical size is approached as material is removed during manufacture. See Figure 7-46A. For example, close-fitting holes and shafts are often given unilateral tolerances.

Bilateral Tolerance System

Bilateral tolerances allow variations in both directions from a design size. Bilateral variations are usually given with locating dimensions. They are also used with any dimensions that can be allowed to vary in either direction, as shown in Figure 7-46B. Angle tolerances are usually bilateral.

Placing Tolerances

A tolerance numeral is placed to the right of the dimension numeral and in line with it. It may also be placed below the dimension numeral with the dimension line between them. Figure 7-47 shows both arrangements.

Limit System

A limit system shows only the largest and smallest dimensions allowed, as shown in Figure 7-48. The tolerance is the difference between the limits.

The amount of variation permitted when dimensioning a drawing can be given in several ways. For both linear and angular tolerances, the ways recommended in this book are as follows:

■ If the plus tolerance is different from the minus tolerance, two tolerance numbers are used, one plus and one minus. Refer again to Figure 7-46. *Note:* Two tolerances in the same direction should not be called for.

■ When the plus tolerance is equal to the minus tolerance, use the combined plus-and-minus symbol (±) followed by a single tolerance number. See Figure 7-49.

■ Show the maximum and minimum limits of size. For both location and size dimensions given directly (not by note), place the high-limit number (maximum dimension) above. Place the low-limit number (minimum dimension) below. See Figure 7-48A. Where the limits are given in note form, as shown in Figure 7-48B, place the minimum number first and the maximum number second.

■ You do not always have to give both limits.
  – A unilateral tolerance is sometimes given without stating that the tolerance in the other direction is zero, as shown in Figure 7-50A.

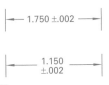

Figure 7-49
Using a combined plus-and-minus sign.

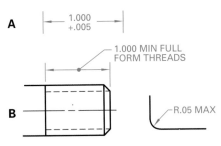

Figure 7-50
Expressing a single tolerance or limit.

  – MIN or MAX is often placed after a number when the other limit is not important. Depths of holes, lengths of threads, chamfers, etc., are often limited in this way, as shown in Figure 7-50B.

■ The number of decimal places shown in the tolerances should always be the same as the number of decimal places in the basic dimension.

Dimensioning for Fits

The tolerances on the dimensions of interchangeable parts must allow these parts to fit together at assembly, as shown in Figure 7-51. When mating parts do not need to be interchangeable, you can dimension as shown in Figure 7-52. The size of one part does not need to be held to a close tolerance. It is to be made the proper size at assembly for the desired fit. For further information about limits and fits, see ANSI B4.1.

To calculate dimensions and tolerances of cylindrical parts that must fit well together, you must first decide which dimension you will use for the basic size. You can use either the minimum hole size or the maximum shaft size as the basic size.

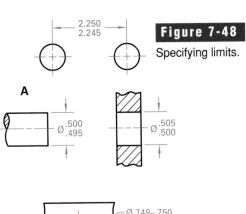

Figure 7-48
Specifying limits.

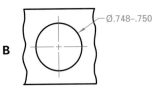

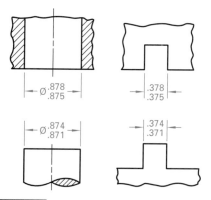

Figure 7-51

Indicating dimensions or surfaces that must fit closely.

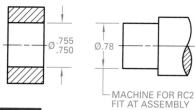

Figure 7-52

Dimensioning noninterchangeable parts that must fit closely.

Basic Hole System

A **basic hole system** is one in which the design size of the hole is the basic size and the allowance is applied to the shaft. To determine the limits for a fit in the basic hole system, follow these steps:

1. Give the minimum hole size.
2. For a clearance fit (a condition in which there is always positive clearance), find the desired allowance, or minimum clearance, from the minimum hole size. For an interference fit (a condition in which there is always negative clearance), add the desired allowance, or maximum interference.
3. Adjust the hole and shaft tolerances to get the desired maximum clearance or minimum interference, as shown in Figure 7-53.

By using the basic hole system, you can often keep tooling costs down. This is possible because standard tools such as a reamer or broach can be used for machining.

Basic Shaft System

A **basic shaft system** is one in which the design size of the shaft is the basic size and the allowance is applied to the hole. To figure out the limits for a fit in the basic shaft system, follow these steps:

1. Give the maximum shaft size.
2. For a clearance fit, find the minimum hole size by adding the desired allowance (minimum clearance) to the maximum shaft size. Subtract for an interference fit.
3. Adjust the hole and shaft tolerances to get the desired maximum clearance or minimum interference. See Figure 7-54.

Use the basic shaft method only if there is a good reason for it, such as when a standard-size shaft is necessary. For additional information on American National Standard limits and fits, see Appendix tables C-9 through C-16.

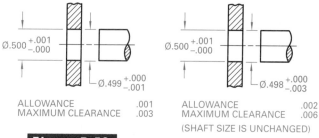

Figure 7-53

Fits in the basic hole system.

Figure 7-54

Fits in the basic shaft system.

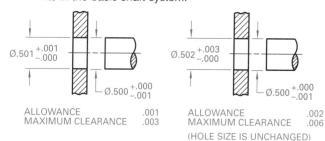

Geometric Dimensioning and Tolerancing

An engineering drawing of a manufactured part is intended to convey information from the designer to the manufacturer and inspector. It must contain all information necessary for the part to be correctly manufactured. It must also enable an inspector to make a precise determination of whether the parts are acceptable. Therefore, each drawing must convey three essential types of information:

- material to be used
- size or dimensions of the part
- shape or geometric characteristics

The drawing must also specify permissible variations for each of these aspects, in the form of tolerance and limits. The addition of this material, size, and shape information to an engineering drawing is known as **geometric dimensioning and tolerancing** (GD&T).

Geometric dimensioning and tolerancing can be one of the most important subjects learned by those who will be entering the manufacturing workplace. It is a very flexible communication system that can help designers specify the intent of the design throughout the entire manufacturing process. Geometric dimensioning and tolerancing is used on a daily basis by engineers, tool makers, manufacturers, inspectors, assemblers, and others in many different manufacturing industries. If geometric dimensioning and tolerancing is applied properly, and employees actually follow the geometric specifications on the drawing, the probability of making better parts increases significantly.

Modern systems of tolerancing include geometric and positional tolerancing, use of datum and datum targets, and precise linear and angular tolerances. These systems provide designers and drafters with a means of expressing permissible variations in a very precise manner. This section covers the application of geometric dimensioning and tolerancing methods to technical drawings.

It is not necessary to use geometric tolerances for every feature on a part drawing. In most cases, if each feature meets all dimensional tolerances, form variations will be adequately controlled by the accuracy of the manufacturing processes and the equipment used.

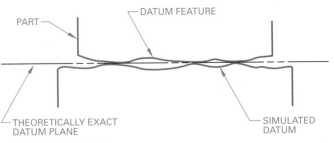

Figure 7-55

A theoretical datum.

Datums

Prior to a discussion of the geometric dimensioning language, you need to understand how parts are positioned for machining, inspection, and assembly. Parts are positioned on datums. A **datum** is a theoretically exact point, line, axis, or area from which geometric measurements are taken. A **theoretical datum** is established by the contact of a datum feature and a simulated datum.

The **datum feature** is any physical portion of a part. The **simulated datum** is what the datum feature contacts and should imitate the mating part in the assembly. See Figure 7-55. A simulated datum may be a mounting surface of a machine tool, a surface of an assembly fixture, or a surface of an inspection holding fixture.

The role of an engineering drawing is to specify what the part should be like after machining or assembly. Therefore, finished surfaces are generally selected as datum features. However, this is not always possible. In many cases specific points, lines, or areas of a surface are defined as datum targets. **Datum targets** are specified where datum features are rough, uneven, or on different levels, such as on castings, forgings, or weldments. It is very common for a part to be supported by one or more machined surfaces and one or more datum targets.

Datum Reference Frame

It may not be possible to make parts exactly the same, but it is possible to design a reliable and repeatable support structure while they are being machined. Parts may move up and down, in and out, and from side to side. They may also rotate. These movements are known as **degrees of freedom**. See

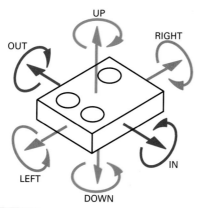

Figure 7-56

The six degrees of freedom. Limiting a part's movement in one or more of these directions during manufacture improves the accuracy of the part.

Specifying Datum Features

Datum features may be identified by the methods shown in Figure 7-57. The datum symbol may be attached to a visible line representing the datum feature, as shown in Figure 7-57A, or to an extension line, as shown in Figure 7-57B. If the datum feature is a rectangular size feature, the straight line connecting the square and the triangle must be in line with the dimension line that states the size of the feature. See Figure 7-57C.

The letters used in the square box do not have to be in alphabetical order. The important thing is how each letter is used on the rest of the drawing. Choose letters that will not be misunderstood due to their appearance elsewhere in a different context.

Specifying Datum Targets

If specific portions of a feature will be used to establish the theoretical datums, they are identified with datum target symbols. Datum targets are of three types: points, lines, and areas. Figure 7-58 illustrates how each type is shown on a drawing.

Figure 7-58A shows an example of a datum target point. A large X is placed where the part will rest on the tooling. Figure 7-58B shows an example of a datum target line. A phantom line is used to show where the line of contact will be. Figure 7-58C shows

Figure 7-56. Each direction in which a part can move during manufacture decreases the accuracy of the finished part. Therefore, movement in each direction must be restrained.

The restrictive environment created to hold the parts is called a **datum reference frame**. One of the major tasks involved in designing and machining parts is figuring out exactly what the datum reference frame should be.

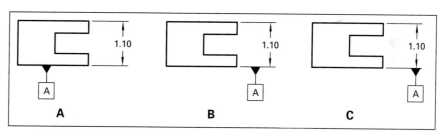

Figure 7-57

Placement of the datum feature symbol.

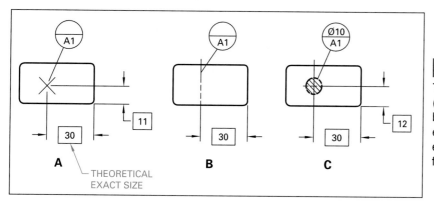

Figure 7-58

The datum target may be (A) a point; (B) a line; or (C) an area. Notice that basic dimensions are used in these examples. Basic dimensions are enclosed in a box to distinguish them from general dimensions.

Table 7-2. Geometric Characteristic Symbols

Symbol	Symbol Name	Symbol	Symbol Name
—	Straightness	◎	Concentricity
▱	Flatness	⊕	Position
○	Circularity	═	Symmetry
⌭	Cylindricity	∅	Diameter
⌒	Profile of a Line	Ⓜ	Maximum Material Condition
⌓	Profile of a Surface	Ⓣ	Tangent Plane
//	Parallelism	Ⓛ	Least Material Condition
∠	Angularity	Ⓟ	Projected Tolerance Zone
⊥	Perpendicularity	Ⓕ	Free State
↗	Circular Runout	⟨ST⟩	Statistical Tolerance
↗↗	Total Runout	↔	Between

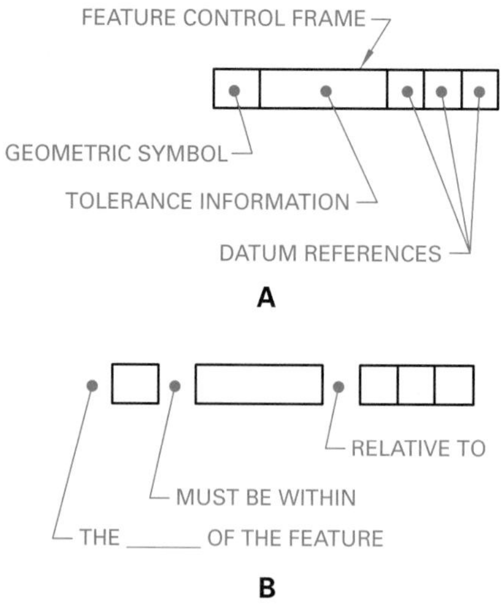

A

B

Figure 7-59

The structure and use of a feature control frame.

an example of a datum target area. The area is shown with a phantom line that has been crosshatched. The area may be any shape.

In all cases, the datum target is identified with a letter and number placed in the bottom half of a circle. The letter identifies the datum, while the number identifies the specific target. The datum target area size is placed in the upper half of the datum target symbol.

The dimensions for datum targets may be basic dimensions, as shown, or general toleranced dimensions. If basic dimensions are used, the actual location tolerances for the datum targets are determined by the employees who make the tooling.

Geometric Dimensioning Sentence Structure

The ASME Y14.5M standard defines fourteen main geometric symbols used to describe geometric conditions. Several other symbols may also be used. The feature control symbols and their names are shown in Table 7-2. The sizes of these symbols are shown in Appendix C-18.

These symbols, along with numbers, are placed in a rectangular box called a **feature control frame**, which is divided into two or more compartments. Figure 7-59A shows that the first compartment contains the geometric characteristic symbol. The second compartment contains the tolerance information. Additional compartments can be added to contain datum references. These are the variables within the basic sentence structure.

The information contained in the feature control frame may be read like a sentence. The first words spoken include the geometric characteristic name. For example, an introductory phrase for Position would be "The Position of the feature." The term *axis, axes,* or *center plane* is added when the control is related to the size features.

The lines dividing the compartments are where the connecting phrases are spoken. The first connecting phrase is "must be within." See Figure 7-59B. The second connecting phrase is "relative to." These connecting phrases can remain the same for all geometric specifications. English translations are provided for each specification in this chapter.

Table 7-3. Types of Tolerance Zones

Parallel Lines	Parallel Planes	Cylinders
Straightness	Flatness	Straightness
Circularity	Parallelism	Parallelism
Circular Runout	Perpendicularity	Perpendicularity
Profile of a Line	Angularity	Angularity
Parallelism	Cylindricity	Position
Perpendicularity	Total Runout	Concentricity
Angularity	Profile of a Surface	
	Straightness	
	Position	
	Symmetry	

Tolerance Zones

This section shows logical relationships between the characteristics by examining their common attributes. The most common attribute is the type of **tolerance zone** they use. The characteristics may be divided into three different tolerance zone types: parallel lines, parallel planes, and cylinders. Table 7-3 shows the characteristics that fall into each type of tolerance zone.

Parallel Lines

The parallel lines tolerance zone type is a two-dimensional area. The distance between the parallel lines is the tolerance zone. It is specified by the geometric tolerance in the feature control frame. In each of the parallel lines cases presented below, the tolerance zone may be at any or all positions on the surface. Each individual trace of the surface is separate from all others.

Figure 7-60 illustrates four examples of how a parallel lines tolerance zone may be applied to a plane surface. Figure 7-60A shows the geometric characteristic Straightness. Because Straightness is not related to a datum, it is considered a refinement of the size dimension. The tolerance zone will always remain within the size zone. It is placed in the view where the inspection will take place.

Other geometric controls that use two parallel lines as a tolerance zone are Parallelism, Perpendicularity, and Angularity, as shown in Figure 7-60B, C, and D. Notice that each feature control frame has a note below it that reads, "EACH ELEMENT." This note is added to specify a two-dimensional inspection.

The definition of parallel lines used in this section is, "A line extending in the same direction with and equidistant at all points from another line." This definition may

include circles or cylinders that are concentric. The Circularity and Circular Runout examples shown in Figure 7-61 display two parallel lines about a common center point.

The main difference between Circularity and Circular Runout is that Circularity is considered a refinement of the size dimension, so it requires no datum. The Circular Runout of a surface is controlled

Figure 7-60

Examples of the parallel lines type of tolerance.

A
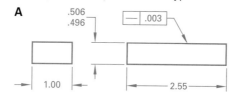
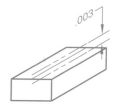

TRANSLATION: THE STRAIGHTNESS OF THE FEATURE MUST BE WITHIN THREE THOUSANDTHS.

B
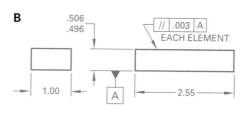
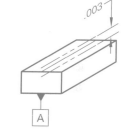

TRANSLATION: THE PARALLELISM OF EACH FEATURE ELEMENT MUST BE WITHIN THREE THOUSANDTHS RELATIVE TO DATUM FEATURE A.

C
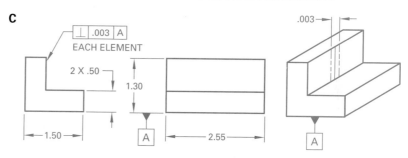

TRANSLATION: THE PERPENDICULARITY OF THE FEATURE ELEMENT MUST BE WITHIN THREE THOUSANDTHS RELATIVE TO DATUM FEATURE A.

D
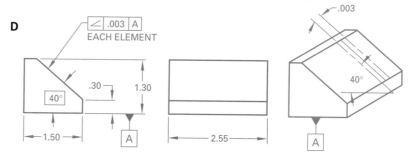

TRANSLATION: THE ANGULARITY OF EACH FEATURE ELEMENT MUST BE WITHIN THREE THOUSANDTHS RELATIVE TO DATUM FEATURE A.

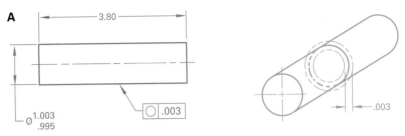

TRANSLATION: THE CIRCULARITY OF THE FEATURE MUST BE
WITHIN THREE THOUSANDTHS.

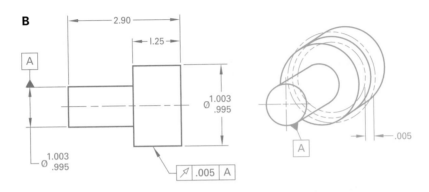

TRANSLATION: THE CIRCULAR RUNOUT OF THE FEATURE MUST BE WITHIN
FIVE THOUSANDTHS RELATIVE TO DATUM FEATURE A.

Figure 7-61

Two parallel lines can define the tolerance zone for concentric cylinders.

relative to an axis derived from a diameter that is different from the one being controlled. It is considered a surface-to-axis control. The tolerance zone for Circularity must remain within the size tolerance, but in Circular Runout the tolerance may exceed the size tolerance if required.

The example shown in Figure 7-62 is of Profile of a Line. Profile uses basic dimensions to define a true profile. The area between two parallel splines, or curved lines, defines the tolerance zone. In the example, the splines are an equal distance from and on either side of the true profile. Profile is the only parallel lines specification that may control size as well as form.

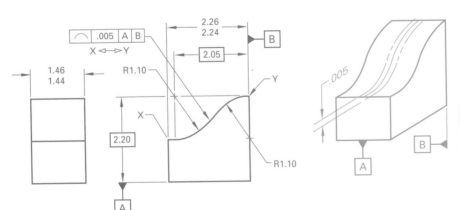

Figure 7-62

Two parallel splines can define the tolerance zone for the true profile of a line.

TRANSLATION: THE LINE PROFILE OF THE FEATURE MUST BE WITHIN FIVE THOUSANDTHS
RELATIVE TO DATUM FEATURES A AND B BETWEEN POINTS X AND Y.

Parallel Planes

The parallel planes tolerance zone types are very similar to the parallel lines examples, except that they are three-dimensional volumes instead of two-dimensional areas. The parallel planes tolerance zone is the space between two parallel surfaces. The distance between the surfaces is specified on a drawing by the geometric tolerance in the feature control frame.

Figure 7-63 illustrates four different geometric characteristics that use the distance between two parallel planes as their tolerance zone. Flatness and Parallelism, shown in Figure 7-63A and B, are the same except that Parallelism is related to another surface, but Flatness is not. Because Flatness is not related to a datum, it is considered a refinement of the size dimension. Even though Parallelism is related to a datum, it is nevertheless considered a refinement of the size dimension because it is a control of opposing surfaces.

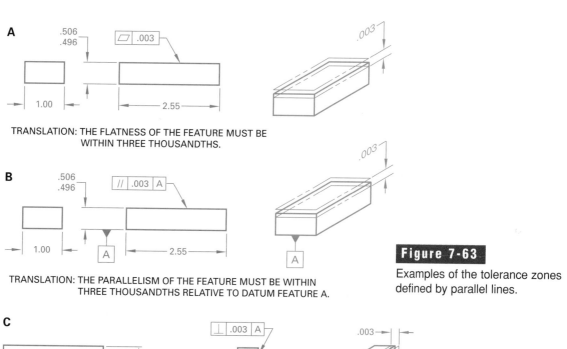

A

TRANSLATION: THE FLATNESS OF THE FEATURE MUST BE WITHIN THREE THOUSANDTHS.

B

TRANSLATION: THE PARALLELISM OF THE FEATURE MUST BE WITHIN THREE THOUSANDTHS RELATIVE TO DATUM FEATURE A.

Figure 7-63

Examples of the tolerance zones defined by parallel lines.

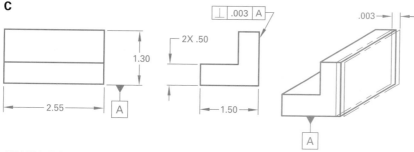

C

TRANSLATION: THE PERPENDICULARITY OF THE FEATURE MUST BE WITHIN THREE THOUSANDTHS RELATIVE TO DATUM FEATURE A.

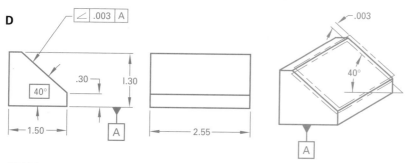

D

TRANSLATION: THE ANGULARITY OF THE FEATURE MUST BE WITHIN THREE THOUSANDTHS RELATIVE TO DATUM FEATURE A.

Perpendicularity and Angularity, shown in Figure 7-63C and D, are the same except the tolerance zone in Perpendicularity is always oriented at a basic 90° angle to the datum surface. The basic angle in Angularity must be specified.

Figure 7-64 parts A and B show examples of Cylindricity and Total Runout, respectively. Cylindricity may be thought of as a combination of Straightness and Circularity. When combined, they form two concentric cylinders around a common axis. Because Cylindricity is not related to a datum, it is considered a refinement of the size dimension. Total Runout uses the same tolerance zone type but, like Circular Runout, the tolerance zone is relative to a datum axis that is derived from a different diameter than the one being controlled.

The true profile in Profile of a Surface must be specified with basic dimensions. The tolerance zone in the example shown in Figure 7-65 is equally distributed on either side of the true profile. If Profile of a Surface is applied with datums, it may control size, form, orientation, and position.

Figure 7-66 illustrates three examples of how to control the center plane of a size feature. The Straightness example in Figure 7-66A allows the form of the part to bow or warp out-side the maximum size dimension. Straightness is the only geometric characteristic that will allow this to happen. The tolerance zone controls the center plane of the part. Because of this, the tolerance is applied to the size of the part and not to a surface.

The Position example in Figure 7-66B uses the same tolerance zone type as Straightness, but it is a control that specifies a centering of the slot relative

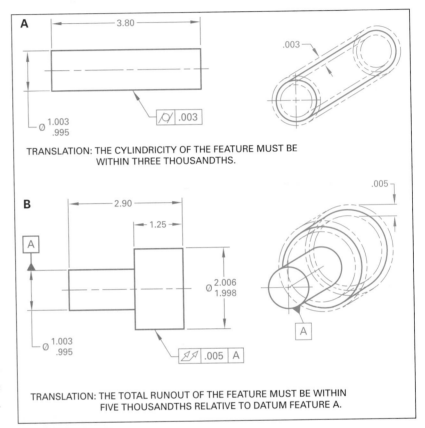

TRANSLATION: THE CYLINDRICITY OF THE FEATURE MUST BE WITHIN THREE THOUSANDTHS.

TRANSLATION: THE TOTAL RUNOUT OF THE FEATURE MUST BE WITHIN FIVE THOUSANDTHS RELATIVE TO DATUM FEATURE A.

Figure 7-64

Cylindricity and Total Runout tolerance zones may be defined using two parallel planes.

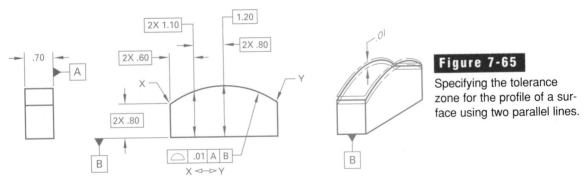

TRANSLATION: THE SURFACE PROFILE OF THE FEATURE MUST BE WITHIN ONE ONE–HUNDREDTH INCH RELATIVE TO DATUM FEATURES A AND B BETWEEN LINES X AND Y.

Figure 7-65

Specifying the tolerance zone for the profile of a surface using two parallel lines.

A

TRANSLATION: THE STRAIGHTNESS OF THE FEATURE CENTER PLANE MUST BE
WITHIN SEVEN THOUSANDTHS AT MAXIMUM MATERIAL CONDITION.

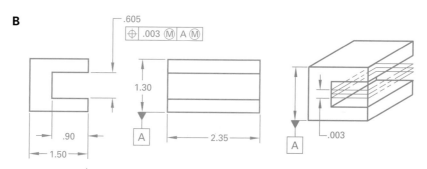

B

TRANSLATION: THE POSITION OF THE FEATURE CENTER PLANE MUST BE WITHIN
THREE THOUSANDTHS AT MAXIMUM MATERIAL CONDITION
RELATIVE TO DATUM FEATURE A AT MAXIMUM MATERIAL CONDITION.

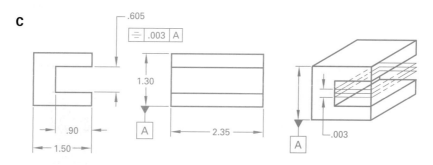

C

TRANSLATION: THE SYMMETRY OF THE FEATURE CENTER PLANE MUST BE WITHIN
THREE THOUSANDTHS RELATIVE TO DATUM FEATURE A.

Figure 7-66

Controlling the center plane of a size feature.

to the outside surfaces of the part. Position is intended for interchangeable fits.

The Symmetry example in Figure 7-66C is similar to Position. Their difference lies in how the tolerance is applied. Symmetry can only be applied on a "regardless of feature size" basis. It may be applied in noninterchangeable situations.

Cylinders

The cylindrical tolerance zone is the most used of the three tolerance zone types. It is a control of the axis of a hole or cylinder. All geometric characteristics that use this type of zone must have a diameter symbol placed in front of the tolerance value.

Letters enclosed in circles after the tolerance value or any size datum references indicate that the tolerance applies at a specified size condition. If there are no size condition symbols, it means that the tolerance applies regardless of feature size.

The Straightness control in Figure 7-67 is similar to the example in Figure 7-66, except for the shape of the tolerance zone. This tolerance also allows the form of the part to bow or warp outside the maximum size dimension.

Figure 7-67

The tolerance zone for Straightness can be defined using a cylinder.

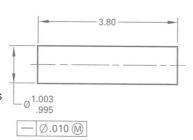

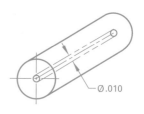

Ø 1.003
.995

— | Ø.010 Ⓜ

TRANSLATION: THE STRAIGHTNESS OF THE FEATURE AXIS MUST BE WITHIN
TEN THOUSANDTHS AT MAXIMUM MATERIAL CONDITION.

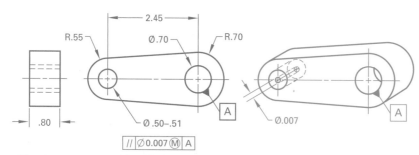

R.55   Ø.70   R.70

2.45

.80

Ø.50–.51

// | Ø0.007 Ⓜ | A

Ø.007

Figure 7-68

The Parallelism control feature can be used to control the parallelism of one hole to another.

TRANSLATION: THE PARALLELISM OF THE FEATURE AXIS MUST BE WITHIN
SEVEN THOUSANDTHS AT MAXIMUM MATERIAL CONDITION
RELATIVE TO DATUM FEATURE A.

Figure 7-69

Using Perpendicularity to control a cylinder at right angles to the datum surface.

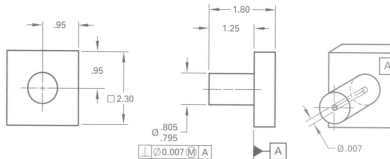

.95

.95

□ 2.30

1.80

1.25

Ø .805
.795

⊥ | Ø0.007 Ⓜ | A

A

Ø.007

TRANSLATION: THE PERPENDICULARITY OF THE FEATURE AXIS MUST BE
WITHIN SEVEN THOUSANDTHS AT MAXIMUM MATERIAL
CONDITION RELATIVE TO DATUM FEATURE A.

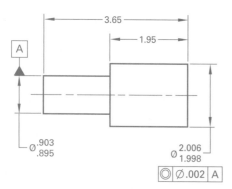

3.65

1.95

A

Ø .903
.895

Ø 2.006
1.998

◎ | Ø.002 | A

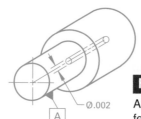

Ø.002

A

Figure 7-70

An example of the Position control feature.

TRANSLATION: THE CONCENTRICITY OF THE FEATURE AXIS MUST BE WITHIN
TWO THOUSANDTHS RELATIVE TO DATUM FEATURE A.

The example in Figure 7-68 illustrates how the parallelism of one hole may be controlled using another Parallelism control. The established cylinder for the controlled hole must be parallel to the axis defined by the datum hole.

A Perpendicularity example is shown in Figure 7-69. The controlling cylinder is oriented at a basic 90° angle relative to the datum surface.

The Position example shown in Figure 7-70 can be thought of as a combination of Parallelism and Perpendicularity with location. The centers of the tolerance cylinders are located with basic dimensions from the datum surfaces.

Figure 7-71 shows a Concentricity example. The cylinder tolerance zone is aligned with the axis of the datum diameter. It is referred to as an *axis-axis control*.

Tolerance Zone Combinations

The previous examples have illustrated single tolerance zones only. It is not unusual for different geometric characteristics to be used in combination. Usually, the lower segment of the geometric control is considered a refinement of the upper segment. Two examples are presented to show this concept.

Figure 7-72 shows Parallelism and Flatness used together. The Parallelism control is a refinement of the size dimension, and the Flatness control is a refinement of the Parallelism.

A very common combination is Position and Perpendicularity. Depending on the arrangement, Position may include Perpendicularity. The Perpendicularity control shown in Figure 7-73 further refines the Position control.

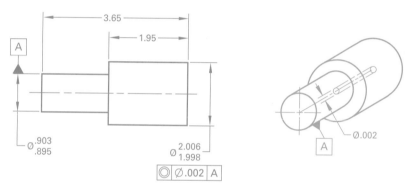

TRANSLATION: THE CONCENTRICITY OF THE FEATURE AXIS MUST BE WITHIN TWO THOUSANDTHS RELATIVE TO DATUM FEATURE A.

Figure 7-71

An example of the Concentricity control feature.

Figure 7-72

Using more than one geometric characteristic.

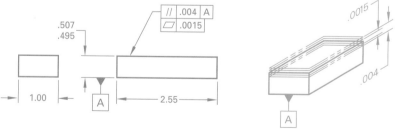

TRANSLATION: THE PARALLELISM OF THE FEATURE MUST BE WITHIN FOUR THOUSANDTHS RELATIVE TO DATUM FEATURE A AND FLAT WITHIN FIFTEEN TEN-THOUSANDTHS.

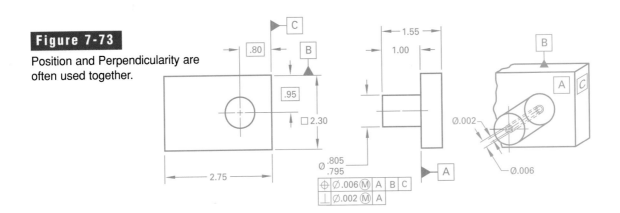

Figure 7-73

Position and Perpendicularity are often used together.

TRANSLATION: THE POSITION OF THE FEATURE AXIS MUST BE WITHIN SIX THOUSANDTHS AT MAXIMUM MATERIAL CONDITION RELATIVE TO DATUM FEATURES A, B, AND C AND PERPENDICULAR WITHIN TWO THOUSANDTHS TO DATUM FEATURE A.

Surface Texture

There is no such thing as a perfectly smooth surface. All surfaces have irregularities. Sometimes a drafter has to tell how much roughness and waviness the surface of a material can have and the lay direction of both. There are standards for these characteristics, and each characteristic has a symbol to represent it.

Refer to ANSI B46.1 for a complete discussion of surface texture. The following paragraphs about surface texture are adapted from that text. They are included here with the permission of the publisher, The American Society of Mechanical Engineers (ASME).

Surfaces, in general, are very complex in character. This standard deals only with the height, width, and direction of the surface irregularities. These are of practical importance in specific applications.

Definitions of Terms

A working knowledge of the relevant terms will help you understand surface texture designations more clearly. These terms have to do with surfaces made by various means. Among these are machining, abrading, extruding, casting, molding, forging, rolling, coating, plating, blasting, burnishing, and others. Study the definitions in this section before attempting to use surface texture characteristics.

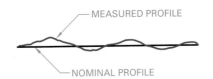

MEASURED PROFILE

NOMINAL PROFILE

Figure 7-74

An enlarged profile shows that a surface is not as it appears.

- **surface texture** Surface texture includes roughness, waviness, lay, and flaws. It includes repetitive or random differences from the nominal surface that forms the pattern of the surface.
- **profile** The profile is the contour, or shape, of a surface in a plane that is perpendicular to it. Sometimes an angle other than a perpendicular one is specified.
- **measured profile** The measured profile is a representation of the profile obtained by instruments or other means. See Figure 7-74.
- **microinch** A microinch is one millionth of an inch (.000 0001"). Microinches may be abbreviated μin.
- **micrometer** A micrometer is one millionth of a meter (.000 0001 m). Micrometers may be abbreviated μm.

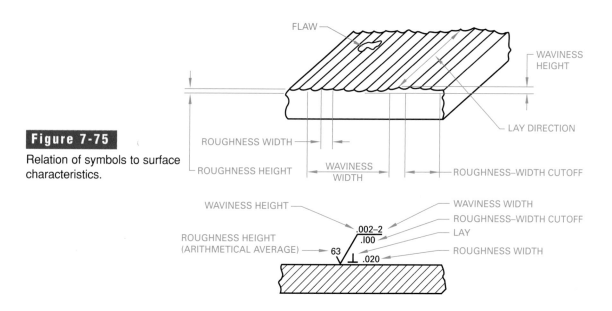

Figure 7-75

Relation of symbols to surface characteristics.

FLAW

WAVINESS HEIGHT

LAY DIRECTION

ROUGHNESS WIDTH

ROUGHNESS HEIGHT

WAVINESS WIDTH

ROUGHNESS–WIDTH CUTOFF

WAVINESS HEIGHT

WAVINESS WIDTH
ROUGHNESS–WIDTH CUTOFF

ROUGHNESS HEIGHT
(ARITHMETICAL AVERAGE)

.002–2
.100

LAY

63

.020

ROUGHNESS WIDTH

Table 7-4. Preferred Series Roughness		
Roughness	**Values**	**Grade**
50	2000	12
25	1000	11
12.5	500	10
6.3	250	9
3.2	125	8
1.6	63	7
0.8	32	6
0.4	16	5
0.2	8	4
0.1	4	3
0.05	2	2
0.025	1	1

Table 7-5. Standard Roughness-Width Cutoff Values*						
mm	0.075	0.250	0.750	2.500	7.500	25.000
inches	.003	.010	.030	.100	.300	1.000

*When no value is specified, the value .030" (0.750 mm) is assumed.

- **roughness** Roughness is the finer irregularities in the surface texture. Roughness usually includes irregularities caused by the production process. Among these are traverse feed marks and other irregularities within the limits of the roughness-width cutoff. See Figure 7-75.
- **roughness height** For the purpose of this book, roughness height is the arithmetical average deviation. It is expressed in microinches or micrometers measured normal to the centerline. The preferred series of roughness-height values is shown in Table 7-4.

- **roughness width** Roughness width is the distance between two peaks or ridges that make up the pattern of the roughness. Roughness width is given in inches or millimeters.
- **roughness-width cutoff** This is the greatest spacing of repetitive surface irregularities to be included in the measurement of average roughness height. Roughness-width cutoff is rated in inches or millimeters. Standard values are shown in Table 7-5. Roughness-width cutoff must always be greater than the roughness width in order to obtain the total roughness-height rating.
- **waviness** Waviness is covered by surface-texture standards. Geometric tolerancing now covers this surface condition under flatness. Flatness is a condition in which all surface elements are in a single plane. Flatness tolerances are applied to surfaces to control variations in surface texture. Waviness results from factors such as machine or work deflections, vibration, chatter, heat treatment, or warping strains. Roughness may be thought of as superimposed on a "wavy" surface.

Table 7-6. Lay Symbols

Symbol	Designation	Example
‖	Lay parallel to the line representing the surface to which the symbol is applied.	DIRECTION OF TOOL MARKS
⊥	Lay perpendicular to the line representing the surface to which the symbol is applied.	DIRECTION OF TOOL MARKS
X	Lay angular in both directions to the line representing the surface to which the symbol is applied.	DIRECTION OF TOOL MARKS
M	Lay multidirectional.	
C	Lay approximately circular relative to the center of the surface to which the symbol is applied.	
R	Lay approximately radial relative to the center of the surface to which the symbol is applied.	

Figure 7-76

Applications of surface texture symbols and ratings.

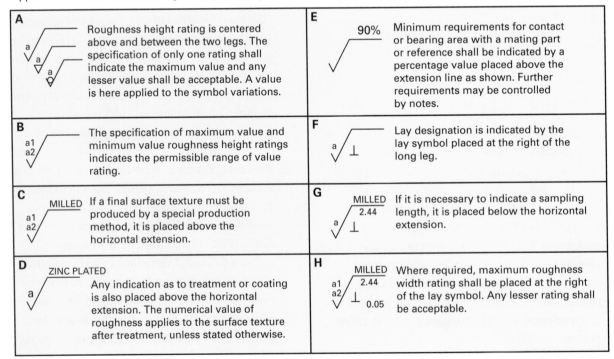

A Roughness height rating is centered above and between the two legs. The specification of only one rating shall indicate the maximum value and any lesser value shall be acceptable. A value is here applied to the symbol variations.

B The specification of maximum value and minimum value roughness height ratings indicates the permissible range of value rating.

C If a final surface texture must be produced by a special production method, it is placed above the horizontal extension.

D Any indication as to treatment or coating is also placed above the horizontal extension. The numerical value of roughness applies to the surface texture after treatment, unless stated otherwise.

E Minimum requirements for contact or bearing area with a mating part or reference shall be indicated by a percentage value placed above the extension line as shown. Further requirements may be controlled by notes.

F Lay designation is indicated by the lay symbol placed at the right of the long leg.

G If it is necessary to indicate a sampling length, it is placed below the horizontal extension.

H Where required, maximum roughness width rating shall be placed at the right of the lay symbol. Any lesser rating shall be acceptable.

- **waviness height** Waviness height is rated in inches as the peak-to-valley distance.
- **waviness width** Waviness width is rated in inches or millimeters as the spacing of successive wave peaks or successive wave valleys. When specified, the values are the maximum amounts permissible.
- **lay** Lay is the direction of predominant surface pattern. Ordinarily, it is determined by the production method used. Lay symbols are shown in Table 7-6.
- **flaws** Flaws are irregularities that occur at one place or at relatively infrequent or widely varying intervals in a surface. Flaws include defects such as cracks, blowholes, checks, ridges, and scratches. The effect of flaws is not included in the roughness-height measurements unless otherwise specified.
- **contact area** Contact area is the amount of area of the surface required to be in contact with its mating surface. Contact area should be distributed over the surface with approximate uniformity. Contact area is specified as shown in Figure 7-76.

Designation of Surface Characteristics

Where no surface control is specified, you can assume that the surface produced by the operation will be satisfactory. If the surface is critical, the quality of the surface needed should be shown.

The symbol used to designate surface irregularities is the check mark with horizontal extension, as shown in Figure 7-77A. The point of the symbol must touch the line to indicate which surface is meant. It may also touch the extension line, or a leader pointing to the surface. The long leg and extension are drawn to the right as the drawing is read. When only roughness height is shown, the horizontal extension may be left off. Figure 7-77B shows the typical use of the symbol on a drawing.

When the symbol is used with a dimension, it affects all surfaces defined by the dimension. Areas of transition, such as chamfers and fillets, should usually be the same as the roughest finished area next to them. Surface-roughness symbols always apply to the completed surface unless otherwise indicated. Drawings or specifications for plated or coated parts must tell whether the surface-roughness symbols apply before, after, or both before and after coating or plating.

Figure 7-76 shows how roughness, waviness, and lay are called for on the surface symbol. Only those ratings necessary to specify the desired surface need to be shown on the symbol.

Symbols for lay are shown in Table 7-6. Roughness ratings usually apply in a direction that gives the maximum reading. This is normally across the lay.

This is the end of the material extracted and adjusted from Surface Texture, ANSI B46.1. For more information, use the complete ISO and ANSI standards.

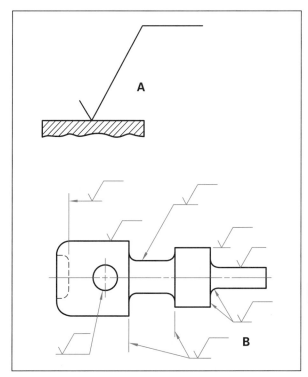

Figure 7-77

(A) The surface symbol; (B) applying the surface symbol on a drawing.

Board Drafting Techniques

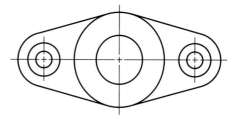

The process of dimensioning a drawing involves several steps. In many instances, drafters prefer first to prepare a freehand sketch with dimensions and notes and then to do the final drawing. This allows the drafter to determine the amount of space required for dimensions and notes before costly errors are made on the final drawing. Either way, the process of adding dimensions is exactly the same.

Figure 7-78 is a two-view drawing of a post socket. Follow the steps below to create the dimensioned drawing.

1. Prepare a freehand sketch of the post socket views to determine how much space will be required for dimensions and notes. Figure 7-79 shows the two views with adequate space allowed between and around them.
2. Prepare an instrument drawing of the two views, complete with centerlines.
3. Study the shape and details of the views before beginning the dimensioning process. Once you are sure about which dimensions are needed and where they should be placed, draw all necessary

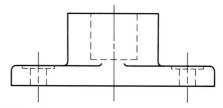

Figure 7-79

Two complete views of the post socket spaced to provide room for dimensions and notes.

extension lines, extend centerlines, and add leaders as needed. See Figure 7-80.

4. Finally, add dimensions, arrowheads, specific and general notes, and any other details necessary to complete the dimensioning process. Remember, when the drawing is finished, it should tell the production workers the exact shape and size of the finished part. Your drawing should look like the one in Figure 7-78.

Figure 7-78

Finished drawing of the post socket.

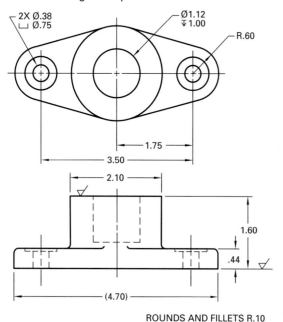

ROUNDS AND FILLETS R.10

Figure 7-80

The post socket with extension lines and leader lines added.

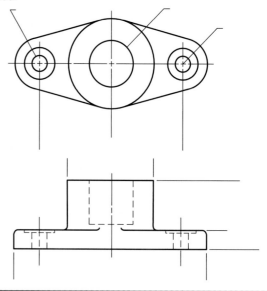

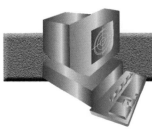
I n this section, you will create a CAD drawing of the post socket shown in Figure 7-81. Notice that this is the same drawing shown in Figure 7-78, but with subtle differences. The differences are due to the use of the AutoCAD software. The differences are minor, and both versions are acceptable by ANSI and ISO standards.

The CAD procedures are slightly different from the board procedures. For example, you don't need to draw a freehand sketch to determine how much space will be needed for dimensions and notes. Instead, you can estimate how far apart to place the views. If you find later that you have allowed too little or too much space, you can move an entire view, complete with dimensions and notes, to space them properly.

To dimension a drawing using AutoCAD, you will use a series of dimensioning commands. The names of the commands are long, so the easiest way to use these commands is to display the Dimension toolbar. To display the toolbar, pick Toolbars… from the View menu at the top of the screen. In the dialog box that appears, pick the check box next to Dimension to turn on the toolbar.

►CADTIP

Working with Toolbars

When you turn on a toolbar, it appears in a separate window on the screen. Therefore, you can move the toolbars around as necessary to get them out of the way of your drawing activities. You can also "dock" them by dragging them to the top or either side of the drawing area.

Creating the Drawing

Follow these steps to create a CAD drawing of the post socket:

1. Set up the drawing file. Study the drawing in Figure 7-81 to determine the appropriate sheet size. Create a new drawing using the corresponding drawing template, and set the grid and snap to convenient intervals. Create layers for objects, centerlines, hidden lines, and dimensions, being sure to use the appropriate linetypes. Set the units to decimal inches with 2 decimal places.

2. Decide which view to draw first and approximately where in the drawing area to place the view. In this case, create the primary centerlines for the top view first, because they will help define the features of both views. Then offset the primary vertical centerline 1.75 to both sides to create the secondary centerlines.

3. Use the LINE, CIRCLE, and TRIM commands with the appropriate object snaps to create both views of the post socket, complete with center

Figure 7-81

The appearance of the post socket drawing when completed on a CAD system. Notice that the CAD operator placed the dimensions, centerlines, and hidden lines on different, color-coded layers.

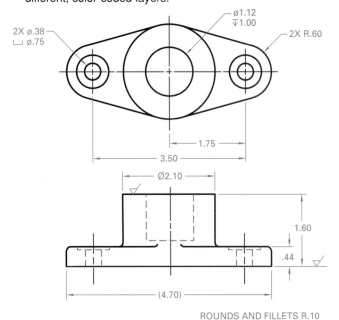

ROUNDS AND FILLETS R.10

lines. Ignore the fillets and rounds when you create the front view. After the basic view is in place, use the FILLET command set to a radius of .10 to create the fillets and rounds. You may choose to use construction lines to create the sides of the front view from the end radii in the top view. See Figure 7-82.

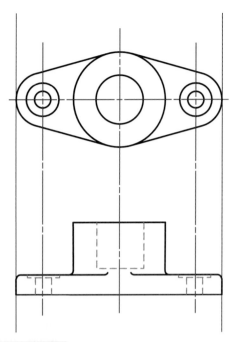

Figure 7-82

Use the centerlines and construction lines to align the views properly.

Setting the Dimension Style

AutoCAD provides a standard dimension style that controls the appearance of dimensions in a drawing. The settings in this style usually need to be changed before you can dimension a specific drawing correctly. To set up the dimension style, enter the DIMSTYLE command or pick the Dimension Style icon on the Dimension toolbar. When the Dimension Style Manager dialog box appears, choose to modify the standard style. The dialog box that appears has several tabs, as shown in Figure 7-83.

For this drawing, you will need to make changes to the first four tabs. However, you should be familiar with all of them. The tabs are listed below. Read through all of the information, and set the values and options described for the Lines and Arrows, Text, Fit, and Primary Units tabs. When you have finished selecting the appropriate settings, pick the OK button to close the dialog box; then pick Close to close the Dimension Style Manager and save your settings.

■ **Lines and Arrows tab.** Use the down arrows and scrollbars to set the dimension and extension line color and line weight to BYLAYER. Enter a value of .12 for the baseline spacing and for "Extend beyond dim lines." Set the arrow

size to .12. Leave the rest of the options at their default values.

■ **Text tab.** Set the text style to Roman. If Roman does not appear as an option, pick the … button to the side of the text style name and choose romans.shx from the list of font names. Then return to the text tab and choose Roman or romans for the text style. Set the text height at .12.

■ **Fit tab.** In the Fit Options portion of the dialog box, pick Arrows to be the first element to be placed outside the extension lines if both text and arrows will not fit.

■ **Primary Units tab.** Choose decimal dimensions with two decimal places. Note that this is set separately from the units for the drawing in general. If you do not specify two decimal places here, AutoCAD will display its default of four decimal places, regardless of your Units settings for the drawing. In Zero Suppression, pick the Leading check box to suppress leading zeros. Pick the Leading box in the Angular Measurements section to suppress leading zeros in angular dimensions. (Remember to leave these boxes unchecked for metric drawings.)

■ **Alternate Units tab.** This tab applies only for drawings that require dual dimensioning. You may safely ignore the Alternate Units tab for all of your drawings related to this book.

- **Tolerance tab.** Even though the Tolerance tab is not needed for this drawing, you should pick the tab and review the options for tolerance method. Pick each of the options and observe the sample drawing in the dialog box to see its effects. You will need to use this tab for toleranced drawing problems at the end of this chapter. For now, leave the Method set to None.

Dimensioning the Drawing

Now that the dimension style is set up correctly, you can begin to dimension the drawing. Follow these steps:

1. Study the shape and details of the views before beginning the dimensioning process. Be sure you understand which dimensions are needed and where they should be placed. They are determined for you in this case, but you will need to decide for yourself in drawings you create on the job.
2. Place the horizontal dimensions first. Both horizontal and vertical dimensions are created in AutoCAD using the DIMLINEAR command. Enter this command by picking the Linear Dimension button on the Dimension toolbar. Press Enter or click the right mouse button to choose to select an object. Then pick the top line of the front view. The dimension line, extension lines, and dimension value appear. Move the cursor to place them approximately .38″ from the view line, and pick the left mouse button to set the dimension in place.

3. Repeat the process in step 2 to set the bottom dimension of the front view. However, notice that this is a reference dimension. You will need to add parentheses around the value. To do this, use the mouse to select the dimension. Then right-click to see a menu of options. Pick Properties. In the Properties dialog box that appears at the left of the screen, pick Text Override. In the box next to Text Override, enter (4.70). Then pick a point anywhere in the drawing area to activate it, and press the Escape key to deselect the dimension. The overriding text appears at the dimension.

4. The remaining vertical and horizontal dimensions require a slightly different approach. The distances being dimensioned are not defined by a single line. Therefore, you cannot use the Select Objects option of the DIMLINEAR command. Instead, enter the DIMLINEAR command, and then pick the endpoints of the dimension to be placed. Be sure the endpoint object snap is turned on to pick the endpoints exactly. The red boxes in Figure 7-84 show where to select the endpoints for the .44 dimension in the front view. Place the remaining horizontal and vertical dimensions using this method.

5. Radial dimensions are created in AutoCAD using the DIMRADIUS command. Pick the Radius Dimension button on the Dimension toolbar to activate the command. Then pick the R.60 arc on the right side of the top view. Move the cursor and pick the left mouse button to place the dimension.

Figure 7-83

The Modify Dimension Style: Standard dialog box allows you to change the settings of the Standard dimension style.

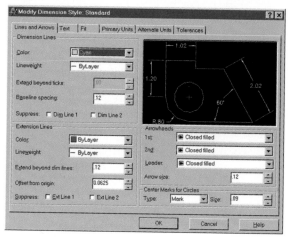

Figure 7-84

Because the fillet is a separate object in AutoCAD, to dimension the height correctly you must use the right endpoint of the horizontal line at the top of the base, as shown by the upper red box.

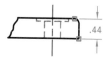

6. Diameters are dimensioned in AutoCAD using the DIMDIAMETER command. Pick the Diameter Dimension button on the Dimension toolbar to activate the command. Then pick one of the holes to be dimensioned and place the dimension as shown in Figure 7-81. Notice that the diameter symbol appears automatically. However, the second line of text for both diameter dimensions must be created separately. AutoCAD does not provide the standard symbols used in this drawing. You must create them yourself. Place them as shown in Figure 7-81. Then use the MTEXT command to create the text. To use this command, enter M at the keyboard just before placing the dimension. An editor appears, allowing you to type in the text and insert the appropriate symbols.

7. Add the finish symbols as shown in Figure 7-81.

8. Use the DTEXT command to add the note at the bottom of the drawing. Your drawing should now look like the one in Figure 7-81.

9. Check the relative positions of the views and adjust them if necessary. Be sure to keep the views aligned exactly.

Using GD&T in AutoCAD

Geometric dimensioning and tolerancing in AutoCAD requires an extra step. However, the software makes it fairly easy to create GD&T "sentences." To demonstrate the use of geometric dimensioning and tolerancing in AutoCAD, use the object in Figure 7-85. Follow these steps.

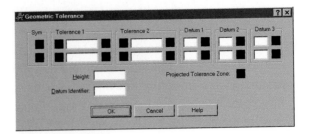

Figure 7-86

The Geometric Tolerance dialog box provides a quick way to set up geometric tolerances.

1. Draw all of the views and dimension them normally.

2. Enter the TOLERANCE command at the keyboard. The Geometric Tolerance dialog box appears, as shown in Figure 7-86.

3. Pick the black box labeled Sym in the first row of the dialog box. The Symbol dialog box appears. Pick the Perpendicularity symbol. The Perpendicularity symbol appears in the black box.

4. The next black box to the right, in the Tolerance 1 area, toggles the diameter symbol on and off. Pick in the box to make the diameter symbol appear.

5. The next (white) box is a text box in which you enter the tolerance value. Pick in the box to activate it, and enter .007 for the value.

6. The black box to the right of the text box controls the material condition. Pick in the box and choose maximum material condition (the M inside a circle).

7. The next three boxes in the row allow you to add a second tolerance, but no further tolerancing is needed for this example, so you can skip them. Move to the white box in the Datum 1 area and enter a capital A for the datum feature.

8. Pick OK. The Geometric Tolerance dialog box disappears, and a feature control frame containing the specifications you have just defined appears at the cursor. Move the cursor to the location shown in Figure 7-85 and pick a point to place it on the drawing.

Figure 7-85

Object for practicing geometric dimensioning and tolerancing procedures in AutoCAD.

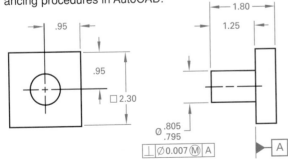

Chapter Summary

- The lines and symbols used for dimensioning a technical drawing are closely controlled by ANSI and ISO standards.

- The unidirectional system of dimensioning is now the preferred system according to ANSI.

- To describe a part for manufacturing, both size and location dimensions are needed.

- ANSI identifies several acceptable methods for dimensioning chamfers, tapers, and curves.

- To specify permissible variations in parts to be manufactured, technical drawings include geometric dimensions and tolerances.

- Parts are positioned on datums for machining, inspection, and assembly.

- When the roughness, waviness, and lay direction of a surface must be controlled, drafters use standard surface texture characteristic symbols and ratings.

- Before you can dimension a drawing in AutoCAD, you must set up the dimension style by either creating a custom style or modifying AutoCAD's Standard style.

Review Questions

1. Name the line used to show where a measurement begins and ends.

2. What is another name for size description?

3. In what dimensioning system are the numerals placed in line with the dimension lines?

4. Size dimensions are one of the two basic types of dimensions. What is the other basic type?

5. What name is given to points, lines, and surfaces that are assumed to be exact?

6. In what dimensioning system are all dimensions and notes read from the bottom of the sheet?

7. Name or sketch the abbreviations or symbols for each of the following terms: diameter, inch, centerline, millimeter, radius, tolerance, and countersink.

8. Sketch various ways in which large and small arcs can be dimensioned.

9. What name is given to the amount by which the accuracy of a part may vary from the absolute measurement?

10. What is the difference between unilateral and bilateral tolerances?

11. Explain the purpose of geometric dimensioning and tolerancing.

12. Name three types of tolerance zones used in geometric dimensioning and tolerancing.

13. What is a feature control frame?

14. Explain the difference between a datum feature and a simulated datum.

15. What two dimensions can be used to describe a cylinder fully in a single view?

16. What units are preferred for technical drawings of machine parts according to ANSI? What are the preferred ISO units?

17. What note must be added to a technical drawing if a hole is to be made in a piece after assembly with its mating piece?

18. Explain how to display the Dimension toolbar in AutoCAD.

19. What is the purpose of the Dimension Style Manager in AutoCAD?

20. What is AutoCAD's text code for the diameter symbol?

Drafting Problems

The drafting problems in this chapter are designed to be completed using board drafting techniques or CAD.

1. For each object shown in Figure 7-87, create a multiview drawing. Take dimensions from the printed scales at the bottom of Figure 7-88. Include dimensions and notes.

Figure 7-87

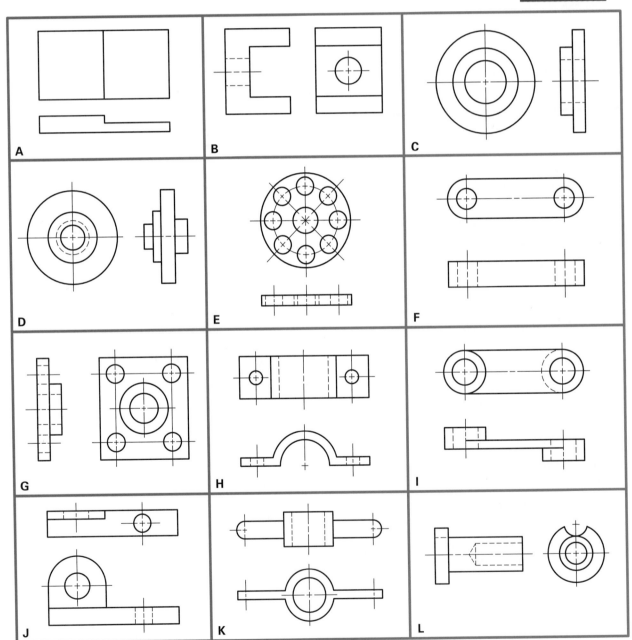

2. Create a multiview drawing of each object shown in Figure 7-88. Take dimensions from the printed scales at the bottom of the illustration. Include dimensions and notes.

For problems 3 through 12, first determine the necessary views. If you are using board drafting instruments, create a freehand sketch and dimension it. Decide on a scale and draw the views. Add all necessary dimensions and notes.

Figure 7-88

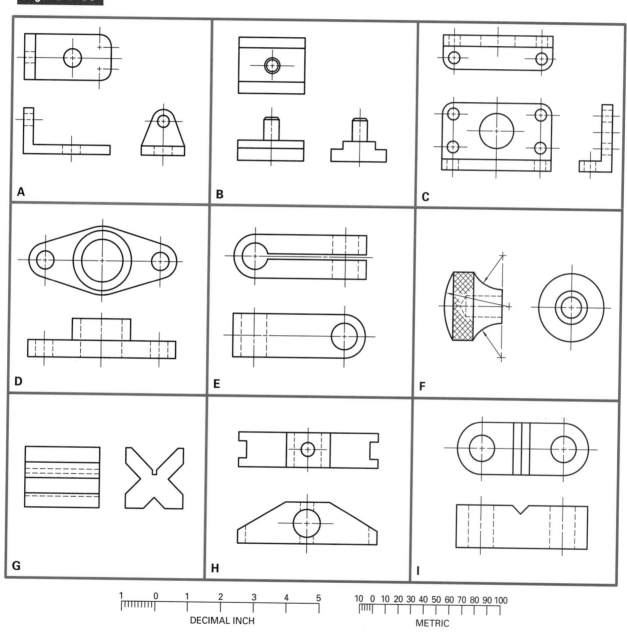

3. Square guide (Figure 7-89). A = 5 mm thick × 44 mm square; B = 30 mm square × 30 mm high; hole = 20 mm square.

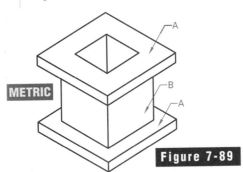

METRIC

Figure 7-89

4. Locator (Figure 7-90). AB = 40 mm; BC = 60 mm; CD = 5 mm; DE = 12 mm; EF = 36 mm; EG = 18 mm; H = 8 mm; hole = ∅10 mm through, 18 mm counterbore, 2 mm deep.

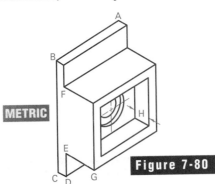

METRIC

Figure 7-80

5. Double-shaft support (Figure 7-91). A = 67 mm; B = 7 mm; C = R21 mm; D = 10 mm; E = ∅6 mm through, 10 mm counterbore, 2 mm deep, 2 holes; F = 43 mm; G = R12 mm; H = 14 mm.

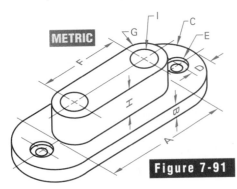

METRIC

Figure 7-91

6. Cradle slide (Figure 7-92). AB = 2.38; BC = 3.56; CD = 5.12; E = 1.50; F = 2.12; G = .88; H = R1.62.

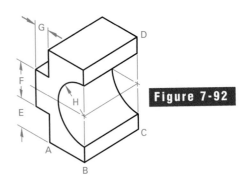

Figure 7-92

7. Pipe support (Figure 7-93). Base plate = .50 thick × 4.50 wide × 6.50 long; A = 2.38; B = R1.50; C = R1.12; D = .50; E = 3.00; F = ∅.38 hole through, countersink to ∅.75, 3 holes; G = 1.00; H = .75; I = 2.25.

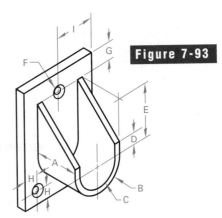

Figure 7-93

8. Stop plate (Figure 7-94). Overall sizes: L = 4.25, W = 2.00, H = .75. AB = .38; AC = 1.00; AE = 2.75; AD = 1.00; JN = .50; M = 1.00; F = ∅.44, 2 holes; G = Boss: ∅1.25 × 50 high, ∅.50 through, .88 counterbore = .12 deep.

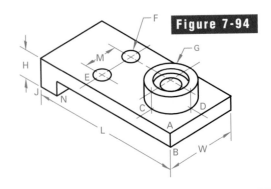

Figure 7-94

9. Idler pulley (Figure 7-95).

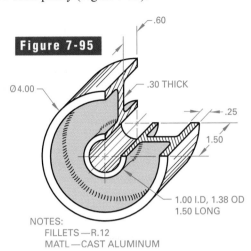

Figure 7-95

.60

Ø4.00

.30 THICK

.25

1.50

1.00 I.D, 1.38 OD
1.50 LONG

NOTES:
FILLETS—R.12
MATL—CAST ALUMINUM

10. Connecting rod (Figure 7-96).

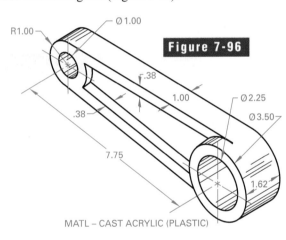

R1.00

Ø1.00

Figure 7-96

.38

1.00

Ø2.25

Ø3.50

.38

7.75

1.62

MATL – CAST ACRYLIC (PLASTIC)

11. Single V-pulley (Figure 7-97).

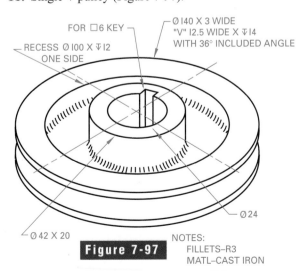

FOR ☐6 KEY

Ø140 X 3 WIDE
"V" 12.5 WIDE X ⤓14
WITH 36° INCLUDED ANGLE

RECESS Ø100 X ⤓12
ONE SIDE

Ø24

Ø42 X 20

Figure 7-97

NOTES:
FILLETS–R3
MATL–CAST IRON

12. Rotor (Figure 7-98).

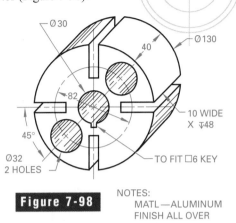

Ø30

40

Ø130

82

10 WIDE
X ⤓48

45°

Ø32
2 HOLES

TO FIT ☐6 KEY

Figure 7-98

NOTES:
MATL—ALUMINUM
FINISH ALL OVER

13. Pin and ring (geometric dimensioning and tolerancing). The pin and ring shown in Figure 7-99 are mating parts. For these drawings to be complete, the diameter features need to be related. The diameter features appear to be drawn relative to the center lines. However, there are no dimensional requirements for the position or orientation of the diameter features. The dimensional objectives are to establish a geometric relationship between the:

- ∅1.613 – 1.616 and ∅2.70
- small diameters and 1.6xx diameters
- larger diameters and two flat mating surfaces

The datums for the following assignments are:

PIN
∅1.607 – 1.610 = datum A
∅.905 – .907 = datum B

RING
2.70 = datum A
∅1.613 – 1.616 = datum B
∅.090 – .913 = datum C

Draw and dimension the pin and ring as shown, and add the geometric dimensions that will translate the following information into geometric symbols and numbers:

a. Ring: The position of the feature axis (∅1.613 – 1.616) must be within seven thousandths at maximum material condition relative to datum feature A.

b. Pin: The position of the feature axis (∅.905 – .907) must be within two thousandths at maximum material condition relative to the datum feature A at maximum material condition.

Ring: The position of the feature axis ($\varnothing$.909 – .913) must be within three thousandths at maximum material condition relative to the datum feature B at maximum material condition.

c. Pin: The Perpendicularity of the feature (left face) of the 1.50 dimension must be within .001 relative to datum feature C.

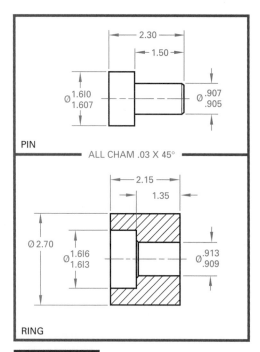

PIN

ALL CHAM .03 X 45°

RING

Figure 7-99

14. Shim and block (geometric dimensioning and tolerancing). The shim and block shown in Figure 7-100 are mating parts. For the drawings to be complete, the width features need to be related. The width features appear to be drawn relative to the center planes. However, there are no dimensional requirements for the position or orientation of the width features. Also, the shim is a stamped part. It is unlikely that the form of the part will fit within the boundary of perfect form at maximum material condition (.055).

The dimensional objectives are to establish a geometric relationship between the:
• reality that the shim will be warped and the maximum material condition size of .055

• slot in the block and the outside surface of the part
• small width features and larger width features

The datums for the following assignments are:

SHIM
1.392 – 1.395 = datum A
bottom of 2.40 = datum B
left side of 3.30 = datum C
1.398 – 1.403 = datum D

BLOCK
left side of 3.40 = datum A

Draw and dimension the shim and block as shown, and add the geometric dimension that will translate the following information into symbols and numbers:

a. Shim: The Straightness of the feature center plane (.051 – .055) must be within ten thousandths at maximum material condition.

b. Block: The position of the feature center plane (1.398 – 1.403) must be within five thousandths at maximum material condition relative to the datum features A, B, and C.

c. Shim: The position of the feature center plane (.607 – .614) must be within two thousandths at maximum material condition relative to datum feature D at maximum material condition.

Figure 7-100

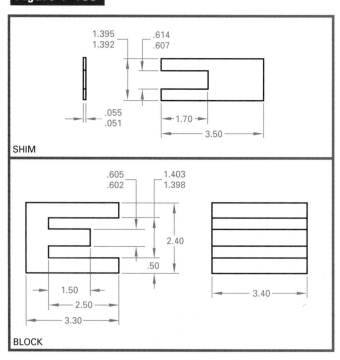

SHIM

BLOCK

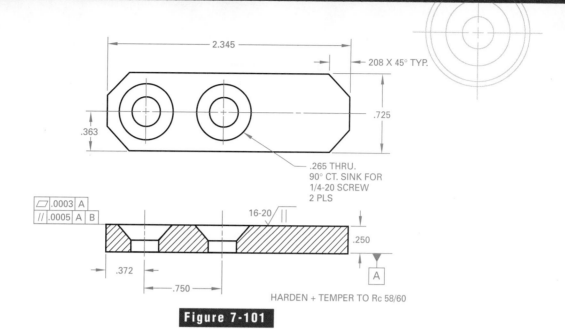

Figure 7-101

15. A designer has submitted a first draft of a CAD drawing of a roller way, as shown in Figure 7-101. Certain features of the drawing do not meet current ANSI standards. Create a second draft of the drawing to meet ANSI standards. Round general dimensions to two decimal places.

Design Problems

Design problems have been prepared to challenge individual students or teams of students. In these problems, you are to apply skills learned mainly in this chapter but also in other chapters throughout the text. The problems are designed to be completed using board drafting, CAD, or a combination of the two. Be creative and have fun!

1. **TEAMWORK** Work as a team to design a two-wheeled scooter. It should be designed to fold flat for easy carrying and storage. Specify lightweight materials. Wheels and bearings should be selected for extreme durability. Share responsibility in the preparation of a complete set of plans. Be sure to include a list of materials. Use general tolerancing and geometric dimensioning and tolerancing where appropriate. Begin with design sketches.

2. Design an adjustable arm for holding your drafting book at a convenient location on your drafting table or at your computer. Begin with design sketches. Make a complete set of plans and include a list of materials.

3. Design a desk or shelf clock. The clock mechanism fits a $\varnothing72$ mm × 20 mm deep hole. The clock should not exceed 64 mm × 152 mm × 254 mm. Materials optional. Prepare a set of drawings and add a materials list.

CHAPTER

Sectional Views

KEY TERMS

associative hatch

broken-out section

crosshatching

cutting plane

full section

half section

offset section

phantom section

removed section

revolved section

section lining

OBJECTIVES

Upon completion of this chapter, you should be able to:

■ Describe the purpose of a sectional view.

■ Select the appropriate type of sectional view to show the hidden feature.

■ Show ribs, webs, fasteners, and similar features in section.

■ Rotate selected features into the cutting plane.

■ Describe and use conventional breaks and symbols.

■ Prepare a drawing with sectional views using both board drafting techniques and CAD.

276

Technical drawings must show all parts of an object, including the insides and other parts not easily seen. These hidden details can be drawn with hidden lines, but this method works well only if the hidden part has a rather simple shape. If the shape is complicated, dashed lines may show it poorly. They can also be confusing, as shown in Figure 8-1A. Instead of dashed lines, a special view called a **section** or **sectional view** should be drawn in these cases. A sectional view shows an object as if part of it were cut away to expose the insides, as shown in Figure 8-1B.

Understanding Sectional Views

To best understand sectional views, imagine that a wide-blade knife has cut through the object. Call the path of this knife a **cutting plane**. Then imagine that everything in front of the cutting plane has been taken away, so that the cut surface and whatever is inside can be seen, as shown in Figure 8-1B. On a normal view, a special line called a **cutting-plane line** shows where the cutting plane passes through the object. See Figure 8-1C. On the sectional view, the cut surface is marked with thin, evenly spaced lines. This is called **section lining** or **crosshatching**. These concepts will be discussed in greater detail later in this chapter.

The basic section lining pattern described above is the ANSI 31 symbol. It is not the only pattern, but it is the one that is used in most cases. ANSI 31 is a general-purpose symbol; you can use it for objects made of any material. It is used especially when more than one kind of material need not be shown, such as on a drawing of a single part.

However, special section lining symbols can be used to show what materials are to be used. ANSI provides for many symbols to stand for different materials, as shown in Figure 8-2. Under this system,

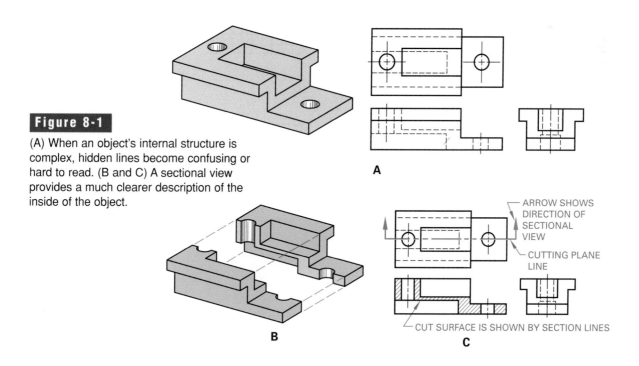

Figure 8-1

(A) When an object's internal structure is complex, hidden lines become confusing or hard to read. (B and C) A sectional view provides a much clearer description of the inside of the object.

A

ARROW SHOWS DIRECTION OF SECTIONAL VIEW

CUTTING PLANE LINE

CUT SURFACE IS SHOWN BY SECTION LINES

B

C

the general-purpose symbol can also mean that an object is made of cast iron. These special symbols are most useful on a drawing that shows several objects made of different materials. These would be used, for example, in an assembly drawing that shows how different parts fit together. However, do not depend on these symbols alone to describe the materials to be used. Specify the exact materials needed in a note or in a list of materials.

The Cutting-Plane Line

The cutting-plane line represents the cutting plane as viewed from an edge, as shown in Figure 8-3. ANSI specifies two forms for cutting-plane lines, as shown in Figure 8-4. The first form is more commonly used. The second shows up well on complicated drawings. At each end of the line, draw a short arrow to show the direction for looking at the section. Make the arrows at right angles to the line. Place bold capital letters at the corners as shown, if needed for reference to the section.

Figure 8-2

ANSI symbols for section lining.

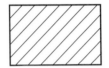

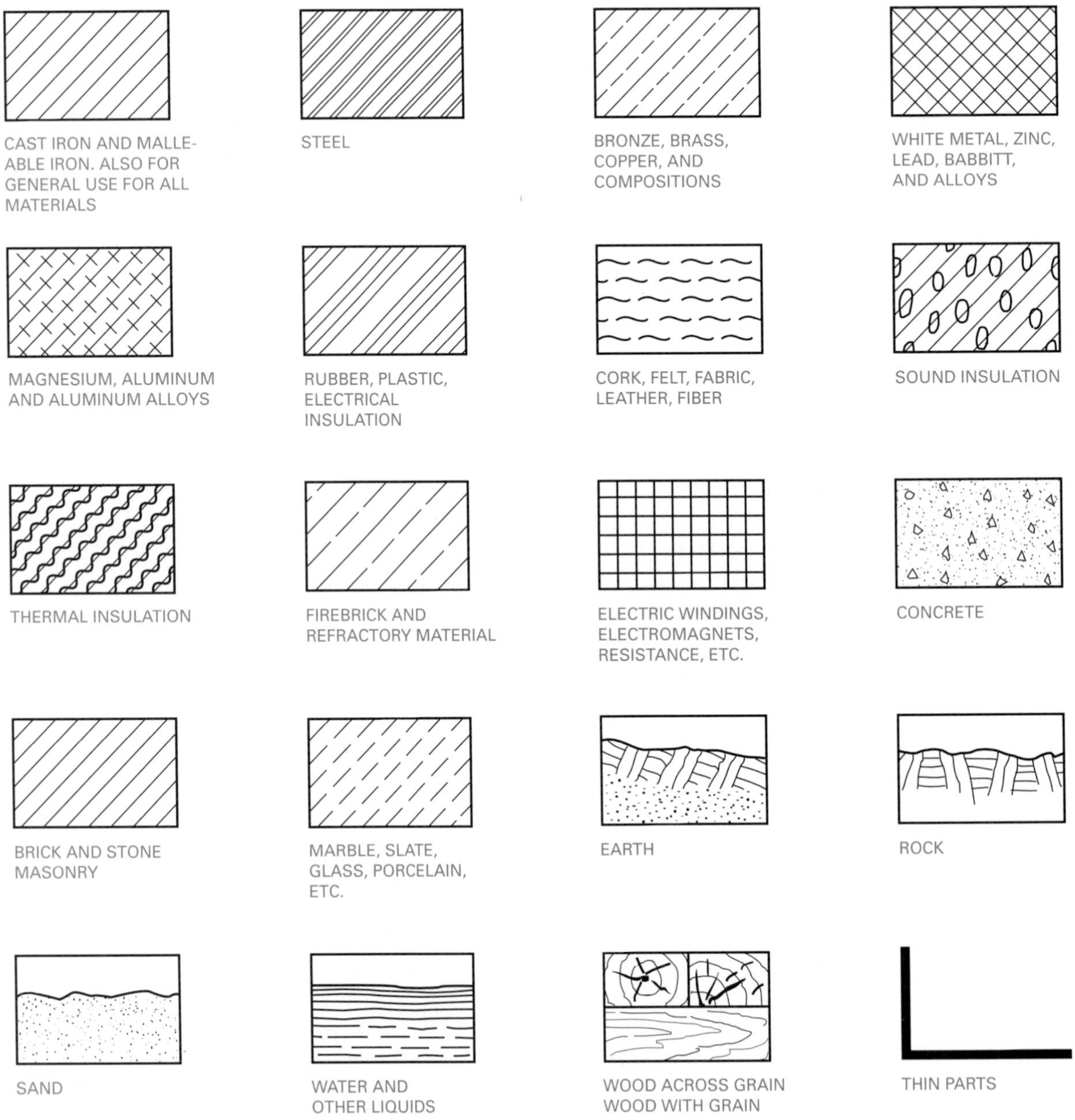

CAST IRON AND MALLE-
ABLE IRON. ALSO FOR
GENERAL USE FOR ALL
MATERIALS

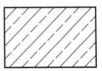

STEEL

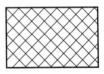

BRONZE, BRASS,
COPPER, AND
COMPOSITIONS

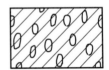

WHITE METAL, ZINC,
LEAD, BABBITT,
AND ALLOYS

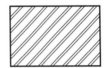

MAGNESIUM, ALUMINUM
AND ALUMINUM ALLOYS

RUBBER, PLASTIC,
ELECTRICAL
INSULATION

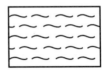

CORK, FELT, FABRIC,
LEATHER, FIBER

SOUND INSULATION

THERMAL INSULATION

FIREBRICK AND
REFRACTORY MATERIAL

ELECTRIC WINDINGS,
ELECTROMAGNETS,
RESISTANCE, ETC.

CONCRETE

BRICK AND STONE
MASONRY

MARBLE, SLATE,
GLASS, PORCELAIN,
ETC.

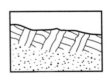

EARTH

ROCK

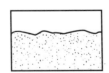

SAND

WATER AND
OTHER LIQUIDS

WOOD ACROSS GRAIN
WOOD WITH GRAIN

THIN PARTS

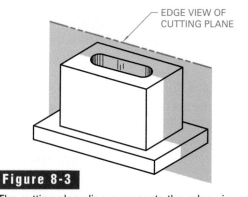

EDGE VIEW OF CUTTING PLANE

Figure 8-3

The cutting-plane line represents the edge view of the cutting plane.

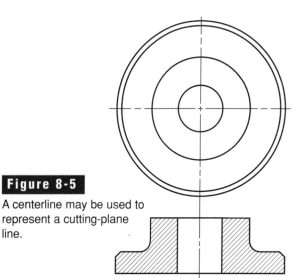

Figure 8-5

A centerline may be used to represent a cutting-plane line.

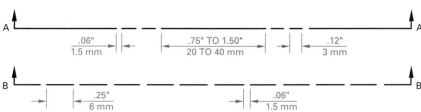

NOTE: ALL SIZES ARE ESTIMATED, NOT MEASURED.

Figure 8-4

Cutting-plane lines.

A cutting-plane line is not needed when it is clear that the section is taken along an object's main centerline or at some other obvious place. Figure 8-5 shows an example of a symmetrical object in which the centerline serves as the cutting-plane line.

Sections Through Assembled Pieces

If a drawing shows more than one piece in section, place the section lines in a different direction on each piece using common angles of 30°, 45°, or 60°, as shown in Figure 8-6A and B. Avoid using horizontal or vertical section lining. Remember, however, that any piece can show several cut surfaces. Make sure that all the cut surfaces of any one piece have section lines in the same direction, as shown in Figure 8-6B.

Figure 8-6

When a sectional view contains more than one assembled piece in section, the section lines must be drawn at different angles for each piece. (A) Two pieces in section require two different angles. (B) Three pieces require three angles.

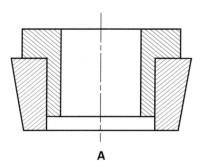

A

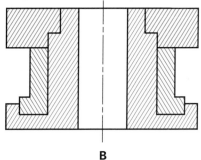

B

Types of Sectional Views

Sectional views can be drawn in many different ways to make the internal features as clear as possible, while keeping the drawing as simple as possible. The various types of sectional views are described here.

Full Sections

A **full section** is a sectional view that shows an object as if it were cut completely apart from one end or side to the other, as shown in Figure 8-7. Such views are usually just called *sections.* The two most common types of full sections are **vertical sections** and **profile sections**, as shown in Figures 8-8 and 8-9, respectively.

Offset Sections

In sections, the cutting plane is usually taken straight through the object. But it can also be offset, or shifted, at one or more places to show a detail or

to miss a part. This type of section, known as an **offset section**, is shown in Figure 8-10. In this figure, a cutting plane is offset to pass through the two bolt holes. If the plane were not offset, the bolt holes would not show in the sectional view. Show an offset section by drawing it on the cutting-plane line in a normal view. Do not show the offset on the sectional view.

Figure 8-7

Full section.

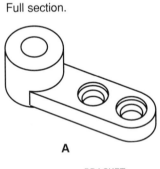

A

BRACKET

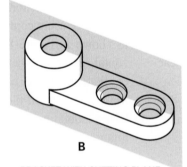

B

BRACKET WITH CUTTING PLANE

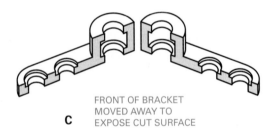

C

FRONT OF BRACKET MOVED AWAY TO EXPOSE CUT SURFACE

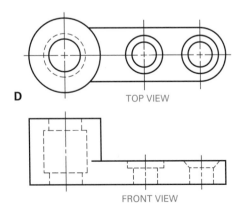

D

TOP VIEW

FRONT VIEW

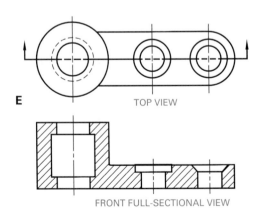

E

TOP VIEW

FRONT FULL-SECTIONAL VIEW

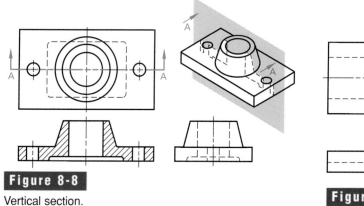

Figure 8-8

Vertical section.

Figure 8-9

Profile section.

Half Sections

A **half section** is one half of a full section. Remember, a full section makes an object look as if half of it has been cut away. A half section looks as if one quarter of the original object has been cut away. Imagine that two cutting planes at right angles to each other slice through the object to cut away one quarter of it, as shown in Figure 8-11A through C. Figure 8-11D shows the exterior of the object (not in section). The half section shows one half of the front view in section, as shown in Figure 8-11E.

Half sections are useful when you are drawing a symmetrical object. Both the inside and the outside can be shown in one view. Use a centerline where the exterior and half-sectional views meet, since the object is not actually cut. In the top view, show the complete object, since no part is actually removed. If the direction of viewing is needed, use only one arrow, as shown in Figure 8-11E. In the top view, the cutting-plane line could have been left out, because there is no doubt where the section is taken.

Figure 8-10

Offset section.

A VIEW OF BEARING FLANGE

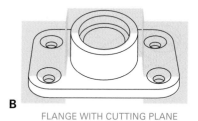

B FLANGE WITH CUTTING PLANE

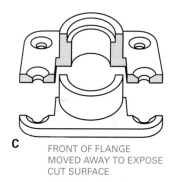

C FRONT OF FLANGE MOVED AWAY TO EXPOSE CUT SURFACE

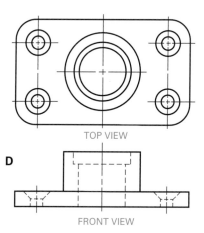

D

TOP VIEW

FRONT VIEW

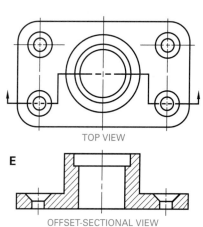

E

TOP VIEW

OFFSET-SECTIONAL VIEW

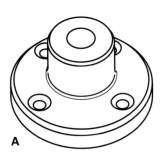

A

VIEW OF PACKING GLAND

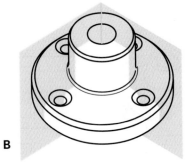

B

PACKING GLAND WITH CUTTING PLANE

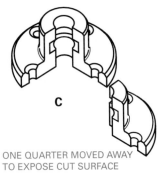

C

ONE QUARTER MOVED AWAY
TO EXPOSE CUT SURFACE

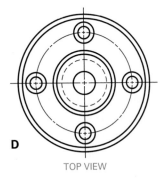

D

TOP VIEW

FRONT VIEW

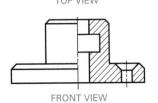

E

TOP VIEW

FRONT VIEW

Figure 8-11

Half section.

Figure 8-12

Broken-out section.

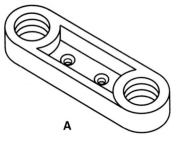

A

VIEW OF DOUBLE PACKING GLAND

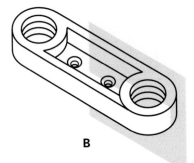

B

GLAND WITH CUTTING PLANE

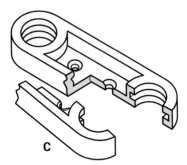

C

PART OF GLAND MOVED
AWAY TO EXPOSE CUT SURFACE

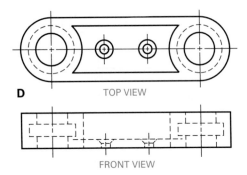

D

TOP VIEW

FRONT VIEW

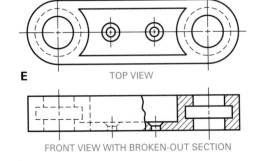

E

TOP VIEW

FRONT VIEW WITH BROKEN-OUT SECTION

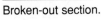

Broken-Out Sections

A view with a **broken-out section** shows an object as it would look if a portion of it were cut partly away from the rest by a cutting plane and then "broken off" to reveal the cut surface and insides. See Figure 8-12. This view shows some inside detail without drawing a full or half section.

Note that a broken-out section is bounded by a short-break line drawn freehand the same thickness as visible lines. Figure 8-13 shows two more examples of broken-out sections.

Revolved Sections

Think of a cutting plane passing through a part of an object, as shown in Figure 8-14. Now think of that cut surface as revolved 90°, so that its shape can be seen clearly, as shown in Figure 8-15. The result is a **revolved section** (also called a **rotated section**).

Use a revolved section when the part is long and thin and when its shape in cross section is the same throughout, as shown in Figure 8-16. In such cases, the view may be shortened, but the full length of the part must be given by a dimension. This lets you draw a large part with a revolved section in a shorter space.

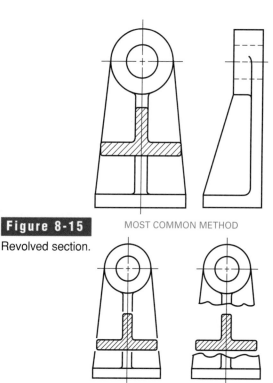

Figure 8-15
Revolved section.

MOST COMMON METHOD

OTHER ACCEPTABLE METHODS

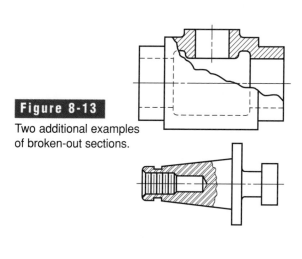

Figure 8-13
Two additional examples of broken-out sections.

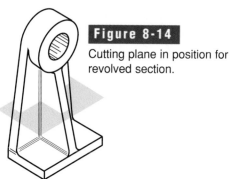

Figure 8-14
Cutting plane in position for revolved section.

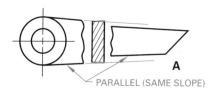

A

PARALLEL (SAME SLOPE)

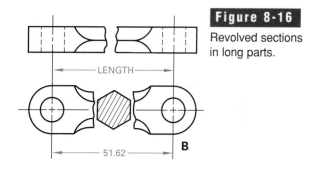

Figure 8-16
Revolved sections in long parts.

LENGTH

51.62

B

Removed Sections

When a sectional view is taken from its normal place on the view and moved somewhere else on the drawing sheet, the result is a **removed section**. Remember, however, that the removed section will be easier to understand if it is positioned to look just as it would if it were in its normal place on the view. In other words, do not rotate it in just any direction. Figure 8-17 shows correct and incorrect ways to position removed sections. Use bold letters to identify a removed section and its corresponding cutting plane on the regular view, as shown in Figure 8-17.

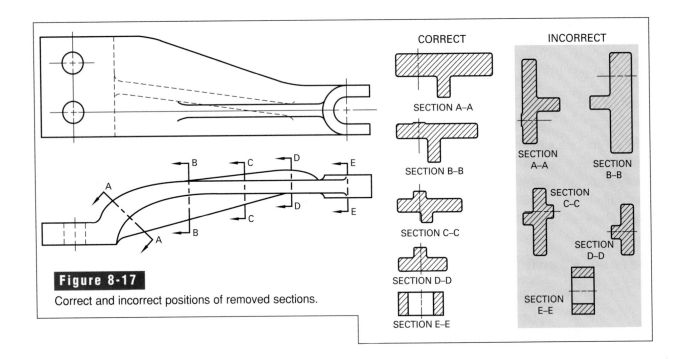

Figure 8-17

Correct and incorrect positions of removed sections.

CORRECT

SECTION A–A

SECTION B–B

SECTION C–C

SECTION D–D

SECTION E–E

INCORRECT

SECTION A–A

SECTION B–B

SECTION C–C

SECTION D–D

SECTION E–E

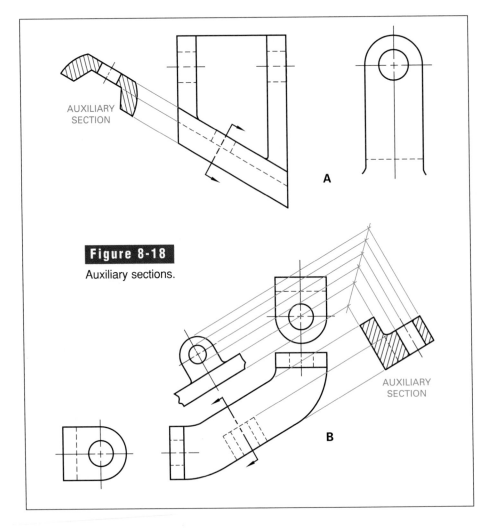

AUXILIARY SECTION

A

Figure 8-18

Auxiliary sections.

AUXILIARY SECTION

B

A removed section can be a sliced section (the same as a revolved section), or it can show additional detail visible beyond the cutting plane. You can draw it at the same scale as the regular views or at a larger scale to show details clearly.

Besides removed sections, you can also draw removed views of the exterior of an object. These too can be made at the same scale or at a larger one. They can also be complete views or partial views.

Auxiliary Sections

When a cutting plane is passed through an object at an angle, as shown in Figure 8-18A, the resulting sectional view is taken at the angle of that plane. It is called an **auxiliary section**, and it is drawn like any other auxiliary view. See Chapter 9 for more information about drawing auxiliary views.

Usually, on working drawings, only the auxiliary section is shown on the cut surface. However, if needed, any or all background features or parts beyond the auxiliary cutting plane may be shown. In Figure 8-18B, notice that the auxiliary section contains hidden lines. It also contains three incomplete views.

Phantom Sections

Use a **phantom section**, or hidden section, to show in one view both the inside and the outside of an object that is not completely symmetrical. Figure 8-19 shows an object with a circular boss on one side. Since the object is not symmetrical, the inside cannot be shown with a half section. A phantom section is used instead. A partial phantom section can sometimes be better than a broken-out section to show interior detail on an exterior view.

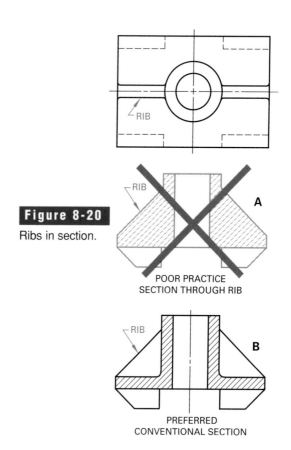

Figure 8-20

Ribs in section.

POOR PRACTICE
SECTION THROUGH RIB

PREFERRED
CONVENTIONAL SECTION

Special Cases

In practice, drafters make exceptions to the general sectioning rules in certain situations. These exceptions have become standard practice in industry. For example, while it is considered undesirable to show hidden lines in a section view, it may be done in special cases to improve accuracy and clarity.

Ribs and Webs in Section

Ribs and **webs** are thin, flat parts of an object that are used to brace or strengthen another part of the object. Often, a true section of an object that contains ribs or a web structure does not appear to show a true description of the part. For example, the section shown in Figure 8-20A would give the idea of a very heavy, solid piece. This would not be a true description of the part. Therefore, when a cutting plane passes through a rib or web parallel to the flat side, do not draw section lining for that part. Instead, draw the part as shown in Figure 8-20B. Think of the plane passing just in front of the rib.

Figure 8-19

Phantom section.

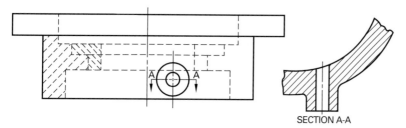

SECTION A-A

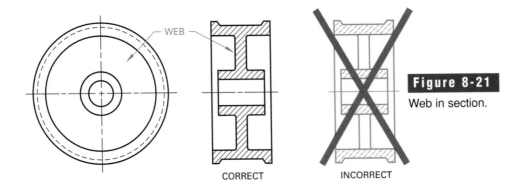

WEB

Figure 8-21

Web in section.

CORRECT INCORRECT

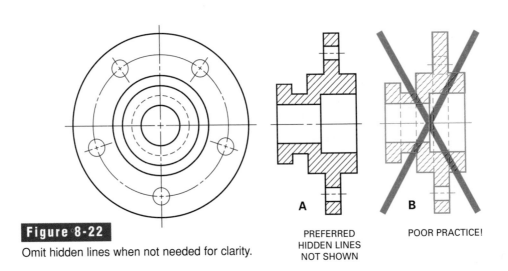

Figure 8-22

Omit hidden lines when not needed for clarity.

A

PREFERRED
HIDDEN LINES
NOT SHOWN

B

POOR PRACTICE!

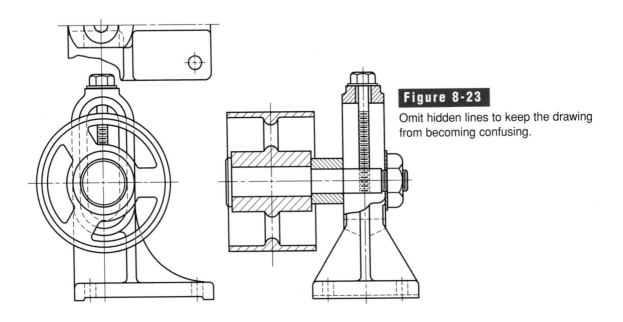

Figure 8-23

Omit hidden lines to keep the drawing
from becoming confusing.

If a cutting plane passes through a rib, a web, or any other thin, flat part at right angles to the flat side, draw in section lines for that part. Figure 8-21 shows an example.

Hidden and Visible Lines

Do not draw hidden lines on sectional views unless they are needed for dimensioning or for clearly describing the shape. In Figure 8-22A, a hub is described clearly using no hidden lines. Compare it with the incorrectly drawn section in Figure 8-22B. On sectional assembly drawings, or sectional views of how parts fit together, hidden lines are generally omitted. This keeps the drawing from becoming cluttered and hard to read. See Figure 8-23. Sometimes a good way to avoid using hidden lines is to draw a half section or partial section.

Normally, in a sectional view, include all the lines that would be visible on or beyond the plane of the section. In Figure 8-24, for example, the section drawing in part A correctly includes the numbered lines, which match the lines on the drawing in part B. A drawing without these lines, as shown in part C, would have little value. Do not draw sectional views in this manner. Figure 8-25 provides another example of how to draw visible lines beyond the plane of the section.

Alternate Section Lining

Alternate section lining is a pattern made by leaving out every other section line. It can be used to show a rib or another flat part in a sectional view when that part otherwise would not show clearly. In Figure 8-26A, an eccentric piece is drawn in section. (An eccentric piece is one that has two or more circular shapes that do not use the same centerlines.) A rib is visible in the top view, but in the sectional view, it is not shown by any section lining.

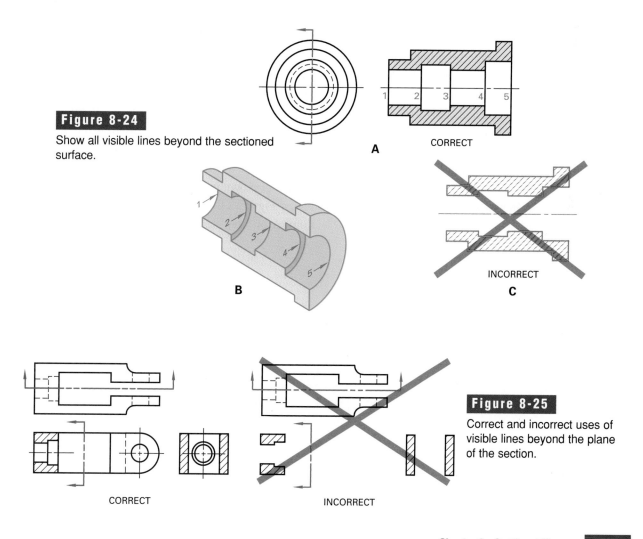

Figure 8-24
Show all visible lines beyond the sectioned surface.

A CORRECT

B

C INCORRECT

Figure 8-25
Correct and incorrect uses of visible lines beyond the plane of the section.

CORRECT

INCORRECT

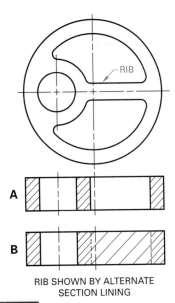

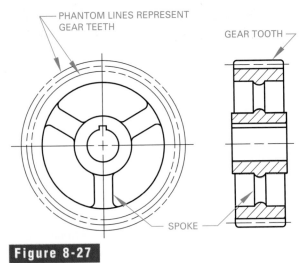

A

B

RIB SHOWN BY ALTERNATE
SECTION LINING

Figure 8-26

Alternate section lining.

SPOKE

Figure 8-27

Spokes and gear teeth should not be sectioned.

Other Parts Usually Not Sectioned

The omission of section lining is standard practice with a flat part such as a rib. But there are no visible lines to represent the rib either, because its top and bottom are both even with the surfaces they join. In fact, without the top view, you might not know that the rib was there. A drawing of an eccentric piece without a rib would look exactly the same. The problem is solved in Figure 8-26B. Here, alternate section lining is used with hidden lines to show the extent of the rib. Alternate section lines are useful to show ribs and other thin, flat pieces in one-view drawings of parts or in assembly drawings.

Do not draw section lines on spokes and gear teeth when the cutting plane passes through them. Leave them as shown in Figure 8-27. Do not draw section lines either on shafts, bolts, pins, rivets, or similar items when the cutting plane passes through them *lengthwise* (through the axis), as shown in Figure 8-28. These objects are not sectioned because they have no inside details. Also, sectioning might give a wrong idea of the part. A drawing showing them in full is easier to read. However, when such parts are cut *across* the axis, they should be sectioned, as shown in Figure 8-29. The sectional assembly in Figure 8-30 shows names and drawings of a number of other items that should not be sectioned.

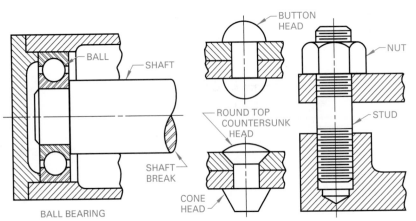

BALL — SHAFT

BUTTON HEAD

NUT

ROUND TOP
COUNTERSUNK
HEAD

STUD

SHAFT
BREAK

CONE
HEAD

BALL BEARING

Figure 8-28

Shafts, bolts, screws, rivets, and similar parts are usually not sectioned if the cutting plane passes through them lengthwise (along the axis).

Rotated Features in Section

A section or an elevation (side, front, or rear view) of a symmetrical piece can sometimes be hard to read if drawn in true projection. Figure 8-31A shows a true projection of a symmetrical piece with ribs and lugs. Notice that it does not show the true shape of the ribs and lugs. When drawing such a view, follow the example in Figure 8-31B. In this drawing, the ribs and lugs have been rotated on the vertical axis until they appear as mirror images of each other on either side of the centerline. Their true shape can now be shown. This is the correct way to draw this type of object. Note that only the parts that extend all the way around the vertical axis are drawn with section lining. Figure 8-32 shows another example in which the lugs are rotated to show true shape. Note that they are not drawn with section lines.

When a section passes through spokes, do not draw section lines on the spokes. Leave them as shown in the section drawing in Figure 8-33A. Compare this drawing with the section drawing for a solid web in Figure 8-33B. It is the section lining that shows that the web is solid rather than made with spokes.

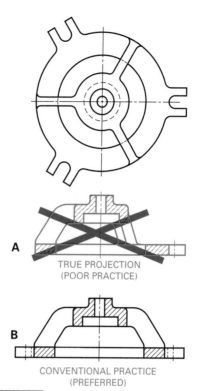

A TRUE PROJECTION (POOR PRACTICE)

B CONVENTIONAL PRACTICE (PREFERRED)

Figure 8-31

Some features should be rotated to show true shape.

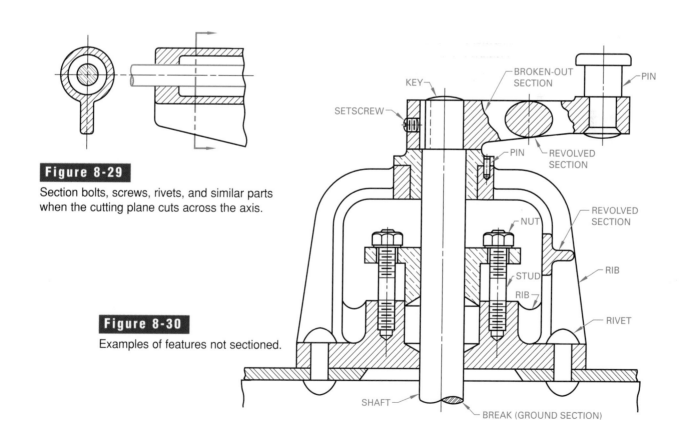

Figure 8-29

Section bolts, screws, rivets, and similar parts when the cutting plane cuts across the axis.

Figure 8-30

Examples of features not sectioned.

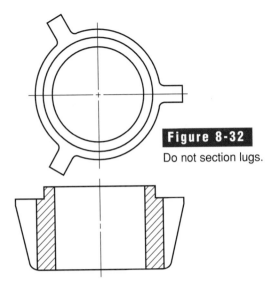

Figure 8-32

Do not section lugs.

Conventional Breaks and Symbols

Conventional breaks and symbols are used to show that a uniform part of a very long object has been cut out of the drawing. This makes some details easier to draw and easier to understand. Figure 8-36 shows methods used to draw long, evenly shaped parts and to break out the drawing of parts. Using a break lets you draw a view to a larger scale. Since the break shows how the part looks in cross section, an end view usually need not be drawn. Give the length by a dimension. The symbols for conventional breaks are usually drawn freehand by board drafters. However, on larger drawings, conventional breaks are often drawn with instruments to give a neat

When drawing a section or elevation of a part with holes arranged in a circle, follow the *preferred* examples in Figure 8-34. In these examples, the holes have been rotated for the section drawing until two of them lie squarely on the cutting plane. These views then show the true distance of the holes from the center, whereas a true projection would not.

Rotating features in drawings is very useful when you want to show true conditions or distances that would not show in a true projection. Moreover, for some objects, only part of the view should be rotated, as shown by the bent lever in Figure 8-35.

Figure 8-34

Preferred and poor practice for showing holes.

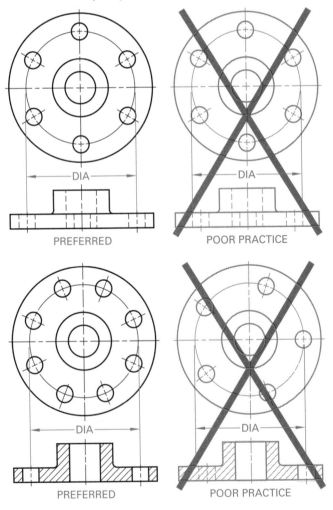

PREFERRED   POOR PRACTICE

PREFERRED   POOR PRACTICE

Figure 8-33

A section through spokes.

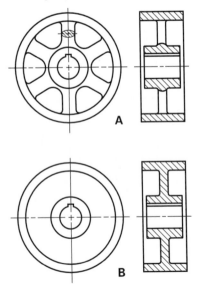

A

B

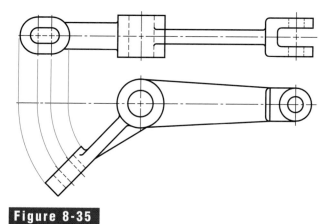

Figure 8-35

Rotation of a part of a view to show true shape.

appearance. CAD operators generally use polylines for breaks on both large and small drawings. Figure 8-37 shows how to draw the break for cylinders and pipes.

Intersections in Section

An intersection, in this discussion, is a point where two parts join, as shown in Figure 8-38. Drawing a true projection of an intersection is difficult and takes too much time. Also, such accuracy of detail is of little or no use to a print reader. Therefore, approximated or conventional sections are usually drawn, as shown in Figure 8-39.

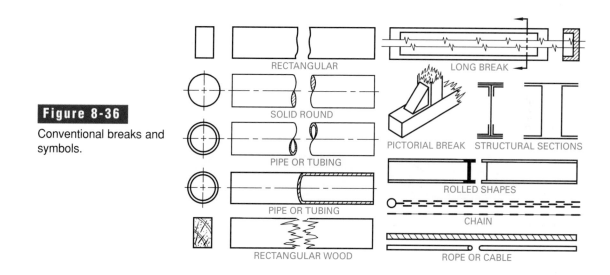

Figure 8-36

Conventional breaks and symbols.

RECTANGULAR

SOLID ROUND

PIPE OR TUBING

PIPE OR TUBING

RECTANGULAR WOOD

LONG BREAK

PICTORIAL BREAK   STRUCTURAL SECTIONS

ROLLED SHAPES

CHAIN

ROPE OR CABLE

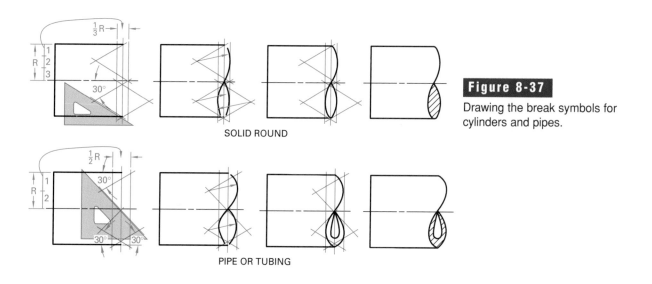

SOLID ROUND

PIPE OR TUBING

Figure 8-37

Drawing the break symbols for cylinders and pipes.

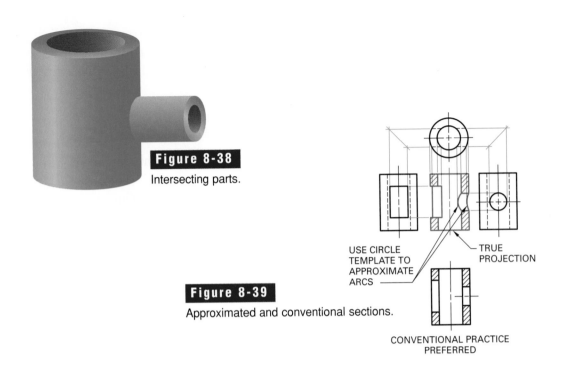

Figure 8-38

Intersecting parts.

Figure 8-39

Approximated and conventional sections.

USE CIRCLE TEMPLATE TO APPROXIMATE ARCS

TRUE PROJECTION

CONVENTIONAL PRACTICE PREFERRED

Board Drafting Techniques

Creating a sectional view using board drafting techniques is similar to creating any other type of view. However, some special considerations are involved with sectional views. Because each line is drawn individually, creating a sectional view can easily become a very time-consuming task. Techniques to help you create good sectional views with a minimum of effort and drawing time are discussed below. Then a practice drawing is presented to help you gain the experience needed to create good sectional views.

Techniques for Sectioning

The spacing of section lines must be adapted for various drawing situations. The size and shape of the object being drawn often determine both the line spacing and the sectioning style.

Spacing of Section Lines

Section lines are spaced close together or far apart, depending on how much space must be filled. See Figure 8-40. According to ANSI, section lines can be spaced from about .03″ (1.0 mm) to .12″ (3.0 mm) apart. However, they must be evenly spaced, and they are usually slanted at a 45° angle. The drawing will be neater if you do not space section lines extremely close together. This will also save time. In most cases, the lines will look best spaced about .10″ apart.

The distance between section lines need not be measured; you may space them by eye. If the area to be covered is large, space the lines farther apart. If the area is small, space the lines closer together. If the area is very small, as for thin plates, sheets, and structural shapes, blacked-in, or solid black, sections

Metric Conversions

The metric standard of measure is the meter, which is subdivided into 100 centimeters, 1,000 millimeters, etc. The meter is defined by international agreement as the distance between two scratches on a platinum-iridium bar kept near Paris. The English standard is the yard, subdivided into 3 feet or 36 inches. The yard is defined as the distance between two scratches on a bar kept in London.

In drafting, the primary working unit of measure in the metric system is the millimeter. It is often necessary for the drafter to convert inches to millimeters or from millimeters to inches. This is especially true in industries involved in international trade. Here is how it is done:

Conversion Factor: $1'' = 25.4$ mm

Example 1: Convert $2''$ to millimeters.
$2'' \times 25.4$ mm $= 50.8$ mm

Example 2: Convert $3.25''$ to millimeters.
$3.25'' \times 25.4$ mm $= 82.55$ mm

In drafting, millimeters are generally given to one decimal place, so 82.55 mm becomes 82.6 mm.

Example 3: Convert 50.8 millimeters to inches.
50.8 mm $\div 25.4 = 2''$

Example 4: Convert 82.6 millimeters to inches.
82.6 mm $\div 25.4 = 3.25196850394''$

In drafting, inches are generally rounded to two decimal places, so 3.25196850394'' becomes 3.25''.

---

may be used. An example of the use of blacked-in sections is shown in Figure 8-41. Note the white space between the parts.

Sectioning Style

When you are drawing a large sectioned area, one way to save time is to use outline sectioning. This method is shown in Figure 8-42. Drafters who use it often draw the section lines freehand and spaced widely apart. You can also shade the sectioned area as shown in Figure 8-43. Shade only along its outline as shown in Figure 8-44, or rub pencil dust over it. Apply a fixative to prevent smudging.

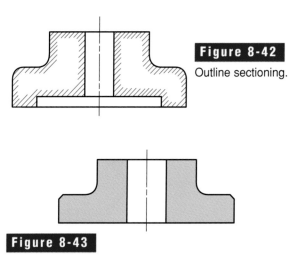

Figure 8-41

You may blacken in the entire sectioned area instead of using section lines when the area is very small.

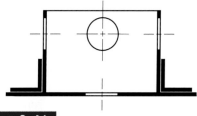

Figure 8-42
Outline sectioning.

Figure 8-40

Space section lines by eye. The distance between section lines varies according to the size of the space to be sectioned.

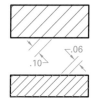 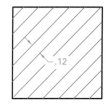

Figure 8-43
A cut surface may be shaded.

Figure 8-44

A cut surface may have a shaded outline.

Do not draw section lines parallel to or at right angles to an important visible line. See Figure 8-45. You may, however, draw them at any other suitable angle and space them at any width. Section lines at different spacing and angles are commonly used to identify different sectioned parts.

Practice Drawing

To develop a sectional-view drawing, first determine which normal views are necessary. Once this is determined, you can decide what type of sectional view is needed to show interior detail clearly with few, if any, hidden lines. As for any multiview drawing, you may then find it useful to prepare a freehand sketch with dimensions and notes. This will allow you to determine the amount of space required for views, dimensions, and notes before starting on the final drawing.

Figure 8-46 is a pictorial drawing of a flat-belt pulley. It will serve as the basis for the drawing developed in Figure 8-47. A circular front view and a side or profile view will be required. Since it is a symmetrical object, a half-section on the profile view will be sufficient to show the interior detail.

Begin the drawing by blocking in the two views. For the circular view, use centerlines to establish the

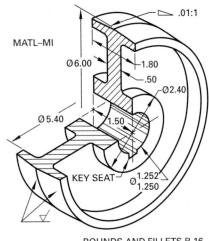

ROUNDS AND FILLETS R.16

Figure 8-46

The flat-belt pulley developed in the practice drawing.

location of the center point and draw a light circle (R3.00) to lay out the view. Project light lines from the circle to block in the profile view, as shown in Figure 8-47A.

Next, use light construction lines to finish blocking in both views as shown in Figure 8-47B. Darken visible lines and add section lining. Finally, complete the drawing by adding dimensions and notes as shown in Figure 8-47C.

The drawing process is essentially the same for all sectional-view drawings. It is important to remember that careful planning and freehand sketching will generally reduce the amount of time it takes to produce the final drawing.

Figure 8-45

Do not draw section lines parallel to or perpendicular to a main line of the view.

POOR PRACTICE

A

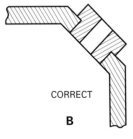

CORRECT

B

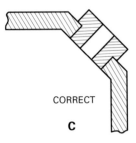

CORRECT

C

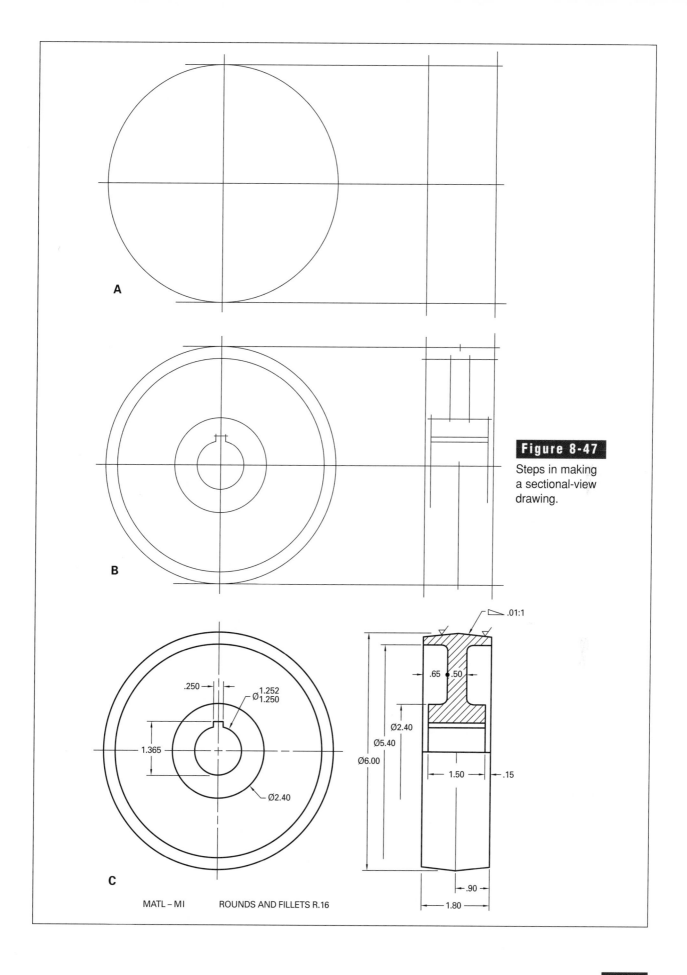

A

B

Figure 8-47

Steps in making a sectional-view drawing.

C

MATL – MI   ROUNDS AND FILLETS R.16

.250

Ø1.252
1.250

1.365

Ø2.40

.01:1

.65   .50

Ø2.40

Ø5.40

Ø6.00

1.50   .15

.90

1.80

CAD Techniques

Among CAD operators, the more common term for section lining is *crosshatching*, or simply *hatching*. AutoCAD and most other CAD programs have commands that provide an easy method for creating a hatch within the boundaries you specify.

Boundary Hatching

In AutoCAD, you use the Boundary Hatch dialog box (BHATCH command) to specify the type of hatch, its scale, and its location on the drawing. See Figure 8-48. Notice that the default pattern is ANSI 31—the general-purpose symbol. The swatch just below the pattern window allows you to see at a glance which pattern is selected.

Hatch Angle

The default angle of 0 results in a hatch that looks exactly like the swatch, with lines slanted at 45°. This is considered 0 because it is the most common angle. However, recall that for drawings in which two or more pieces must be sectioned, you will need to distinguish among them by placing the lines at different angles. Refer again to Figure 8-6B. To do this, set the angle to a value other than zero. Pick the arrow to the right of the Angle text box to choose from a list of other common angles.

Line Spacing

The scale of the hatch determines the line spacing. Section lines are spaced close together or far apart, depending on how much space must be filled. For small areas, they should be closer together, and for larger areas, they should be farther apart. Generally, you can start with the default scale of 1.0000. AutoCAD allows you to preview the result on your actual drawing. Therefore, if the scale does not seem right, you can return to the dialog box and change the scale as many times as necessary to achieve the right effect before you actually apply the hatch to the drawing. See Figure 8-49.

Defining Hatch Boundaries

The Pick Points and Select Objects buttons return you to the drawing so that you can specify the area to be hatched. Pick Points allows you to pick a point anywhere within the areas to be hatched and let AutoCAD calculate the boundaries. This allows you to hatch more than one area at one time. See Figure 8-50. Select Objects allows you to pick an object to be hatched. This is helpful when centerlines or other lines are present within the object that AutoCAD might interpret as a hatch boundary.

Figure 8-48

AutoCAD's Boundary Hatch dialog box.

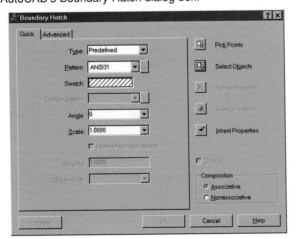

Figure 8-49

The scale of a hatch controls the spacing of the section lines.

HATCH SCALE .5   HATCH SCALE 1.0   HATCH SCALE 2.0

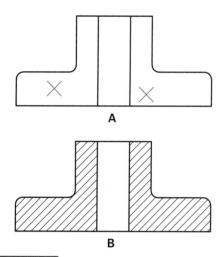

(A) Pick a point inside each area to be hatched. (B) AutoCAD hatches both areas simultaneously. The two hatched areas are created as a single AutoCAD object.

Associativity

The Composition area at the bottom right of the dialog box allows you to choose an associative or nonassociative hatch. An **associative hatch** is one that changes when you change the objects that form its boundaries. For example, suppose you hatch a ∅3.00 circle, as shown in Figure 8-51A. Later, the design changes, and you need to change the diameter of the circle to ∅2.25. When you change the circle, an associative hatch updates automatically to reflect the new diameter, as shown in Figure 8-51B. If you use a nonassociative hatch, the hatch does not update, as shown in Figure 8-51C.

After you have set up the hatch, pick the Preview button in the lower left corner to see the hatch in place on your drawing. Press Enter or right-click to return to the dialog box and make any necessary changes. You can preview the hatch as many times as necessary. When you are satisfied with the hatch, pick OK to apply the hatch to the drawing.

Erasing a Hatch

All of the lines that make up a hatch in AutoCAD are considered a single object. Therefore, to remove the hatch, you can enter the ERASE command (or press E and Enter) to remove the entire hatch pattern from a drawing.

►CADTIP

Advanced Hatching

AutoCAD provides several options for removing hatches from holes in an object and from around dimension text. Experiment with the options in the Boundary Hatch dialog box to become familiar with these options.

Practice Drawing

Figure 8-52 is a pictorial drawing of a flat-belt pulley. It will serve as the basis for the drawing developed in Figure 8-53. A circular front view and a side or profile view will be required. Since it is a symmetrical object, a half-section on the profile view will be sufficient to show the interior detail.

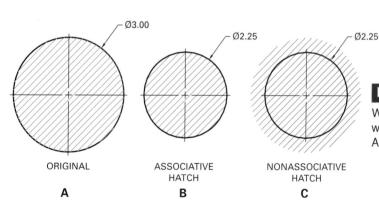

ORIGINAL   ASSOCIATIVE HATCH   NONASSOCIATIVE HATCH

A   B   C

When a hatch is associative (A), it changes when the boundaries that define it change (B). A nonassociative hatch does not change (C).

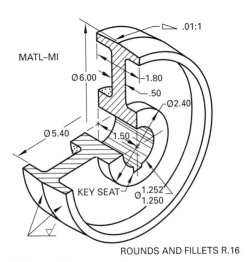

MATL-MI

Ø6.00 — 1.80
— .50
Ø2.40
Ø5.40 1.50
.01:1

KEY SEAT — Ø1.252
1.250

ROUNDS AND FILLETS R.16

Figure 8-52

The flat-belt pulley developed in the practice drawing.

Planning

Study the views in Figure 8-53 carefully before you begin. It is important to plan how to use AutoCAD most efficiently. For this drawing, you should be able to see that the views are easiest to draw by making liberal use of construction lines and the OFFSET command. If you draw the circles for the front view first, you can snap horizontal construction lines to the quadrants of the circles to identify key points on the profile view. Using those lines and the vertical centerline of the profile view, you can use OFFSET to place most of the features in the profile view. For example, offset the vertical centerline .90 to the right and left to locate the edges of the profile view.

Notice the incline noted at the top right of the profile view: .01:1. Using a CAD program, it is possible to draw this incline exactly, and it is tempting to do so. However, drawn at its true proportions, this incline is far too subtle to be seen by print readers. Therefore, common practice is to exaggerate the incline in the drawing and place the true angle in a note, as shown in the illustration. It is important, however, to be consistent. Use the same angle for the inclines at both the top and the bottom of the view. In Figure 8-53, construction lines were used to transfer the top and bottom points from the top and bottom quadrants of the largest circle in the front view. Then these lines were offset .10″ to the inside, as shown in Figure 8-54, and the inclines were constructed using the Intersection object snap to pick the points shown by X's.

Setting Up the Drawing

After you have determined the best way to draw the views, create a new drawing file and set it up for a B-size sheet. Set up the units and layers. In addition to the usual Objects, Hidden Lines, and Centerlines layers, create a Hatch layer to hold the hatches. Specify a thin line width, such as .13 mm, for the Hatch layer.

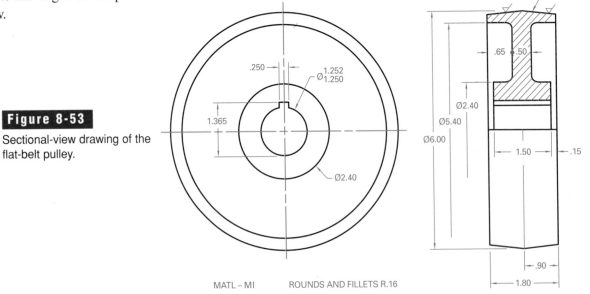

Figure 8-53

Sectional-view drawing of the flat-belt pulley.

MATL – MI   ROUNDS AND FILLETS R.16

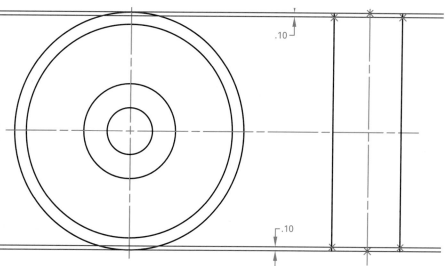

Figure 8-54

Offset the top and bottom construction lines by .10 to establish the incline for the profile view.

.10

.10

The dimensions on this drawing require special attention. Notice that some of the dimensions are given to two decimal places, and others are given to three decimal places. One of the dimensions is a limit dimension. It is possible to use a single dimension style for the entire drawing and simply override the attributes for individual dimensions. However, this is poor practice, and it takes a significant amount of extra time. Instead, you should set up three distinct dimension styles to use in this drawing. Follow these steps:

1. Enter DIMSTYLE or pick the Dimension Styles button on the Dimension toolbar to open the Dimension Style Manager.
2. Instead of modifying the Standard style, pick the New… button to begin a new dimension style. Name the style Two Decimal Places, as shown in Figure 8-55. Accept the defaults to base the style on the Standard style and apply it to all dimensions. Pick Continue to display the same tabbed dialog box you used in Chapter 7. Set up this dimension style exactly as you set up the Standard style in Chapter 7, including all of the text and line attributes. Be sure to set the primary units to two decimal places. Pick OK to return to the Dimension Style Manager.
3. Pick the New… button of the Dimension Style Manager again to create another new style. Name this one Three Decimal Places. If you base the style on Standard, you will have to specify them all over again. To avoid this, base this new style on the Two Decimal Places style

you created in step 2. Now the only change you will need to make is to set the primary units to three decimal places. Then pick OK to return to the Dimension Style Manager.
4. Pick the New… button a third time to create the third new style. Name this style Limits. Since the limit dimension in the flat-belt pulley drawing has three decimal places, base the Limits style on the Three Decimal Places style you created in step 3. Now you will only have to change the tolerance. Pick the Tolerances tab and set Method to Limits. Then set the upper value and lower value to .001. By doing this, you can later create the hole with an actual radius of 1.251. When you dimension the hole using the Limits dimension style, the upper and lower tolerances of .001 will produce the required dimension. Pick the box next to Zero suppression to activate it, and then pick OK to return to the Dimension Style Manager.

Figure 8-55

Create a new style named Two Decimal Places.

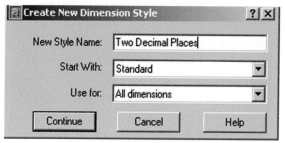

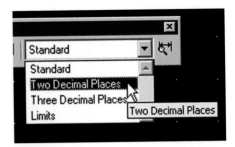

Figure 8-56

The Dimension Style Control dropdown box on the Dimension toolbar allows you to change dimension styles quickly and easily.

5. Pick Close to close the Dimension Style Manager.

On the Dimension toolbar, the Dim Style Control dropdown box says Standard. This refers to the Standard dimension style. To change the style, pick the down arrow and select the style you want. See Figure 8-56. All of the styles you have created will appear in this list.

Drawing the Views

Follow these steps to create the two-view drawing of the flat-belt pulley:

1. Draw both of the views shown in Figure 8-53, using the discussion in the previous sections for guidance.
2. From the Dimension Style Control box on the Dimension toolbar, select Three Decimal Places. Then place the two three-place dimensions in the front view of the pulley.
3. From the Dimension Style Control box, select the Limits dimension style. Then place the diameter dimension for the hole in the front view.
4. Activate the Two Decimal Places dimension style. Create the overall ⌀6.00 dimension and the .90 and 1.80 dimensions at the bottom of the profile view.
5. To create the partial dimensions in the profile view (⌀5.40 and ⌀2.40), offset temporary lines from the horizontal centerline to place the lower end of the dimension. Create the dimensions, and then delete the two temporary lines. The text of the two dimensions may overlap the ⌀6.00 dimension. If so, pick one of the dimensions to activate it. Notice the blue grip box on the

dimension text. Pick the grip so that it becomes solid red. Then drag the text straight up (Ortho will help) so that it no longer interferes with the other text, as shown in Figure 8-53.
6. Notice that the ⌀5.40 and ⌀2.40 dimensions have an incomplete lower dimension line, no lower arrow, and no lower extension line. To achieve this effect, doubleclick the ⌀5.40 dimension to display its properties in the Properties dialog box at the left of the drawing area. Pick Lines & Arrows to see a list of properties of the lines and arrows that make up the dimension. From the list, pick Arrow 2 so that a down arrow appears next to the property, as shown in Figure 8-57. Pick the down arrow, scroll almost all the way to the end of the list of available arrowheads, and select None. This removes the second arrowhead. Then, in the Lines & Arrows section of the Properties dialog box, pick Ext line 2 and pick the down arrow that appears. Select Off to turn the second extension line off. Repeat this procedure for the ⌀2.40 dimension.

►CADTIP

First and Second Arrowheads
In the Properties list, the "first" arrowhead is the arrowhead at the end of the dimension that you picked first when you created the dimension. The procedure in step 6 assumes that you picked the upper points first when you created these two dimensions. If you picked the lower points first, then you should change the values of Arrow 1 and Ext line 1 instead of Arrow 2 and Ext line 2.

7. Create the 1.50 interior dimension in the profile view. Then pick the Continue Dimension button on the Dimension toolbar to add the .15 dimension in exact alignment with the 1.50 dimension. Notice that AutoCAD doesn't ask you for a first point; the first point is automatically set to the end of the 1.50 dimension. Use the Perpendicular object snap to snap the second point of the dimension to the exterior line of the profile view and press Enter twice to leave the Continue Dimension mode.

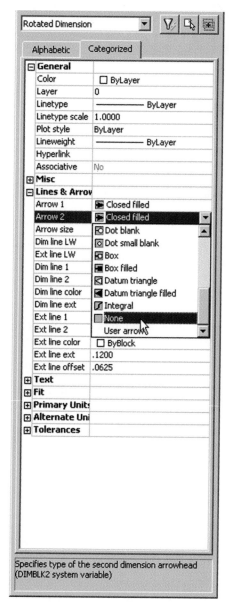

Figure 8-57

Scroll to the next-to-last item in the Arrow 2 dropdown list to select None for the second arrowhead.

9. Use the Quick Leader button on the Dimension toolbar to place the leader for the incline note. Use the Nearest object snap to snap to the top of the profile view, and pick a second point up and to the right for the second point of the leader. Then keep pressing Enter to accept all the prompts. When the text editing box appears, pick OK without adding any text. Pick OK to place the leader on the drawing without any text. Then use lines or a polyline to create the small triangle, and position it as shown in Figure 8-53, next to the leader. Use the DTEXT command, set to the Roman text style and a height of .12, to create the text, and position it as shown.

10. Add the notes using DTEXT.

Creating the Hatch

Finally, you can add the hatch to the sectioned area of the drawing. Enter the BHATCH command by picking the Hatch icon in the toolbar to the left of the drawing area. Use the ANSI 31 general-purpose pattern with an angle of 0 and a scale of 1. Select Pick Points, and pick a point anywhere inside the area to be hatched. Preview the hatch, and if it looks like the hatch in Figure 8-53, pick OK to set the hatch in place.

8. Create the .65 and .50 interior dimensions. Notice the dot that takes the place of arrowheads between the dimensions. To create the dot, display the properties of the .65 dimension (doubleclick it), and change the second arrowhead to None. Then display the properties of the .50 dimension, and change the first arrowhead to Dot.

Success on the Job

Getting Information
Like most software, CAD programs have Help features and manuals that provide more information about how to use the software. Being able to locate and use these reference sources is a skill that will increase your chances for success on the job.

Chapter Summary

- Sectional views are used to show complex interior details.

- Special section lining patterns, or symbols, are used to represent specific materials.

- Cutting-plane lines or centerlines may be used to show where the section is to be taken.

- Section lining on adjacent parts should be drawn at different angles.

- Each of the various types of sectional views has a specific purpose; these views should not be used interchangeably.

- Hidden lines are used on sectional views only if they are needed for clarity.

- Bolts, shafts, pins, and other similar parts are usually not sectioned even if the cutting plane passes through them.

Review Questions

1. What is another term for sectional view?

2. What type of line is used to show where the cutting plane passes through an object?

3. What is another name for the patterns used in section lining?

4. In inches, how far apart does ANSI recommend spacing general-purpose section lining?

5. What is another name for crosshatching?

6. What kind of sectional view results when the cutting plane passes through the entire view?

7. What type of sectional view results when the cutting plane changes direction to pass through specific details?

8. What type of sectional view is most often used to show a symmetrical view in section?

9. What type of sectional view would you most likely use to show the cross-sectional shape of a spoke?

10. What is the section lining called when you leave out every other one for clarity on a rib or web in section?

11. In AutoCAD, if you specify a hatch angle of 0, at what angle will the section lines appear?

12. What determines the spacing of section lines in AutoCAD?

13. What is an associative hatch?

14. How can you change the appearance of an arrowhead or remove it from a dimension?

Drafting Problems

The drafting problems in this chapter are designed to be completed using board drafting techniques or CAD.

1. Parts A and B of Figure 8-58 show examples of half and full sections. In the half sections, the hidden line is optional. Study these examples carefully before attempting any of the drawing assignments in this chapter. For parts C through L of Figure 8-58, take dimensions from the printed scales at the bottom of the page. Make a full or half section as assigned. Add dimensions if required by your instructor. Estimate the sizes of fillets and rounds.

Figure 8-58

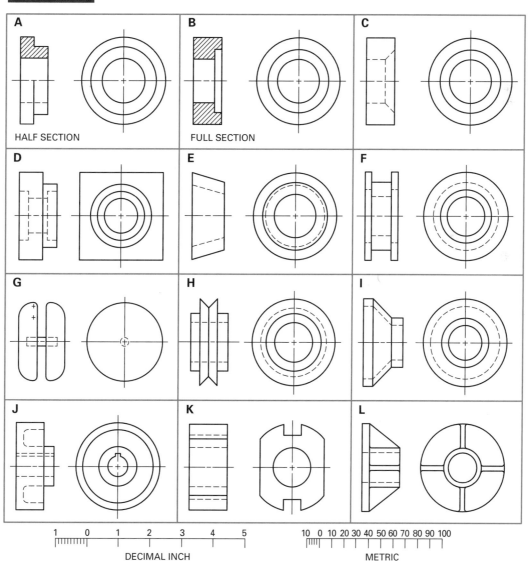

2. Take dimensions from the printed scale for each of the two-view drawings in Figure 8-59. Draw both views. Make a sectional view as assigned. Add dimensions if required by your instructor. Estimate the sizes of fillets and rounds.

3. Take dimensions from the printed scale for each of the drawings in Figure 8-60. Draw both views and section the view indicated by the cutting-plane line. Add dimensions if required by your instructor. Estimate the sizes of fillets and rounds.

Figure 8-59

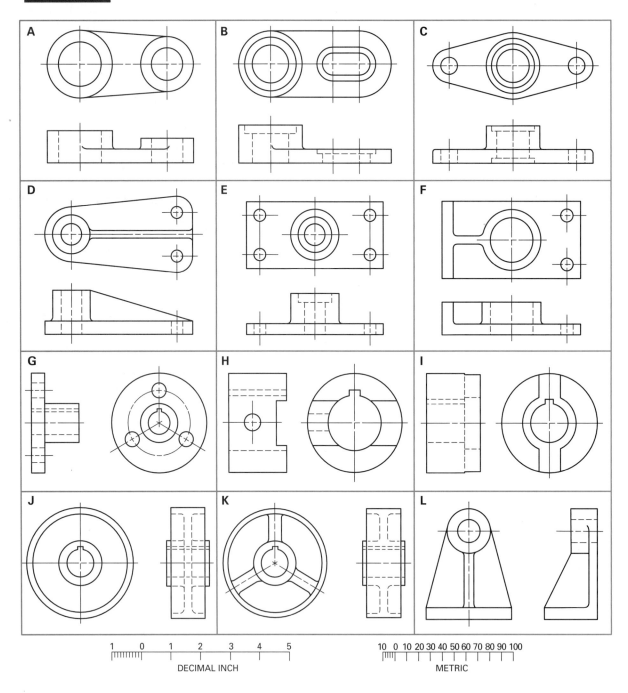

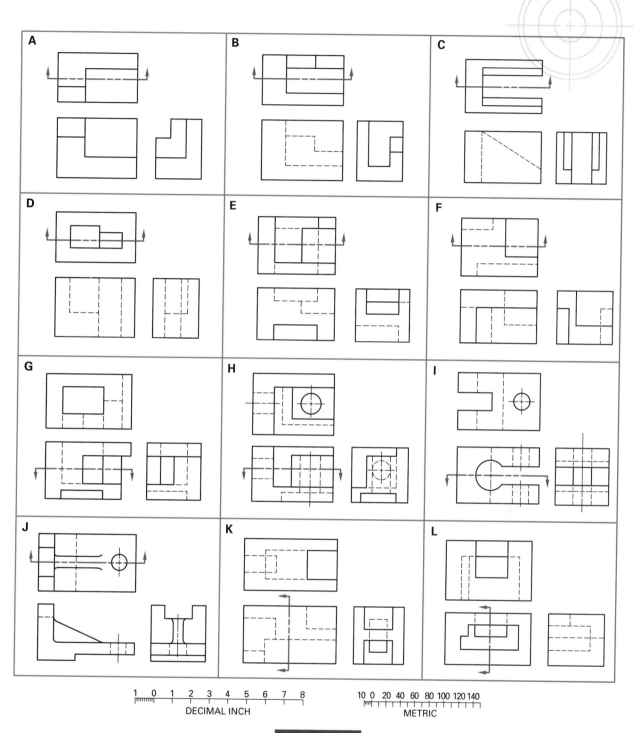

1 0 1 2 3 4 5 6 7 8
DECIMAL INCH

10 0 20 40 60 80 100 120 140
METRIC

Figure 8-60

4. Make a two-view drawing of the collar shown in Figure 8-61. Show a full or half section as assigned.

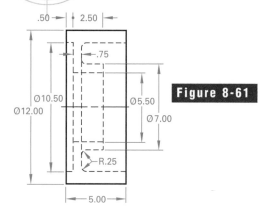

Figure 8-61

5. Make a two-view drawing of the steam piston shown in Figure 8-62. Show a full or half section as assigned.

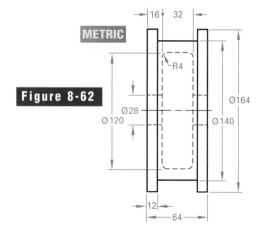

Figure 8-62

METRIC

6. Make a two-view drawing of the shaft cap shown in Figure 8-63. Show a full or half section as assigned.

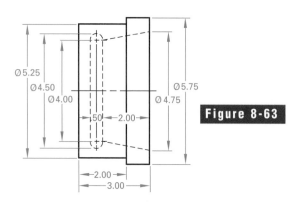

Figure 8-63

7. Make a two-view drawing of the protected bearing shown in Figure 8-64. Show a full or half section as assigned.

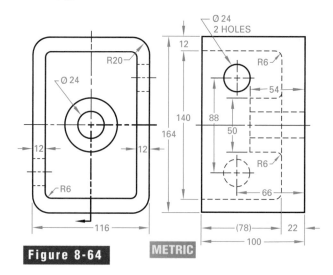

Figure 8-64   METRIC

8. Make a two-view drawing of the water-piston body shown in Figure 8-65. Show a full or half section as assigned.

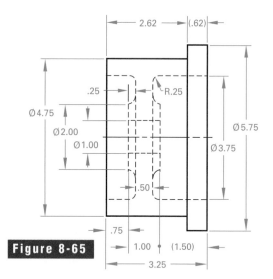

Figure 8-65

9. Make a two-view drawing of the cylinder head shown in Figure 8-66. Show a full or half section as assigned.

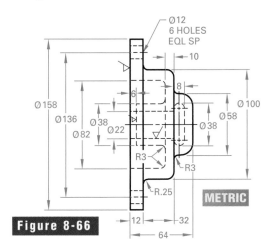

Figure 8-66

10. Make a two-view drawing of the cylinder cap shown in Figure 8-67. Show a full or half section as assigned.

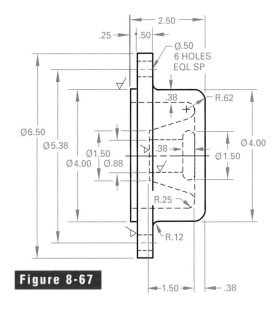

Figure 8-67

11. Make a two-view drawing of the cone spacer shown in Figure 8-68. Show a full or half section.

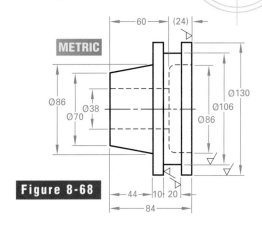

Figure 8-68

12. Draw the rod guide shown in Figure 8-69, using the scale shown in the figure. Make top and front views. Show a broken-out section as indicated by the colored screen.

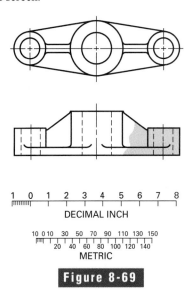

Figure 8-69

13. Draw the chisel shown in Figure 8-70, using the scale from problem 12. Make revolved or removed sections on the colored centerlines. A is a .25″ × 3.00″ (6.3 × 76 mm) rectangle; B is a 1.25″ (32 mm) octagon (measured across the flats); and C and D are circular cross sections. Determine the scale for the drawing before you begin.

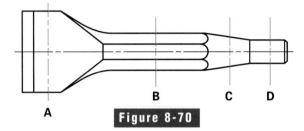

A   **Figure 8-70**

14. Draw the structural joint shown in Figure 8-71, using the scale shown in the figure. Make a full-sectional view of the joint with rivets moved into their proper positions on the centerlines.

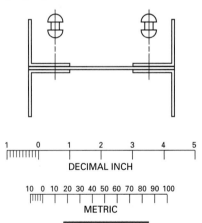

Figure 8-71

15. Draw the adjusting plate shown in Figure 8-72, using the scale from problem 14. Draw front and top views. Make the broken-out section as indicated by the colored screen.

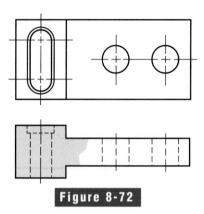

Figure 8-72

16. Draw the grease cap shown in Figure 8-73, using the scale from problem 14. Make front and right full or half sections as assigned.

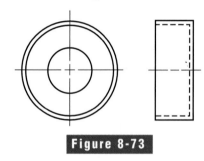

Figure 8-73

17. Draw the rotator shown in Figure 8-74, using the scale from problem 14. Complete the right-side view and make a full or half section.

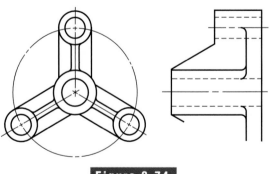

Figure 8-74

18. Draw the screwdriver shown in Figure 8-75. Use the scale from problem 14, and draw the screwdriver twice the size shown. Add removed or revolved sections on the colored centerlines. The overall length is 6.60″.

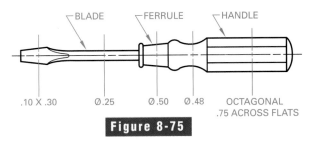

Figure 8-75

For problems 19 through 27, draw the required views at a suitable scale. Make one of the views a sectional view, as appropriate.

19. Base plate. See Figure 8-76. Material: Cast iron.

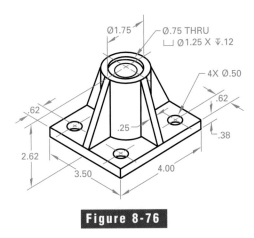

Figure 8-76

20. Shaft base. See Figure 8-77. Material: Cast iron.

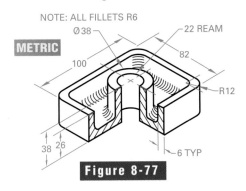

Figure 8-77

21. Step pulley. See Figure 8-78. Material: Cast iron.

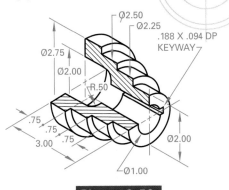

Figure 8-78

22. Lever bracket. See Figure 8-79. Material: Cast iron.

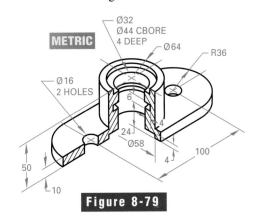

Figure 8-79

23. Idler pulley. See Figure 8-80. Material: Cast iron.

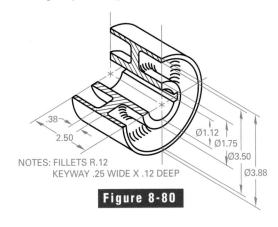

Figure 8-80

24. Retainer. See Figure 8-81. Material: Cast aluminum.

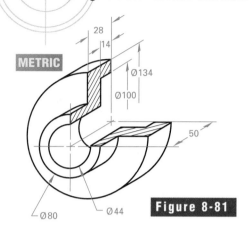

Figure 8-81

26. End cap. See Figure 8-83. Material: Cast iron.

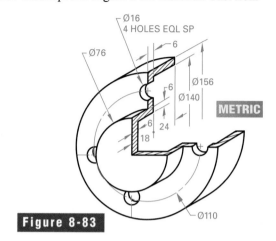

Figure 8-83

25. Rest. See Figure 8-82. Material: Cast aluminum.

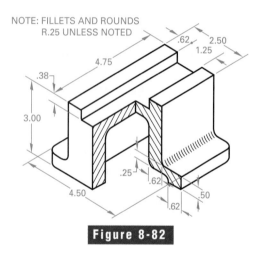

Figure 8-82

27. Flange. See Figure 8-84. Material: Cast aluminum.

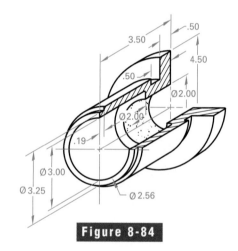

Figure 8-84

For problems 28 through 34, determine a suitable scale and sheet size, and follow the directions to complete each problem.

28. Draw three views of the yoke shown in Figure 8-85. Draw the front view in section. There are two pieces: the yoke and the bushing. Do not copy the illustration.

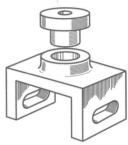

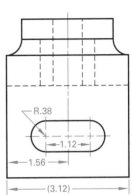

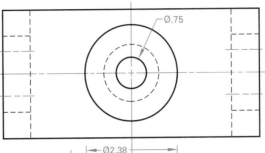

Figure 8-85

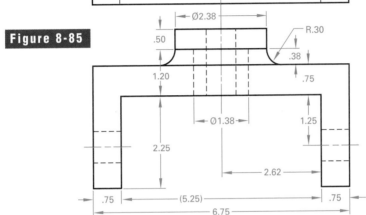

29. Draw three views of the swivel base shown in Figure 8-86. Draw the front view in section.

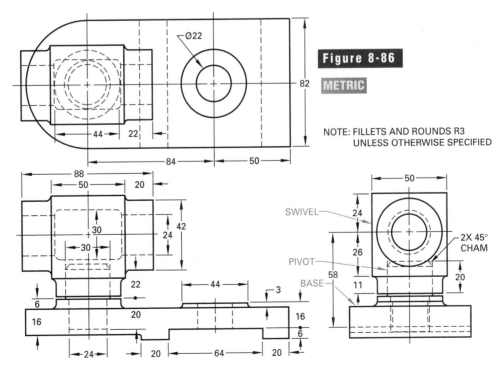

Figure 8-86

METRIC

NOTE: FILLETS AND ROUNDS R3
UNLESS OTHERWISE SPECIFIED

30. Draw three views of the swivel hanger shown in Figure 8-87. Draw the right-side view in section. There are two pieces: the hanger and the bearing.

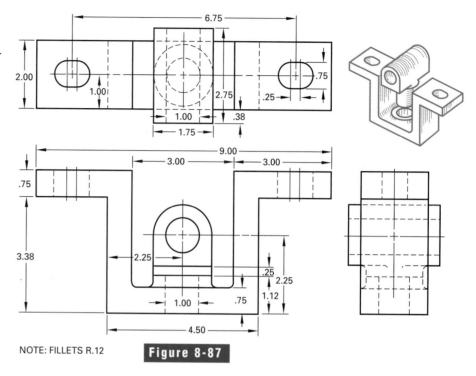

NOTE: FILLETS R.12   **Figure 8-87**

31. Draw three views of the thrust bearing shown in Figure 8-88. Draw the right-hand view in section. There are three parts: the shaft, the hub, and the base.

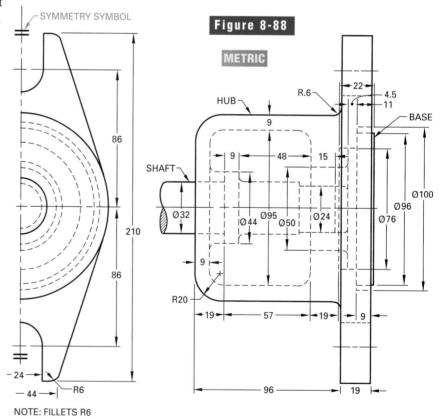

Figure 8-88

METRIC

NOTE: FILLETS R6

32. Make a three-view drawing of the jacket shown in Figure 8-89. Show the front and right-side views in section to improve clarity.

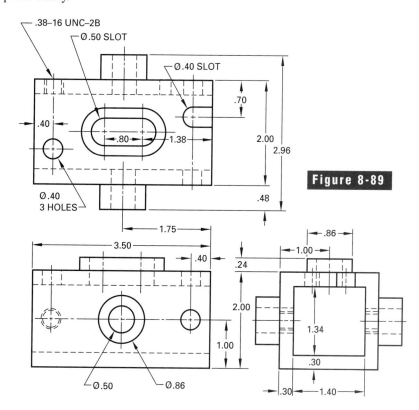

Figure 8-89

33. Make a three-view drawing of the guide block shown in Figure 8-90. Show the front and right-side views in section to improve clarity.

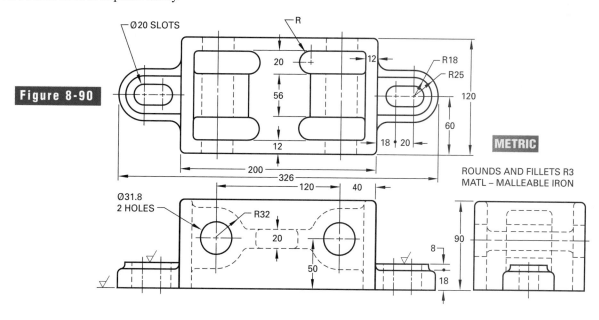

Figure 8-90

METRIC

ROUNDS AND FILLETS R3
MATL – MALLEABLE IRON

34. Prepare a working drawing of the bearing bracket shown in Figure 8-91. Show three views, one in section. Make all changes (shown in color) as specified by the design engineer. Refer to the appendix tables on limits and fits and dimension the precision holes accordingly. Add geometric dimensioning and tolerancing symbols to specify the following:

a. Datum A to be parallel to datum B to within .003″ at MMC.

b. Datum C to be perpendicular to datum A to within .002″ at MMC.

c. Datum C to be perpendicular to datum B to within .002″ at MMC.

d. Datum C to be flat to within .001″.

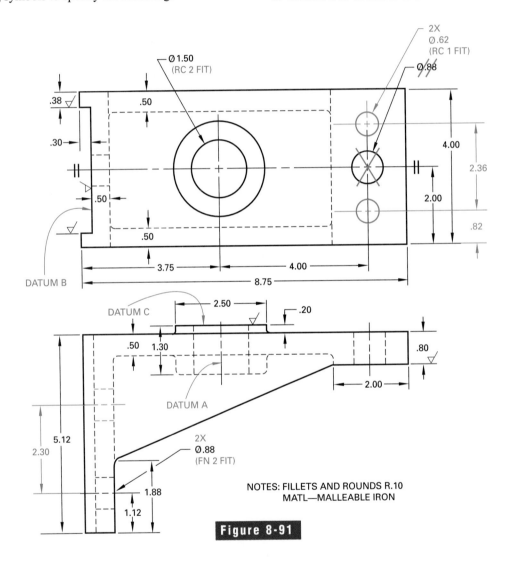

Figure 8-91

NOTES: FILLETS AND ROUNDS R.10
MATL—MALLEABLE IRON

Design Problems

Design problems have been prepared to challenge individual students or teams of students. In these problems, you are to apply skills learned mainly in this chapter but also in other chapters throughout the text. The problems are designed to be completed using board drafting, CAD, or a combination of the two. Be creative and have fun!

1. TEAMWORK Work as a team to design an organizer for your school lockers. It should include specially designed compartments for books, tablets, pencils and pens, and other items that the team decides are important. Material: optional. Include various types of sectional views. Begin with design sketches.

2. Design a golf caddy that attaches easily to the handle of a golf pullcart. It must hold three golf balls, ten tees, a divot repair tool, a scorecard, and a pencil. Each must be easily removed. Material: optional. Include various types of sectional views. Begin with design sketches.

3. Design a pencil and pen caddy for your board drafting or CAD station. It must hold a minimum of five pencils or pens. Material: optional. Include various types of sectional views. Begin with design sketches.

Auxiliary Views and Revolutions

OBJECTIVES

Upon completion of this chapter, you should be able to:

- Determine when a full auxiliary view is required.
- Determine when a partial auxiliary view is required.
- Develop a primary or secondary auxiliary view using board drafting or CAD techniques.
- Project and draw an auxiliary sectional view using board drafting or CAD techniques.
- Develop revolutions using board drafting or CAD techniques.
- Use the concept of revolutions to determine the true size and shape of an inclined surface.

In Chapter 6, you learned how to describe an object with views on the three normal planes of projection: the top, front, and right-side planes. With these planes, you can solve many graphic problems. However, to solve problems involving inclined or oblique surfaces, you must draw views on other planes of projection. This chapter explains how to draw these views on **reference planes** that are parallel to the inclined or oblique surfaces, as shown in Figure 9-1.

Most objects that are designed and manufactured do not conform to convenient rectangular or cylindrical shapes that require only the normal views. Many objects have surfaces that are slanted in one or more directions. You can show the true size of an inclined surface by either an **auxiliary** (additional) **view**, as shown in Figure 9-2A, or a **revolution**, or revolved view, as shown in Figure 9-2B and C. In a revolved view, the inclined surface is turned until it is parallel to one of the principal planes.

In the auxiliary view, it is as if the observer has changed position to look at the object from a new direction. Conversely, in the revolved view, it is as if the object has changed position. Both auxiliaries and revolutions help you visualize things better. They also work equally well in solving problems. This chapter explains how to create auxiliary views and revolutions to show the true size and shape of inclined surfaces.

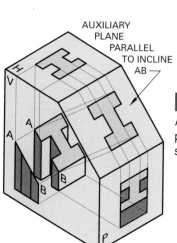

AUXILIARY PLANE PARALLEL TO INCLINE AB

Figure 9-1

An auxiliary view is parallel to the inclined surface it defines.

Success on the Job

Communication Skills

Communication is a means of transferring information through pictures or words. In drafting, we communicate primarily through technical drawings. However, it is also important for the drafter to develop and maintain good oral communication skills. The ability to exchange ideas with fellow drafters—in person or on the telephone—adds considerably to your success on the job.

Figure 9-2

You can determine the true size and shape of an object (A) by creating an auxiliary view or (B and C) by revolving the object.

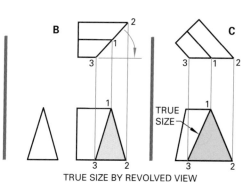

TRUE SIZE BY AUXILIARY VIEW

TRUE SIZE BY REVOLVED VIEW

Auxiliary Views

When an object has an inclined surface, none of the normal views shows the inclined part in its true size and shape. See Figure 9-3A. However, a view on a plane parallel to the inclined surface does show its true size and shape, as shown in Figure 9-3B. An auxiliary view is a projection on an auxiliary plane that is parallel to an inclined surface, as shown in Figure 9-4. It is a view that looks directly at the inclined surface in a direction perpendicular to it. Auxiliary views provide a clear image of the inclined surfaces on an object.

An anchor with a slanting surface is pictured in Figure 9-5A. The three normal views are shown in Figure 9-5B. Not only are these views hard to draw and understand, but they also show three circular features of the anchor as ellipses. In Figure 9-5C, the anchor is described completely in two views, one of which is an auxiliary view.

In Figure 9-6A, a simple inclined wedge block is shown in the normal views. In none of these views does the slanted surface (surface A) appear in its true shape. In the front view, all that shows is its edge line MN. In the side view, surface A appears, but it is foreshortened. Surface A is also foreshortened in the top view. Line MN also appears in both views, but looking shorter than its true length, which shows only in the front view. To show surface A in its true size and shape, you need to imagine an **auxiliary plane** parallel to it, as shown in Figure 9-6B. Figure 9-6C shows the auxiliary view revolved to align with the plane of the paper. By following this method, you can show the true size and shape of any inclined surface.

Primary Auxiliary Views

Auxiliary views are classified according to their origin and which of the three normal planes they are developed from. A **primary auxiliary view** is one that is developed directly from the normal views. There are three primary auxiliary views. Each is developed by projecting as a primary reference the

Figure 9-3

Compare the information given in the normal views (A) with that given in the auxiliary views (B).

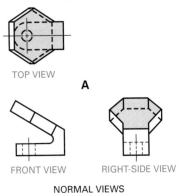

TOP VIEW

A

FRONT VIEW   RIGHT-SIDE VIEW

NORMAL VIEWS

AUXILIARY VIEW IS PREFERRED
TRUE SIZE AND TRUE SHAPE

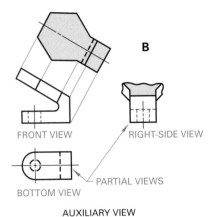

B

FRONT VIEW   RIGHT-SIDE VIEW

BOTTOM VIEW   PARTIAL VIEWS

AUXILIARY VIEW

Figure 9-4

Basic relationship of the auxiliary plane to the normal planes.

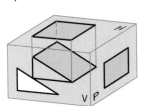

THREE NORMAL PLANES OF
PROJECTION HINGED TOGETHER
A

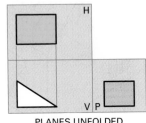

PLANES UNFOLDED
B

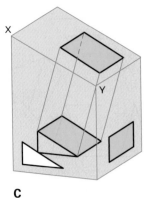

C   VERTICAL PLANE Y PROFILE PLANE

D

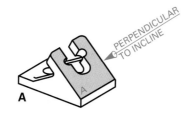

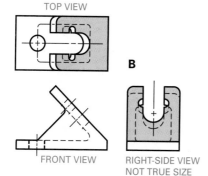

TOP VIEW

B

FRONT VIEW

RIGHT-SIDE VIEW
NOT TRUE SIZE

THREE CIRCULAR FEATURES
IN TRUE SIZE AND
TRUE SHAPE

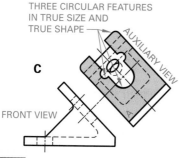

C

FRONT VIEW

Figure 9-5

The pictorial view (A) and the three-view drawing (B) are difficult to draw. The auxiliary view (C) is easier to draw and describes the inclined surface completely.

height, width, or depth obtained from a normal view. Figure 9-7 shows the three primary auxiliary views.

When an auxiliary view is hinged on the front view, the view is a **front auxiliary view**. The primary reference of the front auxiliary view is depth. An example is shown in Figure 9-7A. An auxiliary view that is hinged on the top view is a **top auxiliary view**, as shown in Figure 9-7B. The primary reference of the top auxiliary view is the height of the object. Finally, a view hinged on the right-side view is a **right-side auxiliary view**. Its primary reference is the width of the object, as shown in Figure 9-7C.

Partial Auxiliary Views

If you use break lines and centerlines properly, you can leave out complex curves while still describing the object completely, as shown in Figure 9-8. An auxiliary view in which some elements have been left out is known as a **partial auxiliary view**. In Figure 9-8, a half view is sufficient because the symmetrical object is presented in a way that is easy to understand.

Auxiliary Sections

Sometimes it is useful to show a sectional view of an object. When the cutting plane is not parallel to any of the normal views, the section is known as an **auxiliary section**. In Figure 9-9, the auxiliary section was located by using the cutting plane represented in **edge view** by line AA. See Chapter 8 for more information about sections.

Figure 9-6

Basic relationship of the auxiliary view to the three-view drawing.

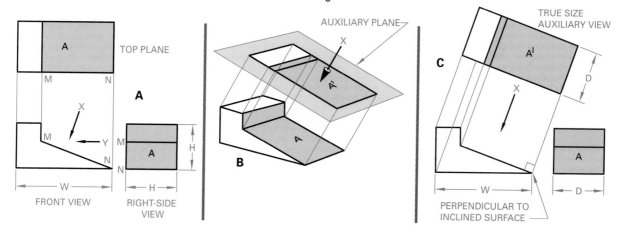

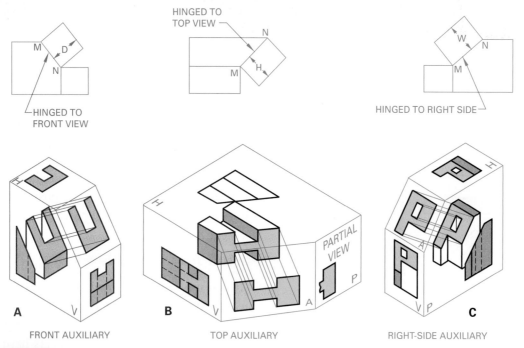

HINGED TO
TOP VIEW

M   D
N

HINGED TO
FRONT VIEW

N
M   H

W   N
M

HINGED TO RIGHT SIDE

H
V
A
FRONT AUXILIARY

H
PARTIAL
VIEW
V   A
P
B
TOP AUXILIARY

H
A
V   P
C
RIGHT-SIDE AUXILIARY

Figure 9-7

Three kinds of auxiliary views, showing how the auxiliary plane is hinged.

Secondary Auxiliary Views

A view projected from a primary auxiliary view is called a **secondary auxiliary view**. Secondary auxiliary views are used to find the true size and shape of a surface that lies along an oblique plane. An **oblique plane** is one that is inclined to all three of the normal planes.

In Figure 9-10, surface 1-2-3-4 is inclined to the three normal planes. In Figure 9-10A, a first auxiliary view has been drawn. It is on a plane perpendi-

cular to the inclined surface. Note that, in this view, points 1, 2, 3, and 4 appear as a line or edge view of the plane. In Figure 9-10B, a secondary auxiliary view has been drawn from the first. It is on a plane parallel to surface 1-2-3-4. This view shows the true shape of the surface.

Figure 9-11 shows another example. In this case, an octahedron (eight triangles making a regular solid) is shown in three views. Triangle surface 0-1-2 is inclined to all three. In Figure 9-11A, a first aux-

Figure 9-8

Partial auxiliary views provide a practical method for explaining details.

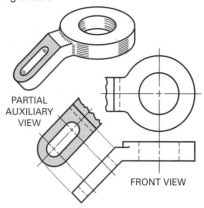

PARTIAL
AUXILIARY
VIEW

FRONT VIEW

Figure 9-9

Drawing an auxiliary section is another way to explain details clearly.

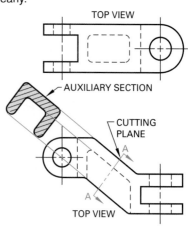

TOP VIEW

AUXILIARY SECTION

CUTTING
PLANE

A

A

TOP VIEW

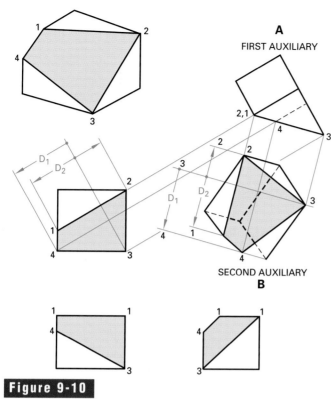

Figure 9-10

A secondary auxiliary view assists in finding the true shape of surface 1-2-3-4.

iliary view has been drawn. It is on a plane perpendicular to triangle surface 0-1-2. Note that line 1-2 in the top view appears as point 1-2 in this auxiliary view and that the triangle now appears as an edge line 0'-1-2. In Figure 9-11B, a secondary auxiliary view has been drawn. It is on a plane parallel to the edge view of triangle surface 0'-1-2 in the first auxiliary view. This secondary auxiliary view shows the true shape of triangle 0-1-2.

Revolutions

When the true size and shape of an inclined surface do not show on a drawing, one solution, as we have seen, is to make an auxiliary view. Remember, in auxiliary views, you set up new reference planes to look at objects from new directions. Another solution is to revolve the object. The resulting drawing is called a revolution. In a revolution, you use the normal reference planes while imagining that the object has been revolved to an angle that places its primary features parallel to the reference planes.

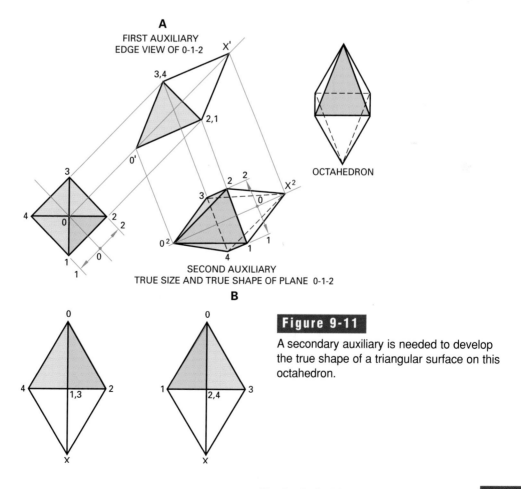

Figure 9-11

A secondary auxiliary is needed to develop the true shape of a triangular surface on this octahedron.

The Axis of Revolution

An easy way to picture an object being revolved is to imagine that a shaft or axis has been passed through it. This imaginary axis is perpendicular to one of the three principal planes. In Figure 9-12, the three principal planes are shown with an axis passing through each one and through the object beyond. When the object is revolved about one of these axes, the axis is called the **axis of revolution**.

An object can be revolved to the right (clockwise) or to the left (counterclockwise) about an axis perpendicular to either the vertical or the horizontal plane. The object can be revolved backward (clockwise) or forward (counterclockwise) about an axis perpendicular to the profile plane.

The Rule of Revolution

The rule of revolution has two parts. See Figure 9-13.
1. *The view that is perpendicular to the axis of revolution stays the same except in position.* This is true because the axis is perpendicular to the plane on which it is projected.
2. *Distances parallel to the axis of revolution stay the same.* This is true because they are parallel to the plane or planes on which they are projected.

Single Revolution

As you have seen, an axis of revolution can be perpendicular to the vertical, horizontal, or profile plane. This section describes the characteristics of each type of revolution.

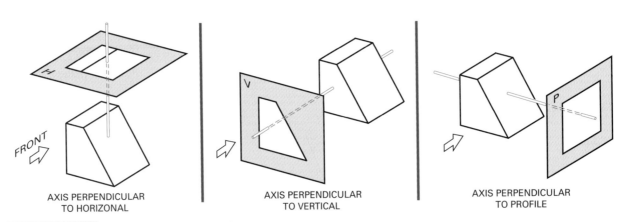

AXIS PERPENDICULAR TO HORIZONAL

AXIS PERPENDICULAR TO VERTICAL

AXIS PERPENDICULAR TO PROFILE

Figure 9-12

Three positions for the axis of revolution. The axis is perpendicular to the principal planes.

Figure 9-13

The rule of revolution. Note that the H shape in the front view has changed only position, not shape.

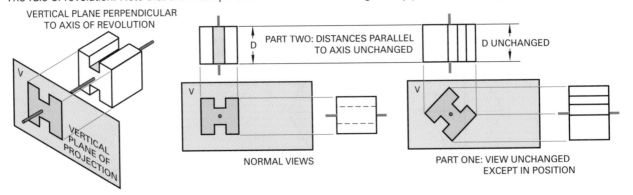

VERTICAL PLANE PERPENDICULAR TO AXIS OF REVOLUTION

VERTICAL PLANE OF PROJECTION

PART TWO: DISTANCES PARALLEL TO AXIS UNCHANGED

D UNCHANGED

NORMAL VIEWS

PART ONE: VIEW UNCHANGED EXCEPT IN POSITION

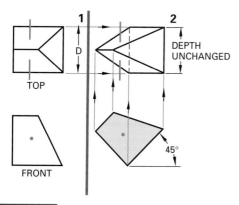

Figure 9-14

Revolution about an axis perpendicular to the vertical plane.

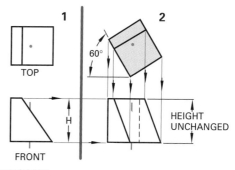

Figure 9-15

Revolution about an axis perpendicular to the horizontal plane.

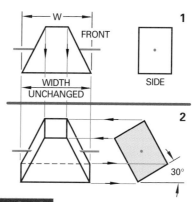

Figure 9-16

Revolution about an axis perpendicular to the profile plane.

Axis Perpendicular to the Vertical Plane

In Figure 9-14, the usual front and top views of an object are shown on the left. On the right, similar views of the same object are shown after the object has been revolved 45° counterclockwise about an axis perpendicular to the vertical plane. Notice that the front view is the same in size and shape, except that it has a new position. The new top view has been made by projecting up from the new front view and across from the old top view. Note that the depth remains the same from one top view to the other.

Axis Perpendicular to the Horizontal Plane

In Figure 9-15, an object is shown on the left in the usual top and front views. The views on the right were drawn after the object had been rotated 60° clockwise about an axis perpendicular to the horizontal plane. The new top view is the same in size and shape as the old top view. The new front view has been made by projecting down from the new top view and across from the old front view. Note that the height remains the same from the original front view to the revolved front view.

Axis Perpendicular to the Profile Plane

At the top of Figure 9-16, another object is shown in the usual front and side views. Below it, the same views show the object after it has been revolved counterclockwise 30° about an axis perpendicular to the profile plane. The new front view has been made

by projecting across from the new side view and down from the old front view. Note that the width remains the same from one front view to the other.

Partial Revolved Views

In a working drawing, you can show an inclined surface by drawing a full or partial view with the object in a revolved position. In Figure 9-17A, the top view shows the angle of a V-shaped part. In the front view, the part is revolved to show its true shape.

In Figure 9-17B, the front view shows the angles at which surfaces of a part are inclined. The inclined surfaces are then revolved in the front view. Next, their dimensions are transferred to the top view. There, the surfaces appear in true size and shape.

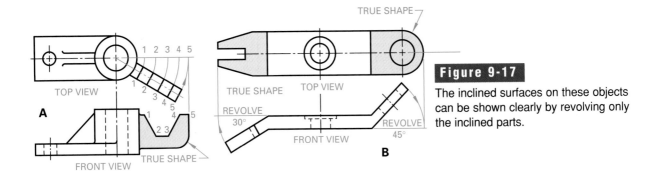

Figure 9-17

The inclined surfaces on these objects can be shown clearly by revolving only the inclined parts.

Successive Revolutions

After an object has been revolved around an axis perpendicular to one plane, it can be revolved again about an axis perpendicular to another plane. This process, known as **successive revolutions**, is shown in Figure 9-18. In part A of the figure, an object is shown in the normal views. In part B, the object has been revolved 30° clockwise about an axis perpendicular to the horizontal plane. In part C, the front view from part B has been revolved 45° clockwise about an axis perpendicular to the vertical plane. In part D, the side view from part C has been revolved about an axis perpendicular to the profile plane until line 3-4 appears as a horizontal line. In all, three successive revolutions occurred, one in each of the three principal planes of projection.

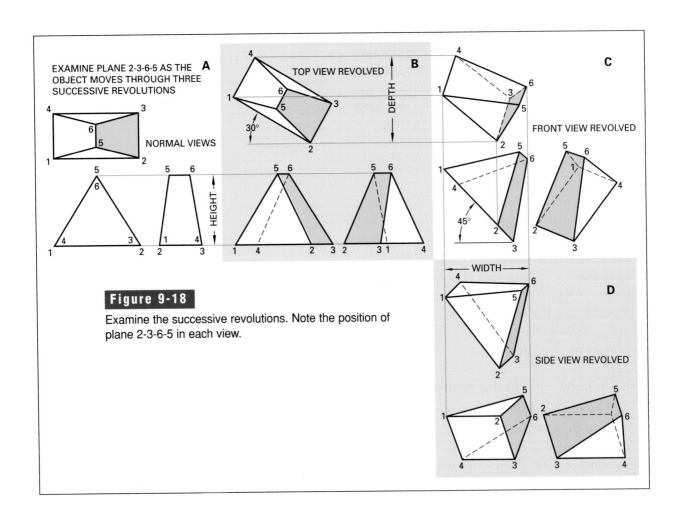

Figure 9-18

Examine the successive revolutions. Note the position of plane 2-3-6-5 in each view.

Other Applications

Revolutions are useful for more than just determining the size and shape of inclined surfaces. For example, revolution can be used to show several possible positions of a movable part. Figure 9-19 shows a front-end loader with various movable parts. In the top view, the loading bucket is revolved to show the maximum angles to which it can be turned. In the front view, revolutions and partial views are given to show lift positions.

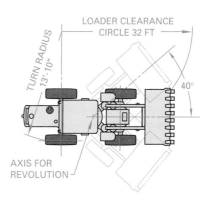

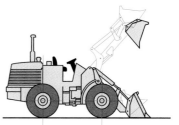

Figure 9-19
The profile of a front-end loader shows several positions of the loading bucket. The plan view shows horizontal rotation.

Board Drafting Techniques

In the following sections, you will construct primary and secondary auxiliary views. Be sure to keep the principles discussed earlier in this chapter in mind as you work through these constructions.

Constructing a Primary Auxiliary View

To construct any primary auxiliary view, use the following steps, as illustrated in Figure 9-20. The method shown in Figure 9-20 is for a front auxiliary view.

Figure 9-20
Steps to construct an auxiliary view.

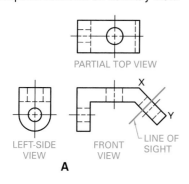

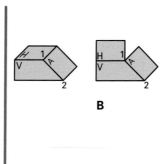

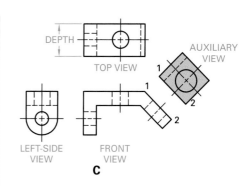

1. Examine and draw the normal views given for an inclined surface, as shown in Figure 9-20A.
2. Find the view that shows the edge view given for an inclined surface. The inclined plane appears as a line in this view. The plane associated with this view is the reference plane, from which the auxiliary plane will be developed.
3. In this view, draw a light construction line at right angles to the inclined surface, as shown in magenta in Figure 9-20A. This is the **line of sight**.
4. Think of the auxiliary plane as being attached by hinges to the view from which it is developed, as shown in Figure 9-20B.
5. From all important points on the reference view, draw projection lines at right angles to the inclined surface (parallel to the line of sight). In Figure 9-20, the important points are labeled 1 and 2.
6. Draw a reference line parallel to the edge view of the inclined surface and at a convenient distance from it. The reference line is shown in magenta in Figure 9-20C.
7. Transfer the depth dimension to the reference line.
8. Project the important points and connect them in sequence to form the auxiliary view, as shown in Figure 9-20C. If you labeled the points for reference, do not leave the labels on the final drawing.

Symmetrical Objects

Figure 9-21 shows how to make a primary auxiliary view of a symmetrical object. In Figure 9-21A, the object is shown in a pictorial view. Follow these steps:

1. Use a center plane as a reference plane, as shown in Figure 9-21B. This is **center-plane construction**.
2. Find the edge view of the inclined plane. In Figure 9-21B, the edge view of this plane appears as a centerline, line XY, on the top view.
3. Label the points on the top view for reference.
4. Transfer these points to the edge view of the inclined surface on the front view, as shown in Figure 9-21B.
5. Parallel to this edge view and at a convenient distance from it, draw the line X'Y', as shown in Figure 9-21C.
6. In the top view, find the distances from the numbered points to the centerline. These are the depth measurement. Transfer them onto the corresponding construction lines you just drew, measuring them off on either side of line X'Y', as shown in Figure 9-21D. The result will be a set of points on the construction lines.
7. Connect and number the points on the construction lines, as shown in Figure 9-21E, to finish the front auxiliary view of the inclined surface.
8. If desired, project the rest of the object from the center reference plane.

Figure 9-21

Steps to draw an auxiliary view using the center-plane reference method.

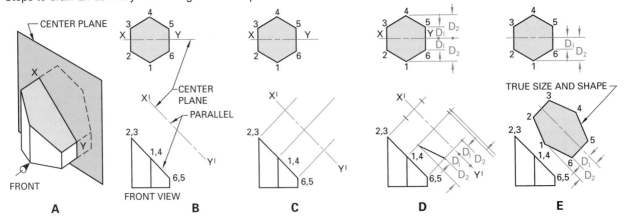

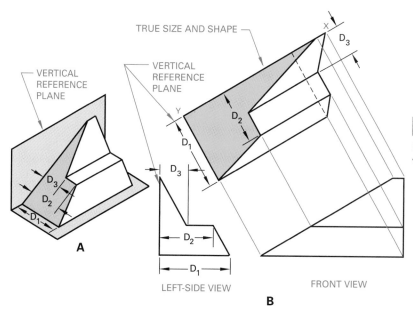

Figure 9-22

Drawing a front auxiliary view using a vertical reference plane.

Using a Vertical Reference Plane

Figure 9-22 shows how to draw a front auxiliary view of a nonsymmetrical object by using a vertical reference plane. Follow these steps:

1. Place the object on reference planes, as shown in Figure 9-22A. These planes are located strictly for convenience in taking reference measurements. The vertical plane can be in front or in back of the object. In this case, it is shown in back.
2. Construct the view as described for symmetrical objects, except lay off the depth measurements D_1, D_2, and D_3 in front of the vertical plane. Figure 9-22B shows the entire object projected onto the front auxiliary plane.

Using a Horizontal Reference Plane

Figure 9-23 shows how to draw a top auxiliary reference plane. The object is a molding cut at a 30° angle. Follow these steps:

1. Imagine a reference plane XY under the molding, as shown in Figure 9-23A.
2. Find points 1 through 6 in the top and left-side views. In the top view, find the edge line of the slanted surface.
3. Draw reference line X'Y' parallel to it and at a convenient distance away.
4. From every point in the top view, project a line out to the line X'Y' and at right angles to it.

Figure 9-23

Drawing (A) a top auxiliary view with a horizontal reference plane; and (B) a front auxiliary view using a vertical reference plane.

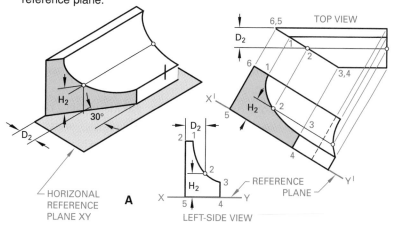

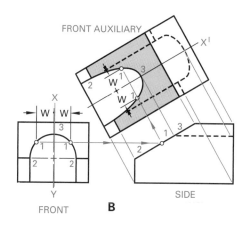

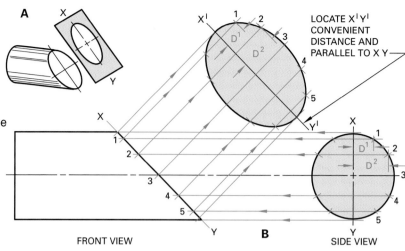

Figure 9-24

Drawing a curve (in this case, an ellipse) in an auxiliary view about the centerline of the cut surface of a cylinder.

FRONT VIEW

SIDE VIEW

LOCATE X'Y' CONVENIENT DISTANCE AND PARALLEL TO X Y

5. In the side view, find height measurements for the various numbered points by measuring up from XY. Lay off these same measurements up from X'Y' along the lines leading to the corresponding points in the top view. Locate more points on the curve as needed in order to draw it accurately. (See the following section, "Curves on Auxiliary Views.") The result is a top auxiliary view with its base on line X'Y', as shown in Figure 9-23A.

Figure 9-23B shows the same process. However, in this illustration, a vertical reference plane is used.

Curves on Auxiliary Views

To draw an auxiliary view of a curved line, locate a number of points along that line. Figure 9-24 shows how to make an auxiliary view of the curved cut surface of a cylinder. The cylinder is shown in a horizontal position. It has been cut at an angle, so the true shape of the slanting cut surface is an ellipse.

This auxiliary view is a front auxiliary view with the depth as its primary reference. To draw it, follow these steps:

1. Begin by locating the vertical centerline XY in the side view, as shown in Figure 9-24A. This line represents the edge of a center reference plane.
2. Locate a number of points along the rim of the side view, as shown in Figure 9-24B. The more points you locate, the more accurate your curve will be.

3. Project lines from these points over to the edge view of the cut surface in the front view.
4. Parallel to this edge view and at a convenient distance from it, draw the new centerline X'Y'.
5. From the points you have located on the edge view, project lines out to line X'Y' and perpendicular to it. Continue these lines beyond X'Y'.
6. Find the depth measurements in the side view by measuring off the distances D_1, D_2, etc., between the centerline XY and the points located along the rim. Take these distances and measure them off on either side of X'Y'.
7. Draw a smooth curve through the points marked to form the ellipse, as shown.

Constructing a Secondary Auxiliary View

Before you begin to draw a secondary auxiliary view, you must first clearly understand the development of a primary auxiliary view. As mentioned earlier in the chapter, a secondary auxiliary view is projected from a primary auxiliary view. Notice in Figure 9-25 that only partial front and top views are drawn initially. The auxiliary views will be used to complete these views.

To construct a secondary auxiliary view of the part shown in Figure 9-25A, follow these steps:

1. Draw partial front and top views as shown in Figure 9-25B. Be sure to allow sufficient space to complete these views later.

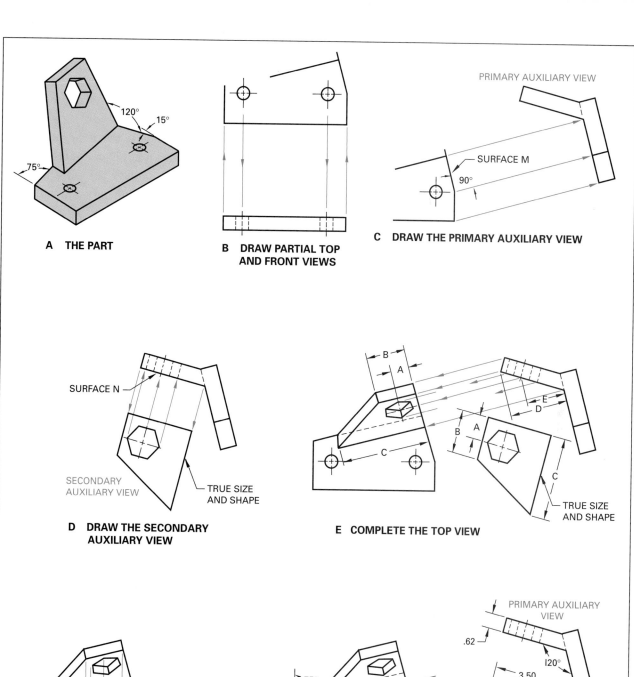

A THE PART

120°

15°

75°

**B DRAW PARTIAL TOP
AND FRONT VIEWS**

PRIMARY AUXILIARY VIEW

SURFACE M

90°

C DRAW THE PRIMARY AUXILIARY VIEW

SURFACE N

SECONDARY
AUXILIARY VIEW

TRUE SIZE
AND SHAPE

**D DRAW THE SECONDARY
AUXILIARY VIEW**

B
A

E
D

B
A

C

C

TRUE SIZE
AND SHAPE

E COMPLETE THE TOP VIEW

D

E

F COMPLETE THE FRONT VIEW

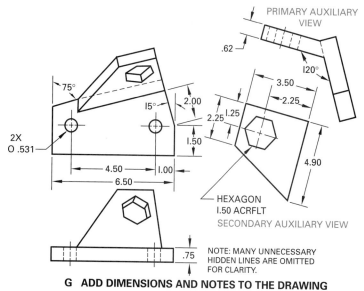

PRIMARY AUXILIARY
VIEW

.62

120°

75°

15°

2.00

2X
Ø .531

3.50

2.25

2.25   1.25

1.50

4.50   1.00

6.50

4.90

HEXAGON
1.50 ACRFLT

SECONDARY AUXILIARY VIEW

.75

NOTE: MANY UNNECESSARY
HIDDEN LINES ARE OMITTED
FOR CLARITY.

G ADD DIMENSIONS AND NOTES TO THE DRAWING

Figure 9-25

Steps in drawing a secondary auxiliary view.

2. Project lines perpendicular to the inclined line in the top view and draw the primary auxiliary view, as shown in Figure 9-25C.

3. Project lines perpendicular to the auxiliary surface of the primary auxiliary view and draw the second auxiliary view, as shown in Figure 9-25D. Notice that the top edge of the secondary auxiliary view is parallel to surface N.

4. Complete the top view by projecting lines from the primary auxiliary view, as shown in Figure 9-25E.

5. Complete the front view by projecting lines from the top view and distances, such as D and E, from the primary auxiliary view. See Figure 9-25F.

6. Darken all lines and add dimensions and notes to complete the drawing, as shown in Figure 9-25G.

Developing Revolutions

As mentioned earlier in this chapter, views can be revolved about axes perpendicular to any of the three primary planes. The practice drawings in this section provide experience in developing revolutions.

Axis Perpendicular to the Vertical Plane

Figure 9-26 shows how to draw a primary revolution perpendicular to the vertical plane. Follow these steps:

1. Imagine an axis AX passed horizontally through a truncated right octagonal prism, as shown in Figure 9-26A. In the front view, it shows as a point. In the top view, and later in the side view, it shows as a line.

2. Revolve the prism clockwise about the axis into a new position, as shown in Figure 9-26B. You can see that the new front view has the same size and shape as the old. Only its position has changed. However, the side view now shows the true size and shape of the truncated surface.

3. Create the side view by projecting across from the new front view and transferring the depth from the top view.

Axis Perpendicular to the Horizontal Plane

As you may recall, you can rotate objects in either a clockwise or a counterclockwise direction. Figure 9-27 shows how to draw an object that is revolved clockwise about an imaginary vertical axis AX. Follow these steps:

1. Draw the three normal views of the object, as shown in Figure 9-27A.

2. Revolve the top view 30° clockwise about the axis AX.

3. Since this is a vertical axis, revolution does not change the height of the object. Therefore, you can project points from the old vertical plane (shown in Figure 9-27A) to make the new one shown in Figure 9-27B.

4. Make the new side view from the front and top views in the usual way.

Figure 9-28 shows how to draw an object through 45°. Notice that the procedure is the same regardless of the direction or degree of rotation.

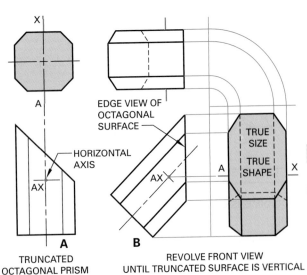

Figure 9-26

Revolution about an axis perpendicular to the vertical plane.

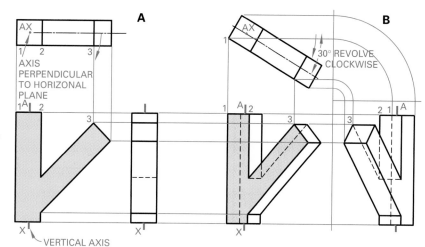

Figure 9-27

Clockwise revolution about an axis perpendicular to the horizontal plane.

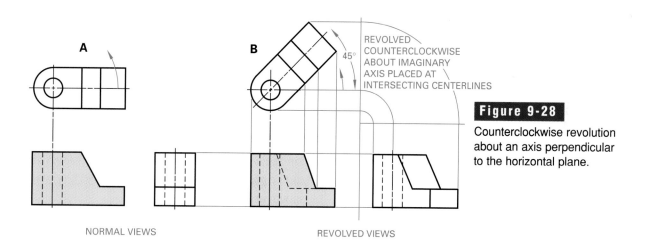

Figure 9-28

Counterclockwise revolution about an axis perpendicular to the horizontal plane.

NORMAL VIEWS

REVOLVED VIEWS

True Shape and Size

A surface shows its true shape when it is parallel to a plane. Since an auxiliary view shows the true size and shape of an inclined surface, it can also be used to find the true length of a line.

True Shape of an Oblique Plane

Figure 9-29 shows how successive revolutions can be used to find the true shape of a surface. Figure 9-29D shows an object on which surface 1-2-3-4 is an oblique plane. It is oblique because it is inclined to all three of the normal views. Follow these steps to develop the true shape of this surface.

1. Draw the object in its normal position, as shown in Figure 9-29A.
2. Revolve it about an axis perpendicular to the horizontal plane until surface 1-2-3-4 is perpendicular to the vertical plane. Now, in the front

view, all you see of this surface is its edge line, as shown in Figure 9-29B.
3. Revolve the object about an axis perpendicular to the horizontal plane until surface 1-2-3-4 is perpendicular to the vertical plane. Now, in the front view, all you see of this surface is its edge line. Surface 1-2-3-4 is now parallel to the profile plane, which shows its true shape. See Figure 9-29C.

True Length of a Line

Both auxiliary views and revolutions can be used to find the true length of a line. In Figure 9-30A, the true length of line OA is not apparent in the top, front, or right-side view because it is inclined to all three normal planes. The auxiliary plane shown in Figure 9-30B does show the true length (TL), because the auxiliary plane is parallel to surface OAB.

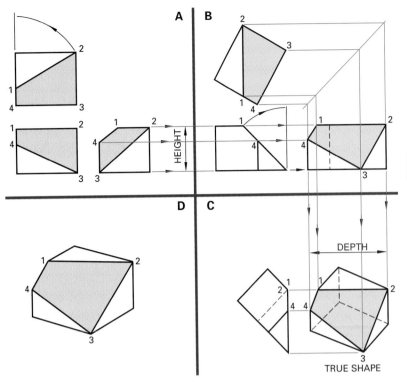

Figure 9-29

To find the true size of oblique plane 1-2-3-4, you must perform successive revolutions.

Figure 9-30C shows another way to show the true length of OA. Revolve the object about an axis perpendicular to the vertical plane until surface OAB is parallel to the profile plane. The side view then shows the true size of OAB and also the true length of OA. A shorter method of showing the true length of OA is to revolve only the surface OAB, as shown in Figure 9-30D.

Figure 9-30E shows the object revolved in the top view until line OA is horizontal in that view. The front view now shows OA in its true length because this line is now parallel to the vertical plane.

In Figure 9-30F and G, still another method is shown. In this case, instead of the whole

object being revolved, just line OA is turned in the top view until it is horizontal at OA′. Point A′ then can be projected to the front view. There, OA′ will be shown at its true length.

You can revolve a line in any view to make it parallel to any one of the three principal planes. Projecting the line on the plane to which it is parallel shows its true length. In Figure 9-30H, the line has been revolved parallel to the horizontal plane. The true length then shows in the top view.

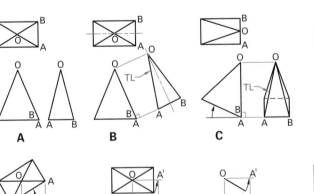

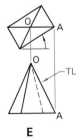
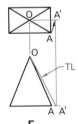
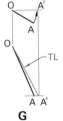
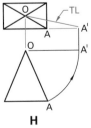

Figure 9-30

Typical true-length problems examined and solved.

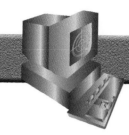

CAD Techniques

The following discussion applies specifically to computer-aided drafting (CAD). Although CAD techniques may differ in their approach from board drafting techniques, it is important to keep the basic drafting principles covered in the first portion of this chapter in mind. Unless specifically stated otherwise, all drafting principles apply equally to board and CAD drawings.

CAD Auxiliary Views

The process of creating an auxiliary view in a two-dimensional CAD drawing is similar to that used in board drafting. However, auxiliary views can usually be drawn in less time because the CAD software provides commands that automate many of the more time-consuming tasks.

Figure 9-31 shows the procedure for creating an auxiliary view using AutoCAD. Follow these steps:

1. Create the front and side views and a partial top view, as shown in Figure 9-31A. Do not dimension the views.
2. Create a construction line (XLINE command) perpendicular to the line that represents the inclined plane at the lower end of the inclined line in the front view. To do this, pick the two endpoints of the short end line, as shown in Figure 9-31B.
3. Copy the construction line to each important point in the front view. See Figure 9-31C.
4. Copy the inclined line to another location on the construction lines, as shown in Figure 9-31D. Use the Nearest object snap to ensure that the endpoint of the inclined line is exactly on the lowest construction line.
5. Offset the line you created in step 4 to the right by the depth dimension, .76, as shown in Figure 9-31D. This defines the depth of the object in the auxiliary view.

6. Trim the construction lines to form the other boundaries of the auxiliary view. Use the Layer Control above the drawing area to move the lines to their appropriate layers: Hidden, Centerline, etc. See Figure 9-31E.
7. Add the other centerline, the hole, and other details based on the dimensions given in Figure 9-31A. The finished drawing should look like the one in Figure 9-31F.

CAD Revolutions

The ability of CAD software to rotate objects allows CAD operators to revolve views quickly and easily on screen. AutoCAD's ROTATE command allows you to revolve any object or group of objects in two or three dimensions.

Revolving Two-Dimensional Views

Figure 9-32 shows the procedure for revolving a two-dimensional view using AutoCAD. Follow these steps:

1. Draw the front and top views, as shown in Figure 9-32A. Do not dimension.
2. With Ortho on, copy the entire front view to the right, as shown in Figure 9-32B.
3. Determine the angle to which you should revolve the view to place the inclined line in a vertical position. See the Tech Math on page 336 for more information about determining the angle.
4. Enter the ROTATE command, use a selection window to select all of the objects in the view you want to revolve, and press Enter. Select the base point by snapping to the right side of the bottom object line, as shown in Figure 9-32C. Then enter the rotation angle you determined in step 3.

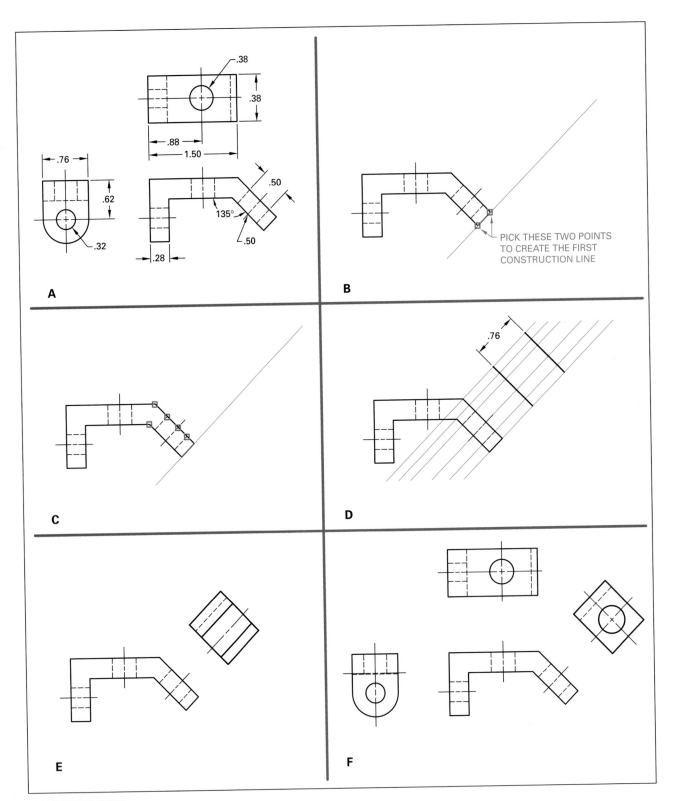

Figure 9-31

Steps to draw an auxiliary view using AutoCAD.

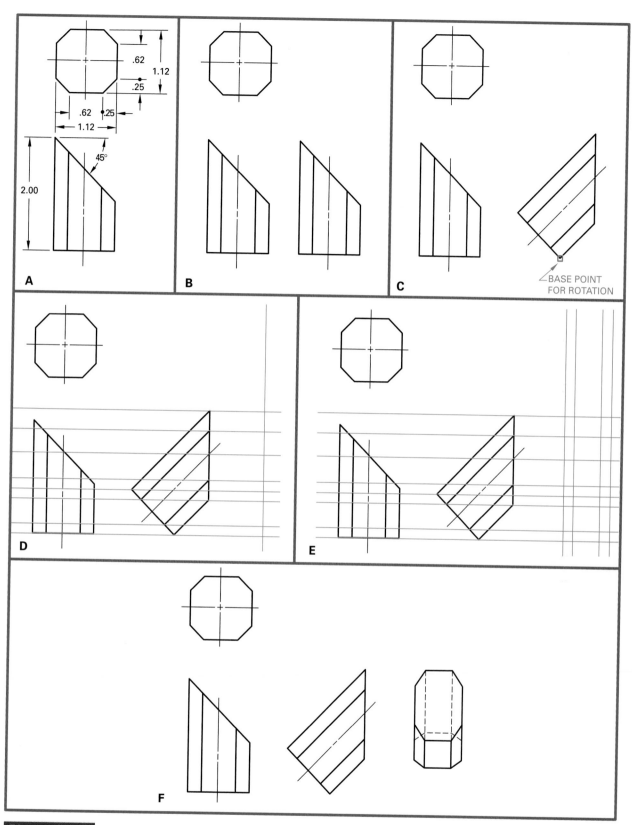

Figure 9-32

Steps to draw a revolution using AutoCAD.

Angles of Rotation

In step 4 of the revolution procedure (page 333), you should be able to determine that the object should be revolved, or rotated, 45° clockwise. By default, AutoCAD rotates objects counter-clockwise.

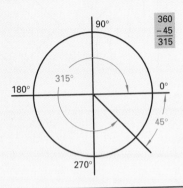

Recall that angles in AutoCAD can be specified up to 360°, as shown in the figure. Therefore, you can determine the correct angle to enter by subtracting 45 from 360 to get 315. If you enter a rotation value of 315°, AutoCAD will rotate the view correctly.

There is an easier way, however. To specify an angle of rotation in a clockwise direction, simply change the sign of the angle. Instead of entering 45°, enter −45°. The effect is the same as entering a positive angle of 315°.

5. Draw horizontal construction lines (XLINE command) through the key points on the revolved view. Add one vertical construction line to serve as the basis for a new front view, as shown in Figure 9-32D.

6. Use the OFFSET command to transfer the depths and distances from the top view, as shown in Figure 9-32E. Refer to Figure 9-32A as necessary for offset distances.

7. Trim the lines and change their layers to finish the new front view showing the true size and shape of the inclined surface. The finished drawing should look like the one in Figure 9-32F.

►CADTIP

Center-Plane Construction

One way to use the vertical construction line you created in step 5 is to use it as a centerline. Notice in the top view that the object is vertically and horizontally symmetrical. Therefore, you can use the center plane, represented by a centerline, as the basis for constructing the revolved view. Simply offset the centerline to the right and left at the appropriate distances to create the object lines. This method is known as **center-plane construction**. However, in CAD, it is usually easier to use the construction line as the left edge of the object and offset the other distances from it.

Revolving Solid Models

In practice, many companies create solid models instead of orthographic drawings of parts or objects that contain oblique surfaces. These models can be revolved using AutoCAD's ROTATE or ROTATE3D command to show the true size and shape of an oblique surface. The advantage of using solid models is that they do not require you to draw extra views. You can revolve the model to show all of the necessary views, including normal views and auxiliary views that show the true size and shape of an oblique surface.

Chapter Summary

■ Auxiliary views are used to solve problems involving inclined surfaces.

■ An auxiliary view is a projection on an auxiliary plane that is parallel to an inclined surface on an object.

■ Primary auxiliary views are projected directly from the normal views.

■ Curved lines are developed on an auxiliary view by projecting reference points onto the reference plane and then connecting the points.

■ A partial auxiliary view, such as a half view, is as easily understood as a full view and can be developed in less time.

■ Secondary auxiliary views are projected from primary auxiliary views.

■ An oblique plane is one that is inclined to all three of the normal planes.

■ Auxiliary views are developed by revolving the plane of projection. Revolutions are developed by revolving the object around an axis of revolution.

Review Questions

1. Under what circumstances do you need to draw auxiliary views of an object?

2. How should you place an auxiliary plane in relation to the inclined surface it describes?

3. What is the difference between a primary auxiliary view and a secondary auxiliary view?

4. Name the three primary auxiliary views.

5. What is center-plane construction?

6. What is a partial auxiliary view? Why is it sometimes preferred over a full auxiliary view in board drafting?

7. What is the basic reason for revolving a view of an object?

8. What is an axis of revolution?

9. What are successive revolutions? When are they needed?

10. Name the three basic single revolutions.

11. List the two parts of the rule of revolution.

12. What two AutoCAD commands are most useful in placing auxiliary views and transferring distances from other views?

13. Briefly describe how to rotate a two-dimensional view using AutoCAD.

14. Explain why some companies use solid models instead of orthographic drawings to represent objects that have inclined or oblique surfaces.

Problems

CHAPTER 9

Drafting Problems

The problems in this chapter can be performed using board drafting or CAD techniques. The problems are presented in order of difficulty, from least to most difficult.

1. For each object in Figure 9-33, only the top view is given. Draw the top and front views and either complete the auxiliary view or just the inclined surface, as directed by the instructor. Figure 9-34 has been developed as an example. The angle X in Figure 9-34 may be 45° or 60°, as assigned. The total height of the front view is 3.75″ for Figure 9-33A through 9-33N.

Figure 9-33

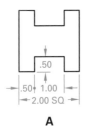

A

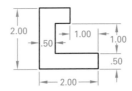

B

C

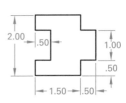

D

E

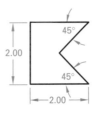

F

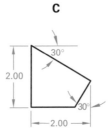

G

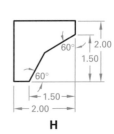

H

I

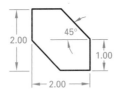

J

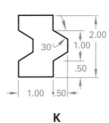

K

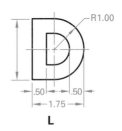

L

M

N

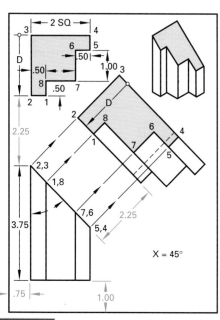

Example for problem 1.

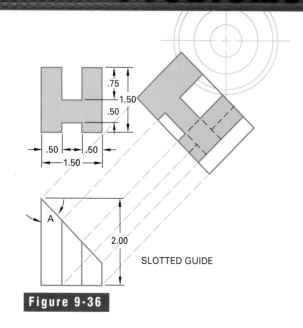

SLOTTED GUIDE

Figure 9-36

4. Draw the front, top, and side views and the front auxiliary projection of the object shown in Figure 9-37. Then change the angle of the inclined surface to 45° and redraw the problem.

2. For each object in Figure 9-35, only the top view is given. Draw the top and front views and either complete the auxiliary view or just the inclined surface, as directed by the instructor. The height of each object is 95 mm.

For problems 3 through 6, draw the views according to the instructions given for each problem.

3. Draw the front, top, and side views and the front auxiliary projection of the object shown in Figure 9-36. Then change the angle of the inclined surface to 30° and redraw the problem.

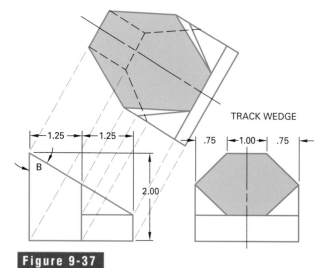

TRACK WEDGE

Figure 9-37

Figure 9-35 METRIC

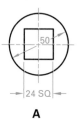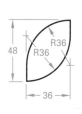

A   B   C   D   E

5. Draw the front, top, and complete auxiliary views, as shown in Figure 9-38. Then change the angle of incline to 30°. Is there any difference in the solution?

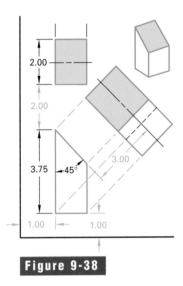

Figure 9-38

6. Draw the front, top, and complete auxiliary views, as shown in Figure 9-39. Then change the angle of incline to 45°. Is there any difference in the solution?

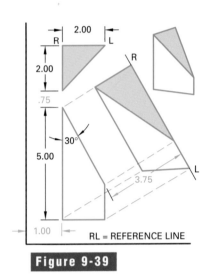

RL = REFERENCE LINE

Figure 9-39

7. Draw the front, top, and auxiliary views for each object shown in Figure 9-40.

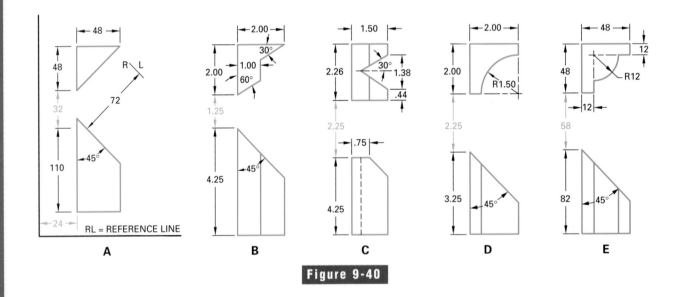

A   B   C   D   E

Figure 9-40

For problems 8 through 12, draw the views according to the instructions given for each problem.

8. Develop the three views of the object shown in Figure 9-41. Draw angle A at 45° and develop the top auxiliary projection.

9. Determine the views necessary to complete the front auxiliary view of the object shown in Figure 9-42. Develop views with angle A = 60°. Alternate problems may be assigned with angle A at 45° or 75°.

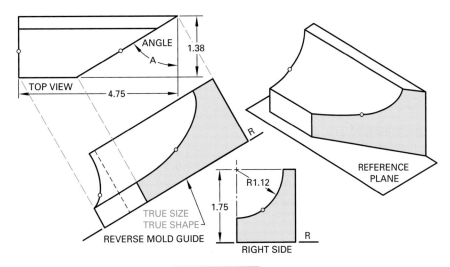

Figure 9-41

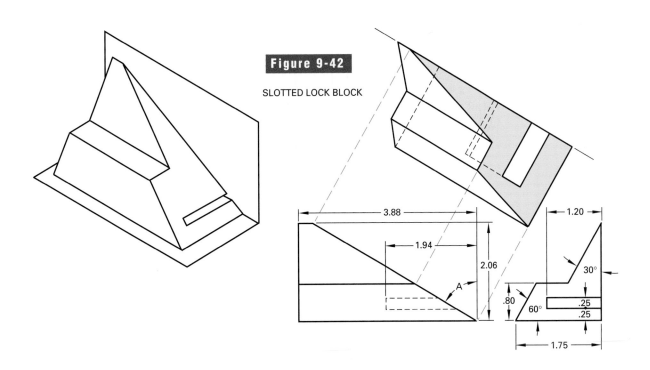

Figure 9-42

SLOTTED LOCK BLOCK

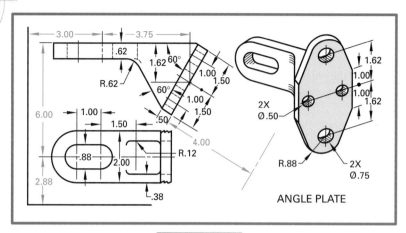

ANGLE PLATE

Figure 9-43

10. A layout and pictorial view of an angle plate are shown in Figure 9-43. Draw the top view and the partial front view as shown. Draw a partial auxiliary view where indicated on the layout. Note that this is an auxiliary elevation. It is made on a plane perpendicular to the horizontal plane.

11. A partial front view, a right-side view, and a partial auxiliary view of an angle cap are shown in Figure 9-44. Draw the views given and another auxiliary view where indicated on the layout. This will be a rear auxiliary view. Dimensioning is required.

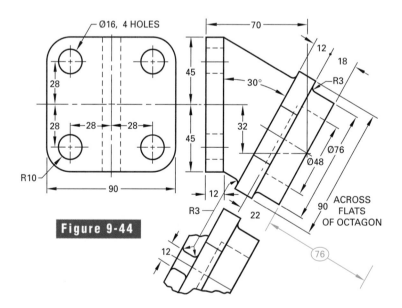

Figure 9-44

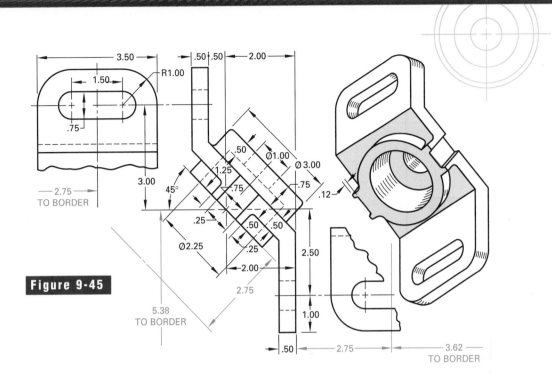

Figure 9-45

12. Figure 9-45 shows a pictorial and layout of an inclined locator. The complete view in the middle is the right-side view. Draw the complete view and the partial views as necessary. Draw an auxiliary view of the inclined locator, as indicated in the layout.

13. Figure 9-46 shows the complete problem. It is given for comparison and is not to be copied. Perform the following tasks:

a. Draw the three-view drawing in its simplest position, as shown in Layout 1.

b. Revolve the front view 45° about an axis perpendicular to the frontal plane, as shown in Layout 2.

c. Revolve the block 30° about an axis perpendicular to the horizontal plane, as shown in Layout 3.

d. Rotate the block 30° from its position in Layout 2 about an axis perpendicular to the side plane, as shown in Layout 4.

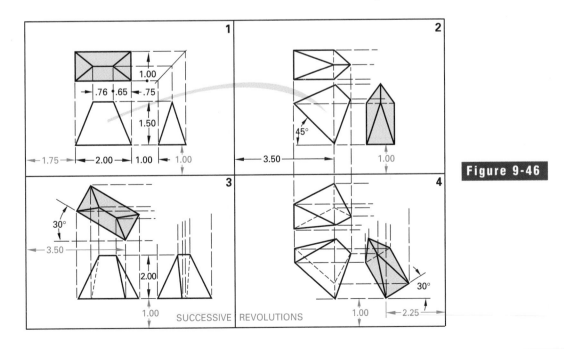

Figure 9-46

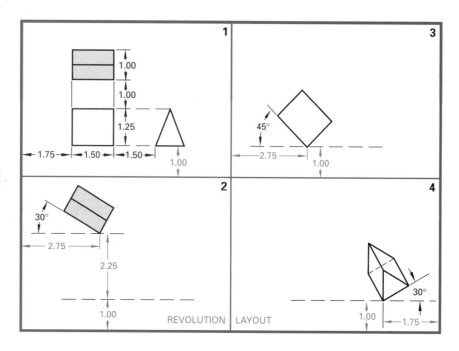

Figure 9-47

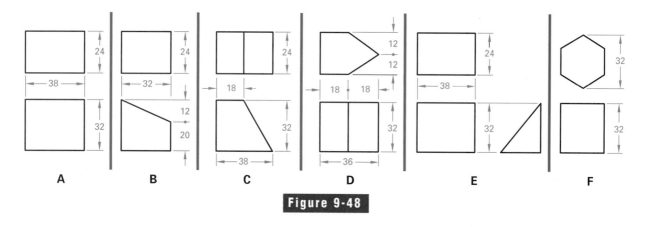

Figure 9-48

Figure 9-49

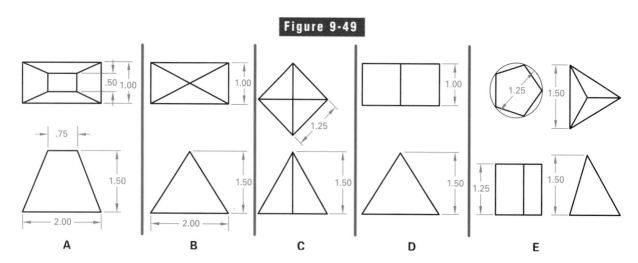

14. Draw the revolved views of the wedge shown in Layout 1 of Figure 9-47. In Layout 2, revolve 45° clockwise about an axis perpendicular to the frontal plane and draw three views. In Layout 3, revolve 30° counterclockwise about an axis perpendicular to the horizontal plane. Finish the revolution as shown in Layout 4.

15. Follow the directions for problem 13 for the objects in Figure 9-48A through F.

16. Follow the directions for problem 13 for the objects in Figure 9-49A through E. Optional assignments may change revolutions to determine true sizes of inclined surfaces.

17. The engineers in your company have redesigned the connecting bar shown in Figure 9-50 to be used on a new tractor hitch. Use the drawing setup shown and make all the design changes the engineer marked in color on the pictorial drawing. Draw front, top, and two auxiliary views. Include all shape and size information necessary for the manufacture of the part. Scale: 1:1.

Design Problems

Design problems have been prepared to challenge individual students or teams of students. In these problems, you should apply skills learned mainly in this chapter but also in other chapters throughout the text. The problems are designed to be completed using board drafting, CAD, or a combination of the two. Be creative and have fun!

1. **TEAMWORK** Work as a team to design an automatic pet food dispenser. Provide all the information necessary to manufacture the dispenser.

2. Design a desk caddy to hold a hand-held calculator, paper clips, 77 mm diameter insert clock, and a 75 mm × 125 mm notepad. The clock should be set at an angle for easy reading.

3. **TEAMWORK** Design a device to store and dispense up to 50 plastic grocery bags. The bags should be inserted at the top and removed from the bottom, one at a time. Design it to hang on a wall.

4. Design a cup rack to display up to 20 collectable cups. The rack can be either a wall unit or a countertop unit.

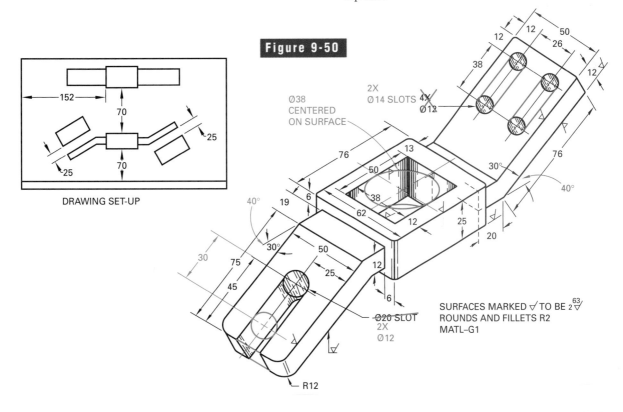

Figure 9-50

DRAWING SET-UP

SURFACES MARKED ⩝ TO BE 2 ⩝
ROUNDS AND FILLETS R2
MATL–G1

Basic Descriptive Geometry

OBJECTIVES

Upon completion of this chapter, you should be able to:

■ Identify points in three-dimensional (3D) space.

■ Identify and describe the three basic types of lines.

■ Identify and describe the three basic types of planes.

■ Establish the relationship among points, lines, and planes in 3D space.

■ Manipulate points, lines, and planes in space to establish true positions, true sizes, and true shapes of features.

■ Establish the true length of an oblique line.

■ Determine the shortest distance between two geometric objects.

■ Determine the true angle between lines or planes.

■ Create points, lines, planes, and solids in 3D space using AutoCAD.

■ Solve descriptive geometry problems using AutoCAD.

The designer who works with an engineering team can help solve problems by producing drawings made up of geometric elements. Geometric elements are points, lines, and planes defined according to the rules of geometry. Every structure has a three-dimensional (3D) form made up of geometric elements. See Figure 10-1. In order to draw three-dimensional forms, you must understand how points, lines, and planes relate to each other in space to form a certain shape. Problems that you might think need mathematical solutions can often be solved through drawings that make manufacturing and construction possible.

Success on the Job

Attitude

Approach your work in a positive and constructive manner. The attitude you display toward your associates and toward the task at hand will generally be reflected in the quality of your work. A positive attitude will yield positive results which, in turn, will enhance your chances for success on the job.

Descriptive geometry is one of the methods a designer uses to solve problems. It is a graphic process for solving three-dimensional problems in engineering and engineering design. In the eighteenth century a French Mathematician, Gaspard Monge, developed a system of descriptive geometry called the Mongean method. The purpose of the system was to solve spatial problems related to military structures. Claude Crozet brought descriptive geometry to the U.S. Military Academy at West Point in 1816. While the Mongean method has changed over the years, its basic principles are still taught in engineering schools throughout the world. By studying descriptive geometry, you develop a reasoning ability that helps you solve problems through drawing.

Most structures designed by people are shaped like a rectangle. This is because it is easy to plan

This bridge shows the result of combining geometric elements.

and build a structure with this shape. This chapter presents a way of drawing that lets you analyze *all* geometric elements in 3D space. Learning to see geometric elements makes it possible for you to describe a structure of any shape. Figure 10-2 shows the basic geometric elements. It also shows some of the geometric features commonly found in engineering designs.

Points

Points are used to identify the intersection of two lines or the corners on an object. A point can be thought of as having an actual physical existence. On a drawing, you can locate a point with a small dot or a small cross. Normally, a point is identified using two or more projections. In Figure 10-3, the normal reference planes are shown in a pictorial view, with point 1 projected to all three planes. The reference planes are shown again in Figure 10-4. When the three planes are unfolded, a flat two-dimensional

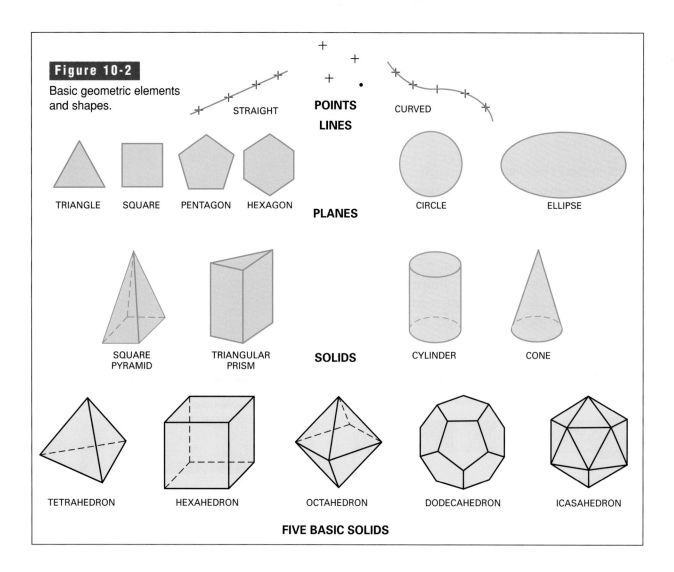

Figure 10-2

Basic geometric elements and shapes.

POINTS

LINES

STRAIGHT   CURVED

PLANES

TRIANGLE   SQUARE   PENTAGON   HEXAGON   CIRCLE   ELLIPSE

SOLIDS

SQUARE PYRAMID   TRIANGULAR PRISM   CYLINDER   CONE

TETRAHEDRON   HEXAHEDRON   OCTAHEDRON   DODECAHEDRON   ICASAHEDRON

FIVE BASIC SOLIDS

(2D) surface is formed. V stands for the vertical (front) view; H stands for the horizontal (top) view; and P stands for the profile (right-side) view.

Points are related to each other by distance and direction, as measured on the reference planes.

Figure 10-5 shows a group of points on the reference planes. You can see the height dimensions in the front and side views, the width dimensions in the front and top views, and the depth dimensions in the top and side views.

Figure 10-3

Locating and identifying a point in space.

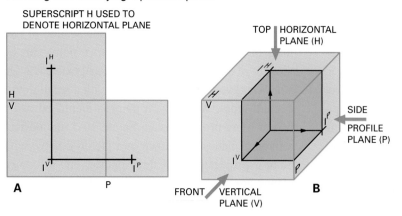

SUPERSCRIPT H USED TO
DENOTE HORIZONTAL PLANE

H
V

A   P

TOP   HORIZONTAL
PLANE (H)

SIDE

PROFILE
PLANE (P)

FRONT   VERTICAL
PLANE (V)

B

Figure 10-4

The point from Figure 10-3 identified on the unfolded reference planes.

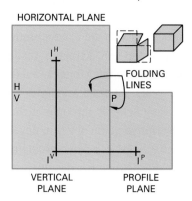

HORIZONTAL PLANE

FOLDING
LINES

H
V   P

VERTICAL
PLANE

PROFILE
PLANE

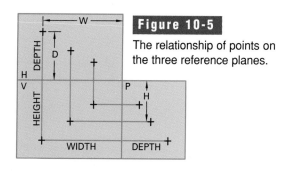

Figure 10-5

The relationship of points on the three reference planes.

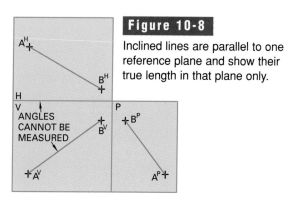

Figure 10-8

Inclined lines are parallel to one reference plane and show their true length in that plane only.

Lines

If a point moves away from a fixed place, its path forms a line. Whereas a point has location only, a line has location, direction, and length. It is easy to draw circular and straight lines. However, plotting irregular curves is somewhat more difficult and must be done very carefully. You can determine a straight line by specifying two points or by specifying one point and a fixed direction.

Basic Lines

Lines are classified according to how they relate to the three normal reference planes. The three basic types of lines are normal, inclined, and oblique.

Normal Lines

A **normal line** is one that is perpendicular to one of the three reference planes. It projects onto that plane as a point, as shown in Figure 10-6. If a normal line is parallel to the other two reference planes, as shown in Figure 10-7, it is shown at its true length (TL).

Inclined Lines

An **inclined line**, like a normal line, is perpendicular to one of the three reference planes. However, it does not appear as a point in that plane. Rather, it appears at its true length. See Figure 10-8. In all other planes, it appears foreshortened.

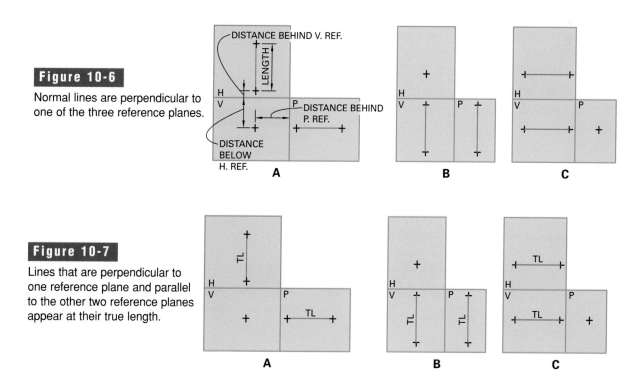

Figure 10-6

Normal lines are perpendicular to one of the three reference planes.

Figure 10-7

Lines that are perpendicular to one reference plane and parallel to the other two reference planes appear at their true length.

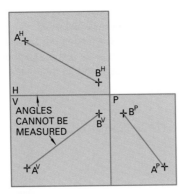

Figure 10-9

Oblique lines appear inclined in all projections, so their true length cannot be determined from the normal reference planes.

Oblique Lines

An **oblique line** appears inclined in all three reference planes, as shown in Figure 10-9. It forms an angle other than a right angle with all three planes. In other words, it is not perpendicular or parallel to any of the three planes. The true length is not shown in any of these views. Also, the angles of direction cannot be measured on the normal reference planes.

True Length of Oblique Lines

Normal lines and inclined lines project parallel to at least one of the normal reference planes. A line parallel to a reference plane shows true length in that plane. Since an oblique line is not parallel to any of the three normal reference planes, you must use an auxiliary reference plane that is parallel to the oblique line to show its true length. See Figure 10-10. The auxiliary reference plane must be perpendicular to its normal reference plane, as shown in Figure 10-11.

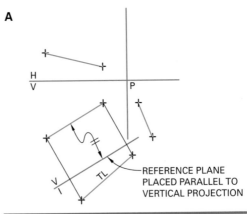

A

REFERENCE PLANE PLACED PARALLEL TO VERTICAL PROJECTION

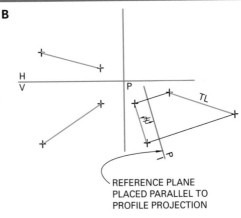

B

REFERENCE PLANE PLACED PARALLEL TO PROFILE PROJECTION

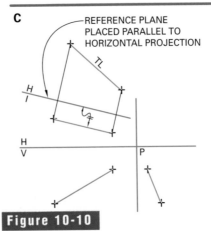

C

REFERENCE PLANE PLACED PARALLEL TO HORIZONTAL PROJECTION

Figure 10-10

The true length of an oblique line can be found by auxiliary projection. Reference planes can be placed parallel to the vertical projection (A), the profile projection (B), or the horizontal projection (C).

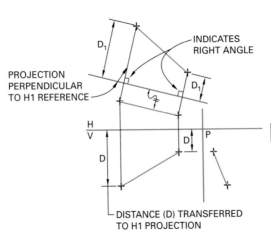

Figure 10-11

An auxiliary reference plane must be perpendicular to its normal reference plane.

Parallel Lines

Figure 10-12 shows the relationship of parallel lines in a three-view study. Line projections are parallel if they appear parallel in all three reference planes. Figure 10-13 shows an example in which the lines seem parallel in the front and top views, but are not parallel in the side view.

Intersecting Lines

If two lines intersect, they have at least one point in common. Figure 10-14 shows the alignment needed to check the point of intersection of two straight lines. Lines 1-2 and 3-4 intersect at point O because point O aligns vertically in the H and V projections.

Now look at lines 5-6 and 7-8 in Figure 10-15. Do the two lines intersect? No, because the points of intersection in the H and V projections are not aligned. The intersection appears to be at point X in the V projection, but it appears to be at point Y in the H projection. Thus, the two lines do not intersect in 3D space.

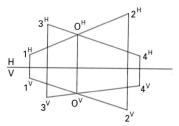

POINT O INDICATES ALIGNED INTERSECTION

Figure 10-14

The point of intersection aligns in the vertical and horizontal reference planes, indicating that the lines do in fact intersect.

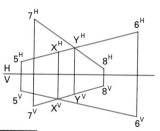

Figure 10-15

The apparent point of intersection of lines 5-6 and 7-8 does not align in the horizontal and vertical reference planes, indicating that the lines do not really intersect.

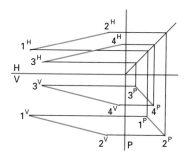

ALL LINES ARE PARALLEL

Figure 10-12

Lines 1-2 and 3-4 are parallel because they appear parallel in all three of the normal reference planes.

Figure 10-13

Lines must appear parallel in all three normal planes to be truly parallel. Lines 5-6 and 7-8 are not parallel because they do not appear parallel in the profile plane.

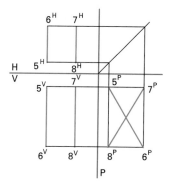

Perpendicular Lines

To find out if two lines are perpendicular, you must find the true length of one of the lines. In the projection in which one of the lines is in true length, the angle between the lines appears at its true size. Therefore, you can see whether the angle between the lines is actually a right angle.

For example, in Figure 10-16A, line 1-2 is parallel to two principal reference planes, so it appears in true length in the vertical projection. In this projection, you can see that line 1-2 and line 3-4 are indeed perpendicular.

In Figure 10-16B, lines 1-2 and 2-3 are oblique in the H and V projections. You must use an auxiliary projection to view one of the lines in true length. In Figure 10-16B, line 2-3 is shown in true length in an auxiliary projection. In the auxiliary, you can see that the two lines are truly perpendicular.

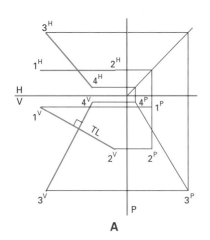

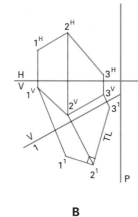

A

B

Industrial Applications of Lines

It may seem that lines drawn on paper mean little and are worth little. However, they do reflect real things. Therefore, they are used in all aspects of industry every day. For example, in areas such as navigational, architectural, and civil engineering, drafters refer to the slope, bearing, azimuth, or grade of a line.

Slope

A line's **slope** is its angle from the horizontal reference plane. Slope is measured in degrees. In Figure 10-17, the true slope of a line is shown in the front view when the line is in true length. To find the slope of an oblique line in true length, use an auxiliary projection perpendicular to a horizontal reference plane. Slope is often used to describe nonvertical or nonhorizontal walls and other features that are not parallel to the normal reference planes.

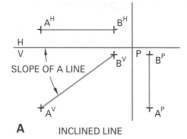

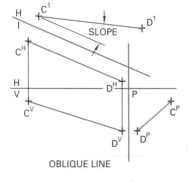

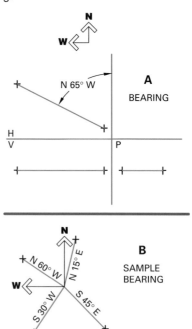

Azimuth and Bearing

You may have heard the terms *bearing* and *azimuth* in connection with aviation. Navigational instruments use these terms to describe the position and direction of aircraft in the air. The angle a line makes in the top view with a north-south line is its **bearing**. See Figure 10-18A. The north-south line is generally vertical, with north at the top. Therefore, right is east and left is west. Make the measurement in the horizontal projection. Dimension it in degrees, as shown in Figure 10-18B.

A measurement that defines the direction of a line off due north is the **azimuth**. The azimuth is always measured off the north-south line in the horizontal plane. It is dimensioned in a clockwise direction, as shown in Figure 10-19.

Grade

The percentage by which land slopes is known as its **grade**. Architectural, civil, and construction engineers specify the grade of land prepared for specific purposes. For example, civil engineers must make certain that the grade of roads that are built in mountainous areas is not too steep. Figure 10-20 shows the scale for constructing a highway with a +12% grade. The grade rises 12′ (3.6 m) in every 100′ of horizontal distance.

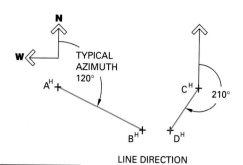

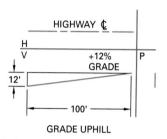

Figure 10-19

Azimuth readings are related to due north.

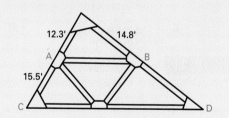

Figure 10-20

Grade is measured in the vertical projection.

TECH MATH

Triangle Proportions

A line parallel to one side of a triangle divides the other two sides proportionately, as shown by the formula next to the illustration below. The formula is stated "*a* is to *b* as *c* is to *d*."

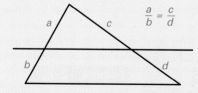

$$\frac{a}{b} = \frac{c}{d}$$

In the illustration on the right, a roof truss is used to illustrate the formula. In this case, AB is parallel to CD. To find the length of BD, substitute available numbers for the letters and solve the equation.

Solution:

$$\frac{BD}{14.8} = \frac{15.5}{12.3}$$

$$BD = \frac{14.8 \times 15.5}{12.3}$$

$$BD = 18.7'$$

Planes

As a line moves away from a fixed place, its path forms a plane. In drawings, planes are thought of as having no thickness. Like true lines, they are also infinite—they extend forever in each direction.

A plane can be determined by any of the following combinations:

■ intersecting lines
■ two parallel lines
■ a line and a point
■ three points not in a straight line
■ a triangle or any other planar (2D) surface, such as a 2D polygon

The Basic Planes

Planes are classified according to their relation to the three normal reference planes. The three basic types of planes are normal, inclined, and oblique planes. As you read the descriptions below, notice that they closely parallel the descriptions of normal, inclined, and oblique lines.

Normal Planes

A **normal plane** is parallel to one of the normal reference planes and perpendicular to the other two. Planes appear as edge views when they are perpendicular to a reference plane. Recall that the edge view of a plane appears as a line.

Figure 10-21 shows three examples of normal planes. In Figure 10-21A, plane 1-2-3 is parallel to the vertical reference plane and perpendicular to the horizontal and profile planes. In Figure 10-21B, the plane is parallel to the horizontal reference plane and perpendicular to the vertical and profile planes. In Figure 10-21C, the plane is parallel to the profile reference plane and perpendicular to the vertical and horizontal planes.

Inclined Planes

An **inclined plane** is perpendicular to one reference plane and inclined to the other two. It is perpendicular to the reference plane in which it shows as an edge view. In the other two reference planes, it appears as a foreshortened surface.

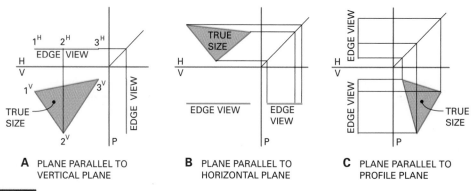

A PLANE PARALLEL TO VERTICAL PLANE
B PLANE PARALLEL TO HORIZONTAL PLANE
C PLANE PARALLEL TO PROFILE PLANE

Figure 10-21

Normal planes parallel to the vertical plane (A), horizontal plane (B), and profile plane (C).

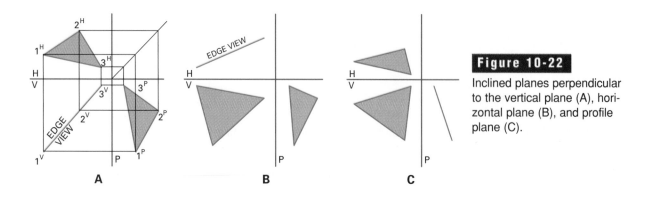

A   **B**   **C**

Figure 10-22

Inclined planes perpendicular to the vertical plane (A), horizontal plane (B), and profile plane (C).

Figure 10-22 shows examples of inclined planes. In Figure 10-22A, inclined plane 1-2-3 is perpendicular to the vertical reference plane. It is inclined to the horizontal and profile planes, where it is foreshortened. Figure 10-22B shows an inclined plane that is perpendicular to the horizontal reference plane, where it shows as a line. The other two reference planes show the plane foreshortened. In Figure 10-22C, the inclined plane is perpendicular to the profile reference plane, where it shows as a line. The plane shows as a foreshortened surface in the other two reference planes.

Oblique Planes

An **oblique plane** is inclined to all three reference planes. An example is shown in Figure 10-23A. Since the oblique plane is not perpendicular to any of the three main reference planes, by definition it cannot be parallel to any of those planes. Thus, it shows as a foreshortened plane in each of the three regular views. Figure 10-23B shows the same oblique plane in a 3D pictorial rendering.

Figure 10-23

An oblique plane in the three-view projection (A) and pictorial (B).

Solving Descriptive Geometry Problems

Now that the basic geometric constructions have been described, you may concentrate on how to use the geometry to solve problems. The board drafting techniques for solving problems in descriptive geometry are much different from the CAD techniques. The difference is due to the CAD software's ability to work in three dimensions. However, it is important to be able to solve 3D problems without the aid of a computer. Therefore, you should study both the "Board Drafting Techniques" and the "CAD Techniques" segments of this chapter carefully.

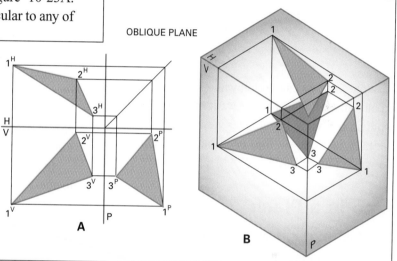

Board Drafting Techniques

This section begins with rather simple operations and proceeds to describe the solutions to more complex problems. It should become clear as you work through the problems that Chapters 9 and 10 are closely related. Almost all problems in descriptive geometry can be worked out using auxiliary planes. You can solve problems by knowing how to find the:

- true length of a line
- point projection of a line
- edge view of a plane
- true size of a plane figure

The ability to understand and solve these problems will build the visual powers necessary for moving on to design problems.

Point on a Line

In Figure 10-24A, line AB on the vertical plane has a point X. To place the point on the line in the other two reference planes, project construction lines perpendicular to the folding lines, as shown in Figure 10-24B.

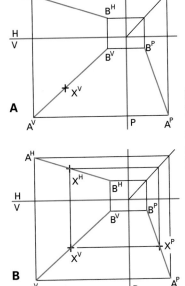

A

B

Figure 10-24

To transfer the location of point X from the vertical reference plane to the other two reference planes, draw straight lines from the point parallel to the fold lines to intersect the line in the other two planes.

Note that you cannot tell whether a point is located on a line using just one view. It may seem to be on a line in one view, but another view may show that it is really in front, on top, or in back of the line, as shown in Figure 10-25.

Line in a Plane

A line lies in a plane if it intersects:

■ two lines of the plane
■ one line of the plane and is parallel to another line of that plane

In Figure 10-26A, line RS must be a part of plane ABC because R is on line AB and S is on BC in all three reference planes. You know that line RS is an oblique line because it is not parallel to any of the normal reference planes and is clearly not perpendicular to the reference planes.

In Figure 10-26B, horizontal line MN is constructed in the vertical projection of plane ABC. A line that is horizontal in the vertical projection is known as a **level line**. Projecting MN to the other reference planes shows that it is an inclined line. The top view shows the true length.

In Figure 10-26C, line XY is constructed parallel to the H/V folding line in the horizontal reference plane. Projected into the vertical plane, it shows as an inclined line in true length. This line is called a **frontal line** because it is parallel to the vertical plane.

Figure 10-26D shows vertical line EF constructed within the plane ABC. It is called a **profile line** because it is parallel to the profile reference plane. Projecting line EF to the profile reference shows the line in true length.

Figure 10-25

Points that fall on a line in all projections are actually on the line. Points 1, 2, and 3 in this illustration are not on line AB, even though they appear to be in some views.

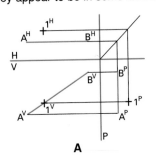

A

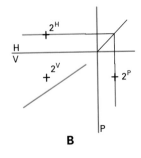

B

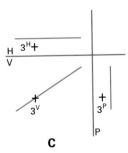

C

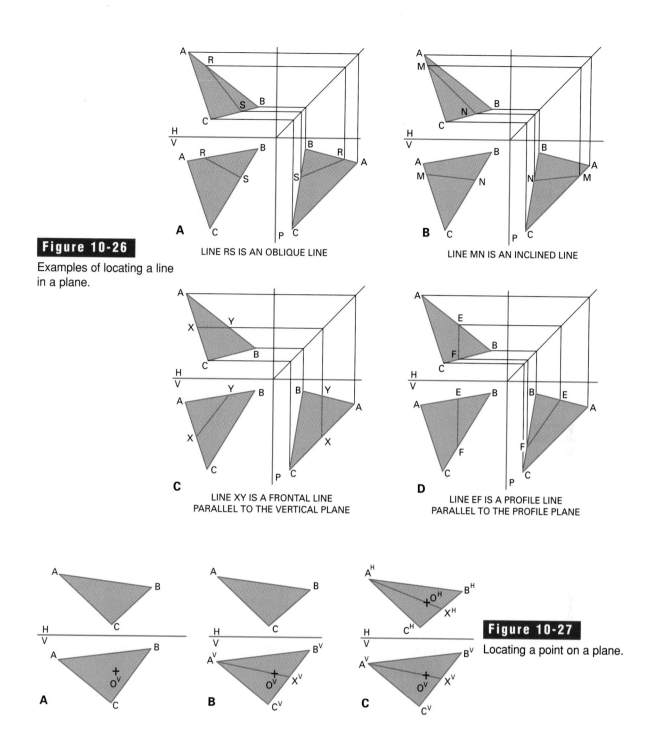

Figure 10-26

Examples of locating a line in a plane.

A — LINE RS IS AN OBLIQUE LINE

B — LINE MN IS AN INCLINED LINE

C — LINE XY IS A FRONTAL LINE PARALLEL TO THE VERTICAL PLANE

D — LINE EF IS A PROFILE LINE PARALLEL TO THE PROFILE PLANE

Figure 10-27

Locating a point on a plane.

Point in a Plane

To locate a point in a plane, project a line from the point to the edges of the plane in which it lies. In Figure 10-27A, point O is on plane ABC. Project line AX, which contains point O, as shown in Figure 10-27B. Then project line AX to ABC in the horizontal reference plane, as shown in Figure 10-27C. Locate point O on the line by drawing a vertical projection to line AX in the horizontal reference plane.

Point View of a Line

A normal line projects as a point on the plane to which it is perpendicular. In Figure 10-28A, line AB is a normal line—it is parallel to the horizontal and profile reference planes. It therefore shows as a point in the vertical reference plane. In Figure 10-28B and C, the same conditions exist. The line projects as a point on the horizontal plane (B) and in the profile plane (C).

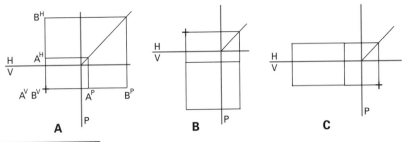

A

B

C

Figure 10-28

A normal line appears as a point in the plane to which it is perpendicular.

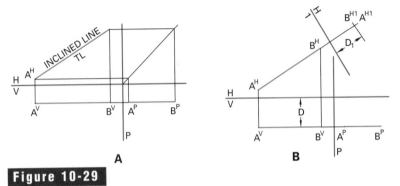

A

B

Figure 10-29

A point projection of an inclined line can be found in an auxiliary projection.

An inclined line projects as a point to an auxiliary plane, as shown in Figure 10-29A. Place a reference plane perpendicular to the inclined line at a chosen distance and label it H/1, as shown in Figure 10-29B. Transfer distance D as shown for a vertical or a horizontal auxiliary projection.

To project an oblique line as a point, use two auxiliary projections. Set up the first auxiliary reference plane parallel to the oblique line, as shown in Figure 10-30A. Then find the true length. Place the secondary auxiliary reference plane perpendicular to the true-length line of the first auxiliary. Locate the point projection by transferring distance X, as shown in Figure 10-30B.

Distance Between Parallel Lines

Point projection is one way to show the true distance between two parallel lines. In Figure 10-31, the parallel lines MN and RS are oblique. Two auxiliary projections are needed to find the point projections. The first auxiliary reference plane H/1 is parallel to MN and RS. In it, lines MN and RS are shown in true length. The second auxiliary reference plane H/2 is perpendicular to the true-length lines in the first auxiliary. The distance between the point projections of the lines is a true distance.

A second way of finding the distance between two parallel lines is shown in Figure 10-32. Think of lines AB and CD as parts of a plane. Connect points A, B, C, and D to form the plane. Draw a horizontal line DX in the top view and project point X into the vertical view. Then draw line DX in the vertical plane. Draw the first auxiliary plane V/1 perpendicular to DX in the vertical view. Find the edge view of plane ABCD by transferring distances 1, 2, 3, and 4, as shown. The secondary auxiliary V/2 shows the true lengths of AB and CD

Figure 10-30

Point projection of an oblique line.

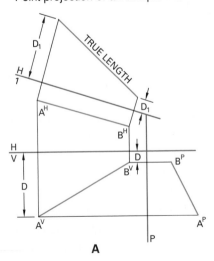

A

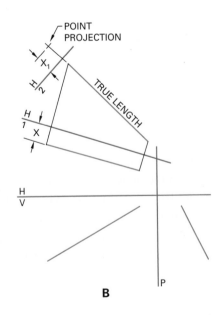

B

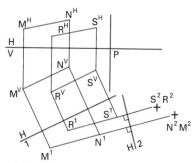

Figure 10-31

Construct the point projection of two parallel lines to find the true distance between them. In this illustration, the two lines are oblique, so two auxiliary planes must be used to achieve the point projection.

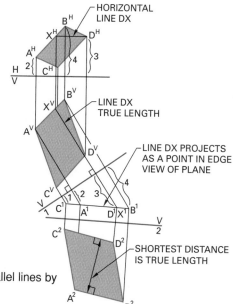

Figure 10-32

Find the distance between two parallel lines by forming a plane.

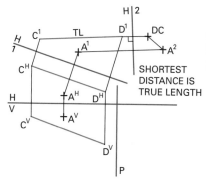

Figure 10-33

Finding the shortest distance from a point to a line.

because plane ABCD is in true size in this view. Measure the true distance between the lines perpendicularly from AB to CD, as shown.

Distance Between a Point and a Line

To find the shortest distance from a point to a line, project the line as a point. In Figure 10-33, project point A and oblique line CD into the first auxiliary projection H/1. In H/1, label the true length of line CD. Place the secondary auxiliary H/2 perpendicular to line CD, and project line CD as a point in this plane. As shown, the distance between points in this projection is true length.

Shortest Distance Between Skew Lines

In Figure 10-34A, lines AB and CD are **skew lines**. That is, the two lines are not parallel and do not intersect. They are both oblique. The shortest distance between these two lines is a perpendicular line between one line and the point view of the other line.

To find the shortest distance between lines AB and CD in Figure 10-34, first find the true length of CD in the first auxiliary. Do this by placing a V/1 reference line parallel to line CD. See Figure 10-34B. Place the secondary auxiliary reference 1/2 perpendicular to the true length of line CD. Find the point projection of line CD and extend line AB as shown. Then construct a perpendicular line from the point projection of CD to line AB. Extend line AB so that it intersects the perpendicular line at point X. Then transfer the intersecting projection back to the first auxiliary, as shown on the extension of line AB.

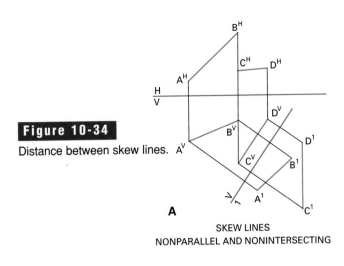

Figure 10-34

Distance between skew lines.

A

SKEW LINES
NONPARALLEL AND NONINTERSECTING

B

LINE AB EXTENDED

SHORTEST DISTANCE

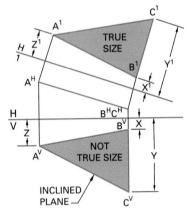

INCLINED PLANE

TRUE SIZE

NOT TRUE SIZE

Figure 10-35

True size of an inclined plane.

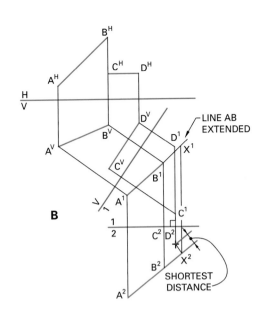

LINE BX PLACED PARALLEL TO H/V REFERENCE

OBLIQUE PLANE

A

EDGE VIEW

TRUE SIZE

B

Figure 10-36

True size of an oblique plane.

True Size of an Inclined Plane

In Figure 10-35, plane ABC shows as an edge view in the top view. Place the auxiliary reference plane H/1 parallel to the edge view and make perpendicular projections. Transfer the distances X, Y, and Z as shown to find the true size of the plane in the first auxiliary projection.

True Size of an Oblique Plane

When plane ABC in Figure 10-36A is projected onto a plane perpendicular to any line in the figure, it shows an edge view in the first auxiliary. In the top view, draw a line BX parallel to the reference plane. Place reference line V/1 perpendicular to the front view of BX. Project the front view of BX into a point projection in the first auxiliary. The point projection is in the edge view of plane ABC, as shown. Place the second auxiliary reference line V/2 parallel to the edge view, as shown in Figure 10-36B. The projection of plane ABC in the secondary auxiliary shows the true size.

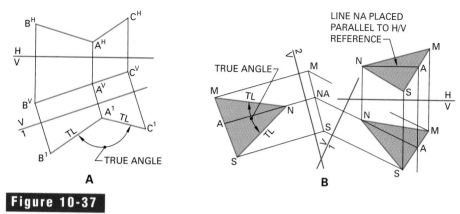

Figure 10-37

Finding the true angle between oblique lines, using two auxiliary planes.

True Angles Between Lines

When two lines show in true length, the angle between them appears in its true value. In Figure 10-37A, the two lines show as an inclined plane. This is because the vertical view shows that lines AB and AC coincide, or lie in a single line. Place the V/1 auxiliary reference parallel to the two lines in the vertical view. The auxiliary view shows the two lines in true length, so it also shows the true angle between the lines.

In Figure 10-37B, oblique lines MS and NS do not show in an edge view. To find the angle between the lines, use two auxiliary planes. The first reference plane is perpendicular to the plane formed by lines NA and MS. The second reference plane is parallel to the first auxiliary view. That is, it is parallel to the edge view of lines MN and NS. The secondary auxiliary view shows MN and NS in true length, so the angle between the two lines is at true size.

Piercing Points

If a line does not lie in a plane and is not parallel to it, the line must intersect the plane. The point of intersection is called a **piercing point**. The line can be thought of as piercing the plane.

Edge-View System

A line crossing the edge view of a plane shows the point at which the line pierces the plane. In Figure 10-38, line RS is neither in plane ABC nor parallel to it. It intersects the plane at a point that is common to both. The edge view of plane ABC is shown in the vertical plane. In the horizontal plane, line RS pierces plane ABC at point P when projected to the vertical plane. If you look at line RS closely in the vertical projection, you will see that element A of the triangle is lower than point R of the piercing line. Therefore, the dashed portion of line RS is invisible.

Figure 10-39 shows another example. In this case, the edge-view system is used to find the point at which line MN intersects oblique plane ABC. The first auxiliary determines the edge view of plane ABC. Carry the piercing point P of line MN back to the vertical and horizontal planes using the projections shown by the arrows.

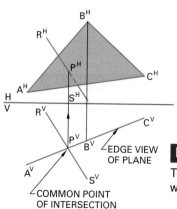

Figure 10-38

The edge-view system of locating the point at which a line pierces a plane.

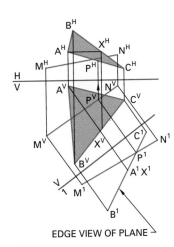

Figure 10-39

Finding the piercing point of a line that intersects an oblique plane.

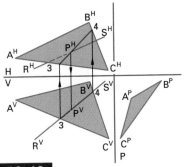

Figure 10-40

The cutting-plane system of locating the point at which a line pierces a plane. Since point S is higher than point R in the vertical projection, point R is below the top projection.

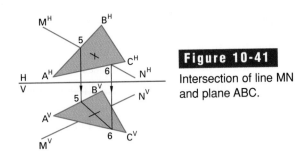

Figure 10-41

Intersection of line MN and plane ABC.

Cutting-Plane System

You can also use a cutting plane containing a line to determine the point at which it pierces a plane. In Figure 10-40, line RS intersects oblique plane ABC. The cutting plane seems to intersect the triangle in the front view at points 3 and 4. Project points 3 and 4 to the top view on lines AC and BC. Connect the points across the plane to find the piercing point P. Then project point P to the front view. Since you now know what parts of line RS are not visible, you can add hidden lines.

Figure 10-41 shows an example in which the cutting plane in the horizontal projection of line MN intersects line AB at point 5 and line AC at point 6. Can the visibility of line MN be determined by inspection?

Angle Between Intersecting Planes

When two planes intersect, they form a **dihedral angle**. A dihedral angle can only be measured in the point view of the intersection.

As you may recall, the intersection of two planes is a straight line. When you find the point view of the line of intersection, the planes in that view are shown as edges, and the angle between the planes is at true

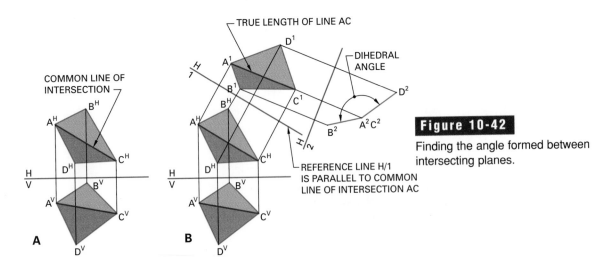

Figure 10-42

Finding the angle formed between intersecting planes.

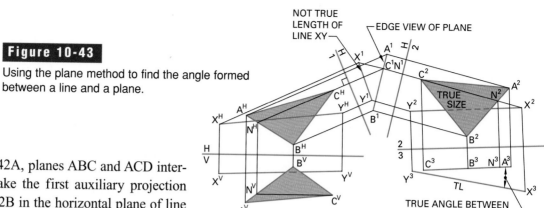

Figure 10-43

Using the plane method to find the angle formed between a line and a plane.

size. In Figure 10-42A, planes ABC and ACD intersect at line AC. Take the first auxiliary projection H/1 in Figure 10-42B in the horizontal plane of line AC. The H/1 auxiliary lets you draw line AC in true length. In the second auxiliary, the point projection of true-length line AC also shows the two given planes as edges. Measure the true angle in the second auxiliary, as shown in Figure 10-42B.

Angle Between a Line and a Plane

You can use a similar concept to find the angle between a line and a plane. If the view shows a plane in the edge view and a true-length line, then it also shows the true angle between the plane and the line.

Plane Method

In Figure 10-43, oblique line XY intersects oblique plane ABC. First find the edge view of oblique plane ABC. Place the first auxiliary H/1 perpendicular to a true-length line in the horizontal projection. Note that line XY is not shown at true length in this first auxiliary view. After projecting the edge view in H/1, place auxiliary plane H/2 parallel to

plane ABC. In the second auxiliary, plot the true size of plane ABC. In a third auxiliary plane 2/3, place the reference line parallel to line XY. The new edge view of the plane and the true-length line form the true angle.

Line Method

In Figure 10-44, oblique plane ABC and oblique line XY intersect in the top view. Place first auxiliary V/1 parallel to line XY to find the true length of XY. Place second auxiliary V/2 perpendicular to the true length of line XY to find the point projection of XY. In the third auxiliary, set reference line 2/3 perpendicular to the true length of B^2O^2 so that plane ABC shows as an edge view. The intersection of the line and plane shows the true size of the angle in this view.

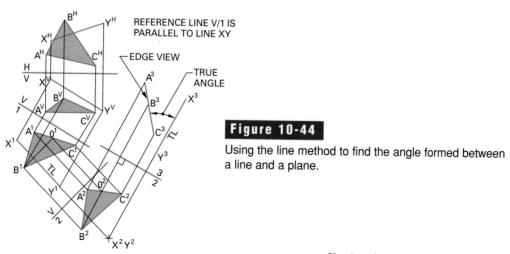

Figure 10-44

Using the line method to find the angle formed between a line and a plane.

True Size of an Oblique Plane by Revolution

Descriptive geometry problems can be solved by revolving an object. Chapter 9, "Auxiliary Views and Revolutions," discussed the rule of revolution. You can use this rule to solve basic problems by changing the position of an object so that the new view shows the needed information.

In Figure 10-45, it is clear that plane ABC is inclined to all the normal reference planes. Place line AX in plane ABC parallel to the horizontal reference line. When you project line AX to the top view, it appears as an inclined line in true length. Project the first auxiliary H/1 in the horizontal reference plane, perpendicular to line AX, to find the point view of AX. Plane ABC appears as an edge view, with X within it. At point AX, edge view B'A'C' is revolved (dashed line) so that it appears parallel to reference line H/1. Projecting points B and C to the horizontal projection allows you to draw the true size of plane ABC in the top view, as shown.

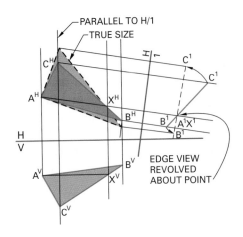

Figure 10-45

Finding the true size of an oblique plane by revolution.

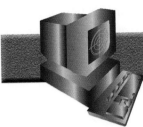

CAD Techniques

CAD can be used in two ways to solve descriptive geometry problems. First, you can use a CAD system to solve problems exactly as you do using board techniques. That is, you can create 2D auxiliary views and revolutions to solve the problems. To use this technique, you can use the commands and techniques you learned in previous chapters. Follow the directions given in the "Board Drafting Techniques" portion of this chapter to solve the problems.

However, it is more practical to use the full power of the CAD system to solve descriptive geometry problems. CAD programs have several commands that allow you to perform location and identification procedures without building elaborate geometric constructions. To solve descriptive geometry problems using CAD, create a 3D model and simply apply the appropriate commands.

3D Coordinate Systems

As you may recall from Chapter 6, working in 3D space requires the addition of a depth axis to the standard 2D Cartesian coordinate system. Figure 10-46 shows the relationship of the X, Y, and Z coordinates used in 3D drawing.

The World Coordinate System

By default in AutoCAD, the computer screen is parallel to the plane formed by the X and Y axes, and the Z axis is perpendicular to the screen. The **origin** (intersection of the axes) is at the bottom left corner of the drawing area. In other words, the top right quadrant (quadrant I) of the Cartesian coordinate system is visible. The shaded portion of Figure 10-46 represents quadrant I. This default viewing configuration is known in AutoCAD as the **world coordinate system (WCS)**.

Other Coordinate Systems

Most CAD programs allow you to define new coordinate systems as necessary. In AutoCAD, a **user coordinate system (UCS)** is a user-defined orientation of the X, Y, and Z axes of the Cartesian coordinate system. Using the UCS command, you can align a new UCS with any planar object, which means that you can create a special UCS to use with any auxiliary plane you may need.

UCS Icon

Notice the X and Y arrows at the bottom left corner of the drawing area. These make up the **UCS icon**. The purpose of the icon is to show the current orientation of the coordinate system. The lines and arrows show the position of the X, Y, and Z axes. When the WCS is the current system, you can't see the Z arrow because it points straight back perpendicular to the screen. However, the UCS icon can be very useful when you have defined one or more user coordinate systems. It helps keep you oriented to the current system.

Drawing in Three Dimensions

CAD programs provide many ways to draw objects in 3D space. In AutoCAD, these include:

- drawing objects with a specified thickness
- extruding 2D objects
- specifying XYZ coordinates
- using solid primitives (solid shapes that are predefined in the software)

Specifying Thickness

AutoCAD's THICKNESS command provides an easy method of creating 3D objects. THICKNESS adds a specific depth to a two-dimensional object, as shown in Figure 10-47. However, it is not strictly a drawing command. Instead, you use it to set the thickness of an object *before* you begin to draw. Then you can use many of the same commands you would use to draw a 2D object.

Figure 10-46

For working in 3D space, you must add the Z, or depth, axis to the familiar Cartesian coordinate system.

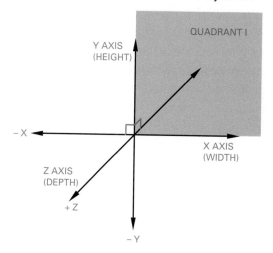

Figure 10-47

Effect of the THICKNESS command. (A) Object drawn without thickness (THICKNESS set to 0). (B) The same object drawn with THICKNESS set to 1.

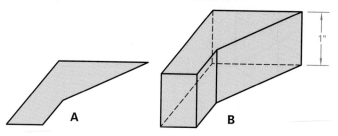

Practice using the THICKNESS command by creating a 5 × 5 × 5-inch cube. Create a practice drawing file, set Snap and Grid to .5, and ZOOM All. Then follow these steps:

1. Enter the THICKNESS command. Notice that the default thickness is 0. This setting creates 2D objects. To create the cube, change the thickness to 5.
2. Use the LINE or PLINE command to draw a 5-inch square.

3. Reenter the THICKNESS command and set the thickness to 0.

Because the thickness is set to 5, you have created a 5-inch cube. It looks like a simple square because you are viewing it from the default viewpoint. In the next section, you will change the viewpoint to see the entire cube.

Setting the Viewpoint

To view the cube from the previous section from a different angle, you can use one of AutoCAD's preset views or create a new viewpoint manually. Follow these steps to explore the preset view options.

1. From the View menu at the top of the screen, select 3D Views and then SE Isometric. This displays the cube from a southeast isometric position, as shown in Figure 10-48. Notice the position of the grid in this view.

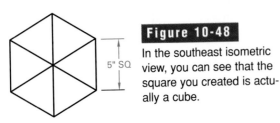

Figure 10-48

In the southeast isometric view, you can see that the square you created is actually a cube.

5" SQ

2. Repeat step 1, but this time select one of the other preset views. Experiment until you are familiar with the various preset opportunities.
3. To return to the default view, enter the PLAN command and select W for WCS. Enter ZOOM All to see the entire drawing area. This view is known as the **plan view**.

The other way to specify a viewpoint is to use the VPOINT command. This command provides more flexibility. You can view the object literally from any point in 3D space. Follow these steps:

1. Enter the VPOINT command and enter 0,1,0 at the prompt. This provides a side view. You will need to ZOOM All to see the entire drawing area.
2. Reenter the VPOINT command and enter 0,0,1. This returns you to the plan view.

This method allows you to set the viewpoint very specifically, but it is not an easy method to use. However, there is more to the VPOINT command.

3. Reenter the VPOINT command, but this time press the Enter key instead of typing in a value. A small tripod and the X, Y, and Z axes appear on the screen.
4. Move the cursor slowly to different locations on the tripod and notice the effect on the axes. The axes represent the viewpoint that will result if you pick the current point. The exact middle of the tripod represents the default view of quadrant I. Start there and move the cursor slowly along the vertical and horizontal lines and then to different locations within the two circles to see the effect on the viewpoint.

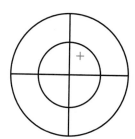

5. Pick a point in the upper, inner circle, as shown in Figure 10-49.

Notice the new position of the UCS icon. By default, the UCS icon follows the origin. This is another helpful reminder of the current viewpoint.

Extruding a 2D Object

It is also possible to extrude a 2D object to give it depth. This method is similar to specifying a thickness, but there are some differences. Unlike the THICKNESS command, AutoCAD's EXTRUDE command works on objects that have already been created. Also, the EXTRUDE command allows you to specify a taper. Figure 10-50 shows the effect of extruding an object with and without a taper. In addition, EXTRUDE works with RECTANGLE and POLYGON as well as LINE and PLINE. This makes the extrusion process more flexible than simply specifying a thickness.

Note: Some versions of AutoCAD may not be capable of extruding objects. If your version does not, use the thickness method or specify the 3D coordinates individually, as described in other sections of this chapter.

The following procedure provides practice in extruding an object and forms the basis for a descriptive geometry problem later in the chapter. Follow these steps:

1. From the View menu, select 3D Views and SE Isometric. This will allow you to see what you are doing in 3D.
2. Use the POLYGON command to create a pentagon inscribed in a $\varnothing.7$ circle.
3. Enter the EXTRUDE command and select the pentagon. Specify an extrusion height of 2 and a taper of 5.

►CADTIP

Removing Hidden Lines

Three-dimensional wireframe views of objects can quickly become confusing. To view the object as it would normally appear, use the HIDE command to remove lines that are hidden in the current viewpoint. To see the hidden lines again, enter REGEN to regenerate the drawing. After removing hidden lines, notice the difference between the pentagonal object and the object you created using the THICKNESS command. The object in part A of the illustration below was created using THICKNESS. It simply adds depth to 2D objects, resulting in a "cookie-cutter" appearance. EXTRUDE creates a solid, as shown in part B of the illustration below. The pentagonal object is actually a simple example of a solid model.

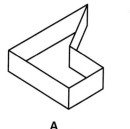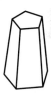

A   B

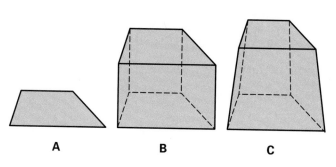

Figure 10-50

The effect of extrusion: (A) the unextruded object; (B) the object extruded without a taper; and (C) the object extruded with a taper of 5.

Specifying Individual Coordinates

Another way to create 3D objects is to figure out the XYZ coordinates of each defining point on the object and then draw the lines individually. This method is extremely time-consuming for complex mechanical assemblies and should be used only if there is a very good reason why a different method cannot be used instead. However, for geometric problem solving, coordinate specification is the perfect way to locate points in a drawing.

The following procedure uses the coordinate method to define a plane that is oblique to the standard view. The plane created here will be used for a descriptive geometry problem in the next section.

1. Enter the LINE command. At the prompt, enter the following sets of absolute XYZ coordinates, pressing Enter after each set. Do not type spaces between the commas and numbers.

   5,1,.5
   4,2,1.5
   7,6,2
   8,5,0
   5,1,.5

2. Use VPOINT or a series of preset views to view the plane from several angles. As you can see, it is oblique to the X, Y, and Z axes.

3. Enter PLAN and W to return to the plan view. The plane looks like a slightly out-of-kilter rectangle in this view.

Descriptive Geometry Problems

Once you have created a 3D object in a CAD system, solving descriptive geometry problems is a fairly easy task. In this section, you will use the pentagonal solid and the plane you created earlier in this chapter to solve some representative problems.

Locating Points

To identify the exact location of a point in 3D space, use the ID command. This command identifies the exact X, Y, and Z coordinates of the point you specify. To demonstrate this, enter the POINT command and use the Endpoint object snap to snap to one of the endpoints of the oblique plane you created earlier. Watch the Command line. The X, Y, and Z values AutoCAD displays should match one of the

points you specified when you created the plane. See Figure 10-51.

Points can exist as single entities in AutoCAD. If you are attempting to locate a single point that is not on a defined line, you must change the setting of PDMODE so that you will be able to see the point. PDMODE is a system variable in AutoCAD that controls how points are displayed on the screen. Follow these steps:

1. Enter the POINT command. At the prompt, enter a coordinate value of −1,5.8,3 to place the point at that location.

2. Enter ZOOM All to be sure you can see the entire drawing. Can you see the point? Probably not. If you can see it at all, it is just a tiny speck on the screen.

3. Enter PDMODE and enter a value of 3. This changes the point display to an X that is more easily visible. Now the position of the point is clear.

4. View the point from several viewpoints. What effect does the negative X value have on the position of the point?

You can use ID to identify the exact location of single points. To do so, be sure to use the Node object snap. (*Node* is another term for *point*.)

True Length of a Line

The true length of any line in 3D space can be determined easily in AutoCAD. Simply use the DIMALIGNED dimensioning command (Aligned Dimension button on the Dimension toolbar). Select the line whose true length you want to find, and place the dimension. The dimension text gives you the true length. After you have determined the true length, you can erase the dimension.

Figure 10-51

The ID command gives the exact location of a point in 3D space. If you pick the point shown on the oblique plane, AutoCAD lists the X, Y, and Z coordinates as 5.000, 1.000, and 0.5000, respectively.

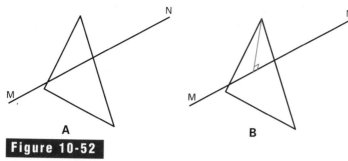

Figure 10-52

(A) An oblique plane and a line that pierces the plane. (B) Draw a line from a point on the plane, and perpendicular to the line. The intersection of this line and line MN is the piercing point.

This method works from any viewpoint, but it is usually easier to see the result if you return to the plan view. Notice that you do not have to create complex auxiliary views to find the true length of a line in AutoCAD.

Determining Distances

AutoCAD's DIST command provides a great deal of information about the relative positions of two points in 3D space. Follow these steps to find the true distance between two points on the oblique plane you created earlier.

1. Enter the DIST command.
2. For the first point, use the Endpoint object snap to snap to one of the endpoints of the plane.
3. For the second point, snap to the endpoint diagonally across the plane from the first point.
   The result is displayed on the Command line. *Note:* If you cannot see the distance, press the F2 key to display a text screen. Review the information that is provided. You now know the exact distance between the two points, the change (delta) on the X, Y, and Z axes, and the angles in and from the XY plane.

Shortest Distance Between Skew Lines

Skew lines are lines that are not parallel and do not intersect. The method for determining the shortest distance between two skew lines is similar to the method used in board drafting. Use the VPOINT command to obtain the point view of one of the lines, and then draw a line from the point view to the other

line. Use the Perpendicular object snap to ensure that the new line is perpendicular to the second line.

Piercing Points

Using AutoCAD, you can find the point at which a line pierces a plane regardless of the orientation of the plane. The procedure that follows creates an oblique plane and a line that passes through the plane. It then demonstrates how to find the piercing point. Refer to Figure 10-52 and follow these steps:

1. Save the first practice drawing as directed by your instructor, and begin a new drawing.
2. Be sure that your UCS is set to World and that you are viewing the plan view. Then create an oblique plane ABC by specifying the following coordinate values:
   3,1,–2
   2,4,1
   1,2,–3
   3,1,–2

► **CADTIP**

Closing a 2D Line

When you are using the LINE command to create a closed 2D figure, such as a triangle, you can create the last segment quickly by entering "c" instead of the last coordinate triple. For example, in step 1 of the procedure for finding a piercing point, instead of entering 3,1,–2 for the last coordinate value, enter the letter "c." (Do not type the quotation marks.) This automatically draws a line segment from the end of the previous segment to close the figure.

3. To create line MN piercing plane ABC, you should first move the UCS icon parallel to plane ABC. To do this, enter the UCS command and then enter 3 to enter the 3point option. Recall that you can use three points to define a plane. Locate the endpoints of plane ABC by snapping to point B, then point C, and then point A. The UCS icon jumps onto the lower left corner of the plane, which is now located at point B.

4. Create a line starting in front of plane ABC at coordinates 1.6,2,2 and ending at 1.6,2,–2. Notice that the only coordinate that changes is the Z coordinate. Drop line AP from the endpoint of A perpendicular to line MN. Enter the ID command and select the intersection of lines MN and AP, or select the point P. The coordinate value of that point should be 1.6,2,0. This is the point at which line MN intersects plane ABC.

5. To return to the World Coordinate System, enter the PLAN command and then W (for World).

Angle Between Intersecting Planes

The procedure for finding the angle between intersecting planes is similar to the procedure for finding the true length of a line. You can use the dimensioning command DIMANGULAR to find the angle directly. There is no need to create auxiliary views.

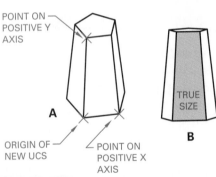

POINT ON POSITIVE Y AXIS

TRUE SIZE

A

B

ORIGIN OF NEW UCS

POINT ON POSITIVE X AXIS

Figure 10-53

Create a new UCS using the 3point method. Specify the points as shown in (A). The plan view of the new UCS is aligned with the screen and shows one face of the solid at its true size and shape (B).

True Shape and Size of a Plane

Using AutoCAD's dimensioning commands, you can dimension a plane correctly without actually seeing the plane in its true size and shape. However, you may find it necessary at times to view the true size and shape of an inclined or oblique plane. The easiest way to accomplish this is to define a user coordinate system that lies on the plane.

None of the five sides of the pentagonal object created earlier in this chapter are parallel to a normal viewing plane. To see the true shape and size of one of the sides, follow these steps:

1. Switch to the NE Isometric view of the drawing.

2. Enter the HIDE command to remove hidden lines. (This is not absolutely necessary, but it makes it easier to see and select the points in the following steps. You may also want to move any interfering objects out of the way before you continue.)

3. Enter the UCS command. Enter N to create a new UCS. When the list of creation options appears, enter 3 (for 3point).

4. For the origin of the new UCS, use the Endpoint object snap to pick the bottom of one of the sides, as shown in Figure 10-53A. For the point on the positive X axis, pick the bottom of the other side of the planar surface. For the point on the positive Y axis, pick the top of the line on which you specified the origin. Notice that the UCS icon moves to the new origin.

5. Enter the PLAN command, and enter C for Current UCS. You can see by the grid that the planar surface is now parallel to the screen. You are now viewing the true size and shape of the surface, as shown in Figure 10-53B.

Note: To return to the plan view of the WCS, enter the PLAN command and enter W for World.

Chapter Summary

- Points can be thought of as having actual physical existence.

- Every object has a three-dimensional form made up of points, lines, and planes.

- The basic types of lines are normal, inclined, and oblique.

- The basic types of planes are normal, inclined, and oblique.

- Descriptive geometry allows the drafter to solve three-dimensional geometric problems with the aid of drawings.

- Using descriptive geometry, the drafter can manipulate points, lines, and planes to determine their true relationship in 3D space.

- Although descriptive geometry problems may seem to have little importance, descriptive geometry is frequently used in industry to solve design and engineering problems.

- CAD programs allow drafters to work directly in 3D space, offering an alternative to traditional descriptive geometry methods.

Review Questions

1. What are the three basic geometric elements?

2. What is the basic shape of most structures designed by people? Why?

3. Explain the purpose of descriptive geometry.

4. How do you locate a point using board drafting techniques? Using AutoCAD?

5. What are the three characteristics of a line?

6. How does a normal line relate to the three normal reference planes of projection?

7. In how many of the normal planes of projection does an inclined line show in true length?

8. How does an oblique line relate to the three main planes of projection?

9. In board drafting, how many auxiliary projections are needed to find the point projection of an oblique line?

10. What are the major characteristics of a plane?

11. Describe how to find out whether two lines really intersect using board techniques. How would you do this using AutoCAD?

12. What is the difference between an inclined plane and an oblique plane?

13. How can you determine the distance between two parallel inclined lines?

14. Using board drafting techniques, how should you place the auxiliary reference plane to find the true size of an inclined line?

15. How can you find the true length of an oblique line using AutoCAD?

16. How can you find the true angle between intersecting lines using board drafting techniques?

17. In what ways is AutoCAD's THICKNESS command different from the EXTRUDE command?

18. How do you find the point at which a line pierces a plane using AutoCAD?

19. Explain how to show an oblique plane or surface at its true shape and size in AutoCAD.

20. What is the purpose of the UCS icon in AutoCAD?

Drafting Problems

The problems in this chapter can be performed using board drafting or CAD techniques. Each problem is laid out on a grid. Assume the size of the larger squares to be .5″. Some of the .5″ squares have been subdivided into .125″ squares. Use this information to complete the problems. If you are using a CAD system, recreate the geometry in the CAD system and then use the appropriate CAD techniques to complete the problems.

1. In Figure 10-54, find the true length of line AB. Determine the true length and slope of line CD.

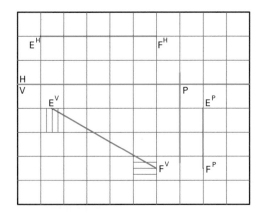

Figure 10-55

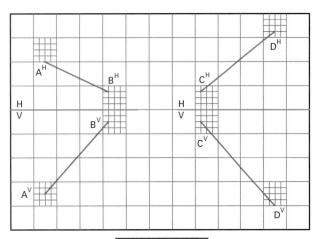

Figure 10-54

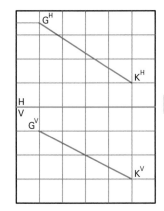

Figure 10-56

2. In Figure 10-55, line EF is the centerline of a pipeline. Scale: 1:500 (1″ = 40′-0). Locate a line X (20′-0) below E on all principal planes of projection. Determine the grade of line EF. Determine the true distance from point E to line X.

3. In Figure 10-56, determine the point projection of line GK.

4. In Figure 10-57, determine the angle LM makes with the vertical plane. What is the bearing?

Figure 10-57

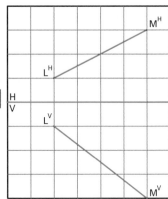

5. In Figure 10-58, determine the slope of line NO. Extend NO to measure 2.25″ (56 mm) long. Draw all three views.

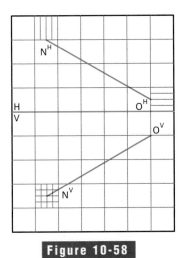

Figure 10-58

7. In Figure 10-60, what is the bearing of line NO located on plane XYZ? Determine the true size of plane XYZ.

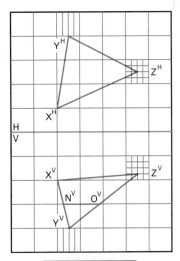

Figure 10-60

6. In Figure 10-59, locate point D in the plan view (horizontal projection). Determine the length of line AD.

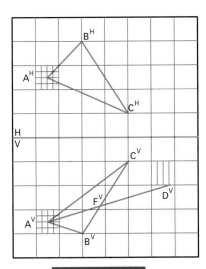

Figure 10-59

8. In Figure 10-61, complete the plan view of plane ABCD and develop a side view.

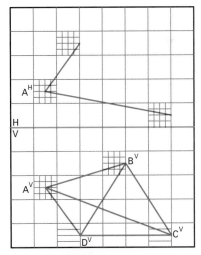

Figure 10-61

9. In Figure 10-62, find the edge view of plane ABCD and determine the angle it makes with the horizontal plane.

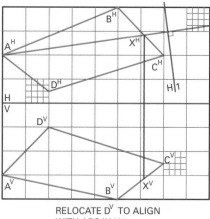

RELOCATE D^V TO ALIGN
WITH ABC IN H/1

Figure 10-62

11. In Figure 10-64, complete the three views showing the intersection of AB and EF.

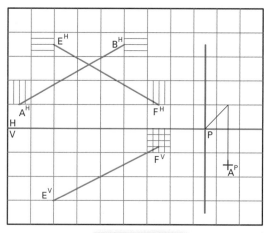

Figure 10-64

10. In Figure 10-63, form the incomplete parallelepiped using points A and L. Determine the proper visibility in all three normal views.

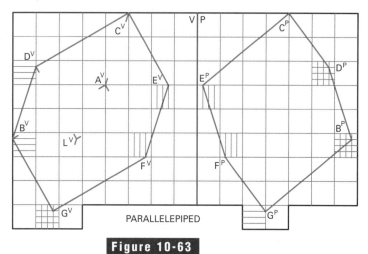

PARALLELEPIPED

Figure 10-63

12. In Figure 10-65, draw the front view of line AB, which intersects line CD. What is the distance from C to A?

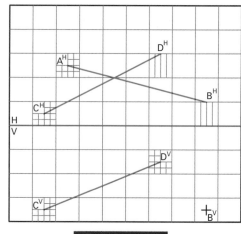

Figure 10-65

13. In Figure 10-66, create a location for plane 1-2-3 in the vertical plane.

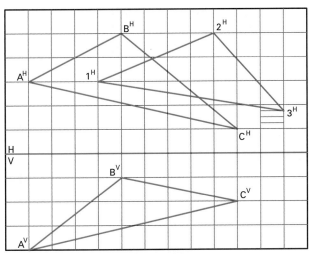

ASSUME THAT POINT 1 TOUCHES PLANE ABC

Figure 10-66

14. In Figure 10-67, construct a plane ABCD parallel to plane 1-2-3.

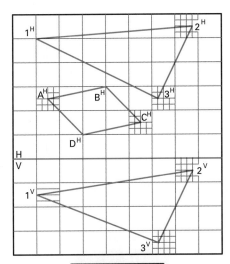

Figure 10-67

15. In Figure 10-68, determine the true size of oblique plane ABC. Draw line XY parallel to plane ABC in the plan view.

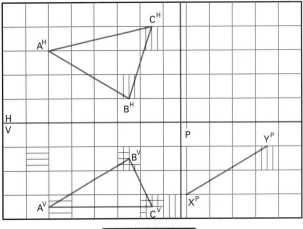

Figure 10-68

16. In Figure 10-69, determine the true size of plane ABC and label its slope.

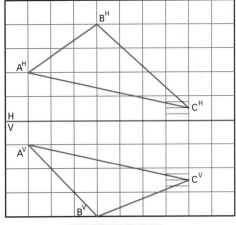

Figure 10-69

17. In Figure 10-70, draw the true size of plane ABC and dimension the three angles of the plane.

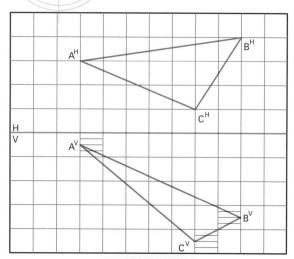

Figure 10-70

18. In Figure 10-71, find the true angle between lines AB and BC.

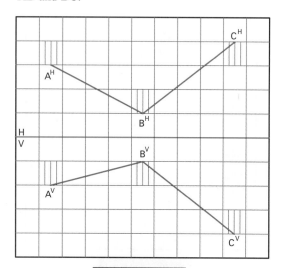

Figure 10-71

19. In Figure 10-72, determine and locate the shortest distance between lines AB and CD.

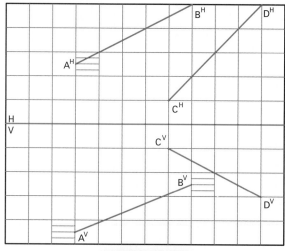

Figure 10-72

20. In Figure 10-73, determine and label the shortest distance between parallel lines AB and MN.

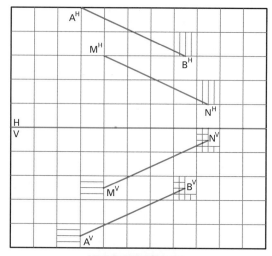

Figure 10-73

21. In Figure 10-74, determine whether line MN pierces the plane.

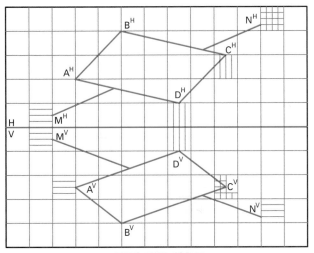

Figure 10-74

22. In Figure 10-75, determine whether line MN pierces plane ABC. Locate line KL so that it pierces the center of plane ABC.

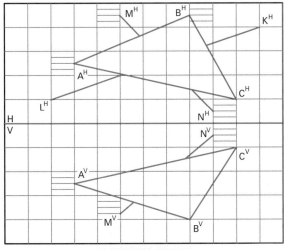

Figure 10-75

23. In Figure 10-76, determine the visibility and angle formed between planes ABC and ABD.

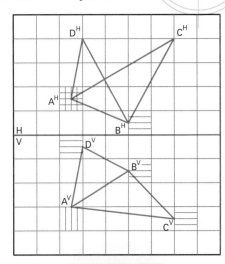

Figure 10-76

24. In Figure 10-77, determine the angle between line BX and plane ABC.

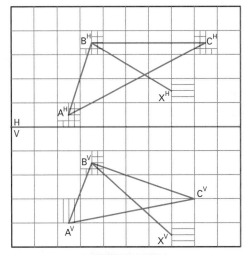

Figure 10-77

25. In Figure 10-78, determine how the two planes intersect and show proper visibility.

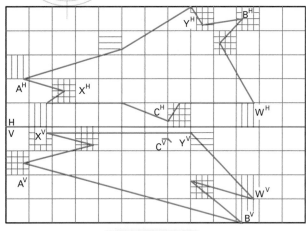

Figure 10-78

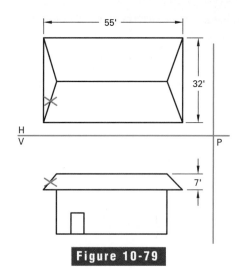

Figure 10-79

Design Problems

Design problems have been prepared to challenge individual students or teams of students. In these problems, you should apply skills learned mainly in this chapter but also in other chapters throughout the text. The problems are designed to be completed using board drafting, CAD, or a combination of the two. Be creative and have fun!

1. Figure 10-79 shows the top and front views of a house. The correct placement for a direct TV dish is at the exact center of the lower left ridge of the roof. Find the true length of the roof ridge of the house, and identify the point at which the dish should be mounted.

2. Design a set of collapsible sawhorses using steel components. They are to be 25.00" (635 mm) high. The top is to be 4.00" (100 mm) wide by 38.00" (965 mm) long. Design the sawhorses to fold into the smallest possible size. Have the legs spread at an oblique angle to the top member. Use adequate bracing to make them stable. Make a complete set of working drawings and a materials list.

3. **TEAMWORK** Work as a team to design a soapbox racecar. It can also be designed as a snow racing machine with runners. Carefully consider both aesthetics and aerodynamics. Design a roll cage using $\varnothing 1.25"$ or square steel tubing. Be sure to add sufficient bracing to provide adequate safety. Each team member should first work independently to develop basic design sketches. The team should then consider all ideas to arrive at the final design. Each member of the team should be responsible for the development of some aspect of the final set of drawings and a materials list.

4. Design a unit of metal scaffolding to be used for construction work or painting. The overall size is to be 8'-0 long by 3'-0 wide by 6'-0 high. Use $\varnothing 1.50"$ aluminum aircraft tubing. Design it to be easy to take apart for transporting, and brace it well for safety. Make a set of working drawings and a materials list.

5. **TEAMWORK** Work as a team to design a piece of playground equipment for children to climb on. The basic design should include round steel or aluminum tubing with welded joints. (See Chapter 15 for more information about welding drafting.) Each team member should first work independently to develop basic design sketches. *Be creative!* Design the apparatus to give children a safe and enjoyable experience. Each member of the team should be responsible for the development of some aspect of the final set of drawings and a materials list.

Fasteners

OBJECTIVES

Upon completion of this chapter, you should be able to:

- ■ Identify and describe various types of fasteners.
- ■ Define common screw-thread terms.
- ■ Specify threads and fasteners on a technical drawing.
- ■ Draw detailed, schematic, and simplified thread representations.
- ■ Name and describe common thread series.
- ■ Describe and specify classes of thread fits.
- ■ Draw various types of threaded fasteners using board drafting or CAD techniques.

A **fastener** is any kind of device or method for holding parts together. Screws, bolts and nuts, rivets, welding, brazing, soldering, adhesives, collars, clutches, and keys are all fasteners. Each of these fastens in a different way. Some fasten parts permanently; others can later be taken apart again or adjusted.

Screws and other fasteners have so many uses and have become so important that engineers, drafters, and technicians must become familiar with their different forms, as shown in Figure 11-1. They must also be able to draw and specify each type correctly.

Unless the drawing is intended for use by a manufacturer of fasteners, the fasteners in technical drawings are usually drawn as symbols. That is, their sizes are drawn by convention that may or may not reflect the exact size of the fastener. However, you must be familiar with the characteristics of the various kinds of fasteners and with the conventions for drawing them.

Screw Threads

The principle of the screw thread has been known for so long that no one knows who discovered it. Archimedes (287-212 B.C.), a Greek mathematician, put the screw thread to practical use. He used it in designing a screw conveyor to raise water. Similar devices are still used today to move flour and sugar in commercial bakeries, to raise wheat in grain elevators, to move coal in stokers, and for many other purposes.

A **screw thread** can be a helical ridge on the external or internal surface of a cylinder, or it can be in the form of a conical spiral on the external or internal surface of a cone or frustum of a cone. All screw threads are shaped basically as a **helix**, or helical curve. Technically, a helix is the curving path that a point would follow if it were to travel in an even spiral around a cylinder and parallel to the axis of that cylinder. In simpler terms, if a wire is

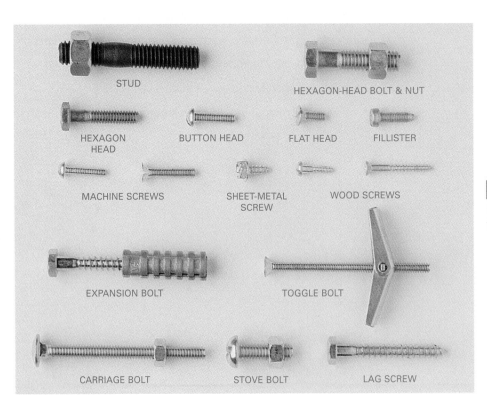

STUD

HEXAGON-HEAD BOLT & NUT

HEXAGON HEAD

BUTTON HEAD

FLAT HEAD

FILLISTER

MACHINE SCREWS

SHEET-METAL SCREW

WOOD SCREWS

EXPANSION BOLT

TOGGLE BOLT

CARRIAGE BOLT

STOVE BOLT

LAG SCREW

Figure 11-1

Examples of threaded fasteners.

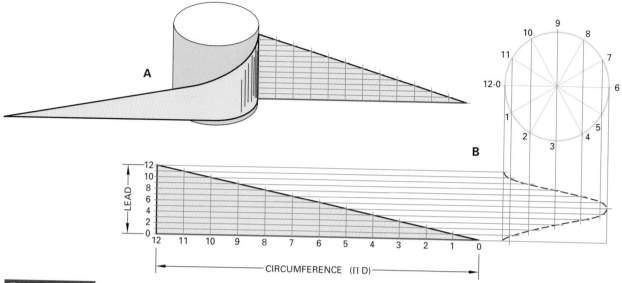

Figure 11-2

Pictorial representation of a helix (A) and a projection of a helix (B).

wrapped around a cylinder in evenly spaced coils, it forms helical curves.

Another way to visualize the shape of screw threads is to cut out a right triangle in paper and wrap it around a cylinder, as shown in Figure 11-2A. If the triangle's base is the same length as the cylinder's circumference, its hypotenuse forms one turn of a helix when wrapped around the cylinder. The triangle's altitude is the pitch of the helix. A right triangle and the projections of the corresponding helix are shown in Figure 11-2B.

The application of the helix is shown in Figure 11-3, the actual projection of a square thread. However, such drawings are seldom made, because they take too much time. Also, they are no more useful than conventional representations.

Figure 11-3

The application of the helix or helical curve.

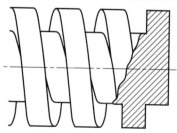

Screw-Thread Standards

The first screws were made for one purpose. They were made without any thought of how anyone else might make one of the same diameter. Later, as industry and mass production developed, goods were produced in quantity by using interchangeable parts. The need then arose for standards for screws and screw threads.

Screw-thread standards in the United States were developed from a system that William Sellers presented to the Franklin Institute in Philadelphia in 1864. Screw-thread standards in England came from a paper presented to the Institution of Civil Engineers in 1841 by Sir Joseph Whitworth. These two standards were not interchangeable.

More and better screw-thread standards have been developed as industrial production has grown more complex. In 1948, standardization committees of Canada, Great Britain, and the United States agreed on the Unified Thread Standards. These standards are now the basic ANSI standards. They are listed in *American National Standard Unified Screw Threads for Screws, Bolts, Nuts, and Other Threaded Parts* (ANSI B1.1) and in Handbook H-28, *Federal Screw Thread Specifications*.

Since 1948, the inch and the metric thread systems have been brought increasingly into line with each other. Prior to 1968, the International Organization for Standardization (ISO) maintained only the metric screw-thread system as its standard. In 1968, ISO adopted the Unified system as its inch screw-thread standard. Since that time, ISO has maintained both the Unified and metric screw thread standards. Fortunately, the two systems are similar.

These similarities are very important as the metric system of measurement comes into worldwide use. To help bring other screw-thread standards into line with this system, ANSI has published the *American National Standard Unified Screw Threads—Metric Translation* (ANSI B1.1a).

Screw-Thread Terminology

Figure 11-4 shows the main terms used to describe screw threads. The Unified and American (National) screw-thread profile shown in Figure 11-5 is the form used for fastening in general. Other forms of threads are used to meet special fastening needs. Some of these threads are shown in Figure 11-6. The sharp V is seldom used today. The square thread and similar forms (worm thread and acme thread) are made especially to transmit motion or power along the line of the screw's axis. The knuckle thread is the thread used in most electric-light sockets. It may also be a "cast" thread. The Dardelet thread automatically locks a screw in place. It was designed by a French military officer but is seldom used today. The former British Standard (Whitworth) has rounded crests and roots. Its profile forms 55° angles. The former United States Standard had flat

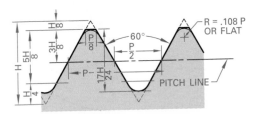

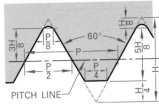

Figure 11-5

Specifications for Unified threads.

crests and roots and 60° angles. The buttress thread takes pressure in one direction only: against the surface at an angle of 7°.

Pitch of a Screw Thread

The **pitch** of a thread is the distance from one point on the thread form to the corresponding point on the next form. It may be defined as a formula as follows:

$$\text{Pitch} = \frac{1}{\text{no. of threads per inch}}$$

Pitch is measured parallel to the thread's axis, as shown in Figure 11-7. The **lead** (pronounced "leed") *L* is the distance along this same axis that the threaded part moves against a fixed mating part when given one full turn. It is the distance a screw enters a threaded hole in one turn.

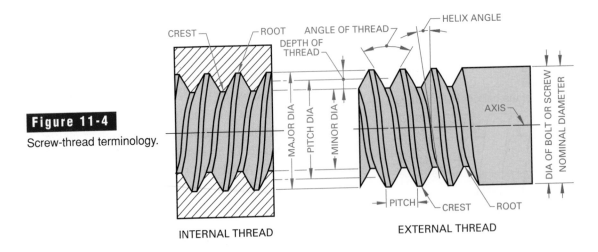

Figure 11-4

Screw-thread terminology.

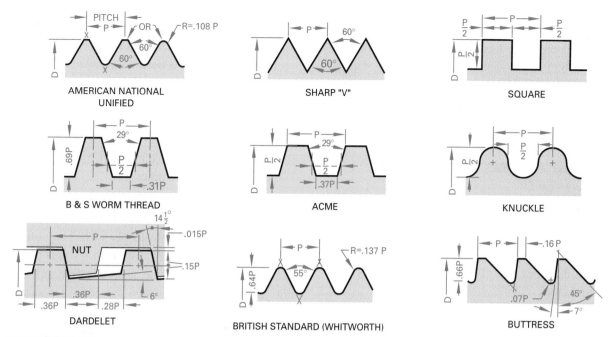

Figure 11-6

Some of the screw-thread profiles used to meet special needs.

Single and Multiple Threads

Most screws have single threads, as shown in Figure 11-7A. A **single thread** is a single ridge in the form of a helix. If you give a single-thread screw one full turn, the distance it advances into the nut (lead) equals the pitch of the thread. A screw has a single thread unless it is marked otherwise.

A **double thread**, shown in Figure 11-7B, consists of two helical ridges side by side. The lead, in this case, is twice the pitch. A **triple thread** is three ridges side by side, as shown in Figure 11-7C. The lead for this thread is three times the pitch. Double and triple threads, which are also called *multiple threads*, are used where parts must screw together quickly. For example, technical pen caps and toothpaste tube caps have multiple threads.

Right- and Left-Hand Threads

A **right-hand thread** screws in when turned clockwise as you view it from the outside end, as shown in Figure 11-8A. A **left-hand thread** screws in when turned counterclockwise. See Figure 11-8B.

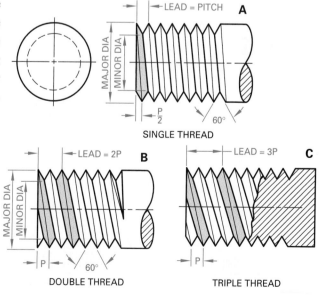

Figure 11-7

Single (A), double (B), and triple threads (C).

Threads are always right-hand threads unless marked with the initials *LH,* meaning "left-hand." Some devices, such as the turnbuckle shown in Figure 11-9, have both right- and left-hand threads. Others, such as bicycle pedals, can have either. A left-hand pedal has left-hand threads; a right-hand pedal has right-hand threads.

Classification of Screw Threads

Screw threads are classified according to several different characteristics. For example, they can be divided according to combinations of diameter and pitch, tolerances and allowances, and various other specifications.

Thread Series

Screws of the same diameter are made with different pitches (numbers of threads per inch) for various uses. In the Unified screw-thread system, the various combinations of diameter and pitch have been grouped in **screw-thread series**. These series are listed in ANSI B1.1. Each is denoted by letter symbols, as follows:

- **Coarse-Thread Series (UNC)** In this series, the pitch for each diameter is relatively large. This series is used for engineering in general.

- **Fine-Thread Series (UNF)** In this series, the pitch for each diameter is smaller (there are more threads per inch) than in the coarse-thread series. This series is used where a finer thread is needed, as in making automobiles and airplanes.

- **Extra-Fine-Thread Series (UNEF)** In this series, the pitch is even smaller than the fine-thread series. This series is used where the thread depth must be very small, as on aircraft parts or thin-walled tubes.

The Unified system also has several constant-pitch-thread (UN) series. These series have 4, 6, 8, 12, 16, 20, 28, or 32 threads per inch. They offer a variety of pitch-diameter combinations that can be used where the coarse, fine, and extra-fine series are not suitable. However, when selecting a constant-pitch series, the first choices should be the 8-, 12-, or 16-thread series. Constant-pitch threads are generally used as a continuation of coarse-, fine-, and extra-fine-thread series in larger diameters.

- **Eight-Thread Series (8UN)** has 8 threads per inch for all diameters.
- **Twelve-Thread Series (12UN)** has 12 threads per inch for all diameters.
- **Sixteen-Thread Series (16UN)** has 16 threads per inch for all diameters.
- **Special Threads (UNS or UN)** are special, non-standard combinations of diameter and pitch.

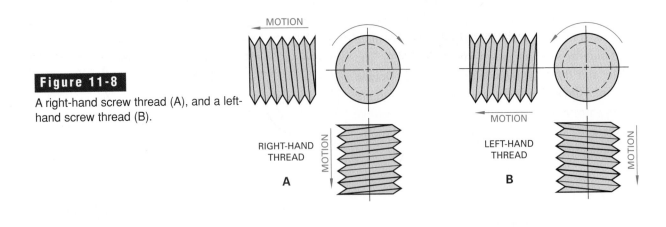

Figure 11-8

A right-hand screw thread (A), and a left-hand screw thread (B).

MOTION

RIGHT-HAND THREAD
MOTION

A

MOTION

LEFT-HAND THREAD
MOTION

B

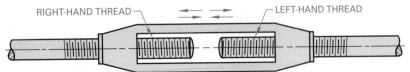

RIGHT-HAND THREAD ———   ⟶   ⎯ LEFT-HAND THREAD

Figure 11-9

A turnbuckle uses right- and left-hand screw threads.

Classes of Fits

Screw threads are also divided into screw-thread classes based on their **tolerances** (amount of size difference from exact size) and **allowances** (how loosely or tightly they fit their mating parts). The exact screw thread needed can be obtained by choosing both a series and a class. In brief, the classes for Unified threads are Classes 1A, 2A, and 3A (external threads), and Classes 1B, 2B, and 3B (internal threads).

- **Classes 1A and 1B** have a large allowance, which allows a loose fit. They are used on parts that must be put together quickly and easily.
- **Classes 2A and 2B** are the thread standards most used for general purposes, such as for bolts, screws, nuts, and similar threaded items.
- **Classes 3A and 3B** are stricter standards for fit and tolerance. They are used where thread size must be more exact.

Thread Specifications

A screw thread is specified by giving its nominal (major) diameter, number of threads per inch, length of thread, initial letters of the series, class of fit, and external (A) or internal (B), as shown in Figure 11-10. Any thread you specify is assumed to be both single and right-hand unless stated otherwise. If the thread is to be left-hand, include the letters LH after the class symbol. If it is to be a double or triple

| Table 11-1. Thread Specification Examples ||
Specification	Meaning
1.25-7UNC-1A	1.25" diameter, 7 threads per inch, Unified threads, coarse threads, Class 1, external
.75-10UNC-2A	.75" diameter, 10 threads per inch, Unified threads, coarse threads, Class 2, external
.88-14UNF-2B	.88" diameter, 14 threads per inch, Unified threads, fine threads, Class 2, internal
1.62-18UNEF-3B-LH	1.62" diameter, 18 threads per inch, Unified threads, extra-fine threads, Class 3, internal, left-hand

thread, include DOUBLE or TRIPLE. Examples of thread specifications and their meanings are shown in Table 11-1.

Sizes of threads and fasteners may be given on a drawing by using either fractional-inch sizes or decimal-inch sizes. Since manufacturers of fasteners often specify them in fractions, it is a generally accepted practice to specify them on the drawing using fractions. However, since many industries have switched entirely to decimal-inch dimensioning, the practice of dimensioning and specifying fasteners in decimal-inch sizes is also now a common practice. Either is correct. Drafters use the method required by the standards and practices of individual companies.

If you are using the traditional dimensioning method shown in Table 7-1, specify tapped (threaded) holes by a note giving the diameter of the tap drill, depth of hole, thread information, internal or external, and depth of thread. For example:

.42 DRILL $\times$ 1.38 DEEP
.50-13UNC-2B $\times$ 1 DEEP

In the more current system, tapped holes are specified only by giving the thread specification and depth. The tap drill diameter and depth of the drilled hole are not given. For example, in the modern system, the specifications given in the preceding paragraph would be reduced to giving only the thread information, thread series, class of fit, internal or external, and depth of thread:

.50-13UNC-2B $\times$ 1.00

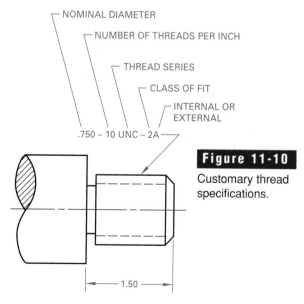

- NOMINAL DIAMETER
- NUMBER OF THREADS PER INCH
- THREAD SERIES
- CLASS OF FIT
- INTERNAL OR EXTERNAL

.750 – 10 UNC – 2A

Figure 11-10

Customary thread specifications.

1.50

CUSTOMARY INCH THREAD CALLOUT

Metric Threads

Specify an ISO metric screw thread by giving its nominal size (basic major diameter) and pitch, both expressed in millimeters. Include an M to denote that the thread is an ISO metric screw thread. Place the M before the nominal size. Use × to separate the nominal size from the pitch.

For the coarse thread series only, do not show the pitch unless you are also giving the length of the thread. If the length is given, separate it from the other designations with an ×. For external threads, the length may be given as a dimension. For example, specify a 10-mm diameter, 1.25 pitch, fine-thread series as M10 × 1.25. Specify a 10-mm diameter, 1.5 pitch, coarse-thread series, however, as M10. Remember, for coarse threads, do not give the pitch unless you also give the length of the thread. If in this example the length were 25 mm, the thread would be shown as M10 × 1.5 × 25.

In addition to the basic designation, you must specify the thread's tolerance class. This is separated by the basic designation by a dash. It consists of two sets of symbols. The first denotes the pitch diameter tolerance. The second, which follows immediately, denotes the crest diameter tolerance. Each of these two sets of symbols is composed of a number giving the grade tolerance, followed by a letter (capital for internal threads and lowercase for external threads) specifying the tolerance position. If the symbols for the pitch diameter tolerance are the same as those for the crest diameter tolerances, you may leave out one set, as shown in Figure 11-11.

Two classes of fits are generally used in the ISO metric system. The first is a general-purpose class and is specified as 6H/6g. The second is for fits that require a closer tolerance and is specified as 6H/5g6g. More detailed information may be obtained from ISO and ANSI standards. However, unless a very specific tolerance class is necessary, many industries have adopted a simplified system for designating thread fits. If the general-purpose fit is acceptable, no tolerance class is specified. If a close fit is necessary, the letter C is placed after the pitch in the thread note. An example is M10 × 1.25C.

Bolts and Nuts

Various types of bolts are shown in Figure 11-12, and the terms used to describe standard bolts and nuts are illustrated in Figure 11-13. In general, bolts and nuts are either regular or heavy (thick), and they are either square or hexagonal. Regular bolts and nuts are used for the general run of work. Heavy bolts and nuts are somewhat larger. They are used where the bearing surface or the hole in the part being held must be larger. The types of bolts and nuts made in regular sizes include square-head, hexagon-head, and semifinished hexagon-head bolts and nuts. Types made in heavy sizes include hexagon-head, semifinished hexagon-head, and finished hexagon-head bolts, and square nuts.

Regular bolts and nuts are not finished on any surface. Semifinished bolts and nuts are processed to have a flat bearing surface, as shown in Figure 11-14. The bearing surface has either a washer face or a face with chamfered corners. Each face has a diameter equal to the distance across the flats (between opposite sides of the bolthead or nut). The thickness of the washer face is about .02″. "Finished" bolts and nuts are so called only because of the quality of their

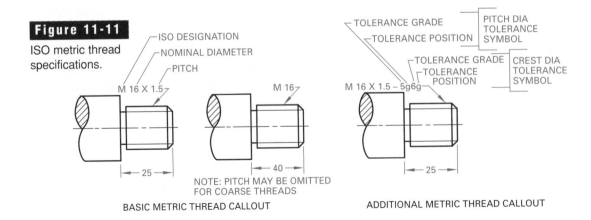

Figure 11-11

ISO metric thread specifications.

BASIC METRIC THREAD CALLOUT

ADDITIONAL METRIC THREAD CALLOUT

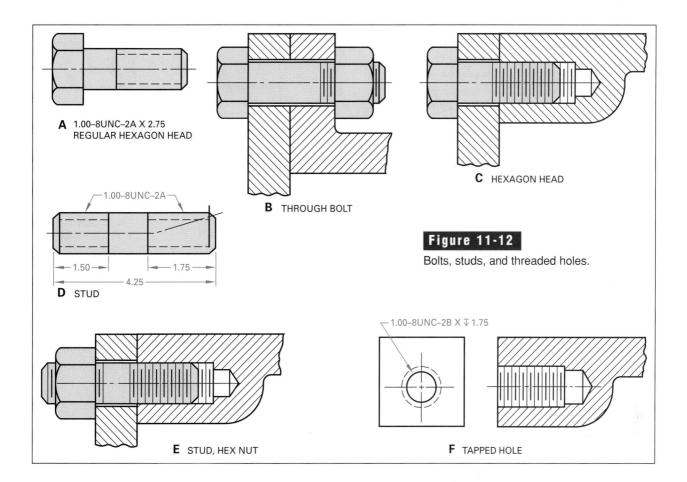

A 1.00–8UNC–2A X 2.75
REGULAR HEXAGON HEAD

B THROUGH BOLT

C HEXAGON HEAD

1.00–8UNC–2A

1.50   1.75
4.25

D STUD

Figure 11-12

Bolts, studs, and threaded holes.

1.00–8UNC–2B X ⤓ 1.75

E STUD, HEX NUT

F TAPPED HOLE

manufacture and the closeness of their tolerance. They do not have completely machined surfaces.

Because boltheads and nuts come in standard sizes, they are generally not dimensioned on drawings. Instead, give the information in a note, as shown in Figure 11-12A. The note specifies a 1.00″ diameter, 8 threads per inch, Unified coarse-thread series, Class 2A fit, 2.75″ long, regular hexagon-head bolt.

A stud or stud bolt has threads on both ends, as shown in Figure 11-12D. It is used where a regular bolt is not suitable and for parts that must be removed often. One way to use a stud is to screw it into a threaded hole permanently. The hole in another part is put over the stud and a nut is screwed onto the end, as shown in Figure 11-12E. Dimension the length of thread from each end as shown. In certain cases, a stud can be passed through two parts and a nut placed on each end. Dimension a tapped hole as shown in Figure 11-12F.

Figure 11-13

Bolt and nut terminology.

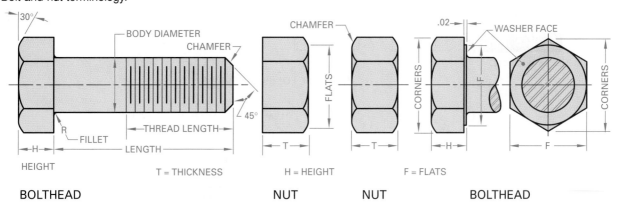

BOLTHEAD

NUT

NUT

BOLTHEAD

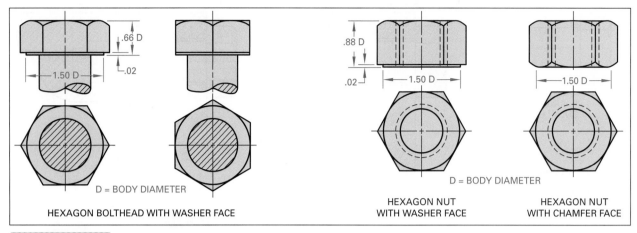

Figure 11-14

Semifinished boltheads and nuts.

Other Threaded Fasteners

Several other kinds of fasteners are in common use for various purposes. Some of them work with nuts, bolts, and screws; others work alone. This section describes several threaded fasteners.

Lock Nuts and Lock Washers

Lock nuts and various devices such as lock washers are used to keep nuts, or bolts and screws, from working loose. Some forms of these locking devices are shown in Figure 11-15. In addition, some self-locking fasteners have threads coated with epoxy that

is combined with a hardening agent before assembly. After assembly, the epoxy hardens and gives a strong, vibration-proof bond. See Figure 11-16.

Cap Screws

A cap screw fastens two parts together by passing through a clearance hole in one and screwing into a tapped hole in the other. Examples of cap screws are shown in Figure 11-17. In most cases, clearance holes are not shown on drawings. Coarse, fine, or 8-thread series may be used on cap screws. Socket-head cap screws may have Class 3A threads. The other head styles have Class 2A threads. Dimensions

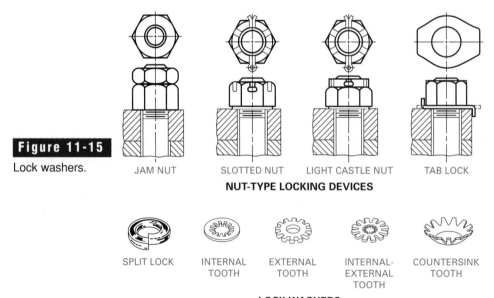

Figure 11-15

Lock washers.

Figure 11-16

Self-locking fastener.

EPOXY

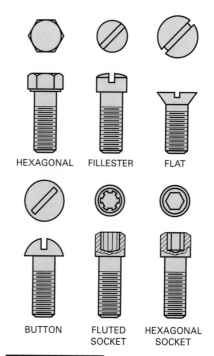

HEXAGONAL   FILLESTER   FLAT

BUTTON   FLUTED SOCKET   HEXAGONAL SOCKET

Figure 11-17

Cap screws.

for drawing various sizes of American National Standard cap screws are given in Appendix C.

Cap screws have a naturally bright finish. This is in keeping with the machined parts with which they are used.

Machine Screws

Figure 11-18 shows examples of machine screws. These screws are used where the fastener must have a small diameter. The finish on machine screws is bright, and the ends are flat, as shown. Sizes below .25″ or 6 mm in diameter are specified by number. Number size machine screws range in size from No. 0, with a diameter of .060″ or 1.524 mm, to No. 12, with a diameter of .216″ or 5.486 mm. Specific sizes for the complete range of number size machine screws can be found in any catalog or technical reference. Machine screws may screw into a tapped hole or extend through a clearance hole and into a square nut. Machine screws up to 2.00″ or 50 mm long are threaded full length. Coarse or fine threads and Class 2 threads may be used on machine screws.

Setscrews

Setscrews, shown in Figure 11-19, are used to hold two parts together in a desired position. They do so by screwing through a threaded hole in one part and bearing, or pushing, against the other. There are two general types: square-head and headless. Headless setscrews can have either a slot or a socket. Any of the points shown can be used on any setscrew. ***SAFETY NOTE:*** Square-head setscrews can cause accidents when used on rotating parts. They may also violate safety codes.

Wood Screws

Figure 11-20 shows various types of wood screws. These screws are made of steel, brass, or aluminum and are finished in various ways. Steel screws can be bright (natural finish), blued, galvanized, or copper-plated. Both steel and brass screws are sometimes nickel-plated. Round-head screws are set with the head above the wood. Flat-head screws are set flush or countersunk.

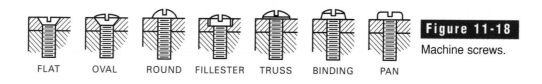

FLAT   OVAL   ROUND   FILLESTER   TRUSS   BINDING   PAN

Figure 11-18

Machine screws.

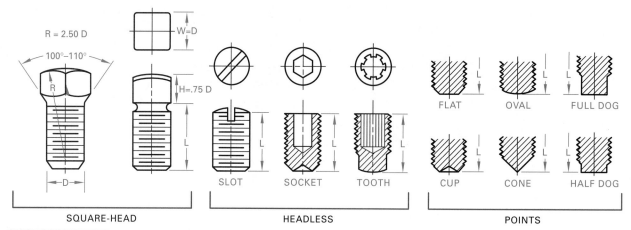

Figure 11-19

Setscrews.

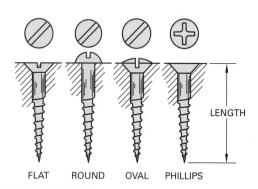

Figure 11-20

Wood screws.

Specify wood screws by number, length, style of head, and finish. For flat-head screws, dimension the overall length. Dimension round-head screws from under the head to the point. Dimension oval-head screws from the largest diameter of the countersink to the point. Sizes and dimensions are listed in Appendix C.

Miscellaneous Threaded Fasteners

Various other threaded fasteners are shown in Figure 11-21. In most cases, the names indicate the ways in which they are used.

Figure 11-21

Miscellaneous threaded fasteners.

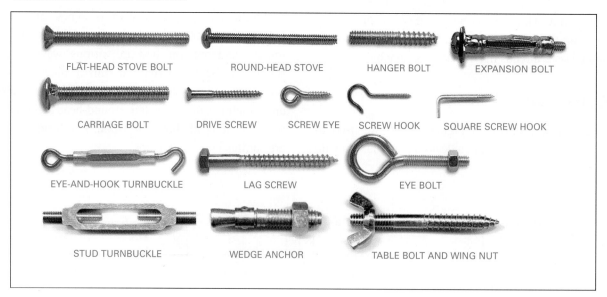

Screw hooks and screw eyes are specified by diameter and overall length. A lag screw, or lag bolt, is used to fasten machinery to wood supports. It is also used in heavy wood constructions when a regular bolt cannot be used. It is similar to a regular bolt but has wood-screw threads. Specify a lag bolt by its diameter and the length from under the head to the point. The head of the lag bolt has the same proportions as a regular bolthead.

Nonthreaded Fasteners

Some applications require fasteners in parts where threading cannot easily be accomplished. In other cases, there is simply no need to remove the fastener once it has been installed, so threads are unnecessary. Nonthreaded fasteners such as keys and rivets are used under these and other circumstances.

Keys

Keys are nonthreaded fasteners used to secure pulleys, gears, cranks, and similar parts to a shaft. Keys are made in different forms for different uses, as shown in Figure 11-22. They range from the saddle key, for light duty, to special forms such as two

square keys, for heavy duty. The common sunk key can have a breadth of about one fourth the shaft diameter. Its thickness can vary from five eighths the breadth to the full breadth.

The Woodruff key is often used in machine-tool work. It is made in standard sizes. Specify it by number (see Appendix C). Special forms of pins have been developed to replace keys for some uses. These pins need only a drilled hole instead of the machining that is needed to make keyseats and keyways.

Rivets

Rivets are rods of metal with a preformed head on one end. They are used to fasten sheet-metal plates, structural steel shapes, boilers, tanks, and many other items permanently. The rivet is first heated red-hot. Then it is placed through the parts to be joined. It is held in place while a head is formed on the projecting end. The rivet is then said to have been "driven."

Large rivets, like those shown in Figure 11-23, have nominal diameters ranging in size from .50″ to 1.75″ (12 mm to 45 mm). Small rivets, shown in Figure 11-24, range from .06″ to .44″ (2 or 3 mm to 11 mm) in diameter. See Appendix C for American National Standard rivet dimensions.

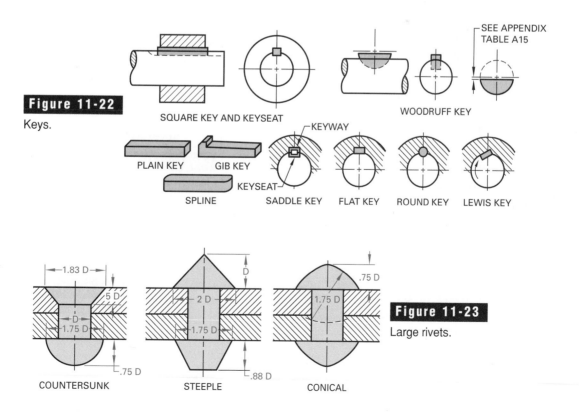

Figure 11-22
Keys.

SEE APPENDIX TABLE A15

SQUARE KEY AND KEYSEAT

WOODRUFF KEY

KEYWAY

PLAIN KEY   GIB KEY

KEYSEAT

SPLINE   SADDLE KEY   FLAT KEY   ROUND KEY   LEWIS KEY

COUNTERSUNK   STEEPLE   CONICAL

Figure 11-23
Large rivets.

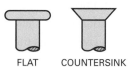

FLAT   COUNTERSINK   BUTTON   PAN   TRUSS OR WAGON BOX

Figure 11-24

Small rivets.

Figure 11-25

Explosive rivets.

Some rivets are made especially for use where one side of the plates cannot be reached or where the space is too small to use a regular rivet. These are called *blind rivets*. One type is the explosive rivet shown in Figure 11-25. It has a small explosive charge in a cavity. After the rivet is inserted, the charge is exploded, forming a head. This rivet is thus excellent for blind riveting, since the head can be formed inside places that are closed or impossible to reach.

Sometimes plates that are riveted together need clear surfaces. This requires flush riveting on one or both sides, as shown in Figure 11-26. These rivets are used on airplanes, automobiles, spacecraft, etc. For some uses, as in tanks or steel buildings, high-strength structural bolts are used. Welding is also in wide use.

Riveted joints are used for joining plates, as shown in Figure 11-27. These joints may be lap or butt joints. They may also have single or multiple riveting.

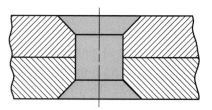

FLUSH BOTH SIDES
THICK PLATES

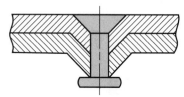

FLUSH ONE SIDE
THIN PLATES DIMPLED

Figure 11-26

Flush rivets.

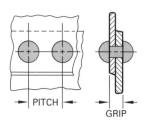

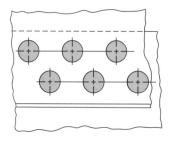

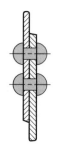

Figure 11-27

Riveted joints.

PITCH   GRIP

SINGLE-RIVETED LAP JOINT   DOUBLE-RIVETED LAP JOINT
STAGGERED RIVETING

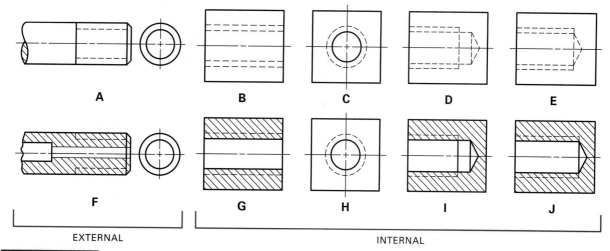

Figure 11-28

Simplified representation of screw threads.

Methods of Representation

Under ANSI rules, screw threads can be drawn in three ways, or representations. These are:

- simplified representation
- schematic representation
- detailed representation

Simplified Representation

A **simplified representation** is much like a schematic representation. For simplified representation, however, draw the crest and root lines as dashed lines, as shown in Figure 11-28B, D, and E, except where any of them would normally show as a visible solid line. See Figure 11-28A, C, F, G, H, I, and J. The drafter saves time by using the simplified representation because it leaves out useless details. Therefore, simplified representation has become the most commonly used in industry.

Schematic Representation

A **schematic representation** shows the threads using symbols, rather than as they really look. For this kind of drawing, leave out the Vs, as shown in Figure 11-29. Also, the pitch need not be drawn to scale. Make it about the right size for the drawing, and then draw the crest and root lines accordingly. Space them by eye to make them look good.

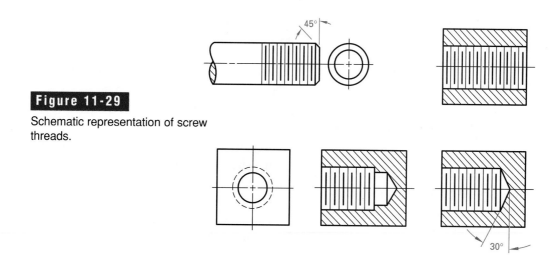

Figure 11-29

Schematic representation of screw threads.

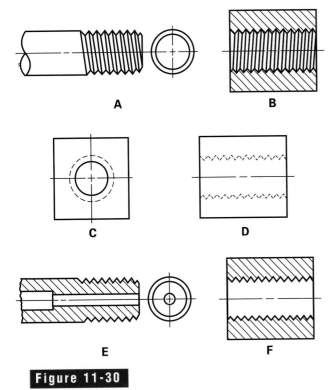

A

B

C

D

E

F

Figure 11-30

Detailed representation of screw threads.

Detailed Representation

A **detailed representation** approximates the real look of threads, as shown in Figure 11-30. For this kind of drawing, it is not necessary to draw the pitch exactly to scale. Instead, estimate the right size for the drawing. In general, detailed representation is not used in working drawings (see Chapter 13), except where needed for clarity. It is also not usually used if the screw is less than 1.00″ (25 mm) in diameter. The realistic, or V-form, thread representation can make a drawing clearer where two or more threaded pieces are shown in section. See Figure 11-31.

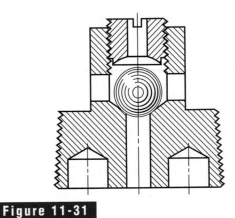

Figure 11-31

Threads in section on assembled pieces.

Board Drafting Techniques

When drawing screw threads, you use special symbols. These are the same whether the threads are coarse or fine, right-hand or left-hand. Use notes to give the necessary information.

Representing Screw Threads

The procedure for drawing screw threads varies depending on the method of representation you choose. Your choice should be based on the intended use of the drawing, as explained in "Methods of Representation."

Drawing a Simplified Representation

To draw a simplified representation, follow the steps in Figure 11-32, as follows:

1. Lay off the outside diameter of the screw, as shown in Figure 11-32A.
2. Lay off the screw-thread depth (estimate) and the chamfer, as shown in Figure 11-32B.
3. Draw the chamfer and a line to show the length of the thread, as shown in Figure 11-32C.
4. Draw dashed lines for the threads, as shown in Figure 11-32D.

Drawing a Schematic Representation

To draw a schematic representation of screw threads, follow the steps in Figure 11-33, as follows:

1. Lay off the outside diameter of the screw thread, as shown in Figure 11-33A.
2. Lay off the thread depth and the chamfer, as shown in Figure 11-33B.
3. Draw thin crest lines at right angles to the axis. See Figure 11-33C.
4. Draw thick root lines parallel to the crest lines, as shown in Figure 11-33D.

The crest and root lines can be at right angles to the axis or slanted to show the helix angle, as shown in Figure 11-34A. To draw slanted threads, give each thread a slope of half the pitch. Notice that the crest lines are thin and root lines are thick. However, you may draw all the lines the same thickness, as shown in Figure 11-34B, to save time, especially on pencil drawings.

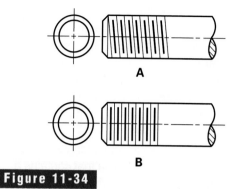

Figure 11-34

(A) Slope-line representation. (B) Uniform-width lines.

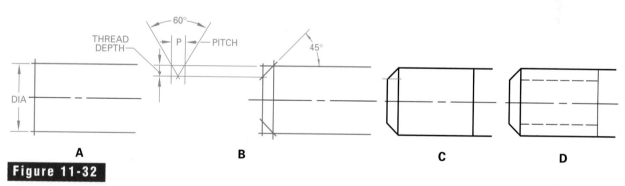

Figure 11-32

Drawing the simplified representation of screw threads.

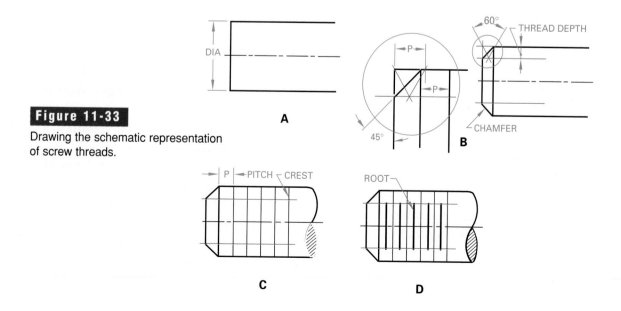

Figure 11-33

Drawing the schematic representation of screw threads.

Calculating the Depth of Threads

The cross-sectional shape of the space between threads of Unified National and ISO metric screw threads is an equilateral triangle, as shown in part A of the illustration below. Remember that an equilateral triangle is a regular figure with three equal sides and three equal angles of 60°. Therefore, to determine the depth of a Unified National or ISO metric thread, simply calculate the altitude of the triangle.

The altitude of an equilateral triangle is calculated by using a simple formula that contains a constant. A *constant* is a known, unchanging value. In this case, the constant is .866 and the formula for calculating the altitude (a) is .866s, as shown in part B of the illustration. This means that every equilateral triangle, regardless of its size, has an altitude that is equal to .866 times or 86.6% of the length of a side of that triangle. For example, an equilateral triangle with sides of 2.00″ would have an altitude of .866 × 2.00 = 1.73″.

Problem: Calculate the depth of .25-20UNC threads.

Solution: The length of side s in part A of the illustration is 1/20″, or .05″.

$$\frac{1}{\text{no. of threads per inch}}$$

a = .866s
a = .866 × .05
a = .0433″ (depth of the .25-20UNC thread)

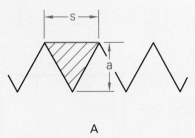

a = ALTITUDE
s = SIDE
a = .866s

A   B

Drawing a Detailed Representation

To draw a detailed representation, draw the screw threads with the sharp-V profile. Use straight lines to represent the helixes of the crest and root lines. To draw the V-form thread, follow the steps in Figure 11-35, as follows:

1. Lay off the pitch *P*, as shown in Figure 11-35A. The pitch need not be drawn to scale. Make it about the right size for the drawing.
2. Lay off the half pitch *P/2* at the end of the thread, as shown.
3. Construct a right triangle with a base of *P/2* and an altitude equal to the outside diameter of the screw. Adjust your triangle, or the ruling arm of your drafting machine, to the slope of this right triangle, and draw all the crest lines with this slope.
4. Use the 30°-60° triangle (or the drafting machine ruling arm set at a 30° angle) to draw one side of the V for the threads, as shown in Figure 11-35B.
5. Reverse the triangle or ruling arm, and complete the Vs.
6. Set the triangle or ruling arm to the slope of the root lines. Draw them as shown in Figure 11-35C. Notice that the root lines do not parallel the crest lines. This is because the **minor diameter** (root diameter) is less than the **major diameter** (crest diameter).
7. Draw 45° chamfer lines using the construction shown in Figure 11-35D.

A hole with an internal right-hand thread is shown as a sectional view in Figure 11-36. Notice that the thread-line slope is in the opposite direction from the external right-hand thread lines on a mating screw. This is because the internal thread lines must match the far side of the screw.

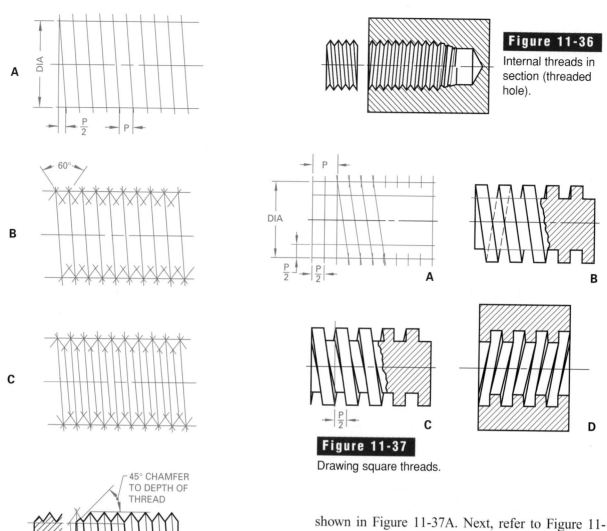

Figure 11-36

Internal threads in section (threaded hole).

Figure 11-37

Drawing square threads.

Figure 11-35

Drawing the detailed representation of screw threads.

Drawing Other Thread Types

Similar procedures are used for drawing other thread types, such as square and acme threads. The procedures for drawing square screw threads and acme threads are presented below.

Square Screw Threads

The square thread has a depth that is one half its pitch. To draw square threads, perform the steps in Figure 11-37. First, lay off the diameter, the pitch P, one-half-pitch spaces, and the depth of the thread, as shown in Figure 11-37A. Next, refer to Figure 11-37B and draw the crest lines. Then draw the root lines, as shown in Figure 11-37C. Figure 11-37D shows an internal square thread drawn in section.

Acme Screw Threads

The **acme thread** has a depth that is one half its pitch. To draw acme threads, follow the method shown in Figure 11-38A. First, lay off the outside diameter, the pitch, and the depth of the thread. Midway between the outside diameter and the depth of the thread is the pitch diameter. Along it, draw the pitch line. On the pitch line, lay off one-half-pitch spaces. Use these to draw the thread profile. Now draw the crest lines and root lines to complete the view. Figure 11-38B shows an enlarged view.

Internal acme threads in section are drawn as shown in Figure 11-38C. They can also be drawn in other ways, including with dashes for the hidden lines or with the outline in section, as shown in Figure 11-38D.

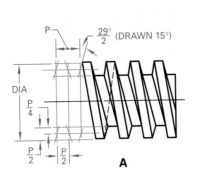

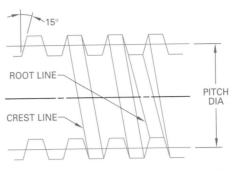

A

B

Figure 11-38

Drawing acme threads.

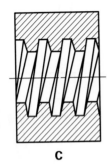

C

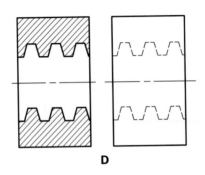

D

Figure 11-39

Regular hexagon bolthead and nut.

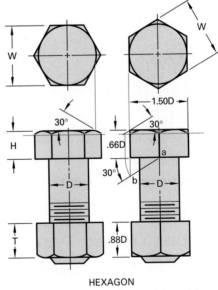

HEXAGON

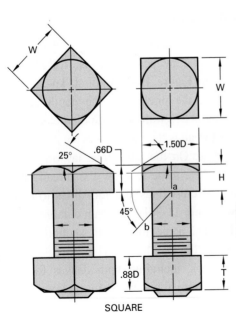

Figure 11-40

Regular square bolthead and nut.

SQUARE

Drawing Boltheads and Nuts

When drawing regular boltheads and nuts, figure the dimensions from the proportions given in Figures 11-39 and 11-40, or consult Appendix C. Draw the chamfer angle at 30° for both the hexagon and the square forms. (The standard for the square form is actually 25°.) Find the radii for the chamfer arcs by trial. Note that you can find one half the distance across the corners, *ab*, by using the construction shown.

Square Bolthead Across Flats

To draw a regular square bolthead across the flats, use the following proportions. If *D* is the major diameter of the bolt and *W* is the width of the bolthead across the flats, then $W = 1.5D$. The height of the head $H = .66D$.

Follow the steps below to draw the bolthead. Refer to Figure 11-41.

1. Draw centerlines and start the top view, as shown in Figure 11-41A.
2. Start the top view by drawing a chamfer circle with a diameter equal to the distance across the flats.
3. Using a 45° triangle, draw a square about the circle.
4. Below the square, start the front view, as shown in Figure 11-41B, by drawing a horizontal line representing the bearing surface of the head. Lay off the height of the head *H*. Then draw the top line of the head.
5. To get the sides of the head, project lines down from the top view.
6. To draw the chamfer, begin by drawing line *ox* (see Figure 11-41A). Revolve the line to make line *oy*, as shown.
7. Run a line from point *y* straight down to the front view, as shown in Figure 11-41B.
8. From point *a* in the front view, draw a 30° chamfer line *ab* out to meet the line from point *y*. This forms the chamfer depth.
9. From point *b*, extend a line horizontally to establish points *c* and *d*, as shown in Figure 11-41C.
10. Draw the chamfer arc through points *c*, *e*, and *d*, using radius *R*. Find the length of *R* by trial, or make it equal to *W*, the width across the flats.
11. Complete the view, as shown in Figure 11-41D.

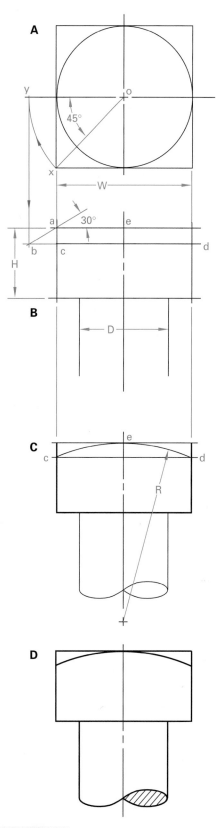

Figure 11-41

Drawing a regular square bolthead across the flats.

Figure 11-42

Drawing a regular square bolthead across the corners.

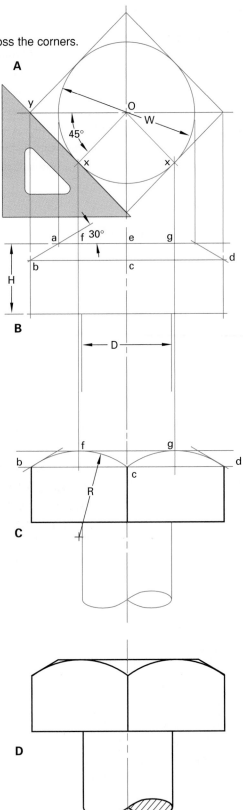

Square Bolthead Across Corners

To draw a regular square bolthead across the corners, use the same proportions you would use to draw it across the flats ($W = 1.5D$, $H = .67D$). Use the following method to create the drawing, as shown in Figure 11-42.

1. Draw centerlines.
2. Start the top view by drawing a chamfer circle with a diameter equal to W, the distance across the flats.
3. Using a 45° triangle, draw a square about the circle.
4. Start the front view by drawing a horizontal line to represent the bearing surface head H.
5. Project lines down from the diameter of the chamfer circle in the top view to meet the top line in the front view.
6. From this point, draw 30° chamfer lines to meet the lines at the sides of the head.
7. Find the two tangent points x shown in Figure 11-42A. Project these points to the top view to get points f and g, as shown in Figure 11-42B and C.
8. Draw line bcd.
9. Draw the chamfer arcs bfc and cgd using radius R. Find radius R by trial, or figure that $R = yo$ = half the distance across the corners.
10. Complete the view, as shown in Figure 11-42D.

Use the same method to draw a nut, except that $T = .88D$ for regular nuts, and $T = D$ for thick nuts. Square bolts are not made in thick sizes.

Boltheads and nuts are usually drawn across the corners on all views of design drawings, no matter what the projection. This is done to show the largest clearance that the bolt or nut must have. It is also done to keep hexagonal heads and nuts from being confused with square heads and nuts.

Hexagon Boltheads

Use the same proportions for hexagon boltheads as for square boltheads. Follow the steps shown in Figure 11-43 to draw a hexagon bolthead across the corners.

1. Start the top view by drawing centerlines and a chamfer circle with a diameter of W, the distance across the flats. See Figure 11-43A.
2. About this circle, draw a hexagon, as shown by lines 1, 2, 3, 4, 5, and 6.
3. Begin the front view, as shown in Figure 11-43B. Draw a horizontal line representing the bearing surface or undersurface of the head.
4. Lay off the height of the head and draw the top line.
5. Project the edges and the chamfer points from the corners of the top view. With these guides, draw in the chamfer line.
6. Draw line *abcd* to locate the chamfer intersections, as shown in Figure 11-43C.
7. Draw arc *bc* using radius R_1. Complete the view by drawing arcs *ab* and *cd* using radius R_2. Find the radius of R_1 and R_2 by trial, as shown in Figure 11-43C and D.

To draw a hexagon bolthead across the flats, proceed as shown in Figure 11-44. Draw hexagon nuts in the same way, but note the difference between the height of the head and the thickness of the nut.

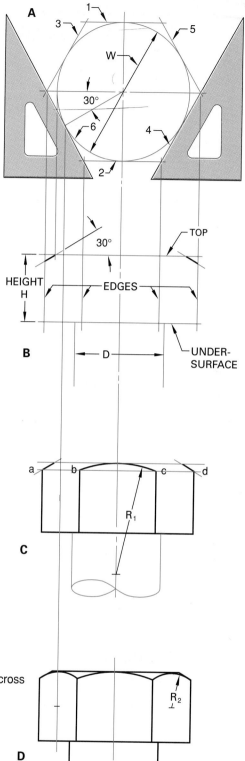

Figure 11-43

Drawing a regular hexagon bolthead across the corners.

Thick Boltheads and Nuts

When drawing thick hexagon boltheads and nuts, figure the dimensions from the proportions given in Figure 11-45. A new standard covers hexagon structural bolts. These bolts are made of high-strength steel, and they are used for structural steel joints.

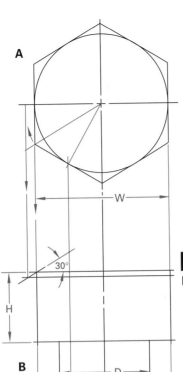

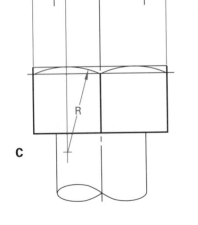

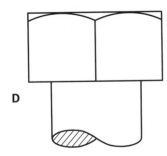

Figure 11-44

Drawing a regular hexagon bolthead across the flats.

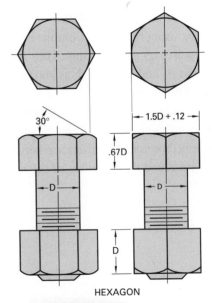

Figure 11-45

Thick hexagon bolthead and nut.

CAD Techniques

One of the advantages of using a CAD system for mechanical drawings is that you only have to draw a complex item, such as a fastener, once. You can then copy the item, or object, as many times as necessary to complete the drawing. In fact, using AutoCAD's blocking ability, you can use the object in other drawings, too.

Fasteners are such common items that many software programs, including the later versions of AutoCAD, provide symbol libraries that contain predrawn fasteners. A **symbol library** is a drawing file that contains a large number of commonly used items, saved as blocks, that you can use in other drawings. See Figure 11-46. From a fastener symbol library, you can select the appropriate fastener, insert it into the drawing at the appropriate size and orientation, and avoid the time needed to create it yourself. Recall that fasteners in most technical drawings are symbolic—they do not have to be drawn to their exact physical size, although they should be drawn according to existing conventions.

There are three ways to incorporate fasteners from symbol libraries into your drawings. You can:

- create your own symbol library
- use AutoCAD's supplied symbol library (if your version of AutoCAD includes it)
- purchase a third-party symbol library

The method you choose depends on several factors. Each method is discussed in this section.

Creating a Symbol Library

There are several reasons why you might want to create your own symbol library. Unless you are using a newer version of AutoCAD, you may not have access to predefined symbol libraries. Your budget, or your company's budget, may not be sufficient to buy third-party software. If the fastener you need is highly specialized, you may not be able to find it in standard symbol libraries. If any of these conditions apply, you may choose to create your own library.

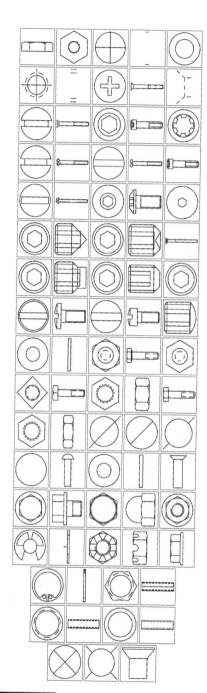

Figure 11-46

A symbol library contains a collection of symbols that you can use in your drawings.

Table 11-2. Estimations for Drawing Threads													
Major Diameter	.25	.31	.38	.44	.50	.56	.62	.69	.75	.81	.88	.94	1.00
Depth	.03	.03	.05	.05	.06	.06	.06	.06	.08	.09	.09	.09	.09
Pitch	.06	.06	.06	.06	.09	.09	.09	.09	.12	.12	.12	.12	.12

As mentioned earlier, a symbol library in AutoCAD is merely an ordinary drawing file that holds a collection of symbols—in this case, fasteners—for use in other drawings. Therefore, the first thing to do is create a new drawing to contain the symbols. Name it "Fasteners Library" so that you will be able to find it quickly when you need it for other drawings.

Many symbol libraries contain hundreds or even thousands of symbols. The symbol library you create in this chapter is intended to be an example only, so you will put only a few fasteners into it. It will provide a basis that you can extend later to meet your needs.

Drawing the Fasteners

Before drawing the fasteners for a symbol library, you should determine whether you want to create simplified, schematic, or detailed representations. All three are used in CAD drawings, depending on the intended purpose of the drawing. Detailed representations are often created as highly accurate 3D solid models so that they can be used with engineering drawings that actually test fit, function, and other characteristics. However, because simplified thread representations are used in most 2D industrial applications, the symbol library you will create in this section will use simplified representation.

TECH●MATH

Determining Pitch and Thread Depth

For screw sizes other than those shown in Table 11-2, you may need to calculate the approximate pitch and thread depth to complete a schematic drawing. First, look up the screw in Appendix C. Find the major diameter and the number of threads per inch. To get the pitch, divide the number of threads per inch by the major diameter of the screw. To approximate the thread depth, subtract the tap drill diameter (also provided in the table) from the major diameter, and divide the result by 2.

Problem: Find the pitch of 1.25-7UNC threads.
Solution: From Appendix C, find the row for 1¼″ and determine that for the UNC threads series, there are 7 threads per inch.

$$\text{Pitch} = \frac{\text{major diameter}}{\text{no. of threads/inch}} = \frac{1.25''}{7} = .18''$$

Problem: Find the approximate thread depth for 1.25-7UNC threads.
Solution: From Appendix C, determine that 1.25-7UNC threads have a tap drill size of 1.109.

$$\text{Approx. Thread Depth} = \frac{\text{major diameter} - \text{tap drill diameter}}{2} = \frac{1.25'' - 1.109''}{2} = \frac{.141''}{2} = .071''$$

Remember that symbols, rather than accurate scale drawings, are used to show fasteners on 2D drawings. Therefore, there is no need, either in board drafting or in CAD, to create the thread symbols using exact sizes. Instead, drafters use fairly standard estimations, or conventions, in their drawings. Table 11-2 shows the major diameter, thread depth, and pitch conventions for drawing threads from .25″ to 1″ in diameter.

Use the CAD commands and techniques discussed in Chapters 2 through 10 to create simplified representations of the following fasteners for your symbol library:

- .500-13UNC by 1.25 hexagon-head bolt
- .500-13UNC hexagon-head nut
- .625-18UNF by 1.50 flat-head cap screw
- Ø.75 by 1.25 button-head rivet

Figure 11-47 shows how each fastener should look in simplified representation. Create each fastener at its true length and height, but remember that you may estimate the threads according to Table 11-2. You will need to refer to the reference tables in Appendix C for some of the sizes needed. You may also wish to review the illustrations in the "Board Drafting Techniques" part of this chapter for appro-priate angles and proportions. However, do not copy the techniques used in board drafting. Instead, think about how best to use AutoCAD commands such as OFFSET, TRIM, and EXTEND to create the fasteners in the most efficient manner. Allow plenty of space around each fastener.

►CADTIP

Saving Your Work
Because you will be doing a large amount of work on the symbol library, you may need to spend a lot of time on it. Be sure to save your work often. That way, if the power suddenly fails, you will lose only the part of the drawing you created since the last time you saved the file.

AutoCAD's SAVETIME command can be used to save the current drawing on a timed schedule. For example, if you set SAVETIME to 5 minutes, AutoCAD will automatically save your drawing every 5 minutes. To use SAVETIME, simply enter the SAVETIME command at the keyboard. Then enter the number of minutes to establish how often AutoCAD automatically saves your drawing.

By default, AutoCAD automatically keeps one backup file of every file you save. If your file corrupts or gets deleted accidentally, search for a file of the same name, but with a "bak" extension instead of a "dwg" extension. Rename the file to give it the "dwg" extension. Then you can open it just like any other AutoCAD drawing. It will contain all of the information that was in the file *before* the last time you saved.

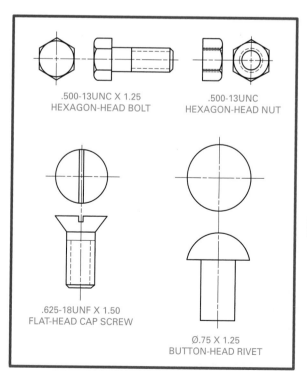

.500-13UNC X 1.25
HEXAGON-HEAD BOLT

.500-13UNC
HEXAGON-HEAD NUT

.625-18UNF X 1.50
FLAT-HEAD CAP SCREW

Ø.75 X 1.25
BUTTON-HEAD RIVET

Figure 11-47

Simplified representations for the custom symbol library.

Creating the Blocks

After you have created all of the symbols for your symbol library, the next step is to create blocks from the symbols. A **block** in AutoCAD consists of one or more objects (lines, arcs, etc.) that you have identified as a single object. To do this, you use the BLOCK or WBLOCK command.

The difference between the BLOCK command and the WBLOCK command is that BLOCK defines the object within the current drawing file. WBLOCK saves the object as a separate DWG file so that you can use it with other drawings more easily. This is especially important in earlier versions of AutoCAD.

To create a block of the hexagon-head bolt, follow these steps:

1. Enter the BLOCK command.

Depending on your version of AutoCAD, you may see a Block Definition dialog box, or you may see prompts at the Command line. In either case, you must enter the same information: the name of the block, the insertion point, and the objects to be included in the block. The **insertion point** is the point at which the block will be attached to the cursor when you insert the block into the drawing. Choosing the insertion point wisely can save time when you are using the block to build a drawing.

2. Enter the following information for this block:
   Name of block: HEX-HD_BOLT
   Insertion point: Use the Intersection object snap to pick the intersection of the top of the bolt head and the centerline.
   Objects to include in block: Use a selection window to select all of the objects that make up the hexagon-head bolt.

3. Press Enter or pick OK to complete the block.

Don't be alarmed if the bolt disappears. In many versions of AutoCAD, the bolt disappears when you block it. Newer versions give you the choice of whether to retain the geometry on the screen. In either case, the bolt now exists in the drawing database as a block, so you can insert it as many times as necessary.

4. Use the same procedure to block the remaining fasteners for your symbol library. Name them HEX-HD_NUT, MACH_SCR, RIVET, and WD_SCR.

5. If the fasteners disappeared when you blocked them, reinsert them into the drawing. Enter the INSERT command and choose to enter the HEX-HD_BOLT block. Notice that it appears on the cursor at the insertion point you specified. Pick a point in the drawing to place the bolt. Notice that you have the opportunity to specify the angle at which the block inserts. Press Enter to insert it at its original orientation. Repeat for each block until all of the symbols are shown in the drawing.

Adding Attributes

To make the symbols in your symbol library even more useful, you can assign attributes to each block. An **attribute** is textual information about the object in a block. For example, you can use an attribute to attach the estimated cost of the object shown in the block. Then, every time you insert the block into the drawing, the cost is recorded in the drawing database. You can later use the attributes to provide a list of materials for the part or assembly shown in the drawing, complete with total estimated cost. You will learn more about materials lists in Chapter 13.

Creating the Attributes

In AutoCAD, the ATTDEF command allows you to create attribute definitions. An attribute definition contains:

- a tag, or name, that identifies the attribute
- a prompt to the user for information (variable attributes only; see below)
- information about the value associated with the attribute

Attributes can be constant or variable. A **constant attribute** is one for which you assign a fixed value. When you insert the block, that value is automatically associated with it. A **variable attribute** is one for which the value can vary. Each time you insert a block that has a variable attribute, AutoCAD prompts you for the value to assign.

Variable attributes are very useful in drafting, especially for the creation of working drawings (see Chapter 13). For example, suppose you have a drawing that contains many bolts. Some are .500-13UNC

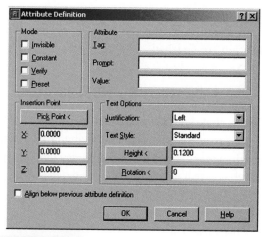

Figure 11-48

The Attribute Definition dialog box.

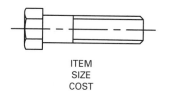

ITEM
SIZE
COST

Figure 11-49

Position the attribute below the symbol, as shown here.

hexagon-head bolts, and others are 1.25-7UNC hexagon-head bolts. You can assign a variable attribute to your HEX-HD_BOLT block to specify its size. This allows you to use one symbol for both types of bolts, while still keeping track of how many of each size you are using in the drawing.

For this symbol library, you will create three attributes for each symbol: Item, Size, and Cost. Follow these steps:

1. Enter the ATTDEF command.

Depending on your version of AutoCAD, you may see a dialog box similar to the one shown in Figure 11-48, or you may see prompts at the Command line. In either case, you will need to enter the same information.

2. Enter the following information:
   Mode: Invisible (In earlier versions of Auto-CAD, press the I key to toggle the Invisible mode on or off.)
   Tag: ITEM
   Prompt: Item name?
   Text height: .12
   Insertion point: Pick a point below the hexagon-head bolt.

3. Press Enter or pick OK to complete the attribute definition. The word ITEM should appear below the hexagon-head bolt.

4. Reenter the ATTDEF command and enter the following information:
   Mode: Invisible

Tag: SIZE
Prompt: What size?
Value: .500-13UNC
Text height: .12
Insertion point: Pick a point just below the ITEM definition.

Notice that the SIZE definition includes a value. This value will appear as the default when you insert the block, but because it is a variable attribute, you can change its value.

5. Reenter the ATTDEF command and enter the following information:
   Mode: Invisible
   Tag: COST
   Prompt: Cost of item?
   Value: .06
   Text height: .12
   Insertion point: Pick a point just below the SIZE definition.

This completes the attributes for the hexagon-head bolt, as shown in Figure 11-49.

6. Repeat steps 1 through 4 to create the definitions for the other blocks.

►CADTIP

Copying and Modifying Attributes

To create attributes quickly for the remaining symbols, copy the attributes for the hexagon-head bolt and place them beneath each of the other symbols. To modify the default values of SIZE and COST, simply modify the properties of those attributes using the Properties dialog box. Choose default values that are typical for the object, or values that you know will be common in the drawing.

Adding Attributes to Blocks

To add the attributes to the symbol blocks, you must redefine the blocks. Follow these steps:

1. Enter the EXPLODE command and pick the hexagon-head bolt symbol. This "explodes" the block back into its original collection of objects. (This does not affect the block in the drawing database.)
2. Enter the BLOCK command. For the name of the block, enter HEX-HD_BOLT. Proceed as you did to create the block originally, but this time include all three of the attributes, as well as the bolt, in the objects to be blocked. When AutoCAD warns you that a block by that name already exists, press Enter or pick OK to redefine it.
3. Repeat steps 1 and 2 to add attributes to each of the other symbols.
4. Use the INSERT command to reinsert the HEX-HD BOLT block into the drawing. Pick an insertion point near the top left corner of the drawing area. Then notice the Command line. As part of the block insertion process now, you can specify the attributes of the block. Enter an item name of Hex Bolt, and press Enter to accept the default values for size and cost.

When the symbol appears on the screen, notice that the attributes do not appear. This is because you specified invisible attributes. Even though they don't appear on the screen, the attributes are attached to the block. Making the attributes invisible allows you to track the attributes of blocks without cluttering up the drawing.

5. Reenter the INSERT command to place each of the remaining symbols back into the drawing.

Finishing the Symbol Library

To finish the symbol library, you may want to add text to identify names of the various blocks. This is not absolutely necessary; it is a courtesy to other people who may use the library. It may also help you when you want to place the symbols into other drawings. Also be sure that all of the blocks are visible in the drawing so that it is easy to see what the symbol library contains. Be sure to save the symbol library when you are finished.

Using a Custom Symbol Library

Now that you have created a custom symbol library, you can use it to make drawing tasks easier in other drawings. To use a symbol library that you have created, you can proceed in two ways, depending on your version of AutoCAD.

Inserting the File

If you are using an older version of AutoCAD, the simplest way to use the blocks from a symbol library is to insert the entire symbol library drawing file into the document in which you want to use them. For example, suppose you have created a new drawing, and you know that you will be needing button-head rivets in this drawing. Follow these steps to make the blocks from your Fasteners Library drawing file available in a new document.

1. Create a new drawing file.
2. Enter the INSERT command. Pick the Browse button, find the Fasteners Library, and double-click it.
3. When AutoCAD prompts you to select the insertion point, press the Escape button to cancel the command.

Canceling the command just before specifying the insertion point has the effect of placing all of the block definitions from Fasteners Library into the current drawing, without actually displaying the blocks on the screen.

4. To be sure the blocks have been inserted into the current drawing, enter the INSERT command. For AutoCAD 2000 and later, pick the down arrow next to Name to see a list of the blocks. If you are using an older version of AutoCAD, enter ? and then Enter to see a list of all of the blocks in the drawing database. The names of the blocks from Fasteners Library should appear.
5. Press Escape or Cancel to leave the INSERT command without inserting the blocks.

Figure 11-50

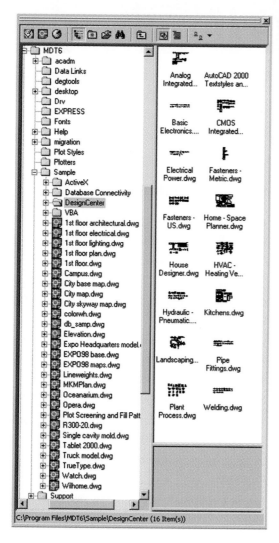

AutoCAD's DesignCenter allows you to drag-and-drop blocks from other files into your drawings.

Browse through the directory tree to see if your version of AutoCAD supplies standard symbol libraries. Some versions of AutoCAD also have a Fasteners library already built into the software. If you have access to the Fasteners library, you can simply pick the item you want and drag it into your current drawing.

The advantage of using default symbol libraries is that you have access to a large number of well-drawn fastener symbols. However, remember that if you use someone else's blocks, you cannot control the attributes associated with the blocks. Therefore, if you will need to use the fasteners later to create a materials list, you might want to either create your own blocks or modify those supplied by AutoCAD. *Note:* **Do not** modify the original AutoCAD files. If you want to modify blocks supplied in a standard library, first save the library using a different name.

Third-Party Symbol Libraries

You can also purchase third-party symbol libraries based on your individual needs. Many companies provide AutoCAD-compatible symbol libraries. Many of them are inexpensive, and some are even free. Consult your instructor about possible sources in your area, or use the Internet to research possible sources. You can usually find this information quickly by using a metasearch engine with keywords such as "CAD symbol library."

Third-party symbol libraries vary in both quality and content. Most of them work with AutoCAD as extensions, so that you can use them from within AutoCAD. If you have DesignCenter, you can also use it to access the symbols in third-party files. However, the procedure for using any third-party product may differ slightly from the method you are using. Read the instructions that come with the software before attempting to use it.

Using DesignCenter

Later versions of AutoCAD come complete with a DesignCenter and several symbol libraries. Check the AutoCAD Help for your version to see if DesignCenter is available to you. DesignCenter is a feature that allows you to access the blocks, linetypes, layers, and other defined items in an existing file without actually opening the file. It displays the items, including a thumbnail or icon, in a dialog box. See Figure 11-50. This is the most efficient way to use symbol libraries.

If DesignCenter is available, pick the AutoCAD DesignCenter icon on the Standard toolbar to open it.

Chapter Summary

- Fasteners include all kinds of devices and methods for holding parts together.

- All screw threads are shaped basically as a helix or helical curve.

- ANSI and ISO maintain drafting standards for threads and fasteners.

- Under ANSI and ISO rules, screw threads can be drawn as detailed, schematic, or simplified representations.

- Screw threads are classified according to several different characteristics, such as diameter and pitch, tolerances, and allowances.

- Complete specifications are required to fully describe threads on a technical drawing.

- Metric thread specifications are somewhat different from those of inch-based threads.

Review Questions

1. What term is used to describe screws, bolts and nuts, rivets, keys, and similar items?

2. Describe the true shape of a screw thread.

3. What do you call the distance from a point on one thread form to a corresponding point on the next thread?

4. What is the distance a threaded part advances with one complete turn called?

5. What types of threads are designed to transmit motion or power along the axis of the screw?

6. Which method of representation is most commonly used in industry?

7. What is another name for double or triple threads?

8. Name the three most common Unified thread series.

9. What is the formula for the pitch of a thread?

10. What do you call a fastener with threads on both ends?

11. What is the difference in designating metric fine- and coarse-thread series?

12. What letter is used in the note for a metric screw thread if a close fit is required?

13. What is the difference between a Class 2A thread and a Class 2B thread?

Drafting Problems

The drafting problems in this chapter are designed to be completed using board drafting techniques or CAD.

1. Use dividers or a scale to take dimensions from the scale below. Draw the views in Figure 11-51 as follows:

   A. Schematic representation of the 1.00-8UNC-2A threads.
   B. End view of (A).
   C. Schematic representation of a section through 1.00-8UNC-2B threads.
   D. Right-side view of (C).
   E. Schematic representation of a section through Ø.88 × 1.50 deep, 1.008UNC-2B × 1.12 deep.
   F. Schematic representation of a section through Ø.88 × 1.50 deep, 1.00-8UNC-2B × 1.50 deep.

2. Use dividers or a scale to take dimensions from the scale below. Draw the views in Figure 11-52 as follows:

   A. Simplified representation of the 1.00-8UNC-2A threads.
   B. End view of (A).
   C. Simplified representation of a section through 1.00-8UNC-2B threads.
   D. Right-side view of (C).
   E. Simplified representation of a section through Ø.88 × 1.50 deep, 1.008UNC-2B × 1.12 deep.
   F. Simplified representation of a section through Ø.88 × 1.50 deep, 1.00-8UNC-2B × 1.50 deep.

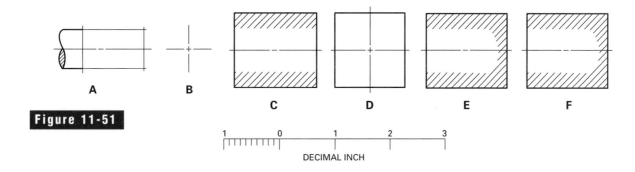

A   B   C   D   E   F

Figure 11-51

1   0   1   2   3
DECIMAL INCH

Figure 11-52

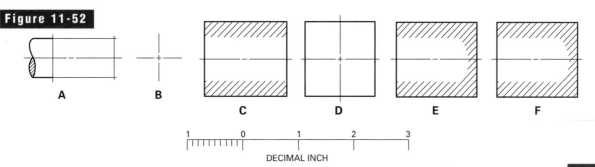

A   B   C   D   E   F

1   0   1   2   3
DECIMAL INCH

For problems 3 through 5, use dividers or a scale to take dimensions from the printed scale. Draw the views as shown and complete them according to the specifications noted on each. Use detailed thread representation.

3. Draw a detailed representation of the screw threads according to the specifications in Figure 11-53.

4. Draw a detailed representation of the double and triple screw threads according to the specifications in Figure 11-54.

5. Draw a detailed representation of the acme and square threads according to the specifications in Figure 11-55.

For problems 6 through 8, draw the views and complete the bolts and nuts in the sectional view. See Appendix C for bolt and nut detail sizes.

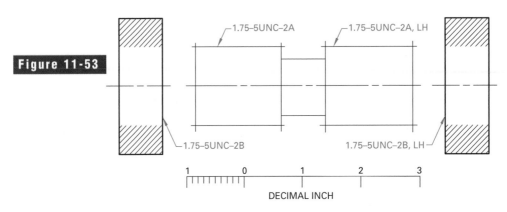

Figure 11-53

1.75–5UNC–2A 1.75–5UNC–2A, LH

1.75–5UNC–2B 1.75–5UNC–2B, LH

DECIMAL INCH

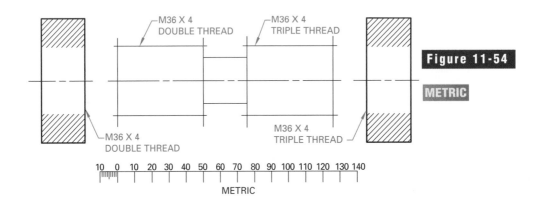

M36 X 4 DOUBLE THREAD M36 X 4 TRIPLE THREAD

Figure 11-54

METRIC

M36 X 4 DOUBLE THREAD M36 X 4 TRIPLE THREAD

METRIC

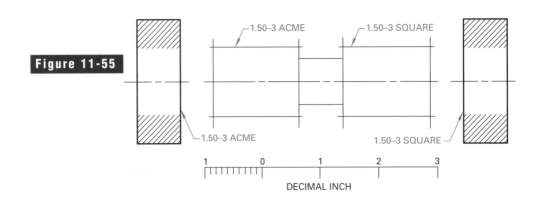

1.50–3 ACME 1.50–3 SQUARE

Figure 11-55

1.50–3 ACME 1.50–3 SQUARE

DECIMAL INCH

6. Draw the sectional view of the regular hexagon bolt and nut. See Figure 11-56.

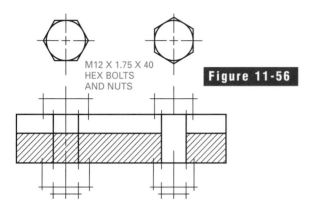

M12 X 1.75 X 40
HEX BOLTS
AND NUTS

Figure 11-56

7. Draw the sectional view of the regular square bolt and nut. See Figure 11-57.

8. Draw the sectional view of the studs shown in Figure 11-58. Complete the Ø12 × 44 studs and hexagon nuts. Other dimensions may be taken from the printed scale at the bottom of the page. Use schematic thread representation.

For problems 9 through 18, take all dimensions from Table 11-3 for the problem assigned, and draw the flange and head plate as shown in Figure

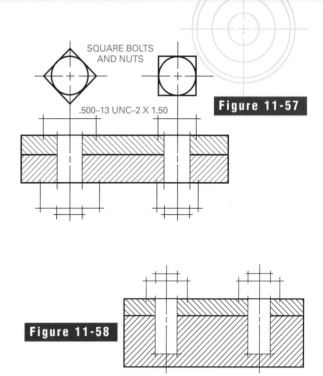

SQUARE BOLTS
AND NUTS

Figure 11-57

.500–13 UNC–2 X 1.50

Figure 11-58

11-59. On the colored centerlines, draw American National Standard hex or square bolts and nuts or metric hex bolts and nuts as specified for the problem. Place the bolthead at the left and show it across the flats. Show the nut across the corners.

Table 11-3. Drafting Problems 9 Through 18										
Problem	Bolt Ø	A	B	C	D	E	F	G	H	I
U.S. Customary										
9.	.250	.25	.32	.62	.50	2.00	.75	.375	.281	.10
10.	.375	.38	.44	.62	.62.	2.25	.75	.375	.406	.12
11.	.500	.50	.56	.88	.62	2.75	1.25	.625	.562	.12
12.	.625	.62	.70	1.12	.75	3.25	1.75	.750	.688	.12
13.	.750	.75	.80	1.25	.88	3.50	2.00	.875	.812	.20
14.	1.000	1.00	1.12	1.50	1.12	4.00	2.50	1.125	1.125	.25
ISO Metric										
15.	6	6	8	16	12	50	20	9.6	6.3	3
16.	10	10	12	20	20	60	20	9.6	10.3	4
17.	16	16	18	24	24	80	40	16.5	16.5	4
18.	24	24	26	32	26	100	60	29.0	25.0	6

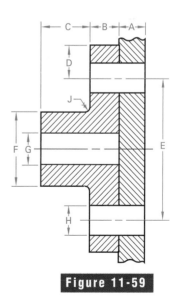

Figure 11-59

For problems 19 through 22, take all dimensions from Table 11-4. On the centerline shown in Figure 11-60, draw a stud with hexagon or square nut, across flats or corners, as directed by the instructor.

Table 11-4. Drafting Problems 19 Through 22

Problem	Bolt Ø	Nut	A	B	C	D	E	F	G
19.	.750	Hex	.80	.88	.75	.812	1.75	1.38	1.00
20.	.875	Sq	.94	1.25	.88	.938	2.00	1.56	1.12
21.	1.000	Hex	1.12	1.12	1.00	1.125	2.25	1.75	1.50
22.	1.125	Sq	1.25	1.50	1.12	1.250	2.75	2.12	1.50

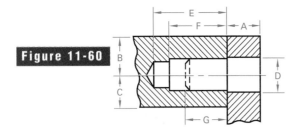

Figure 11-60

For problems 23 through 28, take all dimensions from Table 11-5. Draw the object shown in Figure 11-61. Refer to Appendix C for sizes and draw the assigned style and size of capscrew. Also, draw a top view of the screw head.

Table 11-5. Drafting Problems 23 Through 28

Problem	Bolt Ø	Head Style	A	B	C
		U.S. Customary			
23.	.500	Button (round)	.75	1.75	3.00
24.	.500	Flat	.75	1.75	3.00
25.	.500	Fillister	.75	1.75	3.00
		ISO Metric			
26.	10	Flat	12	40	64
27.	12	Fillister	12	40	64
28.	14	Round	12	40	64

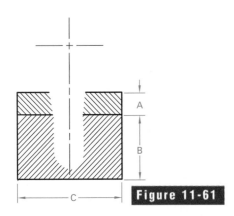

Figure 11-61

For problems 29 through 37, follow the directions to draw the specified fasteners.

29. Refer to Figure 11-62. Draw three setscrews, as described below. Use schematic thread representation, and do not section.
    A. Ø.75 × 1.25 long, square head, flat point.
    B. Ø.75 × 1.25 long, slotted head, oval point.
    C. Ø.75 × 1.25 long, socket head, cup point.

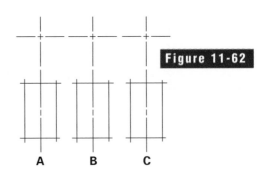

Figure 11-62

30. Draw two views of the spacer shown in Figure 11-63. Use schematic representation to show the threaded holes. A = Ø200; B = 36; C = Ø60; D = M24 (through); E = M10 (through). Add notes and all necessary dimensions.

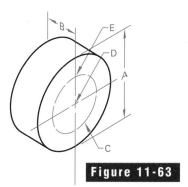

Figure 11-63

31. Draw the Ø1.00 × 3.50 long shaft and Ø1.88 × 1.00 collar, as shown in Figure 11-64. Draw the collar in full section. Add a No. 4 × 2.00 American National Standard taper pin on the vertical centerline. Estimate sizes not given. Materials: shaft—steel; collar—cast iron.

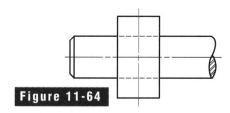

Figure 11-64

32. Draw the necessary views of the shaft support shown in Figure 11-65. At N, draw a Ø.31 setscrew (square head, flat point). At O (four places), draw Ø.38 coarse threads (simplified representation). All fillets and rounds = R.12. AB = 4.50; BC = .50; AD = 3.00; E = 1.50; F = R.88; G = Ø1.00; H = .75; J = .25; K = .62; L = .62; M = .50; P = 1.50.

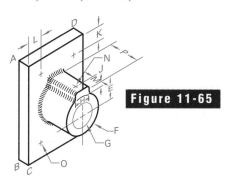

Figure 11-65

33. Draw the Ø1.00 × 3.50 long shaft and Ø2.00 × 1.00 collar, as shown in Figure 11-66. Draw the collar in half section and add an ANSI No. 608 Woodruff key at the top of the shaft. Estimate sizes not given. Materials: shaft—steel; collar—cast aluminum.

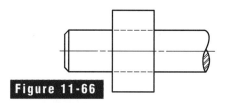

Figure 11-66

34. Draw the view shown in Figure 11-67. Add 2½" #12 wood screws on the colored centerlines. Show the four head types as specified, and draw a top view of each on the center mark above the view.

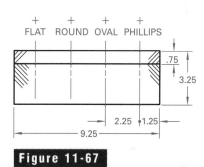

Figure 11-67

35. Draw the view shown in Figure 11-68. On centerline A, draw a Ø.50 × 2.50 fillister-head capscrew (head at top). At B, draw a .38 × 4.00 flat-head capscrew. Show the view in section. Material: steel. Use simplified or schematic thread representation as assigned.

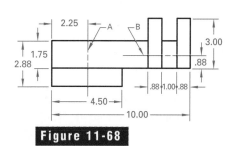

Figure 11-68

36. Draw the figure shown in Figure 11-69. Then draw the rivets specified below. Overall sizes = 1.50 × 7.00. Refer to Chapter 11 and Appendix C and draw the heads (top and bottom) for Ø.50 rivets. Do not dimension.

A. button head

B. high button head

C. cone head

D. flat-top countersunk head

E. round-top countersunk head

F. pan head

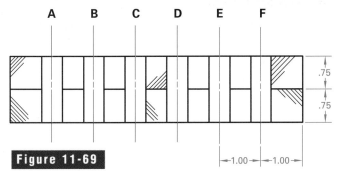

Figure 11-69

37. The thread adapter shown in Figure 11-70 is designed to convert a .68-18UNS-2 spindle thread to a .62-18UNF-2 thread. However, a design change calls for the adapter to convert from Ø1.00″ fine thread series with a Class 3 fit to Ø1.00″ coarse threads with a Class 2 fit. A No. 12 socket-head, cup-point setscrew with fine threads is to be installed at centerline A. You decide the appropriate length. Also, datum A is to be perpendicular to datum B to within .003″ at MMC.

Make a working drawing of the thread adapter using the revised specifications listed above, along with other changes marked on the drawing in color by the design engineer. Use simplified thread representation only. Scale: 1:1.

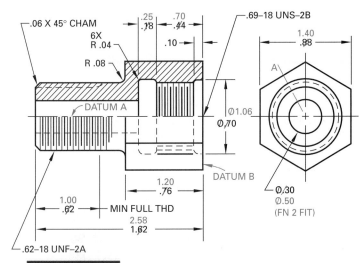

Figure 11-70

Design Problems

Design problems have been prepared to challenge individual students or teams of students. In these problems, you are to apply skills learned mainly in this chapter but also in other chapters throughout the text. They are designed to be completed using board drafting, CAD, or a combination of the two. Be creative and have fun!

1. **TEAMWORK** Work as a team to design a trophy display cabinet. Decide how many and what size trophies you want to display and design the cabinet accordingly. Include as many different types of fasteners as is practical. Remember that fasteners include screws, bolts, adhesives, etc. Material optional.

2. Design a portable cooler that can accommodate six 12-ounce soft-drink cans. Use a wood of your choice for the exterior, a stainless steel liner, and insulation between the box and the liner. Specify epoxy for the wood joints and solder for the metal joints.

3. **TEAMWORK** Design a set of barbecue utensils. Be creative as a team and list as many practical ideas as possible. Have each member of the team produce and present design sketches of a utensil before proceeding with final drawings. Materials and design motifs should be standardized and agreed upon by the team. Use appropriate fasteners to assemble the parts.

4. Design a toddler-age toy used to teach the alphabet. Material optional.

Pictorial Drawing

KEY TERMS

axonometric projection

cabinet oblique

cavalier oblique

dimetric projection

isometric axes

isometric drawing

isoplane

normal oblique

perspective drawing

picture plane

technical illustration

trimetric projection

vanishing point

OBJECTIVES

Upon completion of this chapter, you should be able to:

- List various uses of pictorial drawings.

- Select and draw the most practical type of pictorial for a specific purpose.

- Create isometric drawings with the isometric axes in normal and reversed positions.

- Explain the basic differences in the three types of axonometric projection.

- Make cavalier, normal, and cabinet oblique drawings.

- Develop one-point and two-point perspective drawings.

- Construct irregular curves in pictorial views.

- Select and draw appropriate pictorial sections.

- Manipulate 3D models in AutoCAD to achieve isometric, oblique, and perspective views.

Drafting Principles

Pictorial drawing is an essential part of the graphic language. It is very important in engineering, architecture, science, electronics, technical illustration, and many other professions. It appears in technical literature as well as in catalogs and assembly, service, and operating manuals. Architects use pictorial drawing to show what a finished building will look like. Advertising agencies use pictorial drawings to display new products.

Pictorial drawing is often used in exploded views on production and assembly drawings, as shown in Figure 12-1. These views are made to explain the operation of machines and equipment, to illustrate parts lists, and for many other purposes. See Figure 12-2. In addition, most people use some form of pictorial sketches to help convey ideas that are hard to describe in words.

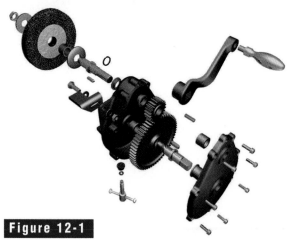

Figure 12-1

An example of a CAD-generated, exploded-view pictorial drawing.

Figure 12-2

An exploded assembly drawing may be used to illustrate a parts list.

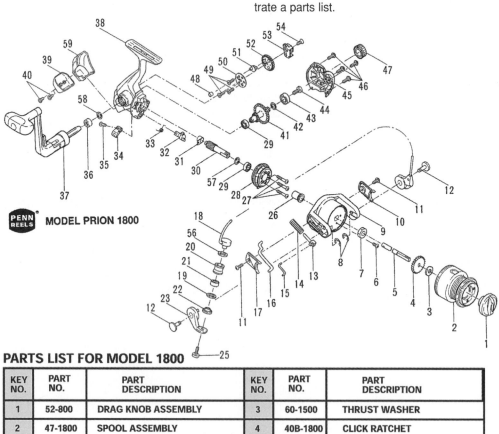

MODEL PRION 1800

PARTS LIST FOR MODEL 1800

KEY NO.	PART NO.	PART DESCRIPTION	KEY NO.	PART NO.	PART DESCRIPTION
1	52-800	DRAG KNOB ASSEMBLY	3	60-1500	THRUST WASHER
2	47-1800	SPOOL ASSEMBLY	4	40B-1800	CLICK RATCHET

Figure 12-3

Types of pictorial drawings.

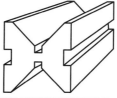

TWO-POINT PERSPECTIVE

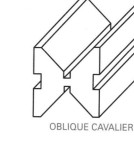

ISOMETRIC

There are three distinct categories of pictorial drawings:

- isometric
- oblique
- perspective

While there are variations of each, the three basic categories are based on how the drawings are constructed and how they appear. Each has its own specific use and is constructed in its own unique way. Figure 12-3 shows a single object drawn using various pictorial techniques.

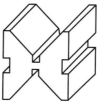

ONE-POINT PERSPECTIVE

OBLIQUE CAVALIER

Isometric Drawing

Chapter 2 discussed techniques for creating isometric sketches. An **isometric drawing** is similar to an isometric sketch except that it is created using instruments. In both cases, the object is aligned with three **isometric axes** at 120° angles from each other, as shown in Figure 12-4A. Several vertical and horizontal positions of the isometric axes are identified in Figure 12-4B and C. You will see how to apply them later in this chapter.

OBLIQUE CABINET

Any line of an object that is parallel to one of the isometric axes is called an **isometric line**. Lines that are not parallel to any of the isometric axes are **nonisometric lines**. See Figure 12-5. An important rule of isometric drawing is:

> *Measurements can be made only on isometric lines.*

Nonisometric lines do not show in their true length, so they cannot be measured.

Figure 12-4

Standard positions for isometric axes.

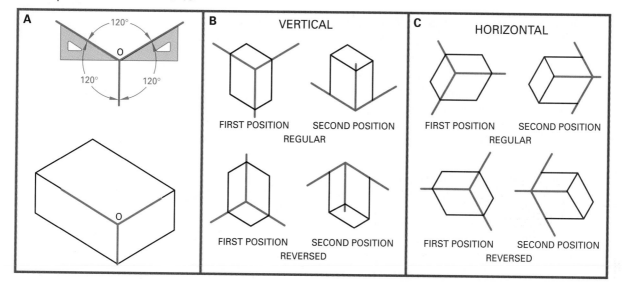

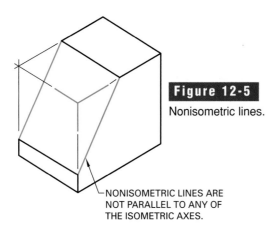

Figure 12-5

Nonisometric lines.

NONISOMETRIC LINES ARE
NOT PARALLEL TO ANY OF
THE ISOMETRIC AXES.

Isometric Projection vs. Isometric Drawing

Isometric projection and isometric drawing are not the same thing, even though many people think they are. This section discusses the differences. While you probably will not have to make pictorial drawings using isometric projection, it is a good idea to understand the theory behind it.

One way to make an isometric projection is by revolution. See Chapter 9 for more information about revolutions. Figure 12-6A shows a cube in the three normal views of a multiview drawing. In Figure 12-6B, each of the three views has been revolved

45°. Notice that the front and side views now show as two equal rectangles. On the side view, a diagonal has been drawn from point O to point B. This is called the **body diagonal**. The body diagonal is the longest straight line that can be drawn in a cube.

In Figure 12-6C, the cube has been revolved upward by 35°16′ to show the body diagonal as a horizontal line in the side view. The front view now forms an **isometric projection**. Its lower edges form an angle of 30° to the horizontal.

Since the front view of the revolved cube is made by projection, its lines are foreshortened—they appear shorter than their true length. In the isometric projection, they appear to be .8165″ instead of their true length of 1.00″.

Figure 12-6D shows an isometric drawing and an isometric projection of the same cube. In the isometric drawing, all edges are drawn their true length instead of the foreshortened length that results from projection. The isometric drawing shows the shape of the cube just as well as the projection. Its advantage is that it is easier to draw, because all its measurements can be made with a regular scale. It can also be drawn without projecting from other views.

For CAD operators, it is important to note that some computer-generated isometric views are isometric projections. Therefore the scale along the axes is not truly a 1:1 scale.

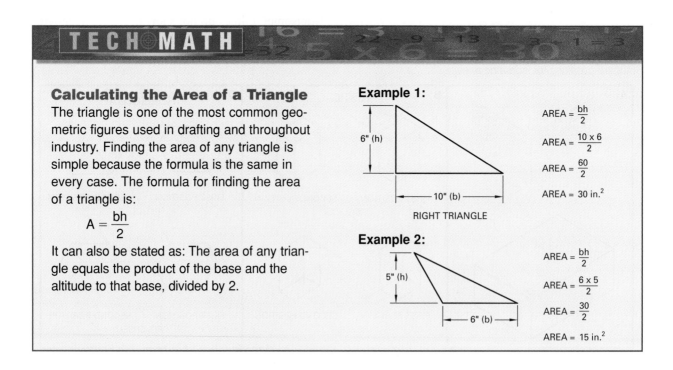

TECH MATH

Calculating the Area of a Triangle

The triangle is one of the most common geometric figures used in drafting and throughout industry. Finding the area of any triangle is simple because the formula is the same in every case. The formula for finding the area of a triangle is:

$$A = \frac{bh}{2}$$

It can also be stated as: The area of any triangle equals the product of the base and the altitude to that base, divided by 2.

Example 1:

6" (h)

10" (b)

RIGHT TRIANGLE

$$AREA = \frac{bh}{2}$$

$$AREA = \frac{10 \times 6}{2}$$

$$AREA = \frac{60}{2}$$

$$AREA = 30 \text{ in.}^2$$

Example 2:

5" (h)

6" (b)

$$AREA = \frac{bh}{2}$$

$$AREA = \frac{6 \times 5}{2}$$

$$AREA = \frac{30}{2}$$

$$AREA = 15 \text{ in.}^2$$

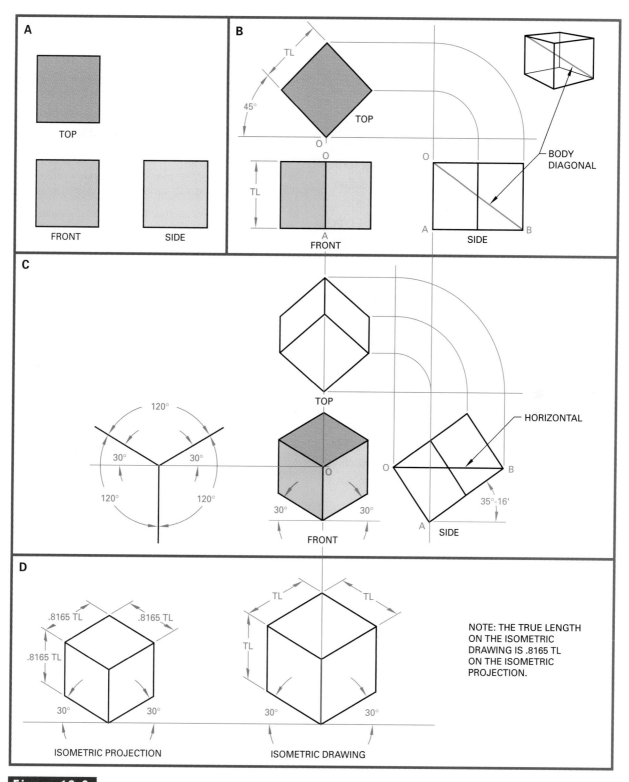

Figure 12-6

Isometric projection and isometric drawing.

Axonometric Projection

Isometric projection is one form of axonometric projection. **Axonometric projection** is projection that uses three axes at angles to show three sides of an object. Other forms are dimetric and trimetric projection. All three are made according to the same theory; the difference is in the angle of projection. See Figure 12-7. In isometric projection, the axes form three equal angles of 120° on the plane of projection. Only one scale is needed for measurements along each axis. Isometric projections are the easiest type of axonometric projection to make. In **dimetric projection**, only two of the angles are equal, and two special foreshortened scales are needed to make measurements. In **trimetric projection**, all three angles are different, and three special foreshortened scales are needed.

Oblique Drawing

Oblique drawings are plotted in the same way as isometric drawings; that is, on three axes. However, in oblique, two axes are parallel to the **picture plane** (the plane on which the view is drawn) rather than just one, as in isometric drawing. These two axes always make right angles with each other, as shown in Figure 12-8. As a result, oblique shows an object as if viewed face on. That is, one side of the object is seen squarely, with no distortion, because it is parallel to the picture plane.

The methods and rules of isometric drawing apply to oblique drawing. However, oblique also has some special rules:

- *Place the object so that the irregular outline or contour faces the front. See Figure 12-9A.*
- *Place the object so that the longest dimension is parallel to the picture plane. See Figure 12-9B.*

Oblique Projection vs. Oblique Drawing

Oblique projection, like isometric projection, is a way of showing depth. See Figure 12-10. Depth is shown by projectors, or lines representing receding edges of the object. These lines are drawn at an angle other than 90° from the picture plane to make the receding planes visible in the front view. As in isometric, lines on these receding planes that are actually parallel to each other are shown parallel. Figure 12-10 shows how an oblique projection is developed. You

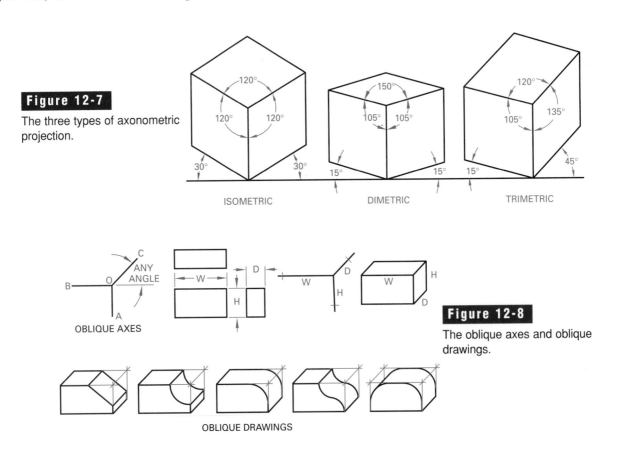

Figure 12-7

The three types of axonometric projection.

ISOMETRIC · DIMETRIC · TRIMETRIC

OBLIQUE AXES

Figure 12-8

The oblique axes and oblique drawings.

OBLIQUE DRAWINGS

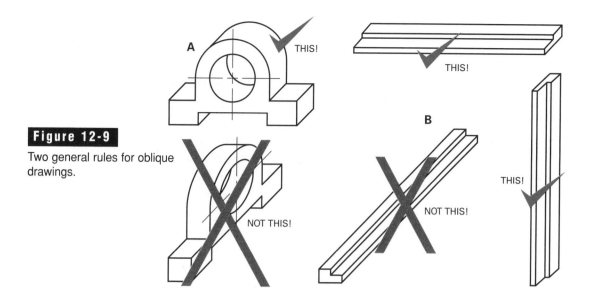

Figure 12-9

Two general rules for oblique drawings.

probably will never have to develop an oblique projection in this way, but as with isometric projection, it is a good idea to understand the theory behind it.

Unlike isometric drawing and projection, no distinction is usually made between oblique projection and oblique drawing. Because oblique drawing can show one face of an object without distortion, it has a distinct advantage over isometric. It is especially useful for showing objects with irregular outlines. Refer again to Figure 12-9A.

Figure 12-11 shows several positions for oblique axes. In all cases, two of the axes, AO and OB, are drawn at right angles. The oblique axis OC can be at any angle to the right, left, up, or down. The best way to draw an object is usually at the angle from which it would normally be viewed.

Types of Oblique Drawings

Oblique drawings are classified according to the length of the receding lines of an object along the oblique axis. Drawings in which the receding lines are drawn full length are known as **cavalier oblique**. Some drafters use three-quarter size. This is sometimes called **normal oblique**, or general oblique. If the receding lines are drawn one-half size, the drawing is **cabinet oblique**. Figure 12-12 shows a bookcase drawn in cavalier, normal, and cabinet drawings. Cabinet drawings are so named because they are often used in the furniture industry.

Perspective Drawing

A **perspective drawing** is a three-dimensional representation of an object as it looks to the eye from a particular point. See Figure 12-13. Of all pictorial drawings, perspective drawings look the most like photographs. The distinctive feature of perspective drawing is that in perspective, lines on the receding planes that are actually parallel are not drawn parallel, as they are in isometric and oblique drawing. Instead, they are drawn as if they were converging, or coming together.

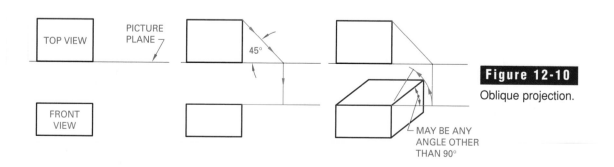

Figure 12-10

Oblique projection.

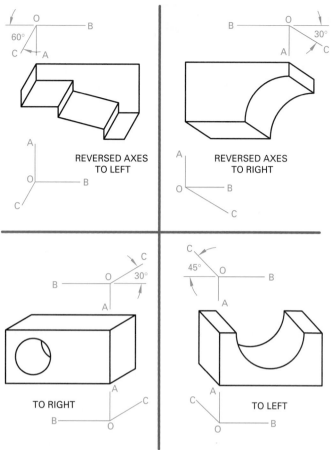

Figure 12-11

Positions for oblique axes.

Definition of Terms

Figure 12-14 illustrates terms used in perspective drawing. The following definitions refer to the card that appears on the picture plane.

- **visual rays** The sight lines leading from points on the card and converging at the eye of the observer.

- **picture plane (PP)** The plane on which the object is drawn. In this case, the object is a card.
- **line of sight (LS)** The visual ray from the eye perpendicular to the picture plane.
- **station point (SP)** The point from which the observer is looking at the card.
- **horizon line (HL)** The line formed where a horizontal plane that passes through the observer's eye meets the picture plane.
- **ground plane** The plane on which the observer stands.
- **ground line (GL)** The line formed where the ground plane meets the picture plane.
- **center of vision (CV)** The point at which the line of sight pierces the picture plane.

Figure 12-15 shows how the projectors, or receding axes, converge in perspective drawing. The point at which they meet is the **vanishing point**. Figure 12-15 also shows how the observer's eye level affects the perspective view. This eye level can be anywhere on, above, or below the ground. If the object is seen from above, the view is an **aerial view**, or bird's-eye view. If the object

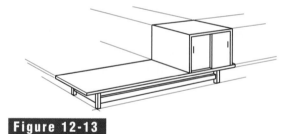

Figure 12-13

Perspective drawing of a music center.

Figure 12-12

Three types of oblique drawings.

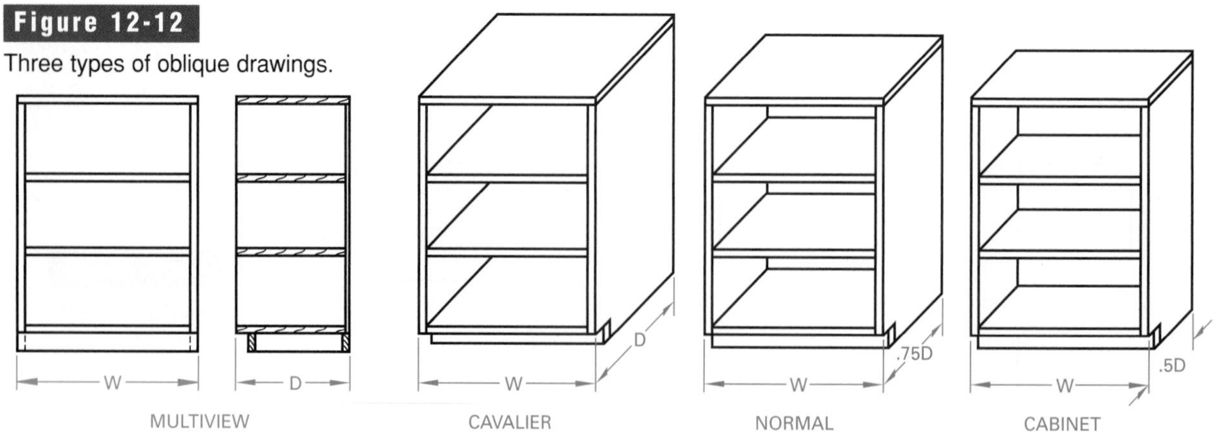

MULTIVIEW   CAVALIER   NORMAL   CABINET

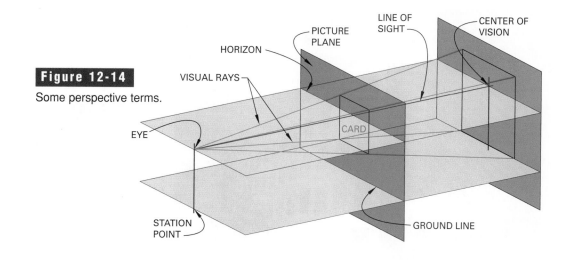

Figure 12-14

Some perspective terms.

LINE OF SIGHT
CENTER OF VISION
PICTURE PLANE
HORIZON
VISUAL RAYS
EYE
CARD
STATION POINT
GROUND LINE

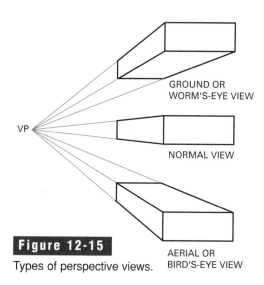

GROUND OR WORM'S-EYE VIEW

VP

NORMAL VIEW

Figure 12-15

Types of perspective views.

AERIAL OR BIRD'S-EYE VIEW

Two-point perspective drawings have two vanishing points. This type is also called *angular perspective*, because none of the faces are drawn parallel to the picture plane. The photograph in Figure 12-17 shows a typical two-point perspective.

Factors That Affect Appearance

Two factors affect how an object looks in perspective. The first is its distance from the viewer, and the second is its position, or angle, in relation to the viewer.

is seen from underneath, the view is a **ground view**, or worm's-eye view. If the object is seen face on, so that the line of sight is directly on it rather than above or below, the view is a **normal view**. The view in Figure 12-14 is a normal view.

Types of Perspective Drawings

Perspective views can have one, two, or even more vanishing points. **One-point perspective**, also called *parallel perspective*, is a perspective view in which there is one vanishing point. See Figure 12-16. Notice that if the lines of the building in Figure 12-16 were extended, they would converge at a single point.

Figure 12-16

The lines of the sidewalk, roof, and building's side appear to converge at a single point in the distance.

Figure 12-17
When a building is viewed at an angle, two sides can be seen. The top and ground lines on each side appear to converge toward points. This is the effect of two-point perspective.

The Effect of Distance

The size of an object seems to change as you move toward or away from it. The farther away from the object you go, the smaller it looks. As you get closer, it seems to grow larger. Figure 12-18 shows a graphic explanation of this effect of distance. An object is placed against a scale at a normal reading distance from the viewer. In that position, it appears to be the size indicated by the scale. However, if it is moved back from the scale to a point twice as far away from the viewer, it looks only half as large. Notice that each time the distance is doubled, the object looks only half as large as before.

The Effect of Position

The shape of an object also seems to change when it is viewed from different angles. This is illustrated in Figure 12-19. If you look at a square from directly in front, the top and bottom edges are parallel. But if the square is rotated so that you see it at an angle, these edges seem to converge. The square also appears to grow narrower. This foreshortening occurs because one side of the square is now farther away from you.

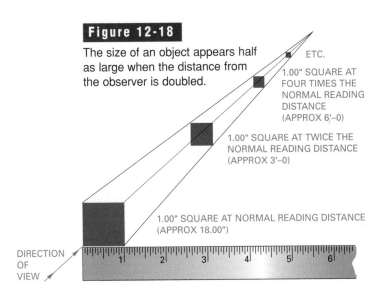

Figure 12-18
The size of an object appears half as large when the distance from the observer is doubled.

ETC.

1.00" SQUARE AT FOUR TIMES THE NORMAL READING DISTANCE (APPROX 6'–0)

1.00" SQUARE AT TWICE THE NORMAL READING DISTANCE (APPROX 3'–0)

1.00" SQUARE AT NORMAL READING DISTANCE (APPROX 18.00")

DIRECTION OF VIEW

Technical Illustration

Generally, **technical illustration** is defined as a pictorial drawing that provides technical information using visual methods. Technical illustrations are used to present complex parts and assemblies graphically in a way that both professionals and the general public can read and understand. They help people understand both the form (shape) and function of parts in assembly. Technical illustrations must show shapes and relative positions in a clear and accurate way.

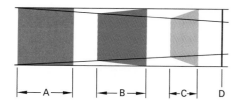

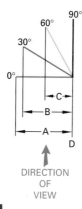

Figure 12-19

The position of the object in relationship to the observer affects its appearance.

Shading may be used to bring out the shape. A technical illustration, however, is not necessarily a work of art. Therefore, the shading must serve a practical purpose, not an artistic one.

Technical illustrations range from sketches to rather detailed shaded drawings and may be prepared using board drafting techniques or sophisticated CAD or modeling software. They may be based on any of the pictorial methods: isometric, perspective, or oblique. The complete project, parts, or groups of parts may be shown. The views may be exterior, interior, sectional, cutaway, or phantom. The purpose in all cases is to provide a clear and easily understood description.

In addition to pictorials, technical illustrations include graphs, charts, schematics, flowcharts, diagrams, and sometimes circuit layouts. Dimensions are not generally a part of technical illustrations because they are not working drawings. However, dimensions are occasionally added to show the relative position of parts or to show the adjustment of parts in an assembly.

Purposes of Technical Illustration

Technical illustration has an important place in all areas of engineering and science. Technical illustrations form a necessary part of the technical and service manuals for machine tools, automobiles, machines, and appliances. In technical illustration, pictorial drawings are used to describe parts in terms of both their form and function. They can also show the steps that need to be followed to complete a product on the assembly line. In fact, they may even be used to set up an assembly line.

Technical illustrations have been used for many years in illustrated parts lists, operation and service manuals, and process manuals. See Figure 12-20. The aircraft and automotive industries in particular have found production illustrations especially valuable. In many industries, pictorial drawings are used when the plan is first designed. These drawings are also used throughout the many phases of production and completion on the assembly line. When the product is delivered to the customer, the service, repair, and operation manuals also include technical illustrations to show the customer assembly and operating procedures.

Figure 12-20

An illustrated parts list.

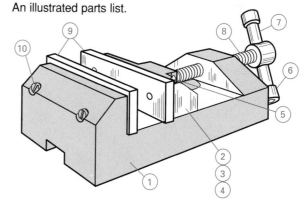

PART NO.	PART NAME	NO. REQ'D
1	BASE	1
2	MOVABLE JAW	1
3	MOVABLE JAW PLATE	1
4	MACHINE SCREW	1
5	LOCKING PIN	1
6	HANDLE STOP	1
7	HANDLE	1
8	CLAMP SCREW	1
9	JAW FACE	2
10	CAP SCREW	2

Choosing a Drawing Type

Most technical illustrations are pictorial line drawings. Therefore, you should have a complete understanding of the various types of pictorial line drawings and their uses. The first half of this chapter describes the various types. Usually, any type of pictorial drawing can be used as the basis for a technical illustration. However, some types are more suitable than others. This is especially true if the illustration is to be **rendered** (shaded).

Figure 12-21 is a V-block shown in several types of pictorial drawing. Notice the difference in the appearance of each. Isometric is the least natural in appearance. Perspective is the most natural. This might suggest, then, that all technical illustrations should be drawn in perspective. This is not necessarily true. While perspective is more natural than isometric in appearance, it takes more time to produce, and it is also more difficult to draw. Thus, it is often a more costly method to use.

The shape of the object also helps to determine the type of pictorial drawing to use. Figure 12-22 shows a pipe bracket drawn in isometric and oblique.

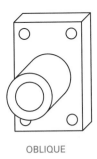 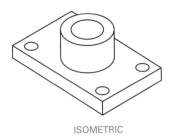

OBLIQUE   ISOMETRIC

Figure 12-22

The shape of an object helps to determine the most suitable type of pictorial drawing to use.

This object can be drawn easily and quickly in oblique. Also, in many cases, the oblique looks more natural than isometric for objects of this shape.

If an illustration is to be used only in-plant, the illustrator usually makes the pictorial drawing in isometric or oblique. These are the quickest and least costly pictorials to make. If the illustration is to be used in a publication such as a journal, operator's manual, or technical publication, dimetric, trimetric, or perspective may be used.

Exploded Views

Perhaps the easiest way to understand an exploded view is this: take an object and separate it into its individual parts, as shown in Figure 12-23. In this illustration, three views are shown in part A, and a pictorial view is shown in part B. In part C, an "explosion" has projected the various parts away from each other. This illustrates the principle of exploded views.

All exploded views are based upon the same principle: projecting the parts from the positions they occupy when put together. Simply, they are just pulled apart. The exterior of a fishing reel is shown in Figure 12-24A. An exploded illustration of the reel is shown in Figure 12-24B. Note that all parts are easily identifiable in the exploded view. Flow lines are generally used to show exactly where each part fits into the assembly.

Identification Illustrations

Pictorial drawings are very useful for identifying parts. They save time when the parts are manufactured or assembled in place. They are also useful for illustrating operating instruction manuals and spare-parts catalogs.

Figure 12-21

A V-block in various types of pictorial drawing.

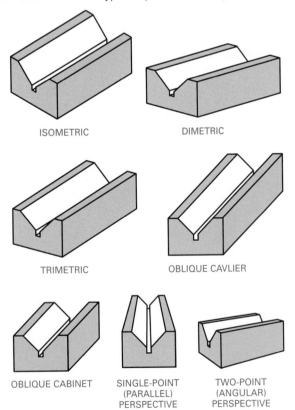

ISOMETRIC   DIMETRIC

TRIMETRIC   OBLIQUE CAVLIER

OBLIQUE CABINET   SINGLE-POINT   TWO-POINT
                    (PARALLEL)   (ANGULAR)
                    PERSPECTIVE   PERSPECTIVE

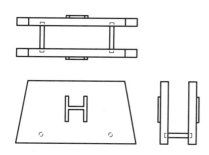

LETTER HOLDER
ORTHOGRAPHIC MULTIVIEW DRAWING

A

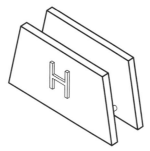

ISOMETRIC ASSEMBLY DRAWING

B

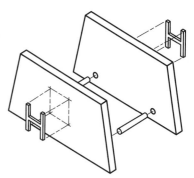

EXPLODED ISOMETRIC DRAWING

C

Figure 12-23

How a view is exploded.

Success on the Job

Pride

Take pride in all that you do. Regardless of the assignment, do it to the very best of your ability. Developing a reputation for consistently turning out top quality work in a timely manner will greatly enhance your chances for career advancement.

Identification illustrations are usually presented in exploded views. If an object consists of only a few parts, identify them by their names as shown in Figure 12-25. If the object contains several parts, number them as shown in Figure 12-24B. In this example, the names of the numbered parts are given in a parts list, a portion of which is shown in Figure 12-24C.

Figure 12-24

A fishing reel (A) with an exploded assembly drawing (B) and partial parts list (C).

A

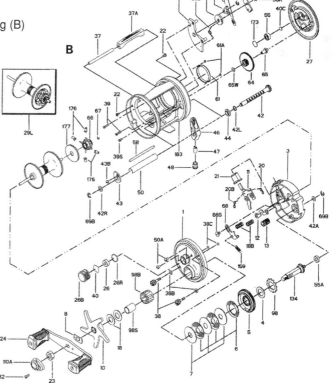

B

 PENN REELS **PARTS LIST FOR MODEL 965**

KEY NO.	DESCRIPTION	MODEL 965 PART NO.
1	Right Side Plate	1N-965
3	Inner R.S. Plate	3-965
4	Fibre Washer	4-60
5	Main Gear	4-60
6	Drag Washer HT-100 (Set of 3)	6C-965
7	Metal Drag Washer (Set of 3)	7C-965
8	Star Tension Washer	8-965
10	Star Drag	10-965

C

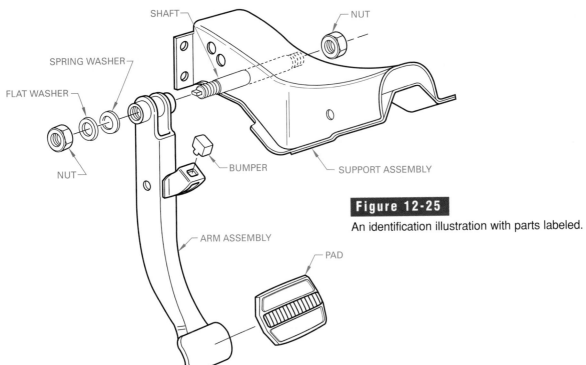

SHAFT

NUT

SPRING WASHER

FLAT WASHER

NUT

BUMPER

SUPPORT ASSEMBLY

ARM ASSEMBLY

PAD

Figure 12-25

An identification illustration with parts labeled.

Rendering

Surface shading, or rendering, may be used when shapes are difficult to read or for aesthetic reasons. For most industrial illustrations, accurate descriptions of shapes and positions are more important than fine artistic effects. You can often achieve satisfactory results without any shading. In general, you should limit surface shading when possible. Shade the least amount necessary to define the shapes that are being illustrated. Shading takes time and is expensive.

In board drafting, different ways of rendering technical illustrations include the use of screen tints, pen and ink, wash, stipple, felt-tip pen and ink, smudge, and edge emphasis, among others. These means of rendering are used in the technical illustrations of aircraft companies, automobile manufacturers, and machine-tool makers. They can even be found in the directions that come with your television.

In CAD drawings, rendering is done in a very different way. Drawings that are to be rendered are created as solid models using 3D drawing techniques. These models can then be rendered using the rendering function of the CAD software. However, many top companies now import the models into high-end, dedicated rendering software. Figure 12-26 shows an example of a part that has been modeled and rendered using a CAD program.

Figure 12-26

CAD programs and third-party rendering software can be used to render solid models.

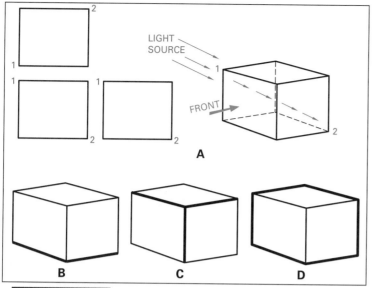

Figure 12-27

Light source and light-shaded cubes.

Outline Shading

Outline shading may be done mechanically or freehand. Sometimes a combination of both methods is used. The light is generally considered to come from behind and above the left shoulder of the observer, across the diagonal of the object, as shown in Figure 12-27A. This is a convention, or standard method, used by board drafters and renderers. In Figure 12-27B, the upper left and top edges are in the light, so they are drawn with thin lines. The lower right and bottom edges are in the shadow. They should be drawn with thick lines. In Figure 12-27C, the edges meeting in the center are made with thick lines to accent the shape. In Figure 12-27D,

Figure 12-28

A maintenance illustration. Notice that only the necessary detail is shown and that just enough shading is used to emphasize and give form to the parts.

the edges meeting at the center are made with thin lines. Thick lines are used on the other edges to bring out the shape. An example of the use of a small amount of line shading is shown in Figure 12-28. In this case, the shading is used to outline important parts of the drawing.

Surface Shading

Surface shading can be done using board drafting techniques or computer software. In either method, the theory of shading is the same. With the light rays coming from the conventional direction, as shown in Figure 12-29A, the top and front surfaces of a cube should be lighted. Therefore, the right-hand surface should be shaded. In Figure 12-29B, the front surface is unshaded and the right surface is lightly shaded using vertical lines. If the front surface has light shading, then the right side should have heavier shading, as shown in Figure 12-29C. Solid shading may sometimes be required to avoid confusion. If the front is shaded, then a darker shade may

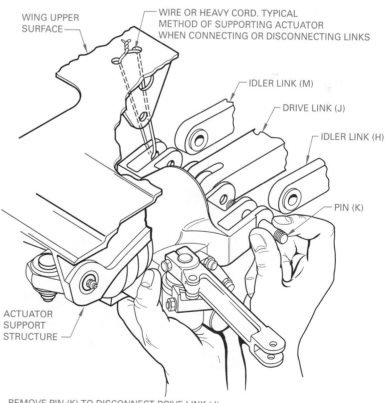

REMOVE PIN (K) TO DISCONNECT DRIVE LINK (J) FROM THE ACTUATOR. PIN (K) IS INSTALLED IN THE SAME MANNER TO CONNECT DRIVE LINK (J) TO THE ACTUATOR

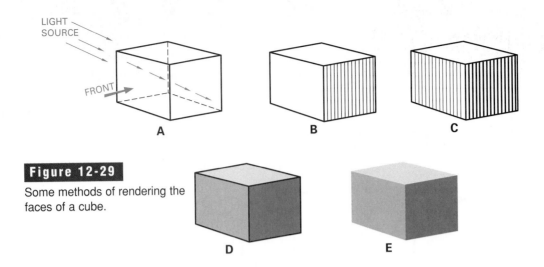

Figure 12-29

Some methods of rendering the faces of a cube.

be used on the right-hand side, as shown in Figure 12-29D. Figure 12-29E was shaded in AutoCAD using the SHADE command. Notice that all three visible sides are shaded differently to define the 3D shape.

Wash Rendering

A wash drawing, or **wash rendering,** is a form of watercolor rendering that has traditionally been done with watercolor and watercolor brushes. CAD drawings can be imported to an illustration program and "painted" to achieve the same effect. Wash rendering is commonly used to render architectural drawings, as shown in Figure 12-30. It is also used for advertising furniture and similar products in newspapers, as shown in Figure 12-31. This technique is highly

specialized and is usually done by a commercial artist. However, some technical illustrators and drafters are occasionally required to do this kind of illustrating.

Photo Retouching

Occasionally, drafters need to change details on a photograph. **Photo retouching** is the process used to accomplish this. It can be done by hand or with special computer software such as Adobe Photoshop®. Details may be added, removed, or simply repaired. This process is often needed in preparing photographs for use in publications. It can be used to change the appearance of specific details or the entire photograph, as shown in Figure 12-32A and B.

Figure 12-30

Wash rendering of an architectural drawing.

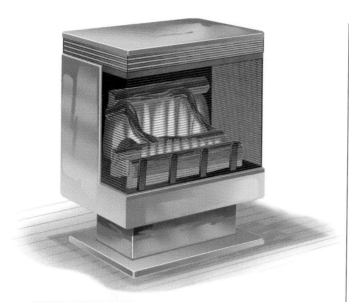

Figure 12-31

This illustration was done electronically. Notice that the computer artist has achieved a wash-rendered effect.

Figure 12-32

Photograph of a table saw, (A) before and (B) after retouching.

A

B

Board Drafting Techniques

In earlier chapters, you learned the basic elements and techniques of drafting. To develop a pictorial drawing, you simply need to develop an understanding of the pictorial concepts and then apply basic drafting skills to the drawing. The various types of pictorial drawings and their applications were discussed earlier in this chapter. This section describes how to apply the principles of pictorial drawing using board drafting techniques.

Isometric Drawing Techniques

In isometric drawing, drawing order is important. For example, you must create the isometric lines—those that are parallel to the isometric axes—before you can begin to draw nonisometric lines. This section explains construction techniques and then steps you through practice exercises for isometric drawing.

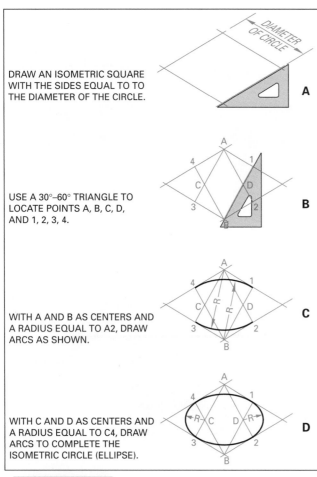

DRAW AN ISOMETRIC SQUARE WITH THE SIDES EQUAL TO TO THE DIAMETER OF THE CIRCLE. **A**

USE A 30°–60° TRIANGLE TO LOCATE POINTS A, B, C, D, AND 1, 2, 3, 4. **B**

WITH A AND B AS CENTERS AND A RADIUS EQUAL TO A2, DRAW ARCS AS SHOWN. **C**

WITH C AND D AS CENTERS AND A RADIUS EQUAL TO C4, DRAW ARCS TO COMPLETE THE ISOMETRIC CIRCLE (ELLIPSE). **D**

Figure 12-33

Steps in drawing an isometric circle.

Isometric Constructions

Before you attempt to create an isometric drawing using drafting instruments, you should understand the techniques used to create various geometric shapes accurately. Several procedures are described in the following sections. Practice these techniques before you attempt to create an isometric drawing.

Isometric Circles

In isometric drawing, circles appear as ellipses. Since it takes a long time to plot a true ellipse, a four-centered approximation is generally drawn, especially for large isometric circles. Isometric circle templates may be used for small ellipses. Figure 12-33 describes how to create a four-centered approximation of an ellipse. Figure 12-34 shows isometric circles drawn on three surfaces of a cube.

Figure 12-35 shows how to make an isometric drawing of the cylinder shown as a multiview drawing in Figure 12-35A. Follow these steps:

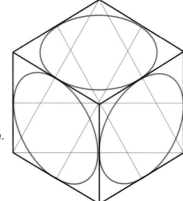

Figure 12-34

Isometric circles on a cube.

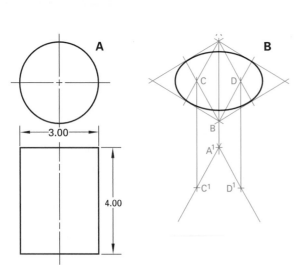

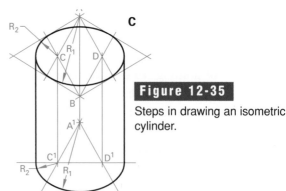

Figure 12-35

Steps in drawing an isometric cylinder.

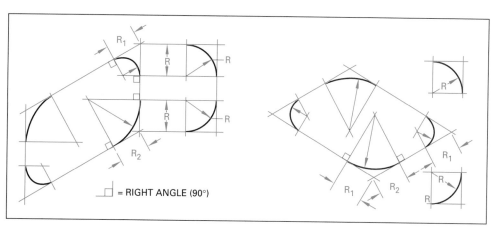

Figure 12-36

Drawing quarter rounds in isometric drawings.

☐ = RIGHT ANGLE (90°)

1. Draw an ellipse of the 3.00″ circle following the procedure shown in Figure 12-33.
2. Drop centers A, C, and D a distance equal to the height of the cylinder (in this case, 4.00″) as shown in Figure 12-35B.
3. Draw lines A′C′ and A′D′.
4. Complete the ellipses as shown in Figure 12-35C. Construct a line through C′D′ to locate the points of tangency. Draw the arcs using the same radii as in the ellipse at the top.
5. Draw the vertical lines to complete the cylinder. Notice that the radii for the arcs at the bottom match those at the top.

To draw quarter rounds, use the method shown in Figure 12-36, or use an ellipse template. In each case, measure the radius along the tangent lines from the corner. Then draw the actual perpendiculars to locate the centers for the isometric arcs. Observe that the method for finding R_1 and R_2 is the same as that for finding the radii of an isometric circle. When an arc is more or less than a quarter circle, you can sometimes plot it by drawing all or part of a complete isometric circle and using as much of the circle as needed.

Figure 12-37 shows how to draw outside and inside corner arcs. Note the tangent points T and the centers 1, 1′, 2, and 2′.

Irregular Curves

Irregular curves in isometric drawings cannot be drawn using the four-center method. To draw irregular curves, you must first plot points and then connect them using a French curve, as shown in Figure 12-38.

Isometric Templates

Isometric templates are made in a variety of forms. They are convenient and can save time when you have to make many isometric drawings. Many of them have openings for drawing ellipses, as well as 60° and 90° guiding edges. Simple homemade guides like those shown in Figure 12-39A are convenient for straight-line work in isometric. These templates can be made to any convenient size. Figure 12-39B shows various ways to position the templates for making an isometric drawing. Ellipse templates like those shown in Figure 12-40 are very convenient for drawing true ellipses. If you use these templates,

Figure 12-37

Constructing outside and inside arcs.

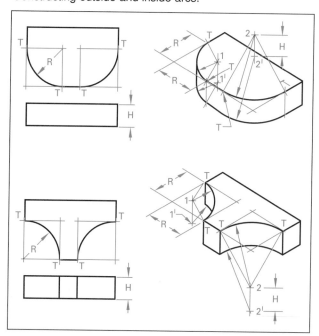

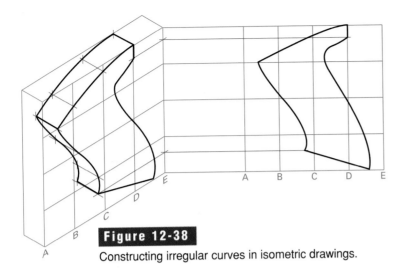

Figure 12-38

Constructing irregular curves in isometric drawings.

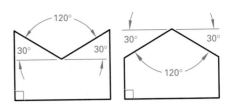

A DESIGN OF HOMEMADE ISOMETRIC TEMPLATES

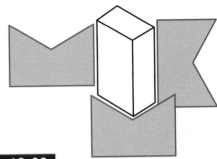

Figure 12-39

Simple isometric templates.

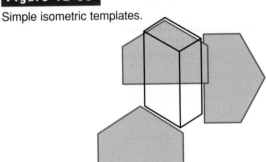

B VARIOUS POSITIONS OF TEMPLATES

your drawings will look better and you will not have to spend time plotting approximate ellipses. See Chapters 3 and 4 for information on templates and how to use them.

Creating an Isometric Drawing

Figure 12-41A shows a multiview drawing of a filler block. To make an isometric drawing of the block, begin by drawing the isometric axes in the first position, as shown in Figure 12-41B. They represent three edges of the block. Draw them to form three equal angles. Draw axis line OA vertically. Then draw axes OB and OC using the 30°-60° triangle. The point at which the three lines meet represents the upper front corner O of the block, as shown in Figure 12-41C.

Measure off the width W, the depth D, and the height H of the block on the three axes. Then draw lines parallel to the axes to make the isometric drawing of each block. To locate the rectangular hole shown in Figure 12-41D, lay off 1.00″ along OC to c. Then from c, lay off 2.00″ to c′. Through c and c′

draw lines parallel to OB. In like manner, locate b and b′ on axis OB and draw lines parallel to OC. Draw a vertical line from corner 3. Darken all necessary lines to complete the drawing, as shown in Figure 12-41E.

Pictorial drawings, in general, are made to show how something looks. Since hidden edges are not "part of the picture," they are normally left out. However, you might need to include them in special cases to show a certain feature.

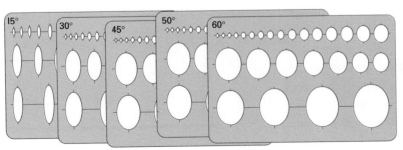

Figure 12-40

Ellipse templates.

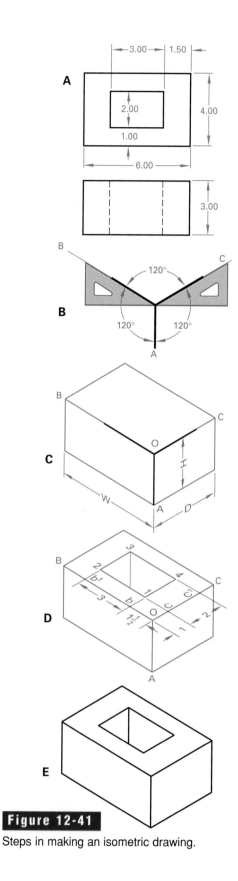

Figure 12-41

Steps in making an isometric drawing.

Figure 12-42 shows how to make an isometric drawing of a guide. The guide is shown in a multi-view drawing in Figure 12-42A. This drawing is more complex because it includes a circular hole and several rounded surfaces. Study the size, shape, and relationship of the views in Figure 12-42A before you proceed. Then follow these steps:

1. Draw the axes AB, AC, and AD in the second position, as shown in Figure 12-42B.

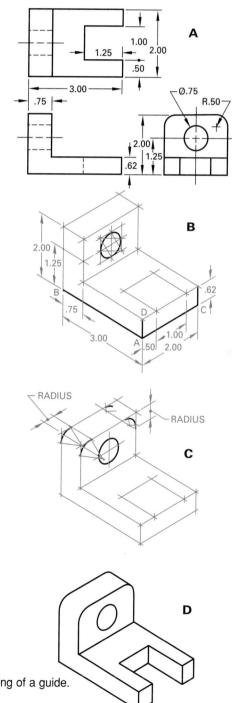

Figure 12-42

Creating an isometric drawing of a guide.

2. Lay off the length, width, and thickness measurements given in Figure 12-42A. That is, starting at point A, measure the length (3.00″) on AB; the width (2.00″) on AC; and the thickness (.62″) on AD.

3. Through the points found, draw isometric lines parallel to the axes. This "blocking in" produces an isometric view of the base.

4. Block in the upright part in the same way, using the measurement given in the top view of Figure 12-42A.

5. Find the center of the hole and draw centerlines, as shown in Figure 12-42B.

6. Block in a .75″ isometric square and draw the hole as an approximate ellipse or use an ellipse template.

7. To make the two quarter rounds, measure the .50″ radius along the tangent lines from both upper corners, as shown in Figure 12-42C. Draw real perpendiculars to find the centers of the quarter circles. See Figure 12-36 for more information about drawing isometric quarter rounds. An ellipse circle template can also be used for this purpose.

8. Darken all necessary lines. Erase all construction lines to complete the isometric drawing, as shown in Figure 12-42D.

Nonisometric Lines

To draw a nonisometric line, first locate its two ends, and then connect the points. Angles on isometric drawings do not show in their true size. Therefore, you cannot measure them in degrees.

Figure 12-43 shows how to locate and draw nonisometric lines in an isometric drawing using the box method. The nonisometric lines are the slanted sides of the packing block shown in the multiview drawing in Figure 12-43A. To make an isometric drawing of the block, use the following procedure.

1. Block in the overall sizes of the packing block to make the isometric box figure as shown in Figure 12-43B.

2. Use dividers or a scale to transfer distances AG and HB from the multiview drawing to the isometric figure. Lay these distances off along line AB to locate points G and H. Then draw the lines connecting point D with point G and point C with point H. This is shown in Figure 12-43C.

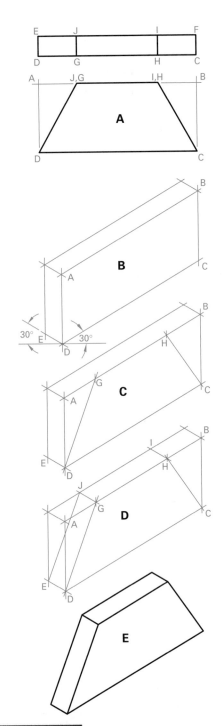

Figure 12-43

Drawing nonisometric lines.

3. Complete the layout by drawing GJ and HI and by connecting points E and J to form a third isometric line, as shown in Figure 12-43D.

4. Erase the construction lines to complete the drawing. See Figure 12-43E.

Angles

To draw the 40° angle shown in Figure 12-44A, use the following procedure.

1. Make AO and AB any convenient length. Draw AB perpendicular to AO at any convenient place. See Figure 12-44A.
2. Transfer AO and AB to the isometric cube shown in Figure 12-44B. Lay off AO along the base of the cube. Draw AB parallel to the vertical axis.
3. Connect points O and B to complete the isometric angle.

Follow the same steps to construct the angle on the top of the isometric cube. This method can be used to lay out any angle on any isometric plane.

Figure 12-45A is a multiview drawing of an object with four oblique surfaces. An isometric view of this object can be made using either the box or the skeleton method. The box method involves the development of a framework, or box, that provides surfaces on which to locate points, as shown in Figure 12-45B. The points are then connected to develop the edges that form the oblique surfaces. In the skeleton method, shown in Figure 12-45C, points are located by taking measurements directly on the base triangle. The points are then connected to develop the edges that form the oblique surfaces.

Reversed Axes

Sometimes you will want to draw an object as if it were being viewed from below. To do so in isometric drawing, reverse the position of the axes. Refer again to Figure 12-4. To draw an object using reversed axes, follow the example shown in Figure 12-46. Consider how an object appears in a multiview drawing, as shown in Figure 12-46A. Then begin the isometric view by drawing the axes in reversed position, as shown in Figure 12-46B. Complete the view with dimensions taken from the multiview drawing, as shown in Figure 12-46C. Darken the lines to finish the drawing.

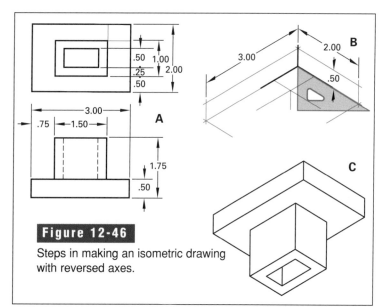

Figure 12-46

Steps in making an isometric drawing with reversed axes.

Figure 12-44

Constructing angles in isometric drawings.

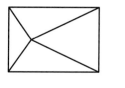

Figure 12-45

Drawing oblique surfaces in isometric drawings.

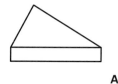

A

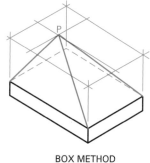

BOX METHOD

B

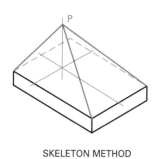

SKELETON METHOD

C

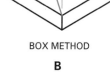

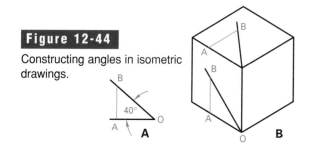

Long Axis Horizontal

When long pieces are drawn in isometric, make the long axis horizontal, as shown in Figure 12-47, or at 30°. For example, a long object is shown in a multiview drawing in Figure 12-47A. The start of an isometric drawing is shown in Figure 12-47B, with the axes shown by thick black lines. In Figure 12-47C, the same object is drawn with the long axis at 30° to the horizontal. The overall size and shape of the object, along with the intended use of the object, will determine which method should be used. Remember, in isometric drawing, draw circles first as isometric squares; then complete them by the four-center method or by using an ellipse template.

Dimensioning Isometric Drawings

There are two general ways to place dimensions on isometric drawings. The older method is to place them in the isometric planes, and to adjust the letters, numerals, and arrowheads to isometric shapes, as shown in Figure 12-48A. The newer unidirectional system, shown in Figure 12-48B, is simpler. In this system, numerals and lettering are read from the bottom of the sheet. However, since isometric drawings are not usually used as working drawings, they are seldom dimensioned at all. Refer to Chapter 7, "Dimensioning," for more information about the aligned and unidirectional methods of dimensioning.

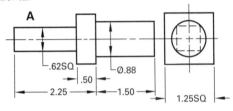

A

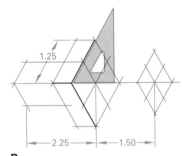

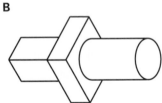

B

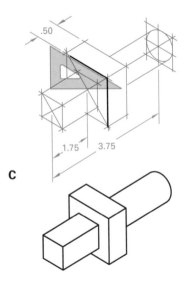

C

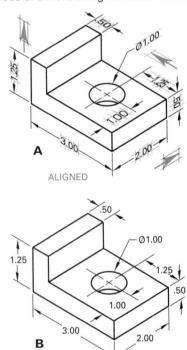

A

ALIGNED

B

UNIDIRECTIONAL

Isometric Sections

Isometric drawings are generally "outside" views. Sometimes, however, a sectional view is needed. To create a sectional view, take a section on an **isometric plane**—a plane that is parallel to one of the faces of the isometric cube. Figure 12-49 shows isometric full sections taken on a different plane for each of three objects. Note the construction lines showing the parts that have been cut away.

Figure 12-50 illustrates an isometric half section. The construction lines in Figure 12-50A are for the complete outside view of the original object. Notice the outlines of the cut surfaces. Figure 12-50B shows how to create the section. Draw the complete outside view as well as the isometric cutting plane. Then erase the part of the view that the cutting plane has cut away.

Oblique Drawing Techniques

In oblique drawing, the front of the object is easy to draw because it is parallel to the picture plane. The rest of the drawing follows rules similar to those for isometric drawings. Lines that are parallel to the axes are drawn first. This section explains construction techniques and then steps you through practice exercises for oblique pictorial drawing.

Oblique Constructions

As with isometric drawing, you should understand how to draw the geometry in an oblique drawing before you begin a complete drawing. The techniques used in oblique drawing are described in the following paragraphs.

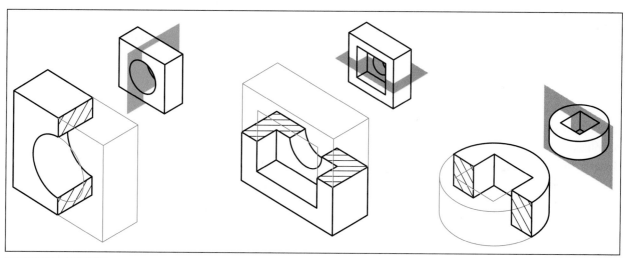

Figure 12-49
Examples of isometric full sections.

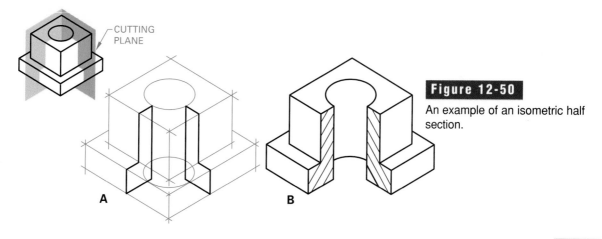

Figure 12-50
An example of an isometric half section.

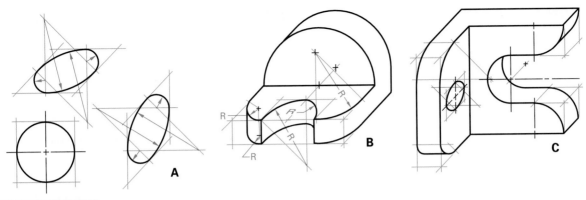

Figure 12-51

Circles parallel to the picture plane are true circles. On other planes, they appear as ellipses.

Oblique Circles

On the front face, circles and curves show in their true shape. On other faces, they show as ellipses. Draw the ellipses using the four-center method or an ellipse template. Figure 12-51A shows a circle as it would be drawn on a front plane, a side plane, and a top plane. Figure 12-51B and C show an oblique drawing with arcs in a horizontal plane and in a profile plane, respectively.

When you draw oblique circles using the four-center method, the results will be satisfactory for some purposes, but they will not be pleasing. Ellipse templates give much better results. If you use a template, first block in the oblique circle as an oblique square. This shows where to place the ellipse. Blocking in the circle first also helps you choose the proper size and shape of the ellipse. If you do not have a template, plot the ellipse as shown in Figure 12-52.

Figure 12-52

Plotting oblique circles.

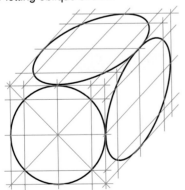

Creating an Oblique Drawing

Figure 12-53 shows the steps in making an oblique drawing. The procedure is much the same as that for creating an isometric drawing. Notice that this oblique drawing can show everything but the two small holes in true shape. Follow these steps:

1. Draw or review the multiview drawing of the object to be drawn in oblique, as shown in Figure 12-53A.

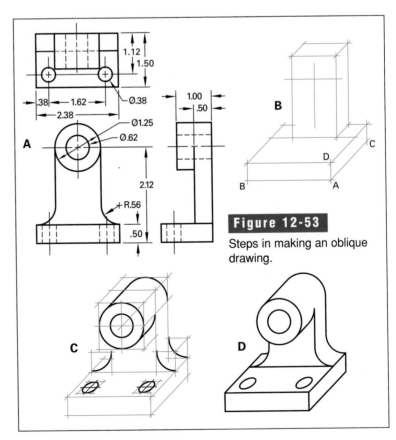

Figure 12-53

Steps in making an oblique drawing.

2. Draw the axes AB, AC, and AD for the base in second position and lay off the length, width, and thickness of the base on the axes, as shown in Figure 12-53B.
3. Draw the base and block in the upright, omitting the projecting boss as shown in Figure 12-53B.
4. Block in the boss and find the centers of all circles and arcs. Draw the circles and arcs, as shown in Figure 12-53C.
5. Darken all necessary lines and erase construction lines to complete the drawing. See Figure 12-53D.

Angles and Inclined Surfaces

Angles that are parallel to the picture plane show in their true size. For all other angles, lay the angle off by locating both ends of the slanting line.

Figure 12-54A is a multiview drawing of a plate with the corners cut off at angles. The rest of the figure shows the plate in oblique drawings. In Figure 12-54B, the angles are parallel to the vertical plane. In Figure 12-54C, they are parallel to the profile plane, and in Figure 12-54D, they are parallel to the horizontal plane. In each case, the angle is laid off by measurements parallel to one of the axes. These measurements are shown by the construction lines.

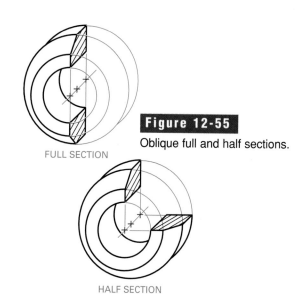

Figure 12-55
Oblique full and half sections.

FULL SECTION

HALF SECTION

Oblique Sections

Like isometric drawings, oblique drawings are generally "outside" views. Sometimes, however, you need to draw a sectional view. To do so, take a section on a plane parallel to one of the faces of an oblique cube. Figure 12-55 shows an oblique full section and an oblique half section. Note the construction lines indicating the parts that have been cut away.

Perspective Drawing Techniques

Perspective drawing involves techniques similar to those used for isometric and oblique drawings. However, perspective drawing is more complex because you must take line of sight, vanishing points, and other items into consideration. This section explains construction techniques and then steps you through practice exercises for one- and two-point perspective drawing.

Perspective Constructions

As with isometric and oblique drawing, you should understand how to draw the geometry in an oblique drawing before you begin a complete drawing. The techniques for inclined surfaces, circles, and arcs are described in the following paragraphs.

Figure 12-54

Angles on oblique drawings.

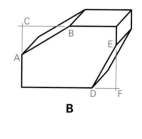

A

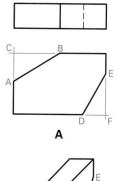

C

B

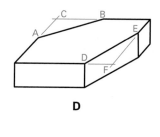

D

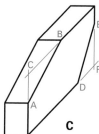

Inclined Surfaces

Plot inclined surfaces in perspective by finding the ends of inclined lines and connecting them. This method of drawing is shown in Figure 12-56.

Circles and Arcs

Figure 12-57 shows how to make a perspective view of an object with a cylindrical surface. First locate points on the front and top views. Then project them to the perspective view. Where the projection lines meet, a path is formed. Draw the perspective arc along the path using a French curve or an ellipse template.

Creating a One-Point Perspective

Figure 12-58 shows an object in multiview and isometric drawings. Figure 12-59 shows how to draw the same object in a one-point, bird's-eye perspective view. Follow these steps:

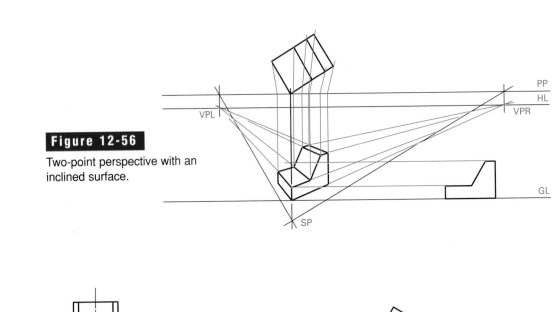

Figure 12-56

Two-point perspective with an inclined surface.

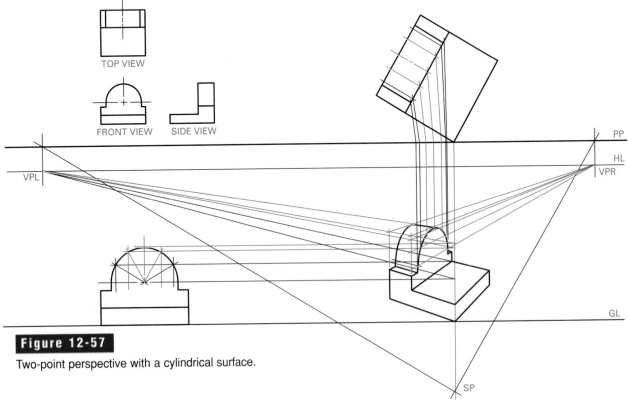

Figure 12-57

Two-point perspective with a cylindrical surface.

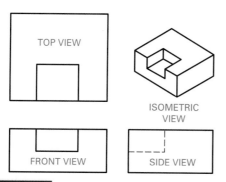

Figure 12-58

Multiview and isometric drawings of an object to be drawn in single-point perspective.

1. Decide on the scale to be used and draw the top view near the top of the drawing sheet, as shown in Figure 12-59A. A more interesting view is obtained if the top view is drawn slightly to the right or left of center.
2. Draw an edge (top) view of the picture plane (PP) through the front edge of the top view.
3. Draw the horizon line (HL). The location will depend on whether you want the object to be viewed from above, on, or below eye level. Draw the ground line. Its location in relation to the horizon line will determine how far above or

below eye level the object will be viewed. See Figure 12-59A.
4. To locate the station point (SP), draw a vertical line (line of sight) from the picture plane toward the bottom of the sheet. Draw the line slightly to the right or left of the top view. Set your dividers at a distance equal to the width of the top view. Begin at the center of vision of the picture plan and step off 2 to 3 times the width of the top view, along the line of sight, to locate the station point. See Figure 12-59B.
5. Project downward from the top view to establish the width of the front view on the ground line. Complete the front view.
6. The vanishing point is the intersection of the line of sight and the horizon line. Project lines from the points on the front view to the vanishing point. See Figure 12-59C. Establish depth dimensions in the following way: Project a line from the back corner of the top view to the station point. At point A on PP, drop a vertical line to the perspective view to establish the back edge. Draw a horizontal line through point B to establish the back top edge.
7. Proceed as in the previous step to lay out the slot detail. See Figure 12-59D.
8. Darken all necessary lines and erase construction lines to complete the drawing.

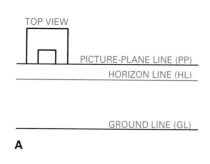

A

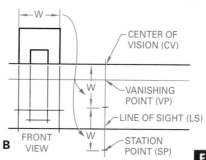

B

Figure 12-59

Procedure for making a one-point perspective drawing (bird's-eye view).

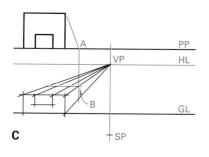

C

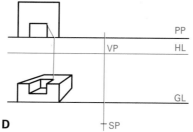

D

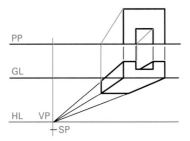

Figure 12-60

One-point perspective (worm's-eye view).

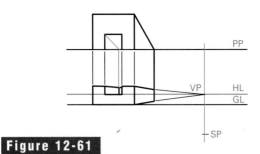

Figure 12-61

One-point perspective (normal view).

In Figures 12-60 and 12-61, the object is drawn in one-point perspective in the other positions. Notice that in all three cases, one face of the object is placed on the picture plane (thus the name *parallel perspective*). Therefore, this face appears in true size and shape. True-scale measurements can be made on it.

Creating a Two-Point Perspective

Figure 12-62 shows an object in multiview and isometric drawings. Figure 12-63 shows how to draw this same object with a two-point bird's-eye perspective view.

1. Draw an edge view of the picture plane (PP). See Figure 12-63A. Allow enough space at the top of the sheet for the top view. Draw the top view with one corner touching the PP. In this case, the front and side of the top view form angles of 30° and 60°, respectively. Other angles may be used, but 30° and 60° seem to give the best appearance on the finished drawing. The side with the most detail is usually placed along the smaller angle for a better view.

2. Draw the horizon line (HL) and the ground line (GL).
3. Draw a vertical line (line of sight) from the center of vision (CV) toward the bottom of the sheet to locate the station point.
4. Draw line SP-B parallel to the end of the top view and line SP-C parallel to the front of the top view. (Use a 30°-60° triangle.) See Figure 12-63B.
5. Drop vertical lines from the picture plane to the horizon line to locate vanishing point left (VPL) and vanishing point right (VPR). Draw the front or side view of the object on the ground line as shown in Figure 12-63B.
6. Begin to block in the perspective view by projecting vertical dimensions from the front view to the line of sight (also called the *measuring line*) and then to the vanishing points. See Figure 12-63C.
7. Finish blocking in the view by projecting lines from points 1 and 2 on the top view to the station point. Where these lines cross the picture plane, drop vertical lines to the perspective view to establish the length and width dimensions. Project point 1' to VPL and point 2' to VPR.
8. Add detail by following the procedure described in steps 6 and 7. See Figure 12-63D.
9. Darken all necessary lines and erase construction lines to finish the drawing.

Figure 12-62

Multiview and isometric drawings of an object to be drawn in two-point perspective.

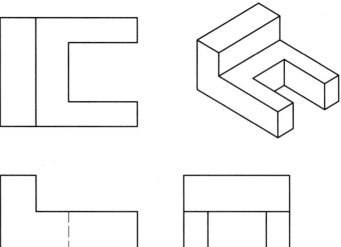

Figure 12-63

Procedure for making a two-point perspective drawing (bird's-eye view).

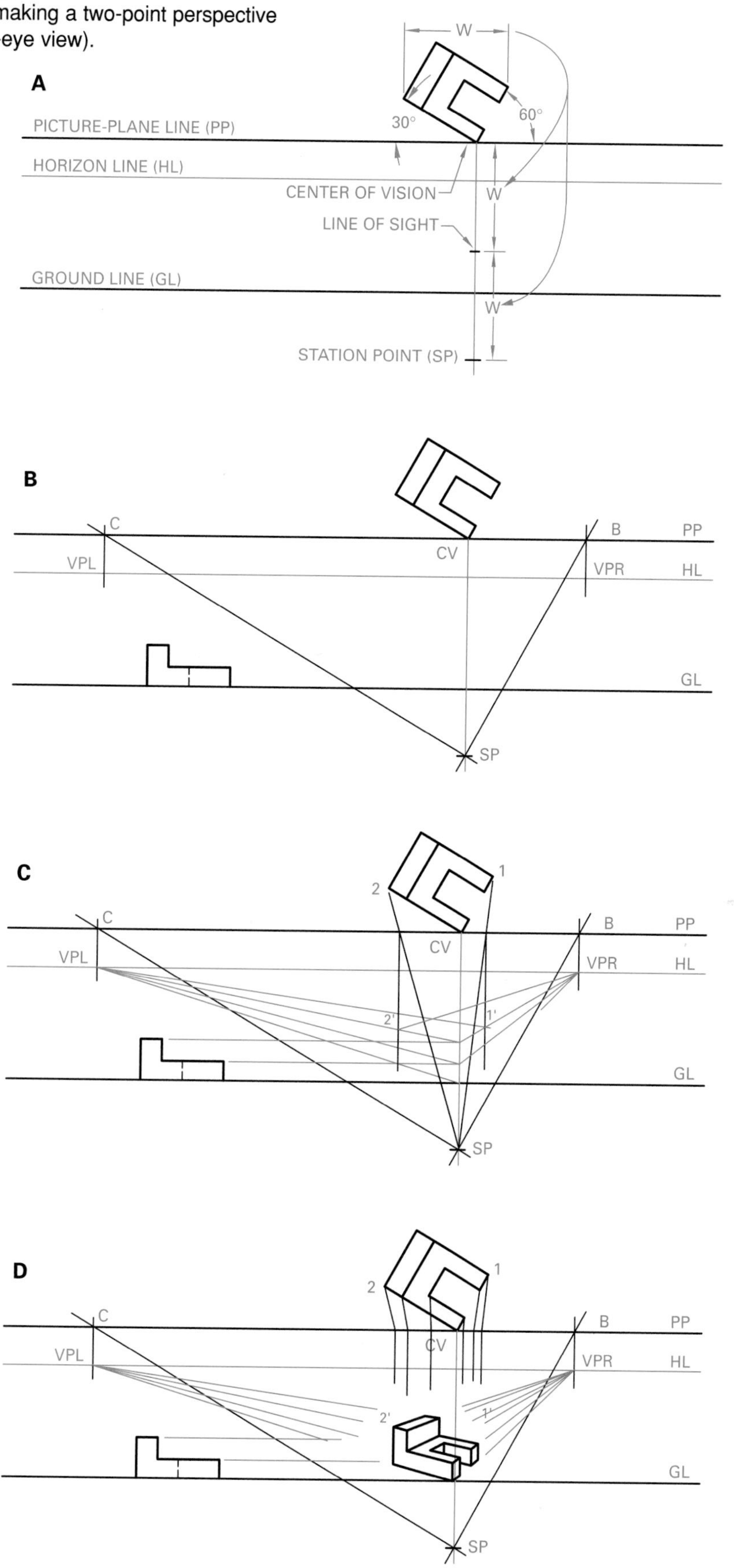

A

PICTURE-PLANE LINE (PP)

HORIZON LINE (HL)

CENTER OF VISION

LINE OF SIGHT

GROUND LINE (GL)

STATION POINT (SP)

30°   60°   W   W   W

B

C   CV   B   PP
VPL   VPR   HL
GL
SP

C

2   1
C   CV   B   PP
VPL   VPR   HL
2'   1'
GL
SP

D

2   1
C   CV   B   PP
VPL   VPR   HL
2'   1'
GL
SP

Figures 12-64 and 12-65 show the same object drawn in the other positions. Notice that none of the faces appear in true size and shape because none of them are on the picture plane.

Perspective Grids

Perspective drawing can take a lot of time. This is because so much layout work is needed before you can start the actual perspective view. Also, a large drawing surface is often needed in order to locate distant points. However, you can offset these disadvantages by using perspective grids. Examples are shown in Figure 12-66. There are many advantages in using grids.

But there is one major disadvantage: a grid cannot show a variety of views. It is limited to one type of view based on one set of points and one view location. However, for the work done in some industrial drafting rooms, this may be all that is needed.

You can buy perspective grids, or you can make your own. Creating your own grids is only practical, however, if you have a number of perspective drawings to make in a special style.

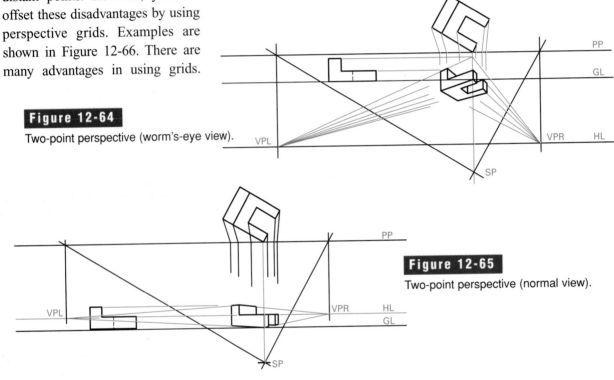

Figure 12-64

Two-point perspective (worm's-eye view).

Figure 12-65

Two-point perspective (normal view).

Figure 12-66

Examples of perspective grids.

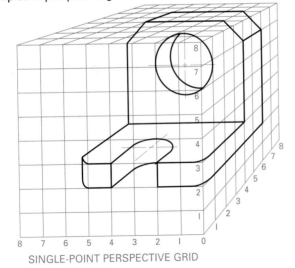

SINGLE-POINT PERSPECTIVE GRID

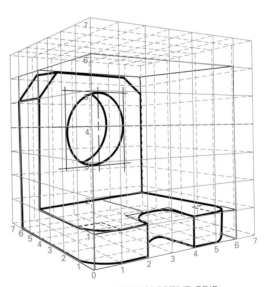

TWO-POINT PERSPECTIVE GRID

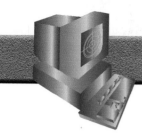

CAD Techniques

In many ways, 2D pictorial drawing is a simpler process if you use a CAD system. For example, you don't have to construct ellipses to represent circles or holes on isometric planes. AutoCAD creates "isocircles" on all three planes. For rendering, however, the CAD work becomes a little more complex, because all rendering is done on 3D models. The sections that follow explain how to use AutoCAD to create various types of pictorial drawings.

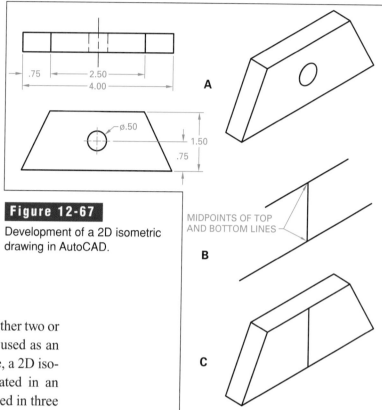

Figure 12-67

Development of a 2D isometric drawing in AutoCAD.

MIDPOINTS OF TOP AND BOTTOM LINES

Isometric Drawing Techniques

You can create isometric drawings in either two or three dimensions. If the drawing is to be used as an isolated illustration or for a single purpose, a 2D isometric is often sufficient. Drawings created in an engineering environment are usually created in three dimensions because 3D models can be used for many different purposes.

2D Isometric Drawings

To create a 2D drawing of the block shown in Figure 12-67A, begin by setting up the grid and snap spacing for an isometric drawing. The rest of the steps are then fairly easy. As in board drafting, you should draw the isometric lines first. Follow these steps:

1. Press the F7 key to turn on the grid, and enter ZOOM All.
2. Enter the SNAP command at the keyboard. Notice the options that appear on the Command line. Enter S (Style), and then enter I (Isometric). Enter a vertical spacing of .25 to finish setting the snap. Notice that the grid and crosshairs (cursor) change to an isometric orientation.

By default, the crosshairs are parallel to the left isometric plane, or **isoplane**. You can change them to

LEFT   TOP   RIGHT

Figure 12-68

Positions of the isometric crosshairs.

the top and right planes using the ISOPLANE command. However, it is faster and more convenient to use one of the shortcut methods. You can toggle through the left, top, and right isoplanes by pressing CTRL+E or simply by pressing the F5 key. Both of these methods work while other commands are active, which can simplify drawing tasks. Figure 12-68 shows the crosshairs in each isometric orientation.

3. Change the crosshairs to the right isoplane.

4. Draw the baseline of the block up and to the right, as shown in Figure 12-67B. Reenter the LINE command and use the Midpoint object snap to snap a second line to the midpoint of the base. Extend the second line vertically up from the midpoint of the base, and make its length equal to the height of the block (1.50″). This temporary vertical line will form the basis for the top line and the isometric circle.

5. At any location on the screen, draw a 2.50″ line parallel to the isometric baseline. This will become the top line of the isometric block. After creating the line, move it into position by selecting the midpoint of the 2.50″ line as the base point for the move and snapping it to the upper end of the temporary vertical line. Refer again to Figure 12-67B.

6. Use the LINE command to connect the ends of the top and bottom lines of the block.

7. Switch the crosshairs to the left isoplane, and draw the .50″ lines at the top and bottom front corner and the top back corner to show the depth of the block. Then connect the .50″ lines to complete the basic shape of the block. See Figure 12-67C.

►CADTIP

The Ortho Mode
Like the snap and grid, the Ortho mode is affected by the isometric orientation. You can therefore use Ortho to keep lines aligned perfectly with the isometric axes. Use Ortho for all isometric lines.

8. Isometric circles, or isocircles, are created using the ELLIPSE command. When the snap is set to Isometric mode, an additional option called Isocircle appears at the Command line when you enter the ELLIPSE command. To create the hole in the block, first make sure the crosshairs are set to the right isoplane. Then enter the ELLIPSE command and select the Isocircle option. Use the Midpoint object snap to snap the center of the isocircle to the midpoint of the temporary vertical line you created in step 4. The command acts exactly like the CIRCLE command. Enter a diameter of .50″ to complete

the isocircle. Erase the temporary vertical line. The completed block should look like the isometric drawing in 12-67A.

3D Isometric Drawings

As you may recall from previous chapters, AutoCAD provides standard, predefined isometric views for 3D objects. Therefore, to create a 3D isometric, you simply build the object in three dimensions, and then change to an isometric view. The following steps use the EXTRUDE command to create a solid model. If your version of AutoCAD does not include EXTRUDE, use the THICKNESS method described in Chapter 10 to create a surface model instead. The basic procedure is otherwise the same.

1. Use the PLINE command to create the front view of the block, just as you would for a 2D orthographic drawing. See Figure 12-69A.

Figure 12-69
Steps to develop a 3D isometric drawing in AutoCAD.

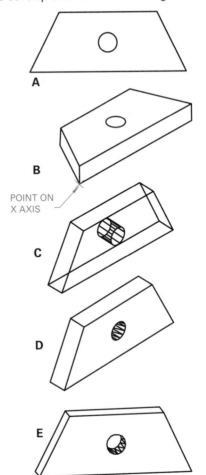

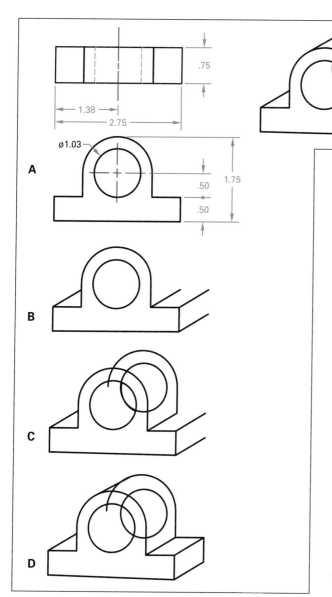

Figure 12-70

Steps to develop a 2D oblique drawing in AutoCAD.

model. Enter X (Xaxis option) to rotate the model around the X axis. At the prompt to specify a point on the X axis, pick the lower corner of the base, as shown in Figure 12-69B. Specify a rotation angle of 90. The model moves into its correct position. See Figure 12-69C. Enter HIDE again to remove hidden lines.

Although this looks like a correct isometric representation of the block, it has one major fault as a solid model. The circle is not truly a hole. It is a solid cylinder in the middle of a solid block. If your version of AutoCAD supports Boolean mathematics, you can subtract the extruded circle from the block to create a real hole. To do this, enter the SUBTRACT command, and pick anywhere on the outside of the block as the solid from which to subtract. Press Enter, and then select the isocircle to subtract. Remove hidden lines again, and notice the difference in the model. You can now see the depth of the hole, as shown in Figure 12-69D. If you were to view the model from another viewpoint, as shown in Figure 12-69E, you would see that the hole extends all the way through the model.

Oblique Drawing Techniques

The procedures for creating an oblique drawing using a CAD system are much like those for creating an isometric drawing. You can create oblique drawings in either two or three dimensions.

2D Oblique Drawings

Because the front view of an oblique drawing is at true size and shape, you can draw the front view as you normally would an orthographic. Then you can change the snap to make the top and side views easier to create. Follow these steps:

1. Create the front view of the pole support shown in Figure 12-70A.

2. Enter the ISOLINES command and set a new density of 10. This is not critical, but it will show the isocircle in better detail in the finished drawing.

3. Enter the EXTRUDE command and select the entire front view. Specify a height equal to the thickness of the block (.50″). Press Enter to specify a taper angle of 0 (no taper).

4. From the View menu, select 3D Views (3D Viewpoint in some versions of AutoCAD) and then SW Isometric. Enter the ZOOM command and enter 1 for the zoom. Then enter the HIDE command to remove hidden lines. Notice that the model seems to be lying on its side, as shown in Figure 12-69B.

5. To orient the model correctly, enter the ROTATE3D command and select the entire

2. Be sure the grid is on, and enter the SNAP command at the keyboard. Enter R (for Rotate) and pick the lower right corner of the front view as the base point. Enter a rotation angle of 30. The grid and crosshairs rotate counterclockwise 30°. Set the snap to .25.

3. Starting at the lower right corner of the base, create the .75″ line that represents the depth of the pole support. Then copy this line to the other key points as shown in Figure 12-70B to define the depth. Use the Endpoint object snap to ensure accuracy.

4. Copy the right upright, hole, and arched top of the front view to the back of the object, as shown in Figure 12-70C.

5. Draw a line tangent to the front and back arcs to connect them, and add the lines to complete the right side of the object, as shown in Figure 12-70D.

6. To complete the oblique drawing, trim away the unwanted lines from construction, as shown in Figure 12-70A.

The procedure given here is for a cavalier oblique drawn at 30°. The same procedure can be used to create an oblique at any angle, simply by changing the angle of the snap rotation. To create a cabinet oblique, divide the depth dimensions by 2 before adding them to the drawing.

3D Oblique Drawings

Like 3D isometric drawings, 3D oblique drawings are created as normal 3D models. The viewpoint is then changed to create the oblique. Follow these steps to create a model of the pole support:

1. Use the PLINE command to draw the front view of the pole support shown in Figure 12-70A. (*Note:* If your version of AutoCAD does not allow extrusion, use the THICKNESS command to set the depth before performing step 1.)

2. Set ISOLINES to 10. Then enter FACETRES and enter a new value of 1. FACETRES is a command variable that controls the appearance of curved objects in 3D views.

3. Use the EXTRUDE command to extrude the object by the depth dimension, .75″.

4. To create the hole, use the SUBTRACT command to subtract the hole from the pole support (if your version of AutoCAD supports this feature).

5. For convenience, select the SW Isometric viewpoint from the View menu. As with the 3D isometric, the drawing is lying on its side, as shown in Figure 12-71A. Use ROTATE3D to rotate it into position. Use the Xaxis option and select the lower front corner as the base point. Rotate it 90° to achieve the position shown in Figure 12-71B.

Using VPOINT

As you may recall from Chapter 10, AutoCAD allows you to view 3D models from any point in space using the VPOINT command. This is an acceptable method of obtaining an oblique view of an object. Figure 12-71B shows an example of an oblique obtained in this way.

Using 3D Orbit

Beginning with AutoCAD 2000, AutoCAD includes a feature known as **3D Orbit**. This feature allows you to rotate a model in 3D space interactively using a spherical orbit algorithm. The advantage of using 3D Orbit is that you can actually see the object as you rotate the view. See Figure 12-72. If this feature is available in your CAD software, activate it using the 3DORBIT command. Move the viewpoint by dragging the mouse. You can also activate the small circles at the quadrants of the green orbit to change the direction of movement.

Perspective Drawing Techniques

Perspective drawing in AutoCAD is done entirely in 3D. Therefore, the first step in drawing any perspective view is to create a normal model of the object. Then you can view the model either in parallel projection (normal) or perspective views.

Figure 12-71

Steps to develop a 3D oblique drawing in AutoCAD.

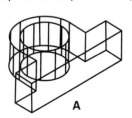
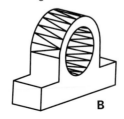

A   B

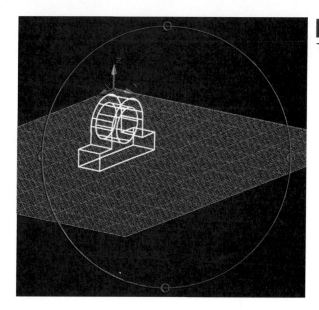

Figure 12-72

The 3D orbit feature.

Figure 12-73

Using DVIEW to create a 3D perspective view.

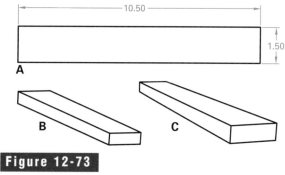

Figure 12-74

The 3DORBIT command offers better control for creating 3D perspective views.

Using DVIEW

In versions prior to AutoCAD 2000, the only method for viewing models in perspective was by using the DVIEW command. This command uses a "camera" and "target" to create the perspective. The line of sight is the line between the camera and the target. To demonstrate the use of the DVIEW command, we will use a simple rectangular block. Follow these steps:

1. Use PLINE to create a rectangular box as shown in Figure 12-73A. Extrude it to a height of .50″, but leave it in the plan view.
2. Enter the DVIEW command and select the box.
3. Enter CA (CAmera) and move the cursor to achieve the view shown in Figure 12-73B. *Note:* You may need to use the Hide option to see the object's position clearly.
4. Enter D to activate the Distance option. A bar appears across the top of the screen. Use the mouse to extend the view to 16x. Then enter H (Hide) again to achieve the view shown in Figure 12-73C. Notice that the box is now viewed in perspective.
5. End the DVIEW command. The box stays in the perspective position even after you leave the DVIEW command.
6. To remove the perspective from the drawing, enter DVIEW again, select the box, and enter O to select the Off option. Press Enter to end the DVIEW command.

Using 3D Orbit

Beginning with AutoCAD 2000, the 3DORBIT command provides a much easier method of adding perspective to a model. Follow these steps:

1. Create the box shown in Figure 12-73A using the PLINE and EXTRUDE commands.
2. For convenience, switch to AutoCAD's preset SW Isometric view.
3. Enter the 3DORBIT command and right-click to present a shortcut menu. From this menu, pick Projection and then Perspective. This places the model into the perspective mode.
4. Use the circular grips at the quadrants of the 3D orbit circle to move the object in the perspective view. Create a view similar to the one shown in Figure 12-74. Press Enter to end the 3DORBIT command.
5. Enter the HIDE command to remove hidden lines. As you can see, the object retains the perspective view even after you end the 3DORBIT command.
6. To remove the perspective view, reenter the 3DORBIT command, right-click, and choose Projection and Parallel.

Chapter Summary

■ Pictorial drawings are useful in engineering, architecture, advertising, and manufacturing.

■ In isometric drawing, measurements can be made only on isometric lines.

■ In isometric drawings, circles appear as ellipses.

■ Isometric drawings can be dimensioned using either aligned or unidirectional dimensioning methods.

■ Isometric, dimetric, and trimetric are the three types of axonometric projections.

■ In oblique pictorial drawing, two axes are parallel to the picture plane.

■ Oblique drawings can be made as cabinet, normal, and cavalier.

■ A perspective drawing is a three-dimensional representation of an object as it appears to the eye from a particular point or location.

■ A technical illustration is a pictorial drawing that provides technical information using visual methods.

■ Isometric and oblique drawings can be created using either 2D or 3D techniques in AutoCAD. Perspective drawings are created in 3D.

Review Questions

1. Describe two uses for pictorial drawings.

2. What are the three most common types of pictorial drawing?

3. Which type of pictorial drawing is most natural in appearance?

4. What do you call a line that is not parallel to any of the three isometric axes?

5. Is an isometric projection larger or smaller than an isometric drawing?

6. Explain why measurements can be taken only along isometric lines in an isometric drawing.

7. Describe a situation in which a reversed isometric drawing might be useful.

8. What is the major practical difference between isometric and dimetric drawings?

9. In what way is an isometric drawing different from an oblique drawing?

10. In an oblique drawing, how many axes are parallel to the picture plane?

11. When creating an oblique drawing, how should you place the object?

12. What is the difference between a cabinet and a cavalier drawing?

13. Describe the general procedure for developing a circle on an oblique drawing using board drafting techniques.

14. What is another name for surface shading?

15. Briefly explain how to create an oblique drawing using AutoCAD.

16. What is the vanishing point in a perspective drawing?

17. What two factors affect how an object looks in perspective?

18. In what way can the grid and snap features of a CAD program help a drafter create an isometric drawing?

19. How do you create isometric circles using AutoCAD?

20. What is the purpose of the DVIEW command in AutoCAD?

Drafting Problems

The drafting problems in this chapter are designed to be completed using board drafting techniques or CAD.

For problems 1 through 15, determine an appropriate scale and create isometric drawings according to the instructions for each problem. Do not dimension.

1. Determine an appropriate scale, and create an isometric drawing of the object(s) assigned from Figure 12-75. *Note:* These objects may also be used for oblique and perspective drawing practice.

Figure 12-75

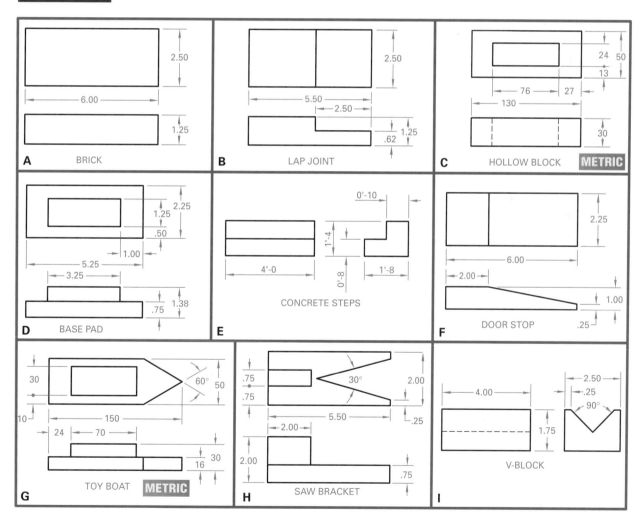

A — BRICK

B — LAP JOINT

C — HOLLOW BLOCK **METRIC**

D — BASE PAD

E — CONCRETE STEPS

F — DOOR STOP

G — TOY BOAT **METRIC**

H — SAW BRACKET

I — V-BLOCK

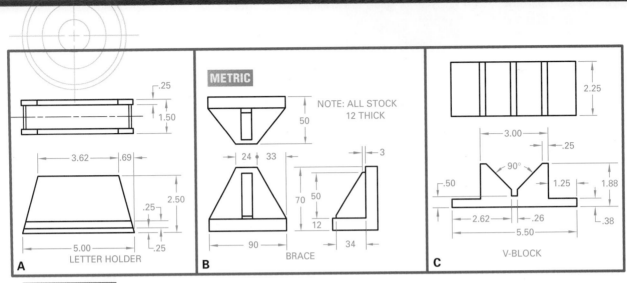

Figure 12-76

2. Determine an appropriate scale, and create an isometric drawing of the object(s) assigned from Figure 12-76. *Note:* These objects may also be used for oblique and perspective drawing practice.

3. Make an isometric drawing of the notched block shown in Figure 12-77. Start at the corner indicated by thick lines.

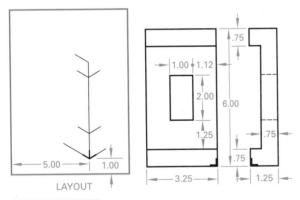

Figure 12-77

4. Make an isometric drawing of the babbitted stop shown in Figure 12-78. Start at the corner indicated by thick lines.

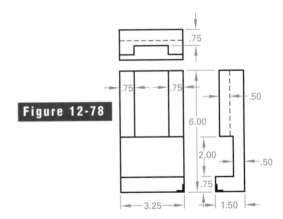

Figure 12-78

5. Make an isometric drawing of the stirrup shown in Figure 12-79. Start the drawing at the lower left. Note the thick starting lines.

6. Make an isometric drawing of the brace shown in Figure 12-80. Start the drawing at the lower right. Note the thick starting lines.

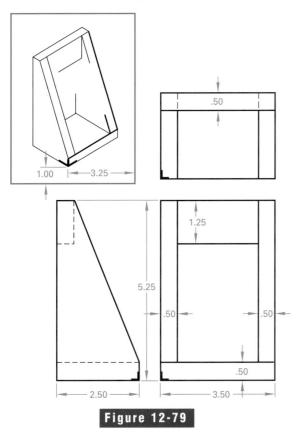

Figure 12-79

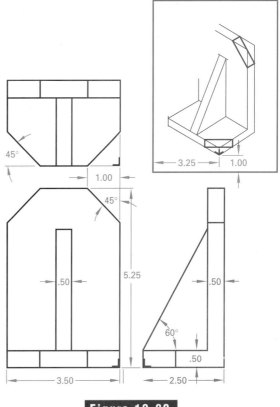

Figure 12-80

7. Make an isometric drawing of the object(s) assigned from Figure 12-81. *Note:* These objects may also be used for oblique and perspective drawing practice.

8. Make an isometric full or half section of the object(s) assigned from Figure 12-81.

Figure 12-81

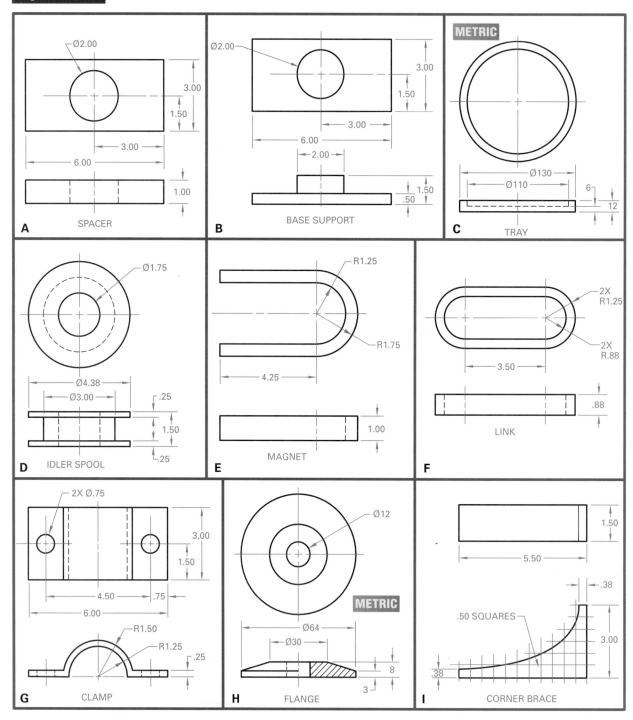

A SPACER

B BASE SUPPORT

C METRIC TRAY

D IDLER SPOOL

E MAGNET

F LINK

G CLAMP

H METRIC FLANGE

I CORNER BRACE

9. Make an isometric drawing of the cross slide shown in Figure 12-82.

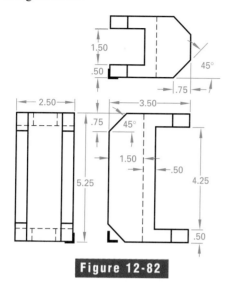

Figure 12-82

11. Make an isometric drawing of a 3″ cube with an isometric circle on each visible side, as shown in Figure 12-84.

Figure 12-84

10. Make an isometric drawing of the ratchet shown in Figure 12-83.

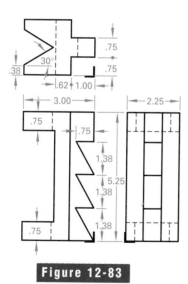

Figure 12-83

12. Make an isometric drawing of a cylinder resting on a square plinth, as shown in Figure 12-85.

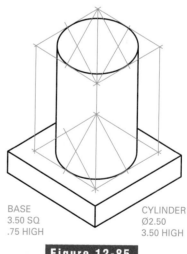

BASE
3.50 SQ
.75 HIGH

CYLINDER
Ø2.50
3.50 HIGH

Figure 12-85

13. Make an isometric drawing of the hung bearing shown in Figure 12-86. Most of the construction is shown on the layout. Make the drawing as though all corners were square, and then construct the curves.

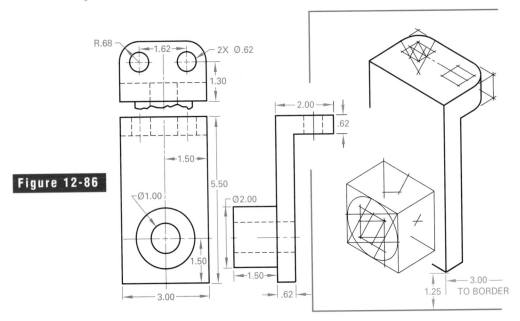

Figure 12-86

14. Make an isometric drawing of the bracket shown in Figure 12-87. Some of the construction is shown on the layout. Make the drawing as though all corners were square, and then construct the curves.

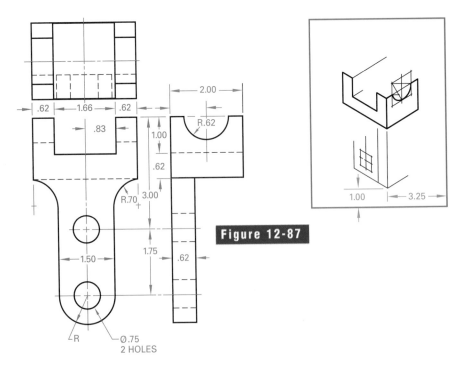

Figure 12-87

15. Make an isometric drawing of the tablet shown in Figure 12-88. Use reversed axes. Refer to the layout on the right.

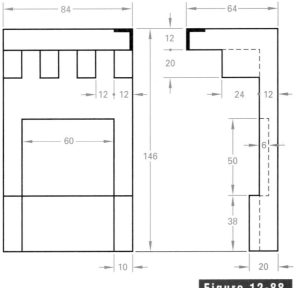
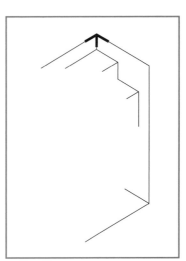

Figure 12-88

For problems 16 through 23, determine an appropriate scale and create oblique drawings according to the instructions for each problem. Do not dimension.

16. Make an oblique drawing of the angle support shown in Figure 12-89.

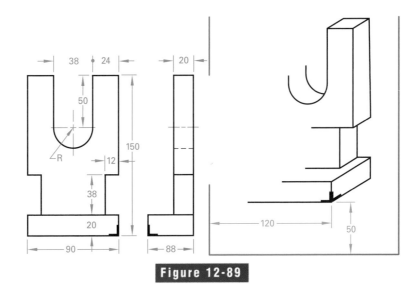

Figure 12-89

17. Make an oblique half or full section of the object(s) assigned from Figure 12-90. *Note:* These objects may also be used for isometric and perspective drawing practice.

Figure 12-90

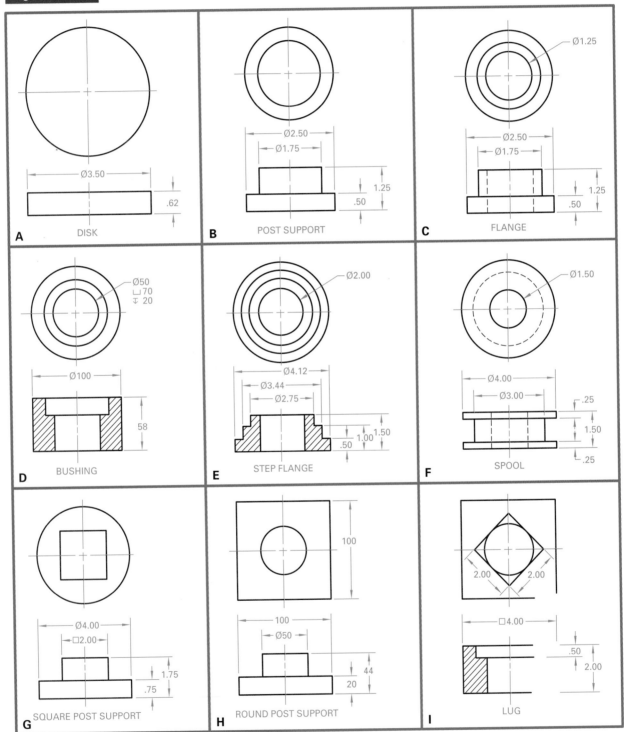

18. Make an oblique drawing of the crank shown in Figure 12-91.

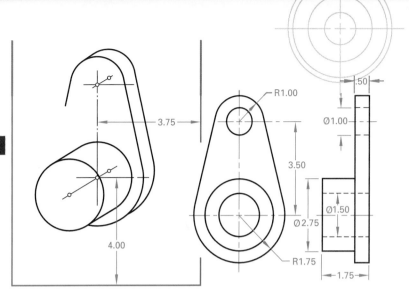

Figure 12-91

19. Make an oblique drawing of the forked guide shown in Figure 12-92.

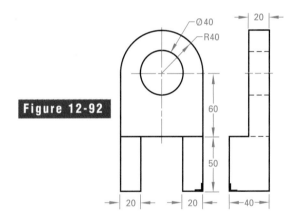

Figure 12-92

20. Make an oblique drawing of the slotted sector shown in Figure 12-93.

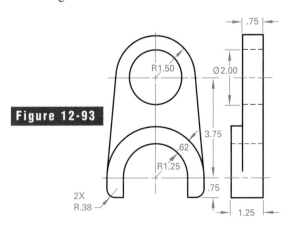

Figure 12-93

21. Make an oblique drawing of the guide link shown in Figure 12-94.

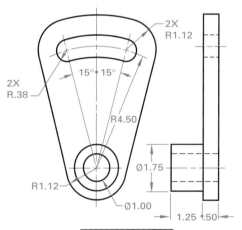

Figure 12-94

22. Make an oblique drawing of the object(s) assigned from Figure 12-95. *Note:* These objects may also be used for isometric and perspective drawing practice.

Figure 12-95

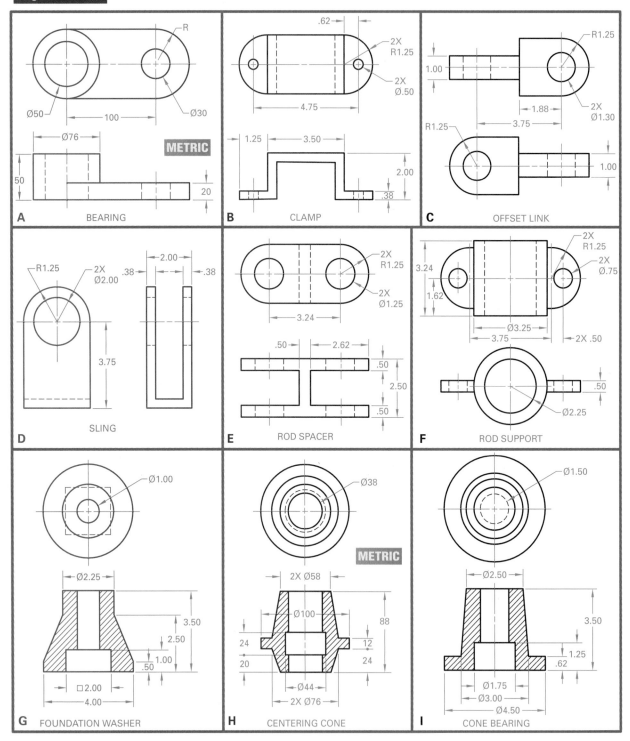

A — BEARING   Ø50   100   Ø30   R   METRIC   Ø76   50   20

B — CLAMP   .62   2X R1.25   2X Ø.50   4.75   1.25   3.50   2.00   .38

C — OFFSET LINK   R1.25   1.00   1.88   3.75   2X Ø1.30   R1.25   1.00

D — SLING   R1.25   2X Ø2.00   2.00   .38   .38   3.75

E — ROD SPACER   2X R1.25   2X Ø1.25   3.24   .50   2.62   .50   2.50   .50

F — ROD SUPPORT   2X R1.25   2X Ø.75   3.24   1.62   Ø3.25   3.75   2X .50   .50   Ø2.25

G — FOUNDATION WASHER   Ø1.00   Ø2.25   3.50   2.50   1.00   .50   □2.00   4.00

H — CENTERING CONE   Ø38   METRIC   2X Ø58   Ø100   88   24   12   20   24   Ø44   2X Ø76

I — CONE BEARING   Ø1.50   Ø2.50   3.50   1.25   .62   Ø1.75   Ø3.00   Ø4.50

23. Make a one-point perspective or two-point perspective drawing of each object assigned. Use any suitable scale.

Figure 12-96

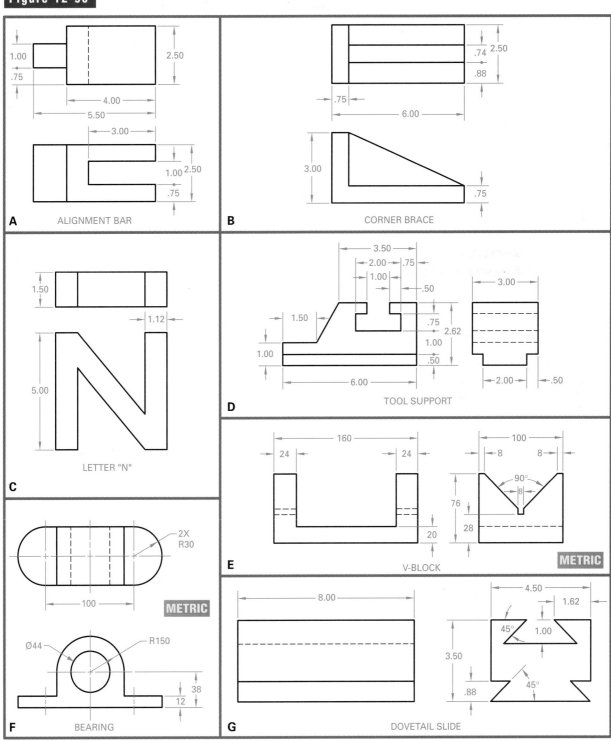

A — ALIGNMENT BAR

B — CORNER BRACE

C — LETTER "N"

D — TOOL SUPPORT

E — V-BLOCK — METRIC

F — BEARING — METRIC

G — DOVETAIL SLIDE

For problems 24 through 28, determine an appropriate scale and create a pictorial drawing according to the problem instructions. Do not dimension unless instructed to do so.

24. Most technical illustrations are basically pictorial line drawings. To develop a good understanding of the relationship of the various types, make an isometric, oblique cavalier, oblique cabinet, one-point perspective, and two-point perspective sketch of the tool support shown in Figure 12-96D.

25. Make instrument or CAD drawings of the same tool support (Figure 12-96D) in isometric, oblique cavalier, oblique cabinet, one-point perspective, and two-point perspective. Compare the sketches with the instrument or CAD drawings. Are they similar? Which type of pictorial drawing gives the most natural appearance?

26. Make a pictorial line drawing of the toy boat shown in Figure 12-75G. Use isometric, oblique, or perspective. Refer to Figure 12-29 and render the boat using the techniques shown in parts C, D, or E.

27. Make an isometric exploded-view drawing of the notebox in Figure 6-65. Add part numbers and a parts list to make an identification illustration. Design your initials as an inlay attached to a circular disk. Redesign the notebox as desired.

28. Make a one-point perspective drawing of one of your own initials. Render it using the technique shown in Figure 12-29C, D, or E.

Design Problems

Design problems have been prepared to challenge individual students or teams of students. In these problems, you are to apply skills learned mainly in this chapter but also in other chapters throughout the text. They are designed to be completed using board drafting, CAD, or a combination of the two. Be creative and have fun!

1. **TEAMWORK** Design an educational toy or game for children ages 3 to 5. Material optional. Carefully consider safety issues. Each team member should develop design sketches and pictorial drawings of his or her design. Include overall dimensions only.

2. Design a portable tool holder to accommodate a cordless electric drill with accessories. The design should also incorporate a means for attaching the tool holder to a tool panel. The accessories include at least a set of drill bits and screwdriver bits. Prepare a list of all items before proceeding with the design. Material optional. Develop design sketches and pictorial drawings with dimensions.

3. **TEAMWORK** Design a device that can be used to convert a portable electric router into a bench-type router. It can be designed as a floor model or a bench-top model. Make it easy and quick to install and remove the router. Material optional. Each team member should develop design sketches of his or her design. Select the best ideas from each to finalize the team design. Prepare a final set of pictorial drawings with dimensions.

4. Design a park bench to be made from at least two different materials. Develop design sketches and pictorial drawings with dimensions.

CHAPTER 13

Working Drawings

KEY TERMS

assembly drawing

assembly working
   drawing

bill of materials

combination drawing

detail drawing

FAO

reference assembly
   drawing

tabulated drawing

title block

working drawing

OBJECTIVES

Upon completion of this chapter, you should be able to:

- Describe the various types of working drawings and explain the purpose of each.

- Set up a working drawing.

- Explain the procedure for checking a set of working drawings.

- Produce detail drawings, assembly drawings, and assembly working drawings.

- Develop and use tabulated drawings for standard parts to be produced in a range of sizes.

- Design and draw a title block, incorporating standard items of information.

- Develop a standard bill of materials.

A **working drawing** gives all the information needed to manufacture or build a single part or a complete machine or structure. It tells precisely the required shape and size of the finished product. It also specifies the kinds of material required, the finish required, and the degree of accuracy required. In addition, it provides all necessary information for assembly of the final product. Nothing must be left to guess. Working drawings are usually multiview drawings with complete dimensions and notes added. An example of a working drawing is shown in Figure 13-1.

Errors on drawings can result in costly mistakes in production. Technical drawings must be complete and clear, and they must conform to standards. Responsibility for these important issues is shared by the drafter, checker, designer, engineer, and other specialists assigned to review the drawings prior to release. Once released from the design and drafting department, they go to the production department. Technically accurate, clearly understood drawings generally contribute to the timely production of good-quality products with a minimum of difficulty on the production line.

"Working drawings" is the generic name for various types of drawings, such as detail drawings, assembly working drawings, reference assembly drawings, tabulated drawings, and other, similar types of drawings. These are the kinds of drawings that become critical working documents for the production of parts and products. Knowledge of the design requirements, manufacturing processes, and drafting practices on the part of the design and drafting team is critical.

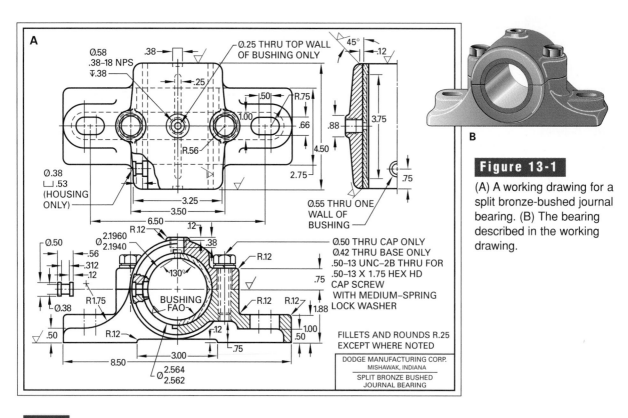

Figure 13-1

(A) A working drawing for a split bronze-bushed journal bearing. (B) The bearing described in the working drawing.

Types of Working Drawings

A good working drawing follows the style and practices of the office or industry in which it is made. Most industries follow the standards recommended by ANSI. That way, plans can be easily read and understood from one industry to another.

No matter what special styles and practices a drafter uses in working drawings, he or she must follow certain basic rules. For example, the drafter must use proper line technique to make the contrast sharp and the drawing easy to read. Also, dimensions and notes must be clear and accurate. The drafter must use standard terms and abbreviations. When the drawing is finished, the drafter must check it carefully to avoid mistakes and to ensure accuracy. The drafter should check the drawing before submitting it to the supervisor or checker for approval. A technical drawing is considered a legal document; to avoid costly mistakes, it must be correct.

A student's drawing may look rough and unfinished next to one made by an experienced drafter or engineer. The finished look of a drawing comes from a thorough knowledge of drafting and a great deal of practice. This chapter discusses the correct order of going about creating a working drawing, as well as some of the procedures that drafters usually follow. You should become thoroughly familiar with these practices. Otherwise, your drawings will not have the style and good form that is demanded by industry today.

Detail Drawings

A drawing of a single piece that gives all the information for making it is called a **detail drawing**. An example is the drawing of a simple part in Figure 13-2. A detail drawing must be a full and exact description of the piece. It should show carefully selected views and include well-placed dimensions. See Figure 13-3.

When a large number of machines are to be produced, detail drawings of each part are often made on separate sheets. This is done especially when some of the parts may also be used on other machines. In some industries, however, several parts of a machine are detailed on the same sheet. Sometimes a separate detail drawing is made for each of several workers involved, such as the patternmaker, machinist, and welder. Such a drawing shows only the dimensions

Figure 13-2

A detail drawing.

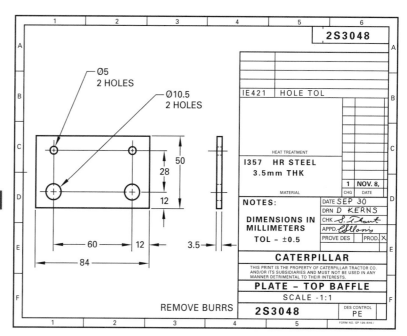

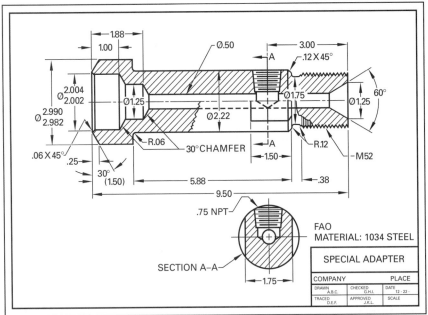

Figure 13-3

A single view and an extra section provide a complete description of this special adapter.

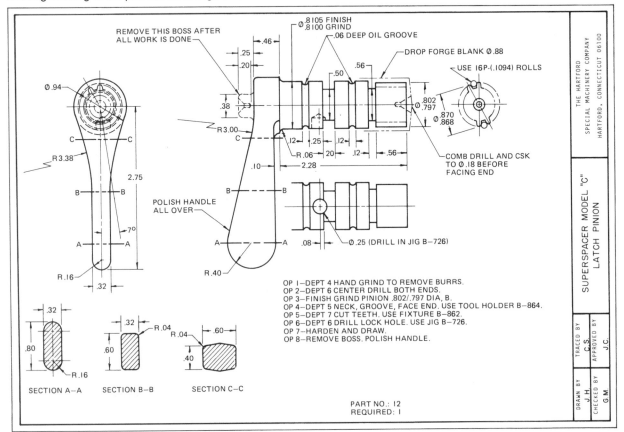

Figure 13-4

(A) A forged index-plunger operating handle. (B) The same handle after machining.

Figure 13-5

Working drawing of the part shown in Figure 13-4.

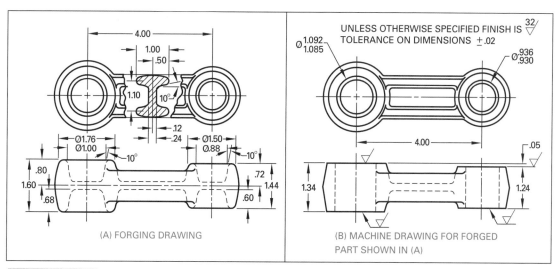

Figure 13-6

Two-part detail drawing showing separate information for forging and machining.

and information needed by the worker for whom it is made. Figure 13-4 shows an index-plunger operating handle. This illustration shows the piece as a forging (A) and after it has been machined (B). Figure 13-5 shows a detail drawing of the same piece. Notice how the drawing shows the part after material has been removed by the machining process. Notice also the detailed list of machine operations.

Figure 13-6 shows a two-part detail drawing. This type of detail is known as a **combination drawing**. One half gives the dimensions for the forging. The other half gives the machining dimensions and notes.

Detail drawings are often made for standard parts that come in a range of sizes. Figure 13-7 is a **tabulated drawing** of a bushing. In this kind of drawing,

the dimensions are identified by letters. A table placed on the drawing tells what each dimension is for different sizes of the part. Either all or some of the dimensions can be given in this way. A similar kind of drawing is one in which the views are drawn with blank spaces for dimensions and notes, as shown in Figure 13-8A. These are then filled in as needed with the required information, as shown in Figure 13-8B. The views in this type of drawing are not to scale, except perhaps for one size.

Assembly Drawings

A drawing of a fully assembled construction is called an **assembly drawing**. Such drawings vary greatly in how completely they show detail and dimensioning. Their special value is that they show how the parts fit together, the look of the construction as a whole, the dimensions needed for installation, the space the construction needs, the foundation, the electrical or water connections, and so forth. When an assembly drawing gives complete information, it can be used as a working drawing. It is then called an **assembly working drawing**. This kind of drawing can be made only when there is little or no complex detail. An example is shown in Figure 13-9.

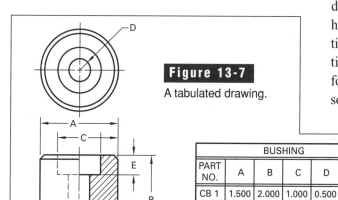

Figure 13-7

A tabulated drawing.

BUSHING					
PART NO.	A	B	C	D	E
CB 1	1.500	2.000	1.000	0.500	0.375
CB 2	1.625	2.125	1.125	0.625	0.437
CB 3	1.750	2.250	1.250	0.750	0.500
CB 4	1.875	2.375	1.375	0.875	0.562
CB 5	2.000	2.500	1.500	1.000	0.625

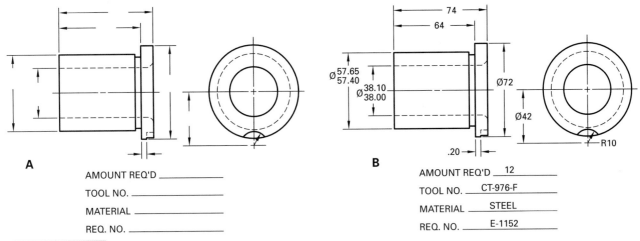

A

AMOUNT REQ'D _____

TOOL NO. _____

MATERIAL _____

REQ. NO. _____

B

AMOUNT REQ'D ___12___

TOOL NO. ___CT-976-F___

MATERIAL ___STEEL___

REQ. NO. ___E-1152___

Figure 13-8

Detail drawing of a standard part with dimensions blank (A) and filled in (B).

You can often show furniture and other wood construction in assembly working drawings by adding enlarged details or partial views as needed. See Figure 13-10. Note that fractional-inch dimensions are used in this drawing. The fractional-inch system is still commonly used in the furniture and woodworking industries. Note also that enlarged views are drawn in pictorial form; they are not regular orthographic views. This method is peculiar to the cabinet-making trade and is not normally used in mechanical drawing.

Figure 13-9

An assembly working drawing for a belt tightener.

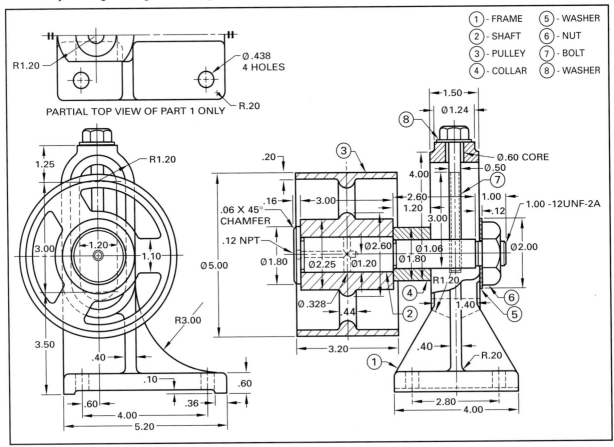

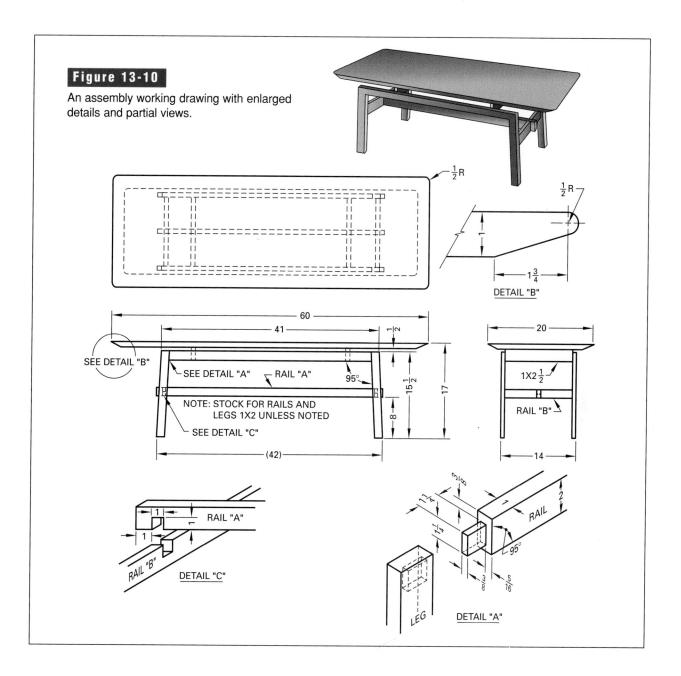

Figure 13-10

An assembly working drawing with enlarged details and partial views.

DETAIL "B"

SEE DETAIL "B"

SEE DETAIL "A"   RAIL "A"   95°

NOTE: STOCK FOR RAILS AND
LEGS 1X2 UNLESS NOTED

SEE DETAIL "C"

RAIL "A"

RAIL "B"   DETAIL "C"

RAIL "B"

1X2 ½

RAIL

95°

LEG   DETAIL "A"

Assembly drawings of machines are generally made to a small scale. Dimensions are chosen to tell overall distances, important center-to-center distances, and local dimensions. All, or almost all, hidden lines may be omitted. Also, if the drawing is made to a very small scale, unnecessary detail may be omitted. This has been done, for example, in the drawing shown in Figure 13-11. Drawings in which details have been omitted are called **outline assembly drawings**. Either exterior or sectional views may be used. When the main purpose is just to show the general look of the construction, only one or two views need to be drawn. Because some assembled constructions are so large, you may need to draw different views on separate sheets. However, you must use the same scale on all sheets.

A **reference assembly drawing** is a special assembly drawing that identifies parts to be assembled. See Figure 13-12. Note the tabular list in the upper right-hand corner. Note also the dimensions shown.

Many other kinds of assembly drawings are made for special purposes. These include part assemblies for groups of parts, drawings for use in assembling or erecting a machine, drawings to give directions for upkeep and use, and so forth.

Figure 13-11

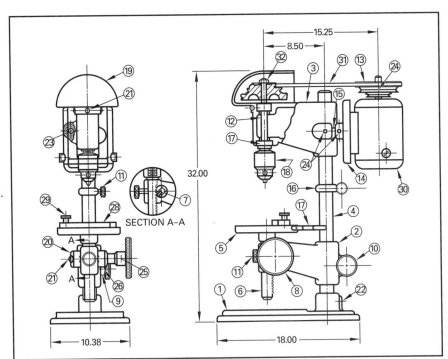

An outline assembly drawing.

Figure 13-12

A reference assembly drawing of the hanger assembly shown in the pictorial view.

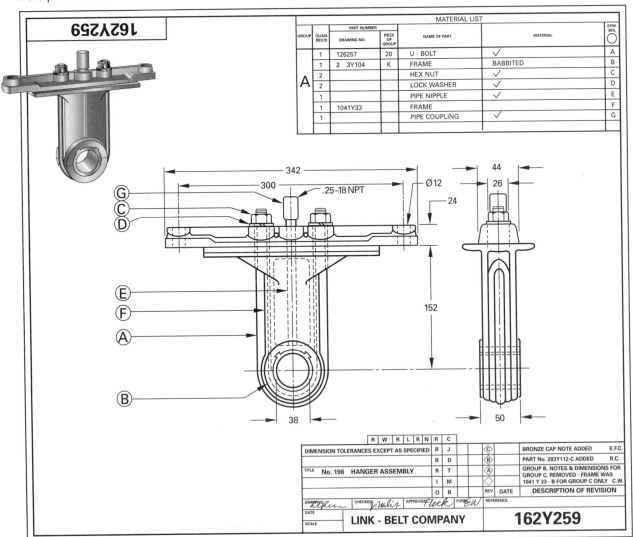

162Y259

GROUP	QUAN-REQ'D	PART NUMBER DRAWING NO.	PIECE OF GROUP	NAME OF PART	MATERIAL	SYM-BOL ◯
A	1	126257	20	U - BOLT	✓	A
	1	2 3Y104	K	FRAME	BABBITED	B
	2			HEX NUT		C
	2			LOCK WASHER	✓	D
	1			PIPE NIPPLE	✓	E
	1	1041Y33		FRAME		F
	1			PIPE COUPLING	✓	G

MATERIAL LIST

		R	W	R	L	R	N	R	C			
DIMENSION TOLERANCES EXCEPT AS SPECIFIED		R	J						◇C◇	BRONZE CAP NOTE ADDED	E.F.C.
		R	D						◇B◇	PART No. 283Y112-C ADDED	R.C.
TITLE No. 198 HANGER ASSEMBLY		R	T						◇A◇	GROUP B, NOTES & DIMENSIONS FOR GROUP C, REMOVED - FRAME WAS 1041 Y 33 - B FOR GROUP C ONLY C.W.
		I	M						◇◇	
		O	R						REV	DATE	DESCRIPTION OF REVISION

DRAWING CHECKED APPROVED FORM

REFERENCE

DATE

SCALE

LINK - BELT COMPANY   **162Y259**

Setting Up a Working Drawing

Regardless of whether you use board or CAD techniques, your drawings will be easier to draw and use if you set them up properly. For example, the views and scales you choose greatly affect the readability of the drawing. Therefore, you must take the time to plan the drawing before you begin.

Choosing Views

To describe an object completely, you generally need at least two views. The views you choose should always be those that are easiest to read. Each view must add information to the description of the object. If it does not, it is not needed and should not be drawn.

In some cases, only one view is needed, as long as the shape and size are clearly understood. Also, you may choose to give information in a note rather than in a second view. For complex pieces, more than three views may be needed. As you plan the views you need, consider whether the additional views should be partial views, auxiliary views, or sectional views.

If you have any question about which views to draw, think about why you are making the drawing. A drawing is judged on how clearly and precisely it gives the information needed for its purpose.

Choosing a Scale

The scale for a detail drawing is chosen according to three factors:

- how large the drawing must be to show details clearly
- how large it must be to include all dimensions without crowding
- the size of the paper

Whenever possible, a drawing should be drawn full size, at a 1:1 scale. Other scales that are commonly used are half, quarter, and eighth. Avoid scales such as 1:6, 1:3, and 1:1.33. If a part is very small, you can sometimes draw it to an enlarged scale, perhaps twice its full size (2:1) or more. See Chapter 4, "Basic Drafting Techniques," for more information about scales.

When you draw a number of details on one sheet, make them all to the same scale if possible. If you must use different scales, note the scale near each drawing. It is often useful to draw a detail, or part detail, to a larger scale on the main drawing. This saves both time and work in making separate detail drawings. For general assembly drawings, choose a scale that shows the details you want and looks good on the size of paper you are using. Sheet-metal pattern drawings are generally made full size for direct transfer to the sheet material in the shop.

For complete assemblies, a small scale is generally used. The scale is often fixed by the size of the paper the company has chosen for assemblies. For part assemblies, choose a scale to suit the purpose of the drawing. This might be to show how parts fit together, to identify the parts, to explain an operation, or to give other information.

Grouping and Placing Parts

Another part of planning a set of working drawings is to plan where each view or piece will be placed on the drawing sheet. When a number of details are used for a single machine, it is often a good idea to group them on a single sheet or set of sheets. A convenient method is to group the forging details together, the material details together, and so on. In general, show the parts in the position they will occupy in the assembled machine. That way, related parts will appear near each other. Long pieces such as shafts and bolts, however, may not work well using this method. They are often drawn with their long dimensions parallel to the long dimension of the sheet. See Figure 13-13.

Title Blocks

A **title block** is an area on a drawing that contains information about the drawing, the company, the drafter, and so on. Every sketch or drawing must have some kind of title block. However, the form, completeness, and location of the title block can vary. On working drawings, it can be placed in a box in the lower right-hand corner, as shown in Figure 13-14A, or it can be included in a record strip running as far as needed across the bottom or end of the sheet, as shown in Figure 13-14B.

The information included in a title block depends on company policy. Most title blocks contain some or all of the following information:

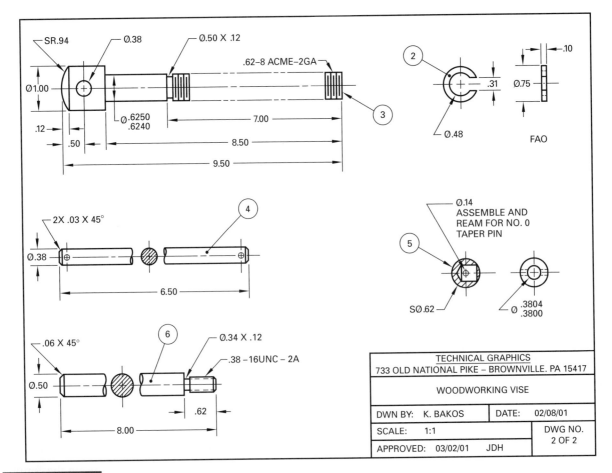

Figure 13-13

Several machine parts drawn on one sheet.

- name of the construction, machine, or project
- name of the part or parts shown, or simple details
- manufacturer, company, or firm name and address
- date (usually the date of completion)
- scale or scales used
- heat treatment, working tolerances, and so forth
- drawing number, shop order number, or customer's order number, according to the system used by the company

- drafting-room record: names and initials, with dates, of drafter and checker; approval of chief drafter, engineer, and so forth

A basic layout for a title block based on ANSI standards is shown in Figure 13-15.

In large drafting rooms, the title block is generally preprinted on the paper, cloth, or film, leaving spaces to be filled in. However, many firms use separate adhesive-backed title blocks. When CAD is being used, the title block is included in the drawing templates.

Figure 13-14

Title blocks: (A) boxed title; (B) strip title.

RIVETS	EXC. NOTED	CENTRAL TEXAS IRON WORKS		
HOLES	EXC. NOTED	WACO　　　　TEXAS　　　　ABILENE		
PAINT		BUILDING _____		
BLUEPRINT RECORD		LOCATION _____		
NO.	DATE	ISSUED FOR	CUSTOMER _____	
		ARCHITECT _____		
		MADE BY	CHK. BY	SHEET NO.
		DATE	REVISIONS	ORDER NO.

A

B

UNIT			NAME OF PIECE		
DR.	DATE	SYMBOL OF MACHINES USED ON	SUPERSEDES DRAW.	STOCK CASTING DROP FORGING	
DR.					
TR.		THE LODGE & SHIPLEY MACHINE TOOL CO.	SUPERSEDED BY DR.	MATERIAL	PIECE NO.
TR. CH.		FORM 795　CINCINNATI, OHIO. U.S.A.			

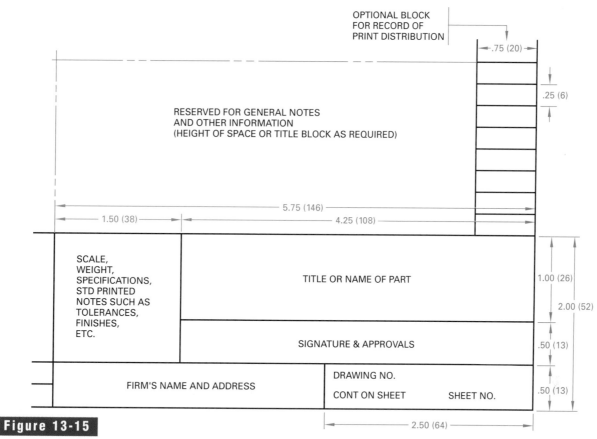

Figure 13-15

Basic layout for a title block. Dimensions shown are ASME/ANSI recommendations and may be varied to accommodate the user's requirements.

Change or Revision Block

A change or revision block is a table that is often attached to the title block, as shown in Figure 13-12. It may also be placed down the right-hand side or along the bottom of the drawing sheet. If the revision block is placed at the top of the sheet, the headings are placed at the top and the revisions are listed and numbered consecutively from the top down. If the revision block is placed at the bottom of the drawing sheet, as shown in Figure 13-12, the headings are placed at the bottom and the revisions are listed and numbered consecutively from the bottom up.

Bill of Materials

Most working drawings include a list of parts, the materials of which they are made, identification numbers, and other important information. This information is especially important on assembly drawings of various kinds. It is also needed on detail drawings where a number of parts are shown on the same sheet.

The names of parts, material, number required, part numbers, and so forth are often given in notes near the views of each part. It is better, however, to place just the part numbers near the views, link them to the views with leaders, and then collect all the other information in tabulated lists. Such a list is called a **bill of materials**. Some companies also call it a "list of materials." ANSI's official term is "parts list." ANSI provides formal specifications for a parts list, but in most cases, less formal formats are used. Figure 13-16 shows an example of a simple bill of materials. The format recommended for use with this book is shown in Figure 13-17. The column widths and names may vary as needed.

For working drawings created using board drafting techniques, the bill of materials is usually placed above the title block or in the upper right corner of the sheet. It can also be written or typed on a separate sheet with a title such as "Bill of Materials for Drawing No. 00" to identify it. This method is preferred by CAD operators.

Figure 13-16

A simple bill of materials.

BILL OF MATERIALS FOR IDLER PULLEY			
NAME	REQD	MATL	NOTES
IDLER PULLEY	1	C1	
IDLER PULLEY FRAME	1	C1	
IDLER PULLEY BUSHING	1	BRO	
IDLER PULLEY SHAFT	1	CRS	
Ø.62 HEX NUT	1		PURCHASED
WOODRUFF KEY #405	1		PURCHASED
.12 OILER	1		PURCHASED

Notes and Specifications

Information that you cannot make clear in a drawing must be given in lettered notes and symbols. For example, special trade information is often given in this way. Notes are used for the following items:

- quantity needed
- material
- kind of finish
- kind of fit
- method of machining
- kinds of screw threads
- kinds of bolts and nuts
- sizes of wire
- thickness of sheet metal

Other, similar information may also be given in notes.

The materials in general use are wood, plastic, cast iron, wrought iron, steel, brass, aluminum, and various alloys. All parts to go together must be of the proper size so that they will fit. Pieces may be left rough, partly finished, or completely finished.

After a part is cast or forged, it must be machined on all surfaces that are to fit with other surfaces. Round surfaces are generally refined on a lathe. Flat surfaces are finished or smoothed on a planer, milling machine, or shaper. Holes are made with drill presses, boring mills, or lathes. Extra metal is allowed for surfaces that are to be finished. To specify such surfaces, place a finish symbol on the lines that represent their edges. If the entire piece is to be finished, write a note such as FINISH ALL OVER, or **FAO**. No other mark is needed.

In the traditional drafting room, the kinds of machining, finish, or other treatment are specified in notes. Examples of these notes are: SPOTFACE, GRIND, POLISH, KNURL, CORE, DRILL, REAM, COUNTERSINK, COUNTERBORE, HARDEN, CASE HARDEN, BLUE, and TEMPER. Often other notes are added for special directions—to explain assembly, for example, or the order of doing work.

In current dimensioning practice, the drafter usually does not specify the method or tool to be used. Only the finished size and shape are given. For additional information, review Chapter 7, "Dimensioning."

Figure 13-17

Recommended form for a bill of materials.

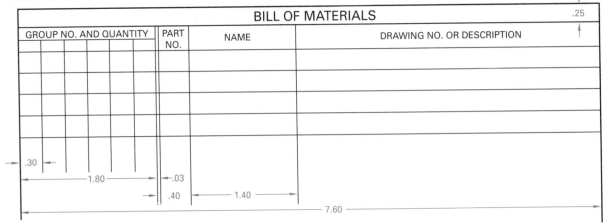

Checking a Drawing

After a drawing is finished, it must be looked over very carefully before it is used. This is called "checking the drawing." It is very important work. A drawing you have made should be checked by someone other than you. A person who has not worked on the drawing will be better able to spot errors. However, you must inspect your own work for errors or omissions before having it further checked.

To make a thorough check, drafters and checkers follow a set order of procedures. The following checks should be made:

1. The views completely describe the shape of each piece.
2. There are no unnecessary views.
3. The scale used is large enough to show all detail clearly.
4. All views are to scale, and the correct dimensions are given.
5. There are no parts that interfere with each other during assembly or operation, and necessary clearance space is provided around all parts that need it.
6. Enough dimensions are given to define the sizes of all parts completely, and no unnecessary or duplicate dimensions are given.
7. All necessary location and positioning dimensions are given with necessary precision.
8. Necessary tolerances and limits and fits are given, along with any other precision information.
9. The kind of material and the quantity needed of each part are specified.
10. The kind of finish is specified, all finished surfaces are marked, and a finished surface is not called for where one is not needed.
11. Standard parts and stock items, such as fasteners, handles, and catches, are used where suitable.
12. All necessary explanatory notes are given and are placed properly on the drawing.

TECH●MATH

Calculating Corner Clearance for a Hex Nut

When designing a product with a hex nut, care must be taken to ensure that there is clearance to tighten the nut in assembly. Calculating corner clearance for a hex nut simply means determining the distance across the corners of a hexagon. This is done by using a simple formula that contains a constant. A **constant** is a known, unchanging value. In this case, the constant is 1.155, and the formula for determining the distance across the corners when you know the distance across the flats is $c = 1.155f$, as shown below.

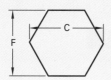

F = FLATS
C = CORNERS
CONSTANT = 1.155

In other words, every hexagon, regardless of its size, has a distance across the corners which is equal to 1.155 times the distance across the flats.

Problem: Find the corner clearance for a hex nut that is .625″ across the flats, as shown below.

.625

Solution: $c = 1.155f$
$c = 1.155 \times .625″$
$c = .722″$

In preparing a working drawing, drafters and students often find it difficult to develop a good layout of the various views on the sheet. Sheet layout for a simple detail drawing is relatively easy. However, sheet layout for a drawing with several parts on one sheet is somewhat more difficult. With experience, you will find it easier and less time-consuming. In the meantime, regardless of how simple or complex the drawing is, some basic steps are necessary to develop a finished working drawing. Refer to Figure 13-18 to practice setting up a working drawing of the marking gage. Follow these steps:

1. Develop a freehand sketch of the views you have chosen for the part or parts given. The sketches need not be to scale, but estimate sizes relatively close to the finished drawing size.

2. Add dimensions and notes to your sketches. This will help you determine the overall space needed for each view or drawing on the sheet.

3. Choose the appropriate scale based on the size and complexity of the part. You may have done this as part of step 1. However, at this point, you can make any necessary changes. For the marking gage, an appropriate scale is 1:1.

4. Determine the amount of space needed on the sheet for each part or view.

5. Select the sheet size that will best accommodate your drawing. Try not to crowd the sheet, but do not select an excessively large sheet. If you are working on a large and complex set of drawings, select one standard sheet size for the entire set of drawings. For the drawing in this exercise, we have chosen an A-size sheet.

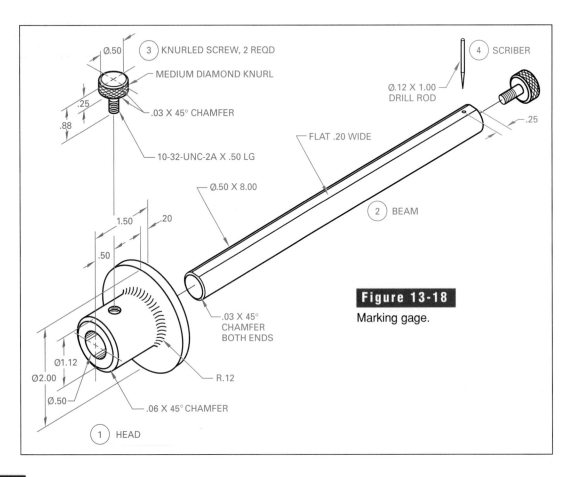

Figure 13-18

Marking gage.

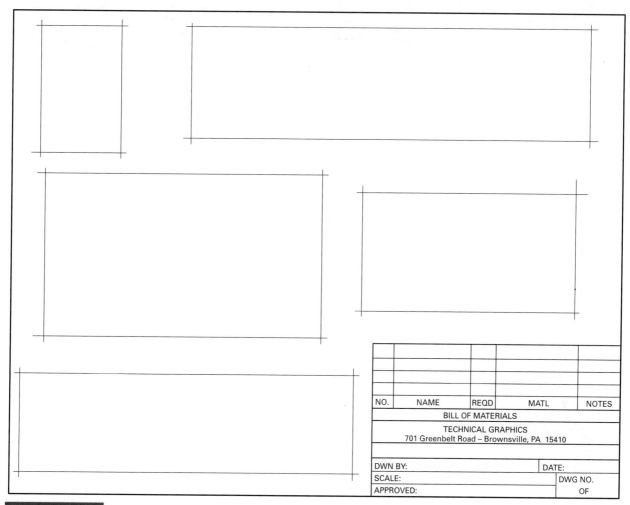

The bill of materials table contains the following columns:

NO.	NAME	REQD	MATL	NOTES

BILL OF MATERIALS

TECHNICAL GRAPHICS
701 Greenbelt Road – Brownsville, PA 15410

DWN BY: | DATE:
SCALE: | DWG NO.
APPROVED: | OF

Figure 13-19

Sheet layout for a working drawing of the marking gage.

6. Draw borderlines, title block, and bill of materials or parts list (if required). See Figure 13-19.

7. Block in spaces for the various views or parts, as shown in Figure 13-19. Use only very light layout lines for this purpose. For drawings of multiple parts, try to group drawings of individual parts according to where and how they fit in the assembly of the final product.

8. Develop the individual drawings in the spaces provided, as shown in Figure 13-20.

9. Darken all lines, add general notes, and fill in the bill of materials as required.

Another method that may be used to develop a good sheet layout is first to prepare rough mechanical drawings of each view or part on separate sheets of paper. Then follow these steps:

1. Select a drawing sheet of appropriate size and rough in the borderlines, title block, and bill of materials or parts list, if required. This sheet will be used for layout purposes only.

2. Place the individual rough drawings on the layout sheet and move them around to develop a pleasing layout. Be sure to allow adequate space for general notes and other details as necessary.

3. Tape the individual rough drawings in place. Carefully review what you have done. Make certain that nothing has been missed in developing the rough draft.

4. Tape a piece of tracing material over the rough drawing. Trace the entire drawing. The result should appear as in Figure 13-20.

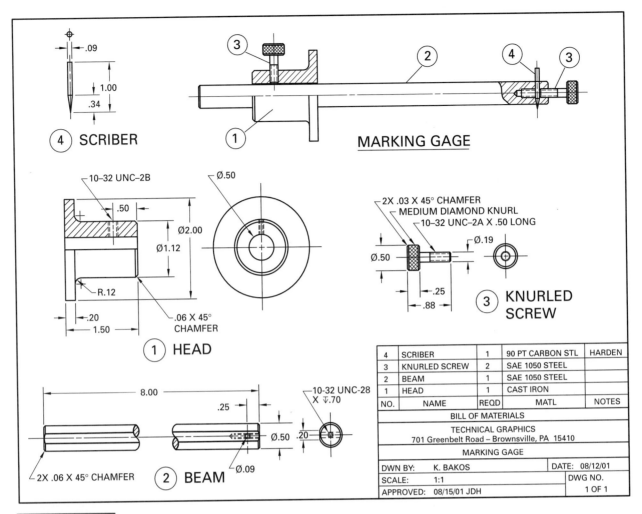

Figure 13-20
Completed working drawing.

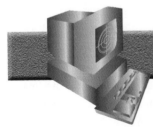

CAD Techniques

In most ways, working drawings are easier to set up using a CAD system than they are using board drafting techniques. For example, as you may recall from Chapter 4, most CAD software provides ANSI and ISO templates, complete with title blocks, for drawing sheets of various sizes. Many companies that use CAD systems also have their own drawing templates. However, you must still take care to set up the drawing correctly. CAD software cannot choose the correct views of an object for you, nor can it select an appropriate scale. You must become proficient at these tasks.

Using Layers

CAD systems greatly reduce the amount of time needed to create a full set of working drawings, particularly when separate drawings are needed for various workers, such as electrical contractors and welders. In a CAD drawing, you can create the basic views only once. Then you place the information needed by each contractor on a separate layer. For example, you might place the basic views on a layer called Machine. Then you could include a Welding layer, an Electrical layer, and as many other layers as necessary to provide the appropriate information to the different contractors. Then, to supply the drawings needed by the electrical contractor, you simply freeze all of the layers for other contractors and either print the drawing or supply it to the contractor electronically. There is no need to create an entirely new drawing for each purpose.

This approach is often beneficial from a record-keeping standpoint, too. By using layers to provide various contractors with information, you are keeping all of this information in a single drawing file. You don't have to keep track of a large number of bulky drawings. Instead, you simply keep backups of a single drawing file. For more information about drawing storage techniques, see Chapter 23.

Multiple Layouts

Beginning with AutoCAD 2000, the AutoCAD software provides the ability to include one or more layout views in a drawing file. These layouts are tabbed at the bottom of the drawing area in a manner similar to the worksheets in Microsoft Excel. By default, each drawing contains two layouts called Layout1 and Layout2. However, you can rename them and add as many layouts as you want. For example, in the drawing in Figure 13-21A, the drafter has included four layouts: one for each part of the assembly, and one for each sectional view. In the closeup of the tabs shown in Figure 13-21B, you can see that the tabs have been given names that make the content of the layouts clear.

All of the layout views are in paper space and include the border and title block specified by the drawing template. However, you can choose what parts of the model-space drawing to include in each layout. For example, in Figure 13-21, all four of the drawings shown in layout views actually exist in model space. The drafter simply chose to show different areas of model space in each layout view.

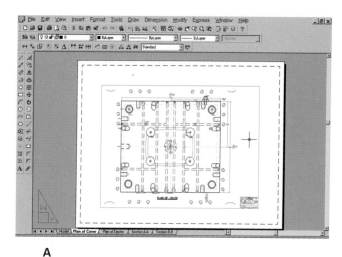

A

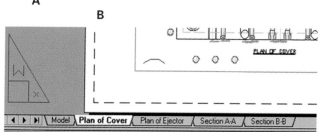

B

Figure 13-21

An entire set of working drawings can be included in a single drawing file by using custom layout views in paper space. (A) One of the layouts for a single-cavity mold drawing. (B) Notice how the drafter has labeled each tab.

Creating a Working Drawing

The initial procedures for creating a working drawing using AutoCAD are basically the same as those for creating any other CAD drawing. However, working drawings require more planning. As in board drafting, you must decide on an appropriate sheet size and scale. However, you should also consider the layers that will be needed and what information each layer will contain. The rest of the procedure will vary depending on the version of AutoCAD you are using.

AutoCAD R14 and Earlier

In the following exercise, you will create a set of working drawings for the exploded assembly drawing of the marking gage shown in Figure 13-22. Use the following steps to begin developing the drawings using AutoCAD Release 14 or earlier.

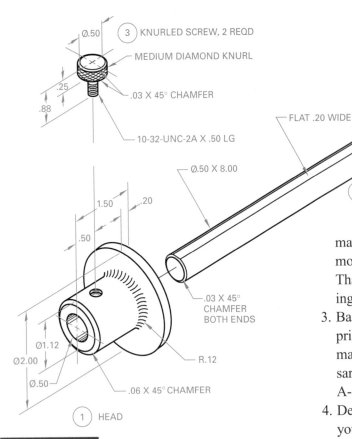

Figure 13-22

Marking gage.

1. Analyze the part or parts carefully to determine the views that will be needed. In the case of Figure 13-22, for example, you will need two-view drawings of the head and beam, a two-view detail drawing of the knurled screw, a detail drawing of the scriber, and an assembly drawing shown in section.

2. Decide how to position the views. To do this, you may want to create a temporary "scratch" file to help you determine the overall space needed for each view or drawing on the sheet. Open a new file and set the drawing limits to a large size to provide plenty of drawing space. Calculate the amount of space needed for each view and draw a rectangle or other polygon to represent the boundaries of that view. See Figure 13-23. Recall that this can be an approximation, because views can be moved in the electronic file for exact placement. Be sure to include space for dimensions and notes in your calculations. You may want to add labels to identify the various views. If the parts are very complex, you

may decide to separate the views into two or more groups for use on different drawing sheets. That is not necessary for the marking gage drawing, however.

3. Based on your work in step 2, select an appropriate size for the drawing sheet, and decide how many drawing sheets you will need (if necessary). For the marking gage, we have selected an A-size sheet.

4. Decide what layers to use for the drawing. As you may have noticed, AutoCAD automatically creates layers for the border and title block. The additional layers you might want to create for the marking gage include Object lines, Dimensions, Centerlines, Hidden lines, and Hatches. (*Note:* In practice, layer names and contents are often specified by the individual companies. In some cases, the layers are even set up in the company's template drawings. You should check company guidelines before completing this step.)

Figure 13-23

Create a new layout using the Layout Wizard.

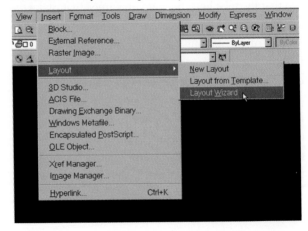

For the marking gage drawing, only a single file is necessary because you will place all of the views on a single drawing sheet. If you decided to use more than one drawing sheet, you would need to complete the steps in this section for each sheet.

Continue to develop the working drawing by following these steps:

1. Create a new drawing using a template file. Call the file Marking Gage. Choose the ANSI A (ansi_a.dwt) drawing template.

2. Set up the layers that you have identified as necessary, and set the text style to Roman Simplex.

3. Using your scratch file as reference, zoom in on the location for each view and develop it as you normally would. The one difference in this case is that you will be grouping more than one two-view drawing with other views on the same drawing sheet.

►CADTIP

Drawing the Screw Efficiently

Notice that a side view of the knurled screw is needed once in the knurled screw detail and twice in the assembly view. The most efficient way to use the CAD software is to draw the screw once, block it, and then reinsert it at the other two locations, rotating it as required.

4. Fill in the title block with the appropriate information by selecting the placeholder text and picking the Properties button.

5. Generate the information needed for the bill of materials. By looking at the exploded assembly drawing in Figure 13-22, you can see that the completed assembly will need one head, one beam, two knurled screws, and one scriber. Create a block for each item, including the knurled screw if you didn't create it earlier. Add invisible, variable attributes as follows:

Attribute Tag	Prompt	Default
Material	What material?	CRS
Notes	Add any notes:	

(Note that you do *not* want to set a default for the Notes tag.) Explode each block on the screen, if necessary, and reblock to include the attributes. Refer to Chapter 11 for more information about adding text attributes to blocks.

6. Before you generate the actual bill of materials, notice that, although you need only two knurled screws to complete the assembly, the knurled screw block is used three times in the drawing. To avoid erroneously including three screws in the bill of materials, explode one of the screws. To do this, pick the Explode button or enter EXPLODE at the keyboard, and select one of the screws. When you explode a block, it loses its associated attributes.

7. Generate the bill of materials. (See "Bill of Materials Generation" later in this chapter.)

8. Check the drawing.

AutoCAD 2000 and Later

If you are using AutoCAD 2000 or later, you should plan the number of layouts to use and the information to be provided on each layout. Keep in mind that you should label each layout so that others who use the file will understand immediately what each layout contains. Refer again to Figure 13-21. For the Marking Gage drawing, you could follow a procedure very similar to that described in the previous section for users of AutoCAD R14 and earlier. You would simply use one of the layout tabs for the entire working drawing. However, some industries require that separate drawing sheets be used for various views. Therefore, it is good to know how to use multiple layouts.

The following procedure presents a second way to lay out the working drawings for the marking gage shown in Figure 13-22. Follow these steps to create a working drawing using AutoCAD 2000 or later:

1. Analyze the part or parts carefully to determine the views that will be needed. In the case of Figure 13-22, you will need two-view drawings of the head and beam, a two-view detail drawing of the knurled screw, a detail drawing of the scriber, and an assembly drawing shown in section.

2. Assume that the company you work for requires that orthographic drawings be drawn on a separate sheet from assembly drawings. Decide how many drawing sheets you will need. For the marking gage, you will need two layout sheets,

one for the four orthographic drawings, and one for the assembly drawing.

3. Select an appropriate size for the drawing sheets. For the marking gage, we have selected an A-size drawing sheet. AutoCAD allows you to set each layout on a different size sheet if you prefer, but this is not good practice. You should keep all the pages of a set of working drawings on drawing sheets of the same size.

4. Decide what layers to use for the drawing. As you may have noticed, AutoCAD automatically creates layers for the border and title block. The additional layers you might want to create for the marking gage include Object lines, Dimensions, Centerlines, Hidden lines, and Hatches. (*Note:* In practice, layer names and contents are often specified by the individual companies. In some cases, the layers are even set up in the company's template drawings. You should check company guidelines before completing this step.)

5. Create a new drawing using the ANSI A template. Call the file Marking Gage.

6. Set up the layers that you have identified as necessary, and set the text style to Roman Simplex.

►CADTIP

AutoCAD Versions and Layout Options

If you are using AutoCAD 2000i or later, the Layout1 and Layout2 tabs do not appear unless you start the drawing from scratch. When you open a new drawing based on the ANSI A template, a single layout view labeled ANSI A Title Block appears. All nine of the steps listed here for creating a layout view have already been performed for this layout. All you have to do is change the name of the tab. To change the name, right-click the tab and select Rename. Name this layout Orthographics. Then right-click the new layout tab, select the From template option, and create your second layout view based on the ANSI A template. Again, all nine steps have been done for you, so you can skip down to the text immediately following step 11.

Now you should set up the layout views. You could use the default Layout1 and Layout2 tabs, but in some versions of AutoCAD it is more efficient simply to set up new layouts from scratch. Follow these steps:

1. From the Insert menu near the top of the screen, choose Layout and then Layout Wizard. See Figure 13-23.

2. Name the new layout Orthographics. Then pick the Next button to proceed to the next step of the Wizard.

3. From the list of printers that appears on the next screen, choose the one you normally use to print CAD problems. Pick the Next button.

4. Select the ANSI A paper size from the drop-down menu. Pick the Next button.

5. Pick the Landscape radio button to specify the page orientation. Pick the Next button.

6. Choose the ANSI A title block from the list. Pick the Next button.

7. Select a single viewport at a viewport scale of 1:1. Pick the Next button.

8. Pick the Select location button and use a window to enclose most of the blank area on the drawing sheet, as shown in Figure 13-24. Pick the Next button.

9. Pick the Finish button to complete the process.

10. Repeat steps 1 through 9 for the other layout view. Name it Assembly.

11. If your version of AutoCAD automatically placed Layout1 and Layout2 tabs in the drawing, right-click on the Layout1 tab and select Delete from the shortcut menu that appears. Pick OK to delete the Layout1 tab. Repeat to delete the Layout2 tab.

You should now have three tabs at the bottom of the drawing area: one for each paper-space layout view, and one for model space. Follow these steps to continue building the working drawing:

1. Pick the Model tab to return to the working area of the drawing. All of the following steps should be done in model space.

2. Create the orthographic views, as shown in Figure 13-25. Be sure to create the views at full size (scale of 1:1). Draw the views in the positions you want them to appear in the layout.

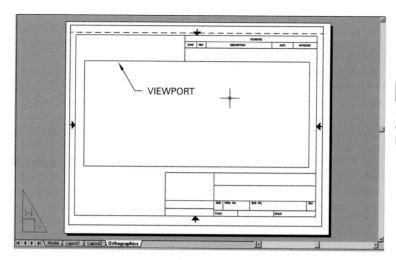

Figure 13-24

Use a window to create a viewport that covers most of the blank space on the layout sheet.

3. Move to another part of the drawing area, well away from the orthographic views, and create the assembly view. Because the orthographics have already been drawn, you can copy parts of them for use in the assembly drawing. For example, you can copy the beam, then delete the interior break, extend the lines to the full

8″, and move the ends to their new positions. Copy the head and rotate the copy 180° to form the basis for the head in the drawing. You will have to rearrange the internal details and reapply the hatch, but that is faster than starting from scratch. The scriber can be used without modification. The screw can simply be copied, rotated, and placed as necessary.

Figure 13-25

Draw the orthographic views in model space.

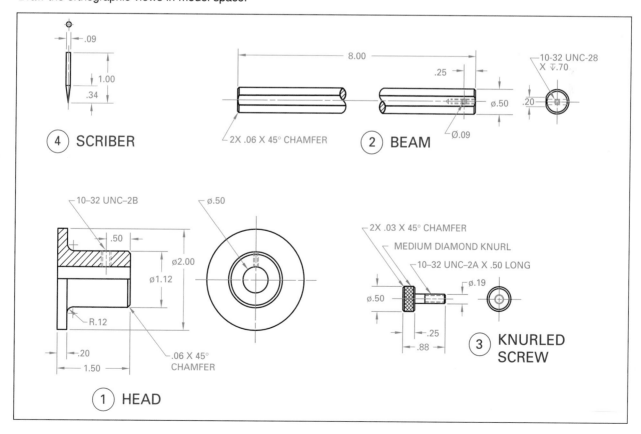

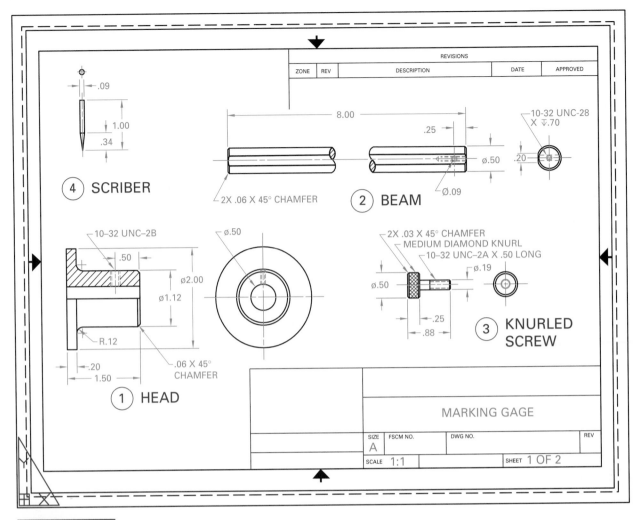

Figure 13-26

Position the drawing so that only the orthographic views appear on the drawing sheet.

4. Set up the orthographic layout view, as shown in Figure 13-26. To do this, pick the Orthographics layout tab. Notice that the entire drawing appears on the layout sheet. To show only the orthographic views, you will need to move them. To do this, pick the PAPER button at the bottom of the screen. The button changes to MODEL, indicating that you are now in model space. This allows you to work directly with the objects you drew in model space. Use the PAN command to move the entire drawing so that the orthographic views are centered on the sheet and the assembly drawing is hidden from view. If you need to move individual views, you can use MOVE or any other drawing command because you are in model space, even though the border and title block are visible.

5. Pick the Assembly layout tab and repeat the procedure in step 4 to center the assembly view on the sheet, as shown in Figure 13-27. The orthographic views should not show on this sheet.

6. Fill in the title block with the appropriate information. *Note:* In some versions of AutoCAD, the title block is blocked, and the text is attribute text. The easiest way to change the text is to delete the attributes entirely and use DTEXT to enter the required information.

7. Generate the information for the bill of materials. By looking at the exploded assembly drawing in Figure 13-22, you can see that the completed assembly will need one head, one beam, two knurled screws, and one scriber. Create a block for each item, including the

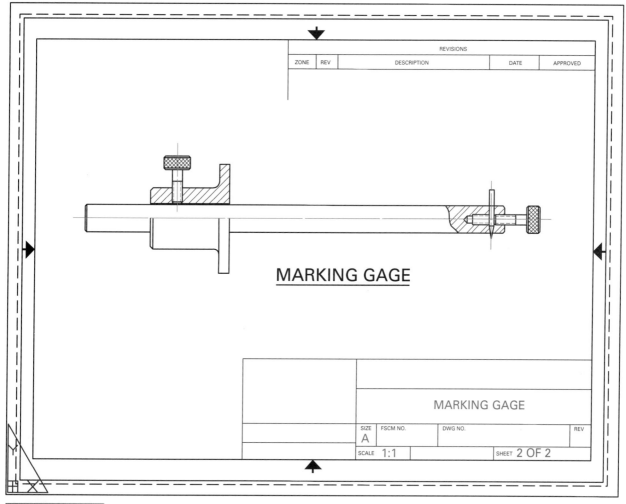

Figure 13-27

The Assembly layout sheet.

knurled screw if you didn't create it earlier. Add invisible, variable attributes to each block as follows:

Attribute Tag	Prompt	Default
Material	What material?	CRS
Notes	Add any notes:	

(Note that you do *not* want to set a default for the Notes tag.) Explode each block on the screen, if necessary, and reblock to include the attributes. Refer to Chapter 11 for more information about adding text attributes to blocks.

8. Before you generate the actual bill of materials, notice that, although you need only two knurled screws to complete the assembly, the knurled screw block is used three times in the drawing. To avoid erroneously including three screws in the bill of materials, explode one of the screws.

To do this, pick the Explode button or enter EXPLODE at the keyboard, and select one of the screws. When you explode a block, it loses its associated attributes.

9. Generate the bill of materials. (See "Bill of Materials Generation" below for details.)
10. Check the drawing.

Bill of Materials Generation

AutoCAD and many other CAD programs allow you to generate a bill of materials automatically, based on the attributes associated with blocks you have inserted into the drawing. The advantage of using this method is that it is very accurate. Each time you insert a block, AutoCAD stores the attribute information in the drawing database. When you want to create the bill of materials, you extract the attribute information into a table in a text or spreadsheet document.

ANSI has approved this method of generating a bill of materials (parts list). According to ANSI, the parts list should be set as a separate page that is included with the working drawings. This is why the bill of materials is not included on the CAD documents shown in Figures 13-24, 13-26, and 13-27.

Creating a Template File

To display attributes as a bill of materials, the first thing you need is an **extraction template file**. This file contains the tags from attributes you want to include in the bill of materials. Only the tags listed in this file will be extracted. This file also defines the type of data each attribute holds, the maximum length of information that can be entered for the attribute, and, for numeric data, the number of decimal places to display. The extraction template file is typically very short. It contains one line of data for each attribute you want to extract. Figure 13-28A shows an extraction template file for use with the marking gage drawing. Notice that the file has only two lines. Create this file using Notepad, and save it using the name attext.txt. Note that the "txt" extension is very important. Be sure to save it into the same folder that contains the Marking Gage.dwg file.

Figure 13-28B explains how to read the extraction template file. As you can see, this kind of file is very easy to create. You simply list the tag name, followed by a seven-digit number that indicates the type of data, field length required, and decimal places. Note that the number must be seven digits long, so if the field length is less than 100, you must insert a 0 in the hundreds place.

Extracting the Attributes

Now you can extract the attributes from the Marking Gage drawing. *Note:* This procedure extracts the attributes from the Marking Gage drawing. To read the resulting file, you must have access to a database program or a spreadsheet program such as Microsoft® Excel.

Follow these steps to extract the attributes from the Marking Gage drawing:

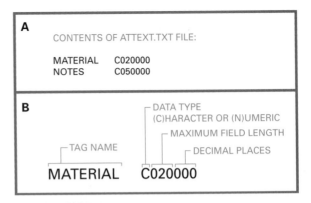

Figure 13-28

(A) Enter the text for the attext.txt file exactly as shown here. Use a single tab to separate the tag name from the number. (B) Explanation of the number used for each tag.

1. Enter the ATTEXT command at the keyboard to display the Attribute Extraction dialog box, as shown in Figure 13-29.
2. Pick the top radio button to choose a comma delimited file format. In this format, the attributes are separated in the output file by commas. We are using this format because most spreadsheets and databases can read it.
3. Pick the Template File… button and navigate through Windows to the location of your attext.txt file. Doubleclick the file name to select it. Attext.txt should then appear in the box to the right of the Template File… box, as shown in Figure 13-29.

Figure 13-29

The Attribute Extraction dialog box.

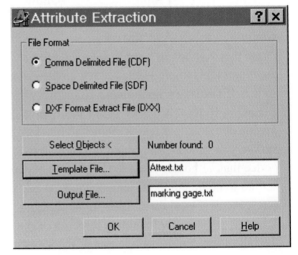

4. By default, the name of the output file is marking gage.txt. That is an appropriate name, but AutoCAD places the file in the AutoCAD directory unless you specify a different location. Pick the Output File… button and change to the folder where you store your AutoCAD drawings.

5. Pick the OK button to complete the extraction.

Completing the Bill of Materials

Now open any available database or spreadsheet program. Then follow the steps below. *Note:* This procedure assumes that Microsoft Excel is being used. If you are using a different program, the steps listed here may vary slightly.

1. Pick the icon to open an existing file, and go to the folder where you store your AutoCAD files. The marking gage.txt file probably won't appear in the selection box. To make it appear, go down to the Files of Type dropdown box and select All Files (*.*). The marking gage.txt file should now appear. Doubleclick to open it.

2. Because txt is not a native Excel file format, Excel presents a Wizard to help you display the text properly. Choose a delimited file, starting at row 1, and pick the Next button. In the next dialog box, choose commas as the delimiting character. Notice that the sample text separates into three columns. Pick the Next button.

3. Choose a general text format for all of the columns, and pick the Finish button to open the file.

4. The text appears, but the columns are not wide enough to display the text correctly. Position the cursor on the column headings over the line between columns A and B. When the double-

Figure 13-30

The marking gage.txt attribute file in Excel after column widths have been adjusted.

'SCRIBER'	'90 PT CARBON STL'	'HARDEN'
'HEAD'	'CASTIRON'	''
'BEAM'	'SAE 1050 STEEL'	''
'KNURLED SCREW'	'SAE 1050 STEEL'	''
'KNURLED SCREW'	'SAE 1050 STEEL'	''

Figure 13-31

Finished bill of materials for the marking gage.

BILL OF MATERIALS			
QUANTITY	ITEM	MATERIAL	NOTES
1	'SCRIBER'	'90 PT CARBON STL'	'HARDEN'
1	'HEAD'	'CASTIRON'	''
1	'BEAM'	'SAE 1050 STEEL'	''
2	'KNURLED SCREW'	'SAE 1050 STEEL'	''

headed arrow appears, move the cursor to widen the column. Repeat to widen the other two columns. When you finish, the contents of the file should look like Figure 13-30.

5. Save the file in the spreadsheet format.

6. Add a new column on the left side of the spreadsheet. (Choose Columns from the Insert menu.)

7. Add two rows at the top of the spreadsheet. (Choose Rows from the Insert menu.)

8. In the new top row, enter the text BILL OF MATERIALS.

9. Highlight the first four cells in the top row. Then pick Cells from the Format menu, pick the Alignment tab, and pick Center Across Selection from the Horizontal dropdown box. This centers the label over the rest of the table.

10. In the second row, enter QUANTITY, ITEM, MATERIAL, and NOTES as headings for the appropriate columns. Highlight the entire second row and pick the Center button to center the text on each column.

11. Notice that KNURLED SCREW appears twice. Delete the second occurrence and enter a 2 in the QUANTITY column for the remaining KNURLED SCREW row.

12. Enter a 1 in the QUANTITY column for each of the other items.

At this point, the bill of materials is in an acceptable format for use with the Marking Gage working drawings, as shown in Figure 13-31. If you wish, however, you may add formatting to make the file more easily readable. For example, you may wish to make the BILL OF MATERIALS label and the column heads bold.

Chapter Summary

■ A good working drawing follows the style and practices recommended by ANSI and/or ISO.

■ A detail drawing gives all necessary information for making a single part.

■ A drawing of a fully assembled product is an assembly drawing.

■ Setting up a working drawing involves choosing views and an appropriate scale, grouping and placing parts, and adding details such as notes, specifications, and a bill of materials.

■ Tabulated drawings are made for standard parts that are to be produced in a range of sizes.

■ A title block contains information about the drawing, the company, the drafter, drawing scale, name of the part, date drawn, etc.

■ A bill of materials lists the part numbers, names, quantity, and other related information.

■ CAD operators use layers efficiently to manage information needed by various contractors or departments using a single drawing.

■ Some CAD programs allow the operator to set up multiple drawing sheet layouts within a single drawing file.

■ Many CAD programs, including AutoCAD, can generate a bill of materials automatically from data that already exists in the drawing file.

Review Questions

1. What kind of drawing shows all the information necessary for manufacturing a part or a complete product?

2. Is Figure 13-1 a detail drawing, an assembly drawing, or an assembly working drawing?

3. Is Figure 13-5 a detail drawing, an assembly drawing, or an assembly working drawing?

4. What factors should you take into consideration when setting up a working drawing?

5. If an entire part is to be machined, what general note should you place on the drawing?

6. What do you call a drawing of a single part with all information needed for manufacturing it?

7. Briefly describe the process of checking a set of working drawings.

8. What is the outside diameter of Part No. CB 4 in Figure 13-7?

9. What three factors help determine the scale used on a drawing?

10. What is another way of specifying the scale used for the drawing in Figure 13-13?

11. What symbol is used to specify a finished surface?

12. What type of drawing might you create to represent a standard part that will be produced in a range of sizes?

13. What is the purpose of a title block?

14. What is ANSI's name for a bill of materials?

15. Explain how a single CAD drawing can be used as both a welder's detail and an electrical detail without becoming cluttered.

16. How many drawing files does a set of working drawings require?

17. What is the advantage of including multiple layouts in a single CAD file?

18. What is the purpose of an extraction template file?

Drafting Problems

The drafting problems in this chapter are designed to be completed using board drafting techniques or CAD.

For problems 1 through 28, follow the instructions for each drawing problem. Determine an appropriate scale if the scale is not given, and dimension the drawings.

1. Make a working drawing of each part of the coupler shown in Figure 13-32. Ends: die-cast aluminum. Spacer: rubber.

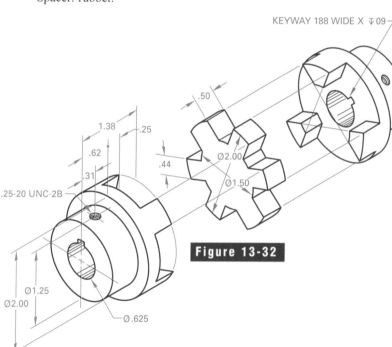

KEYWAY 188 WIDE X ↧ 09

1.38
.25
.62
.50
.44
.31
.25-20 UNC-2B
Ø2.00
Ø1.50
Ø1.25
Ø2.00
Ø.625

Figure 13-32

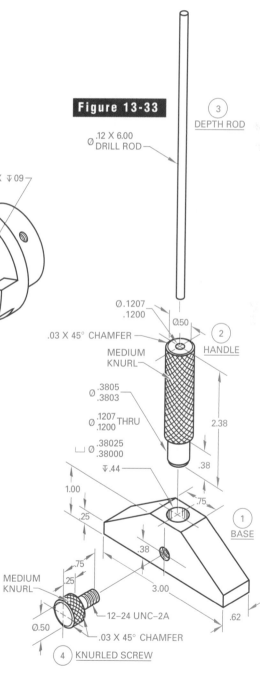

Figure 13-33

Ø .12 X 6.00
DRILL ROD

③ DEPTH ROD

Ø .1207
.1200
.03 X 45° CHAMFER
MEDIUM KNURL
Ø50

② HANDLE

Ø .3805
.3803
Ø .1207 THRU
.1200
⌴ Ø .38025
.38000
↧ .44
2.38
.38

1.00
.25
.75

① BASE

.38
3.00
.62

MEDIUM KNURL
.75
.25
Ø.50
12–24 UNC–2A
.03 X 45° CHAMFER

④ KNURLED SCREW

2. Make an assembly drawing of the coupler shown in Figure 13-32. Estimate all sizes and details not given.

3. Make a working drawing of each part of the depth gage shown in Figure 13-33. All parts: cold-rolled steel.

4. Make an assembly drawing of the depth gage shown in Figure 13-33. Estimate all sizes and details not given.

5. Make a working drawing of each part of the trammel shown in Figure 13-34. Specify "2 REQD" for the point, body, and knurled screw. The point is to be heat-treated after machining.

6. Make an assembly drawing of the trammel shown in Figure 13-34. Estimate all sizes and details not given.

7. Make a working drawing of each part of the arbor shown in Figure 13-35. Flanges: die-cast aluminum. Shaft: cold-rolled steel.

8. Make an assembly drawing of the arbor shown in Figure 13-35 with a Ø6.00 × 1.00 grinding wheel between the flanges. Show sectional views where practical. Draw all fasteners. Estimate all sizes and details not given.

9. Making a working drawing of each part of the power expansion bit shown in Figure 13-36. Cutter: tool steel. Body: cast iron. Use sectional views where necessary.

10. Make an assembly drawing of the power expansion bit shown in Figure 13-36. Estimate all sizes and details not given.

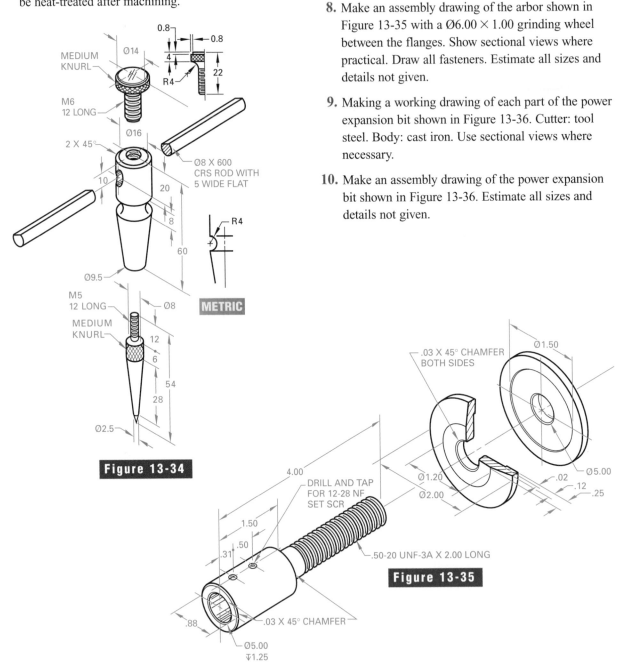

Figure 13-34

Figure 13-35

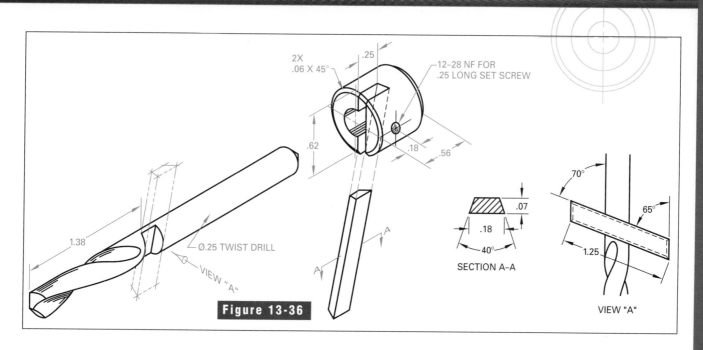

2X
.06 X 45°

.25

12–28 NF FOR
.25 LONG SET SCREW

.62

.18   .56

Ø.25 TWIST DRILL

1.38

VIEW "A"

A

A

.07

.18

40°

SECTION A–A

70°

65°

1.25

VIEW "A"

Figure 13-36

11. Make a working drawing of each part of the level shown in Figure 13-37, except do not draw the level glass. Body: die-cast aluminum. Top plate: cold-rolled steel. Use sectional views where necessary. Fillets = ⅛ R.

12. Make an assembly drawing of the level shown in Figure 13-37. Redesign the level to include vertical and 45° angle level glasses if desired. Estimate all sizes and details not given.

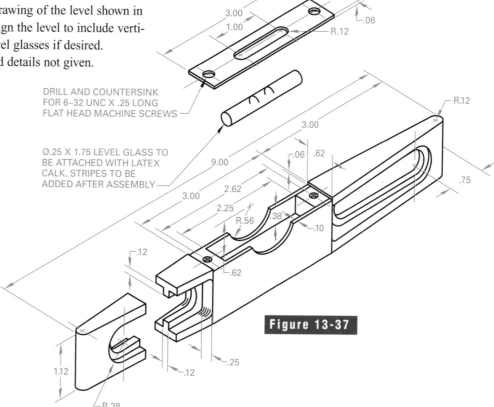

.62

3.00

1.00

.06

R.12

DRILL AND COUNTERSINK
FOR 6–32 UNC X .25 LONG
FLAT HEAD MACHINE SCREWS

Ø.25 X 1.75 LEVEL GLASS TO
BE ATTACHED WITH LATEX
CALK. STRIPES TO BE
ADDED AFTER ASSEMBLY

R.12

3.00

.06   .62

.75

9.00

3.00

2.62

2.25

R.56   38°   .10

.12

.62

1.12

.12   .25

R.38

Figure 13-37

13. Make a working drawing of each part of the circle cutter shown in Figure 13-38. Cutter: tool steel. Body and tool holder: cold-rolled steel.

14. Make an assembly drawing of the circle cutter shown in Figure 13-38. Show sectional views where necessary. Draw all fasteners. Estimate all sizes and details not given.

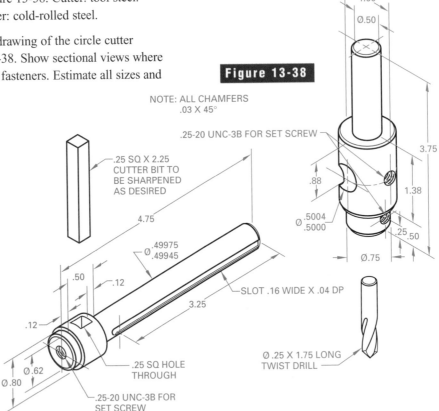

Figure 13-38

NOTE: ALL CHAMFERS
.03 X 45°

.25-20 UNC-3B FOR SET SCREW

1.00

Ø.50

3.75

.88

1.38

Ø .5004
.5000

.25 .50

Ø.75

.25 SQ X 2.25 CUTTER BIT TO BE SHARPENED AS DESIRED

4.75

Ø .49975
.49945

.50

.12

.12

.62

Ø.62

Ø.80

3.25

SLOT .16 WIDE X .04 DP

.25 SQ HOLE THROUGH

.25-20 UNC-3B FOR SET SCREW

Ø .25 X 1.75 LONG TWIST DRILL

15. Make an assembly working drawing of the coffee table shown in Figure 13-39. Include all necessary details and partial views. Choose an appropriate wood and types of joints (see Figure 13-10). Make a complete bill of materials. Estimate all sizes not given.

16. Make an assembly working drawing of the storage cabinet shown in Figure 13-40. Include all necessary details and partial views. Choose an appropriate wood and types of joints (see Figure 13-10). Make a complete bill of materials. Estimate all sizes not given.

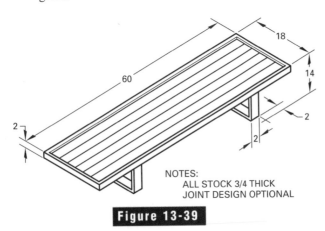

18

14

60

2

2

2

NOTES:
ALL STOCK 3/4 THICK
JOINT DESIGN OPTIONAL

Figure 13-39

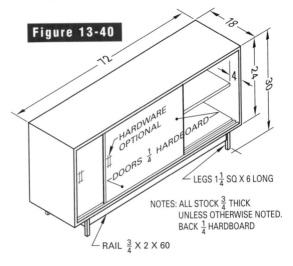

Figure 13-40

18

72

4

24

30

HARDWARE OPTIONAL

DOORS $\frac{1}{4}$ HARDBOARD

LEGS $1\frac{1}{4}$ SQ X 6 LONG

NOTES: ALL STOCK $\frac{3}{4}$ THICK UNLESS OTHERWISE NOTED. BACK $\frac{1}{4}$ HARDBOARD

RAIL $\frac{3}{4}$ X 2 X 60

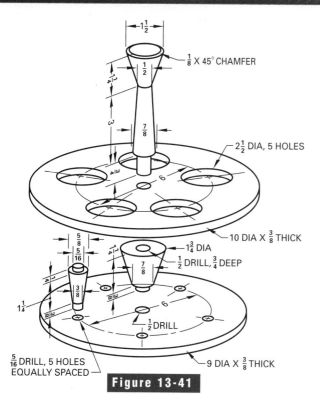

$1\frac{1}{2}$

$\frac{1}{2}$

$1\frac{1}{4}$

3

$\frac{7}{8}$

$\frac{1}{8}$ X 45° CHAMFER

$2\frac{1}{2}$ DIA, 5 HOLES

6

10 DIA X $\frac{3}{8}$ THICK

$\frac{5}{8}$
$\frac{5}{16}$

$1\frac{3}{4}$ DIA

$\frac{1}{2}$ DRILL, $\frac{3}{4}$ DEEP

$\frac{7}{8}$

$\frac{3}{8}$

$1\frac{1}{4}$

6

$\frac{1}{2}$ DRILL

$\frac{5}{16}$ DRILL, 5 HOLES
EQUALLY SPACED

9 DIA X $\frac{3}{8}$ THICK

Figure 13-41

17. Make detail working drawings of each part of the beverage server shown in Figure 13-41. Material: hardwood.

18. Make an assembly drawing of the beverage server shown in Figure 13-41.

19. Make a complete bill of materials for the beverage server shown in Figure 13-41. Based on the bill of materials, estimate the cost to manufacture the server, and prepare an estimated budget.

20. Make an assembly working drawing of the cement float shown in Figure 13-42. Materials: as noted. Use a 1.00″ grid for the curved outline of the handle. Add a bill of materials on the same drawing sheet. (Exception: CAD operators may include a separate bill of materials.)

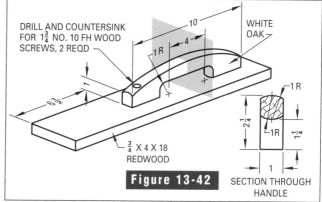

DRILL AND COUNTERSINK
FOR $1\frac{3}{4}$ NO. 10 FH WOOD
SCREWS, 2 REQD

10

4

1R

WHITE
OAK

1

$2\frac{1}{2}$

$\frac{3}{4}$ X 4 X 18
REDWOOD

1R

$2\frac{1}{4}$

1R

$1\frac{1}{4}$

1

Figure 13-42

SECTION THROUGH
HANDLE

21. Make a working drawing of the housing shown in Figure 13-43. Use partial and sectional views where needed. Material: cast iron. Estimate all sizes not given.

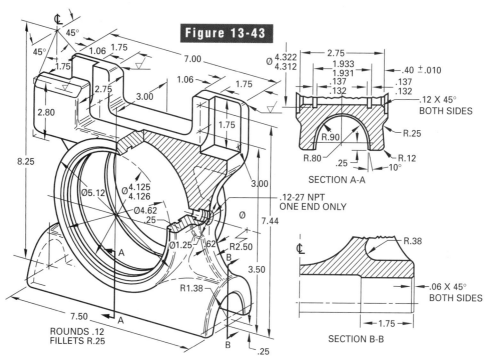

Figure 13-43

45°

45°

1.06   1.75

1.75

2.75

2.80

3.00

7.00

1.06

1.75

1.75

8.25

Ø5.12

Ø$\frac{4.125}{4.126}$

Ø4.62

.25

Ø1.25

.62

R2.50

R1.38

A

A

B

B

7.50

ROUNDS .12
FILLETS R.25

3.00

7.44

3.50

.25

Ø$\frac{4.322}{4.312}$

2.75

$\frac{1.933}{1.931}$

$\frac{.137}{.132}$

.40 ±.010

$\frac{.137}{.132}$

.12 X 45°
BOTH SIDES

R.90

R.25

R.80

.25

R.12

10°

SECTION A-A

.12-27 NPT
ONE END ONLY

R.38

.06 X 45°
BOTH SIDES

1.75

SECTION B-B

22. Make detail working drawings of each part of the hung bearing shown in Figure 13-44. Fully dimension each part. All bolts are 16 mm in diameter. Do not draw bolts and nuts. Estimate sizes not given.

23. Make a working drawing of the end base shown in Figure 13-45. Use partial and sectional views where needed. Material: cast iron. Estimate all sizes not given.

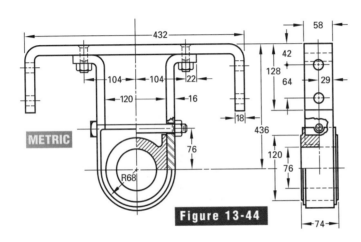

METRIC

Figure 13-44

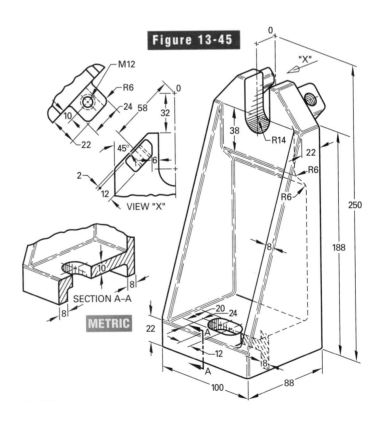

Figure 13-45

VIEW "X"

SECTION A–A

METRIC

24. Make a front view and section of the cushion wheel shown in Figure 13-46 at full size. This type of wheel is used on warehouse or platform trucks to reduce noise and vibration.

25. Make a complete set of detail drawings, full size, with a bill of materials, for the cushion wheel shown in Figure 13-46. Three sheets will be needed. Rivets are purchased and therefore need not be detailed, but they should be listed in the bill of materials.

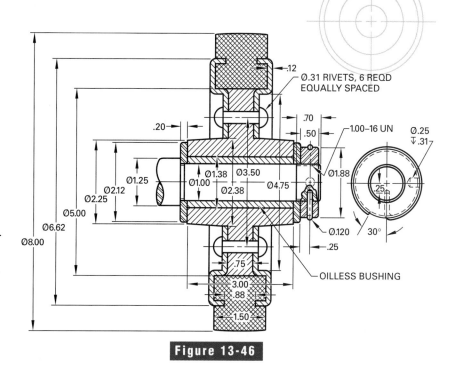

Figure 13-46

26. Make detail drawings of the crane hook parts shown in Figure 13-47.

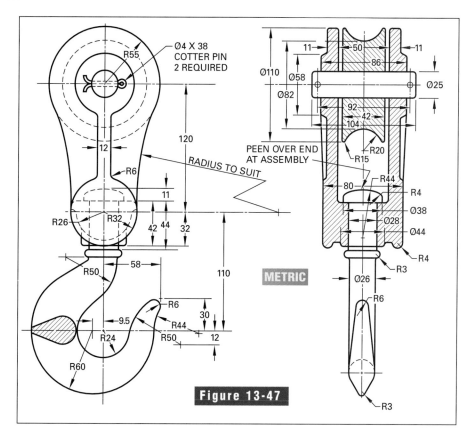

Figure 13-47

27. Make detail drawings of the base, pulley, bushing, and shaft shown in Figure 13-48. Include a bill of materials for the complete pulley-and-stand unit. Scale: Full size. Use three sheets. Top view may be drawn as a half-plan view.

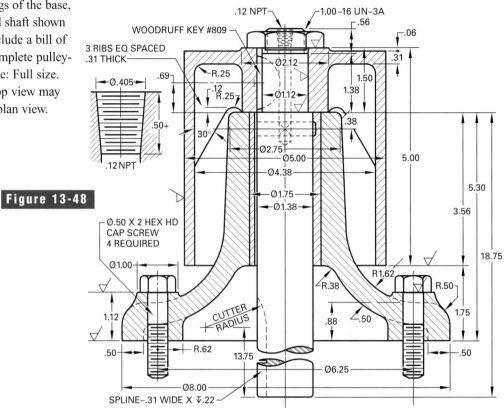

Figure 13-48

28. Make a two-view assembly drawing in section of the universal joint shown in Figure 13-49.

Figure 13-49

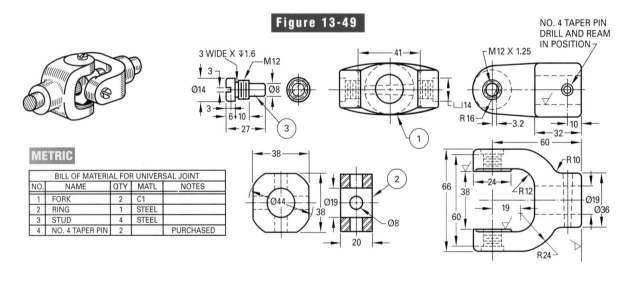

METRIC

BILL OF MATERIAL FOR UNIVERSAL JOINT				
NO.	NAME	QTY	MATL	NOTES
1	FORK	2	C1	
2	RING	1	STEEL	
3	STUD	4	STEEL	
4	NO. 4 TAPER PIN	2		PURCHASED

Problems 29 through 31 refer to Figure 13-50. A jig is a device used to hold a machine part (called the work or production) while it is being machined so that all the parts will be alike within specified limits of accuracy. Notice the work in the upper left corner of the drawing in Figure 13-50. Notice also that the drawing shown does not follow present ANSI standards. Be sure to make necessary changes to update each drawing.

29. Make a detail working drawing of the jig body shown in Figure 13-50.

30. Make a complete set of detail drawings for the jig, including a bill of materials. Use as many sheets as needed.

31. Make a complete assembly drawing of the jig using three views. Give only such dimensions as are needed for putting the parts together and using the jig.

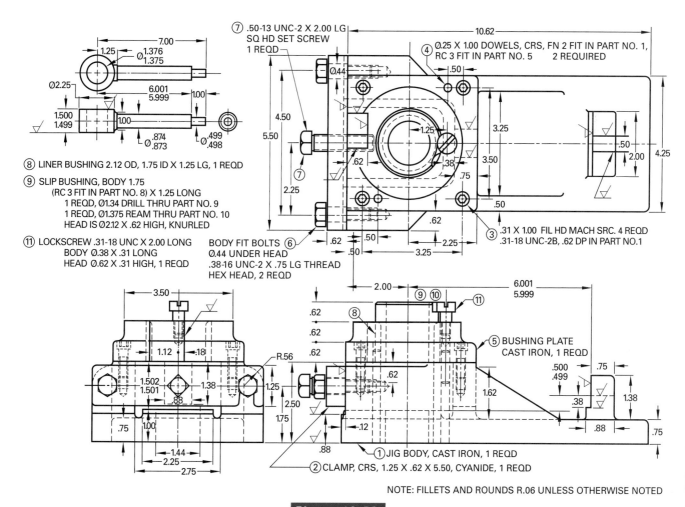

Figure 13-50

For problems 32 through 37, follow the instructions for each drawing problem. Determine an appropriate scale if the scale is not given, and dimension the drawings.

32. Prepare a detail drawing of the journal bearing shown in Figure 13-1. Scale: Full size.

33. Prepare detail drawings of the frame, shaft, and pulley shown for the belt tightener in Figure 13-9. Scale: Full size.

34. Make detail drawings of all of the parts shown in Figure 13-13. Scale: Full size. Convert all dimensions to metric sizes. Use simplified thread representation. Be sure to use current ANSI dimensioning standards.

35. Make a working drawing of the cross slide shown in Figure 13-51. Scale: 1:1.

36. Make a working drawing of the guide bracket shown in Figure 13-52. Scale: 1:1.

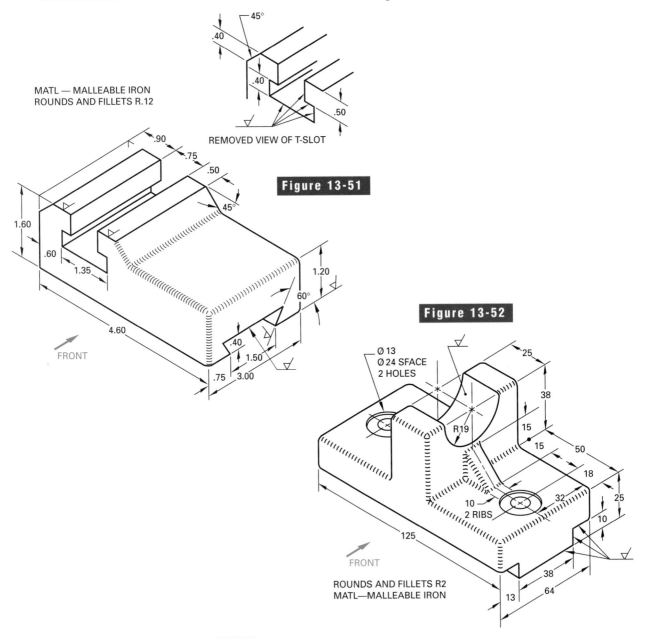

Figure 13-51

Figure 13-52

37. On a B-size sheet, make a set of working drawings of the three parts of the housing shown in Figure 13-53. Scale: 1:1. Use sectional views as appropriate for clarity. Use the appendix to calculate limit dimensions. Include a bill of materials. Add geometric dimensioning and tolerancing symbols based on the following:

- Datum A to be perpendicular to datum B and C within .002 at MMC.
- Datum B to be perpendicular to datum A to within .003 at MMC and parallel to datums B and C within .002 at MMC.
- Datum C to be parallel to datum B to within .001.

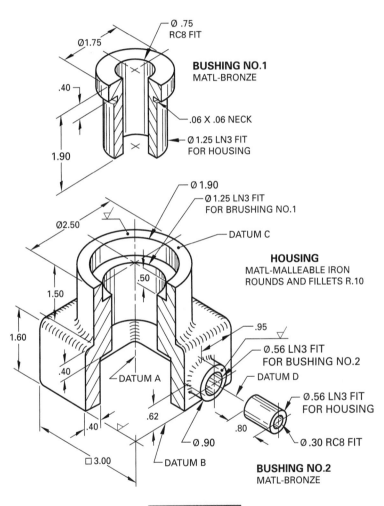

Figure 13-53

38. An assembly drawing of a universal joint is shown in Figure 13-54, along with detail drawings of the yoke and swivel. Standard stock parts such as screws and pins are shown only on the assembly drawing. This product is now ready to be manufactured and sold. However, a pictorial drawing is needed to show the customer the relationship of parts for assembly. Complete the following on an A- or B-size drawing sheet:

• Make an exploded isometric assembly drawing showing all parts in their correct relative positions for assembly. Add part numbers.
• Create a parts list that includes quantity, item name, material, description (material size, etc.), and part numbers.
• Render using the shading technique of your choice.

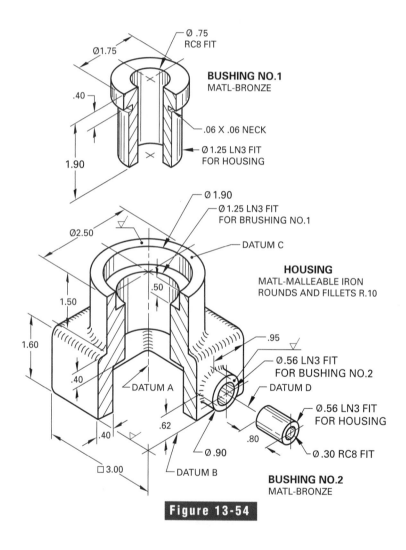

Ø .75
RC8 FIT

Ø1.75

.40

BUSHING NO.1
MATL-BRONZE

.06 X .06 NECK

1.90

Ø 1.25 LN3 FIT
FOR HOUSING

Ø 1.90

Ø 1.25 LN3 FIT
FOR BRUSHING NO.1

Ø2.50

DATUM C

HOUSING
MATL-MALLEABLE IRON
ROUNDS AND FILLETS R.10

1.50

.50

1.60

.95

Ø.56 LN3 FIT
FOR BUSHING NO.2

.40

DATUM A

DATUM D

Ø.56 LN3 FIT
FOR HOUSING

.40

.62

.80

Ø.30 RC8 FIT

□ 3.00

Ø.90

DATUM B

BUSHING NO.2
MATL-BRONZE

Figure 13-54

Design Problems

Design problems have been prepared to challenge individual students or teams of students. In these problems, you are to apply skills learned mainly in this chapter but also in other chapters throughout the text. They are designed to be completed using board drafting, CAD, or a combination of the two. Be creative and have fun!

1. **TEAMWORK** Work as a team to design a line of modular closet-organizing units. Begin by preparing a list of items that are generally stored in a closet. From the list, determine the sizes and types of individual units needed. Develop design sketches for team review. Prepare a complete set of working drawings and a bill of materials for each unit. Finally, prepare an assembly drawing to show how the various units fit together. Materials optional.

2. **METRIC** Design a collapsible sawhorse. Height is 560 mm. Length is 800 mm. Make a set of working drawings, a bill of materials, and an assembly drawing. Material optional.

3. **TEAMWORK** Design a series of tool holders for use in a home workshop. Begin by developing a list of tools you would expect to find in a home workshop. Divide the list among the team members. Each team member is responsible for preparing working drawings for each holder on his or her list. Prepare design sketches for team review. Material optional.

4. **TEAMWORK** Design a portable, collapsible stadium seat. Begin with design sketches. Consider factors such as weight and cost of materials. How can you minimize costs while providing all of the required features? Prepare assembly working drawings with a materials list. Other design considerations:
   - Add storage for a rain poncho.
   - Add a removable cup holder.
   - Create a bill of materials for the stadium seat.
   - Estimate the cost to produce the seat, and prepare a budget. (*Hint:* Remember to include any sales tax in the cost of materials so that your budget will not be underfunded.)
   - If time allows, form a student business to market the stadium seat.

Pattern Development

development

elbow

measuring line

parallel-line
  development

pattern

pattern development

radial-line
  development

stretchout

transition piece

triangulation

Upon completion of this chapter, you should be able to:

■ Explain how pattern development is used in the packaging industry.

■ Describe the general principles of pattern development.

■ Discuss the three main types of pattern development.

■ Explain the purpose of transition pieces and intersections.

■ Prepare patterns using parallel-line development, radial-line development, and triangulation.

■ Prepare patterns for intersecting prisms and cylinders.

A **pattern development** is usually a full-size layout of an object made on a single flat plane. However, very large objects may need to be drawn to a reduced scale. The book cover shown in Figure 14-1 is an example of a pattern development. In part A of the illustration, the cover is laid out flat. In part B, it is wrapped around a book to make a protective covering. Notice that it fits neatly around all surfaces. It does so because each part has been carefully measured and laid out in relation to other parts. Pattern developments are also called **stretchouts**, or simply **developments**.

Making pattern developments is an important part of industrial drafting. Many different industries use pattern developments. Familiar items such as pipes, ducts for hot- or cold-air systems, parts of buildings, aircraft, automobiles, storage tanks, cabinets, boxes and cartons, frozen food packages, and countless other items are designed using pattern development.

To make any such item, a pattern development is first drawn as a **pattern**. This pattern is then cut from flat sheets of material that can be folded, rolled, or otherwise formed into the required shape. Materials used include paper; various cardboards, plastics, and films; metals such as steel, tin, copper, brass, and aluminum; wood; fiberboard; fabrics; and so on.

Whether you are using board drafting techniques or CAD, the drafting principles are the same. For example, the finished pattern or development must usually be done full size. A development that is not full size is not a pattern; it is simply a drawing or representation of the pattern. Understanding the principles and terminology of pattern development will make the drawing task easier.

The Packaging Industry

Packaging is a very large industry that uses pattern developments. Creating packages takes both engineering and artistic skill, because each package design must meet many requirements. For example, packages and containers for industrial goods are designed to be mass-produced at a reasonable cost. However, they must also perform their assigned jobs adequately. Packages for fragile objects, for instance, must be designed so that they protect their contents during shipment. Packages must also look attractive for sales appeal.

Another consideration is the durability of the package. Some packages are meant to be used just briefly and then thrown away. Others are made to last a long time.

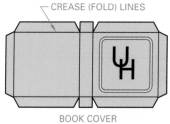

CREASE (FOLD) LINES

ЩН

BOOK COVER

A

Figure 14-1

A book cover is an example of a pattern development.

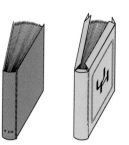

THE COVER FITS TIGHTLY AROUND THE BOOK

B

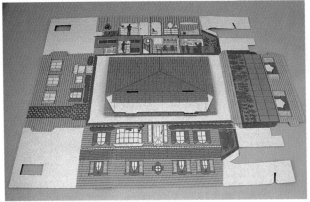

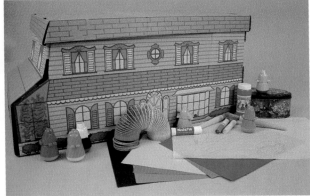

Figure 14-2

This container was made by cutting and folding a flat sheet.

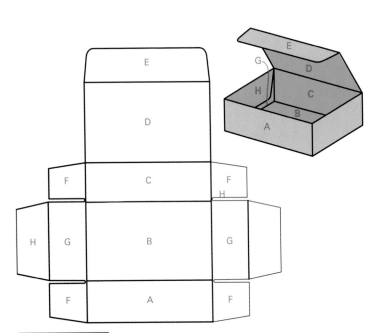

To meet these requirements, designers use many different materials in various thicknesses. Some are made of thin or medium-thick paper stock, as shown in Figure 14-2. This material can also be folded easily into the desired shape or form. Some, like the box in Figure 14-3, are designed so that no glue is needed. Others may need glue on their tabs.

Packages made of cardboard, corrugated board, and many other materials require an allowance for thickness. Examples are boxes made up of a separate container and cover, as shown in Figure 14-4, and a slide-in box like the one shown in Figure 14-5.

Figure 14-3

Pattern for a one-piece package with fold-down tabs.

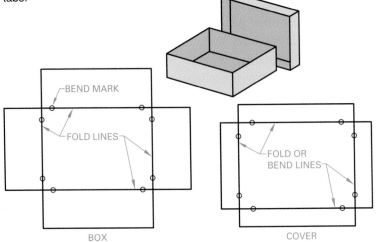

Figure 14-4

Pattern for a box and cover. Notice that the fold lines on the cover are positioned so that the cover will fit over the box after assembly.

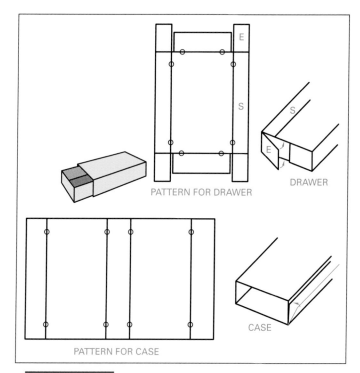

PATTERN FOR DRAWER

DRAWER

PATTERN FOR CASE

CASE

Figure 14-5

A two-part package with a slide-in box. The fold lines on the drawer are positioned so that the box will slide in correctly after assembly.

Sheet-Metal Pattern Drafting

Many different metal objects are made from sheets of metal that are laid out, cut, and formed into the required shape and then fastened together. The objects can be shaped by bending, folding, or rolling. See Figure 14-6. They can be fastened by riveting, seaming, soldering, or welding.

For each sheet-metal object, two drawings are usually made. One is a pictorial drawing of the finished product. The other is a development, or pattern, that shows the shape of the flat sheet that, when rolled or folded or fastened, will form the finished object. See Figure 14-7.

Figure 14-6

The bend in this metal spatula makes it easier to use.

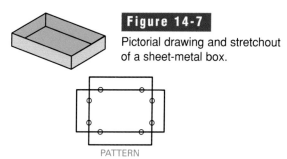

Figure 14-7

Pictorial drawing and stretchout of a sheet-metal box.

PATTERN

A great many thin metal objects without seams are formed by **die stamping**, or pressing a flat sheet into shape under heavy pressure. See Figure 14-8A. Examples range from household utensils to steel wheelbarrows and parts of automobiles and aircraft. Other kinds of thin metal objects are made by spinning. Such objects include some brass-, copper-, pewter-, and aluminumware. See Figure 14-8B. Notice that stamping and spinning stretch the metal out of its original shape and into its new shape.

Principles of Pattern Development

Sheet-metal patterns, like all other patterns, are developed using principles of surface geometry. There are two general classes of surfaces: plane (flat) and curved. The six faces of a cube are plane surfaces. The top and bottom of a cylinder are also plane surfaces. However, the side surface of the cylinder is curved. See Figure 14-9.

Figure 14-8

Examples of products created by die stamping (A) and spinning (B) sheet metal.

A

B

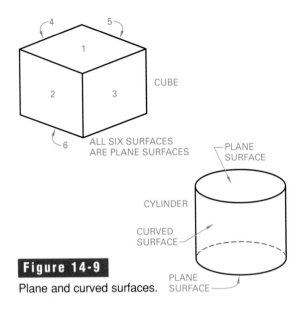

Figure 14-9

Plane and curved surfaces.

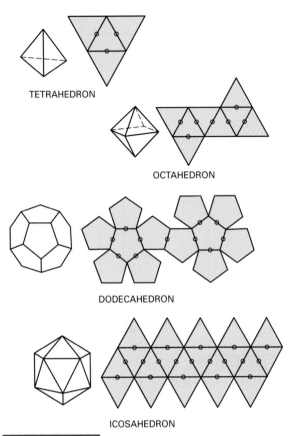

TETRAHEDRON

OCTAHEDRON

DODECAHEDRON

ICOSAHEDRON

Figure 14-11

Patterns for four regular solids.

There are also different kinds of curved surfaces. Those that can be rolled in contact with a plane surface, such as cylinders and cones, are called *single-curved surfaces*. Exact pattern developments can be made for them. Another kind of curved surface is the *double-curved surface*. It is found on spheres and spheroids. Exact pattern developments cannot be made for objects with double-curved surfaces. However, drafters can make approximations.

Figure 14-10 shows how to cut a piece of paper so that it can be folded into a cube. The shape cut out is the pattern of the cube. Figure 14-11 shows the patterns for four other regular solids. To understand pattern development better, lay these patterns out on stiff drawing paper. Then cut them out and fold them to make the solids. Secure the joints with tape. Any solid that has plane surfaces can be made in the same way. Each plane must be drawn in its proper relationship to the others in the development of the pattern.

Finishing a Pattern

Drawing developments is only part of sheet-metal pattern drafting. Drafters must also know about the processes of wiring, hemming, and seaming. In addition, they must know how much material should be added for each. **Wiring** involves reinforcing open ends of articles by enclosing a wire in the edge. See Figure 14-12A. To allow for wiring, a drafter must add a band of material to the pattern equal to 2.5 times the diameter of the wire.

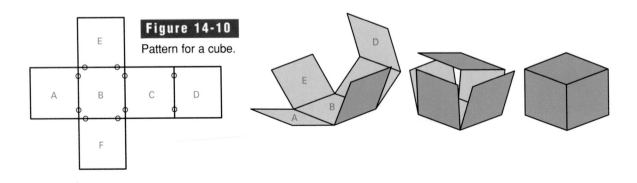

Figure 14-10

Pattern for a cube.

Figure 14-12
Methods of wiring, seaming, and hemming.

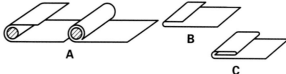

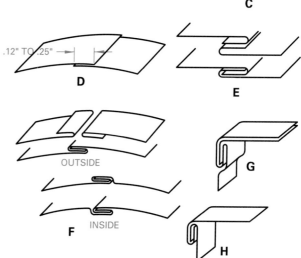

Hemming is another way of stiffening edges. Single- and double-hemmed edges are shown in parts B and C of Figure 14-12. Edges can also be fastened by soldering on lap seams, as shown in part D; flat lock seams, as shown in part E; or grooved seams, as shown in part F. Other types of seams and laps are shown in parts G and H of Figure 14-12. Each has its own general or specific use. How much material is allowed in each case depends on the thickness of the material, method of fastening, and application. In most cases, the corners of the lap are notched to make a neater joint.

.12" TO .25"

D

E

OUTSIDE

G

F   INSIDE

H

TECH•MATH

Calculating Volume

Familiar items such as pipes, storage tanks, cabinets, boxes, and countless other items are designed and patterns are prepared using pattern development. When these items are meant to hold a specific quantity or amount of fluid, solid, or gaseous material, the designer must calculate the volume of the items to make sure they will hold the specified amount of material. For some shapes, calculating the volume is easy. For example, to find the volume of a cube, simply multiply the length times the width times the height. The volume of other shapes requires the use of mathematical formulas.

The volume of a right circular cylinder is determined using the formula:

Volume = (area of base)(height)

For example, the calculations to find the volume of the cylinder shown here are:

Area of base $= \pi r^2$
$= (3.1416)(2^2)$
$= (3.1416)(4)$
$= 12.57$ square inches

Volume $= (12.57)(4)$
$= 50.28$ cubic inches

RIGHT CIRCULAR CYLINDER

The volume of a right circular cone is determined using the formula:

Volume $= \dfrac{\text{(area of base)(height)}}{3}$

Area of base $= \pi r^2$
$= (3.1416)(2^2)$
$= (3.1416)(4)$
$= 12.57$ square inches

Volume $= \dfrac{(12.57)(6)}{3}$
$= 25.14$ cubic inches

Types of Developments

There are three basic types of pattern development: parallel-line development, radial-line development, and triangulation. The type of development needed for an individual object depends on the shape of the object.

Parallel-Line Development

In the patterns for prisms and cylinders, the stretchout line is straight, and the measuring lines, or vertical construction lines, are perpendicular to it and parallel to each other. Making a pattern by drawing the edges of an object as parallel lines is known as **parallel-line development**. The patterns in Figures 14-10 and 14-13 are made in this way.

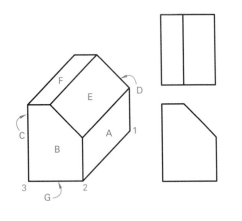

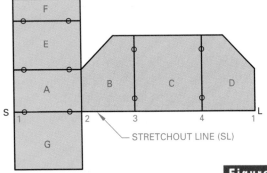

STRETCHOUT LINE (SL)

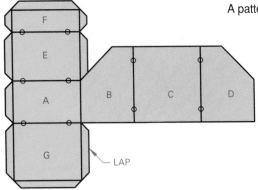

LAP

Radial-Line Development

On cones and pyramids, the edges are not parallel. Therefore, the stretchout line is not a continuous straight line. Also, the measuring lines, instead of being parallel to each other, radiate from a single point. This type of development is called **radial-line development**.

For example, imagine the curved surface of a cone as being made up of an infinite number of triangles, each running the height of the cone. To understand the development of the pattern, imagine rolling out each of these triangles, one after another, on a plane. The resulting pattern would look like a sector of a circle. Its radius would be equal to an element of the cone; that is, a line from the cone's tip to the rim of its base. Its arc would be the length of the rim of the cone's base. The developed pattern of a cone is shown in Figure 14-14.

Triangulation

Some surfaces, such as double-curved surfaces, cannot be developed exactly. The method used to make approximate developments of these surfaces is known as **triangulation**. It involves dividing the surface into triangles, finding the true lengths of the sides, and then constructing the triangles in regular order on a plane. Because the triangles have one short side, on the plane they approximate the curved surface. Figure 14-15 shows examples of products for which the triangulation method must be used. Triangulation may also be used for single-curved surfaces.

Figure 14-13

A pattern for a prism, showing stretchout line and lap.

Figure 14-14

Developed surface of a cone.

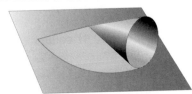

Figure 14-15

Examples of transition pieces.

Transition Pieces

The pieces that are used to connect pipes or openings of different shapes, sizes, or positions are known as **transition pieces**. These pieces have a surface that is a combination of different forms, including planes, curves, or both, and are usually developed by triangulation. A few examples of transition pieces are shown in Figure 14-15.

Intersections

As you may recall from Chapter 10, a line intersects a plane at the piercing point, also called the *point of intersection*. See Figure 14-16. When two plane surfaces meet, the line where one passes through the other is called the *line of intersection*. See Figure 14-17. When a plane surface meets a curved surface, or where two curved surfaces meet, the line of intersection may be either a straight line or a curved line, depending on the surfaces and their relative positions.

Package designers, sheet-metal workers, and machine designers all must be able to find a point at which a line pierces a surface and the line where two surfaces intersect in order to find the true length of the sides. Figure 14-18 shows some ways in which different surfaces intersect.

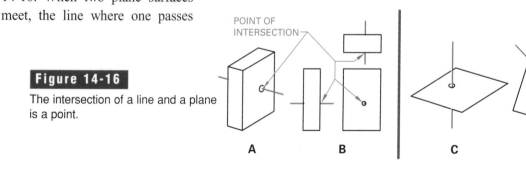

POINT OF INTERSECTION

A   B   C   D

Figure 14-16

The intersection of a line and a plane is a point.

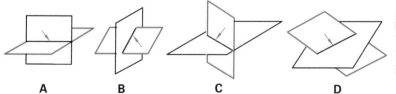

A   B   C   D

Figure 14-17

The intersection of two planes is a line. The arrow points to the line of intersection.

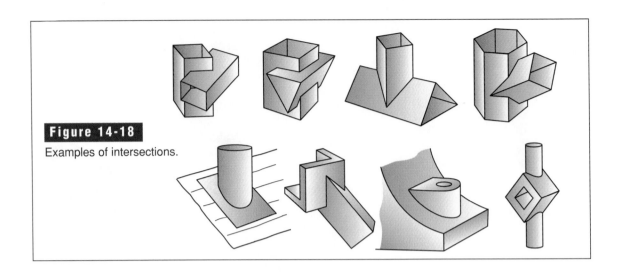

Figure 14-18

Examples of intersections.

Board Drafting Techniques

Pattern development drawings are often done on buff, light green, or light brown paper. They are generally not done on tracing materials such as vellum or film, and they are nearly always done in pencil. Remember that they are done at full size and that accuracy in layout and measurements is of key importance. Only very large objects are drawn to a reduced scale.

Since accuracy is so important, lines are often drawn somewhat thinner than they are for other types of drawings. However, lines still need to be sharp and black. In other words, line technique can be somewhat different, but not of a lesser quality.

Parallel-Line Development

The easiest type of development to create is the parallel-line development. This section describes how to develop patterns for prisms, cylinders, and elbows using parallel-line development.

Prisms

Figure 14-19 is a pictorial view of a rectangular prism. In Figure 14-20, a pattern for this prism is made by parallel-line development. To draw this pattern, proceed as follows:

1. Draw the front and top views full size. Label the points as shown in Figure 14-20A.
2. Draw the stretchout line (SL). This line shows the full length of the pattern when it is completely unfolded. Find the lengths of sides 1-2, 2-3, 3-4, and 4-1 in the top view. Measure off these lengths on SL, as shown in Figure 14-20B.
3. At points 1, 2, 3, 4, and 1 on the stretchout line, draw vertical crease (fold or bend) lines. Make them equal in length to the height of the prism, as shown in Figure 14-20C.

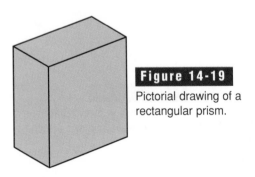

Figure 14-19

Pictorial drawing of a rectangular prism.

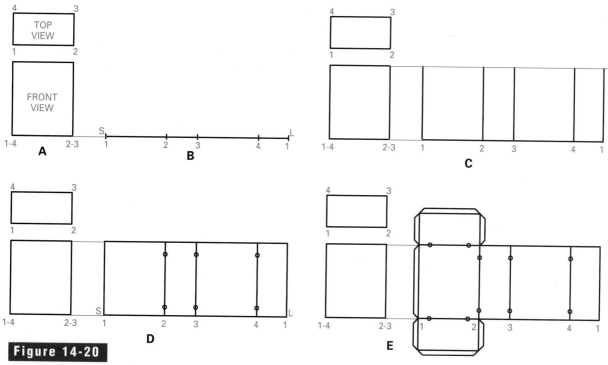

Figure 14-20

Parallel-line development of a rectangular prism.

4. Project the top line of the pattern from the top of the front view. Make it parallel to SL. Darken all outlines until they are sharp and black, as shown in Figure 14-20D. You can use a small circle or X to identify a fold line.

5. Add the top and bottom to the pattern by transferring distances 1-4 and 2-3 from the top view, as shown in Figure 14-20E.

6. Add laps or tabs as necessary for the assembly of the prism. The size of the laps will vary depending on how they are to be fastened and the kind of material used.

A slight variation is the pattern for a truncated prism shown in Figure 14-21. To draw it, first make front, top, and auxiliary views at full size. Label points as shown. The next two steps are the same as steps 2 and 3. Then project horizontal lines from points A-B and C-D on the front view to locate points on the pattern. Connect the points to complete the top line of the pattern. Add the top and bottom as shown. Tabs may be added.

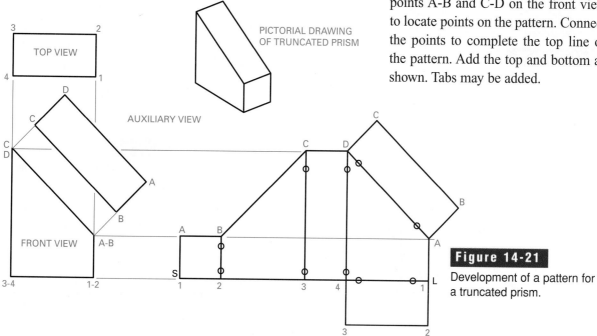

Figure 14-21

Development of a pattern for a truncated prism.

Cylinders

Figure 14-22A shows a pattern development for a cylinder. It is made by rolling the cylinder out on a plane surface.

In the pattern for cylinders, the stretchout line is straight and equal in length to the circumference of the cylinder, as shown in Figure 14-22B. If the base of the cylinder is perpendicular to the axis, its rim will roll out to form the straight line.

To develop a cylinder, imagine that the cylinder is actually a many-sided prism. Each side forms an edge called an **element**. Because there are so many elements, however, they seem to form a smooth curve on the surface of the cylinder. Imagining the cylinder in this way will help you find the length of the stretchout line. This length will equal the total of the distances between all the elements. Technically, of course, the elements are infinite in number. But, for your purposes, you need only mark off elements at convenient equal spaces around the circumference of the cylinder. (Figure 14-23 shows various methods of dividing a circle.) Then add up these spaces to make the stretchout line. This must equal the circumference of the cylinder.

Figure 14-24 is a pictorial view of a truncated right cylinder, showing the imaginary elements. Figure 14-25 shows how to develop a pattern for this cylinder. To draw this pattern, proceed as follows:

1. Draw the front and top views at full size. Divide the top view into a convenient number of equal parts (12 in this case) to locate a set of equally spaced points in the top view.
2. Transfer the points in the top view to the front view to locate points at intervals on the inclined surface.
3. Begin the stretchout line. You will determine its actual length later, when you mark off the elements. Again, the stretchout line must equal the circumference of the cylinder.
4. Using dividers, find the distance between any two consecutive elements in the top view. Then mark off this distance along the stretchout line as many times as there are parts in the top view. Label the points thus found, as shown in Figure 14-25. Then draw a vertical construction line upward from each point. *Note:* In this and subsequent steps, the colored arrows on the figure show the direction in which the various lines are projected.
5. From these intersection points, project horizontal construction lines toward the development.

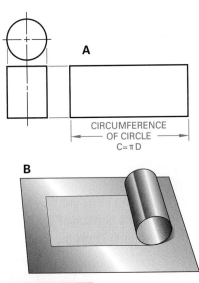

Figure 14-22

Developed surface of a right circular cylinder.

CIRCUMFERENCE OF CIRCLE
C = π D

Figure 14-23

Dividing a circle.

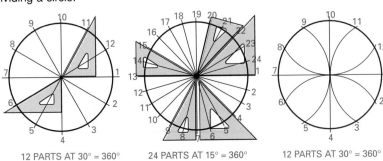

12 PARTS AT 30° = 360°   24 PARTS AT 15° = 360°   12 PARTS AT 30° = 360°

Figure 14-24

Pictorial drawing of a truncated right cylinder.

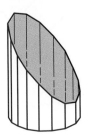

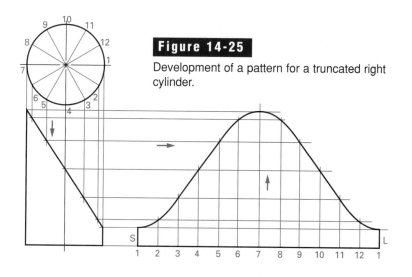

Figure 14-25

Development of a pattern for a truncated right cylinder.

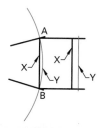

Figure 14-26

The chords used to approximate a cylinder are slightly shorter than the radial distance they represent, because a straight line is the shortest distance between two points.

6. Locate the points where the horizontal construction lines intersect the **measuring lines** (vertical lines) from the stretchout line. Connect these points in a smooth curve.

7. Darken outlines and add laps as necessary.

Since the surface of a cylinder is a smooth curve, your pattern will not be wholly accurate. This is because it was made by measuring distances on a straight line (chord) rather than on a curve. Figure 14-26 represents part of the top view of the cylinder discussed above. It shows that the distance from point to point is slightly less along the chord than along the arc. The difference can be found by figuring the actual length of the arc using the following formula, in which *d* stands for *diameter*:

Circumference = πd

As long as you include enough elements to represent the cylinder adequately, however, the difference is so slight that the inaccuracy is not critical.

The number of elements you choose to use in cylinder development will vary depending on the size of the cylinder, the application, and other factors. However, be sure to avoid using too few or too many elements. If you use too few, the pattern approximation will not be accurate enough; if you use too many, the development becomes much more difficult, and it takes a much longer time to create. See Figure 14-27.

A slightly different method for developing a cylinder is shown in Figure 14-28. In this case, a front and half-bottom view are used. Attaching the two views saves time and increases accuracy. Notice that both methods produce the same result.

Figure 14-27

Choosing a reasonable number of elements is important.

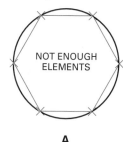

A

ARC LENGTH: 1.0472
CHORD LENGTH: 1.000
DIFFERENCE: .0472

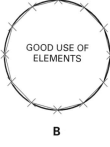

B

ARC LENGTH: .5236
CHORD LENGTH: .5176
DIFFERENCE: .0060

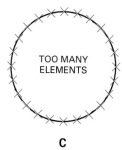

C

ARC LENGTH: .2618
CHORD LENGTH: .2611
DIFFERENCE: .0007

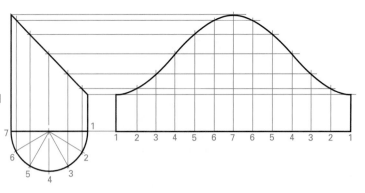

Figure 14-28

Development of a pattern for a truncated right cylinder, using a combined front and half-bottom view.

Elbows

An **elbow** is a joint in a pipe or duct—a place at which two pieces meet at an angle other than 180°. The simplest type of elbow is a square, or two-piece, elbow. Other, more complicated pieces provide a smoother curve.

Square Elbows

A square elbow consists of two cylinders cut off at 45°. Therefore, only one pattern is needed, as shown in Figure 14-29. Allow a lap for the type of seam to be made, if required.

If a lap is not needed on the curved edges, both parts can be developed on one stretchout, as shown in Figure 14-30. Notice that the seam on part A is on the short side, while on part B it is on the long side. In Figure 14-29, the seam on both pieces is on the short side. In most cases, this is not critical.

Four-Piece Elbows

The pattern for a four-piece elbow is shown in Figure 14-31. To draw it, proceed as follows:
1. Draw arcs with the desired inner and outer radii to produce an elbow to fit the desired pipe size, as shown in Figure 14-31A. Divide the outer circle into six equal parts.

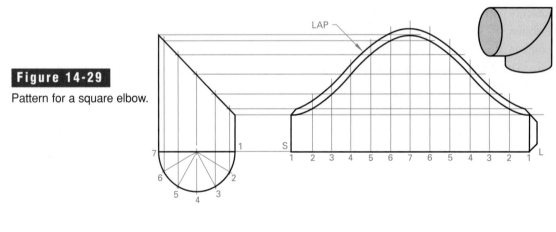

Figure 14-29

Pattern for a square elbow.

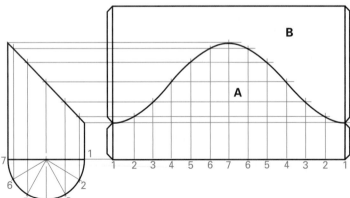

Figure 14-30

Both parts of the pattern may be made on one stretchout.

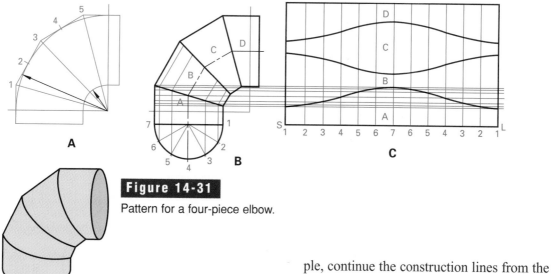

Figure 14-31

Pattern for a four-piece elbow.

2. Draw a half-bottom view and divide it, as shown in Figure 14-31B. Project the elements from this view to the front view.

3. Develop the pattern for part A just as you developed the pattern for the square elbow. If you do not have to allow for laps on the curved edges, you can draw the patterns for all four parts as shown in Figure 14-31C. To find what lengths to mark off on the vertical lines, work from the front view. In the front view of part B, for exam-

ple, continue the construction lines from the half-bottom view, but make them parallel to the surface lines. Then measure all the lines within part B. The curves for all the parts are the same. Therefore, you need plot only one of them. It can then be used as a template for the others.

Radial-Line Development

Items that have conical or pyramidal shapes cannot be developed using parallel-line development because the stretchout line is not a straight line. If a conical or pyramidal object has an axis at 90° to its base, it can be developed using radial-line development. The procedures in this section show how to develop a pattern for right (90°) cones and pyramids.

Right Circular Cone

A right circular cone is one in which the base is a true circle and the tip is directly over the center of the base. See Figure 14-32A. The pattern for a cone is shown in Figure 14-32B. To draw the pattern, proceed as follows:

1. Draw front and half-bottom views to the desired size.

2. Divide the half-bottom view into several equal parts. Label the division points as shown.

3. On the front view, measure the slant height of the cone; that is, the true distance from the apex to the rim of the base (line A1). Using this length as a radius, draw an arc of indefinite length as a measuring arc. Draw a line from apex A to the arc at any point a short distance from the front view.

Figure 14-32

Development of a pattern for a cone.

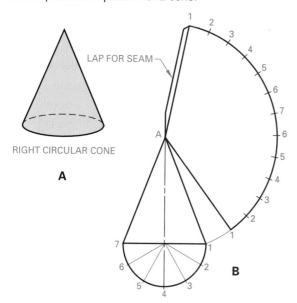

4. With dividers, find the straight-line distance between any two division points on the half-bottom view. Then use this length to mark off spaces 1-2, 2-3, 3-4, and so forth, along the arc. Label the points to be sure none have been missed. Complete the development by drawing line A1 at the far end.

5. Add laps for the seam as required. How much to allow for the seam depends on the size of the development and the type of joint to be made.

A frustum of a cone is created by cutting through the cone on a plane parallel to the base, as shown in Figure 14-33A. To draw the development for a frustum of a cone, proceed as shown in Figure 14-33B. Use the same method as shown in Figure 14-32, but draw a second arc AB from point B on the front view.

Truncated Circular Cone

A circular cone that has been cut along a plane that is not parallel to the base is known as a *truncated circular cone*. See Figure 14-34A. The pattern for such a cone is shown in Figure 14-34B. To draw it, proceed as follows:

1. Draw the front, top, and bottom (or half-bottom) views.

2. Proceed as in Figure 14-32 to develop the overall layout for the pattern.

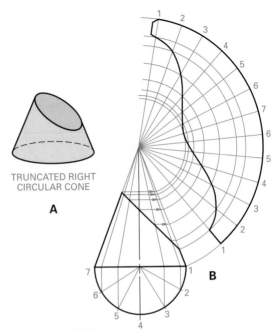

TRUNCATED RIGHT CIRCULAR CONE

A

B

Figure 14-34

Development of a pattern for a truncated right circular cone.

3. Project points 1 through 6 from the bottom view to the front view and then to the apex. Label the points where they intersect the miter line (cut line) to avoid mistakes. These lines, representing elements of the cone, do not show in true length in the front view. Their true length shows only when they are projected horizontally to the points on the arc.

4. Project the elements of the cone from the apex to the points on the arc.

5. In the front view, find the points on the miter line that were located in step 3. Project horizontal lines from them to the edge of the front view. Continue these lines as arcs through the development. Mark the points where they intersect the element lines. Join these points in a smooth curve. Complete the pattern by adding a lap.

Pyramids

Before you can begin to develop a pattern for a pyramid, you must find the true length of its edges. For example, in the pyramid shown in Figure 14-35A, you need to find the true length of OA. Figure 14-35B shows the top and front views of the pyramid. In neither view does the edge OA show in true

Figure 14-33

Development of a pattern for a frustum of a cone.

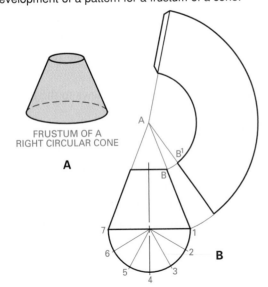

FRUSTUM OF A RIGHT CIRCULAR CONE

A

B

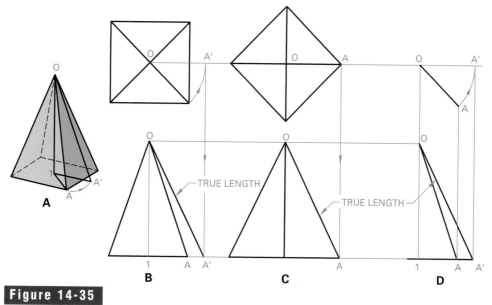

Figure 14-35

Finding the true length of a line.

length. However, if the pyramid were in the position shown in Figure 14-35C, the front view would show OA in true length. In this figure, the pyramid has been revolved about a vertical axis until OA is parallel to the vertical plane. In Figure 14-35D, line OA is shown before and after revolving (OA′).

The construction in Figure 14-35D is a simple way to find the true length of the edge line OA. Revolve this view to make the horizontal line OA′. Project A′ down to meet a base line projected from the original front view. Draw a line from this intersection point to a new front view of O. This line will show the true length of OA.

Right Rectangular Pyramids

Figure 14-36 shows the pattern for a right rectangular pyramid. To draw it, proceed as follows:
1. Find the true length of one of the edges (O1 in this case) by revolving it until it is parallel to the vertical plane (O1′).
2. With the true length as a radius, draw an arc of indefinite length to use as a measuring arc.
3. On the top view, measure the lengths of the four base lines (1-2, 2-3, 3-4, 4-1). Mark these lengths off as the straight-line distances along the arc.
4. Connect the points and draw crease lines. Mark the crease lines.
5. Add base 1-2-3-4 as shown.

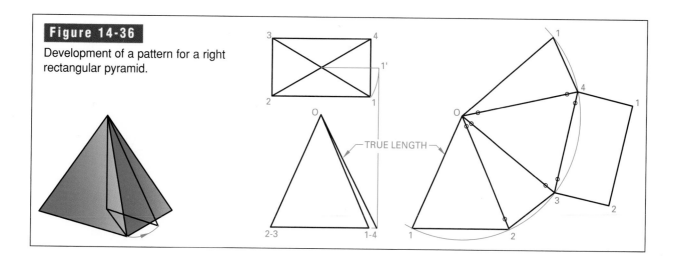

Figure 14-36

Development of a pattern for a right rectangular pyramid.

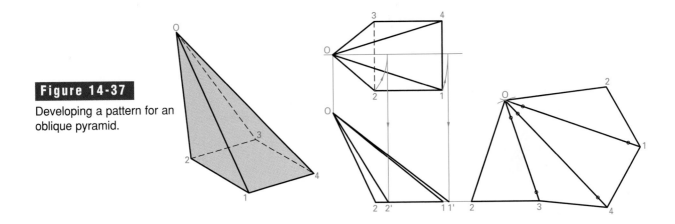

Figure 14-37

Developing a pattern for an oblique pyramid.

Oblique Pyramids

Figure 14-37 shows a development of an oblique pyramid. To draw it, proceed as follows:

1. Find the true lengths of the lateral edges. Do this by revolving them parallel to the vertical plane, as is shown for edges O2 and O1. These edges are both revolved in the top view, then projected

to locate 2' and 1'. Lines O2' and O1' in the front view are the true lengths of edges O2 and O1. Edge O2 = edge O3. Edge O1 = edge O4.

2. Start the development by laying off 2-3. Since edge O2 = edge O3, you can locate point O by plotting arcs centered on 2 and 3 and with radii the true length of O2 (O2'). Point O is where the arcs intersect.

3. Construct triangles O-3-4, O-4-1, and O-1-2 with the true lengths of the sides to complete the development of the pyramid as shown.

Truncated Oblique Pyramids

Figure 14-38 shows a pattern for a truncated oblique pyramid. The pyramid has an inclined surface ABCD. If it were not truncated, it would extend to the apex point O. To draw the pattern, proceed as follows:

1. Find the true lengths of OA, OB, OC, and OD. For this pyramid, OA = OD and OB = OC. To locate these lengths, locate B' and A' in the front view. Lines OB' and OA' will give the true lengths.

Figure 14-38

Developing a pattern for a truncated oblique pyramid.

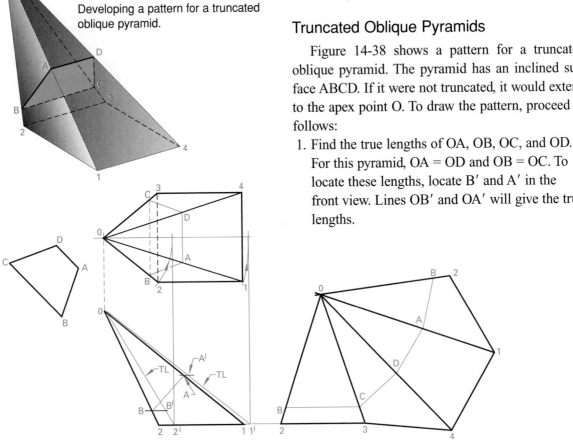

2. Make a development of the pyramid as if it extended to point O and were not truncated.
3. Lay off the true lengths of O-A, O-B, O-C, and O-D on the corresponding edges of this development. Join them to make the pattern for the edge of the frustum.

The inclined surface is shown in the auxiliary view and can be attached to the rest of the pattern as indicated. Also, a base 1-2-3-4 can be attached.

Triangulation

Figure 14-39 shows how triangulation is used in developing an oblique cone. To draw this development, proceed as follows:

1. Draw elements on the top- and front-view surfaces to create a series of triangles, as shown in Figure 14-39A and B. Number the elements 1, 2, and so forth. For a better approximation of the curve, use more triangles than are shown in Figure 14-39.
2. Find the true lengths of the elements by revolving them in the top view until each is horizontal. From the tip of each, project down to the front-view base line to get a new set of points 1, 2, and so forth. Connect these with the front view of point O to make a true-length diagram, as shown in Figure 14-39C.

3. To plot the development shown in Figure 14-39D, construct the triangles in the order in which they occur. Take the distances 1-2, 2-3, etc., from the top view. Take the distances O1, O2, etc., from the true-length diagram. Connect the curve and add tabs if needed.

Developing Transition Pieces

The development of transition pieces incorporates elements of parallel-line development, radial-line development, and triangulation. The exact techniques needed depend on the type of transition to be developed.

Square-to-Square Transition

Figure 14-40 shows a transition piece connecting two square ducts, one of which is at 45° to the other. This piece is made up of eight triangles—four of one size and four of another. To draw the development, proceed as follows:

1. On the top and front views, number the various points as shown. Lines 1-2, 2-3, 3-4, and 4-1 show in their true size in the top view. Lines AB, BC, CD, and DA show in their true size in both views.

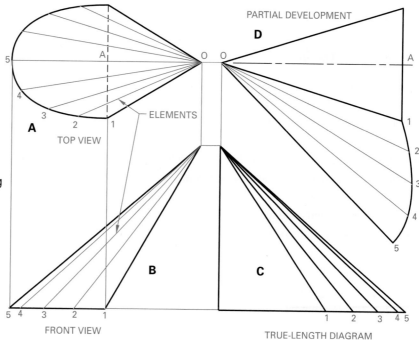

Figure 14-39

Triangulation is used in developing an oblique cone.

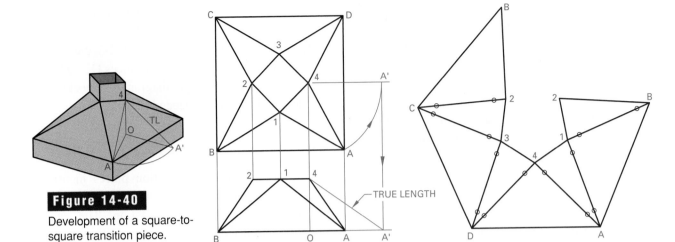

Figure 14-40

Development of a square-to-square transition piece.

2. Find the true length of one of the slanted lines, in this case A4. Do this by revolving it in the top view until it is parallel to the vertical plane. Then project down to the front view, where the true length will show in line 4A′. This will also be the true length of D4, D3, and so forth.

3. Start the development by drawing line DA.

4. Using A and D as centers and a radius equal to the true length of A4, draw intersecting arcs to locate point 4 on the development of the piece.

5. Draw another arc using point 4 as a center and a radius equal to line 4-3. Point 3 is located where this arc intersects the arc drawn in step 4 with D as a center.

6. Proceed to lay off the remaining triangles until the transition piece is completed.

Rectangular-to-Round Transition

Figure 14-41 shows a transition piece made to connect a round pipe with a rectangular cone. This piece contains four large triangles. Between them are four conical parts with apexes at the corners of the rectangular opening and bases, each one quarter of the round opening. To draw the development, proceed as follows:

1. Start with the partial cone whose apex is at A. Divide its base, 1-4, into a number of equal parts, such as 1-2, 2-3, and 3-4. Then draw lines A2 and A3 to give triangles approximating the cone.

2. Find the true length of each of these lines. In practice, this is done by constructing a true-length diagram (diagram I). The construction is

based on the fact that the true length of each line is the hypotenuse of a triangle whose altitude is the altitude of the cone and whose base is the length of the top view of the line.

3. Begin diagram I by drawing the vertical line AE (the altitude of the cone) on the front view. On base EF, lay off distances A1, A2, and so forth, taken from the top view. In the figure, this is done by swinging each distance about point A in the top view, then dropping perpendiculars to base EF.

4. Connect the points thus found with point A in diagram I to find the desired true lengths. Diagram II, constructed in the same way, gives the true lengths of triangle lines B4, B5, and so forth, on the cone whose apex is at B.

5. Start the development with the seam line A1. Draw line A1 equal to the true length of A1 (taken from the true-length diagram I).

6. Draw an arc with point 1 on the development as a center and a radius equal to the distance 1-2 in the top view. Then draw another arc using A as a center and a radius equal to the true length of A2. The two arcs intersect at point 2 on the development.

7. Draw an arc with point 2 as the center and radius 2-3. Then draw another arc with center A and a radius equal to the true length of A3. Where these two arcs intersect will be point 3.

8. Proceed similarly to find point 4. Then draw a smooth curve through points 1, 2, 3, and 4.

9. Attach triangle A4B in its true size. Find point B by drawing one arc from A with radius AB

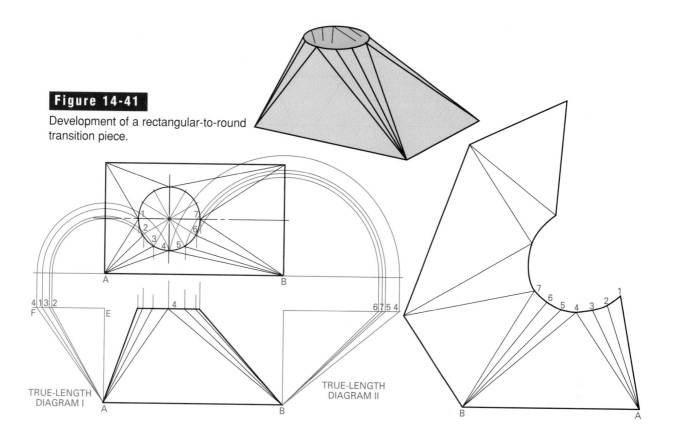

Figure 14-41

Development of a rectangular-to-round transition piece.

TRUE-LENGTH DIAGRAM I

TRUE-LENGTH DIAGRAM II

taken from the top view and another arc from 4 with a radius the true length of B4. Where these arcs intersect will be point B.

10. Continue until the development is completed.

Drawing Intersections

The intersection of two three-dimensional objects requires special attention. The exact location of the line of intersection must be determined before a pattern can be developed.

Intersecting Prisms

Figure 14-42 shows how to draw the intersection of two prisms. To draw the complete front and top views of the intersecting prisms, proceed as follows:

1. Draw the hexagon shape in the top view at 2.00″ across the flats.
2. Project downward from the corners of the hexagon to establish the vertical lines for the front view.
3. Measure the 3.50″ vertical distance to establish the top and bottom of the front view.

4. Locate the exact center of the front view. You can do this easily and quickly by striking diagonals from corner A to corner C and from corner B to corner D.
5. Draw a light construction line through the center point X at 60° to the horizontal (120° to the vertical).
6. Measure 2.00″ along the inclined line in both directions from point X to establish the ends of the square prism.
7. In any convenient location to the right or left, construct an auxiliary view of the square prism. Figure 14-42 shows it to the right of the front view.
8. Project back to the front view to establish the top and bottom edges of the square prism. The lines of intersection on the front view will be added later.
9. Project lines upward from G, F-F′, and G′ on the right side to establish horizontal distances on the diamond shape in the top view. Do likewise on the left side.
10. Draw a horizontal layout line through the center of the top view (G-G′).

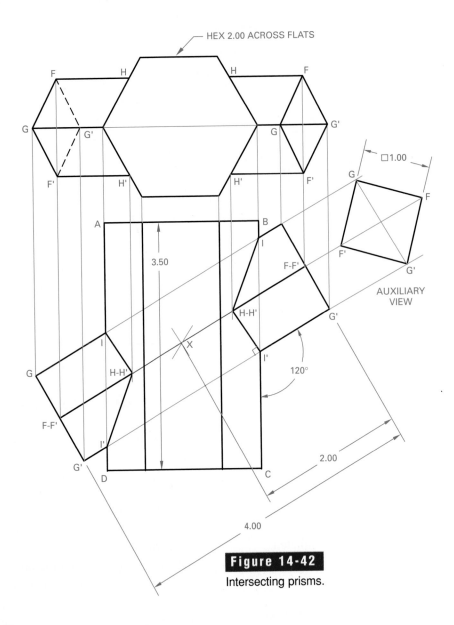

□1.00

AUXILIARY
VIEW

3.50

F-F'

H-H'

120°

2.00

4.00

Figure 14-42

Intersecting prisms.

11. Establish points F and F′ in the top view by projecting both points F-F′ from the front view to the top view. This completes the top view.

12. Add the lines of intersection on the front view by projecting points H-H′ down from the top view to the front view. Connect points I and I′ with points H-H′ to complete the drawing.

A second method for developing complete views of intersecting prisms is shown in Figure 14-43. Proceed as follows:

1. Draw the hexagon shape in the top view and project downward to draw the front view of the hexagon. Measure to locate and draw the top and bottom edges.

2. At any convenient location, construct a miter line (refer to Chapter 6). Use the miter line to locate and draw the vertical lines for the hexagon in the side view. Project horizontal lines from the front view to construct the top and bottom of the hexagon in the side view.

3. Lay out the 25 mm square on the side view and project points upward and across to locate the square prism in the other two views. Measure to locate the ends of the square prism in the top view and project downward to locate its ends in the front view.

4. Lay out the lines of intersection (A-B-C) by projecting points A, B, and C downward from the top view and corresponding points across from the side view. Connect the points as shown.

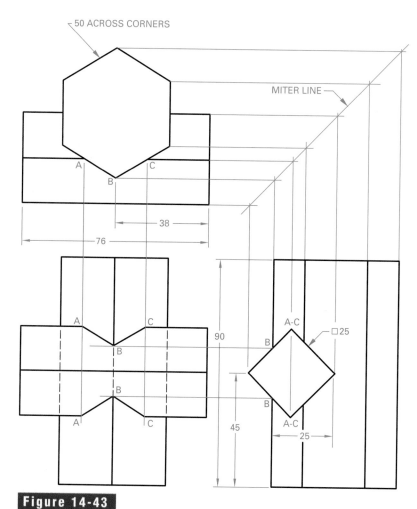

50 ACROSS CORNERS

MITER LINE

A C
B

38

76

A C
B

90

B A-C □25

B B

B A-C

45

25

Figure 14-43

Intersection of a four-sided prism and a six-sided prism.

Intersecting Cylinders

Figure 14-44A shows how to draw the line of intersection of two cylinders. Since there are no edges on the cylinders, you have to assume positions for the cutting planes. Draw plane AA to contain the front line (element) of the vertical cylinder. This plane will also cut a line, or element, on the horizontal cylinder. The intersection of these two lines in the front view identifies a point on the required curve. Similarly, planes BB, CC, and DD cut lines on both cylinders that intersect at points common to both cylinders. Figure 14-44B shows the development of the vertical cylinder.

Figure 14-45 shows how to draw the line of intersection of two cylinders joined at an angle. Here the cutting planes are located by an auxiliary view. To make the development of the inclined cylinder, take the length of the stretchout line from the circumference of the auxiliary view. Choose a cutting plane that divides this circumference into equal parts so that the measuring lines are equally spaced along the stretchout line. Project the lengths of the measuring lines from the front view. Join their ends into a smooth curve.

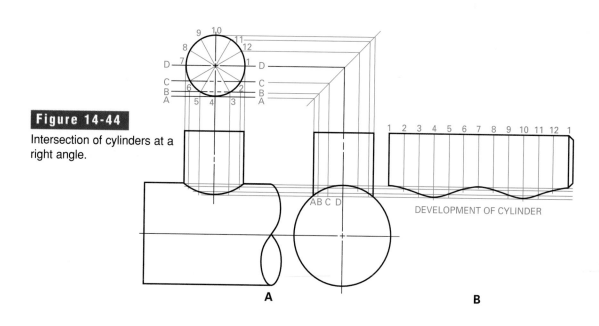

Figure 14-44

Intersection of cylinders at a right angle.

DEVELOPMENT OF CYLINDER

A   B

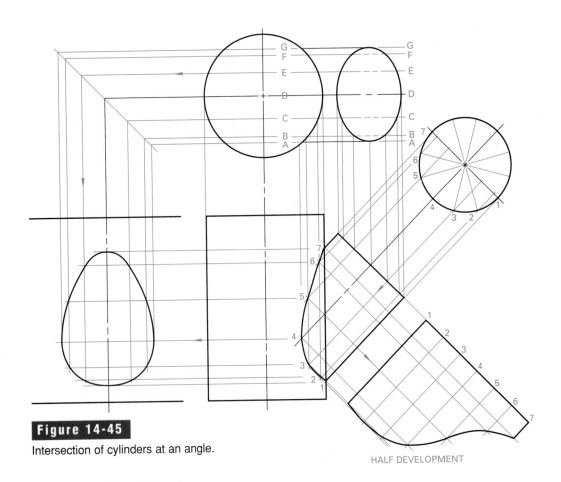

Figure 14-45

Intersection of cylinders at an angle.

HALF DEVELOPMENT

Intersection of Cylinders and Prisms

You can also find the intersection line of a cylinder and a prism by using cutting planes. In Figure 14-46, a triangular prism intersects a cylinder. Planes A, B, C, and D cut lines on both the prism and the cylinder. These lines cross in the front view at points that determine the curve of intersection. To make the development of the triangular prism, take the length of the stretchout line from the top view. Take the lengths of the measuring lines from the front view. Note that one plane of the triangular prism (line 1-5 in the top view) is perpendicular to the axis of the cylinder. Make the curve of the intersection line on that face using the radius of the cylinder. Create the other curve with a French curve.

Figure 14-46

Intersection of a prism and a cylinder.

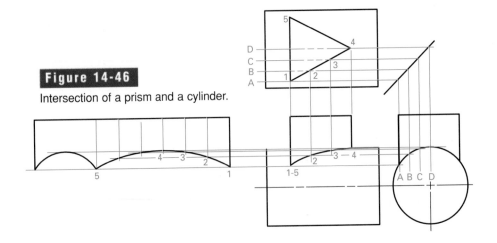

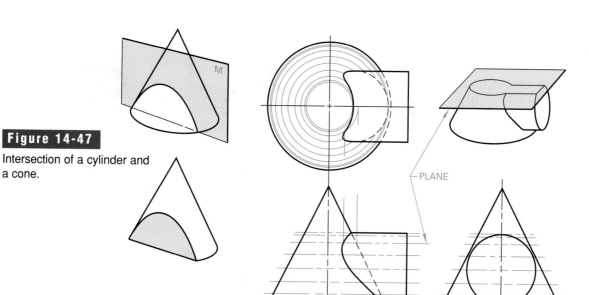

Figure 14-47

Intersection of a cylinder and a cone.

Intersection of Cylinders and Cones

To find the line of intersection of a cylinder and a cone, use horizontal cutting planes, as shown in Figure 14-47. Each plane cuts a circle on the cone and two straight lines on the cylinder. Points of intersection occur where the straight lines of the cylinder cross the circles of the cone in the top view. Project these points onto the front view to get the intersection line. Figure 14-48 shows this construction for a single plane. Use as many planes as are needed to make a smooth curve.

Intersection of Planes and Curved Surfaces

Figure 14-49 shows the intersection of a plane MM and the curved surface of a cone. To find the line of intersection, use horizontal cutting planes A, B, C, and D. Each plane cuts a circle from plane MM. Thus, you can locate points common to MM and the cone, as in the top view. Project these points onto the front view to get the curve of intersection.

Figure 14-48

A cutting plane.

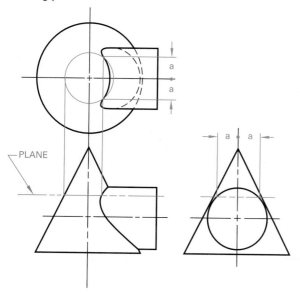

Figure 14-49

Intersection of a plane and a curved surface.

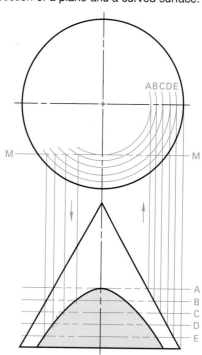

Figure 14-50 shows the intersection at the end of a connecting rod. To find the curve of intersection, use cutting planes perpendicular to the rod's axis. These planes cut circles, as shown in the right-side view. The circles, in turn, cut the "flat" at points that can be projected back as points on the curve.

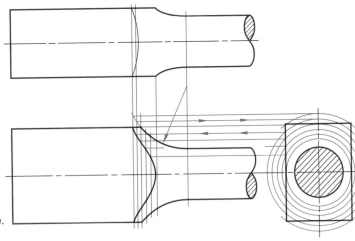

Figure 14-50

Intersection of a plane and a turned surface.

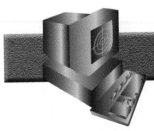

CAD Techniques

In industry, pattern development is generally done using special third-party software. This software may be a standalone product or an add-on to a basic drafting program such as AutoCAD. The advantage of using this software is that it is **parametric**. In other words, it builds the shapes you select according to the parameters you set. For example, in Figure 14-51, the user has specified a truncated circular cone with an upper diameter of 24″, a lower diameter of 36″, and a height of 18″. The software uses those parameters to create a full-size pattern for the cone automatically.

However, you may occasionally need pattern developments that are not available in a parametric program, or the company that hires you may not even use the software. Therefore, you should learn the basic techniques for creating a pattern even if you plan to use parametric software in the future.

The CAD procedures for developing patterns parallel those for board drafting. The basic development is the same—only the tools are different. By now, you should be familiar enough with the basic AutoCAD commands to apply them on your own. A few representative procedures are presented in this section. However, you should practice your CAD skills by returning to the "Board Drafting Techniques" portion of this chapter and working through the remaining procedures on your own.

Figure 14-51

An example of parametric pattern development software. The user specifies the dimensions in the box on the left, and the software automatically creates the pattern development in the drawing window.

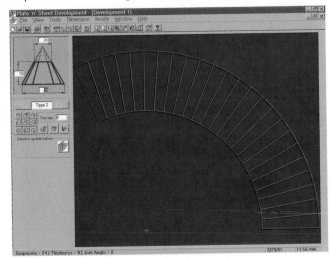

Self-Esteem

A high level of self-esteem is important, both personally and professionally. It is not something that simply happens; it is developed over a period of time and grows with feelings of success. Working toward being precise and thorough in your work, meeting deadlines, being a team player, and other similar positive experiences contribute to your self-esteem. The greater your feeling of self-esteem, the greater are your chances of success on the job.

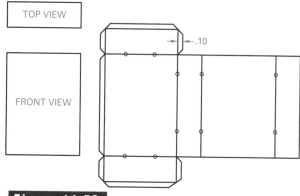

Figure 14-53

CAD development of the rectangular prism from Figure 14-52.

Parallel-Line CAD Development

Figure 14-52 is a pictorial view of a rectangular prism. In Figure 14-53, a pattern for this prism is made by parallel-line development. To draw this pattern, proceed as follows:

1. Use the PLINE command to draw the front and top views at full size.
2. Select the top view and then enter the LIST command. A text window appears listing information about the rectangle. The perimeter equals the length of the stretchout line you need. Draw the stretchout line.
3. At the beginning of the stretchout line, create a 2″ vertical line to represent the beginning of the pattern. Offset this line to the right by 1.5″, .5″, 1.5″, and .5″ (the dimensions of the top view) to create the crease lines and the right end of the development. The last vertical line should lie exactly at the end of the stretchout line. Add the top horizontal across the entire development.
4. Add small circles as shown in Figure 14-53 to identify the crease lines.

5. Add the top and bottom to the pattern by copying the top view and placing it as shown on the development. Use the Multiple option of the COPY command to create both the top and the bottom in a single COPY operation.
6. Add laps or tabs as necessary for the assembly of the prism. The size of the laps will vary depending on how they are to be fastened and the kind of material used. In this case, use the OFFSET command to create .1″ tabs and chamfer the corners of the tabs at 45°.

►CADTIP

Zooming Dynamically

To create the tab chamfers accurately, you will need to zoom in for a closer look. After you have zoomed in to complete one tab corner, enter the ZOOM command and enter D for the Dynamic option. The entire drawing reappears, and a dashed green box shows your current location and level of magnification. A white box of the same size appears at the cursor, with an X at its center. Move the cursor so that the white box is located at the next tab to be chamfered, and right-click to finish the zoom.

If you want to change the zoom magnification, you can do so while the Dynamic option is active by left-clicking. The X at the center of the cursor box changes to an arrow, and moving the cursor changes the size of the white box. The size and location of the box correspond to the size and location of the viewing area when you complete the ZOOM command.

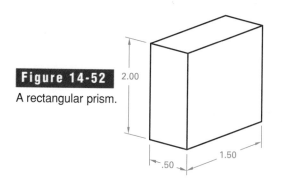

Figure 14-52

A rectangular prism.

Radial-Line CAD Development

AutoCAD provides tools to make radial-line development a fast, accurate process. Figure 14-54 shows a pictorial of a truncated right circular cone. The top radius is .75″, the bottom radius is 1.25″, and the height is 1.12″. Develop the cone as shown in Figure 14-55.

1. Draw front and half-bottom views. Extend a line through the center of the front view to about 1.5″ above the top of the front view. Use the EXTEND command to extend the sides of the front view to find the virtual apex of the truncated cone. See Figure 14-55A.

2. Enter PDMODE and 3 to change the point display to an X that is easily visible. Enter the REGEN command to see the points. Then use the DIVIDE command to divide the half-bottom view into 6 equal parts. See Figure 14-55B.

3. Enter the DIST command and select points A and B to find the true distance from the apex to the rim of the base. Using this length as a radius, create a circle with its center at the apex. Draw line AC from apex A to the circle at point C (any point a short distance from the front view), as shown in Figure 14-55B. Trim away the part of the circle between the front view and line AC to form arc BC.

Figure 14-54

A truncated cone with a top radius of .75″; a bottom radius of 1.25″; and a height of 1.12″.

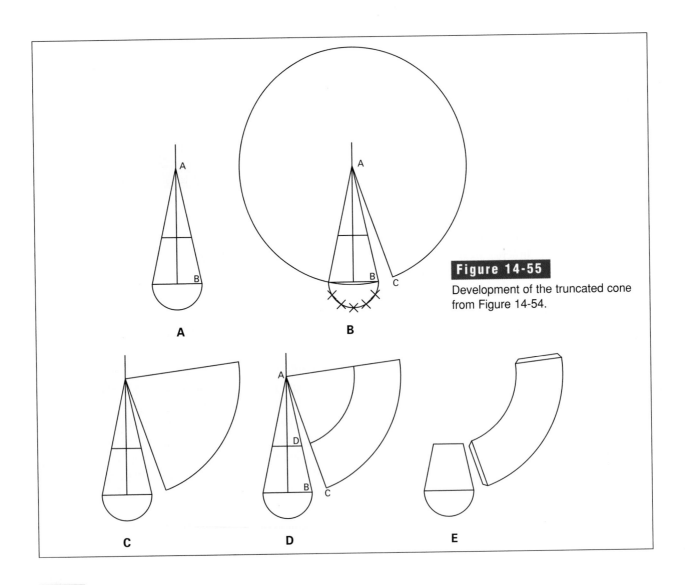

Figure 14-55

Development of the truncated cone from Figure 14-54.

4. Use the DIST command again to find the straight-line distance between any two division points on the half-bottom view. (Use the Node object snap to snap to two of the division points.) Enter the MEASURE command and pick arc BC near line AC. For the segment length, enter the straight-line distance you found using the DIST command. AutoCAD marks points along the arc at the interval you entered.

5. For the development, you need only the first 12 of these intervals. Create a line from the twelfth division mark to the apex (point A). This determines the other end of the development. Trim away the rest of arc BC, and delete the rest of the points. See Figure 14-55C.

6. Enter the DIST command and find the distance from the apex to point D. With this distance as a radius, create another circle with its center at the apex. Trim the circle to the lines that represent the beginning and end of the development, as shown in Figure 14-55D.

7. Clean up the drawing by trimming away unneeded lines. Enter PDMODE and enter a new value of 0 to hide the division points, using REGEN to change the points on-screen. To finish the drawing, add .1″ tabs with 45° chamfers. See Figure 14-55E.

CAD Intersections

Companies that use CAD systems generally create complex solid models of intersecting objects. The pattern development is then created directly from the solid model using third-party software. However, you can develop an intersection in CAD using a procedure similar to that used in board drafting.

Figure 14-56 is a pictorial of the two intersecting prisms from Figure 14-42. Create the intersection shown in Figure 14-57 by following the steps below. *Note:* Construction lines are shown in color in the illustration for clarity only. It is not necessary or desirable to use a different color for these lines, because parts of them are incorporated into the final drawing.

1. Draw the hexagon shape in the top view at 2.00″ across the flats.

2. Use the XLINE command to place construction lines downward from the corners to establish

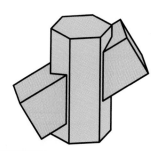

Figure 14-56

Two intersecting prisms.

the vertical lines for the hexagon in the front view.

3. Place line AB horizontally across the construction lines as shown in Figure 14-57A. Offset the line downward by 3.50″ to create the bottom line of the front view. Trim the construction lines to the boundaries of the front view.

4. Locate the exact center of the front view. You can do this easily and quickly by striking a diagonal from corner A to corner C. Start a new line at the midpoint of line AC (the center of the front view), and use polar coordinates to extend it 2.00″ at 30°. Copy the new line using its right endpoint as the base point. Place the copy so that the right endpoint is at the intersection of the diagonals. Enter the PEDIT command and use the Join option to change the two lines into a single polyline that defines the length and location of the rectangular prism.

5. To the right of the front view, construct an auxiliary view of the square prism.

6. Enter the OFFSET command. Instead of selecting the line to offset, enter T to activate the Through option. Select the polyline you created in step 4 as the line to offset. For the through point, choose point D in Figure 14-57B, at the top of the auxiliary view. This creates the top edge of the rectangular prism. Then repeat this operation, choosing point D′ at the bottom of the auxiliary as the through point, to establish the lower edge of the rectangular prism. Connect the ends to complete the front view of the rectangular prism. Erase the diagonal line AC.

7. Use XLINE to draw horizontal construction line DD′ through the center of the top view.

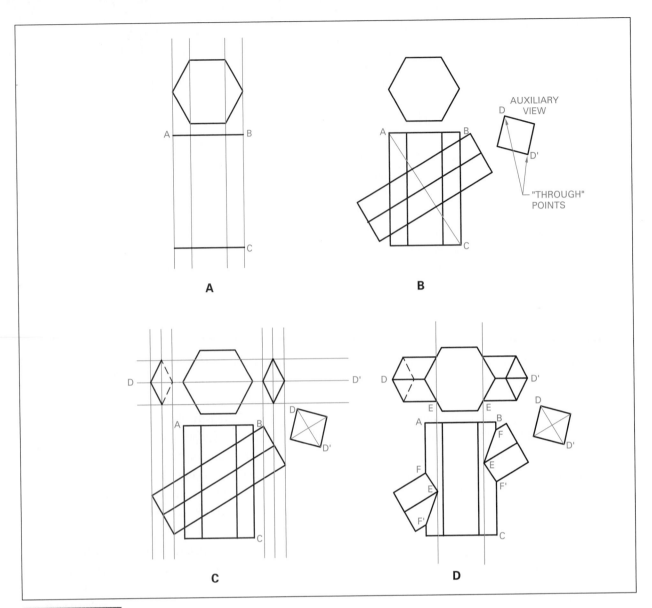

Figure 14-57

Developing intersecting prisms.

8. In the auxiliary view of the rectangular prism, draw both diagonals. Use the DIST command to find the actual distance from the intersection of the diagonals to any corner of the square. Then offset line DD′ by this amount above and below to establish the edges of the rectangular prism in the top view.

9. Establish the right and left ends of the rectangular prism in the top view by creating vertical construction lines from key points on the front view, as shown in Figure 14-57C. Finish the top view by drawing the connecting lines on both sides and trimming the horizontal construction lines. Notice that two of the lines are hidden on the left side of the prism. Clean up the drawing by deleting the vertical construction lines.

10. Add the lines of intersection on the front view by creating vertical construction lines through points E. Connect points F and F′ with points E in the front view and trim away the unneeded portions of the lines, as shown in Figure 14-57D. Delete the two remaining vertical construction lines to complete the drawing.

Chapter Summary

■ A pattern development is a full-size layout of an object made on a single flat plane.

■ Patterns developed for the packaging industry must take into account factors such as duration of product use.

■ Patterns can be developed using any kind of sheet material.

■ Parallel-line development is used to make patterns of objects with parallel edges.

■ Radial-line development is used to make patterns of objects such as cones and pyramids.

■ Triangulation is a method used for making approximate developments of surfaces by using a series of triangles.

■ Exact pattern development can be made only for objects with flat or single-curved surfaces.

■ Transition pieces are used to connect openings of different shapes, sizes, or positions.

■ Intersections are developments of parts that join one another.

Review Questions

1. List two other names for pattern developments.

2. What do you call pressing a flat sheet into a shape under heavy pressure?

3. What must a drafter know, in addition to pattern development techniques, to create a finished sheet-metal pattern?

4. Name the three types of pattern development.

5. Name the two general classes of surfaces.

6. What kind of curved surface is on a sphere?

7. What do you call a development that goes from round to square?

8. What do you call the lines on a stretchout that show where to make a fold?

9. What kind of development uses a series of triangles to produce a pattern?

10. What kind of development is used to produce a pattern for a cone?

11. In terms of pattern development, what is an intersection?

12. Explain the purpose of a transition piece.

13. In what kind of cone is the base a true circle and the apex directly above the center of the base circle?

14. What type of software allows the CAD operator to specify the characteristics of an object and then draws the object according to those specifications?

15. Why do CAD operators usually use third-party software to create pattern developments?

16. Why should drafters learn to create patterns even if they plan to use third-party software?

Problems

Drafting Problems

The drafting problems in this chapter are designed to be completed using board drafting techniques or CAD.

Problems 1 through 10 are planned to fit on an 11.00″ × 17.00″ or 12.00″ × 18.00″ drawing sheet. Draw the front and top views of each problem. Develop the patterns as shown in the example in Figure 14-58. Include dimensions and numbers if required by your instructor. Patterns may be cut out and assembled.

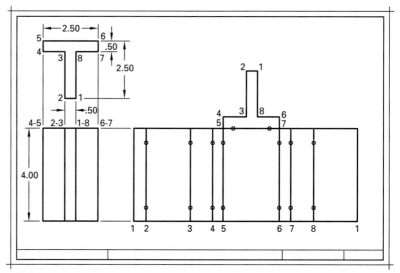

Figure 14-58

Pattern development example.

1. Draw the front and top views of each object shown in Figure 14-59, and then develop the pattern. Add the top in the position it would be drawn for fabrication.

2. Draw the front and top views of each object shown in Figure 14-60, and then develop the pattern. Add the top in the position it would be drawn for fabrication.

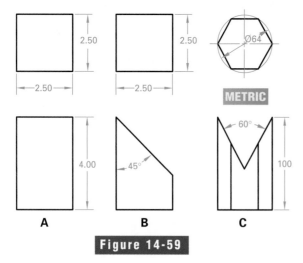

Figure 14-59

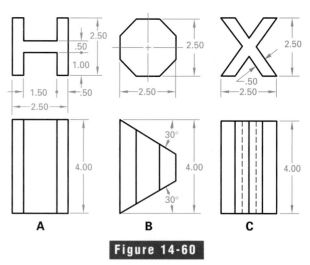

Figure 14-60

3. Draw the front and top views of each object shown in Figure 14-61, and then develop the pattern.

4. Draw the front and top views of each object shown in Figure 14-62, and then develop the pattern.

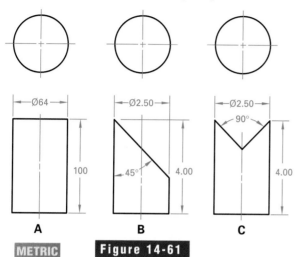

METRIC   **Figure 14-61**

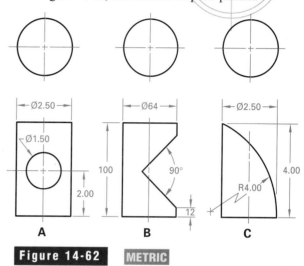

Figure 14-62   METRIC

5. Draw the front and top views and then develop the pattern for each object shown in Figure 14-63.

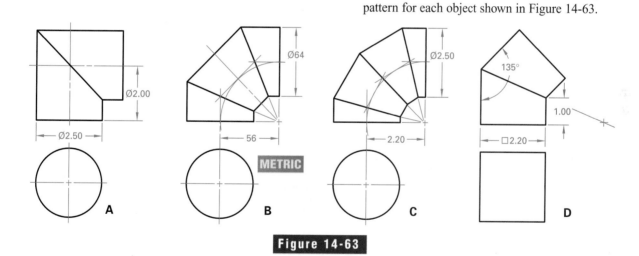

Figure 14-63

6. Draw the front and top views and then develop the pattern for each object shown in Figure 14-64.

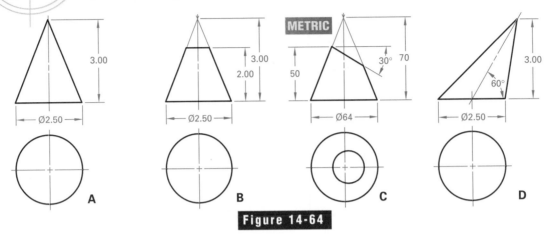

Figure 14-64

7. Draw the front and top views and then develop the pattern for each object shown in Figure 14-65.

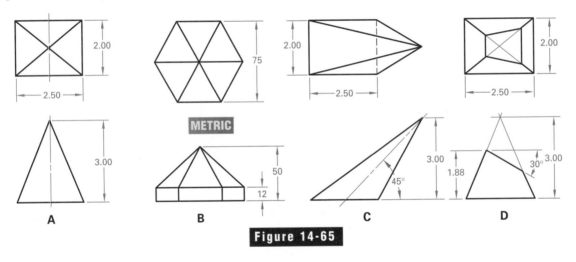

Figure 14-65

8. For the objects in Figure 14-66, create the front views, complete the top views, and develop the patterns.

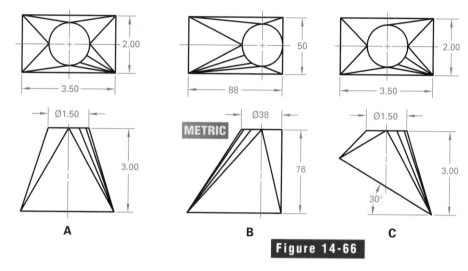
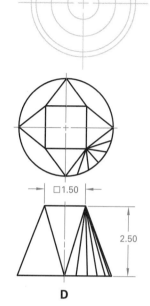

Figure 14-66

9. Draw two views of each pair of objects shown in Figure 14-67, completing the views when necessary by developing the line of intersection and completing the top views. Develop patterns for both parts of each pair.

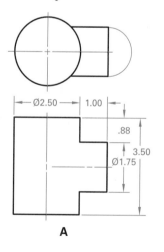
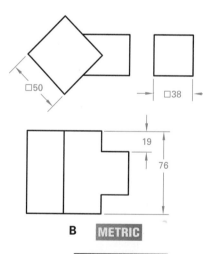
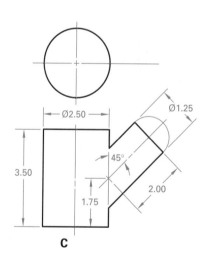

Figure 14-67

10. Draw two views of each pair of objects shown in Figure 14-68. Develop the line of intersection, and complete the top views. Develop patterns for both parts of each pair.

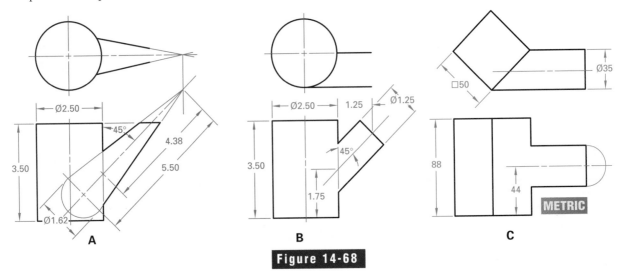

A

B

C

METRIC

Figure 14-68

For problems 11 through 15, follow the directions to create patterns and drawings of the objects as assigned.

11. Make a pattern drawing for the tool tray shown in Figure 14-69. No other views are necessary.

12. Make a pattern drawing of the cookie sheet shown in Figure 14-70. No other views are necessary.

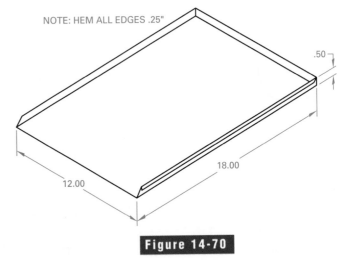

NOTE: HEM ALL EDGES .25"

Figure 14-70

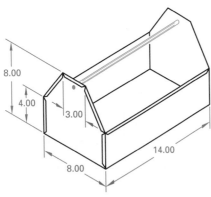

NOTES: 1. ALL HEMS AND TABS .25"
2. HANDLE Ø.50

Figure 14-69

13. Make a pattern drawing of the brass candy tray shown in Figure 14-71. Take dimensions from the printed scale.

14. Make a complete set of working drawings for the candlestick shown in Figure 14-72, including all necessary views and patterns. Take dimensions from the printed scale.

15. Make a complete set of working drawings for the model racer shown in Figure 14-73, including all necessary views and patterns. Take dimensions from the printed scale.

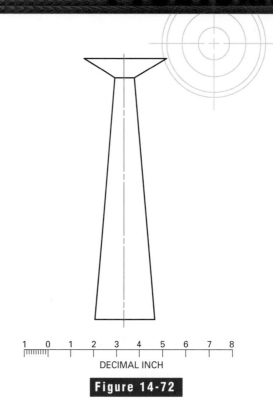

DECIMAL INCH

Figure 14-72

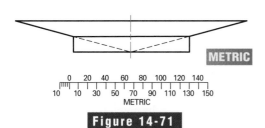

METRIC

METRIC

Figure 14-71

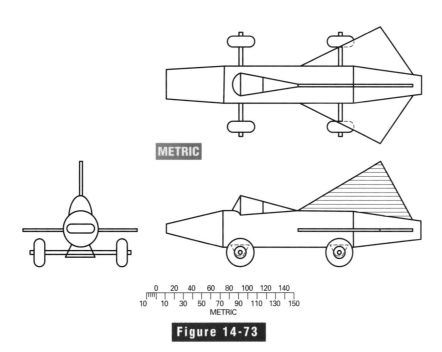

METRIC

METRIC

Figure 14-73

Design Problems

Design problems have been prepared to challenge individual students or teams of students. In these problems, you are to apply skills learned mainly in this chapter but also in other chapters throughout the text. They are designed to be completed using board drafting, CAD, or a combination of the two. Be creative and have fun!

1. Design a carton to be used in shipping the model racer shown in Figure 14-73. The overall dimensions of the racer are $4.50'' \times 5.00'' \times 11.00''$. The carton should be designed as a one-piece development, easily assembled. Design it in a way that will require no adhesive for assembly. Use a computer paint program to design the outside surface of the carton. Use various colors. Trace the pattern onto stiff cardboard, cut it out, and assemble it.

2. **TEAMWORK** Work as a team to design a CO_2 racecar. Be creative and make it an ultramodern concept car. Each team member should develop design sketches for the team to review. As a team, select the final design choice. Prepare a drawing at full size. Each team member should then be assigned the development of one part of the car. Remember, all patterns must be full size. A paper model can then be constructed by cutting out the individual patterns, forming them, and assembling them into the finished model.

3. **TEAMWORK** Design a porch lamp to be installed against an outside wall. The top is to be either a right rectangular pyramid or a right circular cone. The mounting base is to include a right circular cone, a frustum of a cone, or a frustum of a right rectangular pyramid. Material: sheet brass or copper with glass inserts. Prepare a working drawing and all patterns needed for the manufacture of the lamp.

4. Design a carton for the porch lamp designed in design problem 3. Transfer the pattern to stiff cardboard. Cut it out and assemble it.

Welding Drafting

OBJECTIVES

Upon completion of this chapter, you should be able to:

■ Identify the various types of joints and welds.

■ Describe the welding processes classified as fusion welding.

■ Describe the welding processes classified as resistance welding.

■ Use weld symbols correctly in conjunction with other data to develop a complete welding drawing.

■ Determine the appropriate joint preparation for a specific weld application.

■ Convert a drawing for a casting into one that is appropriate for a welded part.

Drafting Principles

Welding is a way of joining metal parts together. The art of welding is very old. In prehistoric times, it was used to make rings, bracelets, and other jewelry. Today it is very important in industry. Welded steel parts are generally lighter, stronger, and longer-lasting than parts made by forging or casting. Figure 15-1A shows a pulley housing made by casting. Compare it with the similar part made by welding, shown in Figure 15-1B.

Welding has become a major assembly method in industries that use steel, aluminum, and magnesium to build cars, trucks, airplanes, ships, or buildings. In a single dump truck, for example, there are hundreds of welds. The basic framework of the truck is assembled by welding.

In order to prepare a welding drawing, the drafter must first understand the basic principles of welding practice. In addition, it is helpful if the drafter understands the advantages and limitations of welding in the design process. For example, welding is not the best method for joining parts that require easy and quick assembly and disassembly. In that case, bolts or screws might be the better choice. In some instances, a combination of fastening methods will best solve a design problem.

The drafter who draws the parts to be welded works with a design engineer who knows what kinds of welding to use with different metals. Nevertheless, the drafter must be familiar with the various welding processes and their associated drafting symbols. Standard drafting symbols for welding have been established by the American Welding Society.

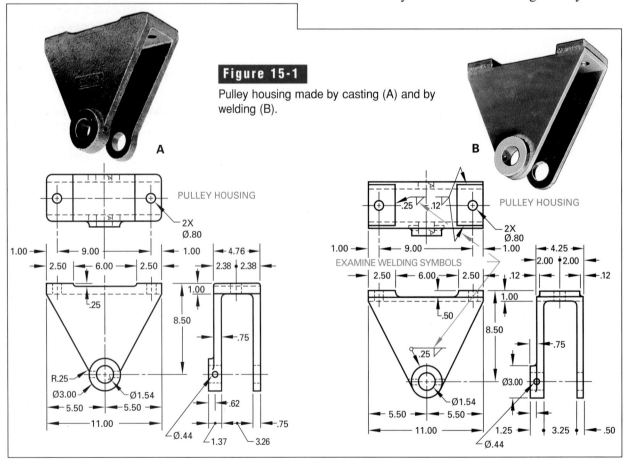

Figure 15-1

Pulley housing made by casting (A) and by welding (B).

Types of Joints and Welds

Industrial processes require a variety of joints and welds because welding is used for so many different purposes. The various types of joints and welds can be combined in many different ways to achieve the desired strength and characteristics for each application. To specify a weld completely, both the type of joint to be achieved and the type of weld must be included.

Joints

The five basic joints used in welding are shown in Figure 15-2. They are:

- butt joints
- lap joints
- corner joints
- edge joints
- T-joints

To detail the right joint, the designer must be familiar with the specified materials, processes, and the conditions of each. Practical welding experience is also helpful in making knowledgeable decisions.

Welds

The right weld for a specific job depends on the type of material specified, the tools to be used, and the cost of preparation. Some of the basic welds are described below.

- **Groove welds** are located in a groove or notch in the work material. The grooves are classified according to their shape, as shown in Figure 15-3. Although Figure 15-3 shows the welds applied to a butt joint, they can also be applied to any other joint. Note that the grooves may be single or double.
- **Fillet welds** are similar to groove welds, except that no groove is made in the material. Instead, the weld rests on top of the joint. Figure 15-4 shows an example of a fillet weld.
- **Plug welds** are welds that fit into a small hole in the work material, as shown in Figure 15-4.
- **Slot welds** are similar to plug welds, except that the shape of the opening for the weld is slot-shaped. The weld symbol for a slot weld is the same as that for a plug weld.

Fusion Welding

In **fusion welding**, heat is applied to create a weld. Fusion welding processes include gas, arc, thermit, and gas-and-shielded-arc welding. Soldering and brazing, although called by their separate names, are also forms of fusion welding.

Figure 15-2

Five basic types of welded joints.

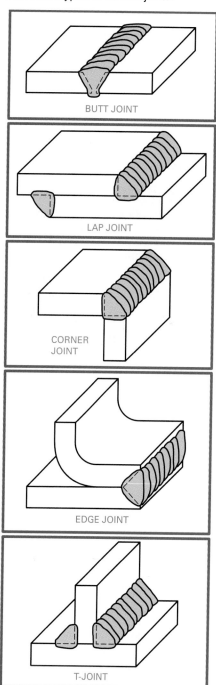

BUTT JOINT

LAP JOINT

CORNER JOINT

EDGE JOINT

T-JOINT

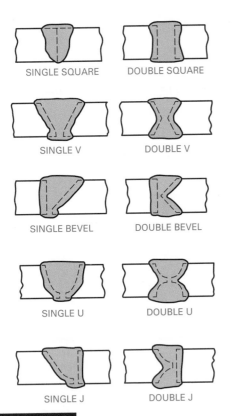

Figure 15-3

Ten basic types of groove welds applied to a butt joint.

The labels in the figure read:
SINGLE SQUARE — DOUBLE SQUARE
SINGLE V — DOUBLE V
SINGLE BEVEL — DOUBLE BEVEL
SINGLE U — DOUBLE U
SINGLE J — DOUBLE J

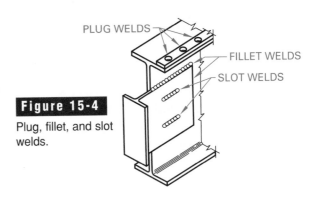

Figure 15-4

Plug, fillet, and slot welds.

Fusion welding requires a welding filler material in the form of a wire or rod. The welder heats the material with a gas flame or a carbon arc. When it melts, it fills in a joint and combines with the metal being welded.

Gas Welding

A combination of gases is used to create the heat for **gas welding**, as shown in Figure 15-5. This process was developed in 1885, when two gases—oxygen, from liquid air, and acetylene, from calcium carbide—were brought into use. Burning acetylene supported by oxygen gives temperatures of between 5000° and 6500°F (2760° and 3595°C).

Arc Welding

In 1881, De Meritens in France first performed a process known as arc welding. **Arc welding** is a form of fusion welding in which an electric arc forms between the work (part to be welded) and an electrode, as shown in Figure 15-6. The arc causes intense heat to develop at the tip of the electrode. This heat is used to melt a spot on the work and on a rod of filler material so that the two can be fused together.

Thermit Welding

The natural chemical reaction of aluminum with oxygen forms the basis for **thermit welding**. A mixture, or charge, made of finely divided aluminum and iron oxide is ignited by a small amount of special ignition powder. The charge burns rapidly, producing a very high temperature. This melts the metal, which then flows into molds and fuses mating parts.

Gas-and-Shielded-Arc Welding

Aluminum, magnesium, low-alloy steels, carbon steels, stainless steel, copper, nickel, and titanium are some of the metals that can be welded using **gas-and-shielded-arc welding**. As its name implies, this process combines arc welding and gas welding. Two

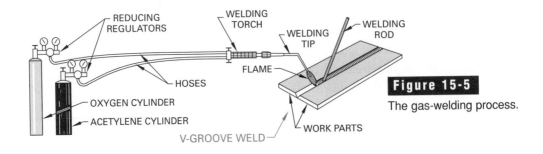

Figure 15-5

The gas-welding process.

Labels in the figure: REDUCING REGULATORS, WELDING TORCH, WELDING TIP, WELDING ROD, FLAME, HOSES, OXYGEN CYLINDER, ACETYLENE CYLINDER, V-GROOVE WELD, WORK PARTS

Temperature Conversions

When a drafter must deal with processes such as welding, forging, casting, soldering, and brazing, it is often necessary to work with the numerical values for the high temperatures needed in these processes. As manufacturers and other companies have begun doing an international business, it has become increasingly necessary for drafters, engineers, and welders to understand and be able to work both in degrees Fahrenheit (F) and in degrees Celsius or centigrade (C). Conversion from one system to another can be accomplished using two simple formulas.

For example, by burning acetylene gas supported by oxygen for gas welding and brazing, temperatures between 5000° and 6500° Fahrenheit can be achieved. These temperatures in Celsius or centigrade are 2760° and 3593°, respectively.

To convert temperatures from degrees Fahrenheit to degrees Celsius, use the formula:

$$C = \frac{5}{9}(F - 32)$$

To apply this formula to the lower limit of temperatures for gas welding and brazing, plug in the numbers as follows:

$$C = \frac{5}{9}(5000 - 32)$$

$$C = \frac{5}{9}(4968)$$

$$C = \frac{5 \times 4968}{9}$$

$$C = 2760°$$

To convert temperatures from degrees Celsius to degrees Fahrenheit, use the formula:

$$F = \frac{9}{5}C + 32$$

To convert the lower limit of temperatures for gas welding and brazing from Celsius to Fahrenheit, use the formula as follows:

$$F = \frac{9 \times 2760}{5} + 32$$

$$F = \frac{24,840}{5} + 32$$

$$F = 4968 + 32$$

$$F = 5000°$$

forms of gas can be used in gas-and-shielded-arc welding: *tungsten-inert gas (TIG)*, and *metallic-inert gas (MIG)*. In tungsten-inert gas welding, the electrode that provides the arc for welding is made of tungsten. Since it provides only the heat for fusion,

some other material must be used with it for filler. In metallic-inert gas welding, the electrode contains a consumable metallic rod. It provides both the filler material and the arc for fusion.

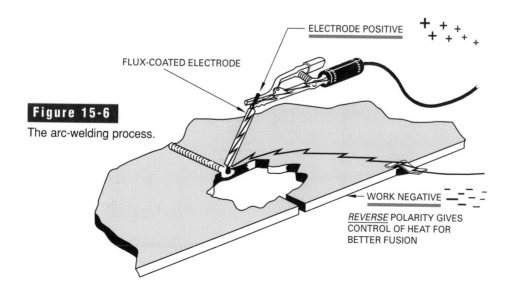

Figure 15-6

The arc-welding process.

Symbols for Fusion Welding

Drafters use special symbols to specify welds on a welding drawing. Figure 15-7 shows an example of the use of welding symbols in technical drawings. In Figure 15-7A, welding symbols are used on a machine drawing. In Figure 15-7B, they are used on a structural drawing.

Figure 15-8 describes the standard welding symbols approved by the American Welding Society. The notes in the illustration explain how to place symbols and data in relation to the reference line. By combining the symbols in Figure 15-8, you can describe any welded joint, from the simplest to the most complex.

Standard Symbols

Welding symbols are a shorthand language. They save time and money in drafting and, if used correctly, ensure understanding and accuracy. They need to be a universal language, and for this reason, the symbols of the American Welding Society, already well established, have been adopted.

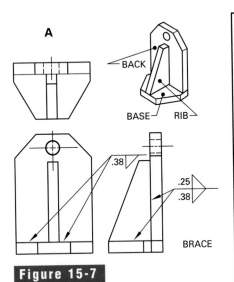

Figure 15-7

The application of welding symbols on a machine part (A) and on a structural drawing (B).

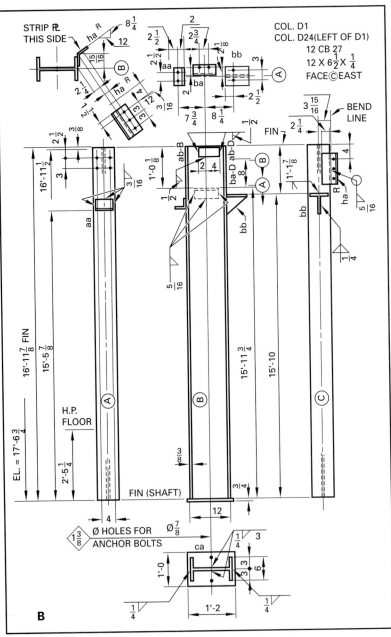

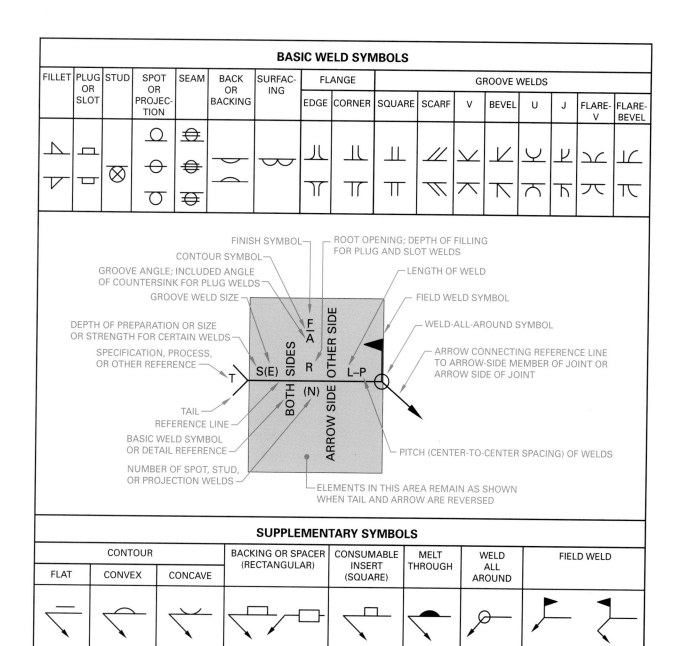

BASIC WELD SYMBOLS

FILLET	PLUG OR SLOT	STUD	SPOT OR PROJEC- TION	SEAM	BACK OR BACKING	SURFAC- ING	FLANGE		GROOVE WELDS							
							EDGE	CORNER	SQUARE	SCARF	V	BEVEL	U	J	FLARE- V	FLARE- BEVEL

SUPPLEMENTARY SYMBOLS

CONTOUR			BACKING OR SPACER (RECTANGULAR)	CONSUMABLE INSERT (SQUARE)	MELT THROUGH	WELD ALL AROUND	FIELD WELD
FLAT	CONVEX	CONCAVE					

NOTE: SIZE, WELD SYMBOL, LENGTH OF WELD, AND SPACING MUST READ IN THAT ORDER FROM LEFT TO RIGHT ALONG THE REFERENCE LINE. NEITHER ORIENTATION OR REFERENCE LINE NOR LOCATION ALTER THIS RULE. THE PERPENDICULAR LEG OR FILLET, BEVEL, J, OR FLARE-BEVEL WELD SYMBOLS MUST BE AT LEFT. ARROW- AND OTHER-SIDE WELDS ARE OF THE SAME SIZE UNLESS OTHERWISE SHOWN. SYMBOLS APPLY BETWEEN ABRUPT CHANGES IN DIRECTION OF WELDING UNLESS GOVERNED BY THE "ALL-AROUND" SYMBOL OR OTHERWISE DIMENSIONED.

Figure 15-8

Weld symbols.

The distinction between the terms *weld symbol* and *welding symbol* should be understood. The weld symbol, explained in Figure 15-8, indicates the type of weld. The welding symbol is a method of representing the weld on drawings. It includes supplementary information and consists of the following elements:

- reference line
- arrow
- basic weld symbol
- dimensions and related data
- supplementary symbols
- finish symbols
- tail
- specifications, process, or other references

SYMBOL   SIGNIFICANCE

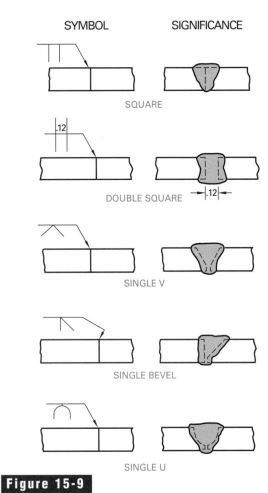

SQUARE

DOUBLE SQUARE

SINGLE V

SINGLE BEVEL

SINGLE U

Figure 15-9

Symbols for five typical groove welds.

Not all elements need be used unless required for clarity. In other words, use only those elements of the welding symbol that are needed to describe the weld completely.

The welding symbol tells what kind of weld to use at a joint and where to place it. Figure 15-9 shows five typical groove welds, along with the symbols used for each. The symbol can be drawn on either side of the joint as space permits. *Note:* When drawing a fillet, bevel, or J-grooved weld symbol, place the perpendicular leg of the symbol to the left, as shown in Figure 15-10.

An arrow leads from the symbol's reference line to the joint. The side of the joint to which the arrow points is called the *arrow side*. The opposite side is called the *other side*. If the weld is to be on the arrow side of the joint, draw the type-of-weld part of the symbol below the reference line, as shown in Figure 15-11A and D. If the weld is to be on the other side, draw the type-of-weld part of the symbol above the reference line, as shown in Figure 15-11B and E. If the weld is to be on both sides of the joint, draw the type-of-weld part of the symbol both above and below the reference line, as shown in Figure 15-11C.

When the weld is to be a J-groove weld, you must place the arrow from the welding symbol correctly to avoid confusion. For example, in part A of Figure 15-12, it is not clear which piece is to be grooved. In part B of the figure, the arrow has been redrawn to show

Figure 15-10

The perpendicular leg on a weld symbol is always drawn to the left.

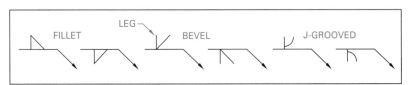

FILLET   LEG   BEVEL   J-GROOVED

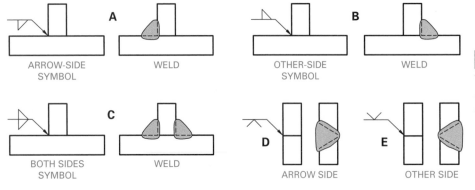

ARROW-SIDE SYMBOL   WELD   OTHER-SIDE SYMBOL   WELD

BOTH SIDES SYMBOL   WELD   ARROW SIDE   OTHER SIDE

A   B   C   D   E

Figure 15-11

Weld symbols show the arrow side and the other side.

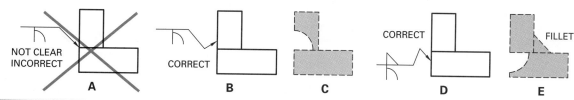

Figure 15-12

Weld symbols for the J-grooved weld.

clearly that the vertical piece is to be grooved, as shown in part C. In part D of Figure 15-12, two welds are called for. The symbol below the reference line includes a J-grooved weld on the arrow side. The arrow shows that it is the horizontal piece that is to be grooved. The symbol above the reference line indicates a fillet weld on the other side. The drawing in Figure 15-12E shows how the completed welds would look.

In Figure 15-13A, the reference dimensions have been included with the welding symbols. Figure 15-13B shows the joint made according to the reference specifications in Figure 15-13A. This joint can be described as follows: a double filleted-welded, partially grooved, double-J T-joint with incomplete penetration. The J-groove is of standard proportion. The radius R is .50″ (13 mm) and the included angle is 20°. The penetration is .75″ (19 mm) for the other side and 1.25″ (32 mm) deep for the arrow side. There is a .50″ (13-mm) fillet weld on the arrow side and a continuous .38″ (10-mm) fillet weld on the other side. The fillet on the arrow side is 2.00″ (50 mm) long. The pitch of 6.00″ (150 mm) indicates that

it is spaced 6.00″ (150 mm) center-to-center. All fillet welds are standard at 45°.

Supplementary Symbols

In Figure 15-13A, notice the solid black dot on the elbow of the reference line. This dot or a black flag is a supplementary symbol for a field weld. This means that the weld is to be made in the field or on the construction site rather than in the shop.

In the tail of the reference line in Figure 15-13A is the typical specification A2. Its meaning is as follows: The work is to be a metal-arc process, using a high-grade, covered, mild-steel electrode; the root is to be unchipped and the welds unpeened, but the joint is to be preheated.

Notice the flush contour symbol over the .50″ (13-mm) fillet weld symbol in Figure 15-13A. This indicates that the contour of this weld is to be flat-faced and unfinished. Over the .38″ (10-mm) fillet weld on the same reference line, there is a convex contour symbol. This indicates that this weld is to be finished to a convex contour. Figure 15-8 shows the supplementary welding symbols to be used for finished welding techniques.

Figure 15-13

Interpretation of weld symbols and specifications.

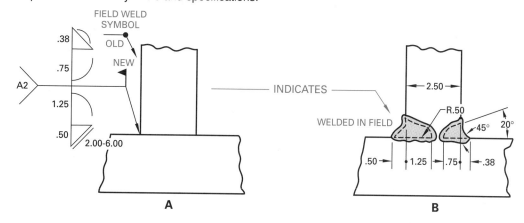

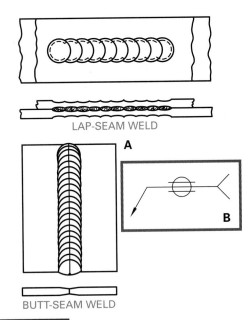

Resistance Welding

In 1857, James Prescott Joule developed **resistance welding**, a form of welding that combines heat and pressure to create a weld. In spite of its early development, however, resistance welding depends on electricity. Therefore, industry did not begin using this process until after the 1880s when electric power became available in large quantities.

Resistance welding is a good method for fusing thin metals. To join two metal pieces, an electric current is passed through the points to be welded. At those points, the resistance to the charge heats the metal to a plastic state. Pressure is then applied to complete the weld.

Projection Welds

A **projection weld** is identified by strength or size. Figure 15-14 parts A and D show pieces set up for such a weld. In each case, one part has a boss pro-

Figure 15-15

Butt-seam and lap-seam welds (A) and the symbol for seam welds (B).

jection. In Figure 15-14B, the reference 700 lb. [3.10 kilonewtons (kN)] means that the acceptable shear strength per weld must be at least 700 lb. (3.10 kN). In Figure 15-14C, the reference data 500 lb. (2.22 kN) specifies the strength of the weld, the 2.00 means that the first weld is located 2.00″ (50 mm) from the left side, and the 5.00 specifies a weld every 5.00″ (125 mm) center to center. In Figure 15-14E, the number .25″ (6.0 mm) specifies the diameter of the weld. In Figure 15-14F, the diameter of the weld is .25″ (6.0 mm), and there is a weld every 2.00″ (50 mm) beginning 1.00″ from the left side. The (5) specifies that there is to be a total of five welds. Notice the arrow-side and other-side indications.

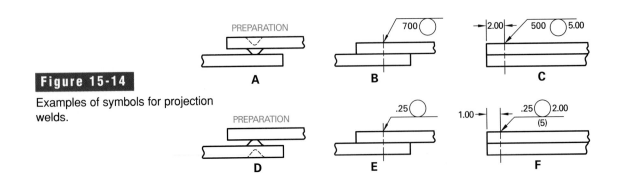

Figure 15-14

Examples of symbols for projection welds.

Seam Welds

A weld along the seam of two adjacent parts is known as a **seam weld**. Figure 15-15A shows butt-seam and lap-seam welds. The symbol for seam welding is shown in Figure 15-15B. The side view shows the two pieces positioned edge to edge for butt-seam welding and overlapping for lap-seam welding, which is done with a series of tangent spot welds.

Flash Welding

A special kind of resistance welding known as **flash welding** is done by placing the parts to be welded in very light contact or by leaving a very small air gap. The electric current then flashes, or arcs. This melts the ends of the parts, and the weld is made.

Spot Welds

When the current and pressure are confined to a small area between electrodes, the resulting weld is called a **spot weld**. Resistance spot welding is done on lapped parts, and the welds are relatively small. Figure 15-16A shows a reference symbol in the top view. The arrow points to the working centerline of the weld. In Figure 15-16A, the minimum diameter of each weld is specified at .30″ (7.6 mm). In Figure 15-16B, the minimum shearing strength of each weld is specified at 800 lb. (3.558 kN). In parts C and D of Figure 15-16, the reference data indicates that the first weld is centered 1.00″ (25 mm) from the left end and the welds are spaced 2.00″ (50 mm) from centerline to centerline.

BASIC RESISTANCE-WELD SYMBOLS			
TYPE OF WELD			
SPOT	PROJECTION	SEAM	FLASH OR UPSET
○	○	⊖	‖

SUPPLEMENTARY SYMBOLS			
WELD ALL AROUND	FIELD WELD	CONTOUR	
		FLUSH	CONVEX
○	⚑	—	⌒

Figure 15-17

Basic resistance-weld symbols.

Symbols for Resistance Welding

Four basic symbols are used in resistance welding. They signify spot, projection, seam, and flash welds, as shown in Figure 15-17. The basic reference line and arrow used for the arc-weld and gas-weld symbols are also used for resistance-weld symbols. However, in general, there is no arrow side or other side. The same supplementary symbols also apply, as Figure 15-17 shows.

Dimensioning Welds

Figure 15-18 shows the typical dimensions for a butt joint with a V-grooved weld. Typically, the manufacturer needs to know the required angle of the joint preparation (A), the root opening (R), and the height of reinforcement (C), specified only if a back or backing weld is to be used on the underside of the

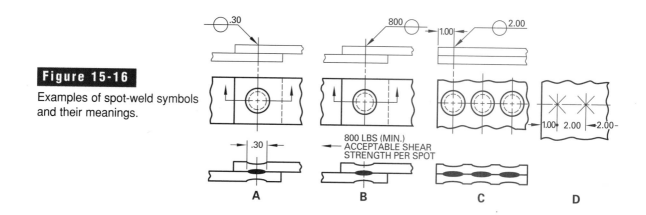

Figure 15-16

Examples of spot-weld symbols and their meanings.

welded joint. Dimensions for a T-joint with a bevel-grooved weld include the bevel angle (B), the root opening (R), and the height of reinforcement (C) if a weld is to be applied to the underside of the joint, as shown in Figure 15-19. The U-grooved joint preparation for a butt joint is given in Figure 15-20. This type of joint is expensive to prepare and is used mainly for joining material 1.00″ or greater in thickness. In this case, dimension the groove angle (A), the radius (R) at the bottom of the groove, the root opening (R), and the height of reinforcement (C), if applicable. The J-grooved joint preparation is generally used on a T-joint, as shown in Figure 15-21. This type of joint requires preparation on one part only, as

shown, and closely resembles the U-grooved joint. Like the U-grooved joint, it is used when joining materials 1.00″ or greater in thickness.

When the design engineer determines that it is not necessary to run a weld the entire length of the joint, the length of the weld is given to the right of the weld symbol, as shown in Figure 15-22. Two or more short welds along a joint are called **intermittent welds** and are specified and dimensioned as shown in Figure 15-23. The pitch (center-to-center spacing) of intermittent welds is shown as the distance between centers of increments on one side of the joint. It is shown to the right of the length dimension with a hyphen between the two dimensions. Staggered intermittent welds are shown with the weld symbols staggered as shown in Figure 15-24.

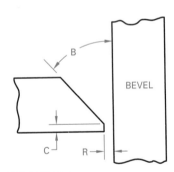

Figure 15-18

Dimensions for a V-joint weld. A = 60° minimum; C = 0 to .12″; R = .12″ to .25″; stock = .50″ to .75″ thick.

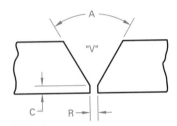

Figure 15-19

Dimensions for a bevel-grooved weld. B = 45° minimum; C = 0 to .12″; R = .12″ to .25″.

Figure 15-20

Dimensions for a U-grooved weld. A = 45° minimum; C = .06″ to .20″; R = 0 to .56″.

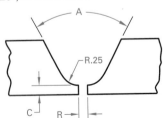

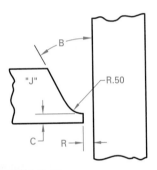

Figure 15-21

Dimensions for a J-grooved weld. B = 25° minimum; C = .06″ to .20″; R = 0 to .56″.

Figure 15-22

Dimensioning the length of a weld.

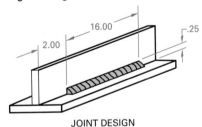

JOINT DESIGN

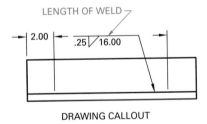

DRAWING CALLOUT

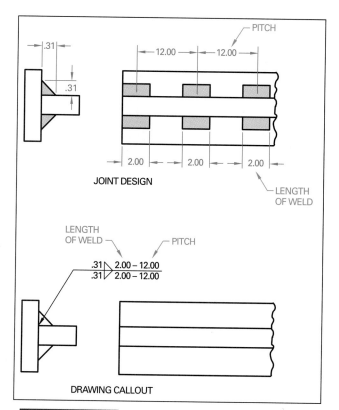

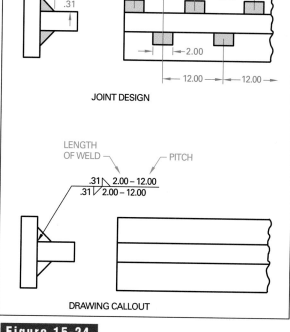

Figure 15-23

Specifying intermittent welds.

Figure 15-24

Staggered intermittent welds.

Board Drafting Techniques

The information covered thus far in this chapter deals with the theoretical aspects of welding and welding drawings. You should now be ready to apply the theory you have learned to the development of a complete welding drawing with dimensions and symbols. The information given on the completed drawing should be sufficient for the manufacture of the part. In order to make this experience more meaningful, we will convert the casting of the connecting link shown in Figure 15-25 into a welded part, or *weldment*.

In the conversion process, several things should be considered:

■ When converting from a casting to a weldment, you can generally reduce the thickness of cast members by one material thickness. For example, the ribs on the casting are .38″ thick. Since steel plate is considerably stronger and tougher than cast iron, the thickness of the replacement steel plate can be .25″ thick. Careful engineering and good judgment become a necessary part of this process.

■ In some cases, changing round features on the casting to square, flat, or rectangular parts on the weldment will reduce joint preparation. This concept will become evident as we continue to convert the connecting link.

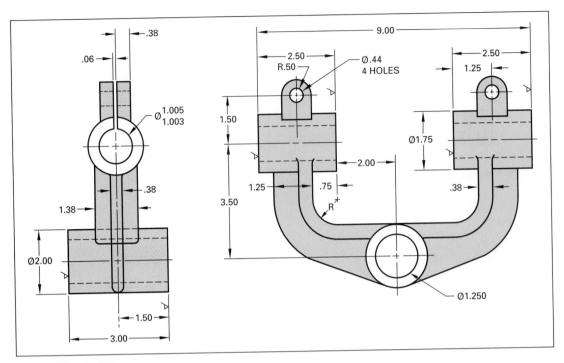

Figure 15-25

Connecting link casting.

QTY	ITEM	MATL	DESCRIPTION	PT NO.
1	SHAFT	STL	☐ 2.00 X 3.00	1
2	SHAFT	STL	☐.75 X 2.50	2
4	SUPPORT	STL	.25 X 1.00 X 1.00	3
2	RIB	STL	.25 X 1.50 X 1.88	4
2	RIB	STL	.25 X 3.00 X 3.38	5
2	RIB	STL	.25 X 1.50 X 2.00	6

Figure 15-26

Welding drawing of the connecting link from Figure 15-25.

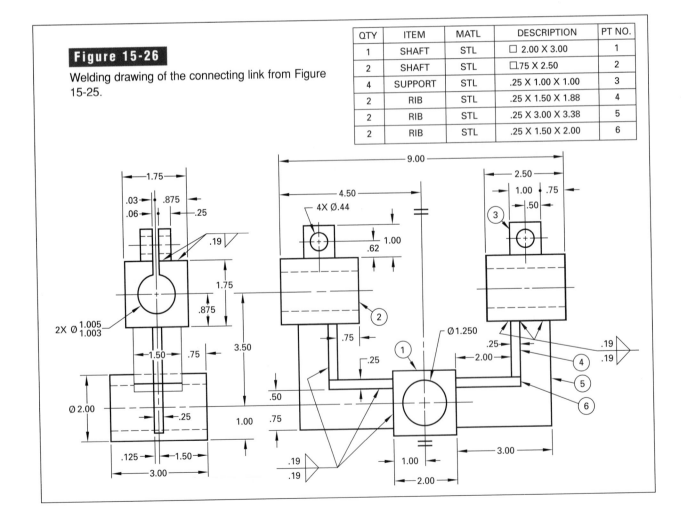

- Since a weldment consists of several individual parts welded together to form an assembly, it is generally useful to number the parts on the welding drawing and prepare a parts list. This saves the manufacturer considerable time in the "tooling up" process.
- Select appropriate weld joints and joint preparations based on the size of the parts to be joined and strength required. In this case, we can use fillet welds throughout with no special joint preparation.
- In most cases, a qualified engineer will determine the size and type of weld to be specified. In this case, the engineer has suggested .19″ fillet welds throughout the part.

To create the welding drawing, follow these steps:

1. Consider the changes required or enabled by converting the connecting link from a casting to a weldment, as described on page 555.

2. Sketch the necessary views, including dimensions and welding symbols, to determine the appropriate scale and drawing sheet size.
3. Create an instrument drawing of the connecting link, incorporating all of the changes you determined in step 1.
4. Add the appropriate dimensions and welding symbols.
5. Number the parts on the welding drawing and create a parts list.

Figure 15-26 is the finished welding drawing. Notice that by converting the round features on the casting to square features on the weldment, we are able to reduce joint preparation and fitting time significantly. This process greatly reduces the cost of manufacturing the part. Notice also that all of the parts except Part No. 5 are stock materials that need only be cut to length, placed in position, and welded. Very limited special cutting and fitting is required.

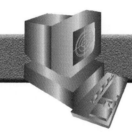

CAD Techniques

The procedure for creating a welding drawing using CAD is similar to that for creating any other CAD drawing. You must, of course, apply the welding theory discussed in this chapter. The feature that sets the CAD techniques apart from board drafting techniques is the use of symbol libraries for the welding symbols. As in other drafting applications, using a symbol library for welding symbols greatly reduces the time and effort required to create a welding drawing.

Welding symbols are available from many different sources. Many CAD programs come with some standard symbol libraries, usually including welding symbols. Many third-party sources also provide welding symbol libraries. Some of these are inexpensive, or even free. You can locate these sources by searching on the Internet using key words such as WELD SYMBOL or CAD WELDING.

Prior to AutoCAD 2000i, AutoCAD did not include symbol libraries as a standard feature. Beginning with AutoCAD 2000i, a welding symbol library was included among the many libraries offered. Figure 15-27 shows an example of the symbols included in AutoCAD's welding library. You can also use AutoCAD 2000i's Point A to find sources for more complex or specialized symbols.

A typical drawing task encountered by a drafter might be to convert the plan for a casting into one for a welded part, or weldment. Refer to Figure 15-25 for a drawing for a casting of a connecting link. To convert this into a welding drawing, you should first consider the practical changes in the construction of

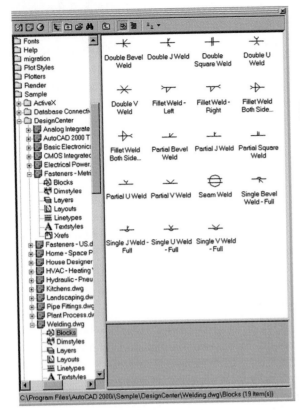

Figure 15-27

Beginning with AutoCAD 2000i, the AutoCAD DesignCenter includes an extensive welding symbol library.

5. Open a welding symbol library. Drag the appropriate symbols from the symbol library into the connecting link drawing. Size and place the symbols as necessary.

6. Create the parts list. If the blocks in the symbol library you are using have embedded attributes, you may be able to use the attributes to create the parts list automatically. See Chapter 11 for more information about using blocks and attributes.

▶CADTIP

Custom Welding Symbols

If you do not have access to a welding symbol library, it may be worth your time to create one. Welding symbols are neither difficult nor time-consuming to create. You can draw them one time, add custom attributes that will make your parts list easier to create, and save the file for use in this and other drawings.

the part. Study the bulleted list of considerations for converting a casting into a weldment that appears in the "Board Drafting Techniques" portion of this chapter. Then follow these steps:

1. Consider the changes required or enabled by converting the connecting link from a casting to a weldment.
2. Determine the appropriate drawing sheet size and the scale at which you will print the finished drawing.
3. Start a new drawing file and create the two-view drawing of the connecting link, incorporating all of the changes you determined in step 1.
4. Add the appropriate dimensions.

Chapter Summary

- Welding is a very old method used to join metal products.

- Industrial processes require a variety of joints and welds because welding is used for so many different purposes.

- Most welding methods can be classified as either fusion or resistance welding.

- The weld symbol indicates the type of weld. The welding symbol is a method of representing the weld on the drawing.

- The complete welding symbol contains as many as eight elements.

- To specify a weld completely, the drafter or engineer must specify both the type of joint preparation and the type of weld to be used, as well as the dimensions appropriate for the specified weld.

Review Questions

1. Which kind of welding was developed first, resistance or fusion? Who was the developer?

2. Name the five basic welding joints.

3. Name nine basic types of groove welds.

4. Describe briefly the difference between gas welding and arc welding.

5. Name the gases that are normally used in gas welding.

6. What is the basic principle of arc welding?

7. What do MIG and TIG stand for?

8. What two types of seams are used in seam welds?

9. What is a spot weld?

10. Explain "arrow side" and "other side."

11. What does a black flag on the welding symbol represent?

12. What does a circle at the intersection of the reference line and the arrow on a welding symbol mean?

13. In general, what dimensions are required to describe a weld completely?

14. List at least two sources for obtaining a welding symbol library.

Drafting Problems

The drafting problems in this chapter are designed to be completed using board drafting techniques or CAD.

1. Make a three-view drawing of the lever stand shown in Figure 15-28. Determine your own dimensions. Provide a support rib for the upright member. Provide dimensions and welding symbols.

2. Make a welding drawing of the double-angle cross bracing shown in Figure 15-29. Select a suitable scale.

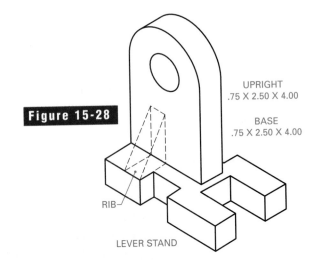

Figure 15-28

UPRIGHT
.75 X 2.50 X 4.00

BASE
.75 X 2.50 X 4.00

RIB

LEVER STAND

Figure 15-29

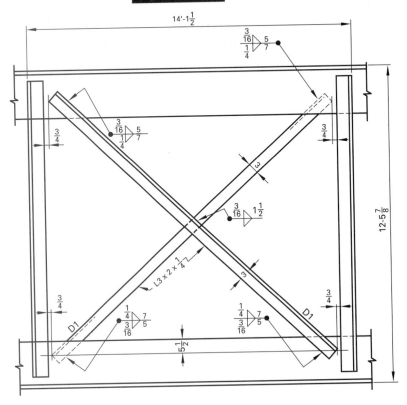

3. Prepare a drawing of each girder section shown in Figure 15-30 and place the symbols for welding in appropriate locations with dimensioning.

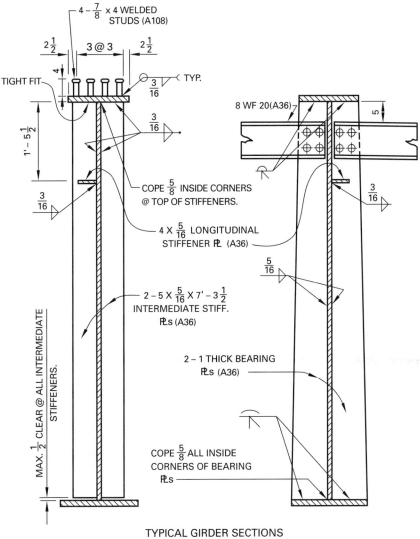

TYPICAL GIRDER SECTIONS

Scale: $\frac{3}{4}$"= 1'-0

Figure 15-30

4. Develop a three-view drawing of the bearing support shown in Figure 15-31. Apply appropriate welding symbols to assemble the five parts that form the bearing support.

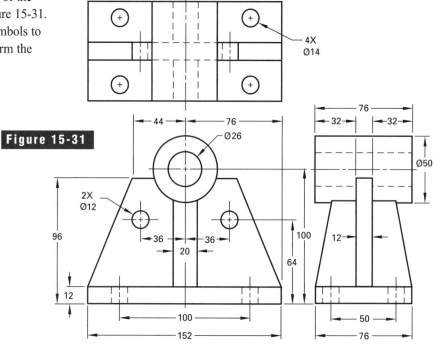

Figure 15-31

5. Prepare three views of the double-bearing swivel support shown in Figure 15-32. Prepare a parts list. Dimension in millimeters.

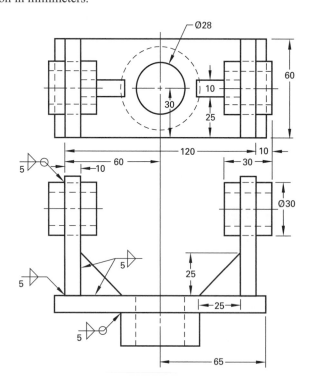

METRIC

Figure 15-32

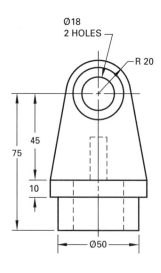

6. Convert the casting drawing for the pin link shown in Figure 15-33 into a weldment drawing. Scale 1:1 on an A-size sheet. Call for .19″ welds and use .25″-thick flat stock to connect the two ends.

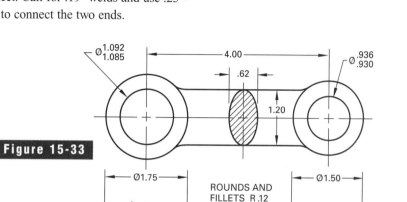

Figure 15-33

ROUNDS AND FILLETS R.12

7. The shaft support shown in Figure 15-34 has been cast in gray iron. This is a rather costly method for producing the part. The engineering division of your company has decided to convert the part from a casting to a weldment as a cost-saving measure. On an A-size sheet, make a welding drawing of the shaft support by converting the casting to a weldment. The thickness of the base plate can be reduced to .25″ thick because steel is tougher and stronger than gray iron. Call for .19″ welds.

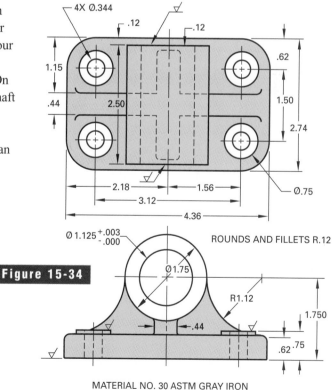

Figure 15-34

MATERIAL NO. 30 ASTM GRAY IRON

Design Problems

Design problems have been prepared to challenge individual students or teams of students. In these problems, you are to apply skills learned mainly in this chapter but also in other chapters throughout the text. They are designed to be completed using board drafting, CAD, or a combination of the two. Be creative and have fun!

1. **TEAMWORK** Design a weightlifting bench that will accommodate at least 500 pounds of free weights. The design should include steel tubing with welded joints. Consider in the design an adjustable inclined bench and an attachment for leg lifts. Each team member should develop sketches of his or her design. Prepare a final set of working drawings with dimensions and welding symbols.

2. Design a stand to support a 1.25″-diameter by 6′-6″ flagpole. It is to be fastened to the floor with No. 14 flathead wood screws. (See Appendix C for screw sizes.) The design should include steel tubing and plate with welded joints. Develop design sketches and working drawings with dimensions and welding symbols.

3. **TEAMWORK** Design a ladder rack for a pickup truck. The design should include square or rectangular steel tubing with welded joints. Each team member should develop sketches of his or her design. Select the best ideas from each to finalize the team design. Prepare a final set of working drawings with dimensions and welding symbols.

4. Design a park bench to be made from at least two different materials. The base should be steel tubing or angle with welded joints. The seat should be wood planks or fiberglass planks. Develop design sketches and working drawings with dimensions and welding symbols.

16

Pipe Drafting

OBJECTIVES

Upon completion of this chapter, you should be able to:

■ Describe the various types of pipe and explain their use in industry.

■ Identify pipe connections and fittings.

■ Explain the differences among gate valves, globe valves, and check valves.

■ Identify various components on a pipe drawing.

■ Identify and use symbols in conjunction with other data to develop a complete pipe drawing.

■ Prepare basic orthographic pipe drawings using board drafting or CAD techniques.

■ Prepare basic pictorial pipe drawings using board drafting or CAD techniques.

■ Prepare a 3D wireframe model of a pipe system using CAD techniques.

Drafting Principles

For centuries, the only thing that was moved from one point to another in pipes and pipe systems was water. Now, a variety of liquid and gaseous materials such as liquid metals, oil, oxygen, nitrogen, steam, acids, and other similar materials are moved through simple to very complex pipe systems. However, pipe materials and fittings have other uses as well. Hydraulic (liquid-based) and pneumatic (air-based) systems are commonly used to apply pressure to controls that activate and move mechanisms on machinery and equipment.

Pipe drafting is a specialized form of drafting that uses lines and symbols to describe the construction of piping components and systems. For example, a plumbing diagram for a residence is an example of a **pipe drawing**. See Figure 16-1. This drawing shows the plumbing contractor precisely where to lay out and install the hot and cold water lines. A more complex plumbing drawing might also show gas lines and sewage lines along with the hot and cold water lines. On a much larger scale, pipe drawings are used to describe the design of oil refineries, complex chemical plants, and liquid materials processing plants. An example of a fuel oil supply system is shown in Figure 16-2.

In general, a design engineer prepares a rough sketch, layout, or specifications for the required pipe system. The drafter, using either board drafting or

Figure 16-1

Pipe (plumbing) drawing for a residence.

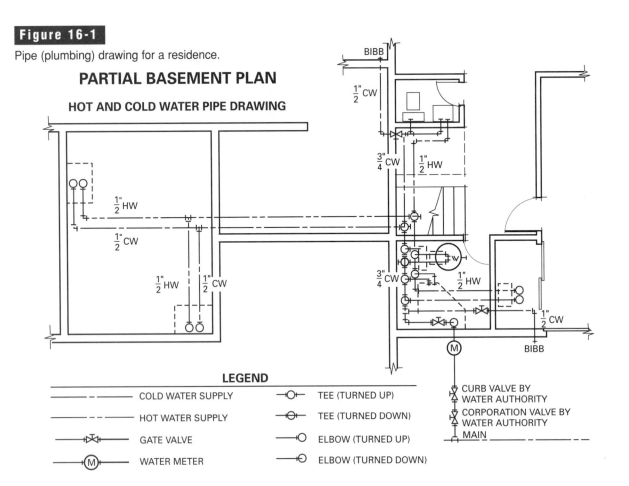

PARTIAL BASEMENT PLAN

HOT AND COLD WATER PIPE DRAWING

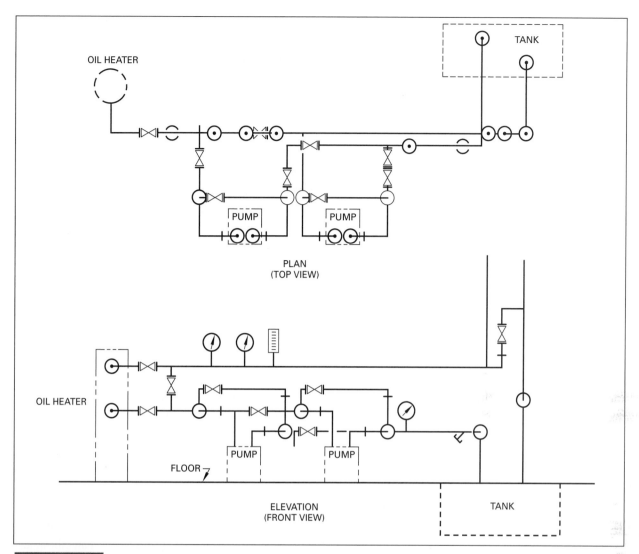

Figure 16-2

Pipe drawings of a fuel oil supply system.

CAD techniques, then prepares the finished pipe drawing for review by engineers and checkers. Pipe drawings range from very simple to extremely complex. For the extremely complex systems, CAD programs can provide virtual models that allow drafters, engineers, and building contractors to "walk through" the finished system visually. This allows them to see pictorially what the finished system will look like while it is still in the design stage. In fact, the more sophisticated CAD systems can make it appear as though you are actually walking through the finished pipe system. Prior to the development of virtual models, design teams often constructed scale models of the proposed system.

Pipe, joints and fittings, and valves are the basic elements of a pipe system. Additional components, including pumps, tanks, furnaces, and boilers, can be added to complete the final integrated system.

Pipe

The pipe used early in history was made of hollow logs fastened end-to-end with tree pitch. This served to transfer water over short distances. Clearly, it could not have been used to transport chemicals and acids such as those transported today through pipe made of stainless steel and plastics.

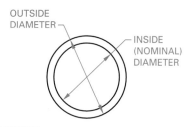

Figure 16-3

The nominal diameter of a pipe is its inside diameter.

In addition to transporting materials, pipe is also used for structural purposes. For example, it is often used as columns to support structural members and for things such as hand railings and the framework for soccer goals.

Pipe is available in various sizes and different types of materials. The size is determined by the required volume and pressure of the substance to be transported. The type of material from which the pipe is made is selected on the basis of the kind of substance to be transported. For example, acids are generally transported through stainless steel or plastic pipe. Water can be transported through steel, wrought-iron, copper, or plastic pipe.

Steel and Wrought-Iron Pipe

Steel or wrought-iron pipes carry material such as water, steam, oil, and gas. They are commonly used when there is a need to accommodate high temperatures and pressures. Standard steel and wrought-iron pipe is specified by the nominal diameter. In pipe sizes, **nominal diameter** refers to the inside diameter (ID). See Figure 16-3.

To accommodate various pressures within the pipe system, several wall thicknesses are available. In fittings for pipe of different wall thicknesses, the outside diameter (OD) of each end of the fitting remains the same. The extra metal is added to the inside to increase the wall thickness. Therefore, the nominal size is simply the general size by which we identify the pipe. For example, the actual inside diameter of 2″ nominal diameter pipe might vary from Ø1.687″ for double extra-strong pipe to Ø2.067″ for standard pipe. In other words, the outside diameter stays the same. The inside diameter changes in order to increase the wall thickness and therefore the strength of the pipe.

Until recently, pipe was available in only three wall thicknesses: standard, extra-strong, and double extra-strong. These thicknesses are designated by schedule, as shown in Figure 16-4. Because of the demand for a greater variety of wall thicknesses in pipe, we now have 10 different wall thicknesses on schedules. Standard pipe is generally referred to as "schedule 40 pipe," and extra-strong pipe is referred to as "schedule 80." Pipe over 12″ is referred to as "OD pipe." This means that the nominal size is the outside diameter of the pipe.

The nominal size of pipe is always given in inches for both the customary inch system and the metric system. However, the actual inside diameter and the actual outside diameter are given in inches for customary inch drawings and in millimeters for metric drawings. Also, threaded fittings are the same for both systems. A 2″ pipe fitting, such as an elbow, has 11.50 threads per inch for both the customary system and the metric system. This allows for a standardized system worldwide. If you look at the pipe charts in Appendixes C-19 through C-23, you will notice that the metric outside diameters and wall thicknesses are given in millimeters. While this appears confusing, it does provide for efficiency in the use of pipe and fittings in international pipe systems.

Figure 16-4

Comparison of wall thicknesses of ½″ nominal-size steel pipe.

A STANDARD SCHEDULE 40

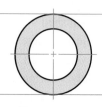

B EXTRA-STRONG SCHEDULE 80

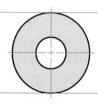

C DOUBLE EXTRA-STRONG SCHEDULE 160

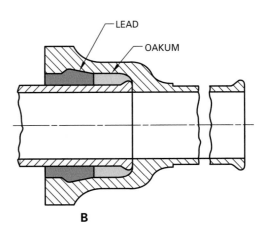

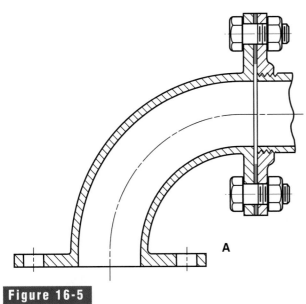

Figure 16-5

Cast-iron pipe joints: (A) flanged; (B) bell and spigot.

Cast-Iron Pipe

Cast-iron pipe has traditionally been used to carry water, natural gas, and sewage. It has been widely used in underground applications because of its resistance to corrosion. Many cities, for example, have underground cast-iron water and waste systems that are more than 100 years old. Due to its much lower cost, plastic pipe is now replacing cast iron for many of these applications.

Cast-iron pipe is specified using the American Water Works Association (AWWA) standards. It is available in a variety of diameters and wall thicknesses. Sections of cast-iron pipe are generally joined using flanged fittings or bell-and-spigot joints, as shown in Figure 16-5.

Seamless Brass and Copper Pipe

Seamless brass and copper pipe are used mainly in plumbing in residential and commercial buildings and in similar applications. They are well suited for these applications because of their ability to withstand corrosion. They have the same nominal diameters as steel or cast-iron pipe, but they have thinner wall sections. Again, due to their much lower purchase and installation costs, plastic pipe and fittings are replacing these materials. However, building codes must be checked before any of these materials are specified.

Copper Tubing

Copper tubing is used in plumbing and heating and where vibration and misalignment are factors, such as in automotive, hydraulic, and pneumatic applications. Copper tubing is relatively expensive, but it is highly corrosion-resistant and transfers heat readily. It is much more flexible than iron pipe, and it can easily be bent when necessary to align with fittings or other tubing. Soft copper tubing is often specified for applications where it is necessary to bend the tubing around corners, fixtures, and appliances. Compression and flare fittings make tubing relatively easy to use.

Plastic Pipe

Plastic pipe is used in residential and small commercial construction for water and sewage. It is made from polyvinyl chloride (PVC). Plastic pipe is inexpensive to purchase and relatively easy to install. Because of its corrosion and chemical resistance, it is used extensively in the chemical industry. It is not generally recommended where high temperatures or high pressures are indicated. However, steel pipe lined with various types of plastic is now in use for high temperature and pressure applications.

Other Types of Pipe

Stainless steel tubing, glass, and vitrified clay are other materials from which pipe is made. Each has very specific uses and is generally specified by the design engineer.

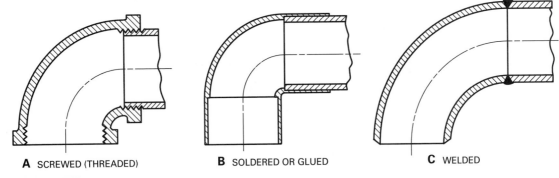

A SCREWED (THREADED)   **B** SOLDERED OR GLUED   **C** WELDED

Figure 16-6

Common types of pipe joints.

Pipe Connections and Fittings

With few exceptions, parts used to join sections of pipe are called **pipe fittings**. However, in some cases, sections of pipe are joined without fittings. For example, when welding joins two sections of steel pipe, this is simply a pipe connection; it does not use a fitting. Sections of plastic pipe are joined using an adhesive or are heat-fused together. Fittings such as elbows and tees are used to make other types of connections and to turn corners.

Pipe connections generally fall into four classes: screw, weld, glue, and flange. Figure 16-5 shows an example of a flange fitting. Figure 16-6 shows a screwed fitting, a soldered or glued fitting, and a welded fitting. The exceptions to these four classes are soldered copper joints and the bell-and-spigot joint used mainly for cast-iron pipe.

Pipe fittings are specified by the nominal pipe size, the name of the fitting, and the material. Some fittings, such as tees, crosses, and elbows, are used to connect different sizes of pipe. These are called **reduced fittings**, and their nominal pipe sizes must be specified. The largest opening of the through run is given first, followed by the opposite end and the outlet. Figure 16-7 illustrates the order in which sizes of reduced fittings are designated.

Screw Fittings

Screw fittings, also called *threaded fittings*, are shown in Figure 16-8. They are generally used on small pipe of Ø2.50″ or less, but they can be used on pipe as large as Ø6.00″. A paste-type joint compound or Teflon® tape is used to lubricate the threads and to close, or seal, any irregularities in the mating threads.

According to ANSI, there are two types of pipe threads: tapered and straight. The most common of these is the tapered thread. It has a 1:16 taper on both the internal thread in the fitting and the external thread on the pipe. This same taper is also used on the external threads of certain fittings such as street elbows and service tees. Refer again to Figure 16-8.

Taper on the threads determines the distance to which the thread of the pipe enters the thread of the fitting. It also ensures a perfectly tight fit between the two parts. Straight pipe threads are used only for very special applications. Information about these may be found in ANSI standards.

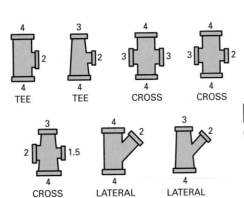

Figure 16-7

Order of specifying the openings of reduced fittings.

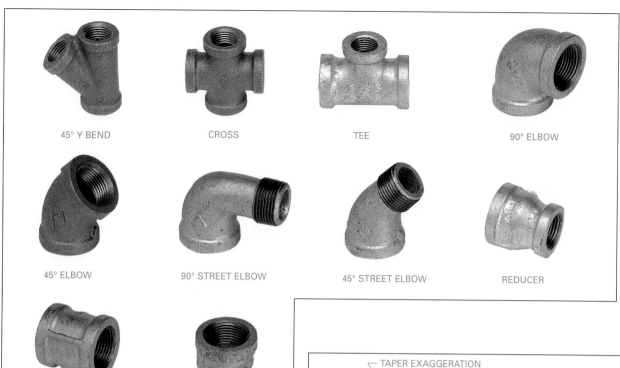

45° Y BEND   CROSS   TEE   90° ELBOW

45° ELBOW   90° STREET ELBOW   45° STREET ELBOW   REDUCER

COUPLING   CAP

Figure 16-8

Screw-type pipe fittings.

Figure 16-9 shows how threads are represented on a drawing and how they are specified. Tapered threads are designated on drawings as NPT (National Pipe, Tapers), which means American Standard Pipe Taper Thread. Notice in Figure 16-9 that pipe threads may be drawn with the taper shown or simply straight. In the thread note, NPT tells you that the thread is tapered. When threads are drawn in tapered form, the taper is exaggerated and the amount of taper is estimated. Straight pipe threads are specified on the drawings as NPS, which means American Standard Pipe Straight Thread. All pipe threads are assumed to be tapered unless otherwise specified. Figure 16-9 shows both schematic and simplified representation. Simplified representation is most commonly used. Refer to Chapter 11, "Fasteners," for specific details of drawing thread symbols.

Figure 16-9

Pipe thread representation.

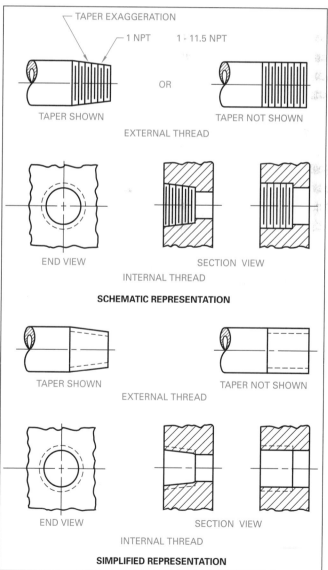

TAPER EXAGGERATION
1 NPT   1 - 11.5 NPT

TAPER SHOWN   OR   TAPER NOT SHOWN
EXTERNAL THREAD

END VIEW   SECTION VIEW
INTERNAL THREAD

SCHEMATIC REPRESENTATION

TAPER SHOWN   TAPER NOT SHOWN
EXTERNAL THREAD

END VIEW   SECTION VIEW
INTERNAL THREAD

SIMPLIFIED REPRESENTATION

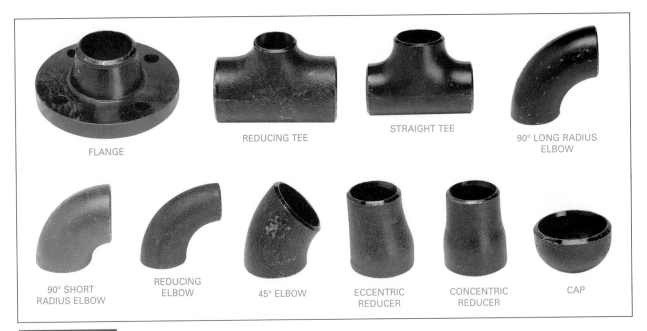

Figure 16-10

Welded pipe fittings.

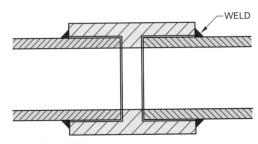

Weld Connections

Weld connections are used mainly on permanent, high-pressure, and high-temperature pipe systems. There are two types of weld connections: butt and socket. Figure 16-10 shows various types of fittings that are butt-welded. Notice that they are beveled on the edges to receive the weld. Butt welds are generally used on pipe Ø2.00″ or larger. Figure 16-11 shows a socket weld joint. This type of weld joint is generally used on pipe Ø2.00″ or smaller.

Figure 16-11

Socket-type weld connection.

Flange Connections

Flange connections, shown in Figure 16-12, provide a quick means of disassembling pipe. Flanges are attached to the pipe ends by welding or threading, as shown in Figure 16-13. The flange faces generally have a gasket between them and are fastened together with bolts.

Figure 16-12

Flanged pipe fittings.

Valves

Valves are used to stop, start, or regulate the flow of fluids and gases in a pipe system. While the design engineer generally specifies the valves to be used, the drafter needs to be aware of the more common types and where they are used. A few of the more common types are described here.

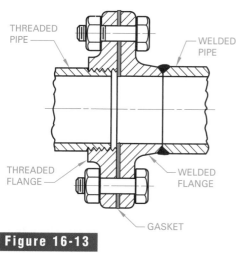

THREADED PIPE

WELDED PIPE

THREADED FLANGE

WELDED FLANGE

GASKET

Figure 16-13

Welded and threaded flanges.

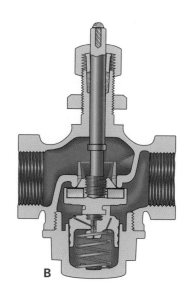

A

B

Figure 16-15

Globe valves.

Gate Valves

Gate valves are used to control the flow of liquids. The wedge, or gate, lifts to allow full, unobstructed flow and lowers to stop it completely. See Figure 16-14. These valves are normally used where the operation of the valve is infrequent. They are not intended to control slight increases or decreases in flow. They are generally used for on/off operation.

Globe Valves

Globe valves, shown in Figure 16-15, are used to control the flow of liquids or gases. They are used primarily for the close, accurate regulation of pressure and volume. If you look carefully at the two types of globe valves, you will notice that there is a slight difference. The

one shown in Figure 16-15A is installed so that pressure is on the disk, which assists the spring in the cap to make a tight closure. It is recommended for the control of air, steam, and other compressibles where instant on/off operation is required. The other type, shown in Figure 16-15B, is recommended for the control of liquids where sudden closure might be objectionable and perhaps dangerous. The cap is fitted with a spring-loaded piston arrangement that slows closure time and helps eliminate shock.

Check Valves

As the name implies, **check valves** permit flow in one direction only. See Figure 16-16. They check, or stop, reverse flow. They are controlled by the pressure and velocity of the material flow alone. They provide no external means of controlling the flow or velocity.

Figure 16-14

Gate valve.

Figure 16-16

Check valve.

Pipe Drawings

The purpose of pipe drawings is to show the size and location of pipes, fittings, valves, and other related components. As in many other aspects of drafting, a set of uniform symbols has been developed to represent individual features on a drawing.

There are two types of pipe drawings: single-line and double-line drawings. See Figure 16-17. **Double-line drawings** take much more time to draw and are therefore much more expensive to produce. As a result, they are generally not used for production or construction work. They are best suited for illustrations in catalogs and other applications in which the visual appearance is important.

Single-Line Drawings

Single-line drawings of pipe systems, also called *simplified drawings*, provide excellent clarity at a greatly reduced cost. As a result, single-line drawings are the method of choice throughout most of industry.

Single-line pipe drawings use a single line to show the arrangement of the pipe, fittings, and related components. The single lines are simply the centerlines of the pipe, regardless of size, drawn as thick black lines to which the symbols for fittings, valves, and other components have been added. The size of the symbols is not critical and is determined by the drafter or the company for which the drafter works.

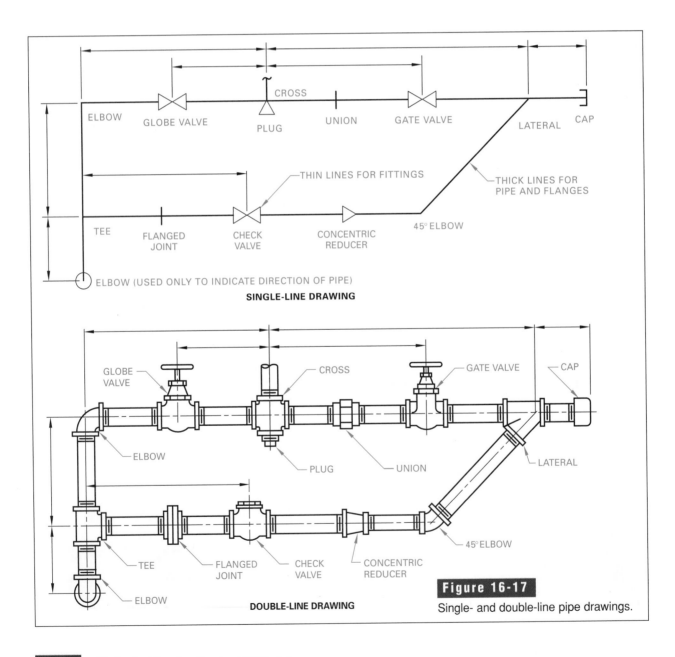

Figure 16-17
Single- and double-line pipe drawings.

Figure 16-18

Orthographic and pictorial pipe drawings.

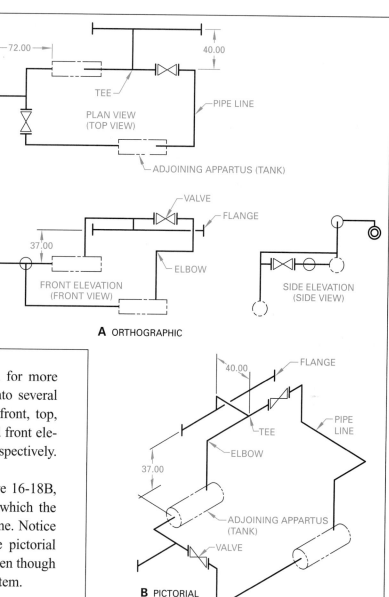

A ORTHOGRAPHIC

B PICTORIAL

When pipe systems carry different liquids, such as hot and cold water, a coded line symbol is often used, as shown in Figure 16-1.

Drawing Projection

Two methods of projection are used in preparing pipe drawings: orthographic and pictorial. Orthographic projection is most suitable for showing single pipes, either straight or bent, in one plane only. See Figure 16-18A. However, this method is also used for more complicated systems, even if they fall into several planes. Notice that in pipe drafting, the front, top, and side views are more commonly called front elevations, plan views, and side elevations, respectively. However, either method is correct.

Pictorial projection, as shown in Figure 16-18B, is recommended for all pipe systems in which the pipes are positioned in more than one plane. Notice how much easier it is to understand the pictorial drawing than the orthographic drawing, even though both drawings represent the same pipe system.

Crossings

When lines representing pipes cross on a drawing, they normally are shown without interruption, as shown in Figure 16-19A. When it is necessary to show that one pipe must pass behind the other, the line representing the pipe farthest from the viewer is shown with a break, or interruption, where the other pipe passes in front of it, as shown in Figure 16-19B.

Figure 16-19

Methods used to show the crossing of pipes: (A) pipe crossings shown without interrupting the pipe passing behind the nearest pipe; (B) using an interrupted line to indicate the pipe farthest away.

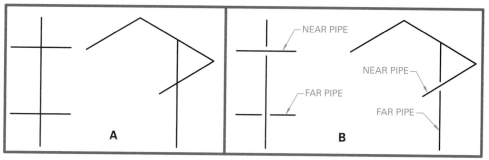

Connections

Permanent connections, whether made by welding or other processes such as gluing and soldering, are shown on the drawing by a heavy dot, as shown in Figure 16-20. A general note or specification may be used to describe the process. A single, thick line instead of a heavy dot may be used to show detachable connections or junctions, as shown in Figures 16-20 and 16-21. The specifications, a general note, or the item list includes the types of fittings, such as flanges, unions, or couplings, and indicates whether the fittings are flanged or threaded.

Figure 16-20

Permanent and detachable pipe connections.

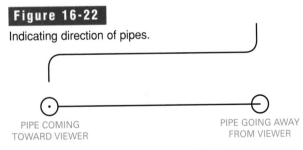

Figure 16-21

Detachable connection symbol used at an apparatus.

Figure 16-22

Indicating direction of pipes.

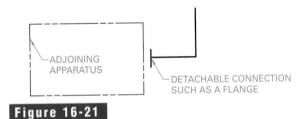

A PIPELINE WITHOUT FLANGES CONNECTED TO ENDS OF PIPE

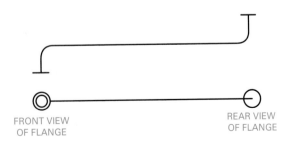

B FLANGES CONNECTED TO ENDS OF PIPELINE

Fittings

Unless there is a special reason to do so, standard fittings such as elbows and tees are not shown on a pipe drawing. They are simply shown, like pipe, as a continuous line. It is the responsibility of the engineers and designers to determine where and what type of fittings to use in building the system. However, the circular symbol for a tee or elbow may be used when it is necessary to show whether the pipe is going away from or coming toward the viewer, as shown in Figure 16-22. Elbows on isometric pipe drawings may be shown without the radius, using square corners. However, the change of direction that the pipe takes should be quite clear if this method is used.

Adjoining Apparatus

If needed, an adjoining apparatus that does not belong to the pipe itself, such as a tank, may be shown by an outline drawn with a thin phantom line, as shown in Figure 16-21. For example, if a boiler is already in place and the drafter simply needs to show the pipe system connected to it, phantom lines are used to represent the boiler.

Dimensioning

Follow these guidelines for dimensioning pipe drawings:

- Dimensions for pipe and pipe fittings are always given from center to center of pipe and to the outer face of the pipe end or flange. See Figure 16-23A.
- Pipe lengths are not normally shown on the drawings. They are left to the installer.
- Pipe and fitting sizes and general notes are placed on the drawing beside the part concerned or, where space is restricted, indicated with a leader.
- An item list is usually provided with the drawing.
- Pipes with bends are dimensioned from vertex to vertex, as shown in Figure 16-23A.
- Radii and angles of bends may be dimensioned as shown in Figure 16-23B. Whenever possible, the smaller of the supplementary angles should be specified.

Figure 16-23

Dimensioning pipes.

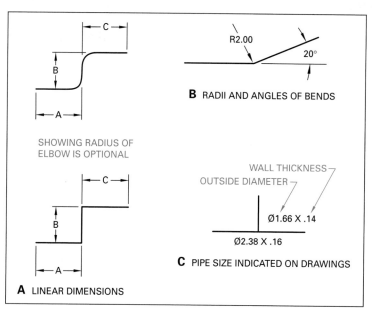

A LINEAR DIMENSIONS

SHOWING RADIUS OF ELBOW IS OPTIONAL

B RADII AND ANGLES OF BENDS

R2.00

20°

WALL THICKNESS
OUTSIDE DIAMETER
Ø1.66 X .14
Ø2.38 X .16

C PIPE SIZE INDICATED ON DRAWINGS

■ The outer diameter of the wall thickness of the pipe may be specified on the line representing the pipe or elsewhere (item list, general note, specification, etc.). See Figure 16-23C.

Orthographic Pipe Symbols

In orthographic projection, if flanges are not required on the ends of pipe, the symbols indicating the direction of the pipe are required. If the pipe is coming toward the viewer, it is shown by two concentric circles, the smaller one being solid, as shown in Figure 16-22A. If the pipe is going away from the viewer, it is shown by one circle. No extra lines are required on any other view.

As shown in Figure 16-22B, flanges are represented by two concentric circles in the front view, by one circle in the rear view, and by a short stroke in the side view. This is the case regardless of their type or size.

Symbols representing valves are drawn with continuous thin lines. The valve spindles should be shown only if it is necessary to indicate their positions. It is assumed that unless otherwise specified, the valve spindle should be in the position shown in Figure 16-24.

TECH MATH

Pipe Length for Bends

It is often necessary to make large-radius bends on pipe or round tubing rather than weld square corners or use fittings. To do so, you may need to know how much pipe is needed to make the bend. An example of this is in the design and construction of the roll cage for a racecar. Follow these steps to find out how much pipe is needed for the bend shown in the illustration.

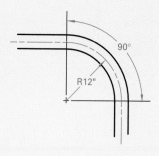

90°

R12"

1. Determine the radius of the bend. In this example, the radius is 12″.
2. Determine the angle of the bend. In this example, the angle is 90°.
3. Calculate the circumference of the entire circle.
   $C = \pi D$
   $C = 3.1416 \times 24″$
   $C = 75.40″$
4. Calculate the percentage of the full circle used for the bend.
   $$\frac{90°}{360°} = \frac{1}{4} = 25\%$$
5. Multiply the percentage by the circumference to determine the length of the bend.
   $75.40 \times 25\% = 18.85″$ (length of bend on centerline)

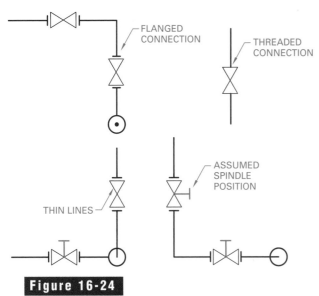

Figure 16-24

Valve symbols. When valve spindles are not shown, they are assumed to be in the positions indicated in this illustration.

Isometric Pipe Drawings

Isometric pipe drawings are pictorial representations of the complete pipe system. They include pictorial drawings of the pipes, fittings, valves, components such as tanks and boilers, and dimensions. In board drafting, orthographic views are generally drawn first and the pictorial drawing follows.

In CAD, the entire pipe system is created in 3D and the isometric drawing is created automatically.

Figure 16-25 is an isometric drawing of a complete pipe system with dimensions. The rules and procedures that apply to the development of this type of drawing are the same as those for producing an isometric drawing of any object. Refer to Chapter 12 for more information about isometric drawings.

Success on the Job

Adapting to Change

Throughout your career, you will be confronted with countless opportunities to make changes. In some cases, the changes will be simple and require little effort. In other cases, they will be more time-consuming and require a great deal of effort. Whatever the case, always show a willingness to adapt to change by readily and willingly accepting new policies and procedures. Continue to learn, and continue to improve upon the manner in which you conduct yourself and the manner in which you do your work. Adapting to change is one of the best ways to guarantee your continuing success on the job.

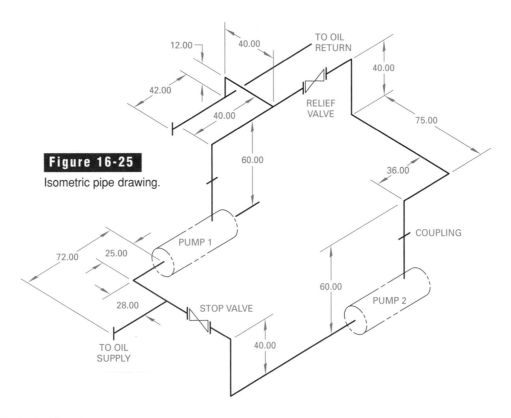

Figure 16-25

Isometric pipe drawing.

Board Drafting Techniques

In many instances, pipe drawings are not made to scale and not completely dimensioned. In this case, the drawing is simply a schematic diagram used to tell the installer, pipe fitter, or manufacturer the relative positions of components. Lengths of individual sections of pipe and specific locations of fittings, valves, and other items are left to the judgment of the installer as long as they fall with the sequence or order described on the print.

In other instances, pipe drawings are made to scale and are fully and accurately dimensioned. Refer again to Figure 16-25. The dimensions are located as if they were locating centerlines on the various components (pipes, fittings, etc.) throughout the system.

Orthographic Pipe Drawing

The following procedure creates a single-line orthographic pipe drawing of the pipe system shown in Figure 16-25. It is drawn to scale, but only critical dimensions are added.

1. Determine the number of views required to describe the pipe system fully. On very simple systems, a single view may be sufficient. On more complex systems, two or more views may be required. For instructional purposes, three views will be used in this case.

2. Begin by laying out the lines representing the centerlines of the sections of pipe for the front elevation (front view) using light construction lines as shown in Figure 16-26.

3. Project from the front elevation to develop the layout for the plan view and right-side elevation. See Figure 16-27.

4. Add fittings, valves, and other components as appropriate. On single-line drawings, the symbols for fittings are generally omitted unless needed for clarity. Darken lines and add dimensions and notes as needed.

5. Add reference letters or numbers and prepare a materials list for all components. If the pipe is the same size and type throughout, a general note can be used for its specification, and it need not be added to the materials list. The finished orthographic drawing is shown in Figure 16-28.

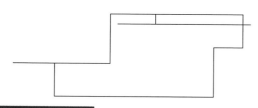

Figure 16-26

Layout of centerlines for pipe sections of the front elevation (front view).

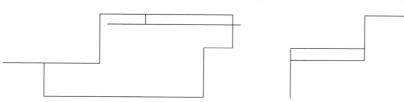

Figure 16-27

Layout of centerlines for all three views.

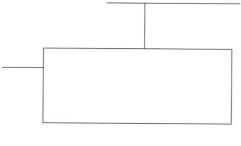

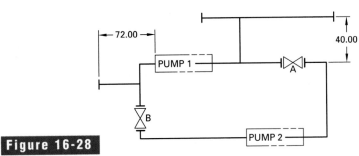

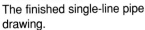

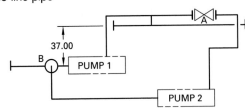

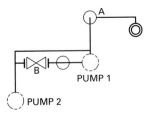

CODE	VALVE	SERVICE
A	RELIEF	PUMPS 1 AND 2
B	STOP	PUMP 2

NOTE: ALL PIPE Ø1.00, SCHEDULE 40

Figure 16-28

The finished single-line pipe drawing.

Isometric Pipe Drawing

Now reverse the process from the previous section to produce an isometric pipe drawing of the same system. Refer to Figure 16-28 for basic information on the design of the system.

1. Begin by laying out the lines representing the centerlines of the sections of pipe in isometric format, as shown in Figure 16-29.

2. Add fittings, valves, and other components as appropriate, as shown in Figure 16-30. The symbols for fittings are omitted unless needed for clarity. Darken lines as shown.

3. Add dimensions, notes, and other symbols as necessary. The finished isometric should look like the one in Figure 16-25.

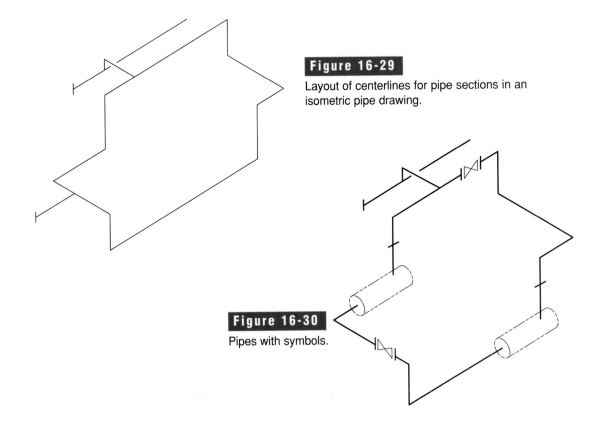

Figure 16-29

Layout of centerlines for pipe sections in an isometric pipe drawing.

Figure 16-30

Pipes with symbols.

CAD Techniques

CAD systems can be used in many ways to create pipe drawings. For simple pipe systems that fall within a single plane, the drafter may create an orthographic drawing. For fairly simple systems that incorporate more than one plane, the drafter may create a 2D isometric drawing, as described in Chapter 12.

For complex pipe systems, CAD operators often create 3D models. The advantage of using a 3D model is that it can be viewed from any point in space. The engineer or designer can "walk through" the system to discover potential conflicts and other problems before the design is finalized.

Orthographic CAD Drawing

The procedure for creating an orthographic pipe drawing using a CAD system is similar to that for creating any other orthographic drawing, and it parallels the board drafting method. For practice, refer to the orthographic drawing developed in the "Board Drafting Techniques" segment of this chapter. Instead of following the steps exactly, use your CAD skills to create the three views. Remember to use the XLINE and OFFSET commands to develop the drawings most efficiently. Also, use the appropriate object snaps to ensure accuracy.

2D Isometric CAD Drawing

The procedure for creating a 2D isometric drawing of a pipe system follows the principles and techniques discussed in Chapter 12, "Pictorial Drawing." Again, you should practice your drawing techniques by completing an isometric drawing of the pipe system developed in the "Board Drafting Techniques" segment. Set the snap style to Isometric before you begin. You will probably want to set the snap interval and grid at 5' or 10'.

When you dimension the drawing, you will notice that many of your dimensions are not neatly lined up with the isometric lines. Figure 16-31A shows an example of one of the dimensions from the pipe system as it appears when you first create it. AutoCAD provides a way to rotate the extension lines into a more natural position, as shown in Figure 16-31B. To do this, pick the Dimension Edit button on the Dimension toolbar, or enter DIMEDIT at the keyboard, and select the dimension to be changed. At the next prompt, enter O to activate the Oblique option. The obliquing angle for the dimension shown here is 30. For other dimensions, you may need to enter different obliquing angles. If the angle doesn't appear as it should, try angles of 30, –30, 15, and –15.

CADTIP

Pipe Symbol Libraries

To expedite both the orthographic drawing and the 2D isometric drawing, you may choose to find and use a library of pipe symbols. These symbols are available within some CAD programs, including AutoCAD 2000i. You can also obtain them free or at low cost from third-party providers or sometimes even from other CAD operators who have developed libraries and have posted them on the Web for general use.

Figure 16-31

Use the Oblique option of the DIMEDIT command to correct the angle of the aligned dimension (A) to place it on the isometric grid (B). Enter –30° when asked for the angle.

75.00

A

75.00

B

3D Model

CAD programs such as AutoCAD are capable of producing two basic types of 3D models. The solid models that were used in some of the previous chapters of this book can have properties: mass, center of gravity, and so on. These properties are generally not needed for pipe drawings.

The other type of 3D model is a simpler construction known as a **wireframe**. Imagine a model that has been built from toothpicks. It has no walls, and therefore it has no interior. These are the characteristics of a wireframe drawing. See Figure 16-32. The only major difference between solid models and wireframes is that wireframes have no intrinsic properties. Single-line isometric pipe drawings are therefore often completed as wireframes. Because they involve a single line, 3D coordinates can be used to create these wireframes quickly and easily.

Drawing the Basic Wireframe

Figure 16-33 shows a pipe system that has been created as a wireframe. Follow these steps to create the wireframe drawing:

1. Set up a new file with drawing limits of 0,0 and 400,300. Set the grid interval to 10 and enter ZOOM All.
2. We will begin the model by drawing the pipe segments on the front plane. To do that, we need to change from the plan (top) view to the elevation (front) view of the drawing. From the View menu, select 3D Viewport and then select Front.

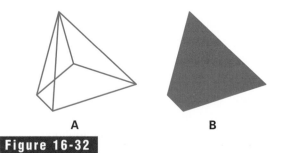

A   **B**

Figure 16-32

Wireframe drawings are like physical models made of toothpicks (A). Like solid models, they can be shaded (B).

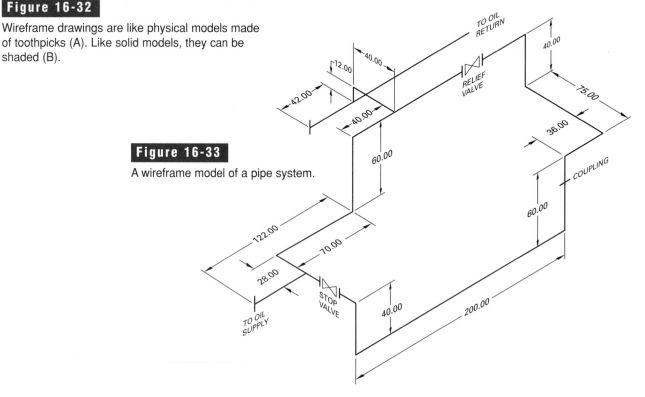

Figure 16-33

A wireframe model of a pipe system.

START POINT
(50,50,0)

@40<-270

@200<0

@60<-90

@36<0

Figure 16-34

Draw the first four segments in the front plane using polar coordinates.

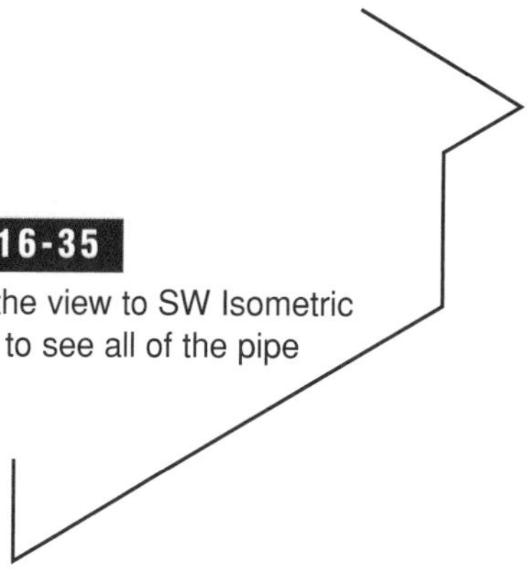

Figure 16-35

Changing the view to SW Isometric allows you to see all of the pipe segments.

3. To draw on the front, you must also change the UCS. Enter the UCS command, enter V to activate the View option, and press Enter. Then reenter the UCS command, enter S to activate the Save option, and name the UCS Front.

4. Now you are ready to begin the drawing. Begin at the top left front corner of the pipe system, as shown in Figure 16-34. Enter the LINE command and enter the coordinates 50,50,0 as the starting point. Because the first four line segments remain the front plane, you can use 2D polar coordinates to define them. Temporarily end the LINE command.

Now you need to draw a line straight back from the front plane of the drawing. For this, you can use XYZ coordinates. However, first you must figure out the exact location of the points you need to specify in 3D space. There are a couple of ways to do this. You could have mapped out all of the points before you began the model. However, it is often easier and quicker to calculate the points as you go. That is the method we will use here.

5. Reenter the LINE command and use the Endpoint object snap to snap to the end of the last segment you drew in step 4. Leave the cursor over this point and read the coordinate values at the bottom left corner of the screen. The coordinate value should be 286,70,0. The next point you need to enter is at exactly the same X and Y coordinates, but at a different depth, or Z coordinate. The Z coordinate equals the distance from the current point to the next point on the pipe system. Therefore, at the To point: prompt, enter 286,70,–75. (The Z value is negative because the line is moving away from you.)

6. Note that you will not be able to see the result of this operation while you are in the default view. To see the 3D model grow as you pro-

ceed, and for easier point selection, choose 3D Viewport and SW Isometric from the View menu. Enter the ZOOM command and then 1 to see the entire drawing. See Figure 16-35.

7. The next point is located 40′ straight up. Therefore, the only value that changes is the Y value. Since the current XYZ coordinate is 286,70,–75, add 40 to the Y value to get the next point: 70 + 40 = 110. Enter the LINE command, snap to the end of the pipe segment you drew in step 4, and enter the next coordinate value: 286,110,–75. Leave the LINE command active from this point on to ensure that the lines connect at their endpoints.

8. A tee occurs along the next segment of pipe. The easiest way to create this part of the drawing is to draw the segment from the current point to the tee, create the side pipes, and then return to the main loop of the pipe system. From the drawing in Figure 16-33, you can determine that the tee is located 126′ to the left of the current point. Therefore, to find the coordinate value, subtract 126 from the current X value: 286 – 126 = 160. The new coordinate is 160,110,–75.

▶ **CADTIP**

Positive and Negative Coordinates

When you manipulate the values of X, Y, and Z coordinates in a wireframe drawing, remember that values are positive when the line moves up, to the right, or toward you. The values are negative when the line moves down, to the left, or away from you.

9. Continue by changing the depth dimension (Z value) to create the short length of pipe that goes back (away from you) 40′. Leave the X and Y values unchanged. The value is 160,110,–115. Draw the short lengths of pipe that extend down 12′ and to the left 42′, and end the LINE command. Add the short line at the end that denotes a detachable connection. Then reenter the LINE command and create the line that goes to the oil return. The length of this line does not matter. The fastest way to create this line is to turn Ortho on to ensure that the line follows the correct angle, and pick a point to end the line at about the point shown in Figure 16-33.

10. Use the BREAK command to trim the oil return line as shown in Figure 16-33 to show that it is behind the other lines. *Note:* Do not use the TRIM command; in 3D views, lines that seem to cross do not actually cross, so the result of the TRIM command may be unpredictable.

11. Return to the tee and continue creating the lines to finish the main body of the pipe system. When you finish, your drawing should look like the one in Figure 16-36.

Adding the Symbols

Now you must add the appropriate symbols to complete the pipe system. Only two symbols are necessary in this drawing: a relief valve and a stop valve. Note that their exact placement along the pipeline is not necessary. However, they must be in line with the appropriate pipes. To do this, you will create two new UCSs, one for each valve.

1. Enter the UCS command, enter N for New, and enter 3 for 3Points.
2. For the new origin, pick the point shown in Figure 16-37. For a point on the positive X axis, turn Ortho on and move the cursor to the right until a line starts to form. Pick a point as shown

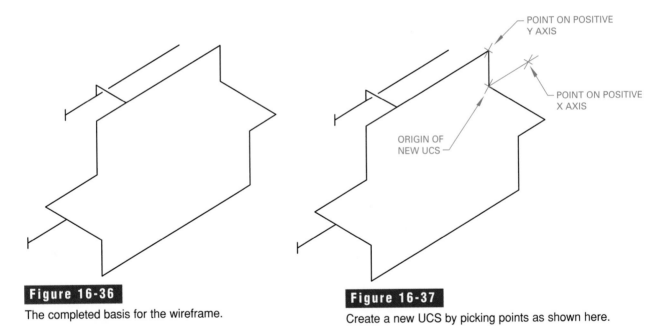

Figure 16-36

The completed basis for the wireframe.

Figure 16-37

Create a new UCS by picking points as shown here.

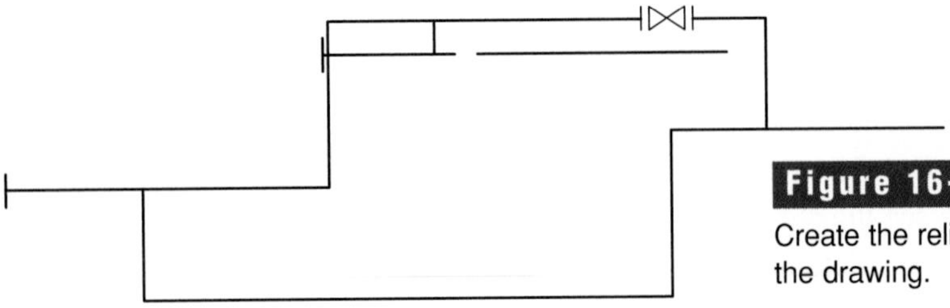

POINT ON POSITIVE Y AXIS

POINT ON POSITIVE X AXIS

ORIGIN OF NEW UCS

Figure 16-38

Create the relief valve at an appropriate size for the drawing.

in Figure 16-37. For a point on the positive Y axis, pick the endpoint as shown in the figure. Save the UCS as Valve1.

3. To draw on the new UCS efficiently, go to its plan view. From the View menu, select 3D Viewpoint, Plan View, and Current UCS. The surface on which you need to draw the relief valve is now parallel to the screen.

4. Draw the relief valve, estimating sizes to make it look appropriate in the drawing, as shown in Figure 16-38.

5. Return temporarily to the SW Isometric view.

6. Create another new UCS using the 3Point option. This time, choose the origin and positive X and Y values shown in Figure 16-39. Save this UCS as Valve2.

7. Go to the plan view of the current UCS and draw the valve, as shown in Figure 16-40.

8. Return to the SW Isometric view. The drawing should now look like the one in Figure 16-41.

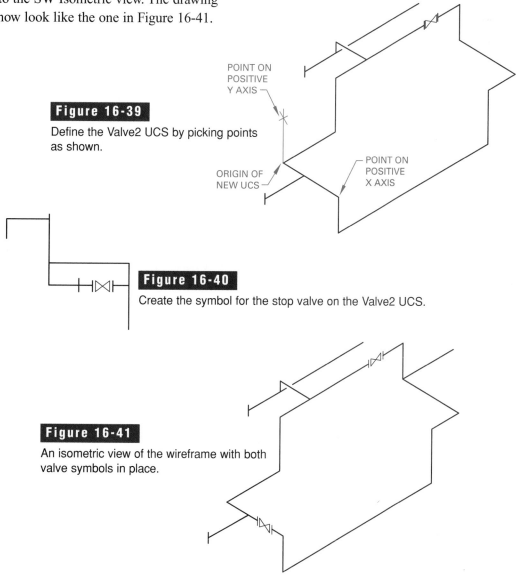

POINT ON POSITIVE Y AXIS

ORIGIN OF NEW UCS

POINT ON POSITIVE X AXIS

Figure 16-39
Define the Valve2 UCS by picking points as shown.

Figure 16-40
Create the symbol for the stop valve on the Valve2 UCS.

Figure 16-41
An isometric view of the wireframe with both valve symbols in place.

Dimensioning the Drawing

Now it is time to dimension the drawing. Look carefully at the isometric drawing in Figure 16-33. Notice that the extension lines run parallel to the isometric axes. This is because the dimensions were created on the appropriate UCSs. Follow these steps to dimension the drawing:

1. Set up the dimension style for this large drawing. Set the text height to 4, for example, and the arrow size to 3. You may need to experiment to find the best settings.

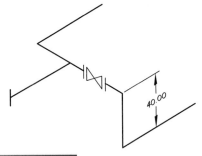

DIMSCALE
To set the size of dimensions, including arrowheads, dimension text, and other features, you can use the DIMSCALE command. The default setting is 1.000 (100%). To make the dimension features half their original size, set DIMSCALE to .5. To make them larger, use a number higher than 1, such as 1.2.

2. Enter the UCS command and R for Restore. Type Front for the name of the UCS to restore.
3. Display the Dimension toolbar. Pick the Linear Dimension button, right-click to select an object to dimension, and pick the lower left leg of the pipe system, as shown in Figure 16-42. Create the dimension as you would for any other drawing. Notice that it lines up automatically with the isometric view. Repeat this procedure for the other two dimensions that lie in the front plane.
4. Restore the Valve1 UCS and complete the dimensions that lie in that plane.
5. On your own, create the appropriate new UCSs for the rest of the dimensions. *Note:* You do not have to save every UCS with a name. However, if you plan to return to a UCS more than once,

Figure 16-42

Place the 40.00 dimension on the front UCS.

you may choose to name it. To work most efficiently, create all of the dimensions that lie on one UCS before moving to another UCS. The finished drawing should look like the one in Figure 16-33.

Note that dimensions are often left off of wireframe models of pipe systems. Because the drawing is made at full size, the engineer or drafter can query the CAD software to find the specific length of a given pipe segment. However, if dimensions are not included, the model must be drawn to scale.

Success on the Job

Clear Documentation
When you create multiple UCSs in a drawing, name them using logical names that others will immediately understand. When many UCSs are needed, you may also want to document them by creating a list either in the file or in a text document associated with the file. This written communication makes it easier for others to access and use your drawing files, which helps ensure your continuing success on the job.

CHAPTER 16 Review

Chapter Summary

- Pipe drafters use lines and symbols to describe the construction of piping components and systems.

- Pipe is used primarily to transport liquids and gases, but it can also be used as a structural member.

- Pipe is available in various wall thicknesses to accommodate different pressures within a pipe system.

- Pipe fittings are used to join sections of pipe in a pipe system.

- Valves are used to control the flow of liquids or gases in a pipe system.

- Pipe drawings can be single-line drawings or double-line drawings.

- Pipe drawings are usually made as orthographics or pictorials.

- When a CAD system is used, a wireframe model is often built.

Review Questions

1. For most pipe, to what does *nominal diameter* refer?
2. What is OD pipe?
3. Name the four kinds of pipe connections.
4. What is a reduced fitting?
5. In what types of pipe systems are weld connections used?
6. What kind of valve would you use in a pipe system in which pressure and volume must be carefully regulated?
7. What is the purpose of a check valve?
8. What kind of pipe drawing is best suited for illustrations in catalogs?
9. Under what circumstances is an orthographic projection usually drawn?
10. When is pictorial projection used for a pipe system?
11. When it is necessary for one pipe to pass behind another, how is this shown on the pipe drawing?
12. How are welded and threaded connections shown on a pipe drawing?
13. How are pipes with bends dimensioned on a pipe drawing?
14. Why is it not always necessary to create a pipe drawing to scale?
15. Briefly explain how to adjust a dimension on a 2D isometric pipe drawing so that it falls on the isometric grid.
16. What is the difference between a wireframe and a solid model?

Drafting Problems

The drafting problems in this chapter are designed to be completed using board drafting techniques or CAD.

1. Refer to Appendix tables C-19 through C-23. Position an A4-size drawing sheet horizontally and make a double-line drawing of a 90° malleable-iron threaded elbow and tee for Ø1.00″ steel pipe. Use metric sizes. Join the two fittings with a piece of Ø1.00″ × 4.00″ steel pipe. Estimate any sizes not given. Scale 1:1.

2. Convert the double-line pipe drawing shown in Figure 16-43 to a single-line drawing. Refer to Appendix B in your textbook for symbols. Pipe and fittings are 2″ except the smaller pipe between the 45° reducing elbow and the reducer, which is 1″. Mark the reducer and the 45° reducing elbow accordingly.

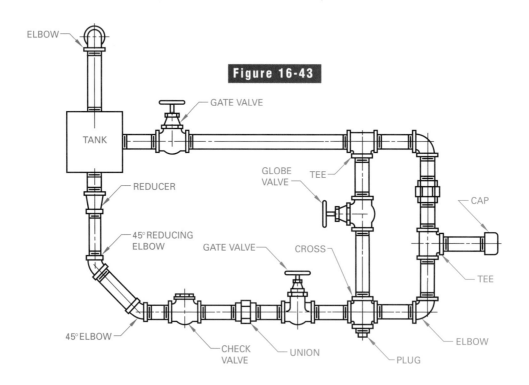

Figure 16-43

3. Convert the single-line pipe drawing in Figure 16-44 to a double-line drawing. Refer to Appendix B in your textbook for symbols. Pipe and fittings are 2″ except the smaller pipe to the left of the reducer, which is 1½″. Label fittings and dimension the drawing if instructed to do so. Scale: ¾″ = 1′-0.

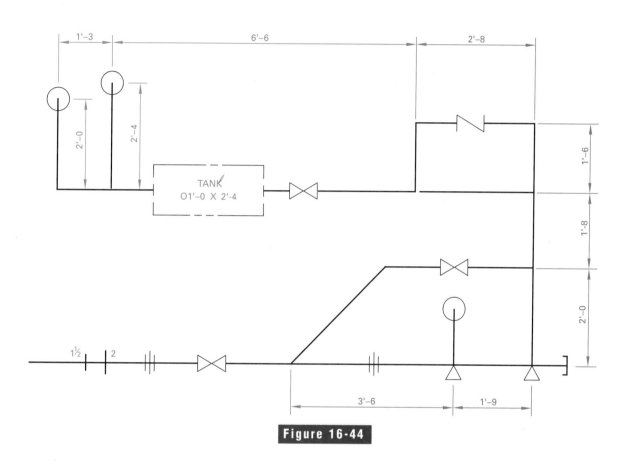

Figure 16-44

4. Convert the three-view single-line pipe drawing in
Figure 16-45 to a single-line isometric pipe drawing.
Pipe and fittings are Ø2.00, schedule 40, threaded.
Label fittings and add dimensions if instructed to do
so. Estimate sizes not given. Scale: ¼″ = 1′-0.

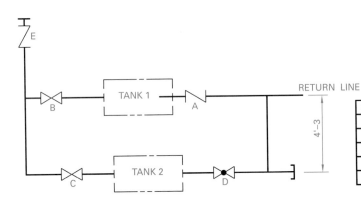

CODE	VALVE	SERVICE
A	CHECK	TANK 1
B	GATE	TANK 1
C	GATE	TANK 2
D	GLOBE	TANK 2
E	CHECK	RETURN LINES

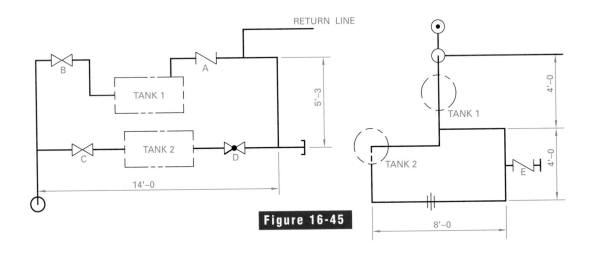

Figure 16-45

5. Figure 16-46 is a drawing of a pipe system assembled using butt-welding fittings and schedule 40 steel pipe. Refer to Appendix table C-23, "American Standard Steel Butt-Welding Fittings." Prepare a fitting schedule and a pipe schedule following the layouts shown below the pipe drawing. When calculating the length of each piece of pipe, be sure to allow for the size of the fittings.

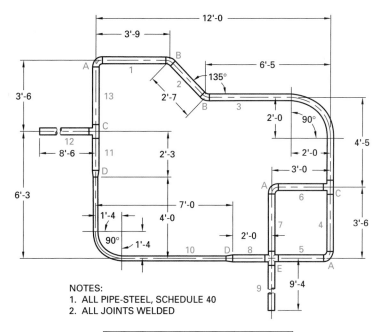

NOTES:
1. ALL PIPE-STEEL, SCHEDULE 40
2. ALL JOINTS WELDED

FITTING SCHEDULE			
CODE	QTY	NAME	SIZE

PIPE SCHEDULE		
CODE	DIAMETER	LENGTH

Figure 16-46

Design Problems

Design problems have been prepared to challenge individual students or teams of students. In these problems, you are to apply skills learned mainly in this chapter but also in other chapters throughout the text. They are designed to be completed using board drafting, CAD, or a combination of the two. Be creative and have fun!

1. **TEAMWORK** Design a soccer goal to be constructed using steel or wrought-iron pipe and steel fencing material. Obtain information on sizes and other specifications from your school's soccer coach. As a team, decide on the best method for connecting the joints of pipe. Make a complete set of orthographic and isometric drawings, including a materials list.

2. Design a seesaw to be constructed and used in your community park. Specify 2″-diameter pipe with threaded fittings for the main frame. The frame is to be anchored in concrete. The seesaw beam is to be made from 2″-diameter pipe with threaded caps on each end and wooden seats bolted in place. Design the beam so that it can be adjusted to balance with varying weight on each seat. The finished design should include the frame and three beams. Consider all safety issues in the design.

3. **TEAMWORK** Design a scaffold for use in the construction and painting of houses. The design should involve Ø1″ pipe with threaded fittings. The overall size of the scaffold is to be 3′-0 deep, 10′-0 long, and 8′-0 high. Design it so that 2 × 12 planks can be placed at a height of 4′-0 or at 8′-0. Use pipe flanges or casters on the base of each vertical member (leg). Each team member should develop design sketches of his or her design. Select the best ideas from each to finalize the team design. Prepare a final set of working drawings with dimensions and pipe symbols.

4. **TEAMWORK** Design an apparatus for bodybuilding. The team should first list the specific types of exercises the apparatus is to accommodate. Design the entire unit around steel pipe. Other materials may be used as necessary. Each team member should work on detailing a portion of the entire unit to accommodate a specific exercise. Then the team should work on integrating all of the designs into one final unit.

5. **TEAMWORK** Design a piece of playground equipment using mainly pipe with threaded fittings. Other materials can be used in a limited amount. Begin by brainstorming a list of activities that children would enjoy doing on this type of apparatus. Each team member should develop solutions for one or more item from the list. The team should then come together to develop the overall design. More than one structure might be necessary to accommodate all activities listed. All units are to be secured in concrete. Prepare a final set of working drawings with dimensions and a materials list.

Cams and Gears

OBJECTIVES

Upon completion of this chapter, you should be able to:

■ Explain the purposes and applications of cams and gears.

■ Develop a displacement diagram.

■ Develop a profile of a cam.

■ Describe the three main types of cam motion.

■ Describe the features of a typical gear drawing.

■ Draw gear teeth using the simplified method.

■ Develop a gear-tooth drawing.

Drafting Principles

Cams and gears are two of the basic elements that work in combination to form a machine. A *mechanism* is a system of mechanical parts connected to a motion or drive source such as a motor. Mechanical parts include things such as cams, gears, belts, linkages, shafts, bearings, and housings. While most of these parts are of little use alone, in combination they serve many purposes in machine design.

This chapter deals specifically with cams and gears. They are complex elements of machine design that can transmit motion, change the direction of motion, or change the speed of motion in a machine.

See Figure 17-1. The ability to draw cams and gears and to understand their function is an important step in becoming a machine designer. This chapter introduces the skills that are basic to understanding and drawing these important machine parts.

Cams

A **cam** is a machine part that usually has an irregular curved outline or a curved groove. When the cam rotates, it transmits a specific, continuous motion to another machine part that is called the **follower**. The

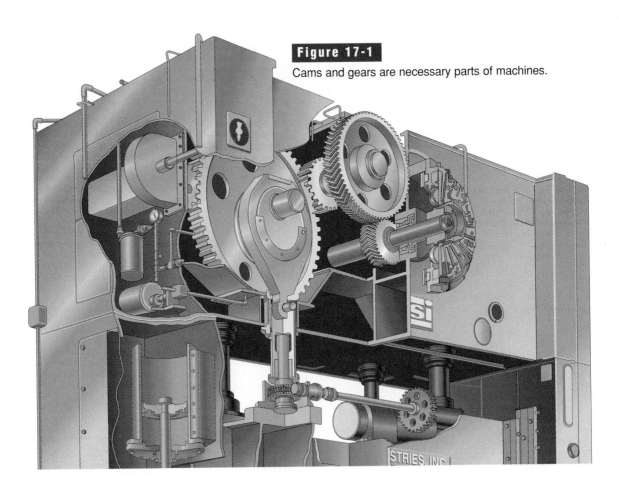

Figure 17-1

Cams and gears are necessary parts of machines.

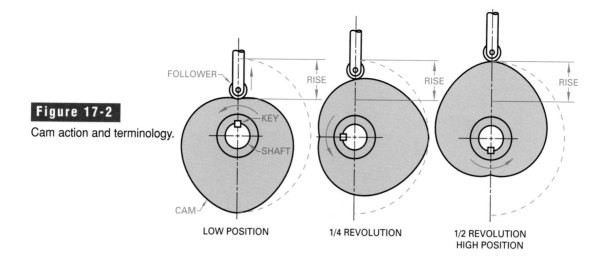

Figure 17-2

Cam action and terminology.

FOLLOWER
RISE
KEY
SHAFT
CAM
RISE
RISE

LOW POSITION   1/4 REVOLUTION   1/2 REVOLUTION
HIGH POSITION

cam and the follower together make up the cam mechanism. The cam drives the follower. The design of the cam determines the motion and path of the follower.

This motion is needed in all automatic machinery. Therefore, cams are important to the automatic control and accurate timing found in many kinds of machinery. The unlimited variety of shapes makes the cam useful to the designer.

The illustrations in Figure 17-2 provide a pictorial description of how the cam works. The stroke, also called the *rise,* takes place within one half of a full revolution, or 180°. The movement is repeated every 360°, or once during each full revolution.

Kinds of Cams

All cams can be thought of as simple inclines that produce or transmit a predetermined motion. Cams of various sizes and shapes transmit motion to the follower in specific ways. For example, a plate cam like that in Figure 17-3 has a follower that moves up and down as the cam turns on the shaft. A cylindrical cam such as the one in Figure 17-4 has a follower that moves back and forth parallel to the axis of the shaft. The grooved cam in Figure 17-5 has a follower that follows the groove, moving in an irregular pattern as the cam turns on a shaft.

Figure 17-4

A cylindrical cam.

Figure 17-3

A plate cam.

Figure 17-5

A grooved cam.

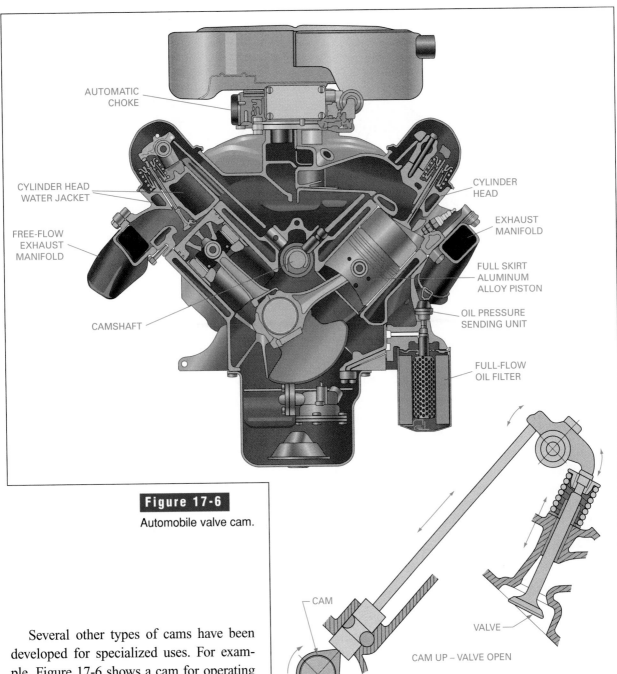

Figure 17-6

Automobile valve cam.

CAM

VALVE

CAM UP – VALVE OPEN

Several other types of cams have been developed for specialized uses. For example, Figure 17-6 shows a cam for operating the valve of an automobile engine. This cam has a flat follower that rests against the face of the plate cam. Figure 17-7 shows several other types of cams:

■ A slider cam moves the follower up and down as the cam moves back and forth.
■ An offset plate cam has a point follower that is off center.
■ A cam with a pivoted roller follower transmits motion at a 90° angle.
■ A cylindrical-edge cam may employ a swinging follower.

Cam Followers

Figure 17-8 shows three types of followers for a plate cam. The roller follower reduces friction to a minimum, so it transmits force at high speeds. The point follower and the flat-surface follower are made with a hardened surface to reduce wear from friction. These followers are generally used with cams that rotate slowly.

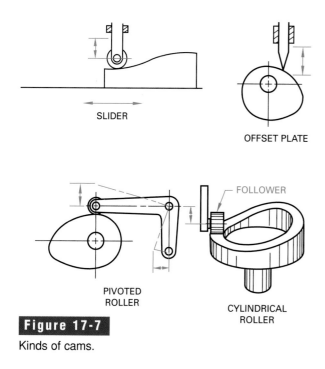

SLIDER

OFFSET PLATE

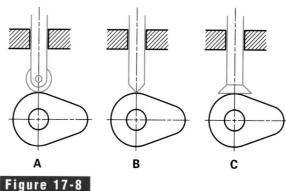

Figure 17-8

Plate cam followers.

PIVOTED
ROLLER

CYLINDRICAL
ROLLER

Figure 17-7

Kinds of cams.

follower. However, the cam and follower are normally not in contact during a drop (fall) or dwell (rest) unless contact is brought about by an outside pressure. Outside pressure is exerted by a spring pushing the follower against the cam to ensure direct contact.

Plate cams require a spring-loaded follower to ensure that contact is made throughout a full revolution. The rise, or lifting, of the follower by the cam is made through direct contact between the cam and

Displacement Diagrams

The shape of the cam determines the direction of motion in the follower, as shown in Figure 17-9A. A **displacement diagram** such as the one in Figure 17-9B shows the motion the cam will produce through one revolution. The length of the displacement diagram represents one revolution of 360°. The height of the diagram represents the total displacement stroke of the follower from its lowest position—in this case, 1.875″.

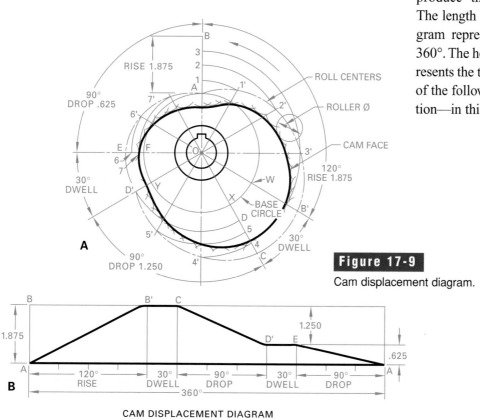

Figure 17-9

Cam displacement diagram.

CAM DISPLACEMENT DIAGRAM

Cam Motion

Cams can be designed so that the followers have three different types of motion. Displacement diagrams are used to plot the different kinds of motion.

Uniform Motion

Cams that follow **uniform motion** (steady motion) are suitable for high-speed operations. Technically, uniform motion is "straight-line" movement in which time and distance are directly proportional. In other words, equal distances on the rise are made for equal distances on the travel. However, to avoid a sudden jar at the beginning and end of the motion, designers use arcs to change it slightly. In Figure 17-10, the magenta line represents uniform motion that has been modified in this way.

The cam in Figure 17-9A displays uniform motion. The length, or time, of a revolution is divided into convenient parts. The parts are proportional to the number of degrees for each action, based on one full 360° revolution. Each part or division is called a *time period*. These proportional parts are identified as A, 1', 2', B', C, 4', 5', D', E, 6', 7', and back to A, as shown. Each proportional part or time period is a 30° angular division of the *base circle* shown in Figure 17-9A.

In Figure 17-9A, point O is the center of the cam shaft, and point A is the lowest position of the center of the roller follower. The center of the roller follower must be raised 1.875" with uniform motion during the first 120° of a revolution of the shaft. It must then dwell for 30°, drop 1.250" for 90°, dwell for another 30°, and drop .625" during the remaining 90°. The shaft is assumed to revolve uniformly (with constant speed).

Figure 17-10

Uniform motion.

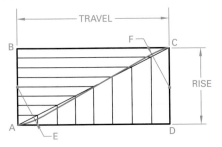

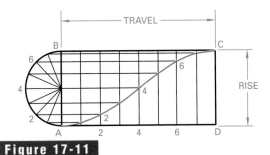

Figure 17-11

Harmonic motion.

Harmonic Motion

Other cams follow **harmonic motion**, which is a smoother-acting motion that follows a harmonic curve. Like uniform-motion cams, harmonic-motion cams are suitable for use with high speeds. They are also used when a smooth start-and-stop motion is needed.

The method for plotting a harmonic curve is shown in Figure 17-11. To draw harmonic motion, first draw a semicircle with the rise as the diameter. Divide the semicircle into eight equal parts, using radial lines. Then divide the travel into the same number of equal parts. Project the eight points horizontally from the semicircle until they intersect the corresponding vertical projections, as shown. Finally, draw a smooth curve through all eight points.

Uniformly Accelerated and Decelerated Motion

Cams that employ a steadily increasing and decreasing speed, rather than a constant speed, have **uniformly accelerated and decelerated motion**. The motion produced by these cams is very smooth. It follows a parabola, or parabolic curve. A parabola is formed when a cone is sliced vertically at any place other than through the center. Refer again to Figure 14-49.

To design a cam with uniformly accelerated and decelerated motion, first divide the rise into parts proportional to 1, 3, 5, . . . , 5, 3, 1, as shown in Figure 17-12. Note that the parts are not of equal size. Then divide the travel into the same number of parts, but divide it into equal parts. Project the points horizontally from the rise until they intersect the corresponding vertical projections. Finish by drawing a smooth curve through all the points of intersection, as shown.

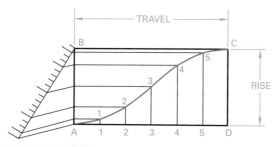

Figure 17-12

Uniformly accelerated and decelerated motion.

Figure 17-14

Friction wheels are a simple means of transmitting rotary motion from one shaft to another.

Gears

A device that transmits motion using a series of teeth is called a **gear**. A few of the many kinds of gears are illustrated in Figure 17-13.

Spur Gears

One of the most practical and dependable machine parts for transmitting rotary motion from one shaft to another is the **spur gear**. The operation of simple spur gears can be explained in this way: If the rims of two wheels are in contact, as shown in Figure 17-14, both will revolve if only one is turned. If the small friction wheel is two thirds the diameter of the larger wheel, it will make 1.50 revolutions for every 1.00 revolution of the larger wheel. This assumes that no slipping occurs. However, when the load on the driven wheel gets larger and the wheel is hard to turn, slipping begins to occur. Therefore, fric-

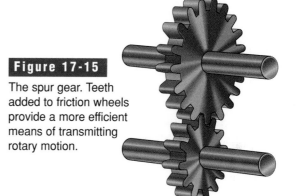

Figure 17-15

The spur gear. Teeth added to friction wheels provide a more efficient means of transmitting rotary motion.

tion wheels cannot be counted on for a smooth transfer of rotary motion. When teeth are added to the wheels in Figure 17-14, they become spur gears, as shown in Figure 17-15. Teeth added to the wheels provide the same kind of motion as rolling friction wheels. Now, however, there is no slipping.

Figure 17-13

Several gears that are used as typical machine elements: (A) rack and pinion; (B) spur gears; (C) planetary (internal) gear set; (D) spiral bevel gears; (E) eccentric spur gears; .

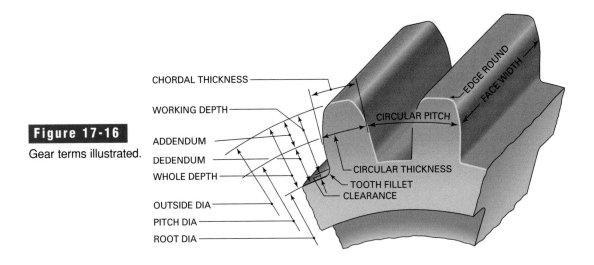

Figure 17-16

Gear terms illustrated.

CHORDAL THICKNESS

WORKING DEPTH

ADDENDUM

DEDENDUM

WHOLE DEPTH

OUTSIDE DIA

PITCH DIA

ROOT DIA

EDGE ROUND

FACE WIDTH

CIRCULAR PITCH

CIRCULAR THICKNESS

TOOTH FILLET

CLEARANCE

Figure 17-16 illustrates the parts of a spur gear. Refer to this illustration as you work with the gear formulas presented in this chapter.

Gear Teeth

The basic forms used for gear teeth are involute and cycloidal curves. These curves are explained in the following paragraphs.

The gears in Figure 17-15 are good examples of involute gears. The small spur gear is called the **pinion**. The teeth on these gears have a special shape that lets them mesh smoothly. This shape is an **involute curve**. Figure 17-17 explains an involute curve that is used in drawing gear teeth. As shown in the illustration, you can think of an involute of a circle as a curve made by taut string as it unwinds from around the circumference of a cylinder.

A cycloidal curve can be thought of as the path of a curve formed by a point on a rolling circle, as shown in Figure 17-18. The information given in this chapter is for the 14½° involute system. It can be used for a 20° involute system as well, since the only practical difference is the number of degrees of the **pressure angle**. The 14½° or 20° refers to the pressure angle, as shown in Figure 17-19. The pressure angle and the distance between the centers of mating spur gears determine the diameters of the base circles. The point of tangency of the gears is their **pitch diameter**. The pitch diameters are equal to the diameters of the rolling friction wheels that are replaced by the gears. The involute is drawn from the base circle, which is smaller than the pitch circle.

In Figure 17-20, R_A is the radius of the pitch circle of the gear with the center at A. R_B is the radius of the pitch circle of the pinion with the center at B. The distance between gear centers is $R_A + R_B$. The line of pressure $T_A T_B$ is drawn through O, which is the point of tangency of the pitch circles. It makes

Figure 17-17

Involute of a circle.

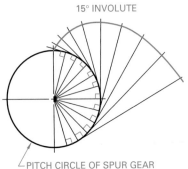

15° INVOLUTE

PITCH CIRCLE OF SPUR GEAR

Figure 17-18

A cycloidal curve.

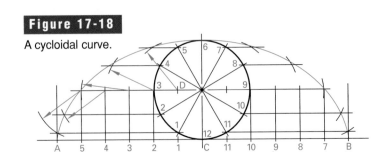

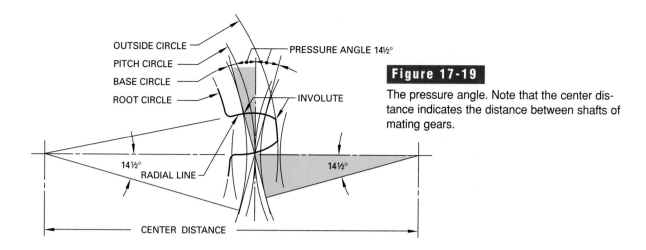

Figure 17-19

The pressure angle. Note that the center distance indicates the distance between shafts of mating gears.

OUTSIDE CIRCLE
PITCH CIRCLE
BASE CIRCLE
ROOT CIRCLE
PRESSURE ANGLE 14½°
INVOLUTE
14½°
RADIAL LINE
14½°
CENTER DISTANCE

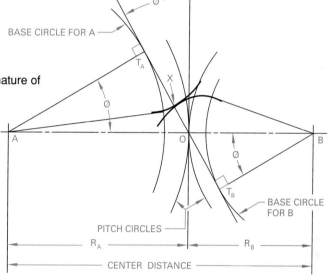

Figure 17-20

Gear tooth interaction; the rolling nature of surface contact.

BASE CIRCLE FOR A
Ø
Ø
T_A
X
A
O
B
Ø
T_B
BASE CIRCLE FOR B
PITCH CIRCLES
R_A
R_B
CENTER DISTANCE

the pressure angle ϕ (Greek letter phi) with the perpendicular to the line of centers. This angle is 14½°. Note that lines AT_A and BT_B are drawn from centers A and B perpendicular to the pressure-angle line $T_A T_B$. A point X on a cord (line of pressure line $T_A T_B$) will describe the points that form the involute curve as the cord unwinds. This represents the outlines of gear teeth outside the base circles. The profile of the gear tooth inside the base circle is a radial line. Figure 17-17 shows the cord unwinding off the surface on the base circle.

Applying Spur-Gear Formulas

Table 17-1 defines terms commonly used in gear formulas, and Table 17-2 presents the actual formulas. Refer to these two tables as you apply the gear formulas.

Drafters apply these formulas to the written descriptions and information they are given. The following information is typical of the instructions a drafter might receive.

A pair of involute gears is to be drawn according to the specifications that follow:

- The pressure angle will be 14½°.
- The distance between parallel shaft centers will be 12.00″.
- The driving shaft will turn at 800 revolutions per minute (RPM).
- The diametral pitch (number of teeth per inch of pitch diameter) equals 4.

To find the rest of the information needed to draw the gears, follow these steps:

1. Find the pitch radius of the pinion and the spur gear. These calculations are based on the ratio of the velocity of the two cylinders. One cylinder drives the other. Thus, the ratio is obtained by

Table 17-1. Spur-Gear Terms and Symbols

Term	Symbol	Definition
	N	number of teeth
	N_G	number of teeth of gear
	N_P	number of teeth of pinion
pitch diameter	D	diameter of pitch circle
	D_G	pitch diameter of gear
	D_P	pitch diameter of pinion
diametral pitch	P	number of teeth per inch of pitch diameter
addendum	a	radial distance the tooth extends above the pitch circle
dedendum	b	radial distance the tooth extends below the pitch circle
outside diameter	D_O	overall gear size: pitch diameter plus twice the addendum
root diameter	D_R	pitch diameter minus twice the dedendum
whole depth	h_t	radial distance from the root diameter to the outside diameter; equal to addendum plus dedendum
circular pitch	p	distance from a point on one tooth to the same point on the next tooth measured along the pitch circle; the distance of one tooth and one space
clearance	c	the distance between the top of a tooth and the bottom of the mating space when gear teeth are meshing; the difference between the addendum and the dedendum
working depth	h_k	the distance a tooth projects into the mating space; twice the radial distance of the addendum
pressure angle	o	the direction of pressure between teeth at the point of contact
base-circle diameter	D_b	the circle from which the involute profile is developed

dividing the velocity of the driver by the velocity of the driven member. The velocities are stated in RPM.

Pitch radius of pinion:

$$R_P = \frac{400}{400 + 800} \times 12.00'' = 4.00'' \text{ radius}$$

Pitch radius of spur gear:

$$R_S = \frac{800}{400 + 800} \times 12.00'' = 8.00'' \text{ radius}$$

The velocity ratio therefore equals ½(4.00″:8.00″).

2. Find the number of teeth on the pinion:
   $N_P = DP = 4 \times 8 = 32$
3. Find the number of teeth on the spur gear:
   $N_S = DP = 4 \times 16 = 64$
4. Find the addendum:
   $a \dfrac{1}{p} = \dfrac{1}{4} = .25''$
5. Find the dedendum:
   $b = \dfrac{1.157}{p} = \dfrac{1}{4} = .289''$

Involute gears are interchangeable as long as they have the same diametral pitch, pressure angle, addendum, and dedendum.

Involute Rack and Pinion

A rack and pinion is shown in Figure 17-21. A rack is a gear with a straight pitch line instead of a circular pitch line. The tooth profiles become straight lines. These lines are perpendicular to the line of action.

Worm and Wheel

The **worm gear** is similar to a screw. It may have single or multiple threads. Figure 17-22 shows how the worm and wheel mesh at right angles. This system is used to transmit motion between two perpendicular, nonintersecting shafts. The wheel is similar to a spur gear in design, except that the teeth must be curved to engage the worm gear.

Table 17-2. Spur-Gear Formulas

number of teeth	$N = DP = \dfrac{\pi D}{p} = DO \times P - 2$
pitch diameter	$D = \dfrac{N}{P} = D_O - 2a$
diametral pitch	$P = \dfrac{N}{D} = \dfrac{\pi}{p}$
addendum	$a = \dfrac{1}{P} = \dfrac{p}{\pi}$
dedendum	$b = \dfrac{1.157}{P} = \dfrac{1.157p}{\pi}$
outside diameter	$D_O = \dfrac{N + 2}{P} = D + 2a = \dfrac{(N + 2)p}{\pi}$
root diameter	$D_R = D - 2b = D_O - 2(a + b)$
whole depth	$h_t = a + b = \dfrac{2.157}{P} = \dfrac{2.157p}{\pi}$
circular pitch	$\dfrac{\pi D}{N} = \dfrac{\pi}{P}$
circular thickness	$t = \dfrac{p}{2}$
chordal thickness	$t_c = D\sin\left(\dfrac{90°}{N}\right)$
clearance	$c = \dfrac{.157}{P} = \dfrac{.157p}{\pi}$
working depth	$h_k = 2 \times a = 2 \times \dfrac{1}{P}$

Figure 17-21

Rack and pinion.

they are in spur gears. The pitch diameter is the diameter of the pitch cone in the bevel-gear design. The symbols for important features, such as the angles, are listed in Table 17-3. The three Greek letters that are used to represent bevel angles are α (alpha), δ (delta), and γ (gamma).

Figure 17-22

Application of a worm and wheel.

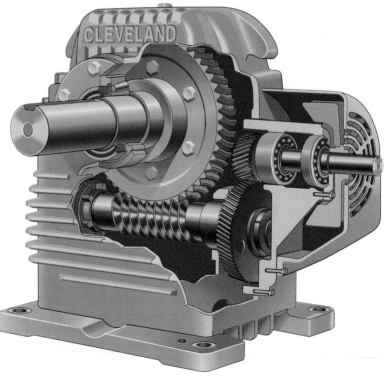

Bevel Gears

When two gear shafts intersect, bevel gears are used to transfer motion. Figure 17-23 shows four sets of rolling friction cones. Think of bevel gears as replacing the friction cones, just as the spur gear replaced the circular friction wheels. Figure 17-24 illustrates mating bevel gears. The smaller gear is called the pinion. When the gears are the same size and the shafts are at right angles, bevel gears are referred to as miter gears.

Some basic information about bevel gears is given in Figure 17-25. Look for similarities and differences in the design of spur and bevel gears. The circular pitch and the diametral pitch are based on the pitch diameter just as

Gear Ratio

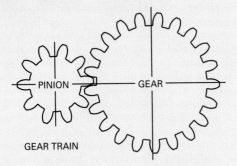

GEAR TRAIN

The diameters of the gear and pinion can be used to determine the gear ratio. Calculating the gear ratio determines the amount of increase or decrease in the speed of a combination of gears, or gear train. If the diameter of the gear is twice that of the pinion, it will have twice as many teeth as the pinion. Therefore, the gear ratio can also be determined by the relationship of the number of teeth on each. For example, the gear in the illustration has 24 teeth; the pinion has 12 teeth. Therefore, the pinion will make two complete turns for every one turn of the gear. This results in a gear ratio of 1:2. If the pinion is attached to a motor that rotates at

1000 revolutions per minute (RPM), the gear will turn at a speed of 500 RPM, because the pinion rotates twice for every revolution of the gear.

The actual gear ratio is found by dividing the number of teeth on the gear by the number of teeth on the pinion. In this case, $24 \div 12 = 2$. The gear ratio is 2:1, and the pinion ratio is 1:2.

Mechanical advantage and gear ratio are closely related. If the pinion (drive gear) with 12 teeth is driving the gear (driven gear) with 24 teeth, the mechanical advantage is 2. In other words, the torque, or turning power, of the gear is doubled. Here is how it works:

$$\text{Mechanical advantage} = \frac{\text{Number of teeth on the driven gear}}{\text{Number of teeth on the drive gear}}$$

$$\text{Mechanical advantage} = \frac{24}{12}$$

$$\text{Mechanical advantage} = 2$$

In other words, by doubling the size of the driven gear, you increase the torque by a ratio of 2:1. Therefore, if a motor is driving the pinion, the gear could carry a load double what it could carry without the gears.

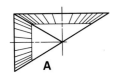

A

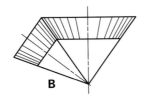

B

C

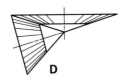

D

Figure 17-23

Rolling cones that represent bevel gearing.

Figure 17-24

Bevel gear.

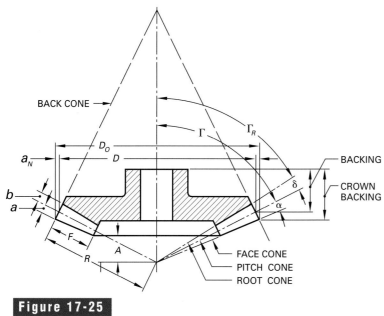

BACK CONE

Γ

Γ_R

D_O

a_N

D

b

a

δ

BACKING

CROWN
BACKING

α

F

A

R

FACE CONE
PITCH CONE
ROOT CONE

Figure 17-25

Bevel-gear terms. Note the cone shape of the gear.

Table 17-3. Symbols for Bevel Gears	
Symbol	**Definition**
α	addendum angle
δ	dedendum angle
Γ	pitch angle
Γ_R	root angle
Γ_O	face angle
a	addendum
b	dedendum
a_n	angular dedendum
A	cone distance
F	face
D	pitch diameter
D_O	outside diameter
N	number of teeth
P	diametral pitch
R	pitch radius

Gear Drawings

It is not necessary to show the teeth on typical gear drawings. A drawing for a cut spur gear is shown in Figure 17-26. The gear blank should be drawn with dimensions for making the pattern and for the machining operations. The spur-gear drawing should include information for cutting the teeth. It should also include information for the tolerances required and a notation of the material to be used. On assembly drawings, a simplified gear may be used, as shown in Figure 17-27. Even though the drawing is simplified, however, it should include all necessary notes for making the gear.

On working drawings of bevel gears, the needed dimensions for machining the blank must be given, as well as all the gear information. An example of a working drawing for a cut bevel gear is shown in Figure 17-28.

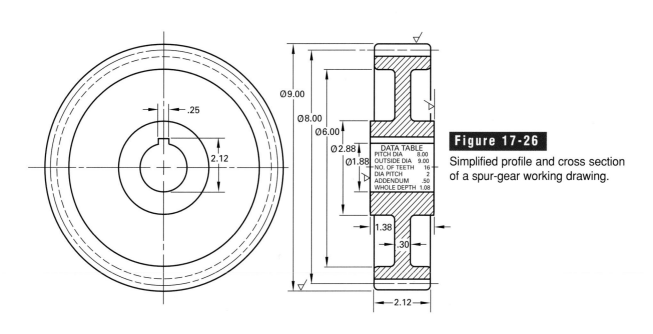

.25

2.12

Ø9.00

Ø8.00

Ø6.00

Ø2.88

Ø1.88

DATA TABLE	
PITCH DIA	8.00
OUTSIDE DIA	9.00
NO. OF TEETH	16
DIA PITCH	2
ADDENDUM	.50
WHOLE DEPTH	1.08

1.38

.30

2.12

Figure 17-26

Simplified profile and cross section of a spur-gear working drawing.

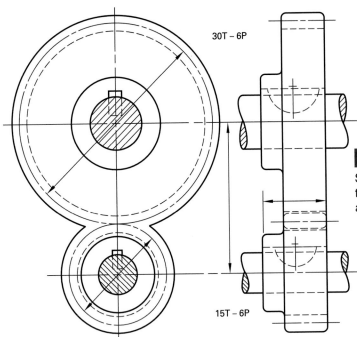

30T – 6P

15T – 6P

Figure 17-27

Simplified drawing of mating spur gears. Note the tangent pitch circles, the number of teeth, and the pitch.

The American National Standards Institute has established standards for satisfactorily detailing gear drawings. ANSI Y14.7.1 and ANSI Y14.7.2 are standard references that can be used for further study of the subject of gear detailing.

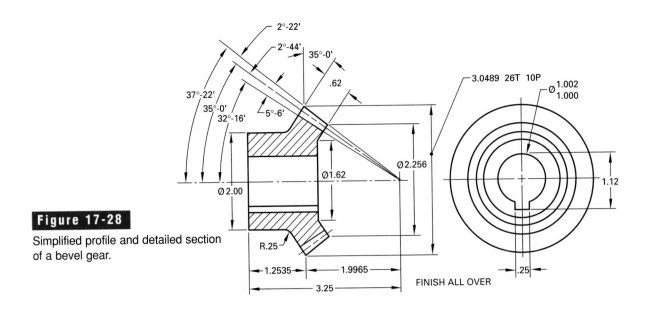

Figure 17-28

Simplified profile and detailed section of a bevel gear.

Board Drafting Techniques

By now, you should have a rather clear understanding of the basic elements of the design of cams and gears. However, before proceeding, you may want to go back to the section on cams and review terminology and cam nomenclature. Be sure you understand these completely, and then proceed.

Drawing Cam Profiles

The profile of the cam in Figure 17-29 has five important features: a rise, dwell, drop, another dwell, and another drop. Follow these steps.

1. **Rise.** Divide the rise AB, or 1.875″, into a number of equal parts. Four parts are used in the illustration, but eight parts would make the layout more accurate. The rise occurs from A to W (120°). Divide it into the same number of equal parts as the rise (four at 30°) with radial lines from O. Using center O, draw arcs with radii O1, O2, O3,

and OB until they locate 1′, 2′, 3′, and B′ on the four radial lines. Use an irregular curve to draw a smooth line through these four points.

2. **Dwell.** First draw an arc B′C (30°) using radius O′B′. This allows the follower to be at rest, because it will stay at the same distance from center O.

3. **Drop.** At C, lay off CD (1.250″) on a radial line from O. Divide it into any number of equal parts (three are shown). Next, divide the arc XY (90°) on the base circle into the same number of equal parts (three). Draw three radial lines from center O every 30°. Then draw arcs with center O and radii O4, O5, and OD to locate points 4′, 5′, and D′ on the three radial lines. Using an irregular curve, draw a smooth line through points 4′, 5′, and D′.

4. **Dwell.** Draw an arc D′E′ (30°) with radius OD. This will provide the constant distance from center O to let the roller follower be at rest.

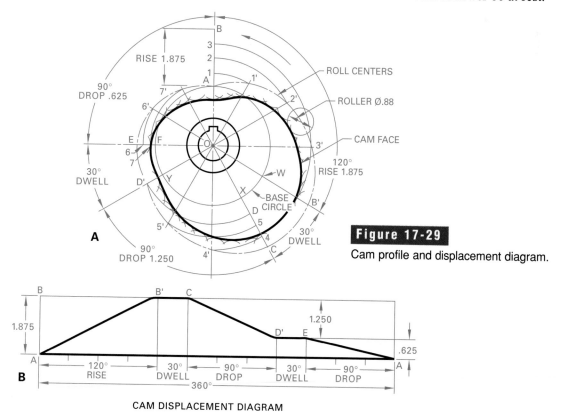

Figure 17-29

Cam profile and displacement diagram.

CAM DISPLACEMENT DIAGRAM

5. **Drop.** In the last 90° of a full revolution, the roller will return to point A. It will move through a distance EF, or 6.25″. First, divide EF into any number of equal parts (three are shown). Then divide arc FA into the same number of equal parts. Next, draw radial lines every 30°. Draw arcs with radii O6 and O7 to locate points 6′ and 7′. Using the irregular curve as a guide, draw a smooth curve through points E, 6′, 7′, and A. This finishes the roll centers.

6. **Finish the Profile.** Using the line-of-roll centers as a centerline, draw successive arcs with the radius of the roller, as shown in Figure 17-29. Then use an irregular curve to draw the cam pro-

file. The profile should be a smooth curve tangent to the arcs you drew representing the roller.

Another drawing for a face, or plate, cam is shown in Figure 17-30. Note that the amount of movement, or rise, is given by showing the radii for the dwells (4.50″ and 7.00″). Harmonic motion is used, and there seem to be two rolls working on this cam. In Figure 17-31, a drawing for a *barrel cam* (cylindrical cam) is shown with a displacement diagram. The diagram shows two dwells and two kinds of motion. Note that the distance traveled from center to center is 1.50″ and is called *throw*. Practice your technique by drawing the displacement diagrams for these two cams.

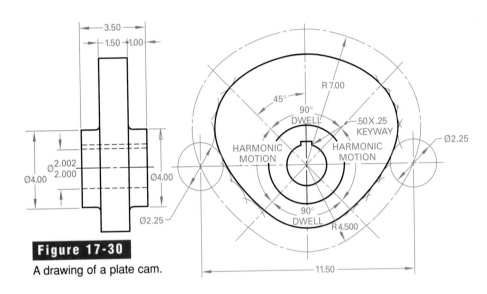

Figure 17-30

A drawing of a plate cam.

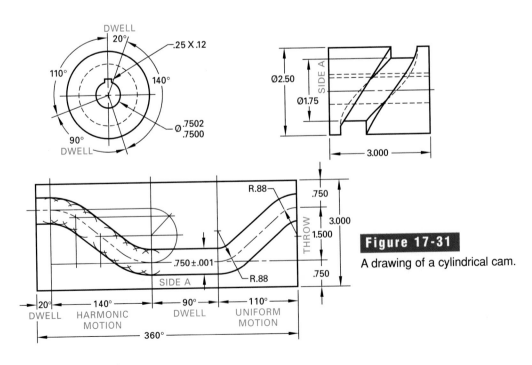

Figure 17-31

A drawing of a cylindrical cam.

Drawing Gear Teeth

Although it is typically not necessary to show the teeth on a gear drawing, there are some occasions when the drafter must include them. Figure 17-32 shows an example of a gear detail drawn using the simplified method. To simplify the drawing of the involute curve, note that the radius in Figure 17-32 is drawn as ⅛ the pitch diameter. The procedure for drawing gear teeth using this method is as follows:

1. Draw the addendum circle, pitch circle, and root circle.
2. Draw a line tangent to the pitch circle at any point on the pitch circle.
3. Draw a line through the intersection of the pitch circle and the tangent line at an angle of 14½° (may be drawn at 15°) to the tangent line.

4. Draw the base circle tangent to the 14½° (15°) line.
5. Calculate or estimate the chordal thickness and step this distance off along the pitch circle.
6. With the compass set at a distance equal to ⅛ of the pitch diameter, place the compass point on the base circle and strike the arcs, as shown in Figure 17-32.
7. Add the short radial lines from the curve of the tooth to the fillets. The size of the fillet is estimated. It can be drawn to any suitable size that appears appropriate.
8. Darken all lines.

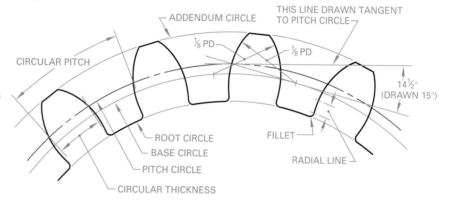

Figure 17-32

Simplified method of drawing a gear tooth.

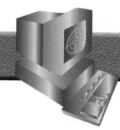

CAD Techniques

The following discussion applies specifically to computer-aided drafting (CAD). Although CAD techniques may differ in their approach from board drafting techniques, it is important to keep the basic drafting principles covered in the first portion of this chapter in mind. Unless specifically stated otherwise, all drafting principles apply equally to board and CAD drawings.

Creating a Cam Profile

The profile of the cam in Figure 17-33 has five important features:

- rise of 1.875″ over 120°
- dwell, or rest period, over 30°
- drop of 1.250″ over 90°
- dwell over 30°
- drop of .625″ over 90°

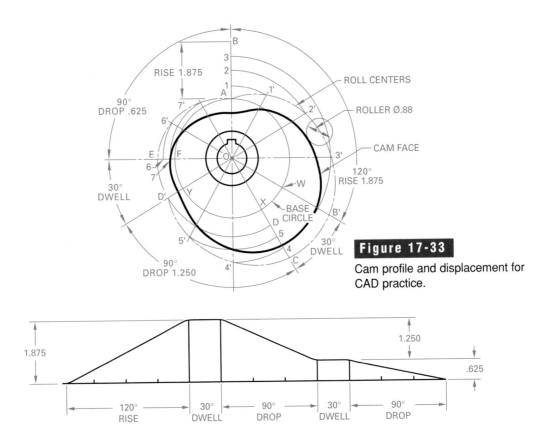

Figure 17-33

Cam profile and displacement for CAD practice.

Development of the cam profile is an interactive process that involves creating an accompanying cam displacement diagram. Follow these steps to create the cam profile and displacement diagram.

1. Set up a drawing file with limits of 0,0 and 15,12.
2. Create the base circle for the cam profile with a diameter of 3.88″.
3. Place the hole in the center of the cam by creating a Ø1.00″ circle using the same center point as the base circle. Remember to use the Center object snap to locate the center.

The next step involves the use of the ARRAY command. AutoCAD provides two types of arrays for 2D drawings: rectangular and polar. A **rectangular array** is one in which an object is copied into a specified number of rows and columns, as shown in Figure 17-34A. A **polar array** copies the object a specified number of times to fill a circle or arc that you specify, as shown in Figure 17-34B.

4. Divide the circle into 30° increments. To do this, first create a vertical 4″ line with its starting point at the center of the base circle, as shown in Figure 17-35. Then enter the ARRAY command and pick the vertical line. Enter P to select a polar array. There are 360° in a circle, and you want to divide it into equal 30° divisions. Therefore, divide 360

Figure 17-34

(A) A rectangular array; (B) a polar array.

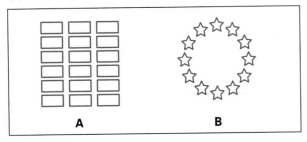

Figure 17-35

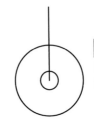

Create a 4″ vertical line up from the center of the base circle.

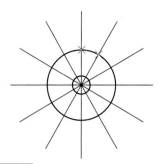

Figure 17-36

Measure the distance between two consecutive radial lines along the base circle by entering the DIST command and picking the points shown here.

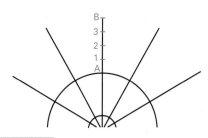

Figure 17-38

Divide line AB into four equal parts. *Note:* In this and the following illustrations, some elements are shown in color for clarity. It is not necessary to use these colors in your CAD drawing.

by 30 to find that the number of items in the array should be 12. Enter 12 as the number of items, and accept the default angle to fill of 360°. Enter Y (Yes) to rotate the arrayed objects. The arrayed lines appear, dividing the circle into perfect 30° increments.

5. Use the DIST command to measure the distance from the intersection of the base circle and one of the 30° division lines to the intersection of the base circle and the next division line, as shown in Figure 17-36. This provides the length for each 30° increment on the displacement diagram. Each increment will be 1.00″.

6. Begin the cam displacement diagram by laying out the base line AA. Begin the line near the lower left corner of the drawing area. It should be equal to 12 segments of 1.00″ each, or 12.00″ total. Enter the PDMODE command and enter a new value of 4. This value creates tic marks at the location of points and divisions. Then enter the DIVIDE command, select the base line, and divide it into 12 parts. The tic marks appear at the required 30° intervals.

7. Using the base line and the bulleted rise, dwell, and drop information at the beginning of this section, complete the cam displacement diagram, as shown in Figure 17-37. Add a small

fillet with a radius of .30 at each end of each segment of the diagram, as shown in the figure. *Note:* For the fillet at the beginning of the base line, you will need to create a temporary line starting at the left end of the base line and extending to the left. Fillet the displacement line to this line to get the appropriate curve. Then delete the temporary line and, if necessary, extend the base line to meet the fillet.

8. Dimension the displacement diagram as shown in Figure 17-37.

Now you can return to the cam profile and continue to develop it. To do this, you will actually develop the path of the follower first. Notice that the follower in this case is a roller. The roller path is the most critical part of the profile, because the purpose of the entire cam development is to create a cam that makes the roller exhibit the specific rise, dwell, and drop behavior stated in the bulleted list at the beginning of this section. Follow these steps:

1. Trim the vertical division line to the top of the base circle. Then extend a new line AB from the intersection of the vertical division line with the top quadrant point of the circle, extending upward by 1.875″ (the specified rise). Use the DIVIDE command to divide this line into four equal segments, corresponding to the four 30° segments in the rise, as shown in Figure 17-38.

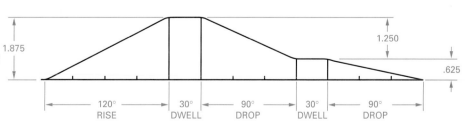

Figure 17-37

The finished displacement diagram.

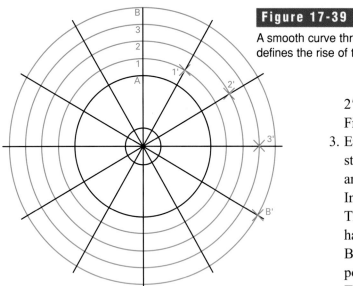

Figure 17-39

A smooth curve through points A, 1′, 2′, 3′, and B′ defines the rise of the roller.

Change PDMODE to 3 to see the division points. *Note:* Figure 17-38 shows the points labeled with letters. These are for your reference only. There is no need to add them to the drawing.

2. Using base circle center O, create circles with radii O1, O2, O3, and OB to locate points 1′,

2′, 3′, and B′ on the four radial lines. See Figure 17-39.

3. Enter the ARC command, and create an arc that starts at point A, has a center point at point 1′, and ends at point 2′. Be sure to use the Intersection object snap to place the arc exactly. Then create a second arc that starts at point 2′, has a center point at point 3′, and ends at point B′. The result is a smooth curve connecting point A with point B′, as shown in Figure 17-40. This line defines the path of the center of the roller through the rise segment of the cam revolution. Delete the circles with radius 1, 2, and 3, but do not delete the circle with radius B.

4. The next phase is a dwell of 30° duration. During a dwell, the roller remains at exactly the same distance from the center O. Therefore, you can use a portion of the circle with radius B to define the path of the roller. Trim the circle so that only arc B′C remains, as shown in Figure 17-41.

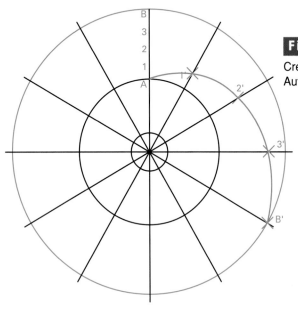

Figure 17-40

Create a smooth curve through the points using AutoCAD's ARC command.

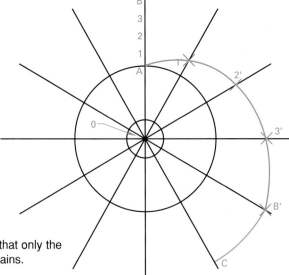

Figure 17-41

Trim the circle with radius B so that only the segment between B′ and C remains.

5. Now the roller needs to drop by 1.250″ over the next 90°. The procedure will be similar to that used to define the rise. First, delete radial line OC. Then create a new line CD. The line should start at point C and extend toward center point O for 1.250″ (the specified drop), which means the line should be at a 120° angle. At the To point prompt, enter @1.25<120.

6. Divide line CD into three equal parts using the DIVIDE command. Create three circles with center O and radii of O4, O5, and OD to locate points 4′, 5′, and D′, as shown in Figure 17-42.

7. Enter the ARC command, and connect points C, 4′, and 5′. Reenter the ARC command and start the second arc at point 5′. Estimate a center point about halfway between point 5′ and point D′, and end the arc at point D′. The arcs should form a smooth curve. If your second arc does not connect smoothly with the first one, pick the arc to select it, and pick the middle grip (blue box) so that it turns red. Move the grip as necessary to change the shape of the arc to provide a smooth curve. This completes the first drop. Delete the circles with radius 4 and 5, but do not delete the circle with radius D.

8. The next phase is another dwell of 30° duration. Trim the circle so that only arc D′E remains, as shown in Figure 17-43.

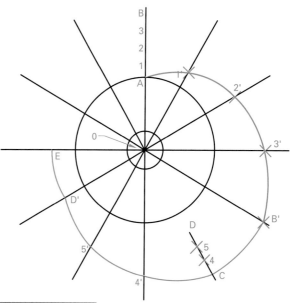

Figure 17-43

Connect points C, 4′ 5′, and D′ to complete the arc for the first drop, and then trim the circle with radius D to create the second dwell.

9. The final drop begins at point E and ends at point A; in other words, it drops by .625″ over 90°. The portion of radial line OE that lies between the base circle and point E is, by definition, .625″. Therefore, trim away the rest of line OE, leaving only segment EF. Then divide segment EF into three equal segments using the DIVIDE command, and create circles with center O and radii 6, 7, and F to create points 6′, 7′, and F′. Following the same procedure you used for the first drop, define the path of the roller through these points.

10. When you have completed the roller path, delete the remaining circles, including the base circle, but do not delete the Ø1.00″ hole at the center of the cam. Also delete the radial lines and the division points. Enter the PEDIT command and pick anywhere on the roller path. When AutoCAD informs you that the selection is not a polyline, choose to convert it to a polyline. Then enter the J (Join) option, and select the remaining segments of the roller path, following a *counterclockwise* path, in order to change the entire path into a single polyline. End the PEDIT command and pick the roller path to make sure that all of the segments were added. *Note:* If the last segment did not add

Figure 17-42

Create three additional circles to identify the path of the roller through the first drop.

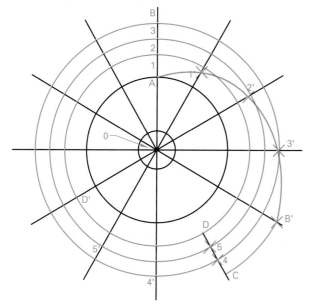

correctly, reenter PEDIT, pick the polyline, and enter C (Close). The closing segment will follow exactly the same path as the underlying segment. Delete the underlying segment. Then select the polyline and change its linetype to Center. With the polyline still selected, enter LTSCALE and change the linetype scale to .500. The finished roller path is shown in Figure 17-44.

11. The actual cam profile can now be created in one easy step. You know from Figure 17-33 that the diameter of the roller is Ø.88″. By definition, then, the path of the center of the roller is .44″ on the outside of the cam profile. In other words, the cam profile is .44″ to the inside of the roller path. Enter the OFFSET command, enter an offset distance of .44, and pick the roller path. Then pick a point inside the roller path to create the actual cam profile. Pick the cam profile and change its linetype to continuous.

12. To finish the cam profile drawing, add the roller with a center point at any point along the roller path, remembering that its diameter is .88. Add vertical and horizontal centerlines. The finished working drawing is shown in Figure 17-45.

Drawing Gear Teeth

Gears are easier to draw using CAD because you can simply create one tooth in detail and then array it to form the entire gear. We will use a polar array to complete a drawing of a gear showing the individual gear teeth. Follow these steps:

1. Draw the addendum circle, the pitch circle, and the root circle, using the dimensions shown in Figure 17-46.

2. Create a 2.00″ line from the top quadrant point of the pitch circle at a downward angle of 14.5°. (Use the Quadrant object snap for the first point, and enter @2<345.5 at the To point prompt.) Then create a circle with the same center point as the other circles. Use the Tangent object snap to snap the circle to the 2.00″ line. This becomes the base circle. Erase the 2.00″ line.

3. This gear will have 12 teeth. To determine the shape of the curved portion on the upper part of

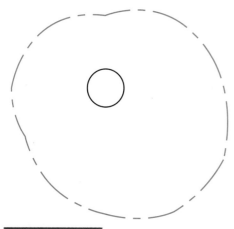

Figure 17-44

The completed roller path.

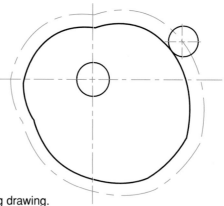

Figure 17-45

The finished working drawing.

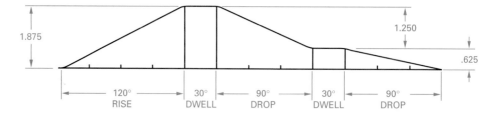

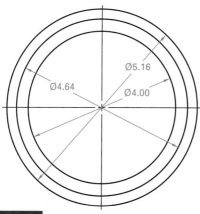

Figure 17-46

Create the addendum circle, the pitch circle, and the root circle.

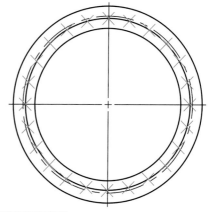

Figure 17-47

Use DIVIDE to create 24 equal segments on the base circle.

the tooth, first divide the base circle into 24 equal segments (because there are two sides for each tooth). Before you do this, enter the PDMODE command and change the point display to 3 so that the points will show up. Then enter the DIVIDE command and pick the base circle. Enter 24 for the number of segments. The drawing should look like the illustration in Figure 17-47.

4. Zoom in on a small portion of the circle. Create a circle with its center at one of the division marks on the base and a radius equal to ⅛ of the

►CADTIP

Smooth Circle Representations

When you zoom in to a small part of a circle, you may notice that the circle looks segmented at the higher zoom level. To fix the circle's appearance, you can enter the REGEN command to regenerate the drawing at the new zoom level. However, if you plan to use several different zoom levels in a drawing, you can avoid the problem by setting VIEWRES to a higher level. VIEWRES is an AutoCAD system variable that controls the resolution of curved lines on a scale of 1 to 20,000. The default value is 100. By entering a new, higher value, you can force the curved lines to be more accurate at high zoom levels. Try setting the zoom level to various levels. For this drawing, you may want to set VIEWRES to a value of 5,000.

pitch diameter, or .58″. Then create an identical circle with its center at the next division mark on the base circle. See Figure 17-48A. The area common to these two circles begins to define a gear tooth. Trim away the excess portions of both circles, as shown in Figure 17-48B.

5. Create a line from the center of the base circle tangent to one of the arcs that forms the tooth. Create a similar line tangent to the arc that forms the other side of the tooth, as shown in Figure 17-49A. These lines are the radial lines that form the relatively straight portions near the bottom of the tooth. Then zoom in on the tooth and trim away the bottom part of the arcs

Figure 17-48

Begin the gear tooth by creating circles with a radius equal to ⅛ the pitch diameter at two consecutive division points on the base circle. Trim away the excess parts of the circle as shown.

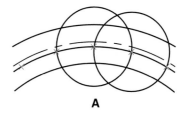

A

B

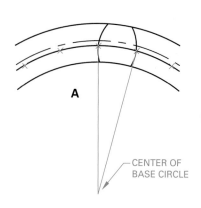

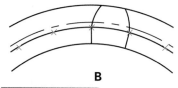

A

B

CENTER OF
BASE CIRCLE

Figure 17-49

Create the bottom sides of the tooth along radial lines from the center of the base circle.

and the part of the lines that extends below the root circle. The result should look like Figure 17-49B.

6. Add the fillets at the bottom of the tooth. These can be estimated to give a good appearance in the drawing. In this case, you might try using a fillet radius of .04″.

7. Trim away the addendum circle, leaving only the portion that forms the top of the gear tooth. See Figure 17-50. This completes one tooth.

8. You can now array the tooth to complete the rest of the gear teeth automatically. Enter the ARRAY command. At the Select objects prompt, carefully pick all of the lines and arcs that make up the tooth, and press Enter. Then enter P to choose a polar array, and use the Center object snap to select the center of the base circle as the center point of the array. Enter 12 for the number of items in the array, and accept the default 360° angle to fill. Enter Y (Yes) to rotate the objects as they are arrayed. Zoom out to see the effect of the ARRAY operation: the remaining teeth appear, perfectly spaced around the gear.

9. Delete the pitch circle and the base circle, and remove the division marks.

10. Before trimming the base circle, use it as a basis for creating the hole at the center of the

> ►**CADTIP**
>
> **Removing Division Marks**
> The fastest way to remove the division marks from the screen is to change PDMODE back to 0. Remember, however, that if you simply change their display, the points still exist in the drawing database. If the drawing is to be used directly to drive certain machinery, the points can create problems. Therefore, if the drawing is intended to be used with a CAM or CIM setup, you should delete the points individually.

gear and the circles that define the indented portion of the gear. Enter the CIRCLE command and use the Center object snap to snap to the center of the base circle. Create the hole with a diameter of 1″.

11. Trim away the base circle from the base of each tooth. The finished gear should look similar to the one in Figure 17-51.

Figure 17-51

Array the tooth and delete the portions of the base circle that overlap the gear teeth to complete the gear.

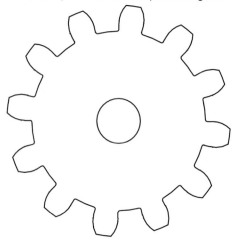

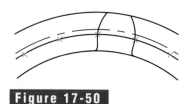

Figure 17-50

The completed gear tooth.

Chapter Summary

■ Cams and gears transmit motion, change the direction of motion, or change the speed of motion in a machine.

■ A cam and follower together make up a cam mechanism.

■ A displacement diagram shows the motion a cam will produce through one revolution.

■ The design of the cam determines the direction of motion and the path of the follower.

■ Cams can be designed to provide uniform motion, harmonic motion, and uniformly accelerated and decelerated motion.

■ Gears transmit motion using a series of teeth.

■ The basic forms used for gear teeth are involute and cycloidal curves.

■ A rack is a gear with a straight pitch line rather than a circular pitch line.

■ Bevel gears are used when two gear shafts intersect.

Review Questions

1. Other than time, what is important in the design of a cam?

2. Describe the purpose of a displacement diagram.

3. List three types of cams and three types of followers.

4. List three uses of a cam.

5. Why is harmonic motion used in a cam?

6. What technical term describes the ratio of the number of teeth to the pitch diameter?

7. Rolling cones are used to describe the meshing of what kind of gears?

8. What two bits of information are needed to find circular pitch?

9. List two applications of bevel gears.

10. Name the four circles used in drawing gears.

11. What is a polar array?

12. Explain why arrays are useful in developing a gear drawing that includes gear teeth.

Drafting Problems

The drafting problems in this chapter are designed to be completed using board drafting techniques or CAD.

1. Make a profile of the radial plate cam shown in Figure 17-52. Prepare a displacement diagram similar to the one in Figure 17-9 to illustrate the travel patterns from the base circle.

2. Prepare a drawing of the displacement diagram and the cam profile in Figure 17-9. Shaft: Ø.62"; roller: Ø.56"; base circle: Ø2.50"; hub: Ø1.12". Diagram line = 2.50 × 3.14.

3. Prepare diagrams similar to those in Figures 17-10 through 17-12 to illustrate the three types of cam motion. Travel: 2.50"; rise: 1.25".

4. Draw the two views of the plate cam shown in Figure 17-30. Prepare a displacement diagram to illustrate the rises.

5. Draw a box (grooved) cam for the conditions given in Figure 17-53. Draw the path of the roll centers and lay off the angles. From A to B, rise from 2.50" to 3.50", with modified uniform motion. From B to C, dwell, radius 3.50". From C to D, drop from 3.50" to 2.50", with harmonic motion. From D to A, dwell, radius 2.50". Draw the groove, using a roller with a 1.00" diameter in enough positions to fix outlines for the groove. Complete the cam drawing. Keyway .50" × .12".

Figure 17-52

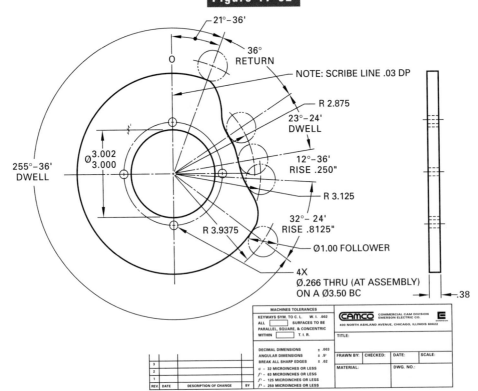

- 21°– 36'
- O
- 36° RETURN
- NOTE: SCRIBE LINE .03 DP
- R 2.875
- 23°– 24' DWELL
- 12°– 36' RISE .250"
- Ø 3.002 / 3.000
- 255°– 36' DWELL
- R 3.125
- 32°– 24' RISE .8125"
- R 3.9375
- Ø1.00 FOLLOWER
- 4X Ø.266 THRU (AT ASSEMBLY) ON A Ø3.50 BC
- .38

MACHINES TOLERANCES			
KEYWAYS SYM. TO C. L.	W. I. .002		
ALL ☐ SURFACES TO BE PARALLEL, SQUARE, & CONCENTRIC			
WITHIN ☐ T. I. R.			

CAMCO COMMERCIAL CAM DIVISION EMERSON ELECTRIC CO.
400 NORTH ASHLAND AVENUE, CHICAGO, ILLINOIS 60622

TITLE:

DECIMAL DIMENSIONS	± .003	FRAWN BY:	CHECKED:	DATE:	SCALE:
ANGULAR DIMENSIONS	± .5°				
BREAK ALL SHARP EDGES	± .02				
⌀ - 32 MICROINCHES OR LESS		MATERIAL:		DWG. NO.:	
f¹ - 63 MICROINCHES OR LESS					
f² - 125 MICROINCHES OR LESS					
f³ - 250 MICROINCHES OR LESS					

REV.	DATE	DESCRIPTION OF CHANGE	BY
3			
2			
1			

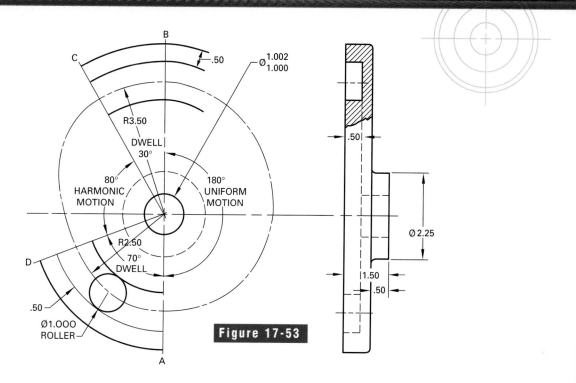

R3.50

DWELL
30°

80°
HARMONIC
MOTION

180°
UNIFORM
MOTION

.50

Ø1.002
1.000

R2.50

70°
DWELL

.50

Ø1.000
ROLLER

.50

Ø2.25

1.50

.50

Figure 17-53

6. Prepare a two-view drawing of the spur gear in Figure 17-26. Make a simplified profile and section as shown. Using the given data, draw in two gear teeth after calculating circular thickness.

7. Complete the gear data for the gears shown in Figure 17-54, using formulas from this chapter. Make an enlarged drawing of a mating spur and pinion as shown. Select a suitable scale. Use an involute to draw the gear tooth profile or ⅛ PD, as directed by the instructor.

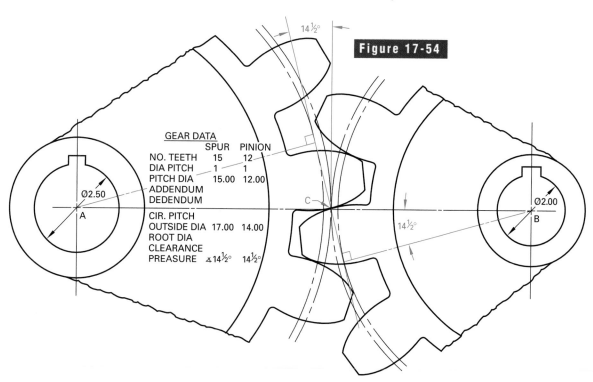

Figure 17-54

14½°

GEAR DATA	SPUR	PINION
NO. TEETH	15	12
DIA PITCH	1	1
PITCH DIA	15.00	12.00
ADDENDUM		
DEDENDUM		
CIR. PITCH		
OUTSIDE DIA	17.00	14.00
ROOT DIA		
CLEARANCE		
PREASURE	∡14½°	14½°

Ø2.50

A

Ø2.00

B

C

14½°

8. Prepare a working drawing of the mating gears in Figure 17-27. Calculate the data necessary to draw the gear teeth and prepare a data table.

9. Prepare a spur-gear detail drawing, using the data given in Figure 17-55 to determine the necessary dimensions. Note that the gear is designed in metric dimensions (using a comma in place of a decimal point) and is dual-dimensioned with decimal inches.

10. Prepare the two views of the bevel gear in Figure 17-28. Review the simplified profile and insert the proper pitch circles as centerlines in the front view. Dimension the drawing.

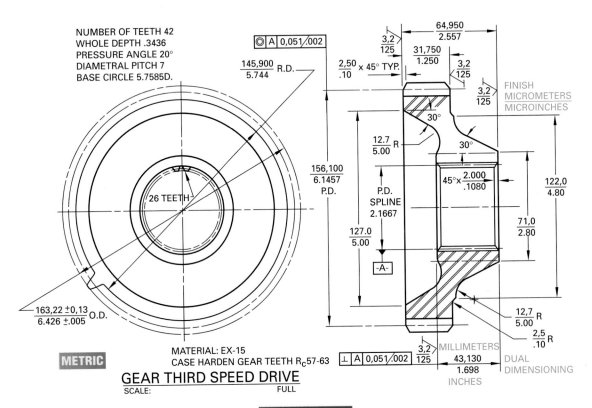

Figure 17-55

Design Problems

Design problems have been prepared to challenge individual students or teams of students. In these problems, you are to apply skills learned mainly in this chapter but also in other chapters throughout the text. They are designed to be completed using board drafting, CAD, or a combination of the two. Be creative and have fun!

1. TEAMWORK Design an automatic bubble-blowing machine that uses commercial bubble solution. It can be designed to use compressed air from a commercial compressor or from a built-in air pump (diaphragm or piston). Use a cam mechanism to open and close the opening where the air and solution form a bubble and exit the machine. Use your most creative imagination in the development of this design.

2. Design a gear mechanism with a cable, powered by a hand crank or electric motor drive, used to raise and lower a 30'-0 steel flagpole for maintenance purposes. The flagpole is hinged at the base. Design the unit to attach to a concrete pad installed 15'-0 from the base of the pole.

3. TEAMWORK Design a device for crushing aluminum beverage cans using a gear mechanism to reduce the effort required to crush the can. The power is to be supplied with a hand crank attached to the gear mechanism. Be aware of safety issues. Design appropriate sheet metal guards to enclose the gears. Add safety devices where needed to protect the operator from flying material.

4. METRIC Design a hold-down device used to hold parts in place on a drill-press table. Use a cam arrangement for holding the part and for quick release. Design the unit to work on stock up to 100 mm thick and have a reach of at least 150 mm. Material optional.

Architectural Drafting

OBJECTIVES

Upon completion of this chapter, you should be able to:

■ Describe and develop a site plan for a single building plot or a small community.

■ Prepare a set of house plans using either board drafting or CAD techniques.

■ Identify and name the parts and materials used in a building.

■ Explain the purpose of each type of plan that is typically included in a set of architectural working drawings.

■ Dimension an architectural drawing.

Drafting Principles

The work that architects do greatly affects the way we live. Our everyday lives take place within an environment designed by architects and city planners. Architects design buildings and the spaces around them to form rural communities, suburbs, or cities. Architecture can be thought of as any type of physical environment, including sprawling suburbs, crowded inner cities, and industrial complexes.

You can easily examine architects' work. Look all around your community. Evaluate some of the new and old styles of housing. Look closely at the apartments, condominiums, churches, banks, stores, and offices in your neighborhood. Are they useful and attractive? Did the architect do a good job of planning? Were durable, attractive materials selected? See Figure 18-1.

A FUNCTIONAL FLOOR PLAN

0 4 8 FT

PERSPECTIVE — PEN & INK RENDERING
LEVEL THREE - PRESENTATION FOR CLIENT – 3 **B**

Figure 18-2

(A) A functional traffic pattern; (B) the architect's visualization of the finished home.

Figure 18-1

Condominiums such as these add geometric design elements as well as a large number of residences to a community.

Those who study and evaluate architectural structures look for three things:

- The structure must be suitable for the activity for which it was designed. A good design takes traffic flow into consideration, as shown in Figure 18-2A.
- The structure must be well engineered. Structural members have to be well constructed, and materials must be suitable.
- The structure must have *aesthetic value*, or beauty. See Figure 18-2B.

In order to understand architectural drafting, it is important that you have a basic understanding of all aspects of this large and complex field. Issues such as career opportunities, community planning, styles of architecture, and various other architectural concepts play a major role in introducing you to this special kind of mechanical drafting.

The Architect's Office

Many different types of jobs are related to the building industry. The rapid growth of small towns and big cities demands the creation of new environments. The team includes:

- architects
- city planners
- landscape architects
- interior designers
- specification writers
- drafting technicians
- illustrators
- construction supervisors
- structural engineers
- mechanical engineers
- office personnel

All of these people work together on large architectural projects. Some employees work with the clients. Others offer design services. Someone has to estimate costs.

The team must also consider new materials and ways of building. For example, one major concern today is energy conservation. The design team is constantly challenged to plan structures that minimize energy use or to use alternative forms of energy. Another challenge is planning and constructing buildings that are compatible with or even beneficial to the environment. See Figure 18-3. As laws are passed requiring stricter measures, the design team must explore new ways to comply. In addition, job estimates and schedules must take local and national trade practices and legal restrictions into consideration. For example, floor plans for residential structures should meet the minimum standards set by the Federal Housing Administration (FHA).

Like other types of businesses, architectural firms can be sole proprietorships, partnerships, or corporations. Sole proprietorships have just one owner. In a typical architectural partnership, there are generally

Figure 18-3

Buildings can be designed to fit into the environment.

two to four main partners. The partners employ six to twelve people. However, some partnerships may have a hundred or more employees. Corporations are owned by people who buy stock in the company. Whatever the form of business organization, an architectural group works as a team. Each member of the architectural team has a well-defined position.

A large firm may offer all the basic architectural services, including architectural design, structural design, mechanical engineering, civil engineering, landscape design, interior design, and urban planning. These services may be used in the design and construction of one building or of an entire city. Smaller firms may provide only one service. They may make subcontracts with other specialized firms in working on a project.

Architectural offices usually have an overall project director to coordinate activity on several different projects. Individual team leaders might help find clients or work on a particular project.

Community Architecture

You can see the effects of the architect's planning by looking carefully at any town or city. Many residential areas have a distinctive architectural style, such as a traditional or contemporary style. Others combine a number of styles. Some styles may be built in various materials to create different effects. A town may have wooden, or frame, cottages and brick bungalows. The larger residential buildings usually have rental apartments. Some may be condominiums, in which the units are individually owned, not rented.

Neighborhoods

Neighborhoods are residential areas that contain housing as well as other buildings and spaces people need to live. Residential buildings are arranged in geometric patterns. The more important, busy streets often form the boundaries of a neighborhood. These streets are frequently lined with small stores, small offices, and apartment buildings. When the pattern is rectangular, the basic unit is called a *block*. Parks or recreation centers are conveniently placed throughout each neighborhood. Figure 18-4 shows a plan for a recreation center designed for a neighborhood that has a lot of space.

Figure 18-4

The plot plan and major features of a recreation center in a proposal to a park board.

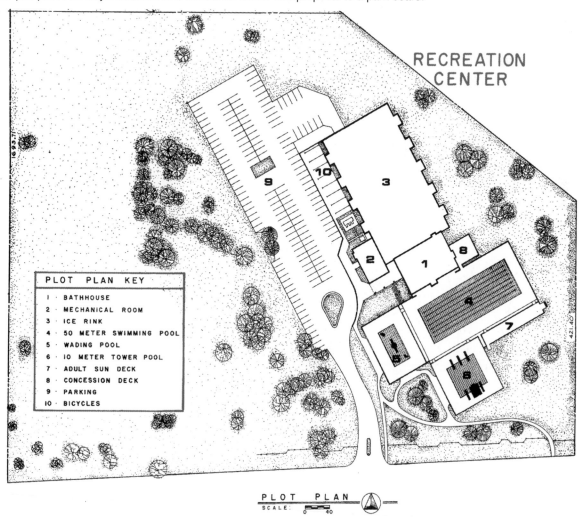

PLOT PLAN KEY

1 · BATHHOUSE
2 · MECHANICAL ROOM
3 · ICE RINK
4 · 50 METER SWIMMING POOL
5 · WADING POOL
6 · 10 METER TOWER POOL
7 · ADULT SUN DECK
8 · CONCESSION DECK
9 · PARKING
10 · BICYCLES

PLOT PLAN
SCALE: 0   40

Urban Development

Neighborhoods combine to make up towns and cities. In the United States, many people have moved from both inner cities and rural areas to suburbs. These are communities on the edges of cities. The growing cities and suburbs need municipal, or public, buildings to serve their residents. Municipal buildings include the city hall and facilities for services such as police, fire protection, water, and sanitation. Health centers such as medical offices and hospitals are also important, as well as cultural facilities such as libraries, museums, and theaters. Architectural services are needed to plan and develop these communities.

Traffic Patterns

Whether a community's streets form rectangular blocks or wandering lanes with "dead ends," they must be connected to major thoroughfares. The major streets are usually lined with large stores, shopping centers, and office buildings. Apartment buildings, banks, and hospitals are also generally located on or near key roads. Building and zoning commissions decide what kinds of buildings will be in different areas. Traffic patterns should be designed to let industrial employees get to and from work easily. The success of a community often relies on good planning for vehicles and pedestrians.

Residential Construction

The main parts of a frame house are shown in Figure 18-5. Not every house has all of these parts. Some parts may be made of different materials. The typical wood framing of an exterior wall is shown in Figure 18-6A. The framing begins on the foundation wall with the sill, header, and sole plate. The stud wall is then erected. Next, sheathing, plywood, or insulation board is put on. Many kinds of facing can be used. Horizontal or vertical siding is typical. In addition, shakes or shingles are often used. Brick veneer, shown in Figure 18-6B, is also common.

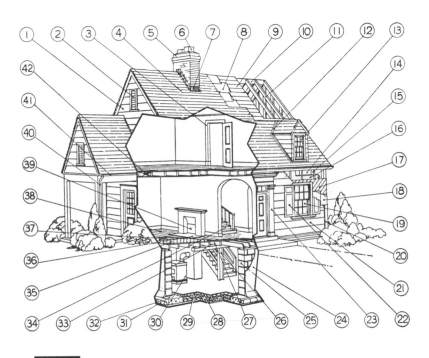

Figure 18-5

Typical parts of a frame house.

1. Gable end	22. Shutters
2. Louver	23. Exterior trim
3. Interior trim	24. Waterproofing
4. Shingles	25. Foundation wall
5. Chimney cap	26. Column
6. Flue linings	27. Joists
7. Flashing	28. Basement floor
8. Roofing felt	29. Gravel fill
9. Roof sheathing	30. Heating plant
10. Ridge board	31. Footing
11. Rafters	32. Drain tile
12. Roof valley	33. Girder
13. Dormer window	34. Stairway
14. Interior wall finish	35. Subfloor
15. Studs	36. Hearth
16. Insulation	37. Building paper
17. Diagonal sheathing	38. Finish floor
18. Sheathing paper	39. Fireplace
19. Window frame and sash	40. Downspout
20. Corner board	41. Gutter
21. Siding	42. Bridging

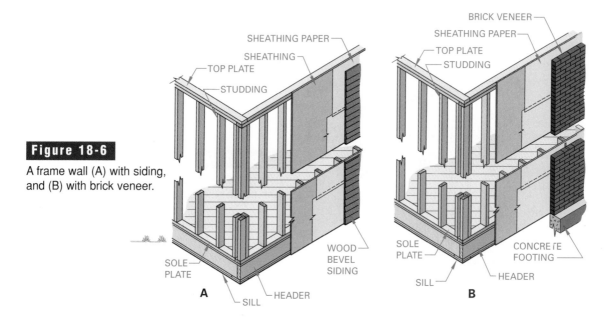

Figure 18-6

A frame wall (A) with siding, and (B) with brick veneer.

SHEATHING PAPER
SHEATHING
TOP PLATE
STUDDING
SOLE PLATE
SILL
HEADER
WOOD BEVEL SIDING

A

BRICK VENEER
SHEATHING PAPER
TOP PLATE
STUDDING
SOLE PLATE
SILL
HEADER
CONCRETE FOOTING

B

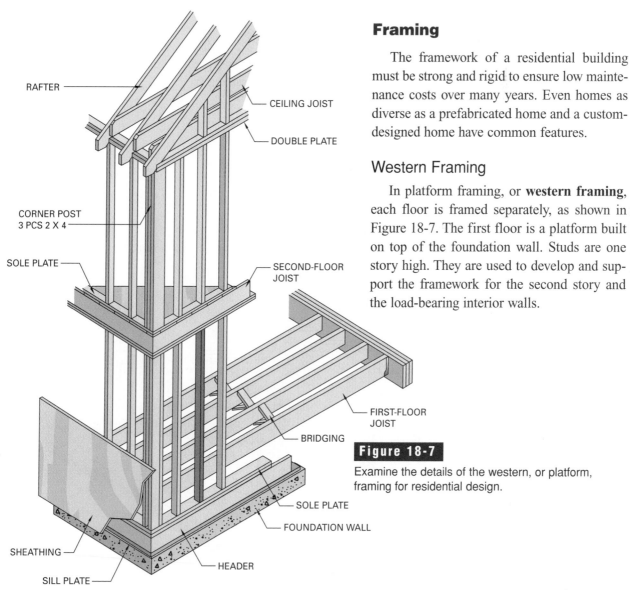

RAFTER
CEILING JOIST
DOUBLE PLATE
CORNER POST 3 PCS 2 X 4
SOLE PLATE
SECOND-FLOOR JOIST
FIRST-FLOOR JOIST
BRIDGING
SOLE PLATE
FOUNDATION WALL
SHEATHING
HEADER
SILL PLATE

Framing

The framework of a residential building must be strong and rigid to ensure low maintenance costs over many years. Even homes as diverse as a prefabricated home and a custom-designed home have common features.

Western Framing

In platform framing, or **western framing**, each floor is framed separately, as shown in Figure 18-7. The first floor is a platform built on top of the foundation wall. Studs are one story high. They are used to develop and support the framework for the second story and the load-bearing interior walls.

Figure 18-7

Examine the details of the western, or platform, framing for residential design.

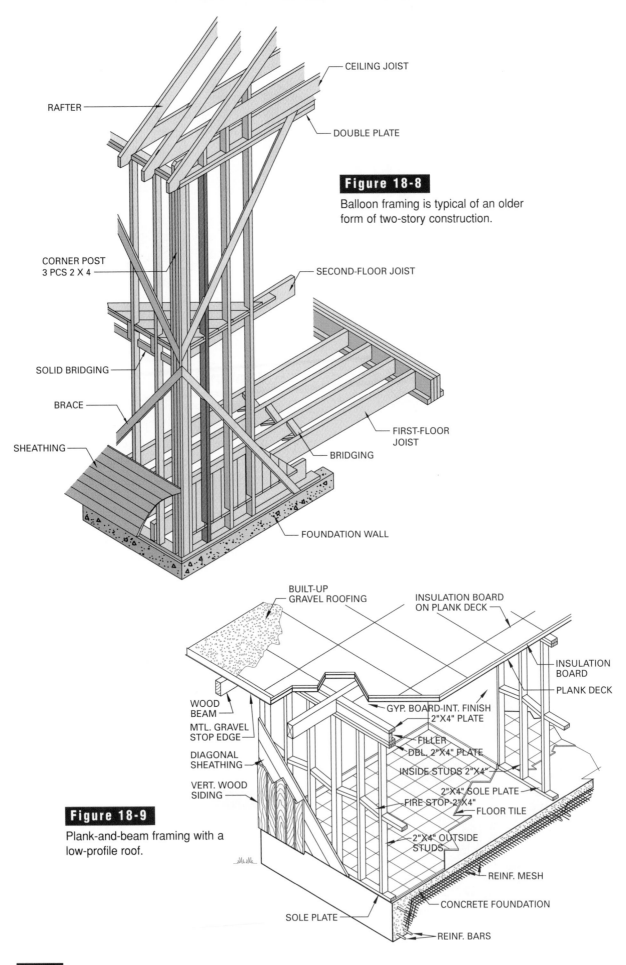

CEILING JOIST

RAFTER

DOUBLE PLATE

Figure 18-8

Balloon framing is typical of an older form of two-story construction.

CORNER POST
3 PCS 2 X 4

SECOND-FLOOR JOIST

SOLID BRIDGING

BRACE

SHEATHING

FIRST-FLOOR
JOIST

BRIDGING

FOUNDATION WALL

BUILT-UP
GRAVEL ROOFING

INSULATION BOARD
ON PLANK DECK

INSULATION
BOARD

PLANK DECK

WOOD
BEAM

GYP. BOARD-INT. FINISH
2"X4" PLATE

MTL. GRAVEL
STOP EDGE

FILLER
DBL. 2"X4" PLATE

DIAGONAL
SHEATHING

INSIDE STUDS 2"X4"

VERT. WOOD
SIDING

2"X4" SOLE PLATE
FIRE STOP-2"X4"

Figure 18-9

FLOOR TILE

Plank-and-beam framing with a
low-profile roof.

2"X4" OUTSIDE
STUDS

REINF. MESH

CONCRETE FOUNDATION

SOLE PLATE

REINF. BARS

Balloon Framing

The studs in a house constructed with **balloon framing** are two stories high, as shown in Figure 18-8. A false girt (ribbon) inserted into the stud wall carries the second-floor joists. A box sill is used. Diagonal bracing brings rigidity to the corners. While this system is not commonly used today, architects must understand it to be able to remodel older homes. Older home remodeling has become a large business in itself.

Plank-and-Beam Framing

Figure 18-9 shows an example of **plank-and-beam framing**. This system uses heavier posts and beams than other systems. These members carry a deck of continuous planking. This kind of framing allows ceilings to be higher and more open, with fewer supporting members, and costs less to build.

Sill Construction

Figure 18-10 shows different types of sill and wall construction. In Figure 18-10A, the frame wall is set up on a box-sill construction. Note the metal termite shield on top of the foundation wall. In Figure 18-10B, a brick veneer starts below the floor line on a stepped foundation wall. Figure 18-10C shows slab construction reinforced with a wire mesh.

Corner Studs and Sheathing

Some typical corner bracing is shown in Figure 18-11. Diagonal sheathing, or covering, was formerly the most common kind of bracing. Today, to save labor and reduce cost, builders use horizontal sheathing and plywood along with modular insulation board. Plywood is used not only on exterior walls, but also for interior subflooring and roof sheathing. Anchor bolts secure the superstructure to the substructure. These are normally ½″ to ¾″ (12 to 19 mm) apart and extend 18″ (457 mm) into the concrete.

Roof Designs

Some basic roof types are shown and named in Figure 18-12. The shape of the roof is often the key to a building's architectural style. The vertical measurement of a roof is called the *rise*. The horizontal

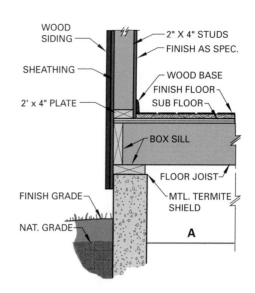

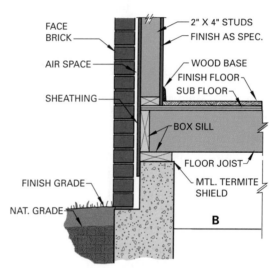

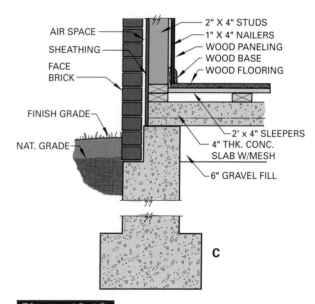

Figure 18-10

Types of sill construction.

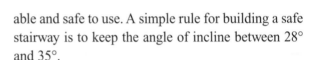

DIAGONAL SHEATHING — CORNER TO NAIL LATH

Figure 18-11

Arrangement of corner studs in forming a frame wall.

dimension is called the *run*. Together, they determine the *roof pitch* (slope).

The most common roof shapes are the gable, hip, flat, and shed shapes. The mansard, gambrel, butterfly, combination, clerestory, and A-frame shapes are used more with specific design styles.

Stairway Framing and Detail

Three types of stairways are the straight run, the platform, and the circular stairways. Stairs are made up of **risers** (the vertical part of each step) and **treads** (the horizontal part of each step). These parts are illustrated in Figure 18-13. The rise of a flight of stairs is the height measured from the top of one floor to the top of the next floor. The run is the horizontal distance from the face of the first riser to the face of the last riser in one stairwell. The run also equals the sum of the width of the treads.

Risers are generally 6½″ to 7½″ (165 to 190 mm) high. The width of the treads is such that the sum of one riser and one tread is about 17″ to 18″ (430 to 480 mm). A 7″ (180-mm) riser and an 11″ (280-mm) tread is considered a general standard. You can easily use a scale to divide the floor-to-floor height into the number of risers. A good stairway feels comfort-able and safe to use. A simple rule for building a safe stairway is to keep the angle of incline between 28° and 35°.

On working drawings, stairways are not usually drawn in their entirety. Instead, break lines are used, and the drawing shows what is on the level beneath the stairs. Refer again to Figure 18-13.

Doors

Doors are usually 6′-8 or 7′-0 (2000 or 2100 mm) high. The width may vary from 2′-0 to 3′-0 (600 to 900 mm), but it is usually 2′-8 or 3′-0 (800 to 900 mm). The thickness varies. Thicknesses of 1⅜″ and 1¾″ (35 to 45 mm) are standard for interior doors. Thicknesses of 1¾″ and 2½″ (45 to 65 mm) are standard for exterior doors. The head, jamb, and sill details may vary depending on whether a swinging, sliding, or folding door is used. The door must fit its frame closely. Yet it must also open and close easily. Figure 18-14 shows various patterns for doors.

Doors to the outside are usually larger than others to allow for heavier use and for bringing in furniture. These doors are usually 3′-0 (900 mm) wide. Bedroom and bathroom doors are usually 2′-6 (792 mm). Bifold and folding accordion doors have special features and sizes.

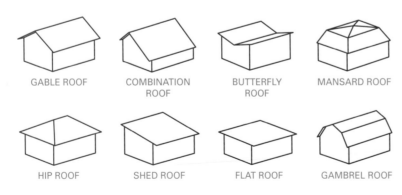

GABLE ROOF   COMBINATION ROOF   BUTTERFLY ROOF   MANSARD ROOF

HIP ROOF   SHED ROOF   FLAT ROOF   GAMBREL ROOF

Figure 18-12

Some typical residential roof types that define style.

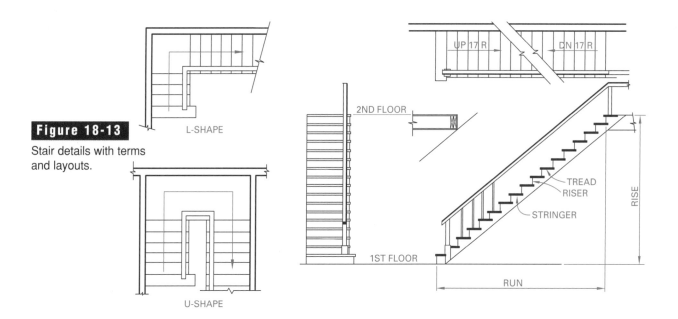

Figure 18-13

Stair details with terms and layouts.

L-SHAPE

U-SHAPE

UP 17 R   DN 17 R

2ND FLOOR

1ST FLOOR

TREAD
RISER

STRINGER

RISE

RUN

Windows

The style of a house determines what style of window is used and how windows are placed. Double-hung and casement windows are practical in most kinds of houses. However, horizontal sliding windows are also very popular. Figure 18-15 shows some of the various types available to the architect.

Casement windows are popular for French and English designs. They are hinged at the sides to swing open (in or out). Hopper windows are hinged at the bottom. Awning windows are also called *projected windows*. Sliding windows move sideways. Double-hung windows are commonly used for Colonial and American-style structures. Each double-hung window contains two independent sashes that can move in a vertical track. A counterbalance, or spring arrangement, holds them at any position desired. Newer windows can have a press-in, spring-loaded track that is very convenient. Fixed windows and jalousies (windows made of adjustable glass slats) are special kinds of windows. Figure 18-16 shows sectional details and some technical terms for a window. Normally, windows are placed in walls so that their tops line up with the top of the door.

Architectural Materials

The new environments designed by architects are created from materials. Detailed drawings and specifications must show how these materials are to be fabricated and constructed. The *Sweets Architectural Catalog* for materials manufacturers is one of the most important tools of the design team. Many materials are available in standard units for the building trades. Others are custom-designed. A few standards are discussed here.

Figure 18-14

Typical front-door patterns.

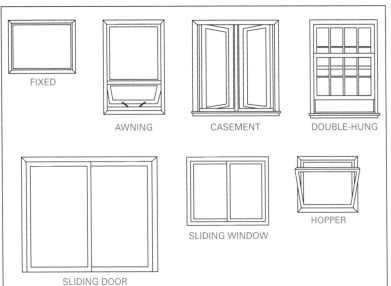

Figure 18-15

Typical windows and sliding doors.

FIXED

AWNING

CASEMENT

DOUBLE-HUNG

SLIDING DOOR

SLIDING WINDOW

HOPPER

Figure 18-16

Double-hung window in a frame wall with technical terms and sectional details.

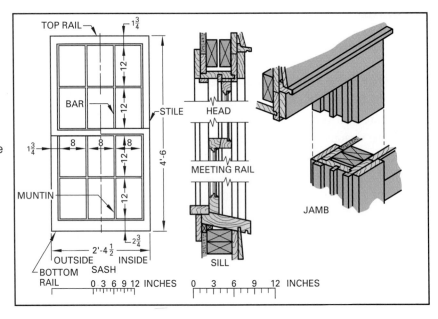

Lumber

Lumber may be specified by its **nominal size**. The nominal dimensions are the dimensions before the wood has been surfaced, or prepared for use. These dimensions differ from the actual dimensions of the surfaced wood. For example, most of the lumber and boards for residential construction are dressed (finished or surfaced) on four sides. The dressed sizes are noted in Table 18-1.

Masonry

Figure 18-17 gives the sizes of brick building materials. The common brick has modular dimensions. Brick, block, stone, and stucco may be used for exterior and interior walls and floors. Most of these materials can serve as structural load-bearing walls for supporting floors and roofs. They generally can be bonded, or overlapped, to increase their structural strength, as shown in Figure 18-18. Bonding also forms decorative patterns. Well-designed masonry needs little upkeep. Its colors and patterns are important to the overall architectural design.

Table 18-1. Standard Sizes of Lumber and Boards

Lumber					
Nominal Size	2 × 4	2 × 6	2 × 8	2 × 10	2 × 12
Dressed Size	$1^1/2'' \times 3^1/2''$	$1^1/2'' \times 5^1/2''$	$1^1/2'' \times 7^1/2''$	$1^1/2'' \times 9^1/2''$	$1^1/2'' \times 11^1/2''$
Nominal Size	4 × 6	4 × 8	4 × 10		
Dressed Size	$3^9/16'' \times 5^1/2''$	$3^9/16'' \times 7^1/2''$	$3^9/16'' \times 9^1/2''$		
Nominal Size	6 × 6	6 × 8	6 × 10		
Dressed Size	$5^1/2'' \times 5^1/2''$	$5^1/2'' \times 7^1/2''$	$5^1/2'' \times 9^1/2''$		
Nominal Size	8 × 8	8 × 10			
Dressed Size	$7^1/2'' \times 7^1/2''$	$7^1/2'' \times 9^1/2''$			
Boards					
Nominal Size	1 × 4	1 × 6	1 × 8	1 × 10	1 × 12
Actual Size, common boards	$3/4'' \times 3^9/16''$	$3/4'' \times 5^9/16''$	$3/4'' \times 7^1/2''$	$3/4'' \times 9^1/2''$	$3/4'' \times 11^1/2''$
Actual Size, shiplap	$3/4'' \times 3''$	$3/4'' \times 4^{15}/16''$	$3/4'' \times 6^7/8''$	$3/4'' \times 8^7/8''$	$3/4'' \times 10^7/8''$
Actual Size, tongue-and-groove	$3/4'' \times 3^1/4''$	$3/4'' \times 5^3/16''$	$3/4'' \times 7^1/8''$	$3/4'' \times 9^1/8''$	$3/4'' \times 11^1/8''$

Figure 18-17

Types and sizes of brick.

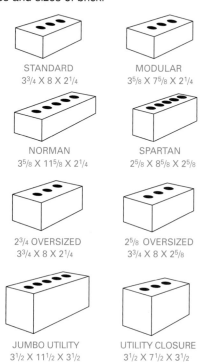

STANDARD
$3^3/4$ X 8 X $2^1/4$

MODULAR
$3^5/8$ X $7^5/8$ X $2^1/4$

NORMAN
$3^5/8$ X $11^5/8$ X $2^1/4$

SPARTAN
$2^5/8$ X $8^5/8$ X $2^5/8$

$2^3/4$ OVERSIZED
$3^3/4$ X 8 X $2^1/4$

$2^5/8$ OVERSIZED
$3^3/4$ X 8 X $2^5/8$

JUMBO UTILITY
$3^1/2$ X $11^1/2$ X $3^1/2$

UTILITY CLOSURE
$3^1/2$ X $7^1/2$ X $3^1/2$

Concrete

Concrete is another type of structural material. Poured into forms, concrete is used as footings and foundation walls to support floor loads. Both interior and exterior walls can be formed from concrete. Concrete can also be precast to give certain finishes. Color, texture, and pattern all work together to create a desired "look" for a concrete wall.

Figure 18-18

Common brick bond for building a wall.

8" ALL ROLOK WALL
COMMON BOND

Calculating Cubic Yards

Concrete is delivered and sold by the cubic yard. To calculate the number of cubic yards needed to pour a footing for a house, proceed as follows:

1. Determine the thickness and width of the footing. For this example, the footing will be 8″ thick by 16″ wide.
2. Determine the lineal feet of footing needed to support all foundation walls. For this example, the number of lineal feet is 264′-0.
3. Convert all three dimensions to yards. (If a dimension is in inches, convert it to yards by dividing by 36. If a dimension is in feet, convert it to yards by dividing by 3.)
4. Calculate cubic yards as follows:

$$\frac{8}{36} \times \frac{16}{36} \times \frac{264}{3} = \frac{33{,}792}{3{,}888} = 8.69$$

or 9 cubic yards.

Modular Coordination

To help simplify and standardize building design, the building industry and ANSI established a standard: Modular Coordination, A62.1. This standard was sponsored by the American Institute of Architects. According to this standard, the basic customary-inch module is 4″ for all U.S. building materials and projects. The ISO has established 100 mm (almost 4″) as the module for countries using the metric system. Modular components are typically designed on a 4″ or 100-mm centerline. All buildings currently designed for customary modular specifications are planned with multiples of 4″. Modular planning reduces building costs and time because materials do not have to be cut to many different sizes, and there is less waste.

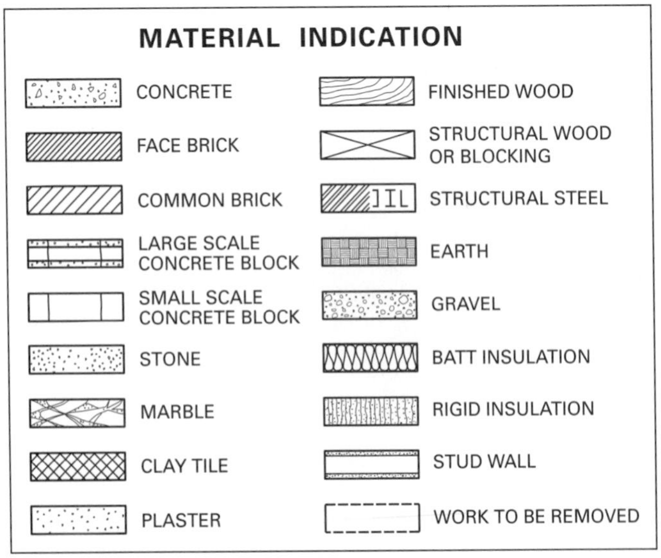

Symbols for Architectural Drawings

Working drawings include many standard symbols that make the drawings less cluttered and easier to read. These symbols are listed in the latest edition of *Architectural Graphic Standards*. The standard graphic symbols include a variety of choices to meet the needs of all types of architectural drawing. Plumbing, electrical, landscaping, and many other types of schematics and symbols have been standardized. These standards are periodically revised by the American Institute of Architects (AIA). CAD symbols approved by the AIA are also available on the Internet in AutoCAD's DWG format.

Material Symbols

Architectural material symbols are similar to section lines or crosshatching in mechanical drafting. However, instead of merely indicating sections, they also indicate the materials used in building. Examples of material symbols are shown in Figure 18-19. These symbols are used in plan, section, and elevation drawings. See Figure 18-20 for examples of how these symbols are used in drawings.

MATERIAL INDICATION

Symbol	Material	Symbol	Material
	CONCRETE		FINISHED WOOD
	FACE BRICK		STRUCTURAL WOOD OR BLOCKING
	COMMON BRICK		STRUCTURAL STEEL
	LARGE SCALE CONCRETE BLOCK		EARTH
	SMALL SCALE CONCRETE BLOCK		GRAVEL
	STONE		BATT INSULATION
	MARBLE		RIGID INSULATION
	CLAY TILE		STUD WALL
	PLASTER		WORK TO BE REMOVED

Figure 18-19

Typical material symbols used in architectural detail.

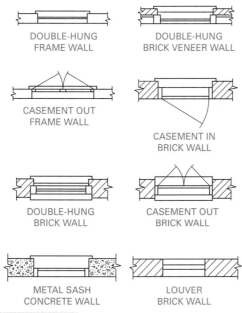

DOUBLE-HUNG FRAME WALL

DOUBLE-HUNG BRICK VENEER WALL

CASEMENT OUT FRAME WALL

CASEMENT IN BRICK WALL

DOUBLE-HUNG BRICK WALL

CASEMENT OUT BRICK WALL

METAL SASH CONCRETE WALL

LOUVER BRICK WALL

Figure 18-20

Window symbols for various types of window openings.

Symbols for Window and Door Openings

Typical windows are shown in Figure 18-20. These standard symbols help the reader interpret window movement. They also show what type of wall material frames the window.

Examples of door symbols are shown in Figure 18-21. The symbol for an outside door shows it ajar in a single line. An arc shows the direction in which the door swings. In residential design, exterior doors generally swing to the interior. In commercial design, exterior doors must swing out for safety reasons. Interior doors are shown in a variety of ways.

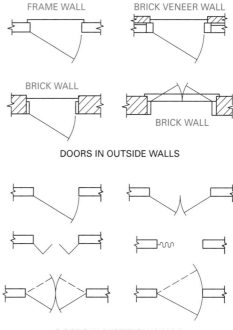

FRAME WALL

BRICK VENEER WALL

BRICK WALL

BRICK WALL

DOORS IN OUTSIDE WALLS

DOORS IN PARTITION WALLS

Figure 18-21

Door openings with swing symbols. These symbols are available on architectural templates for board drafters. CAD operators use architectural symbol libraries.

Working Drawings

Working drawings include plans, elevations, sections, schedules, schematics, and details. Along with the specifications for materials and finish, they are the guides used in the construction and erection of a building. Figure 18-22 is a photograph of a completed home. In the following discussion, we will use this home as an example to demonstrate the various kinds of drawings needed in a complete set of architectural working drawings.

Most residential home designs start as an idea put forth as a sketch of a floor plan, and perhaps a pictorial sketch also, as shown in Figure 18-23. Although they are not formally part of the working drawings, perspective drawings like the one in Figure 18-23 are often created for sales and marketing purposes, as well as for client approval. It is easiest to see a building's architecture in a pictorial presentation drawing. One-, two-, or three-

Figure 18-22

The house built from the plans shown on the following pages.

Figure 18-23

A preliminary, undimensioned study of a house developed for client approval.

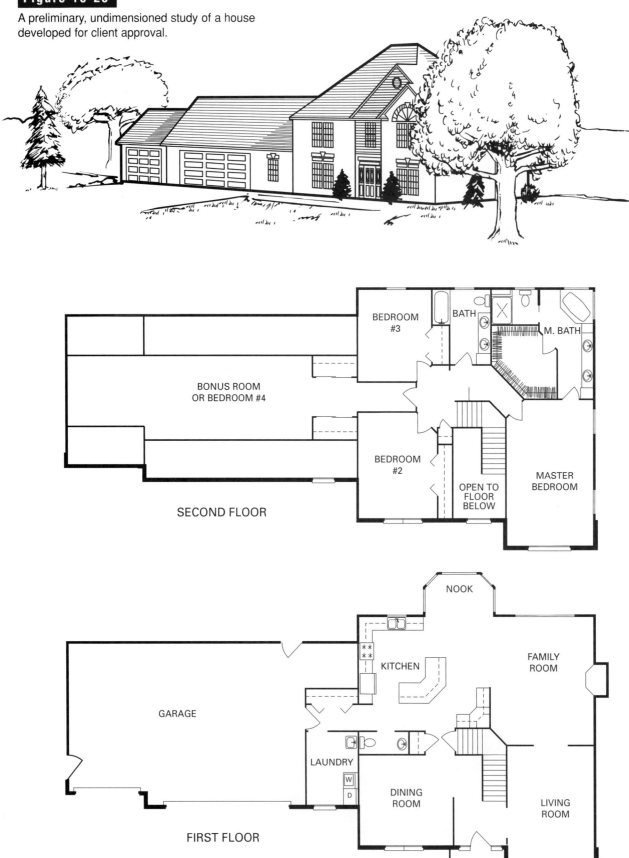

SECOND FLOOR

BEDROOM #3

BATH

M. BATH

BONUS ROOM OR BEDROOM #4

BEDROOM #2

OPEN TO FLOOR BELOW

MASTER BEDROOM

FIRST FLOOR

NOOK

GARAGE

KITCHEN

FAMILY ROOM

LAUNDRY

W
D

DINING ROOM

LIVING ROOM

point perspective drawings show a building in a realistic way. After the basic design has been approved by the client, the architect proceeds to create the working drawings.

Foundation Plan

The foundation plan in Figure 18-24 serves as a guide for constructing the foundation. Therefore, it must be completely dimensioned. It should be checked with the first-floor plan. In fact, you can develop the foundation plan directly from the first-floor plan.

Floor Plan

A floor plan is a section that cuts horizontally through walls and shows room arrangements, as shown in Figure 18-25. In general, a floor plan is made for each floor (story) of a building. It shows all walls, doors, windows, and other structural features. It also shows fixed features such as the cabinets and stairways. In some cases, it may also show heating and plumbing fixtures and lighting outlets. Windows and doors are located accurately, then represented by conventional symbols. Their sizes are listed on schedules that accompany the plans, as shown in Figure 18-26.

The size of a house is generally specified by the total number of square feet in the floor plan. A small house might have only 1200 square feet. A large house might have several thousand square feet.

When specifying the square footage of a house, several variables need to be considered. For example, gross square footage includes the garage, enclosed porches and patios, covered walkways, finished or partially finished basements, etc. When calculating the number of square feet of living space, many of these extras would not be included.

When determining square footage of heated area, consider only the inside to inside dimensions of heated spaces. The square footage of closets is included with the room in which they are located. The basement and garage are included only if they are normally heated. Any finished and heated rooms in the basement would be included.

Figure 18-24

A foundation plan with structural notes and dimensions ready for contractor approval.

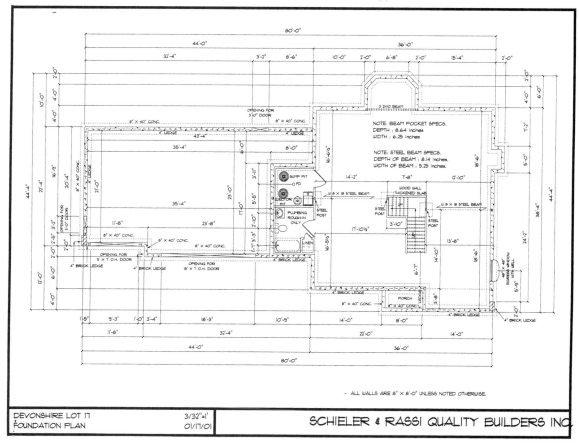

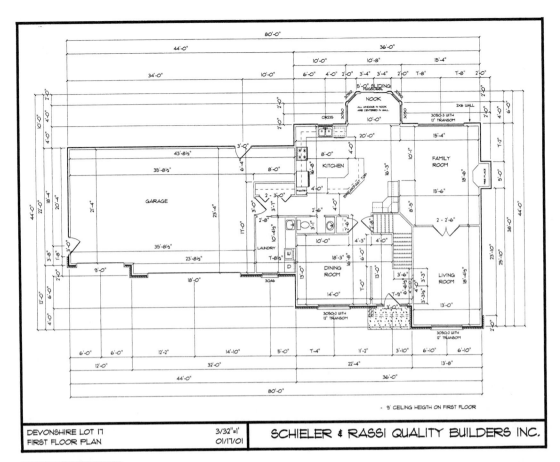

DEVONSHIRE LOT 17
FIRST FLOOR PLAN

3/32"=1'
01/17/01

SCHIELER & RASSI QUALITY BUILDERS INC.

Figure 18-25

The floor plan.

Figure 18-26

Window and door schedules.

SCHEDULES							
WINDOWS				DOORS			
MARK	SIZE	TYPE	REMARKS	MARK	SIZE	TYPE	REMARKS
W1	14-3'-4" x 5'-5"	D.H. VINYL		A	9'-0" x 6'-8"	OVERHEAD	GARAGE
W2	3-3'-4" x 4'-9"	D.H. VINYL		B	18'-0" x 6'-8"	OVERHEAD	GARAGE
W3	1-3'-4" x 4'-6"	D.H. VINYL		C	5'-0" x 6'-8"	FWD GLD PAT	NOOK
W4	2-2'-0" x 5'-5"	D.H. VINYL		D	3'-0" x 6'-8"	1/2 LIGHT	MN ENTR
W5	1- 3'-4" x 2'-9"	D.H. VINYL		E	2-1'-0" x 6'-8"	1/2 LT SIDE LT	
W6	1-3'-0" x 3'-0"	D.H. VINYL		F	3-3'-0" x 6'-8"	6 PANEL	
W7	8-3'-0" x 1'-0"	D.H. VINYL	TRANSOMS	G	2-6'-0" x 6'-8"	BYPASS	
W8	2-4'-0" x 4'-0"	D.H. VINYL	FIXED PICT.	H	2-2'-6" x 6'-8"	BIFOLD	
W9	1-2'-2" x 1'-3"	D.H. VINYL	CIRCLE TOP	I	2-3'-0" x 6'-8"	BIFOLD	
W10	5'-0" x 1'-0"	D.H. VINYL	TRANSOM	J	2'-6" x 6'-8"	SGL POCKET	
				K	2'-4" x 6'-8"	SGL POCKET	
				L	2'-8" x 6'-8"	SGL POCKET	
				M	2-2'-6" x 6'-8"	FRENCH	
				N	2-2'-8" x 6'-8"	FRENCH	
				O	1'-8" x 6'-8"	ST 4 PANEL	
				P	2-2'-6" x 6'-8"	ST 4 PANEL	
				Q	4-2'-8" x 6'-8"	ST 4 PANEL	

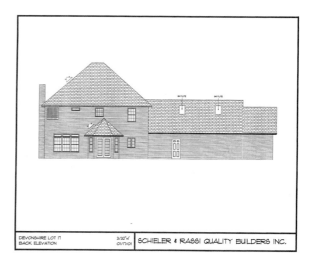

Figure 18-27

Elevations showing foundation walls and roof pitch.

Elevations

The vertical projections of buildings that help define their structural form and architectural style are called **elevations**. The elevations show the exterior look of a house, floor and ceiling heights, openings for windows and doors, roof pitch, and selected materials. The elevations are shown in the architectural style chosen and in relationship to the plan. Elevations allow the architect to study the balance or symmetry of the design and consider the need for changes.

Figure 18-27 shows the elevations for our example home. A complete set of plans includes at least four elevations, one for each side of the building.

Sections

Architects create sections to show internal features that are not visible on the plan view or elevations. The two basic kinds of sections are detail sections and cross sections.

In a **detail section**, a vertical plane cuts through walls to show construction details. Detail sections are generally drawn to a much larger scale than that of the elevation. They show the wall in a much clearer and more detailed view.

Figure 18-28 details a typical wall section from substructure to superstructure. Door and window details are also included. They are developed as the building progresses. That way, the millwork and custom framework will assemble readily.

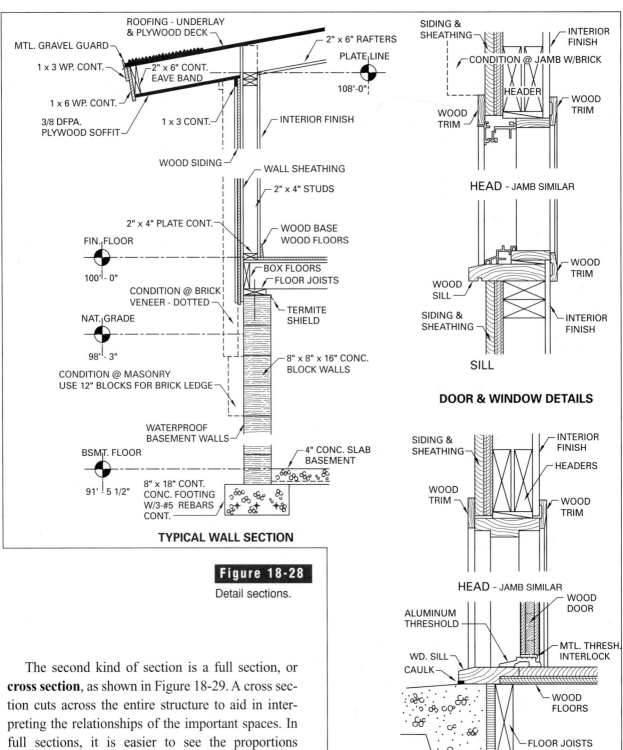

TYPICAL WALL SECTION

- ROOFING - UNDERLAY & PLYWOOD DECK
- MTL. GRAVEL GUARD
- 1 x 3 WP. CONT.
- 2" x 6" CONT. EAVE BAND
- 2" x 6" RAFTERS
- PLATE LINE
- 108'-0"
- 1 x 6 WP. CONT.
- 3/8 DFPA. PLYWOOD SOFFIT
- 1 x 3 CONT.
- INTERIOR FINISH
- WOOD SIDING
- WALL SHEATHING
- 2" x 4" STUDS
- 2" x 4" PLATE CONT.
- WOOD BASE WOOD FLOORS
- FIN. FLOOR
- 100'- 0"
- BOX FLOORS
- FLOOR JOISTS
- CONDITION @ BRICK VENEER - DOTTED
- TERMITE SHIELD
- NAT. GRADE
- 98'- 3"
- 8" x 8" x 16" CONC. BLOCK WALLS
- CONDITION @ MASONRY USE 12" BLOCKS FOR BRICK LEDGE
- WATERPROOF BASEMENT WALLS
- BSMT. FLOOR
- 91'- 5 1/2"
- 8" x 18" CONT. CONC. FOOTING W/3-#5 REBARS CONT.
- 4" CONC. SLAB BASEMENT

Figure 18-28

Detail sections.

DOOR & WINDOW DETAILS

- SIDING & SHEATHING
- INTERIOR FINISH
- CONDITION @ JAMB W/BRICK
- HEADER
- WOOD TRIM
- WOOD TRIM

HEAD - JAMB SIMILAR

- WOOD SILL
- WOOD TRIM
- SIDING & SHEATHING
- INTERIOR FINISH

SILL

- SIDING & SHEATHING
- INTERIOR FINISH
- HEADERS
- WOOD TRIM
- WOOD TRIM

HEAD - JAMB SIMILAR

- ALUMINUM THRESHOLD
- WOOD DOOR
- MTL. THRESH. INTERLOCK
- WD. SILL
- CAULK
- WOOD FLOORS
- FLOOR JOISTS

SILL

The second kind of section is a full section, or **cross section**, as shown in Figure 18-29. A cross section cuts across the entire structure to aid in interpreting the relationships of the important spaces. In full sections, it is easier to see the proportions between spaces and how these relate to construction and use.

Site Plan

Figure 18-30 shows the site plan for the example house. A **site plan**, or plot plan, shows the lot and locates the house on it. It should give complete and accurate dimensions. It should also show all drive-ways, sidewalks, utility easements, and other pertinent information as required by the building inspector. Zoning requirements adopted by the city or community dictate minimum distances the house must be positioned from streets and lot lines on the front, rear, and sides of the lot.

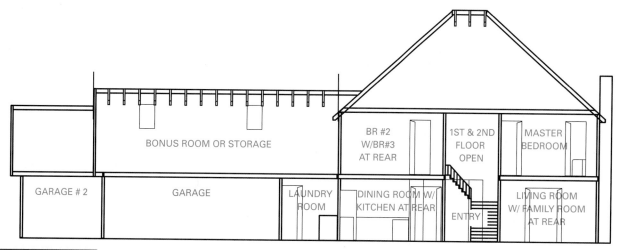

Figure 18-29

The cross section cuts across the entire structure.

BONUS ROOM OR STORAGE

BR #2 W/BR#3 AT REAR

1ST & 2ND FLOOR OPEN

MASTER BEDROOM

GARAGE # 2

GARAGE

LAUNDRY ROOM

DINING ROOM W/ KITCHEN AT REAR

ENTRY

LIVING ROOM W/ FAMILY ROOM AT REAR

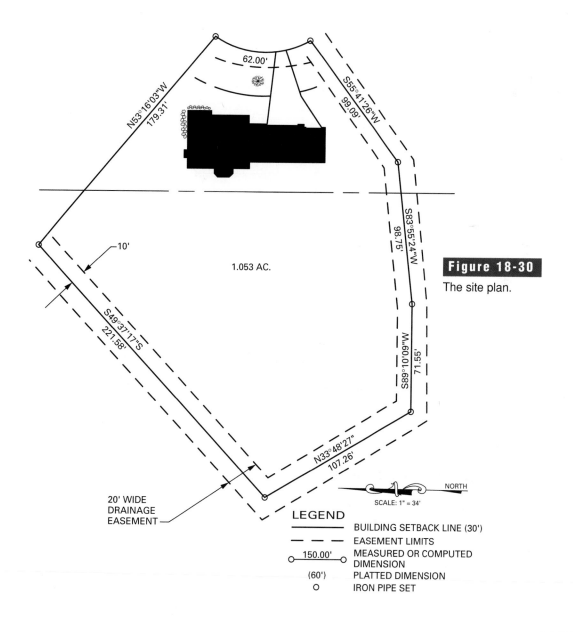

Figure 18-30

The site plan.

62.00'

N53°16'03"W 179.31'

S55°41'26"W 99.09'

S83°55'24"W 98.75'

10'

1.053 AC.

S49°37'17"S 221.58'

S89°01'09"W 71.55'

N33°48'27" 107.26'

20' WIDE DRAINAGE EASEMENT

NORTH
SCALE: 1" = 34'

LEGEND

————	BUILDING SETBACK LINE (30')
– – – –	EASEMENT LIMITS
o——150.00'——o	MEASURED OR COMPUTED DIMENSION
(60')	PLATTED DIMENSION
O	IRON PIPE SET

Other Types of Drawings

Many sets of working drawings also include plans that specify details for various contractors. For example, most sets include an electrical wiring plan. The electrical plan for the example house is shown in Figure 18-31. It shows the circuits for 110V and 220V service. It also shows the electric outlets and switches located for the major appliances.

Other plans that might be included are plumbing plans and heating, ventilation, and air-conditioning (HVAC) plans. However, when the specifications for these specialties are fairly simple and straightforward, these details are sometimes included on the basic floor plan.

Figure 18-31

An electrical plan indicates circuit layout and services.

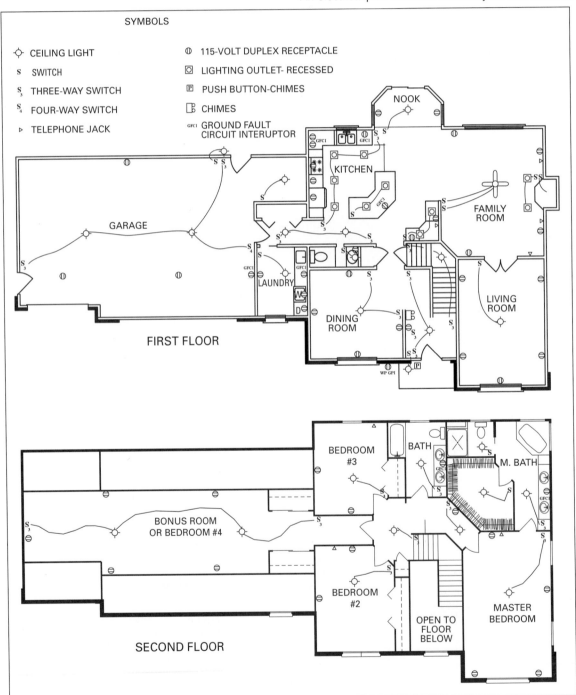

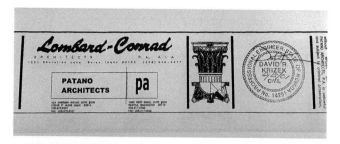

Figure 18-32

Title block and logo.

Schedules and Specifications

Schedules are lists that define and describe details shown by symbols on the actual drawings. A list of all the types of doors and their sizes is an example of a schedule. Refer again to Figure 18-26. The schedules are usually fitted on an unused area of one of the drawing sheets. However, if there are many schedules or if they are particularly complex, they may be placed on a separate drawing sheet.

Architectural **specifications** are detailed written instructions that should accompany every set of plans. They are essential for turning these plans into a building. They note the general conditions of the site. They also specify the materials to the client's and contractor's mutual agreement. Guidelines for specifications have been published by the Construction Specification Institute (CSI) in their *Manual of Practice*. These guidelines are available through CSI.

Preparing Title Blocks

Like other drafting companies, architectural firms usually create a quickly identifiable image with a logo, as shown in Figure 18-32. Some firms use forms with preprinted borders, title blocks, and revision blocks. Others store these features directly in the company's CAD templates.

Dimensioning Techniques

With a few exceptions, the general dimensioning techniques discussed in Chapter 7 are used to dimension architectural drawings. However, the dimension line is unbroken. In addition, the aligned dimensioning method is used in architectural and structural drafting. Dimensions appear above the dimension line and are given in feet and inches, as shown in Figure 18-33.

Figure 18-33

Techniques in architectural dimensioning for a frame wall and a brick wall.

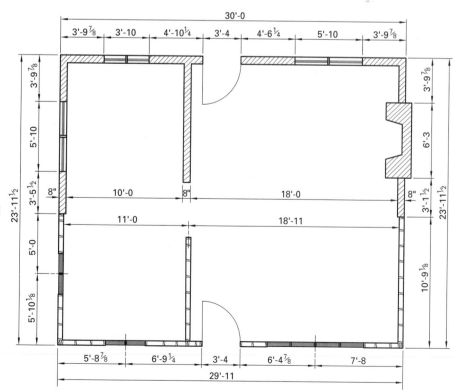

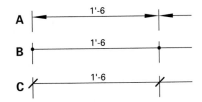

A 1'-6

B 1'-6

C 1'-6

Figure 18-34

Arrowheads, heavy dots, and heavy 45° slashes are used in architectural dimensioning. Dimensioning text is normally ⅛″ or ³⁄₃₆″ in height.

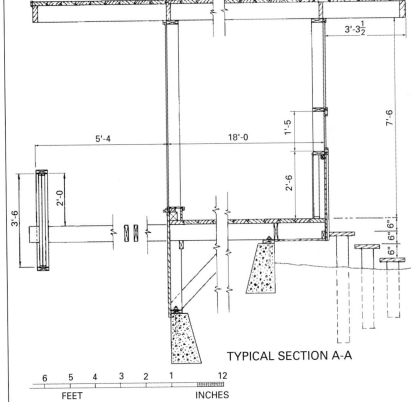

TYPICAL SECTION A-A

FEET   INCHES

Figure 18-35

Dimensions that apply to section and plan drawings.

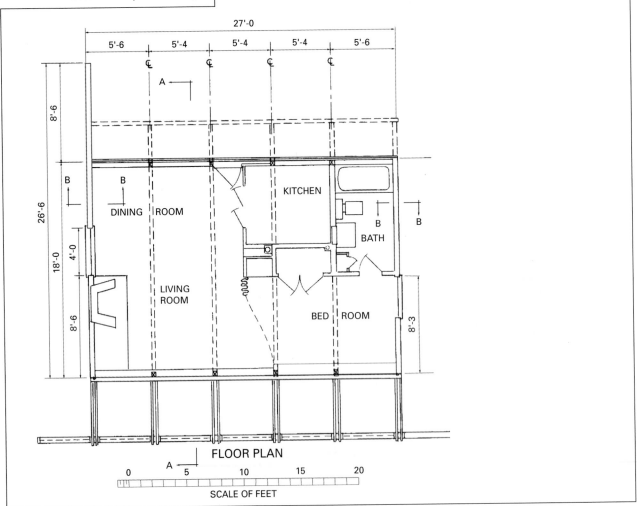

FLOOR PLAN

SCALE OF FEET

The dimension lines are placed about ⅜″ (10 mm) apart and end with the usual arrowhead or with one of several other symbols, as shown in Figure 18-34. Note that dimension lines may cross when sizes of interior rooms are given. The plan has only width and depth dimensions, as shown. Any needed heights are shown as floor-to-ceiling dimensions on one of the elevations or on a suitable wall section, as shown in Figure 18-35.

Construction and Finish Dimensions

The dimensions used on architectural drawings are either finish or construction dimensions. The elevations generally have the expected finished dimensions. The floor plan and sectional details have either construction or finish dimensions, as shown in Figure 18-35.

Dimensioning Exterior Walls

Structural wall openings on the plan drawings for frame and brick-veneer structures are dimensioned in similar ways, but again, there are some differences. Figure 18-36 shows standard dimensioning practice for residential structures with various types

of exterior walls. The crosshatched masonry wall in Figure 18-36A is dimensioned to the outside face of the wall. This is done with solid brick, concrete, and concrete-block walls. The frame wall in Figure 18-36B is dimensioned to the face of the stud. The brick-veneer wall in Figure 18-36C with cross-hatched brick and a sole plate is dimensioned to the face of the stud. The remaining material is assumed to be exterior wall facing.

Architectural Scales

Residential designs and details are usually developed with the aid of reduction scales. The architect's scale is discussed in Chapters 3 and 4. The ¼″ = 1′-0 (1:50 in metric) scale is best suited for plans for houses and small buildings. The usual scale for larger buildings is ⅛″ = 1′-0 (1:100 in metric). Plot plans may be drawn at ⅒″ = 1′-0 or 1/32″ = 1′-0, but it is better to use an engineer's scale at 1″ = 20′-0, 1″ = 30′-0, or 1″ = 40′-0 (1:200, 1:300, or 1:400 in metric). Enlarged details are developed at 1″ = 1′-0 or 1½″ = 1′-0. Sectional details are defined at ½″ = 1′-0 or ¾″ = 1′-0. Some details may require half-size or full-size drawings. Metric scale for enlarged drawings may be 1:5, 1:10, or 1:20.

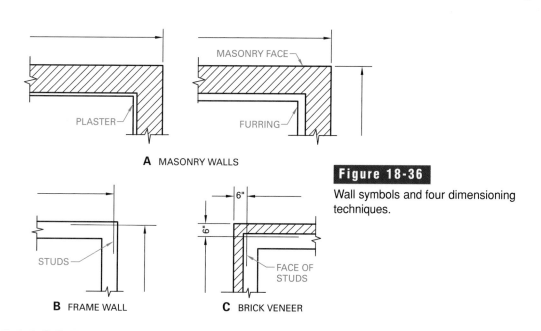

A MASONRY WALLS

B FRAME WALL

C BRICK VENEER

Figure 18-36
Wall symbols and four dimensioning techniques.

Board Drafting Techniques

The board drafting techniques that you have learned throughout the earlier chapters of this textbook apply almost entirely to architectural drafting. The use of instruments, line technique, the projection of views and details, sectional views, dimensioning, and pictorial drawing all play an important role in producing architectural drawings. However, there are a few techniques that are unique to architectural drawings.

Special Techniques and Tools

Like other mechanical drawings, architectural drawings provide the information necessary to build a product. However, this different type of product—buildings and other structures—creates a need for specialized techniques and tools. These will be covered in the next few paragraphs.

Line Technique

Developing good line technique is a very important skill for architectural designing and detailing. In some cases, architectural drafting practices differ slightly from mechanical drafting practices. Figure 18-37 shows the proper technique for drawing lines on an architectural drawing. Notice that the lines intersect and stop after forming a flared corner.

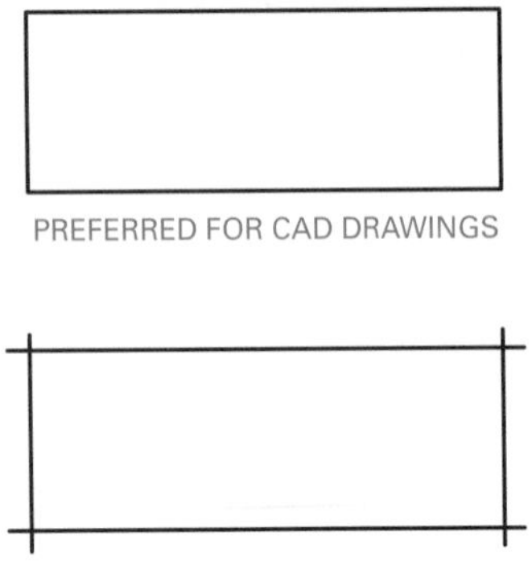

Lettering

Both traditional and contemporary styles of lettering are commonly used by architects. Refer to Chapter 2 for more information about traditional and contemporary lettering styles.

Fractions are always made with a division line. In architectural drafting, two styles are acceptable, as shown in Figure 18-38. Notice that in the style that uses the horizontal division bar, the fraction is twice the height of the whole number. Since the numerals in the fraction never touch the division bar, they are about ¾ the height of the whole number. As described in Chapter 2, guidelines are used for both lettering and dimensioning to aid in producing neat and consistent freehand lettering.

In the second style of fraction—the one that uses a diagonal division bar—the height is only slightly greater than the whole number. In using this style, drafters generally use only the guidelines for whole numbers and letters. The height of the fraction is then

Figure 18-38

Single-stroke freehand architectural-style lettering.

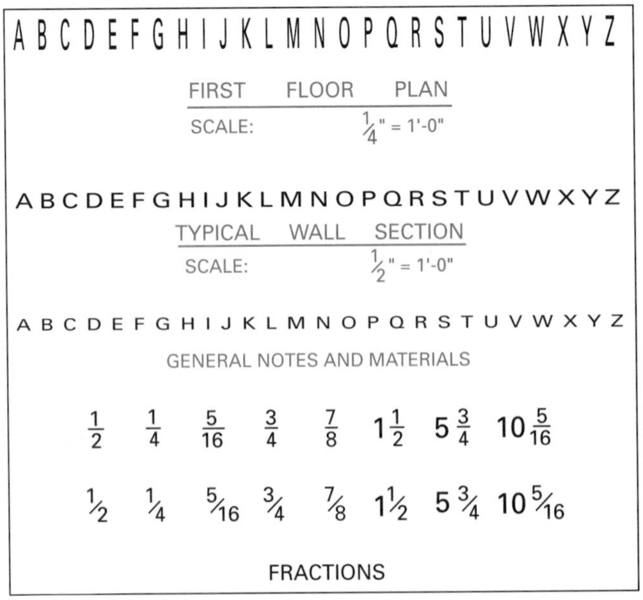

ABCDEFGHIJKLMNOPQRSTUVWXYZ

FIRST   FLOOR   PLAN

SCALE: $\frac{1}{4}$" = 1'-0"

ABCDEFGHIJKLMNOPQRSTUVWXYZ

TYPICAL   WALL   SECTION

SCALE: $\frac{1}{2}$" = 1'-0"

ABCDEFGHIJKLMNOPQRSTUVWXYZ

GENERAL NOTES AND MATERIALS

$\frac{1}{2}$ $\frac{1}{4}$ $\frac{5}{16}$ $\frac{3}{4}$ $\frac{7}{8}$ $1\frac{1}{2}$ $5\frac{3}{4}$ $10\frac{5}{16}$

$\frac{1}{2}$ $\frac{1}{4}$ $\frac{5}{16}$ $\frac{3}{4}$ $\frac{7}{8}$ $1\frac{1}{2}$ $5\frac{3}{4}$ $10\frac{5}{16}$

FRACTIONS

Figure 18-37

Architectural line technique.

PREFERRED FOR CAD DRAWINGS

PREFERRED FOR BOARD DRAFTING

estimated on the lettering guidelines. The angle of the diagonal division bar is approximately 60° and is always estimated. It is never drawn with instruments.

Templates

Templates have improved the quality of drawings and increased efficiency in the architect's office by providing standard symbols that can be drawn more quickly. Templates are usually made of lightweight plastic with openings that form structural shapes or fixtures, as shown in Figure 18-39. You follow the outline of the opening with your pencil or pen to make the graphic image. Architectural templates are available in ⅛″ and ¼″ scales.

Pressure-Sensitive Materials

Many graphic aids are available to help the designer meet professional standards on both plan and elevation drawings. Examples of these heat-resistant and pressure-sensitive symbols are shown in Figure 18-40. Many landscape symbols, such as shrubs, hedges, and trees, are available in both plan and elevation forms. There are also door, furniture, plumbing, and electrical symbols, and many other

Figure 18-40

Examples of pressure-sensitive symbols used for architectural presentation.

Figure 18-39

Examples of templates available for architectural drawings.

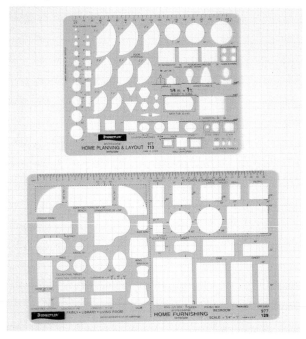

items for presentation and detail drawings. The drafter transfers these symbols to vellum by rubbing them using a smooth, blunt instrument.

Producing a Floor Plan

Throughout this chapter you have learned about the various kinds of construction, construction materials, styles of architecture, and the various drawings that make up a set of house plans. In some respects, a set of house plans is similar to a set of working drawings for a manufactured product. The set must include all of the drawings and all of the necessary information, or specifications, to construct the house.

Figures 18-24 through 18-31 give an example of a set of plans for a residence. The procedure for preparing any of these drawings is essentially the same as the procedure you learned in earlier chapters for preparing a drawing of a machine part. The best practice is to begin with a freehand sketch, add

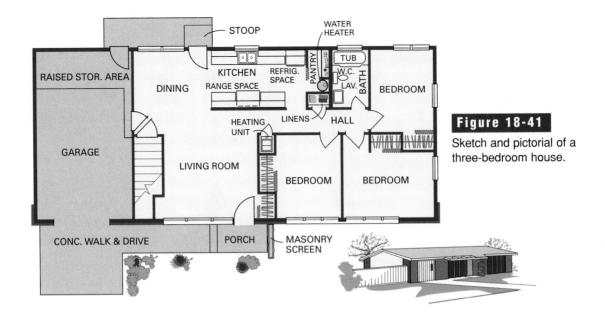

Figure 18-41

Sketch and pictorial of a
three-bedroom house.

details, darken the lines, and add dimensions and notes. We will follow this same procedure in making a drawing of a floor plan for a three-bedroom house. The procedure is as follows:

1. Establish the overall size of the house, the number and types of rooms, and the general layout. For this practice exercise, this information is provided in Figure 18-41.

2. Prepare a freehand sketch of the floor plan. In practice, several sketches may be necessary before you arrive at the precise layout you want. For this exercise, assume that the floor plan

shown in Figure 18-41 is the final design layout. The pictorial sketch is optional for this exercise, but note that a pictorial sketch is often useful in showing a client what the finished residence will look like.

3. In most cases, the scale for the floor plan of a residence is $\frac{1}{4}'' = 1'\text{-}0$. After preparing the drawing sheet, block in the overall size and shape of the floor plan using only light construction lines. Add interior walls and exterior and interior wall openings such as windows and doors. See Figure 18-42.

Figure 18-42

Begin by laying out the exterior and interior walls using light construction lines.

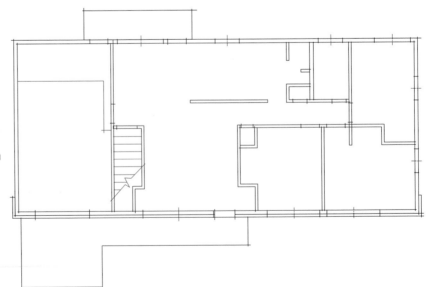

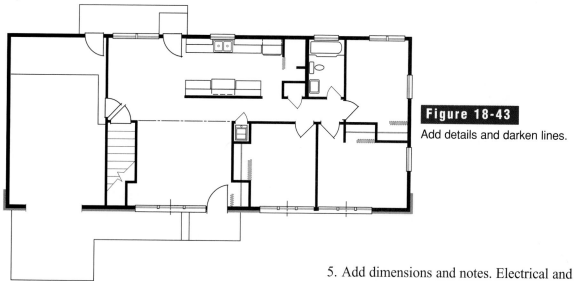

Figure 18-43

Add details and darken lines.

5. Add dimensions and notes. Electrical and plumbing symbols may be added if desired. However, in most cases, separate electrical, plumbing, and heating plans are more practical because they are less cluttered, and therefore easier to read. The finished floor plan is shown in Figure 18-44.

4. Add details such as kitchen cabinets, appliances, bathroom fixtures, door and window symbols, and other details that apply to your particular plan. Darken all lines. Also add any exterior treatments, such as a brick veneer, at this time. See Figure 18-43.

Figure 18-44

The finished floor plan.

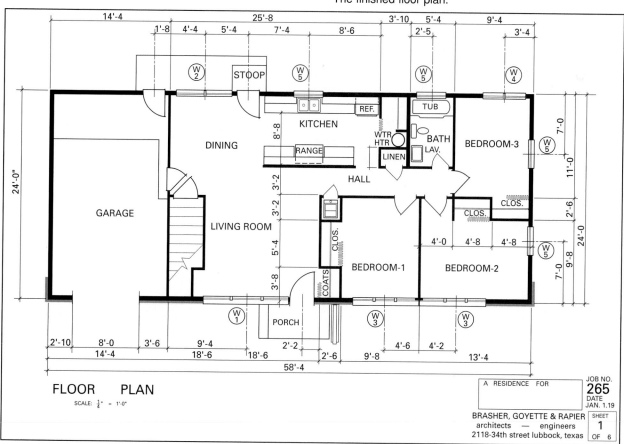

Architectural CADD (computer-aided design and drafting) has grown to include much more than just an electronic means of creating a set of working drawings. Using A/E/C (architectural, engineering, and construction) drawings, architects can now integrate various functions with CADD systems for complete documentation, as shown in Figure 18-45. Advanced systems for the integrated architectural office often include:

- 3D architectural modeling
- architectural production drawings
- space planning/facility management drawings
- engineering production drawings

In addition to A/E/C applications, software companies have developed a variety of architectural applications to meet the demands of different architectural companies. These programs form a distinct subset of the drafting software available for general use. While many architects use general-purpose CAD programs such as AutoCAD and CADKEY, most of them supplement the basic program with add-on software designed especially for architectural design and drafting. Some architectural firms use nothing but specialty software, foregoing the large, expensive, general-purpose packages.

More and more companies are starting to provide advanced three-dimensional visualization of their product designs. These companies are likely to use CADD software designed specifically for 3D architectural drawings, such as DataCAD or ArchiCAD. Figure 18-46 shows a model created in ArchiCAD.

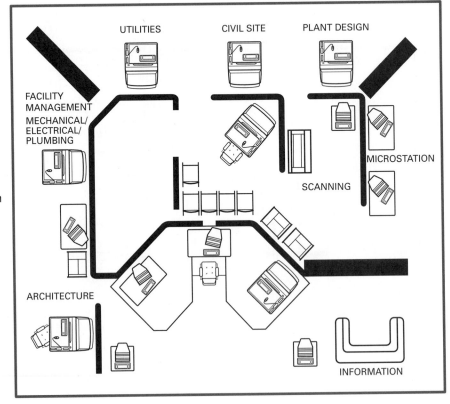

Figure 18-45

An architect's studio with an integrated A/E/C system.

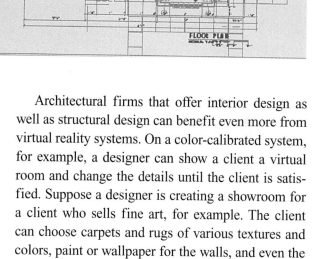

The same plan that was used for construction of this Boston restaurant was used to generate all of the interior and exterior elevations. The architect used ArchiCAD to create 3D views as he drew the 2D plan.

Virtual Evaluation

The advanced architectural and rendering software available today allows specialists in architectural visualization to provide virtual images and effects that help designers and clients "see" what a finished structure will look like—before construction on the structure has even started. For example, a graphic artist or designer can create a photorealistic image of the building based on a 3D model. The artist renders the model using software such as 3D Studio Max, Alias PowerAnimator, or a similar special-effects package. The result is an image in which texture, light and shadow, and balanced proportions make the image look almost like a photograph. See Figure 18-47.

A more radical departure from traditional architectural drafting is the virtual walkthrough. A walkthrough is a "tour" through a 3D model. By viewing the model from the inside, with windows, doors, and other structural features in place, the designer or client can tell if the windows in the room look unbalanced, or if they need to be moved up or down.

The simplest walkthroughs are very structured. The viewer has to look where the "camera" points, without the option of looking right or left. More advanced virtual reality systems allow the viewer to look all around and actually "see" in that direction.

Architectural firms that offer interior design as well as structural design can benefit even more from virtual reality systems. On a color-calibrated system, for example, a designer can show a client a virtual room and change the details until the client is satisfied. Suppose a designer is creating a showroom for a client who sells fine art, for example. The client can choose carpets and rugs of various textures and colors, paint or wallpaper for the walls, and even the type of lighting for the showroom. The designer loads these textures, colors, and lights into the model, and the client "walks through" the showroom to determine the effects.

 Success on the Job

Work Orders

In many companies, work orders are used to communicate specific jobs and tasks between departments. The CAD department receives work orders from other departments that describe exactly what the CAD operators must do to fulfill a job's requirements. In turn, the CAD department may write work orders to departments such as maintenance to request computer repairs, for example. Become familiar with the work order used by your company. Understanding how to fill out a work order completely and accurately will improve your chances for success on the job.

Figure 18-47

This model of a toy store was rendered in Alias PowerAnimator. Notice the special lighting effects and textures that make the model look realistic.

Tools and Commands

For more traditional drafting techniques, Auto CAD provides tools and commands that have special applications in architectural drafting. Study this information before you proceed to the practice example.

Architectural Symbol Libraries

As in other specialty areas of drafting, symbol libraries for architectural drawings can significantly decrease the time and effort required to create a set of working drawings. Symbol libraries are available for many aspects of architectural drawings, including:

■ floor and wall cabinet symbols (plan and elevation views)
■ electrical symbols
■ HVAC symbols
■ window and door symbols (plan and elevation views)
■ furniture (plan and elevation views)
■ landscaping (plan and elevation views)

Double Line Creation

AutoCAD includes an MLINE command that allows you to create multiple parallel lines called *multilines* in a single operation. This command makes it much easier to create the walls on floor and foundation plans. You should practice and experiment with MLINE before you actually use it on a complex floor plan. Note that you can create more than two lines simultaneously with the MLINE command. However, two lines are the default because the command is most commonly used to create double lines.

Figure 18-48

A floor plan for a detached two-car garage with laundry room.

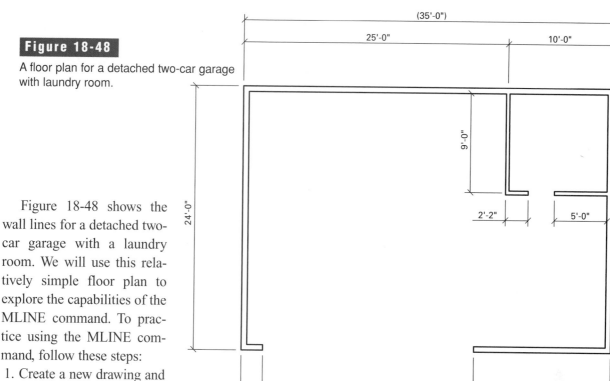

Figure 18-48 shows the wall lines for a detached two-car garage with a laundry room. We will use this relatively simple floor plan to explore the capabilities of the MLINE command. To practice using the MLINE command, follow these steps:

1. Create a new drawing and set it up for an architectural drawing. The fastest way to do this is to start a new drawing by selecting Use a Wizard and Quick Setup. Click Architectural, enter 80′ × 60′, click Finish, and enter ZOOM All to see the entire drawing area.

Before you can use MLINE effectively, you should set up a multiline style that meets your needs. The default multiline style places two lines at an interval of 1 unit apart, with no line caps (short lines at the beginning and end of the multiline connecting the multiple lines). You can change the spacing between the lines using the MLINE options, but you must use the MLSTYLE command to create a new style to handle the line caps.

2. Enter the MLSTYLE command. In the text box next to Name, replace STANDARD with WALLS. Then pick the Save… button. In the Save Multiline Style dialog box, pick Save to save the new style. Then pick the Load… button and select the new WALLS style to load it. This makes it the current multiline style. Pick OK to end the MLSTYLE command. Then reenter the MLSTYLE command and pick the Multiline Properties button. In the Caps portion of the Multiline Properties dialog box, pick the Start and End boxes next to Line to select them, as shown in Figure 18-49. Pick OK to end the MLSTYLE command.

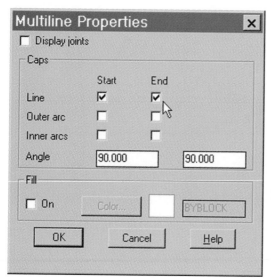

Figure 18-49

Specify starting and ending line caps for the multiline.

3. Enter the MLINE command. Instead of picking a point to begin the line, enter S to activate the Scale option. Scale allows you to set the distance between the lines. The garage in this example is of frame construction, which means that the exterior walls should have a thickness of 6″, so set the scale to 6. (Remember that inches are assumed in architectural units.)

Aligning Multilines

AutoCAD allows you to determine where the points you specify fall on the multilines. Top justification, which is the default, places the specified points at the outermost line. Because wall lines are usually dimensioned to their outside lengths, you should use top justification for architectural floor plans. For interior walls, however, you should use the Zero justification option, which places specified points at the midline of the multiline, because interior walls are dimensioned to their centers. For other applications, you may need to use Bottom justification, which places the innermost line of the multiline at the points you specify.

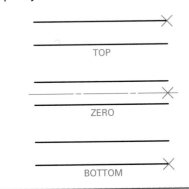

4. With the MLINE command active, pick a point near the middle of the screen, on the left side, as the first point of the multiline. Create the outer shell of the garage using the following relative coordinates:

@2′<180
@24′<90
@35′<0
@24′<270
@13′<180

Press Enter a second time to end the MLINE command. Figure 18-50A shows the result of this operation.

5. Create and load another multiline style, using a name of your choice, that has an ending line cap, but no starting line cap.

6. Reenter the MLINE command, and enter J to activate the Justification option. Set the justification to Zero. Enter the Scale option, and set the wall width to 4″, which is typical of interior walls. Then create the interior partition lines as shown in Figure 18-50B. Start each wall line on the *exterior* surface of the exterior garage wall lines. Use the coordinate display and snap to place the lines accurately.

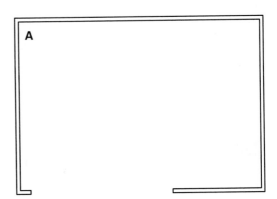

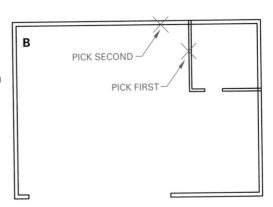

Figure 18-50

Create the walls using the MLINE command, and then clean up the intersections using MLEDIT.

To complete the walls, clean up the intersections between the interior partitions and the exterior wall lines. To do this, use the MLEDIT command. Follow these steps.

1. Zoom in on the intersection of the vertical partition wall and the exterior wall lines.
2. Enter the MLEDIT command. From the visual menu, pick the merged tee. (The words *merged tee* appear at the bottom of the dialog box when you have selected the correct option.) For the first mline, pick the internal wall partition, as shown in Figure 18-50B. It is very important that you pick this wall first to obtain the correct results. Then pick the exterior wall for the second mline. This cleans up the intersection to remove the internal lines.
3. Zoom in on the second intersection that needs to be cleaned up and fix it using the same technique.

Dimension Style

As noted earlier in this chapter, the dimensions on architectural drawings often look different from those on mechanical drawings. For example, the Architectural Tick is often used instead of arrowheads on dimension lines, and dimensions are given in feet and inches instead of decimal inches.

To dimension an architectural drawing, you must therefore set up an architectural dimensioning style. Create a new text style called Cityblue using the cityblueprint.shx font. Then create a new style called Architecture, and specify the following settings:

- arrowheads: Architectural Tick
- units: Architectural
- text style: Cityblueprint
- text height: 6′
- text gap: 3″
- extension lines: 6′
- origin offset: 6′

Be sure to save the new style by picking the Save button before picking OK. With these settings in place, you can use the guidelines given in the "Drafting Principles" section of this chapter to dimension the architectural drawing.

Architectural Dimensions

As you review architectural drawings that were created in AutoCAD, you will notice that the dimensions include the inch mark. For example, a dimension might be 15′-6″. As you may recall, board drafters delete the inch mark, so the same dimension would be written as 15′-6. It is acceptable to include the inch marks on CAD-generated drawings. In fact, some CAD programs, including AutoCAD, set the inch marks automatically.

CAD-Generated Floor Plan

Throughout this chapter you have learned about the various kinds of construction, construction materials, styles of architecture, and the various drawings that make up a set of house plans. In some respects, a set of house plans is similar to a set of working drawings for a manufactured product. The set must include all of the drawings and all of the necessary information, or specifications, to construct the house.

Figures 18-24 through 18-31 give an example of a set of plans for a residence. The procedure for preparing any of these drawings is essentially the same as the procedure you learned in earlier chapters for preparing a drawing of a machine part, with the exceptions noted in this chapter. To create a floor plan using AutoCAD, keep in mind the tools, commands, and techniques discussed in the previous section. To practice the skills necessary to complete a floor plan, refer to the designer's sketch shown in Figure 18-51. The procedure is as follows:

1. Establish the overall size of the house, the number and types of rooms, and the general layout. For this practice exercise, this information is provided in Figure 18-51. Determine the drawing limits, and set up a new drawing using the architectural options described in "Tools and Commands."

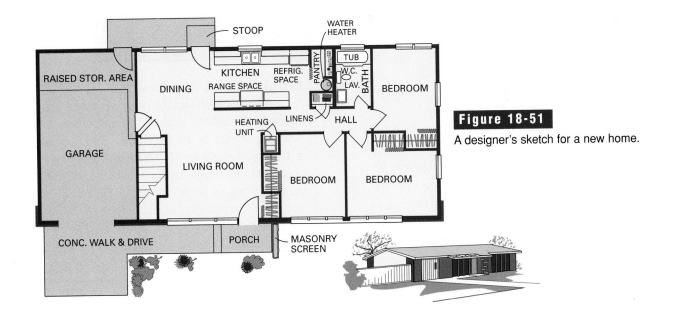

Figure 18-51

A designer's sketch for a new home.

2. Set up and use the MLSTYLE, MLINE, and MLEDIT commands to create the walls and partitions in the house. See Figure 18-52.

3. Using symbol libraries as available and appropriate, add details such as kitchen cabinets, appliances, bathroom fixtures, door and window symbols, and other details. Also add any exterior treatments, such as a brick veneer, at this time. Use AutoCAD's hatch patterns as necessary for these. See Figure 18-53.

4. Set up an appropriate dimensioning style and add the dimensions and notes. The finished floor plan is shown in Figure 18-54.

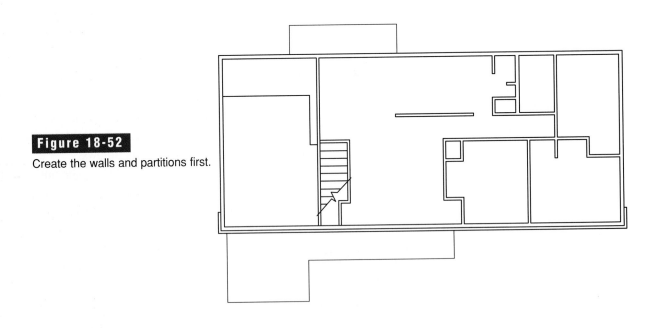

Figure 18-52

Create the walls and partitions first.

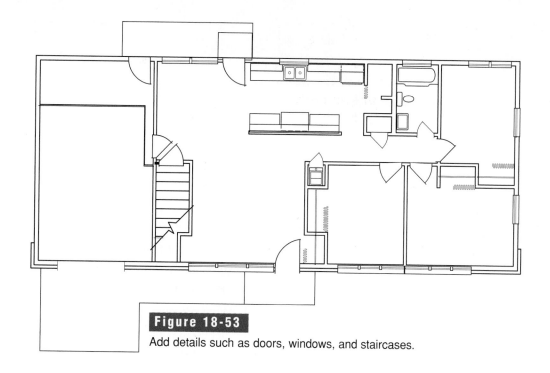

Figure 18-53

Add details such as doors, windows, and staircases.

Figure 18-54

Finish the drawing by adding dimensions, labels, and notes.

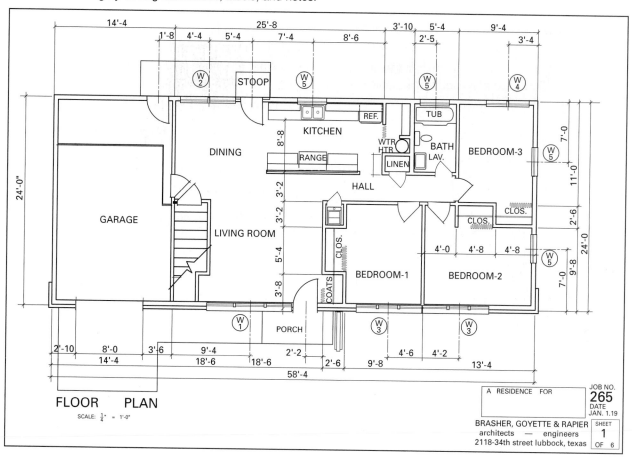

FLOOR PLAN

SCALE: $\frac{1}{4}$" = 1'-0"

A RESIDENCE FOR

JOB NO.
265
DATE
JAN. 1.19

BRASHER, GOYETTE & RAPIER
architects — engineers
2118-34th street lubbock, texas

SHEET
1
OF 6

CHAPTER 18 Review

Chapter Summary

- Architectural firms range in size from one or two to hundreds of employees.

- Community architecture includes the design of neighborhoods, public buildings, and urban development.

- Residential buildings may be of frame, masonry, or concrete construction.

- Common architectural materials include lumber, masonry, and concrete.

- Symbols are used on drawings to reduce drawing time, and to make the drawings clearer and easier to read. Symbols can be thought of as a drafter's shorthand.

- Working drawings are a complete set of drawings and related documents used to guide the construction of a building or other structure.

- The dimensioning techniques for architectural drawings are similar to those for mechanical and other types of drawings, with a few minor differences.

- Tools for creating a CAD-generated architectural drawing include symbol libraries, template overlays, and special commands that ease the drawing process.

Review Questions

1. Define architecture in your own words.
2. What major services does an architect render?
3. What are the three main ways to judge a building's architecture?
4. What are neighborhoods?
5. List three types of framing common in residential design.
6. What determines the pitch of a roof?
7. Draw or sketch four material symbols used by architects.
8. List six types of windows used in houses.
9. What is modular coordination?
10. Name the types of drawings typically found in a set of working drawings.
11. What are the preferred scales for plan and elevation drawings of small buildings?
12. What scales are preferred for enlarging details?
13. Name two styles of lettering used on architectural drawings.
14. What are pressure-sensitive symbols?
15. What is virtual evaluation?
16. Explain how to create double wall lines for a floor plan using AutoCAD.

Drafting Problems

The drafting problems in this chapter are designed to be completed using board drafting techniques or CAD.

For problems 1 through 9, follow the directions to complete each problem. Your instructor will assign an oblique or isometric drawing for problems 1 and 2.

1. Prepare an isometric or oblique drawing of the brick wall section shown in Figure 18-55. Make the section 4'-0 long. You may show the brick bonding, or you may draw it as horizontal lines in pictorial at 30°. Scale: 1½" = 1'-0.

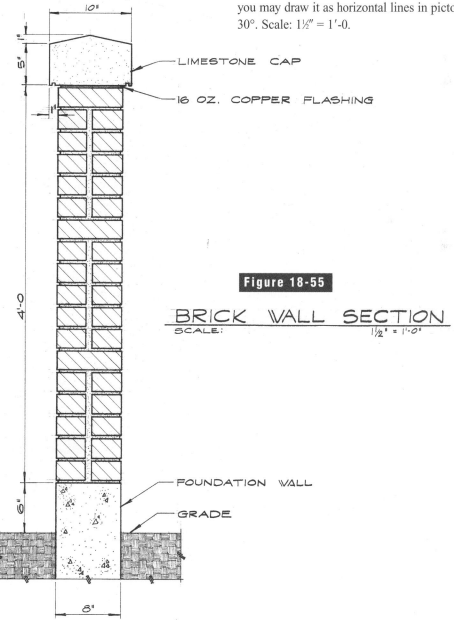

Figure 18-55

BRICK WALL SECTION
SCALE: 1½" = 1'-0"

LIMESTONE CAP

16 OZ. COPPER FLASHING

FOUNDATION WALL

GRADE

2. Prepare an isometric or oblique detail of the concrete wall shown in Figure 18-56. Investigate the sizes of block. Scale: 1½″ = 1′-0.

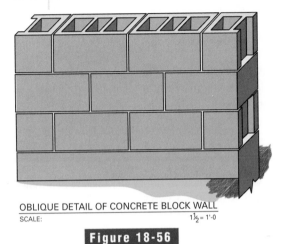

OBLIQUE DETAIL OF CONCRETE BLOCK WALL

SCALE: 1½ = 1′-0

Figure 18-56

3. Draw a footing detail at the scale shown in Figure 18-57, and label the parts.

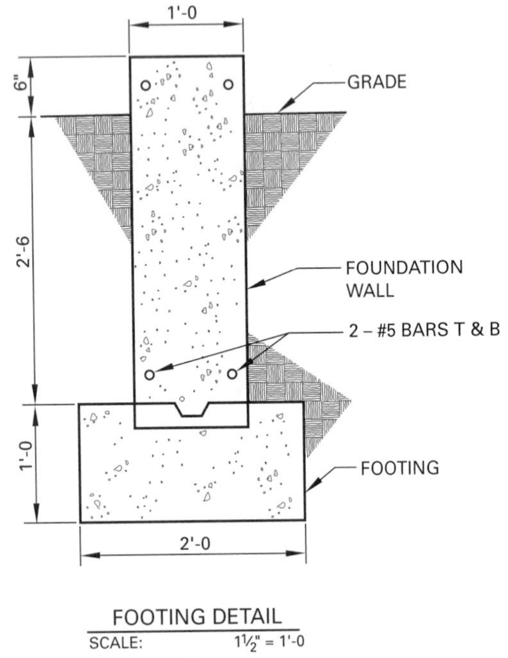

1′-0

6″

2′-6

1′-0

2′-0

GRADE

FOUNDATION WALL

2 – #5 BARS T & B

FOOTING

FOOTING DETAIL

SCALE: 1½″ = 1′-0

Figure 18-57

4. Prepare a section detail of the column footing shown in Figure 18-58, using the given size of the structural items.

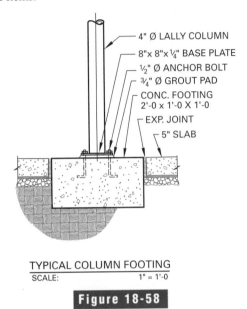

4″ Ø LALLY COLUMN

8″x 8″x ¼″ BASE PLATE

½″ Ø ANCHOR BOLT

¾″ Ø GROUT PAD

CONC. FOOTING 2′-0 x 1′-0 X 1′-0

EXP. JOINT

5″ SLAB

TYPICAL COLUMN FOOTING

SCALE: 1″ = 1′-0

Figure 18-58

5. Prepare a pictorial of the typical roof detail shown in Figure 18-59. Studs, ceiling joists, and rafters are 16″ on center. Joists and rafters: 2″ × 6″. Studs and top plates: 2″ × 4″. Ridge beam: 2″ × 8″. Scale: 1½″ = 1′-0.

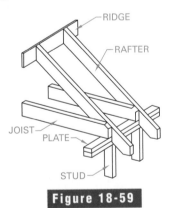

RIDGE

RAFTER

JOIST

PLATE

STUD

Figure 18-59

6. Using the nominal dimension of the materials listed in Figure 18-60, construct the sill detail in section. The lapped 1″ siding has an 8″ exposure. Scale as shown.

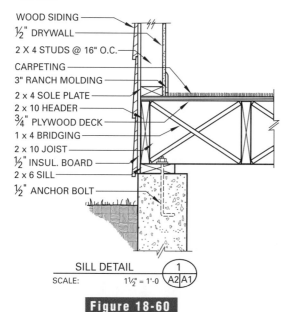

WOOD SIDING
½″ DRYWALL
2 X 4 STUDS @ 16″ O.C.
CARPETING
3″ RANCH MOLDING
2 x 4 SOLE PLATE
2 x 10 HEADER
¾″ PLYWOOD DECK
1 x 4 BRIDGING
2 x 10 JOIST
½″ INSUL. BOARD
2 x 6 SILL
½″ ANCHOR BOLT

SILL DETAIL   1 A2|A1
SCALE: 1½″ = 1′-0

Figure 18-60

7. Prepare a partial girder-column detail as shown in Figure 18-61. The column is 4″ with a 6″ plate welded to the top. The 10″ I-beam is bolted to the plate. The 2 × 4 rests on the 4½″ flange of the beam.

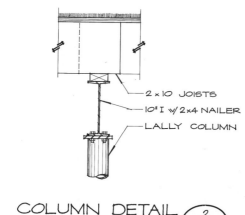

2 x 10 JOISTS
10″ I w/ 2x4 NAILER
LALLY COLUMN

COLUMN DETAIL   2 A2|A1
SCALE: 1½″= 1′-0

Figure 18-61

8. Prepare a plan of the 6″ beam pocket shown in Figure 18-62. The beam is to have a 4″ bearing on the concrete wall.

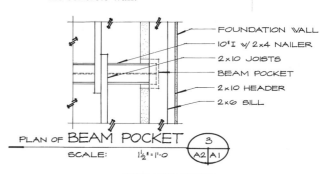

FOUNDATION WALL
10″I w/ 2x4 NAILER
2 x 10 JOISTS
BEAM POCKET
2 x 10 HEADER
2 x 6 SILL

PLAN OF BEAM POCKET   3 A2|A1
SCALE: 1½″ = 1′-0

Figure 18-62

9. Prepare a sectional beam-pocket detail showing the 10″ I-beam on the stepped foundation wall, as shown in Figure 18-63. Allow the beam flange to appear as a nominal ½″ thickness.

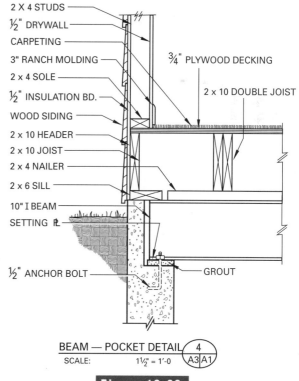

2 X 4 STUDS
½″ DRYWALL
CARPETING
3″ RANCH MOLDING
2 x 4 SOLE
½″ INSULATION BD.
WOOD SIDING
2 x 10 HEADER
2 x 10 JOIST
2 x 4 NAILER
2 x 6 SILL
10″ I BEAM
SETTING ℞
½″ ANCHOR BOLT

¾″ PLYWOOD DECKING
2 x 10 DOUBLE JOIST
GROUT

BEAM — POCKET DETAIL   4 A3|A1
SCALE: 1½″ = 1′-0

Figure 18-63

Problems 10 through 15 are a partial set of plans for a two-story residence. Note that each square on the grids behind these plans is 2'-0 square. Prepare drawings as assigned by the instructor.

10. Draw the front elevation of the two-story residence shown in Figure 18-64. Use the modular grid to establish the sizes. Draw at a scale of ⅛″ = 1'-0. Add windows, shutters, and decorative appointments to suit your own design.

11. Examine the gridded plot plan shown in Figure 18-65, and develop the boundary lines at a scale of 1″ = 20′, on a C-size sheet. Dimension each boundary line. Complete the plan of the house on the site. Landscape with trees and bushes. All radii on the driveway must be 18′.

Figure 18-64

Figure 18-65

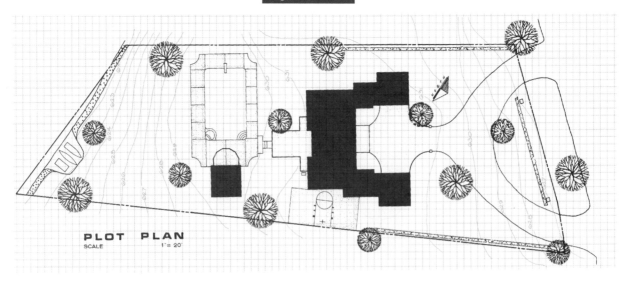

12. Draw a first-floor plan as shown in Figure 18-66 at a scale of ⅛″ = 1′-0. By examining the grid, locate windows on the plan from elevations. Label the rooms and the appropriate sizes for this preliminary architectural study for your client.

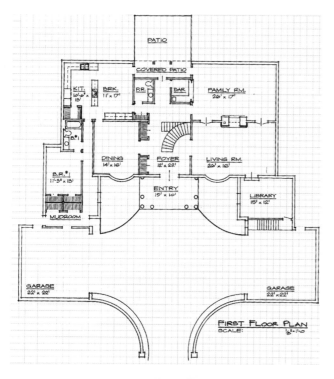

Figure 18-66

13. Prepare a second-floor plan as shown in Figure 18-67. Scale: ⅛″ = 1′-0. Label the rooms with names and dimensions.

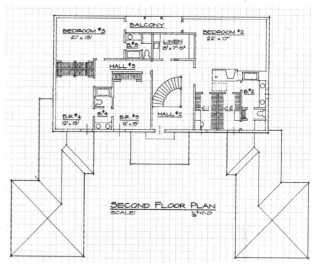

Figure 18-67

14. Draw a right-side elevation by examining the proportions on the modular grid in Figure 18-68. Find the common roof pitch and label it on the elevation.

Figure 18-68

15. Prepare the first-floor plan, as shown in Figure 18-69, at a scale of ½″ = 1′-0. Add the dimensions to the plan. Examine the modular grid for room sizes.

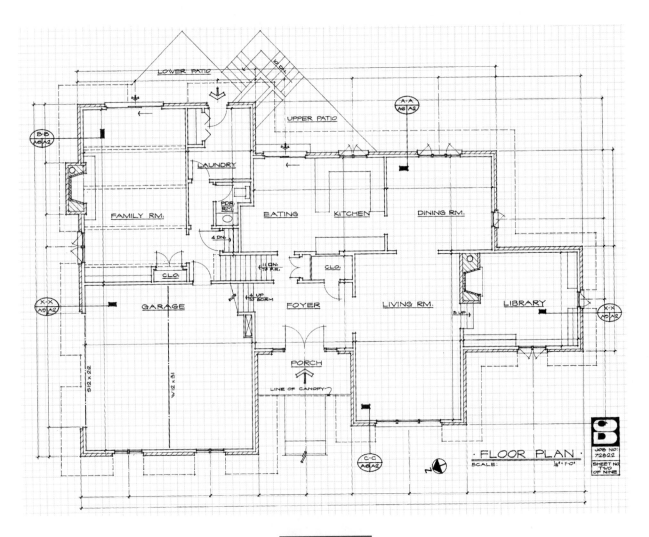

Figure 18-69

For problems 16 through 18, follow the directions to complete each problem.

16. Examine the elevations of the building in Figure 18-70A-D in relation to the floor plan. Draw the plan and lay out the rooms at a scale of ⅛″ = 1′-0. Add appointments and site landscaping. Draw the elevations. This is a preliminary study, so dimensions do not have to be exact at this stage of presentation. Render the elevations.

17. Develop an isometric drawing of the western (platform) framing in Figure 18-7. Rafters: 2″ × 6″; joists: 2″ × 8″; studs: 2″ × 4″.

18. Develop a detailed floor plan of the house shown in Figure 18-25, with complete dimensions. Scale: ¼″ = 1′-0. Calculate the gross square footage and the square footage of the heated area. Optional: Complete a set of plans, elevations, and sections for the house.

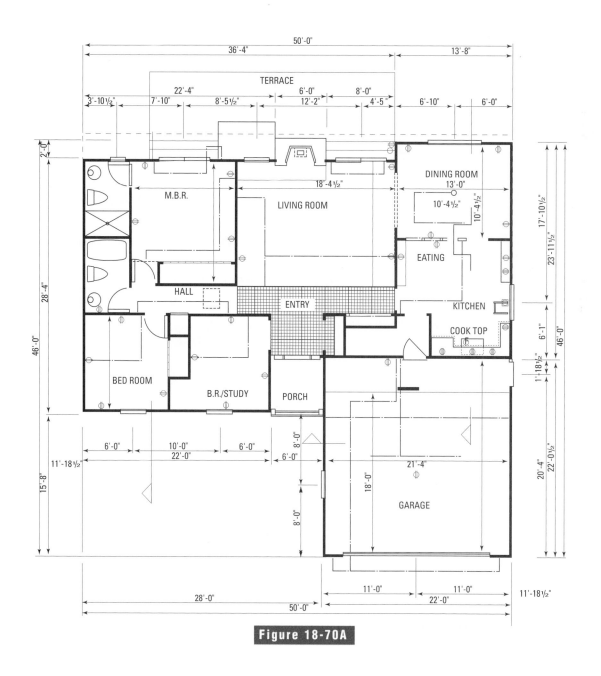

Figure 18-70A

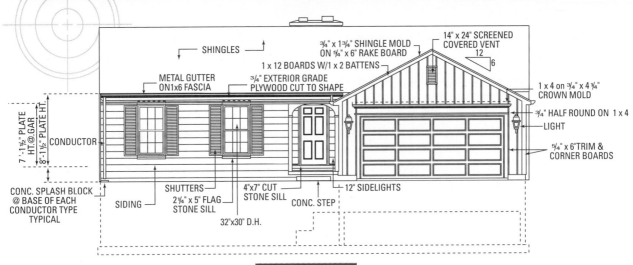

Figure 18-70B

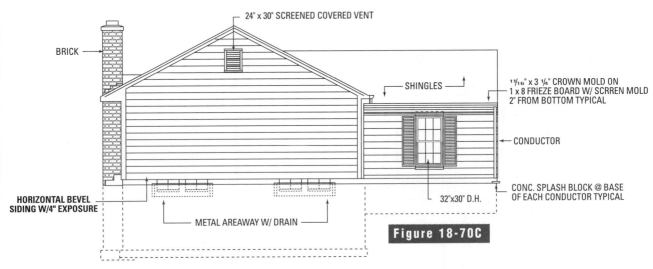

Figure 18-70C

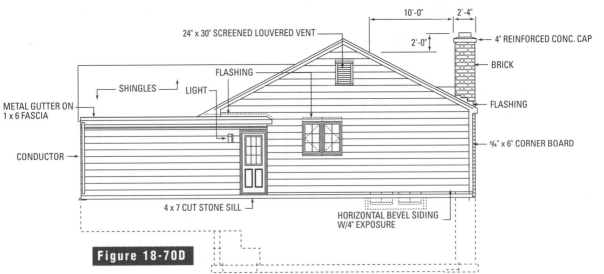

Figure 18-70D

RIGHT SIDE ELEVATION

Design Problems

Design problems have been prepared to challenge individual students or teams of students. In these problems, you are to apply skills learned mainly in this chapter but also in other chapters throughout the text. The problems are designed to be completed using board drafting, CAD, or a combination of the two. Be creative and have fun!

1. TEAMWORK Design an outbuilding in which landscaping and gardening equipment, tools, and supplies can be stored and serviced. The general specifications are as follows:

   Size: Not to exceed 450 square feet

   Base: Concrete slab with floor drains

   Exterior: Rustic design, wood siding, asphalt shingle roof, double exterior doors (minimum 6'-0 opening), and adequate windows for natural light

   Interior: Open space (no partitions), wall and ceiling finish of your choice, fluorescent lighting, 115V electrical outlets on each wall, workbench with tool cabinet for servicing lawn and garden equipment

   The team should agree on an overall design and then individual team members should assume responsibility for developing one of the following:
   • Integral slab and foundation plan
   • Floor plan
   • Elevations
   • Roof plan
   • Typical wall section
   • Electrical plan
   • Window and door schedule
   • Materials list

2. TEAMWORK Design a residence that will be placed on a lot 150'-0 wide × 300'-0 deep. It is a corner lot and is relatively level from side to side but slopes approximately 3'-0 uniformly from front to back. All utilities are underground and run parallel with the front property line. The following are the general specifications:

   Rooms: Kitchen, dining room, living room, family room, laundry, two-car garage, storage for athletic equipment, three bedrooms, two baths, full basement with exercise room

   Exterior: Two-story brick, cedar shingle roof

   Interior: Plaster walls, hardwood floors in all but baths, laundry, kitchen, and basement; ceramic tile in baths and kitchen; vinyl floor covering in laundry; concrete floor in basement

   Size: Not to exceed 3,200 square feet

   The team should agree on an overall design and then individual team members should assume responsibility for developing one of the following:
   • Site plan
   • Foundation plan
   • Floor plans
   • Roof plan
   • Elevations
   • Typical wall sections
   • Electrical plan
   • Plumbing plans
   • HVAC plans
   • Window and door schedules
   • Room finish schedule
   • Details list

   When the design has been completed, prepare a technical presentation and present it to the rest of the class.

3. The architectural firm for which you work is preparing comprehensive plans for a new housing development called Pleasant Hills. For security purposes, there is to be a single entrance drive to allow residents to enter and leave the development. You have been assigned the task of designing an aesthetically pleasing masonry entrance structure that will include the Pleasant Hills sign. Use the following specifications in preparing design and working drawings:

   Size: not to exceed an area 100'-0 wide × 65'-0 deep

   Entrance road: 30'-0 wide

   Material: masonry with cast aluminum letters

   The finished set of plans will include the following:
   • Site plan
   • Foundation plan
   • Elevations
   • Typical sections
   • Structural details
   • Sign details
   • Electrical plan (if applicable)
   • Materials list
   • Landscape plan
   • Pictorial rendering

Structural Drafting

compressive strength

dead load

fabrication shop

finite element analysis
  (FEA)

gage lines

live load

laminating

load

pitch

prestressed concrete

reinforced concrete

truss

Upon completion of this chapter, you should be able to:

■ Describe the responsibilities of the structural drafter.

■ Identify structural steel members that form the framework of
buildings, bridges, and other structures.

■ Explain how lamination increases the usability of wood for structural elements.

■ Describe how concrete is used as a structural component.

■ Describe the use of structural clay systems, such as bricks, as
structural elements.

■ Prepare details of structural steel components using both board
drafting and CAD techniques.

Can you imagine the number of structural drawings required for the construction of a building like the one shown in Figure 19-1? Thousands of drawings and many thousands of hours are involved in the design of buildings, bridges, dams, athletic stadiums, storage tanks, communication towers, and many other kinds of structures. Fortunately, the use of CAD in the design process removes much of the repetitiveness that was once involved in the development of plans for multi-story buildings and other structures. The CAD operator is able to develop a basic structural system and use it as a template. For each level, it is necessary only to add or change those features that are unique to that floor.

Structural drawings are basically working drawings of the framework of a structure. Structural drafters prepare the drawings in cooperation with engineers, planners, and designers from various specialized fields. Involved in this team effort are architectural engineers, structural engineers, designers, and detailers, as well as various other specialists who play a specific role in the overall design.

The Structural Drafter

The structural drafter is usually a member of an engineering team. This team often works under the direction of a project manager or a job superintendent. As an example of this teamwork, struc-

tural and architectural designers often combine their efforts. The architectural designer designs the form of a building based on the function it will have. Then the structural designer designs the frame of the building to fit this form, as shown in Figure 19-2. The construction of buildings, bridges, and other structures depends on the detailed instructions provided by the structural engineer.

A structural drafter usually works at one of the following five jobs:

- Drafting details in an architect's or engineer's office

Figure 19-1

The design of buildings like this requires thousands of pages of structural drawings and specifications.

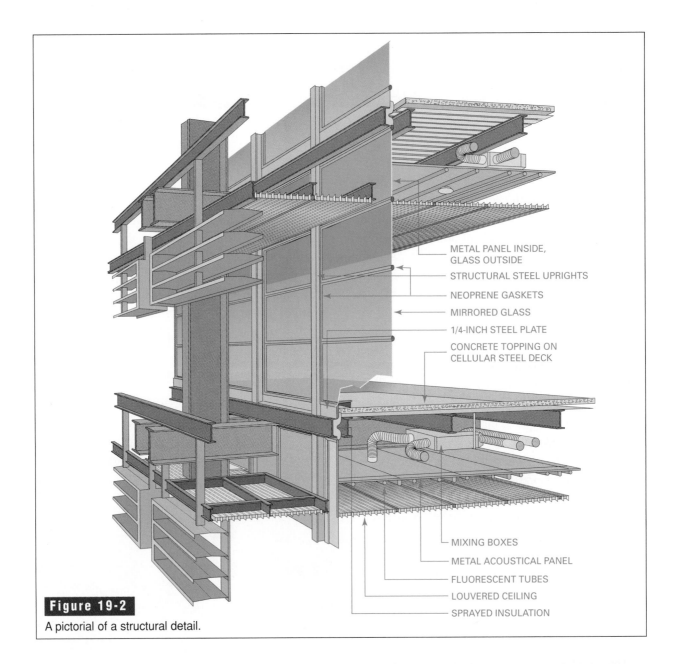

Figure 19-2

A pictorial of a structural detail.

METAL PANEL INSIDE, GLASS OUTSIDE
STRUCTURAL STEEL UPRIGHTS
NEOPRENE GASKETS
MIRRORED GLASS
1/4-INCH STEEL PLATE
CONCRETE TOPPING ON CELLULAR STEEL DECK

MIXING BOXES
METAL ACOUSTICAL PANEL
FLUORESCENT TUBES
LOUVERED CEILING
SPRAYED INSULATION

- Preparing construction details for a contractor (making the shop drawings for a construction company)
- Drafting structural details for a manufacturer of structural materials
- Working for the engineering department of a manufacturing plant that maintains its own engineering operations
- Preparing drawings for government or other agencies that regulate the design and construction of public buildings, bridges, dams, and other structures

 Success on the Job

Short-Term Goals

Setting short-term work goals often means establishing a schedule for the completion of a project. Rather than approach a large project as an insurmountable task, approach it as the sum of its many parts. Prepare a work schedule based on the individual parts, and strive to meet the deadlines you set. Pacing yourself throughout a project will help you to finish on time, which will always improve your chances for success on the job.

Structural Materials

Designers and detail drafters must be familiar with a great many structural materials. In addition, all members of the engineering design team must learn about new construction materials and systems as they are developed. All structural materials have special ways of being assembled or fastened together. These ways must be considered whenever accurate drafting details are prepared.

The basic structural materials used today are steel, wood, concrete, structural clay products, and stone masonry. Different materials have different characteristics. It is the designer's job to choose the right combination of materials to bear the stresses imposed by a structure.

Structural Steel

Steel shapes have framed the skyscrapers of our cities for nearly a century. The American Institute of Steel Construction (AISC) maintains regional offices from coast to coast to help provide guidelines for designing and building steel structures.

Steel makes a good construction material because of the shapes into which it can be formed at the mill. Steel shapes are produced in rolling mills. They are then shipped to **fabrication shops** where they are cut to specific lengths and where connections are prepared.

The AISC publishes the *Manual of Steel Construction (AISC Manual)*. The latest edition of the *AISC Manual*, or any handbook published by a major steel company, lists all the major shapes available and the great variety of their sizes and weights. The *AISC Manual* contains tables for designing and detailing the various shapes in any combination. Figure 19-3 shows cross sections of various plain steel shapes. These shapes are grouped by the AISC, as shown in Table 19-1.

Examples of Steel Structures

Many different kinds of structures are framed in steel. Two examples—a building and a bridge—are examined in this section.

A completed A-frame building is shown in Figure 19-4. Each of the 13 steel A-frame structural members is approximately 220′ (67,000 mm) across at the base, 135′ (41,150 mm) across at the top, and 15′ (4570 mm) across the vertical bents. It is built of

Figure 19-3

Steel shapes in section.

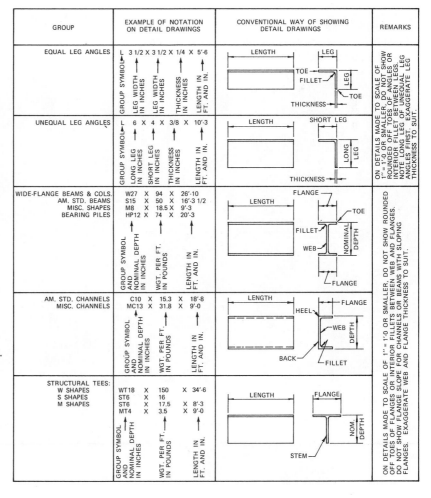

Table 19-1. Structural Steel Shapes

Group Name	Symbol	Comments
American Standard beams	S	Used to be called I-beams.
American Standard channels	C	
Miscellaneous channels	MC	Include special-purpose channels that are not standard.
Wide flange shapes	W	Used both as beams and as columns.
Miscellaneous shapes	M	Lightweight shapes that look in cross section like W shapes.
Structural tees	ST, WT, MT	Made by splitting S, W, and M shapes, usually along the middepth of their webs.
Angles	L	Consist of two legs of equal or unequal widths, at right angles to each other.
Plates and flat bars	PL, Bar	Rectangular in cross section.

tubes, wide-flanged sections, and chords made of 18″ × 26″ (455 × 660 mm) tubes. Each A-frame was assembled on the site and erected in five pieces. All the steel members are connected with high-strength bolts (HSB).

The A-frames were designed with the aid of a computer to hold up under many different combinations of **loads**, the weights or pressures borne by a structure. These include wind loads, temperature changes, the **dead load** (weight of the building), and the **live load** (weight added temporarily, such as furniture). If the building is designed correctly, these loads are carried through the structure to the ground. For the building in Figure 19-4, which is located in Florida, the designer had to allow for wind loads from hurricane winds. Allowance must always be made for loads created by geographical location, the prevailing weather, the function of the building, and the size and shape of the structure.

Figure 19-4

The structural shapes in this A-frame resort hotel are dependent on structural steel detail.

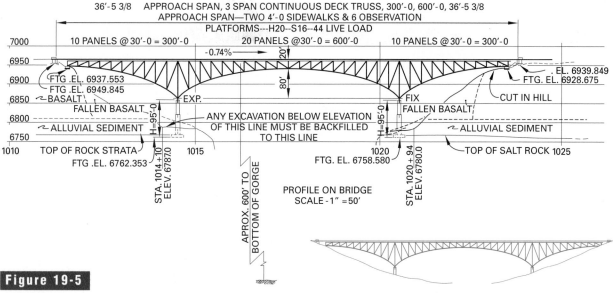

Figure 19-5

A schematic design of the bridge framework for a bridge over the Rio Grande in New Mexico.

Figure 19-5 shows the schematic for the steel framework of a bridge over a deep gorge of the Rio Grande near Taos, New Mexico. The bridge has a rigid structural frame made of high-strength steel fastened with high-strength bolts and welds. More than 1900 customary tons, or 1725 metric tons, of structural steel were used in its construction. The center span is 600' (183 m) long, and the two side spans are each 300' (91.5 m) long. The distance from the canyon floor to the bridge floor is 600' (183 m). A detail drawing of a typical welded-steel member of this bridge appears in Figure 19-6.

Steel Systems

Using steel, engineers have developed a number of new structural systems. The system shown in Figure 19-7 consists of four or five modular units in a geometric pattern. The basic unit is made of four or five parts that are bolted together. The unit can be used in canopies and roofs, as well as integrated wall and roof systems. Note how the geometric pattern of the exposed steel becomes a part of the overall design.

Figure 19-6

A welded beam for the bridge shown in Figure 19-5.

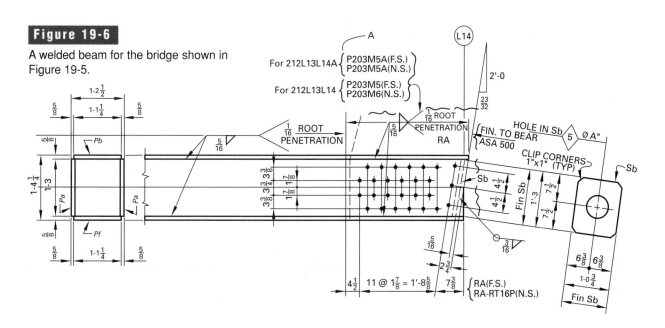

Figure 19-7

Geometrically arranged structural members help shape buildings.

Structural Domes

Figure 19-8 shows a dome over a theater in Reno, Nevada. The geometry used in this structure is geodesic. It is different from other dome geometry in that its strength extends in all directions. The three-dimensional triangulated framing of the dome makes it exceptionally strong, as shown in Figure 19-9.

Trusses

A **truss** is a configuration of structural elements that adds strength to a structure. Roofs and bridges are examples of two structures that often depend on trusses. Figure 19-10 shows diagrams and names for some roof trusses and bridge trusses. The ones shown are only a few of those available. Each type can also be modified to carry specific loads.

Figure 19-8

A flexible geodesic steel dome for a theater.

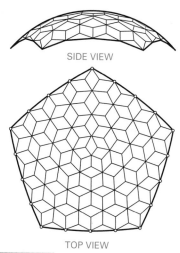

SIDE VIEW

TOP VIEW

Figure 19-9

The dome in Figure 19-8 has geometric characteristics that increase its strength.

Fastening Structural Steel

Structural steel requires high-strength fasteners. Examples of fasteners often used for structural members include rivets, high-strength steel bolts, and welds.

As you may recall from Chapter 11, rivets are permanent fasteners that have a preformed head on one end. After being driven through the materials to be

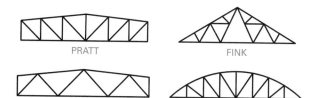

PRATT   FINK

WARREN   BOWSTRING

Figure 19-10

Roof- and bridge-truss diagrams.

fastened, a head is formed on the other end. The standard symbols for rivets are shown in Figure 19-11. A typical buttonhead rivet is shown in Figure 19-11B. If the riveting is to be done in the shop, the drafter uses shop rivet symbols. These are open circles the diameter of the rivet head. If the riveting is to be done in the field (on the construction site), the drafter uses field rivet symbols. These are blacked-in circles the diameter of the rivet hole. Lines on which rivets are spaced are called **gage lines**. The distance between rivet centers on the gage lines is called the **pitch**.

High-strength steel bolts are rated by the American Society for Testing and Materials (ASTM). The bolt can be applied in the field or in the shop. The hole into which it fits is normally 1/16″ (2 mm) larger than the bolt. Figure 19-12 shows two

---

TECH⊕MATH

Roof Pitch Analysis

In the design of a roof truss, the designer often gives the distance from the edge of the roof to the center of the building and the *pitch*, or rise relative to the run. If the pitch of the roof shown in the illustration below is 4/5, it means that the roof rises four units for every five horizontal units.

Problem: Find the length of the rafter (*d*).

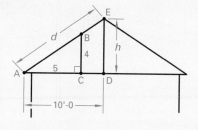

Solution: Since triangle ABC is similar to triangle AED, the proportion is written:

$$\frac{h}{10} = \frac{4}{5}$$

$$h = \frac{4 \times 10}{5}$$

h = 8 feet (total rise)

Using the Pythagorean theorum:

$$d^2 = (8)^2 + (10)^2$$

$$d^2 = 164$$

d = 12.8 feet (length of rafter)

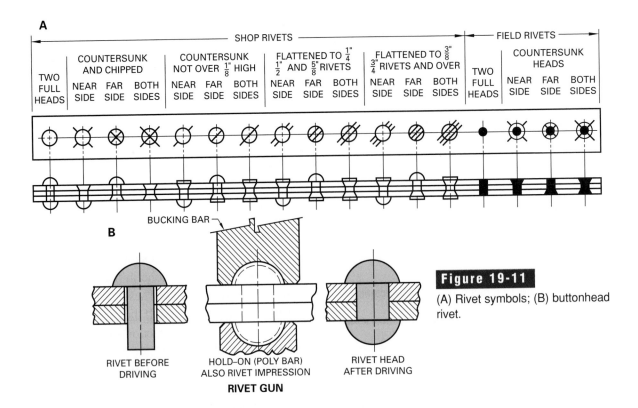

	SHOP RIVETS														FIELD RIVETS				
TWO FULL HEADS	COUNTERSUNK AND CHIPPED			COUNTERSUNK NOT OVER $\frac{1}{8}$ HIGH			FLATTENED TO $\frac{1}{4}$ $\frac{1}{2}$ AND $\frac{5}{8}$ RIVETS			FLATTENED TO $\frac{3}{8}$ $\frac{3}{4}$ RIVETS AND OVER			TWO FULL HEADS	COUNTERSUNK HEADS		
	NEAR SIDE	FAR SIDE	BOTH SIDES	NEAR SIDE	FAR SIDE	BOTH SIDES	NEAR SIDE	FAR SIDE	BOTH SIDES	NEAR SIDE	FAR SIDE	BOTH SIDES		NEAR SIDE	FAR SIDE	BOTH SIDES

Figure 19-11

(A) Rivet symbols; (B) buttonhead rivet.

BUCKING BAR

RIVET BEFORE DRIVING

HOLD–ON (POLY BAR) ALSO RIVET IMPRESSION

RIVET GUN

RIVET HEAD AFTER DRIVING

kinds of bolted connections: frame and seated. The bolt transmits the force of the beam load to the column.

Structural steel members can also be welded together. This is usually accomplished using the metal-arc process. The fillet weld is the most common on structural connectors. See Chapter 15 for a review of the standard welding processes and symbols.

Drafting Structural Details

The plain shapes shown in Table 19-1 and Figure 19-3 are basic to structural detailing. You must be familiar with them in order to make adequate drawings.

Structural details are prepared at a scale of 1″ = 1′-0 for beams up to 21″ (533 mm) in depth. For beams of greater depth, a ¼″ = 1′-0 scale is preferred. The overall length of structural members can be shortened as long as details are shown adequately. Also, very small dimensions, such as a clearance, can be exaggerated to clarify views.

Designers place on their design drawings all the information needed to prepare shop drawings. Figure 19-13 shows a small part of a designed floor plan.

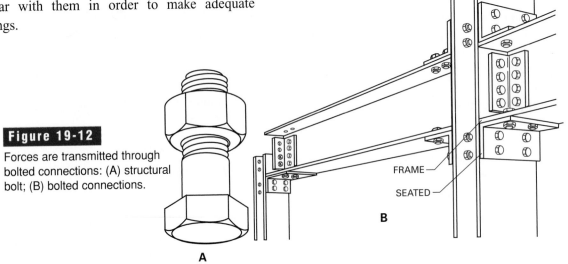

Figure 19-12

Forces are transmitted through bolted connections: (A) structural bolt; (B) bolted connections.

FRAME

SEATED

A

B

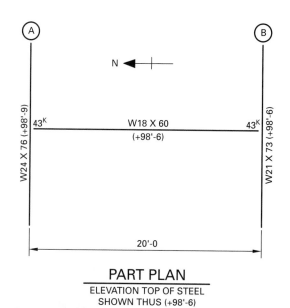

PART PLAN
ELEVATION TOP OF STEEL
SHOWN THUS (+98'-6)

Figure 19-13

A small part of a plan, arranging steel members between beams A and B.

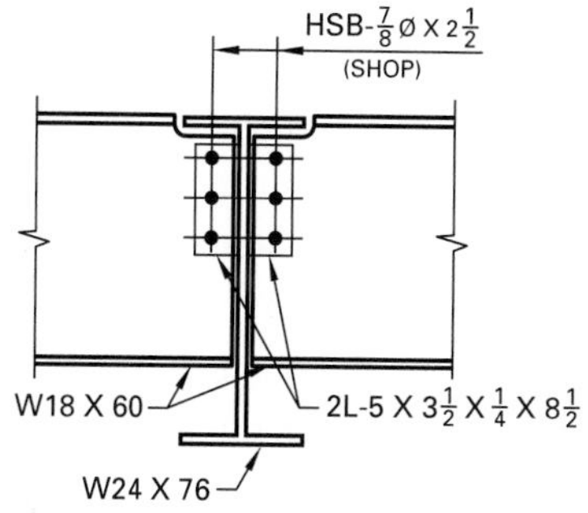

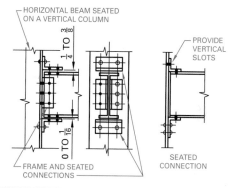

Figure 19-15

Typical connection or framing of mating beams.

Figure 19-16

Frame and seated connections.

The designer has included enough notes and dimensions on the plan to prepare a shop drawing of the wide-flanged beam (W).

The 20'-0 dimension is presumed to be the structural bay, or distance from A to B. The structural members are at right angles to one another unless noted. The height given on the line diagram of a beam is significant. Height elevations are assumed to be level at the figure given. The figure shows two elevations: 98'-6 and 98'-9.

Detail drawings such as the one in Figure 19-14 are essential for making structural pieces. The figure is a detail of a beam. This kind of drawing seldom describes the connections of mating parts. See Figure 19-15. Instead, it just shows features such as connection angles that will be involved when the piece is used in building.

In preparing structural details, the drafter refers to the handbook and the dimensions for detailing. Figure 19-16 shows both frame and seated connections.

Dimensioning Structural Steel Members

In structural drawings, dimensions are given primarily to working points. For beams, give dimensions to the centerline. For angles and (normally) to channels, give dimensions to the backs. Give vertical dimensions on beams and channels to the tops or bottoms.

Generally, do not dimension the edges of flanges and the toes of angles. Make the dimension lines continuous and unbroken. Use the aligned system of dimensioning. When dimensions are in

Figure 19-14

Typical beam detail.

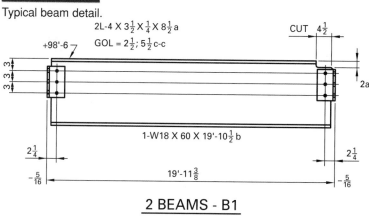

2 BEAMS - B1

feet and inches, use the foot symbol, but not the inch symbol. *Exception:* CAD-generated drawings may use both the foot and inch symbols.

Figure 19-17 is a structural drawing of a small steel roof truss. This symmetrical piece is detailed about the left of a centerline. Study the drawing closely. On the drawing, each member is completely dimensioned or described. In addition, the dimensions shown adequately relate the fixed location of each structural member.

Wood for Construction

Wood is commonly used for the frames of homes and other small structures. Details for wood construction have been developed by the American Forest & Paper Association, 1111 19th St., NW, Suite 800, Washington, DC 20036. They are now used as a standard method of construction.

Structural timber can be manufactured in many forms by cutting wood into thin slabs and gluing them together. This process is known as **laminating**. Builders can buy these "factory-grown" timbers in any size or shape. See Figure 19-18. Some of the forms available include tudor arches, radial arches, parabolic arches, A-frames, and tapered beams. The American Institute of Timber Construction (AITC) has set up guidelines for makers of structural glued, laminated timber.

Figure 20-19 shows some of the construction details that must be used with structural timber. These detail drawings show how timbers are joined together and how structural members are anchored to foundations.

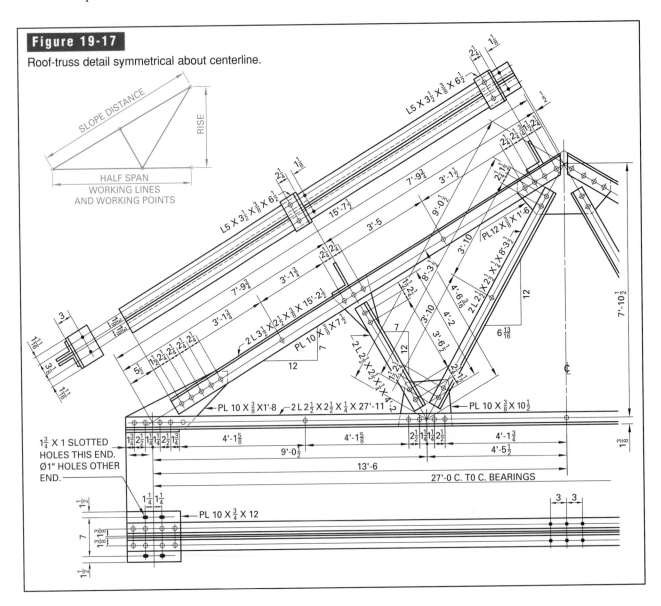

Figure 19-17

Roof-truss detail symmetrical about centerline.

Figure 19-18

Laminated wood forms take many shapes.

Concrete Systems

Many of our buildings, bridges, and dams have been made possible by the use of concrete. The American Concrete Institute has prepared a manual of standard practice for concrete structures.

Concrete has only limited strength unless it is specially prepared. It is made from a mixture of gravel, sand, water, and portland cement. Various grades are produced, depending on the proportions of these ingredients. The concrete can also be reinforced or prestressed.

Reinforced concrete has steel bars embedded in it. These bars are arranged to bear the structural loads that the concrete could not support by itself. When concrete and steel are combined, they can be used in various ways, including the monolithic form shown in Figure 19-20. A typical reinforced concrete detail is shown in Figure 19-21.

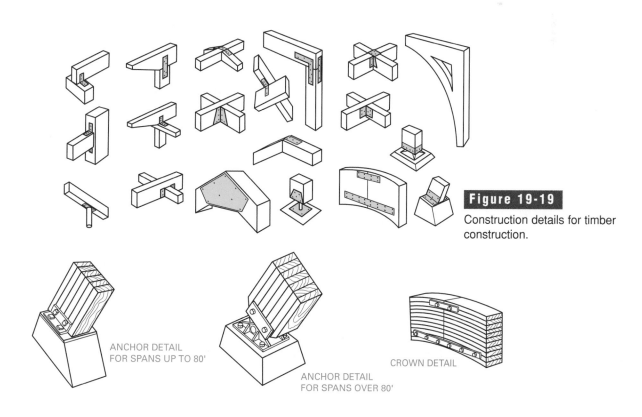

Figure 19-19

Construction details for timber construction.

ANCHOR DETAIL
FOR SPANS UP TO 80'

ANCHOR DETAIL
FOR SPANS OVER 80'

CROWN DETAIL

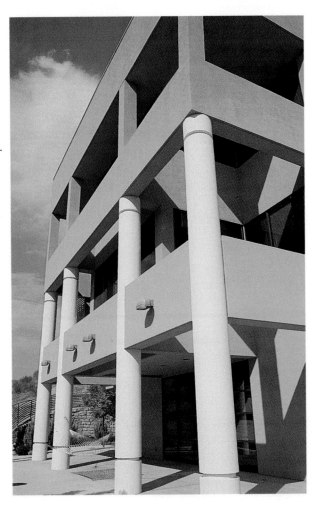

Figure 19-20

Monolithic concrete forms shape new structures.

In **prestressed concrete**, the reinforcing bars are stretched before the concrete is poured over them. The prestressed form will then accept a predefined load. This combination of concrete and stretched steel is stronger than either plain concrete or reinforced concrete.

Concrete forms designed by a structural engineer are drawn for the manufacturer's use only. Construction drawings of these forms are prepared by the manufacturer. These drawings are made to show the contractor the location, placement, and connections, as shown in Figure 19-22.

Structural Clay Systems

The solid brick wall is the oldest type of masonry construction known. Bricks are made from different types of clay in many shapes, forms, and colors. The common brick size is $2\frac{1}{4}'' \times 3\frac{3}{4}'' \times 8''$ ($57 \times 95 \times 203$ mm). Brick walls are made to support floors and roofs. For the vertical walls to be able to carry the horizontal floors and roofs, the bricks must have high **compressive strength**.

Bricks in construction are arranged in overlapping and interlocking patterns and fastened together with connecting mortar joints. This produces a structural assembly that acts as a single structural unit. Figure 19-23 shows some of the common bonds and structural patterns used with bricks. Note the terms applied to the brick.

Figure 19-21

Reinforced concrete detail showing the most efficient use of concrete and steel. The sectioned area represents the concrete that can be eliminated when the concrete is reinforced as shown.

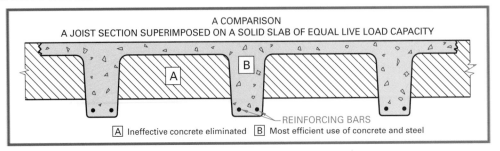

A COMPARISON
A JOIST SECTION SUPERIMPOSED ON A SOLID SLAB OF EQUAL LIVE LOAD CAPACITY

REINFORCING BARS

A Ineffective concrete eliminated   B Most efficient use of concrete and steel

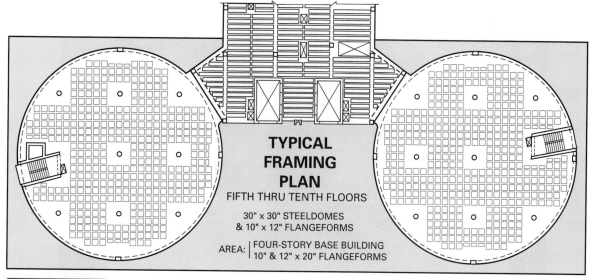

Figure 19-22

Concrete placement plan.

Stone Masonry

The strength of a structural clay system generally depends on the strength of the mortar joints. When the design limits this strength, the designer can call for reinforced masonry. This is masonry with steel rods or wire embedded in the mortar. Brick or concrete masonry can also be used to enclose a structural steel framework. This provides enough fireproofing to meet standard building codes.

For centuries natural stone was used as a major structural material. It was commonly used to construct foundations for buildings or for complete structures such as bridges and walls. Many of the arch-shaped bridges built hundreds of years ago still stand and are still in service (Figure 19-24). Most of these early stone structures were built without the use of mortar and remain in service today because of

Figure 19-23

Brick bonding forms structural walls.

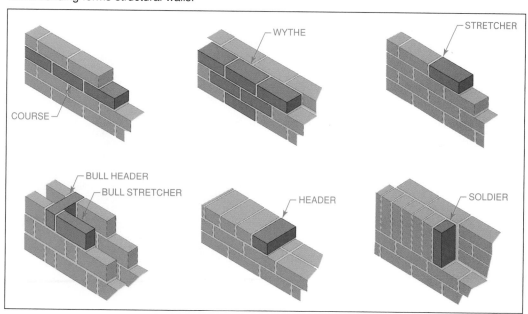

Figure 19-24

Natural stone has been used for centuries to create bridges and other structures.

the careful fitting of the stones to one another. In most cases, the stone used for these structures was simply "field stone," so called because it was used in the size and shape as it was found in the fields. In most cases little or no additional shaping was done unless it was necessary to fit the stone into a particular opening.

More recently, stone has become more decorative than structural. The most common types of stone used in construction are sandstone, limestone, granite, slate, and marble. All of these can be used as they come directly from nature or they can be cut to virtually any desired size or shape. Figure 19-25 is an example of the use of limestone in the construction of a fireplace. Since it does not support other structural members, its use is considered more decorative than structural. Granite, slate, and marble are somewhat more expensive than sandstone or limestone and, therefore, are used more sparingly in most cases.

In many instances, brick and other molded masonry materials, including molded stone, have now replaced natural stone in the construction industry. Brick and other molded materials are less expensive, more uniform in size and shape, and quicker and easier to install. These features result in lower cost without sacrificing strength or aesthetics.

Figure 19-25

The limestone in this fireplace is more decorative than structural.

Board Drafting Techniques

The architect, or architectural designer, first designs the form for a building or other structure based on the customer's needs and budget. The structural designer, or design engineer, then develops the structural design to support the architect's design. Once the design work is completed, it is the structural drafter's job to prepare the final set of working drawings.

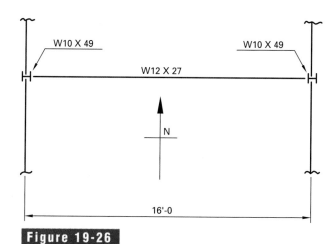

Figure 19-26

Partial design drawing.

Figure 19-26 is a partial design drawing. It represents one very small part of what may be a very large and complex grid of columns and beams that make up the structural plan for a building. The W10 × 49 tells us that the design engineer is specifying a 12″ wide-flange beam that weighs 49 pounds per foot for the columns. The W12 × 27 is calling for a 12″ wide-flange beam that weighs 27 pounds per foot. The 16′-0 dimension is the center-to-center spacing of the two outside beams.

The structural designer has determined that the W12 × 27 beam is to be attached to the W10 × 49 columns using a seated beam connection. The beam is to be attached using bolts to connect the beams to the angles, and the angles are to be welded to the column. When the symbol for the location of bolts is drawn solid, it means that the bolts are to assemble the two parts at the job site, not in the shop. The structural drafter is responsible for preparing this detail. It will consist of two views with all necessary information for final construction. As the structural drafter, you should proceed as follows:

1. Block in the two views consisting of the column, beam, and angles in the front and side views, as shown in Figure 19-27.

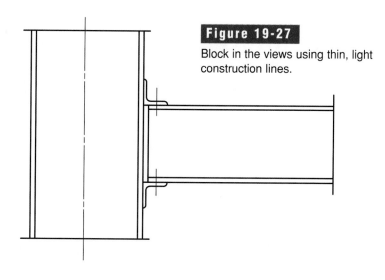

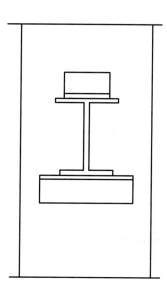

Figure 19-27

Block in the views using thin, light construction lines.

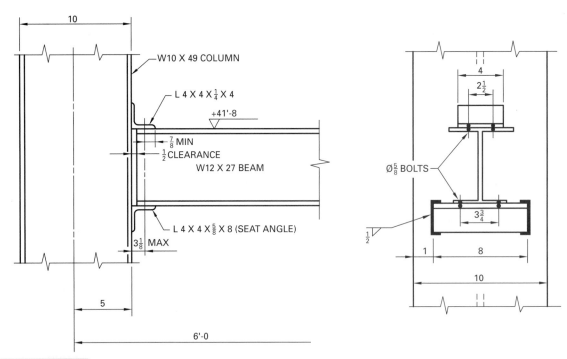

Figure 19-28

Finished detail drawing of a seated beam connection.

2. Darken the lines.
3. Add dimensions, notes, and other details as necessary to finish the drawing, as shown in Figure 19-28.

The triangular-shaped symbol with +41'-8 tells us that the top elevation of the beam when it is installed is 41'-8 above a given reference point or reference elevation.

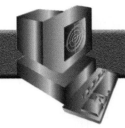

CAD Techniques

CAD software provides tools that drafters can use to create traditional two-dimensional structural details and other drawings that are needed in structural applications. However, as in the architectural field, CAD also opens up new possibilities for structural engineers and designers.

Structural Analysis

Traditional board drafting records design decisions in a way that allows the builder or manufacturer to build the product accurately. The CAD software can help in the actual design of a structure by providing a basis for virtual structural analysis.

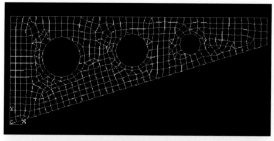

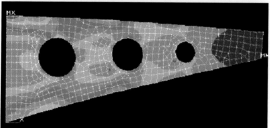

Figure 19-29

The results of a typical computerized finite element analysis, showing stress contours and possible weak spots in a design.

materials, and members, analysis quickly becomes more difficult. In these situations, structural designers use software such as **finite element analysis (FEA)** programs to test computer models of structural members. The user assigns material properties to various parts of a CAD solid model. Then the user applies various "loads" to the model, and the FEA software automatically creates stress contours that show any weak spots in the structural design. See Figure 19-29.

Creating a Structural Detail

After the design work is completed for a project, it is the structural drafter's job to prepare the final set of working drawings. In practice, the process for creating a structural detail using a CAD system is similar to that for creating any other 2D CAD drawing.

Figure 19-30 is a partial design drawing. It represents one very small part of what may be a very large and complex grid of columns and beams that make up the structural plan for a building. The W10 × 49 tells us that the design engineer is specifying a 10″ wide-flange beam that weighs 49 pounds per foot for

Structural engineers can test a proposed design without building a prototype.

Intended use and the environment in which the structure will stand are among the many factors that can affect the weight or force that a structure must withstand. Because the effects of failure can be disastrous, a structure's load-bearing capacity and structural strength are critical in structural design.

Many factors, such as the types of material and the method of fastening used, affect the strength of the structure. Obviously, it is impossible to test the structure directly before it is built. For straightforward analysis of simple structural elements, the structural designers use calculators and tables in engineering texts. However, in complicated structures that may contain several interacting forces,

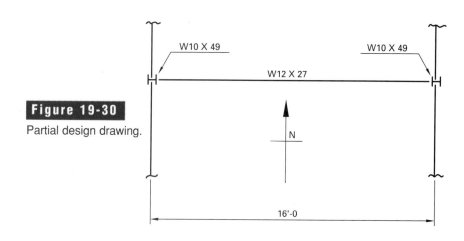

Figure 19-30

Partial design drawing.

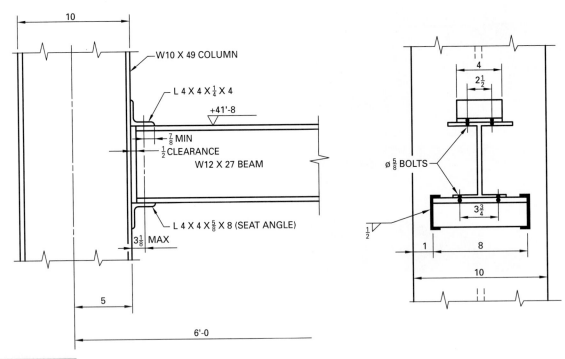

Figure 19-31

The finished CAD-generated detail drawing of a seated beam connection.

the columns. The W12 × 27 is calling for a 12″ wide-flange beam that weighs 27 pounds per foot. The 16′-0 dimension is the center-to-center spacing of the two outside beams.

The structural designer has determined that the W12 × 27 beam is to be attached to the W10 × 49 columns using a seated beam connection. The beam is to be attached using bolts to connect the beams to the angles, and the angles are to be welded to the column. When the symbol for the location of bolts is drawn solid, it means that the bolts are to assemble the two parts at the job site, not in the shop. The structural drafter is responsible for preparing this detail. It will consist of two views with all necessary information for final construction. As the structural drafter, you should proceed as follows:

1. Determine the limits needed for the two-view drawing.
2. Create a new drawing, and set up the units and limits according to your calculations in step 1.
3. Create the object lines for the front (left) view.
4. Use the RAY command to extend key lines from the front view to the side view.
5. Trim the ray lines and add the other object lines to complete the side view.
6. Add dimensions, notes, and other details as necessary to finish the drawing, as shown in Figure 19-31. The triangular symbol with +41′-8 tells us that the top elevation of the beam when it is installed is 41′-8 above a given reference point or reference elevation.

To achieve the dimension style suggested in this chapter for structural drawings, you will need to change the Standard dimension style or create a new one in AutoCAD. Although the specific steps vary among versions of AutoCAD, all versions allow you to specify an unbroken dimension line with the dimension set above the line.

Chapter Summary

■ The structural drafter is generally a member of a rather large and complex engineering team.

■ Structural drafters prepare the drawings used to construct the framework of a building or other structure.

■ While there are many different materials used in construction, the basic ones are steel, wood, concrete, structural clay products, and stone masonry.

■ The design of the framework of a structure must include consideration of many different kinds of loads in addition to the weight of the structure itself.

■ Structural steel members are generally fastened together by welding, bolting, or riveting.

■ Structural timber can be manufactured in many forms by laminating.

■ The load-carrying capacity of concrete can be increased by reinforcing or prestressing it.

■ Many structures are possible only through the use of concrete.

■ The strength of a structural clay system depends on the strength of the mortar joints.

Review Questions

1. Name five types of jobs that are commonly done by structural drafters.

2. Name five basic structural steel shapes.

3. What is the major advantage of a geodesic dome?

4. Name four types of trusses.

5. What is the difference between a frame and a seated steel-beam connection?

6. How do you dimension beams, angles, and channels?

7. How do you draw dimension lines on structural drawings?

8. What symbols are used in structural dimensions, and when do you use them?

9. In what way does lamination increase the usability of wood for structural elements?

10. What are the ingredients in concrete?

11. What is the difference between reinforced concrete and prestressed concrete?

12. What characteristic of vertical walls made of structural clay systems allows them to carry the weight of floors and other loads?

13. What are the metric dimensions of a common brick?

14. In addition to preparing 2D structural drawings efficiently, what capability does CAD software add for the structural engineer or designer?

15. What must you do to achieve the preferred dimension style for structural drawings using AutoCAD?

Problems

Drafting Problems

The drafting problems in this chapter are designed to be completed using board drafting techniques or CAD.

1. Draw the detail of a frame connection on an S-beam, as shown in Figure 19-32. The S-beam is 8″ × 18.4 lb., the flange is 4″ wide, the web is .270″, and the 3″ × 3″ connection angle is 6″ long. Pitch = 1½″; gage = 1¼″; rivet = ½″. Prepare the detail at half scale.

2. Prepare a partial detail of the frame connection of an 180-wide flanged beam with a 240-wide flanged beam, as shown in Figure 19-33. Scale: 30 = 10-0. Prepare a partial part-plan diagram about column center C, with noted members, and elevations.

3. Prepare a detail drawing of the standard 20″ × 64.5 lb. S-beam shown in Figure 19-34. The flange is 6¼″ wide with web thickness of ½″. Scale: ¾″ = 1′-0.

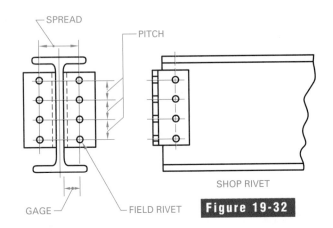

SHOP RIVET

Figure 19-32

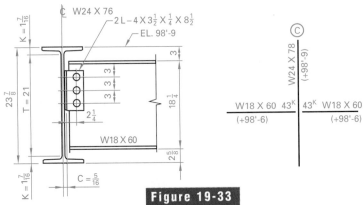

Figure 19-33

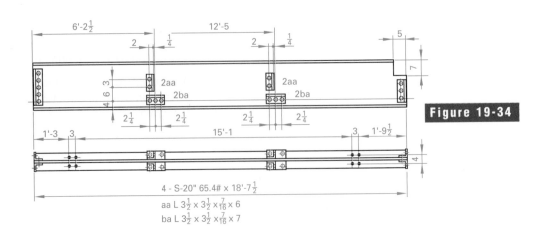

Figure 19-34

4 - S-20″ 65.4# × 18′-7½
aa L 3½ × 3½ × 7⁄16 × 6
ba L 3½ × 3½ × 7⁄16 × 7

4. Prepare a drawing of the reinforced concrete detail shown in Figure 19-35. Scale: 1½″ = 1′-0.

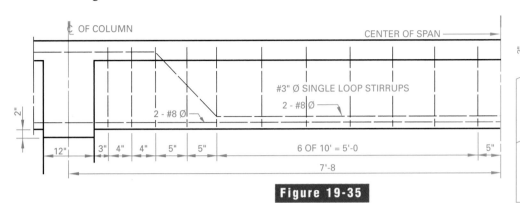

Figure 19-35

5. Prepare a truss detail of Figure 19-17 at a scale of 1″ = 1′-0 (1:12). Prepare the left half of the symmetrical beam as shown about the centerline.

6. Detail the standard steel shapes shown in Figure 19-36 at an appropriate scale, and show sectioning.

7. Prepare a detail for the girder shown in Figure 19-37. The girder is 22″ high with a 6″ flange. Develop the girder showing the conventional lines (shown in magenta) with heavy line weights. Identify angles, welds, stiffeners, and field bolts. With the instructor's guidance, fill in missing dimensions. Scale: 1½″ = 1′-0.

Figure 19-36

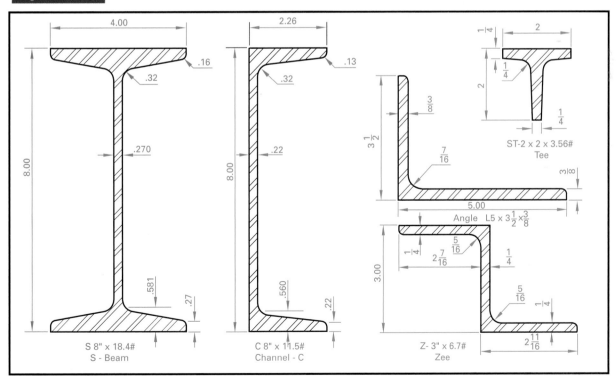

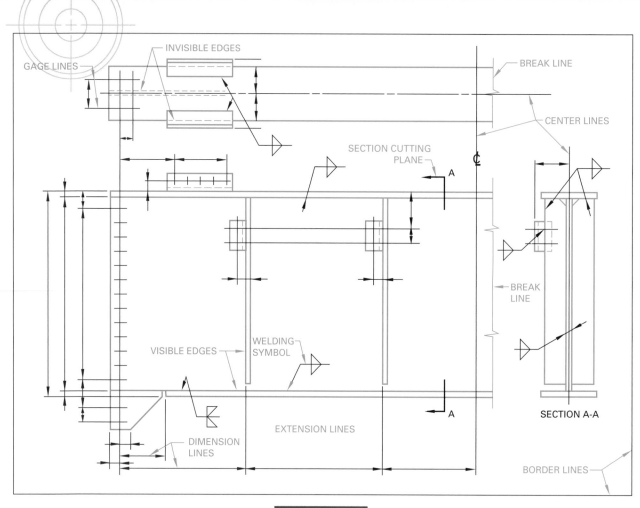

Figure 19-37

Design Problems

Design problems have been prepared to challenge individual students or teams of students. In these problems, you are to apply skills learned mainly in this chapter but also in other chapters throughout the text. The problems are designed to be completed using board drafting, CAD, or a combination of the two. Be creative and have fun!

1. Design a welding workbench for use in a fabricating shop using steel components. It is to be 34″ (864 mm) high. The top surface, ¾″ steel plate, is to be 48″ × 72″ (1200 mm × 1800 mm). Make a complete set of working drawings, including connection details and a materials list.

2. **TEAMWORK** Work as a team to design a truss for a gable roof. The truss should extend across an unsupported span of 30′-0 (9000 mm). The rise is 8′-0 (2400 mm). Each team member is to make a scale model (1″ = 1′-0 or 1:10 metric scale) of the truss using a different material for each. Test the comparative strengths of the scale models using free weights. To test the scale models, set them upright and support the ends on sawhorses or strong tables. Hang free weights from the underside and record the amount of weight supported before the trusses fracture. Compare the strength of the various materials. *Work safely!*

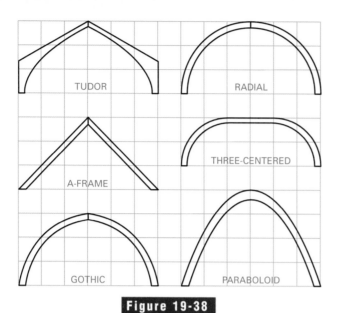

Figure 19-38

3. Design a creative, artistic welded sculpture using structural steel components. Design it to sit in a reinforced concrete base or to be bolted to a reinforced concrete slab. Make a working drawing of your creation and a materials list.

4. **TEAMWORK** Assign each member of the design team one of the laminated wood trusses shown in Figure 19-38. Have each team member design and draw a structural steel connection to join the tops of the two members of the truss assigned. Prepare a connection detail, including fasteners if appropriate. Design the connections as fabricated steel parts. Each grid block on Figure 19-38 represents a 8'-0 (2400 mm) square. The thickness of the components is 10" (254 mm). Estimate sizes as necessary.

Map Drafting

KEY TERMS

block diagram

cartographer

cartography

contour

contour interval

dip

fault

geology

operations map

plat

spline

stratum

strike

topographic maps

OBJECTIVES

Upon completion of this chapter, you should be able to:

- Describe the work and career opportunities of a cartographer.

- Explain how scales are determined and shown on maps.

- Identify various types of maps, such as plats, structural maps, geological sections, block diagrams, and contour maps.

- Explain the purposes of geological maps, geological sections, and subsurface maps.

- Produce a contour map with standard symbols using both board drafting and CAD techniques.

- Produce a plat using both board drafting and CAD techniques.

apmakers are people who have been trained to gather the necessary information and prepare maps. Mapmakers are also called **cartographers**. Skilled map drafters prepare maps in detail for government agencies, civil engineers, scientists, geographers, and geologists.

Mapmaking, or **cartography**, is a graphic method of representing facts about the surface of the earth or other bodies in the solar system. Aircraft and satellites take high-altitude photographs that can help mapmakers work efficiently and accurately. See Figure 20-1. Cartographers use a method called **photogrammetry** to make three-dimensional measurements from the resulting photographs. Computers bring speed and efficiency to nearly every stage in the process, enabling cartographers to create specialized maps from existing measurements rather than through extensive and costly fieldwork.

Cartography is a highly specialized field of drafting. It has its own set of rules, regulations, techniques, and methods. As a result, much of this chapter deals with the theory and principles behind map drafting and concepts that need to be understood before even the simplest maps can be drawn.

Career Opportunities

Many jobs are available for those who are skilled in map preparation. The field of civil engineering is always expanding. Railroads, highways, harbor facilities, airports, and space stations are just a few areas in this broad industry in which map planning is being used.

The drafter may prepare maps and charts under the direction of the design engineer or cartographer. There may be opportunities for advancement in areas such as photogrammetry, surveying, or research and development projects with the geographer.

Success on the Job

Feedback
Any feedback, good or bad, about your work or personal qualities can be useful information. Write down all the comments you receive and review them periodically. This will help you to spot trends in yourself and in the quality of your work. Any improvements you make as a result of this practice will enhance your chances for success on the job.

Figure 20-1
Satellites can take high-resolution photographs from which scientists make map surveys of the earth and other planets in our solar system.

Map Scales

Some maps that describe ownership of property, such as city plats, must be very accurate. They may be drawn to a large scale in order to note all the information of physical property. Other maps, such as those showing the geography of states or countries, may need to show boundary lines, streams, lakes, or coastlines over large areas. The scale on these maps must be quite small. In fact, some maps of very large areas may use a scale of several miles to the inch.

Satellite photos divide the United States into quadrangle maps no larger than 71 miles, drawn at a scale of 1:24,000. At that level, 1″ on the map equals 2000 feet on the ground. This allows the map to show buildings, roads, and other physical features.

The civil engineer's scale is used for map drawings. Distances are given in decimals of a foot, mile, meter, or kilometer. See the maps in geography and history books for examples.

A map cannot be read accurately unless the scale is known. Therefore, the scale of a map must be shown as part of the basic map information. The scale of a map is generally noted as 1″ = 500′, or 1 part = 6000 parts, noted as 1:6000. The scale of 1″ = 1 mile can also be shown as 1:63,360.

Types of Maps

Maps that are made for different reasons have distinct features and notations. Examples of several kinds of maps are illustrated and explained in this section.

Plats

A map used to show the boundaries of a piece of land and to identify it is called a **plat**. The amount and kind of information presented on a plat depend on its purpose.

The plat of a survey that was made to accompany the legal description of a property is shown in Figure 20-2. Accuracy of information on a plat is all-important. It must agree with the legal description exactly.

Other types of plats are *planimetric* and *cadastral*. Planimetric plats are made to keep a record of street improvements and to show the location of utilities in towns and cities. Cadastral plats are used to record sizes and location of property for tax purposes. A plat is shown in Figure 20-3. Notice the numbering of the lots and the location of streets, sidewalks, and other details important to city officials.

Operations Maps

Maps that show the relationship between the land's physical features and the operation that is to be performed are known as operations maps. An example of an **operations map** is shown in Figure 20-4. Well-executed operations maps help engineering, management, or government groups in the presentation. A presentation well done aids in the selling of a program.

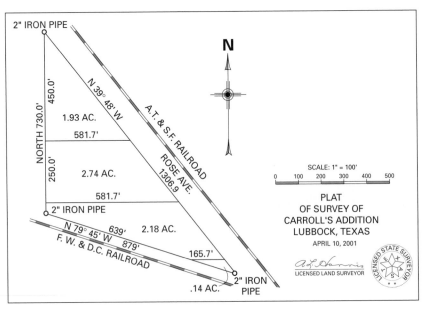

Figure 20-2

Plat of a survey. Note the parts that make up the plat: the acreage of each part, the iron pipe locating the corners of the graphic scale, the signature of the surveyor, and the official seal. One acre = 43,560 square feet.

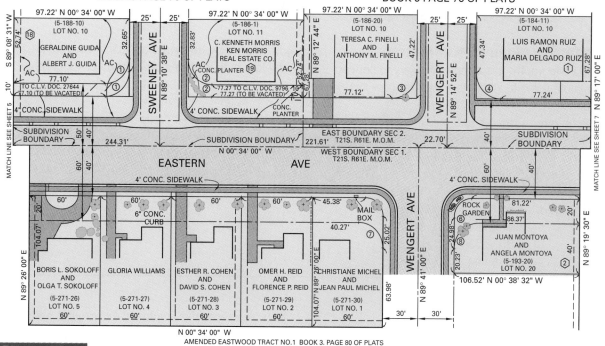

Figure 20-3
Part of a city map.

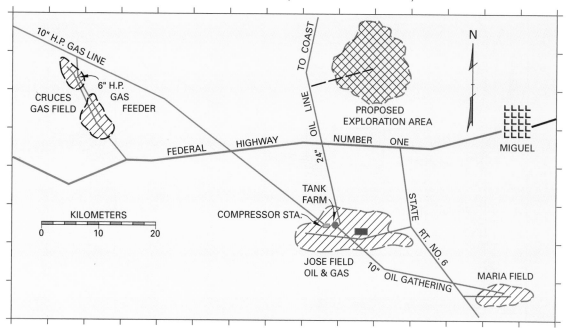

Figure 20-4
An oil-field operations map. Note that the scale is in kilometers (1 km = .621 statute mile).

Statute Miles to Kilometers

In cartography, it is often necessary to convert statute miles to kilometers. Here is how it is done:

Problem: Convert 10.5 miles to kilometers

Solution: 1 statute mile = 1.609 kilometers

$10.5 \times 1.609 = 16.9$ kilometers

Problem: Convert 6.8 square miles to square kilometers

Solution: 1 square mile = 2.590 square kilometers

$6.8 \times 2.590 = 17.6$ square kilometers

The following are some additional conversion factors that are commonly used in cartography:

1 kilometer = .6214 statute mile

1 meter = 39.37 inches

1 meter = 1.094 yards

1 acre = 4047 square meters

1 cubic yard = .7646 cubic meter

1 cubic meter = 35.31 cubic feet

Contour Maps

Since maps are one-view drawings, vertical distances in ground level do not show. They can, however, be shown by lines of constant level called **contours**, as shown in Figure 20-5. The contour lines represent the height of the ground above sea level. Contour lines that are close together indicate a steeper slope than lines that are far apart. This can be seen by projecting the intersections of the horizontal level lines with the profile section, as shown in Figure 20-5.

Note that the contour map and the profile correspond to the plan and section of an ordinary drawing.

Figure 20-5

A contour map with a profile at section A-A.

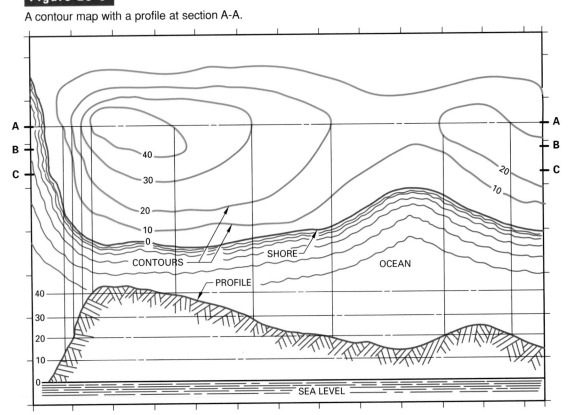

Figure 20-6

A contour map with intermittent streams indicated by the lines with three dashes.

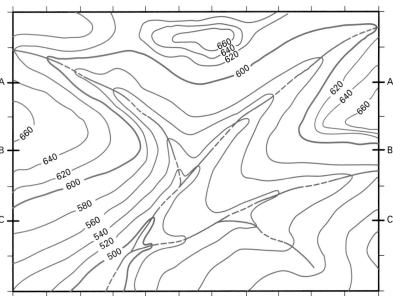

The horizontal cutting plane *AA* on the contour map shows exactly where the profile is taken. You can see how the profile would change if the cutting plane were moved toward the ocean or to some other new position.

The **contour interval** (vertical distance between contour lines) can be adjusted according to the cartographer's needs and the scale of the map. A 10′ distance may be quite satisfactory for some purposes, while a 5′ distance may be used on maps that require a larger scale. For close detail work, such as an irrigation project, the contour distances may be reduced to 2′, 1′, or even .5′. On small-scale maps with a high degree of relief, distances may be increased from 20′ to 200′ or more. As an aid to reading the map, every fifth contour is usually emphasized by drawing a much heavier line.

Another example of a contour map is shown in Figure 20-6. This map uses a contour distance of 20′. The elevation in feet is marked in a break in each contour line. Notice that the drainage is shown in intermittent streams.

Before a contour map can be drawn, elevations must be obtained in the field for several key points that control the drawing of the contours. The following methods are used:

- a grid system in which all intersection elevations are obtained, along with important elevations on grid lines
- points located by transit and stadia rod, with the corresponding elevations figured by plane table
- aerial photographic surveys similar to the one shown in Figure 20-7

Experience in surveying is needed in all of these mapping methods.

Figure 20-7

Portions of an aerial photo and a topographic map compiled by photogrammetric methods.

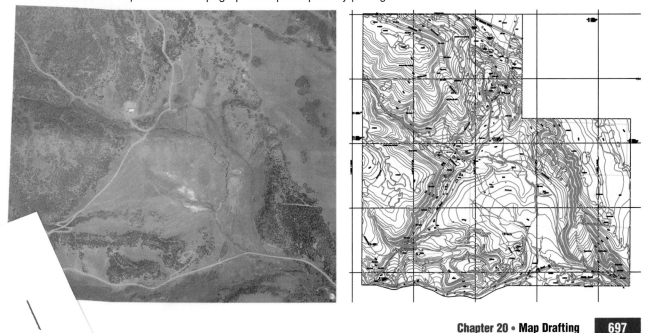

Topographic Maps

Maps that present a complete pictorial description of the areas shown are called **topographic maps**. These maps show information such as boundaries, natural features, structures, vegetation, and relief (elevations and depressions).

Symbols are used for many of the features shown on topographic maps. Some of these are shown in Figure 20-8. Maps using topographic symbols can be obtained at a low cost from the Director, U.S. Geological Survey, Department of the Interior, or from the U.S. Coast and Geodetic Survey, Department of Commerce, Washington, D.C. Naval charts can be ordered from the Hydrographic Office, Bureau of Navigation, Department of the Navy.

Block Diagrams

Discussion thus far has concentrated on mapping in the horizontal plane and the vertical plane by profiles or sections. To help people see three-dimensionally, drafters can create a **block diagram** (a three-dimensional projection using the isometric view). The block diagram in Figure 20-9 was developed from the contour map in Figure 20-6. Keep in mind that each contour represents a level plane, similar to a card in a deck of cards. As in all isometric drawings, true lengths are measured on the isometric axes.

Geological Mapping

To understand geological mapping, you must first understand a little about geology. **Geology** is the science that deals with the makeup and structure of the earth's surface and interior depths. The crust of the earth is made up of three groups of rock: igneous, sedimentary, and metamorphic.

Igneous rocks, for purposes of this discussion, are the basic crystalline materials that make up the earth's crustal ring. This rock was once molten. It has cooled, but it has not been eroded, nor has its makeup changed. See Figure 20-10.

Sedimentary rocks, as a rule, are deposited in water in layers of different thicknesses similar to the layers of an onion. If you cut an onion perpendicular to its axis, you will notice a series of concentric rings. In a slice of the earth's crust made in a sedimentary area, a similar pattern can be seen. A series of layers that can be identified by texture, color, and

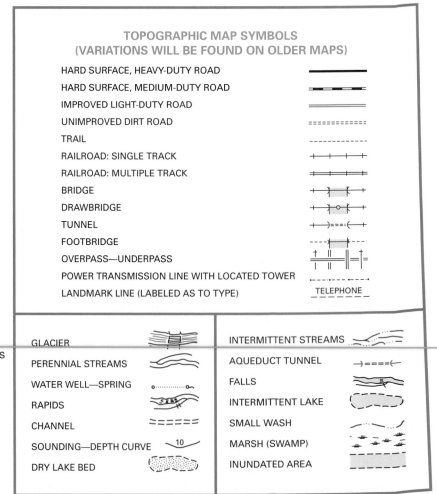

Figure 20-8

Some conventional symbols used on maps. Note the intermittent streams as used in Figure 20-6.

Figure 20-9

A block diagram of the area shown in the contour map in Figure 20-6.

material is visible. Pictures of the Grand Canyon provide a good example of sedimentary layers, as shown in Figure 20-11.

Sedimentary rocks that have been deeply buried, heated to high temperatures, and subjected to great pressure become *metamorphic rocks*. The combined heat and pressure recompose the rocks so that they can no longer be identified as sedimentary. See Figure 20-12.

Nature, being ever-changing, folds, tips, and slices sedimentary layers in a number of ways. Sometimes sedimentary layers include large areas of igneous rock. Often, the whole mass may be tipped, perhaps for miles. In some places, this tipping raises the mass of rock high above sea level. In other places, the mass drops thousands of feet. The geologist making investigations has the problem of representing what has happened or what a particular area looks like.

Figure 20-13 is part of a geological surface map. The red lines represent the line of surface exposure of the contact between two rock formations, like the line of two contacting layers of a cut onion. The geologist locates this in the field and notes the observation point with the T symbol. The **strike**, or direction of contact, is shown by the top of the T. The **dip**, or slope of the contact, is shown by the figures at the T, such as 23° on the left side of Figure 20-13, below section line *XX*. Since the stem of the T, in this case,

Figure 20-10

Igneous rock.

SEDIMENTARY ROCK LAYERS

Figure 20-11

The Grand Canyon provides an excellent example of sedimentary rock.

is pointing to the east, or to the right, the dip is 23° to the east. In other words, this formation slopes 23° below the horizontal and to the east. Another T symbol on the left side shows a 30° dip to the west. The heavy, broken line near the left edge represents a fault. A **fault** is the line along which a rock layer has broken.

Geological Sections

The purpose of geological sections is to help in the interpretation of geological surface maps. The example in Figure 20-14 shows what the geologist believes the area below the surface is like. This is a

section along line *XX* of Figure 20-13. The dips that the geologist noted are used in developing the curvature of the folds. By means of a typical section of the region, the geologist can determine the various normal thicknesses of each **stratum** (plural *strata*), or distinct layer. These values are used in making the section. The fault, as indicated, shows the area at the right to be upthrown. The displacement is easily seen by comparing the position of formation *A* on either side of the fault.

Subsurface Mapping

To show details of strata lying below the surface of the earth, subsurface mapping is used. Subsurface mapping can show the top or bottom of a given formation, or possibly an assumed horizon. Information for constructing such a map is obtained from many sources. These sources may include core holes, electrically recorded logs, seismograph surveys, and so forth.

Figure 20-13
Part of a geological surface map.

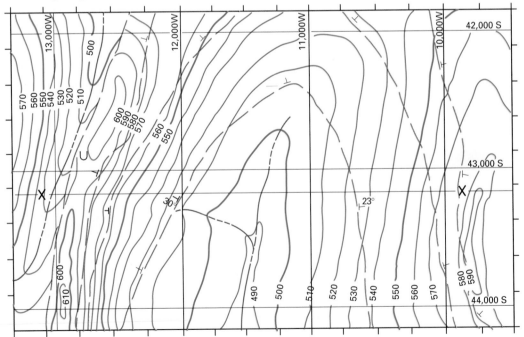

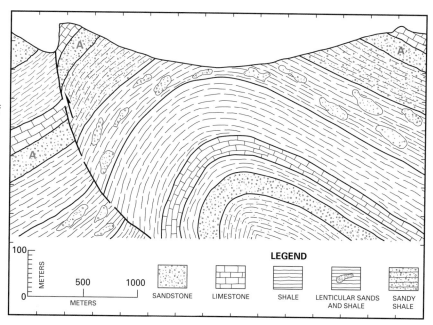

A geological section along line XX of Figure 20-13.

An example of information that was obtained from electrically recorded logs taken in a series of oil wells is shown in Figure 20-15. The wells are located on a grid pattern. Producing wells are represented by a solid black circle. Dry holes are represented by an open circle with outward-extending rays. Notice that the contours are numbered with negative values, or depths that are below sea level. The greater the value, the deeper the point below sea level. Section *XX* shows the thickness of the sand and the level of the oil-water contact.

The readability of geological maps and sections can be greatly improved by the use of colors. In Figure 20-13, colors may be applied to each of the formations between the red formation-contact lines. This can also be done in Figure 20-14 by applying colors to the corresponding formations. The use of colors helps bring out the three-dimensional relationship of the surface and the shape of the structure. Color also helps people understand the geology of the area. In board drafting, the drafter creates a paper print of the tracing, adds color, and then rubs the colored areas of the print carefully to give smooth, even color texture. Color is usually used in CAD-generated drawings as a matter of course.

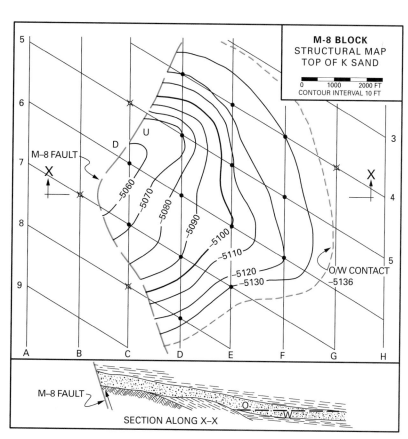

Figure 20-15

A structural map showing strata details below the surface of an oil field.

The creation of geological maps and drawings is an important part of the extractive minerals industry, particularly the petroleum industry. With the aid of maps, it is possible to keep proper records and information so that activity in this economic field can be continued. Standards for records are different from company to company. However, general standards are well covered in technical literature such as publications of the Petroleum Branch of AIME (American Institute of Mining, Metallurgical, and Petroleum Engineers), the AAG (Association of American Geographers), U.S. Geological Survey, U.S. Bureau of Mines, and others.

Board Drafting Techniques

As you have learned, there are many different kinds of maps that are produced in various ways. The map used in this board drafting example is first developed as a contour map, shown in Figure 20-16, and then is further developed into a plat or plot plan, as shown in Figure 20-17. The initial contour map shows the undeveloped land. It is developed around carefully surveyed points, through which the contour lines are drawn.

Figure 20-16

Contour map with profile.

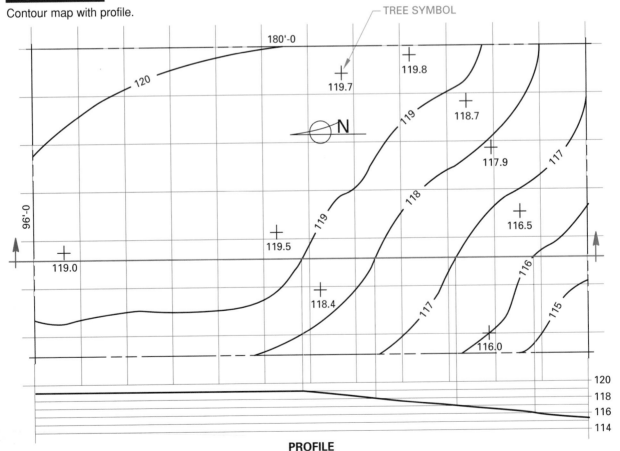

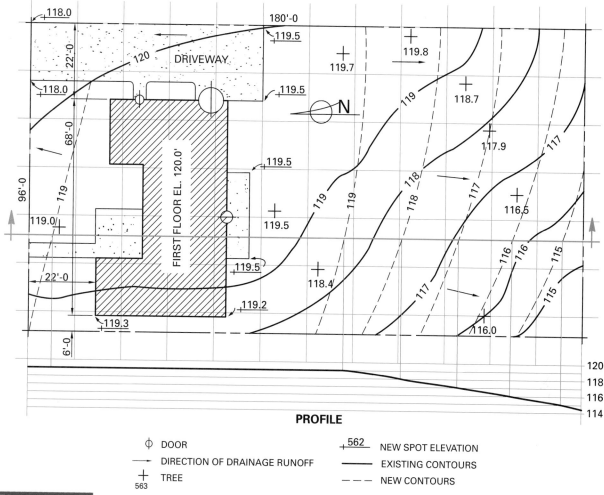

PROFILE

φ DOOR

→ DIRECTION OF DRAINAGE RUNOFF

+ TREE
563

+562 NEW SPOT ELEVATION

—— EXISTING CONTOURS

- - - NEW CONTOURS

Figure 20-17

Finished plat (plot plan) with profile.

Creating a Contour Map

In the past, drafters used contour pens to ink contour lines. Contour pens have blades that swivel so that complex contour lines can be followed easily. However, a technical pen is now preferred; it has replaced the contour pen for nearly all board drafting applications.

The curved lines on a contour map must be drawn with great accuracy because each line represents a precise elevation. To complete the curves on a contour map accurately, use one of the many types of flexible curves that can be bent to match the curves to be drawn on the map.

For the purpose of this exercise, a ½″ grid has been added in the background of Figures 20-16 and 20-17 to help you locate the contours accurately. In

practice, the contours would be developed as discussed earlier in this chapter. Follow these steps:

1. Rough in the contour lines on an A-size sheet at a scale of 1″ = 20′. Use the ½″ grid as reference to place the lines accurately. (*Note:* Do not include the grid on your final drawing.)

2. Develop a profile from the contour map. The profile can be taken through any desired location to show the slopes of the plot. In fact, several profiles are often drawn on plots that are much larger or on which the contour is erratic. Profiles can be taken horizontally, vertically, or at any other angle that best describes the shape of the plot graphically. In this case, create a horizontal profile at a location of your choice.

3. Darken the lines and add notes and text as necessary to finish the drawing. (Do not add the TREE SYMBOL label and leader.)

Creating a Plat

Figure 20-17 shows the proposed improvements to the property shown in the contour map in Figure 20-16. In Figure 20-17, the contour has been revised to show how the property should be graded. While the profile in Figure 20-16 is nearly identical to the one in Figure 20-17, a careful study of the revised contour lines will show that the finished contour will be somewhat smoother and more uniform than the original contour.

Notice that the house is carefully and precisely located in relationship to the property lines. Also, notice that new spot elevations have been added to establish the elevations at critical locations, such as the corners of the house, patio, and driveway. This gives the landscaper precise information regarding elevations and contours.

Create the plat as an overlay to the contour map you drew in the previous section. Follow these steps:

1. Using tracing paper, set up an overlay for the contour map. Trace the property lines and the contour lines carefully.
2. Add the revised contour lines and the spot elevations, using the grid in Figure 20-17 as a guide. (*Note:* Do not include the grid on your final drawing.)
3. Place the house, driveway, and other structures accurately according to the dimensions shown in Figure 20-17.
4. Develop a new profile using the revised contour lines.
5. Darken the lines and add notes and text as necessary to finish the overlay. (Do not add the SPOT ELEVATION label and leader.)

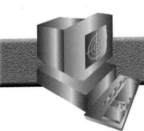

CAD Techniques

One of the conveniences CAD brings to the drafting table is the ability to create a single drawing and use it for more than one purpose. The needs of various contractors can be met by using layers effectively and then controlling their display to show only the information needed for a particular purpose. See Chapter 4 for more information about using layers in this manner. In this exercise, you will create a contour map that shows an undeveloped piece of property, as shown in Figure 20-18. Then, on another layer (or set of layers), you will complete a plot plan to show the finished grading and property improvements, as shown in Figure 20-19.

Creating a Contour Map

The curved lines on a contour map must be drawn with great accuracy because each line represents a precise elevation. To complete the curves on a contour map accurately, first identify the known elevation points and enter them into the drawing. Then use the SPLINE command to connect the points in a smooth line. If necessary, you can use the PEDIT or SPLINEDIT command to adjust the curves. Follow these steps:

1. Create a new drawing. Study the dimensions given in Figure 20-19 to determine the drawing limits. Set the units to architectural, and set the

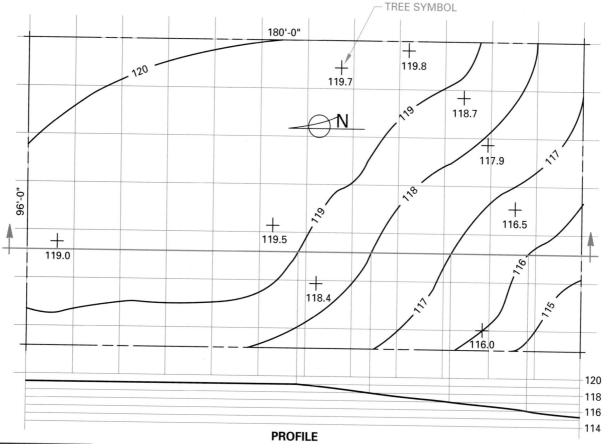

Figure 20-18

CAD-generated contour map.

dimension style for aligned dimensions with the text set above the unbroken dimension line. You may also want to set grid and snap to convenient intervals. (A half-inch grid would match the one in Figures 20-18 and 20-19.)

2. Set PDMODE to 3.

3. Then use the POINT command to place key elevation points for the 120′ contour line. In practice, these points would have been determined in the field. For this exercise, a ½-inch grid has been added behind the contour to help you place the contour lines accurately. Create your elevation points at locations where the contour lines in Figure 20-18 cross the grid. You do not have to create a point at every intersection, but create enough so that your contour line will be accu-

rate. Be sure to include the points at the property lines that define where the contour lines will begin and end.

The points you created in step 3 become the control points for a spline. A **spline** is simply a curved line that passes through or is controlled by a series of points. To create a spline in AutoCAD, you can use the SPLINE command.

4. Enter the OSNAP command to set the running object snaps. Turn on the Node object snap. You may want to disable the other object snaps temporarily.

5. Enter the SPLINE command. Select the point at one end of the contour line. Then snap to each of the other points, in order. The spline will begin to develop as you select points. After you

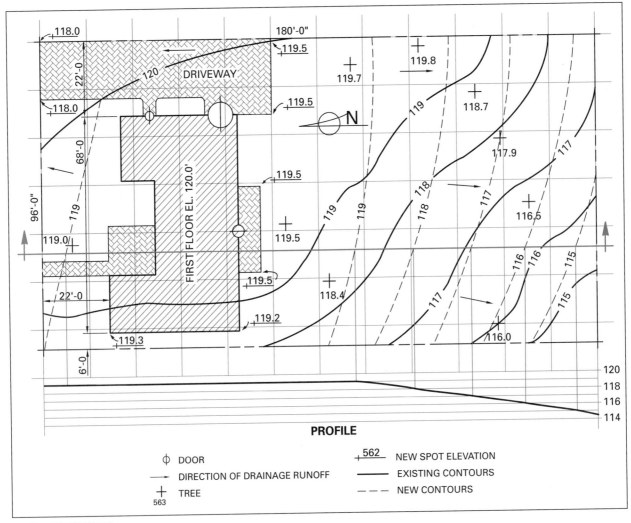

Figure 20-19

CAD-generated plat.

have selected the last point, you will need to press the Enter key three times to complete the spline correctly.

6. Repeat steps 3 through 5 for each of the other contour lines.

7. Create a new layer called OriginalProfile, and make this the current layer.

8. On the OriginalProfile layer, develop a profile from the contour map. The profile can be taken through any desired location to show the slopes of the plot. In fact, several profiles are often

drawn on plots that are much larger or on which the contour is erratic. Profiles can be taken horizontally, vertically, or at any other angle that best describes the shape of the plot graphically. In this case, create a horizontal profile at a location of your choice. Use a construction line for the first magenta grid line, and offset it to create the other grid lines. If you disabled object snaps such as Endpoint and Intersection, you may want to turn them back on before creating the extension lines from the contour map to the profile.

9. Change back to the layer on which you created the contour map. Add the tree symbols, notes, and text as necessary to finish the drawing. (Do not add the TREE SYMBOL label and leader from Figure 20-18.)

Creating a Plat

Figure 20-19 shows the proposed improvements to the property shown in the contour map in Figure 20-18. In Figure 20-19, the contour has been revised to show how the property should be graded. While the profile in Figure 20-18 is nearly identical to the one in Figure 20-19, a careful study of the revised contour lines versus the original contour lines will show that the finished contour will be somewhat smoother and more uniform than the original contour.

Notice that the house is carefully and precisely located in relationship to the property lines. Also, notice that new spot elevations have been added to establish the elevations at critical locations, such as the corners of the house, patio, and driveway. This gives the landscaper precise information regarding elevations and contours.

Because the plot plan contains much information that is not needed for a contour map, you should place the improvements on a separate layer or layers that can be turned off to show only the contour map. Follow these steps:

1. Create new layers called Grading and Improvements. Make Grading the current layer.
2. Add the revised contour lines and the spot elevations, using the grid in Figure 20-19 as a guide. (*Note:* Do not include the grid on your final drawing.)
3. Make Improvements the current layer. Place the house, driveway, and other structures accurately according to the dimensions shown in Figure 20-19.
4. Enter the BHATCH command and choose the AR-HBONE hatch pattern. Pick points in the driveway, front walk, and patio areas to hatch. You will need to use a small scale to make the hatch look appropriate on the drawing. Preview the hatch before picking OK, and adjust it if necessary. Be sure to pick the Remove Islands button to remove the hatch from around text automatically.

►CADTIP

Layers for Hatches
Before you hatch the drawing, you may want to create layers named Brick and SectionLines. Place the hatches on these layers so that they can be removed temporarily for clarity if necessary. You may also want to create the hatches in contrasting colors, as shown in Figure 20-19, to make the drawing easier to read.

5. Use the ANSI31 section lining hatch pattern to hatch the house.
6. Create a new layer called NewProfile, and make this the current layer. Freeze the OriginalProfile layer.
7. On the NewProfile layer, develop a new profile using the revised contour lines.
8. Add tree symbols, notes, and text as necessary to finish the overlay. (Do not add the SPOT ELEVATION label and leader.)

CAD-Automated Cartography

Mapping requires a tremendous amount of information, complicated calculations, and precise drafting skills. CAD programs, along with special third-party mapping software, can ease the process of creating clear, accurate, and up-to-date maps. Third-party packages include software that can import data directly from global positioning satellite (GPS) systems. This allows field workers to use satellite data to specify positions and elevations very accurately. The software imports the data and builds a block diagram based on the individual points. This can be extremely helpful in cases in which the data changes frequently, as in mining and some civil engineering applications. For example, using CAD and a third-party data import system, the block diagram shown in Figure 20-20A can be updated automatically to reflect a cut-and-fill design for land development, as shown in Figure 20-20B. (A *cut* is earth to be removed to obtain a desired level or slope in preparation for construction. A *fill* is earth to be supplied and put in place for the same reason.) Both the original and the revised block diagram would have taken many hours to complete without the computerized tools. In addition, the CAD-generated files have the advantage of being true 3D files, so they can be viewed from any direction.

Some cities also use CAD-based plats to track facilities such as utility poles and fire hydrants automatically. They use CAD's block attributes to assign information such as the individual hydrant numbers, the date the hydrant was last inspected, and the date of the next scheduled inspection. CAD programs can be configured, either with or without third-party software, to alert the CAD operator when an inspection should occur. In this example, the notification might include a list of hydrants to be inspected, as well as flashing icons on the plat at the location of each hydrant to be inspected.

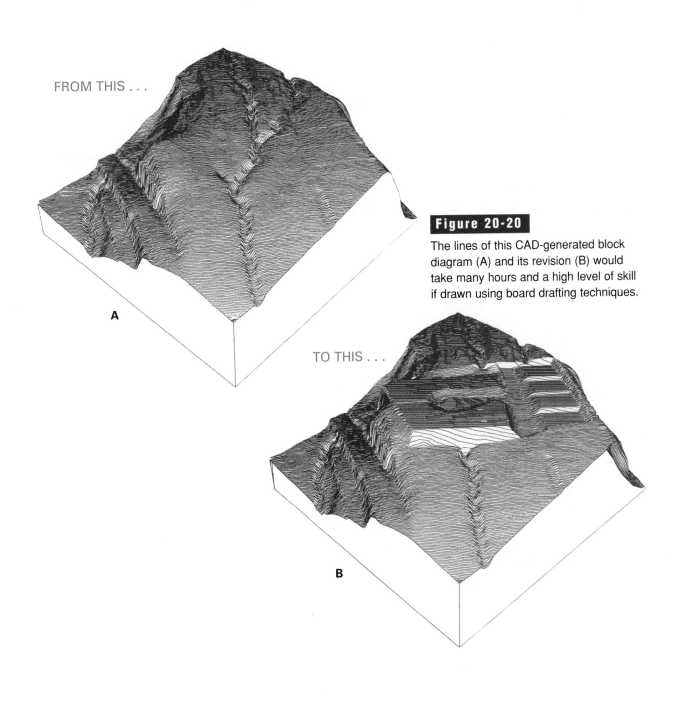

FROM THIS . . .

A

Figure 20-20

The lines of this CAD-generated block diagram (A) and its revision (B) would take many hours and a high level of skill if drawn using board drafting techniques.

TO THIS . . .

B

Chapter Summary

■ Career opportunities in the field of mapmaking include areas such as surveying and research and development projects as well as drafting.

■ Mapmaking is a pictorial method of representing facts about the surface or subsurface of the earth or other bodies in the solar system.

■ The scale of a map depends on its intended use.

■ Accuracy in plotting detail is of extreme importance in preparing maps.

■ Plats, operations maps, contour maps, topographic maps, and block diagrams are but a few of the many types of maps in use today.

■ Geological maps show surface formations, and geological sections help with the interpretation of geological maps. Subsurface maps show the strata below the earth's surface.

■ A large amount of printed material about maps and mapping is available from various government agencies.

■ CAD techniques have now become the most common way to prepare maps.

Review Questions

1. What is another name for mapmaking?

2. Which field of engineering is most generally associated with maps and mapmaking?

3. Is the scale of a map large or small if it covers a large area?

4. Is the scale of a map large or small if it covers a small area?

5. What is a plat?

6. What kind of map shows the relationship between the land's physical features and the purpose for which the land is to be used?

7. What is the purpose of a contour map?

8. What are contour lines?

9. What do you call the vertical distance between contour lines?

10. What kind of information is shown by a topographic map?

11. What do you call a pictorial drawing, generally isometric, of a block of earth showing profiles and contours?

12. What is another word for *dip*?

13. What do you call the line along which a layer of rock has broken?

14. What is a geological map?

15. Name the three groups of rocks that make up the crust of the earth.

16. How are contour lines generated using a CAD program?

Drafting Problems

The drafting problems in this chapter are designed to be completed using board drafting techniques or CAD.

1. Plot the map of Chicago, as shown in Figure 20-21, on a C-size sheet. Scale: ¾″ = 1 mile, or other suitable scale. Calculate the number of square miles, acres, or square kilometers that make up this city. (1 square mile = 2,589,988 square meters)

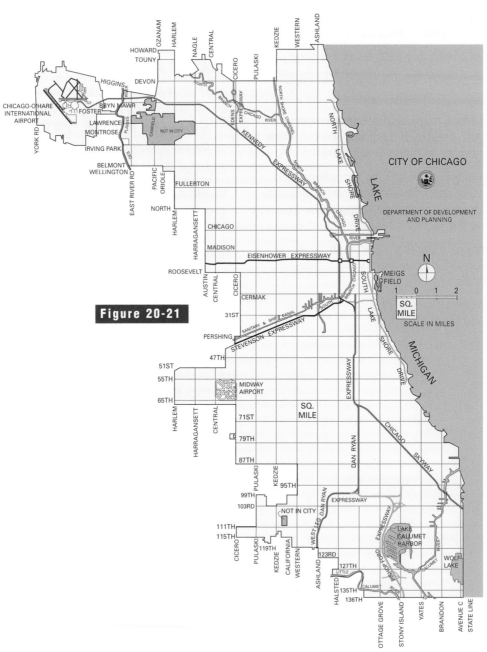

Figure 20-21

2. Make a drawing of a plat survey, as shown in Figure 20-2. Use a working space of 11″ × 15″. Scale: 1″ = 100′. Start the lowest point 8.25″ from the left border and 1.50″ up from the bottom border.

3. Using the drawing from problem 2 as a basis, create an overlay (or additional layers on a CAD system) to lay out a residential tract of land and divide it into lots. Plan most lots with 75′ frontage and 150′ or more in depth. Rose Avenue is a main street.

4. Prepare a drawing of a contour, as shown in Figure 20-5. Use a working space of 8.50″ × 11.00″. Draw 2 times the size shown in the figure. Use dividers, a grid, or a digitizer to enlarge the figure (vertical profile scale 1″ = 20′; horizontal scale: 1″ = 500′). Draw the profile on line *CC*.

5. Make a city map, using the data provided in Figure 20-3, showing streets, sidewalks, and lots with dimensions.

6. Make an operations map, as shown in Figure 20-4. Draw a grid sheet over the map. Short marks along the border are to assist in drawing the grid. Redraw on an 8.00″ × 15.00″ working space. How long is the 10.00″ gas line? How long is the 24.00″ oil line?

7. Make a contour map, as shown in Figure 20-6. Use the grid on the border to enlarge the contours. Scale: grid = 1″ squares. Working space: 8.50″ × 11.00″.

8. Prepare a contour map of Figure 20-15 and a vertical profile through *XX*. The grids are 1.00″ apart. Scale: 1″ = 100′.

9. A client has purchased the building lot shown in Figure 20-22, on which she plans to construct a house (see crosshatched area). Contours need to be changed slightly to provide for a more ideal building lot. As a drafter, your job is to do the following:
   a. On a B-size sheet, use a grid (.25″ squares) to draw the plot plan as shown.
   b. Draw profiles as indicated by cutting-plane lines *AA* and *BB*.
   c. Using colored lines or dashed lines, revise the contour to result in a gentle slope in all directions away from the outline of the house.
   d. Draw revised profiles of *AA* and *BB* using colored lines or dashed lines.
   e. Add landscaping symbols for shrubs, trees, etc., if assigned.

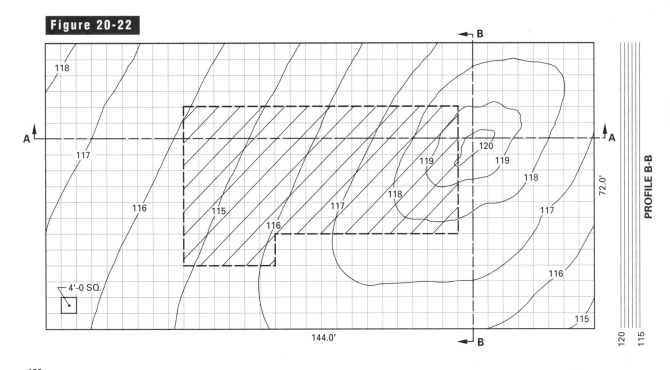

Figure 20-22

PROFILE A-A

Design Problems

Design problems have been prepared to challenge individual students or teams of students. In these problems, you are to apply skills learned mainly in this chapter but also in other chapters throughout the text. The problems are designed to be completed using board drafting, CAD, or a combination of the two. Be creative and have fun!

1. Develop a plat for an ideal building lot. Show contour lines and horizontal and vertical profiles at the highest elevation. Show the location of a house, sidewalks, driveway, trees, and utility lines.

2. **TEAMWORK** Design a housing development consisting of 20 building lots and additional space for public service features such as a mini-park with a playground. Show streets, sidewalks, structures, trees, utility lines, and other significant details. The lots should have approximately 100′ (30 meters) frontage and be 120′ (36.5 meters) deep. The team should do all preliminary design work as a sketch on a chalkboard. The final drawing may be done using board drafting or CAD techniques.

Electricity/ Electronics Drafting

OBJECTIVES

Upon completion of this chapter, you should be able to:

■ Define basic electrical and electronic terminology.

■ Differentiate between electricity and electronics.

■ Explain the concept of electricity.

■ Use standard ANSI symbols in the development of electrical and electronic diagrams.

■ Draw series, parallel, and series-parallel circuits.

■ Differentiate between block diagrams and schematic diagrams.

Progress in making electrical power started with Thomas Alva Edison's electricity generator in New York City in 1882. Since then, electricity has completely changed the communication, manufacturing, and utility industries. Electricity and electronics are a powerful team for space shuttles, computers, communications systems, automated machinery, and everyday appliances. Robots have been common in industrial applications for many years. Cordless and cellular telephones are supplementing standard corded telephones in the home as well as on the job. Compact discs and DVDs carry high-quality audio and video signals.

Computers are everywhere, not only as tools themselves, but as adjuncts to automobiles, airplanes, appliances, and toys. "Smart homes" have computerized systems that control indoor and outdoor lights, garage doors, security systems, and even appliances such as coffeemakers. Some homes even have "home theaters" like the one shown in Figure 21-1. The light level and sound level, as well as the actual VCR or DVD controls, are handled electronically using a computerized control system.

Electricity and Electronics

The flow of electrons through wires or other metal conductors is known as **electricity**; it is passive, and it doesn't change. A common example is the wiring in a house. Electricity controls the lights and appliances. It also supplies the source for electronic equipment.

The field of **electronics** originated in the 1950s with the invention of the electron tube (vacuum tube). In general, an electric item is considered electronic if it incorporates electron tubes or semiconductors. *Semiconductors* are materials that are neither good conductors nor good insulators. You

Figure 21-1

Home theaters incorporate intricate electronic control systems.

Figure 21-2

The electron in the structure of an atom.

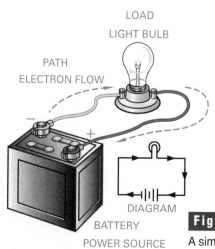

will learn more about the use of semiconductors later in this chapter. For now, you should know simply that the field of electronics is based on principles of electricity. Therefore, to understand electronics, you must first understand electricity.

Electricity

The source of electrical energy is the tiny **atom**. All atoms are made up of many kinds of particles. One of these particles is the **electron**. The electron is the most important particle in the study of electricity and electronics.

The electrons in an atom move around the nucleus, or center, of the atom, as shown in Figure 21-2. All electrons in all atoms are the same. Each electron has a negative charge. This charge forms the basis of electricity.

Voltage and Current

Sometimes electrons can be made to leave their "parent" atoms, the atoms of which they were originally part. This happens, for example, when a piece of wire is connected across the terminals of a battery. The battery produces an electrical pressure called **voltage**. Voltage, which is represented in formulas by a capital E, is measured in volts (V).

The voltage causes a stream of electrons to flow through the wire. This electron flow is called a **current**. When you connect a light bulb to the wire, electrons move through the lamp filament (the thin wire in the bulb) from the battery, as shown in Figure 21-3. The energy of the moving electrons is changed into heat energy as the filament becomes white-hot. The glow of the filament produces light.

The electron pathway is formed by the battery, the wire, and a *load*, in this case the lamp filament. This is a simple form of electric circuit. In other circuits, electrical energy is changed into other kinds of energy. Examples of these kinds of energy include magnetism, sound, and light. Current, which is often represented in formulas by a capital I, is measured in amperes (A).

Direct current (DC) is a flow of electrons through a circuit in one direction only. The current may be fixed at a steady level, or it may vary, as shown in Figure 21-4, but the current value never drops below zero.

Alternating current (AC) is a flow of electrons in one direction during a fixed time period, and then in the opposite direction during a similar time period, as shown in Figure 21-5. One complete AC alternation is called a **cycle**. The number of times this cycle is repeated in one second is known as the **frequency** of the alternating current. Frequency is measured in hertz (Hz). For example, a current in which there are 60 complete cycles per second may be referred to as 60-cycle AC or 60 Hz.

LOAD
LIGHT BULB

PATH
ELECTRON FLOW

DIAGRAM

BATTERY

POWER SOURCE

Figure 21-3

A simple electric circuit.

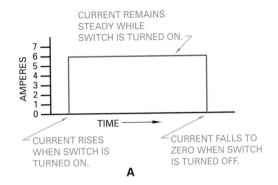

CURRENT REMAINS
STEADY WHILE
SWITCH IS TURNED ON.

CURRENT RISES
WHEN SWITCH IS
TURNED ON.

CURRENT FALLS TO
ZERO WHEN SWITCH
IS TURNED OFF.

A

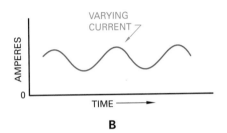

VARYING
CURRENT

B

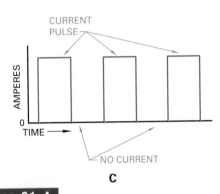

CURRENT
PULSE

NO CURRENT

C

Figure 21-4
Examples of direct current.

Resistance

Electrons can move through some materials more easily than others. For example, electric current flows more easily through a copper wire than through a steel wire of the same size. We say that the steel offers more **resistance** than the copper. We measure resistance in *ohms* (Ω). The symbol for resistance is R.

Materials with small resistance to the flow of electrons are called **conductors**. Silver is the best conductor known. However, it costs too much for general use. Copper and aluminum are also good conductors. They are the most widely used. Materials through which electrons will not flow easily are called **insulators**. The insulators used most often are glass, porcelain, plastics, and rubber compounds.

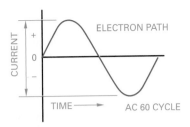

ELECTRON PATH

AC 60 CYCLE

Figure 21-5

Alternating current builds up from zero to a maximum current flow in one direction, falls to zero, then builds to the maximum current flow in the other direction, and again returns to zero.

Ohm's Law

The amounts of voltage, current, and resistance in a circuit are related. This relationship, called *Ohm's law*, may be expressed as a formula. Basically, this formula states that the voltage (E) needed to force a specific amount of current through a circuit is equal to the product of the current (I) and the resistance (R) of the circuit, or $E = I \times R$.

This formula can be rearranged algebraically as necessary to solve for voltage, current, or resistance. As long as you know two of the values, you can find the third. The Tech Math in this chapter explains how to manipulate the formula algebraically. Figure 21-6 shows graphically how these relationships work. By placing the three values in a circle, as shown, and covering the unknown value, you can easily see how to use the equation to find the unknown.

Figure 21-6

Placing the values in a circle makes their relationship easier to remember.

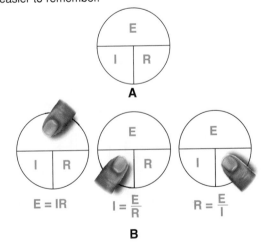

$E = IR$ $I = \dfrac{E}{R}$ $R = \dfrac{E}{I}$

B

Algebra and Ohm's Law

If you know the value of any two of the variables in the formula for Ohm's law, you can find the value of the third by manipulating the formula algebraically. For example, if you know the voltage and the resistance of a circuit, you can find the current. According to the rules of algebra, the value of an equation doesn't change as long as you do the same operation on both sides of the equal sign.

Problem: In a circuit in which the voltage (*E*) is 120V and the resistance (*R*) is 20Ω, what is the current?

Solution: To solve for current (*I*), first divide both sides of the formula by *R*:

$$\frac{E}{R} = \frac{I \times R}{R}$$

The *R*s on the right side cancel out, so the equation can be rewritten as:

$$\frac{E}{R} = I$$

So, you can find the current by dividing the voltage by the resistance:

$$\frac{120V}{20\Omega} = 6 \text{ amps}$$

Power

In general, **power** is the rate at which work is done. In terms of electric circuits, power is used to measure the rate at which electric energy is delivered to a circuit. It can also be used to measure the rate at which an electric circuit converts the energy of moving electrons into another form, such as heat or light energy.

The basic formula for power is $P = E \times I$. Recall that *E* stands for voltage, and *I* stands for current. *P* stands for the power in watts (*W*). Notice that the formula for power looks much like the formula used for Ohm's law. You can use the same techniques discussed in the Tech Math and shown in Figure 21-6 to find the value of any variable in the power formula.

Types of Electric Circuits

The basic types of electric circuits include series circuits, parallel circuits, and series-parallel circuits. These three terms are explained in the following paragraphs.

Series Circuit

Some circuits are wired so that the current flows from the source (battery, generator, etc.) through each load sequentially, one after the other. This type of circuit is known as a **series circuit**.

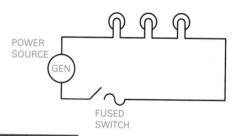

Figure 21-7

A series circuit for three lamps.

An example of a series circuit is shown in Figure 21-7. In this circuit, three lamps get power from a generator when the fused switch is closed. All of the lights must be on. If any one light burns out, the circuit will be open, so none of the lights will light.

Parallel Circuit

Circuits that are wired so that the current can flow through more than one path simultaneously are known as **parallel circuits**. Figure 21-8 shows an example of a parallel circuit. Notice that each lamp is on a separate path or branch of the circuit. If one lamp burns out, the others still work.

A circuit for a siren is shown in Figure 21-9. It can be turned on by any of the four pushbuttons because the buttons are connected in parallel. A good use of this would be an alarm system to warn of an attempted holdup at a store. The pushbuttons, connected in parallel, would be under the counters and in the cashier's office. The symbol used for the siren is the same one that is used for a loudspeaker. The symbol is labeled SIREN to specify that this particular loudspeaker is to be a siren.

Series-Parallel Circuit

A circuit that contains both series and parallel components is called a **series-parallel circuit**. Combining series and parallel circuits permits many different arrangements. In Figure 21-10, lamps C and D are in series. Lamps E and F are in parallel. Both lamps C and D must be on if switch A is closed, since they are in series. When switches A and B are closed, all the lamps are lighted. Lamps E and F will work separately. If one fails, the other stays lighted because they are in parallel. However, because lamps C and D are in series, when one fails, the other does not light.

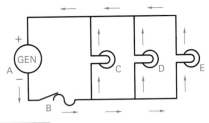

Figure 21-8

A parallel circuit for three lamps.

Figure 21-9

A circuit diagram showing a siren connected to four pushbuttons connected in parallel.

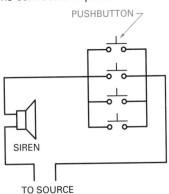

Electronics

As you may recall, electronic devices are those that incorporate electron tubes or semiconductors. Vacuum tubes are now obsolete, although you may still see them in old, collectable radios and televisions. The term *electronics* is now used interchangeably with the term *digital electronics*. Digital electronics involves the use of pulses of low voltage to operate semiconductor devices such as diodes and transistors in equipment such as computers and sound systems. Different voltage levels or currents, called *signals*, control the operation of the equipment by affecting the operation of the semiconductors.

Semiconductors

The simplest semiconductor device is the **diode**. See Figure 21-11. Diodes are usually made of silicon, and they work by acting as a gate. They allow current to pass through in one direction only. The polarity (+ or −) and value of the voltage applied determines whether the diode will conduct current. Diodes can be used singly or together in circuitry to control current flow.

The **transistor** has single-handedly changed the face of electronics by miniaturizing the circuitry needed to control current and amplify input voltages

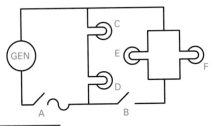

Figure 21-10

A series-parallel circuit.

Figure 21-11

A diode.

Figure 21-12

A transistor.

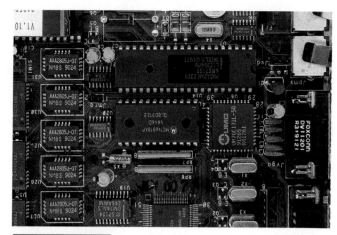

Figure 21-13

Integrated-circuit chips.

or currents. See Figure 21-12. Electronically, it is the equivalent of two diodes placed back-to-back. This arrangement forms three parts: an emitter, a collector, and a base. The emitter is the part of the transistor that is common to both the base and the collector. The base is the input junction of the transistor, and the collector is the output junction. Each of these three parts is connected to wire leads or pins that connect the transistor to the circuit.

In a continuation of the miniaturization of electronics, the **integrated circuit** incorporates components such as resistors, diodes, and transistors into a single substrate so that they can be handled as a unit. Two integrated-circuit "chips" are shown in Figure 21-13 (the large, black items). The development of integrated circuits has enabled electronic devices to become smaller and faster as new products and uses are developed.

Logic Circuits

In digital electronics, a *logic circuit* is one that provides the specified output conditions when one or more specific input conditions are satisfied. Most logic circuits are based on switching techniques using binary (two-part) voltage or current levels. A *logic gate* is an electronic circuit that compares signal inputs logically, and then outputs a result based on the inputs. Logic gates have only two values: on and off. These values are generally assigned the binary equivalents of 1 and 0, respectively.

The three basic types of logic gates are as follows:

- **AND gates** produce a 1 result only if all inputs are 1.
- **OR gates** produce a 1 result if any one or more of the inputs is 1.
- **NOT gates**, or inverters, produce a 1 result if the input is 0; they produce a 0 result if the input is 1.

Electronic circuit designers often find it useful to combine the basic types to simplify the structure of a logic circuit. A NAND gate is a combination of an AND gate and a NOT gate. If both of the inputs to a NAND gate are 1, the output will be 1. If both of the inputs are 0, the output will still be 1. If the two inputs are different, however, the output will be 0. A NOR gate is a combination of an OR gate and a NOT gate. In this gate, if both inputs are 1, the result is 0. If both inputs are 0, the result is 1. If the two inputs are different, the output is 0.

The behavior of the various logic gates is often summarized in *truth tables* that show their output for any given combination of inputs. The schematic symbols and truth tables for the logic gates discussed here are shown in Figure 21-14.

Logic circuits are very useful in controlling electronic devices. For example, thermostats are often controlled using logic circuits. Even devices as complex as computers rely almost entirely on logic circuits.

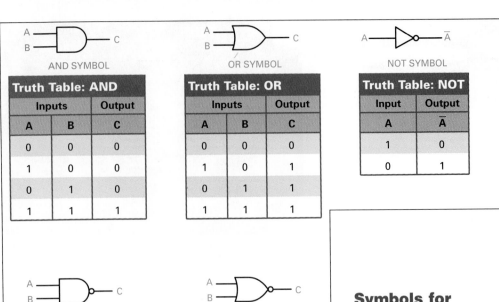

AND SYMBOL

Truth Table: AND		
Inputs		Output
A	B	C
0	0	0
1	0	0
0	1	0
1	1	1

OR SYMBOL

Truth Table: OR		
Inputs		Output
A	B	C
0	0	0
1	0	1
0	1	1
1	1	1

NOT SYMBOL

Truth Table: NOT	
Input	Output
A	$\overline{A}$
1	0
0	1

NAND SYMBOL

Truth Table: NAND		
Inputs		Output
A	B	C
1	1	0
0	0	1
1	0	1
0	1	1

NOR SYMBOL

Truth Table: NOR		
Inputs		Output
A	B	C
1	1	0
0	0	1
1	0	0
0	1	0

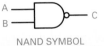

Graphic symbols and truth tables for logic gates.

Symbols for Electrical/Electronic Diagrams

Electrical/electronic drafting requires the use of symbols that are dedicated to this field. Symbols can be drawn quickly and easily on board drawings using templates like the one shown in Figure 21-15. In CAD drawings, the symbols are usually inserted from a symbol library. To avoid confusion, drafters use standard symbols defined by ANSI and the Institute of Electrical and Electronics Engineers (IEEE). Figure 21-16 shows some of the basic symbols used for electrical and electronics drafting. Refer to ANSI/IEEE Y32E, "Electrical and Electronics Graphics Symbols and Reference Designations," for a complete collection of standard symbols.

Electrical/Electronic Diagrams

There are many kinds of electrical and electronic diagrams, used for many different purposes. The major types are discussed in the paragraphs that follow. For additional information about national standards for these diagrams, see ANSI Y14.24, "Engineering Drawing and Related Documentation Practices," part 12, "Electrical/Electronic Diagrams."

As you read the information in the following paragraphs, study the accompanying illustrations carefully. Try to trace the signal through each diagram. Careful study of these drawings will help you understand the process of creating new drawings.

Figure 21-15

Template for electrical and electronic circuits.

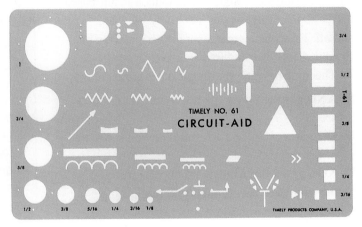

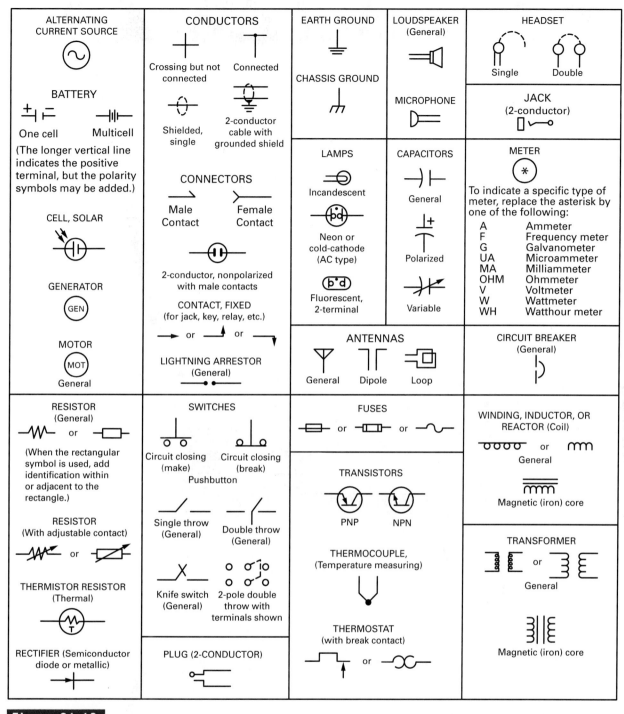

Figure 21-16

Standard electricity/electronics symbols.

Schematic Diagram

A circuit diagram, or **schematic diagram**, shows how a circuit is connected and what it does. As you can see in Figure 21-17, a schematic diagram does not show the physical size or shape of the parts of the circuit. It does not show where the parts of the circuit actually are. Instead, it is used to illustrate the details of the circuit design and to help troubleshooters trace the circuit and its functions.

Layout

The layout of a schematic diagram should follow the circuit, signal, or transmission path from input to output, from power source to load, or in the order that the equipment works. In general, schematic dia-

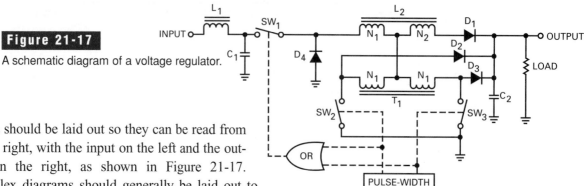

Figure 21-17

A schematic diagram of a voltage regulator.

grams should be laid out so they can be read from left to right, with the input on the left and the output on the right, as shown in Figure 21-17. Complex diagrams should generally be laid out to read from upper left to lower right. They may be laid out in two or more layers. Each layer should be read from left to right. Where possible, include endpoints for outside connections at the outer edges of the circuit layout. Functional groups are often outlined with dashed lines to make the schematic easier to read.

Connecting Lines

The lines that connect the components of a circuit should be drawn as horizontal or vertical lines when possible. Use as few bends and crossovers as possible. Do not connect four or more lines at one point if they can just as easily be drawn another way.

When you draw connecting lines parallel to each other, the spacing between the lines should be no less than .06″ at the final drawing size. Group parallel lines according to what they do. It is best to draw them in groups of three. Allow double spacing between groups of lines.

Interrupted Group Lines

When interrupted lines are grouped and bracketed, identify the lines as shown in Figure 21-18. You can show at the brackets where the lines are meant to go or where they are meant to be connected. Do this using notes outside the brackets, as shown in Figure 21-18A, or by using a dashed line between brackets, as shown in Figure 21-18B. When using a dashed line to connect brackets, draw it so that it will not be mistaken for part of one of the bracketed lines. Begin the dashed line in one bracket and end it in no more than two brackets.

Figure 21-18

Typical arrangement of line identifications and circuit destinations.

Interrupted Single Lines

When a single line is interrupted on a schematic diagram, show where the line is going in the same place you identify it. Identify single interrupted lines in the same way as grouped and bracketed lines.

Single-Line Diagram

Figure 21-19 is an example of a single-line diagram. This type of diagram is a schematic diagram that shows the course of an electric circuit and the parts of the circuit using single lines and symbols.

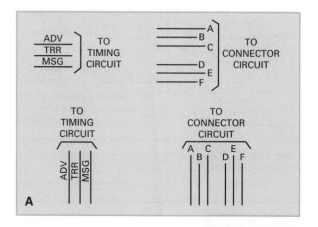

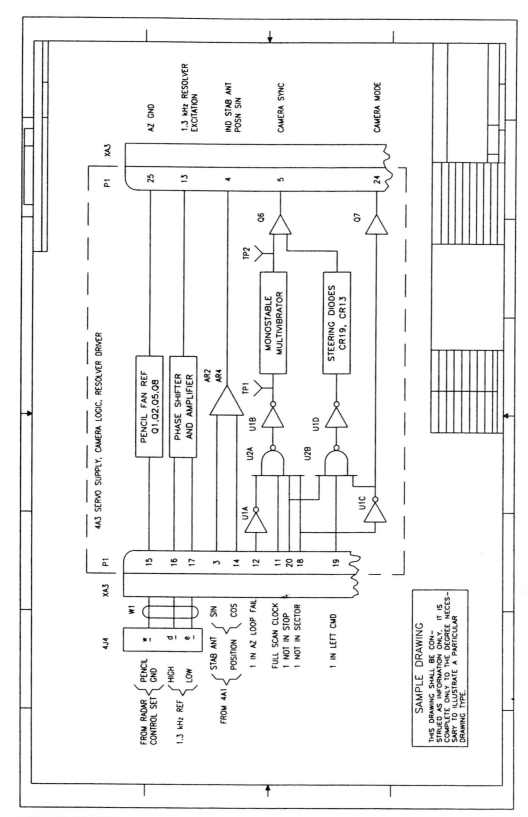

Figure 21-19

Single-line diagram.

It tells in a basic way how a circuit works, leaving out much of the detailed information shown on other types of diagrams. Single-line diagrams make it possible to draw complex circuits in a simple way. For example, in a single-line diagram of a communications or power system, a single line may stand for a multiconductor communication or power circuit.

Functional Block Diagram

Another type of circuit drawing is the **functional block diagram**, which shows the functions of major elements of a circuit or system in a greatly simplified form. The drawing consists of rectangular blocks that represent the elements of the circuit, joined by single lines, and clarified by explanatory notes. See Figure 21-20. The blocks show how the components or stages are related when the circuit is working. Note the arrowheads at the terminal ends of the lines. These arrowheads show which way the signal path travels from input to output, reading the diagram from left to right. The size of a block is generally determined by the amount of text that needs to be included. Graphic symbols are not generally used on functional block diagrams.

Engineers often draw or sketch functional block diagrams as a first step in designing new equipment. Because blocks are easy to sketch, the engineer can try many different layouts before deciding which to use. Block diagrams are also used in catalogs, descriptive folders, and advertisements for electrical equipment. In technical service literature, functional block diagrams aid in the troubleshooting and repair of equipment.

Connection Diagram

A diagram that shows how the components of a circuit are connected is a **connection diagram**. These are often referred to as *wiring diagrams*. A connection diagram shows in detail the physical arrangements of electrical connections and wires between elements within a circuit, as shown in Figure 21-21. Ordinarily, only the wiring for internal connections and connections that have one internal terminal are shown in the diagram. External wiring is omitted.

Interconnection Diagram

An interconnection diagram is similar to a connection diagram, but it shows only external connections between components. The connections inside each component are left out. Figure 21-22 shows an example of an interconnection diagram. As you can see, each component is shown on the diagram by a broken rectangle. The name of the component is included for clarity.

Logic Circuit Diagram

A general explanation of logic circuits was included earlier in this chapter. Logic circuit diagrams are simply diagrams that show the logic functions of a circuit. In addition to the logic symbols discussed earlier, a logic circuit diagram usually includes the pin numbers on integrated circuits, test points, boundaries of the assembly, and any nonlogic functions that might be necessary to describe the circuit completely. Figure 21-23 shows an example of a logic circuit diagram.

Figure 21-20

Functional block diagram of a 20,000W broadcast transmitter.

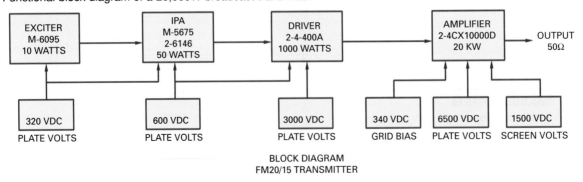

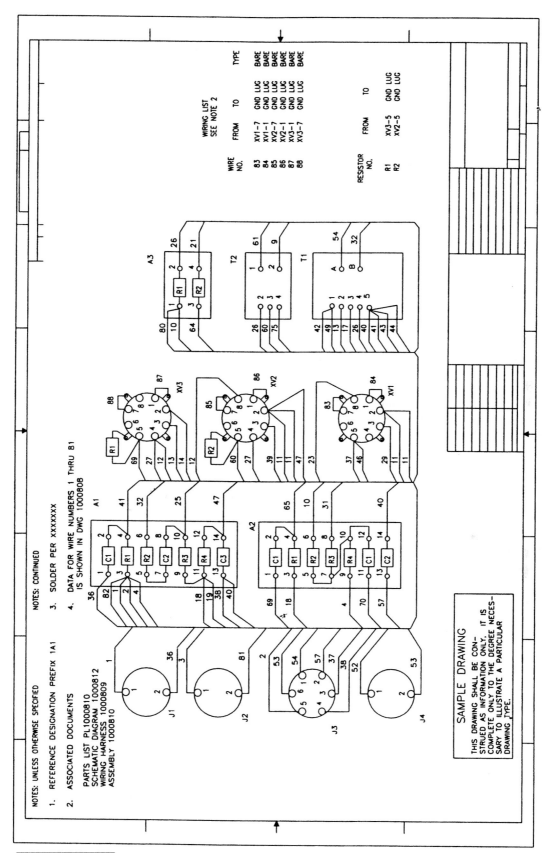

Figure 21-21

Connection or wiring diagram.

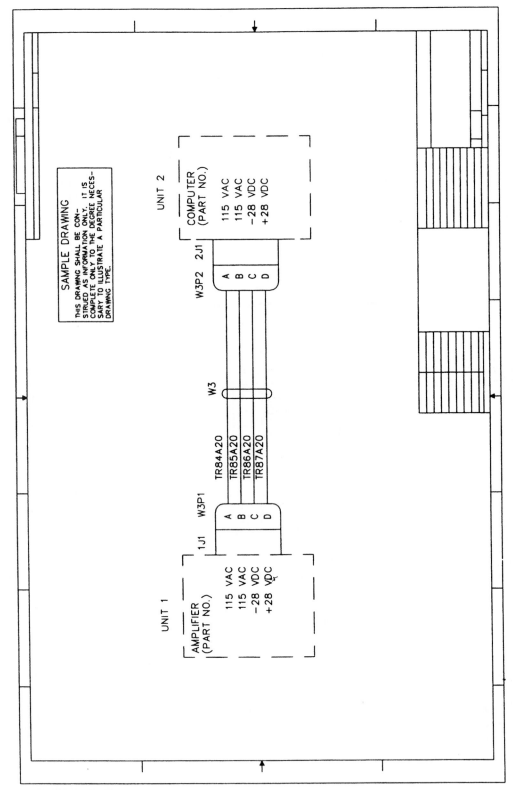

SAMPLE DRAWING

THIS DRAWING SHALL BE CON-
STRUED AS INFORMATION ONLY. IT IS
COMPLETE ONLY TO THE DEGREE NECES-
SARY TO ILLUSTRATE A PARTICULAR
DRAWING TYPE.

UNIT 2

COMPUTER
(PART NO.)

115 VAC
115 VAC
−28 VDC
+28 VDC

W3P2   2J1

A
B
C
D

W3

TR84A20
TR85A20
TR86A20
TR87A20

W3P1

1J1

A
B
C
D

UNIT 1

AMPLIFIER
(PART NO.)

115 VAC
115 VAC
−28 VDC
+28 VDC

Figure 21-22

Interconnection diagram.

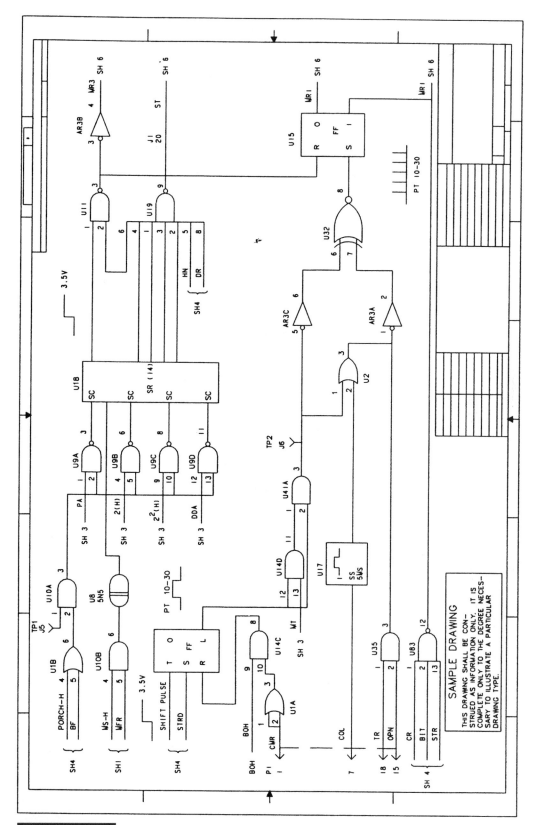

Figure 21-23

Logic circuit diagram.

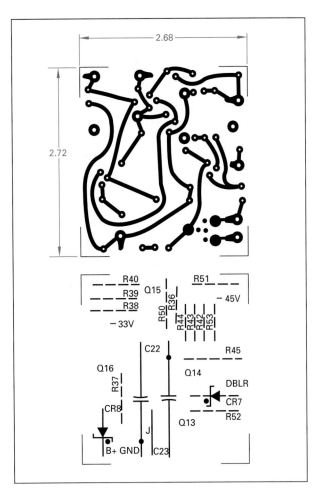

Figure 21-24

A printed circuit layout with a component identification overlay.

Printed Circuit Drawings

Many electronic devices incorporate printed circuit boards into the design of the product. A *printed circuit board* is a rigid or flexible board made of a dieletric base material onto which circuitry can be added. The term *printed* refers to the process used to deposit the circuitry on the board. Common processes include etching, screen printing, and bonding.

Printed circuit drawings are used in making printed circuit boards. Each drawing is an exact layout of the pattern of the circuit needed. The drawing is made actual size or larger. If drawn larger, it can be made smaller by photography. The conductor lines on the pattern should be at least .03″ (1 mm) wide. They should be spaced at least .03″ apart. See Figure 21-24.

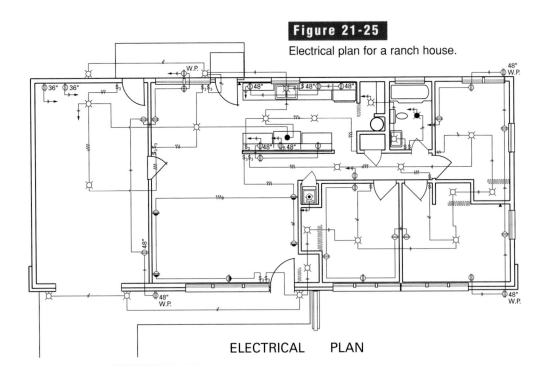

Figure 21-25

Electrical plan for a ranch house.

ELECTRICAL   PLAN

Electrical Layouts for Buildings

Figure 21-25 shows the usual way in which an architect locates electric outlets and switches in a building. This plan only shows where the lights, base plugs, and switches are to be placed. A list of the symbols used in architectural electrical drawings is shown in Figure 21-26.

Note that you cannot build a good electrical system using the diagram in Figure 21-25. A complete and detailed set of electrical drawings is needed to create the actual system. These drawings must be made by someone who knows the engineering needs of the system.

Drawing Conventions

Drafters follow standard conventions when drawing electrical or electronic diagrams. These standards make it easier for people to read and understand the diagrams.

Figure 21-26

Electrical wiring symbols for architectural design and floor plan layout.

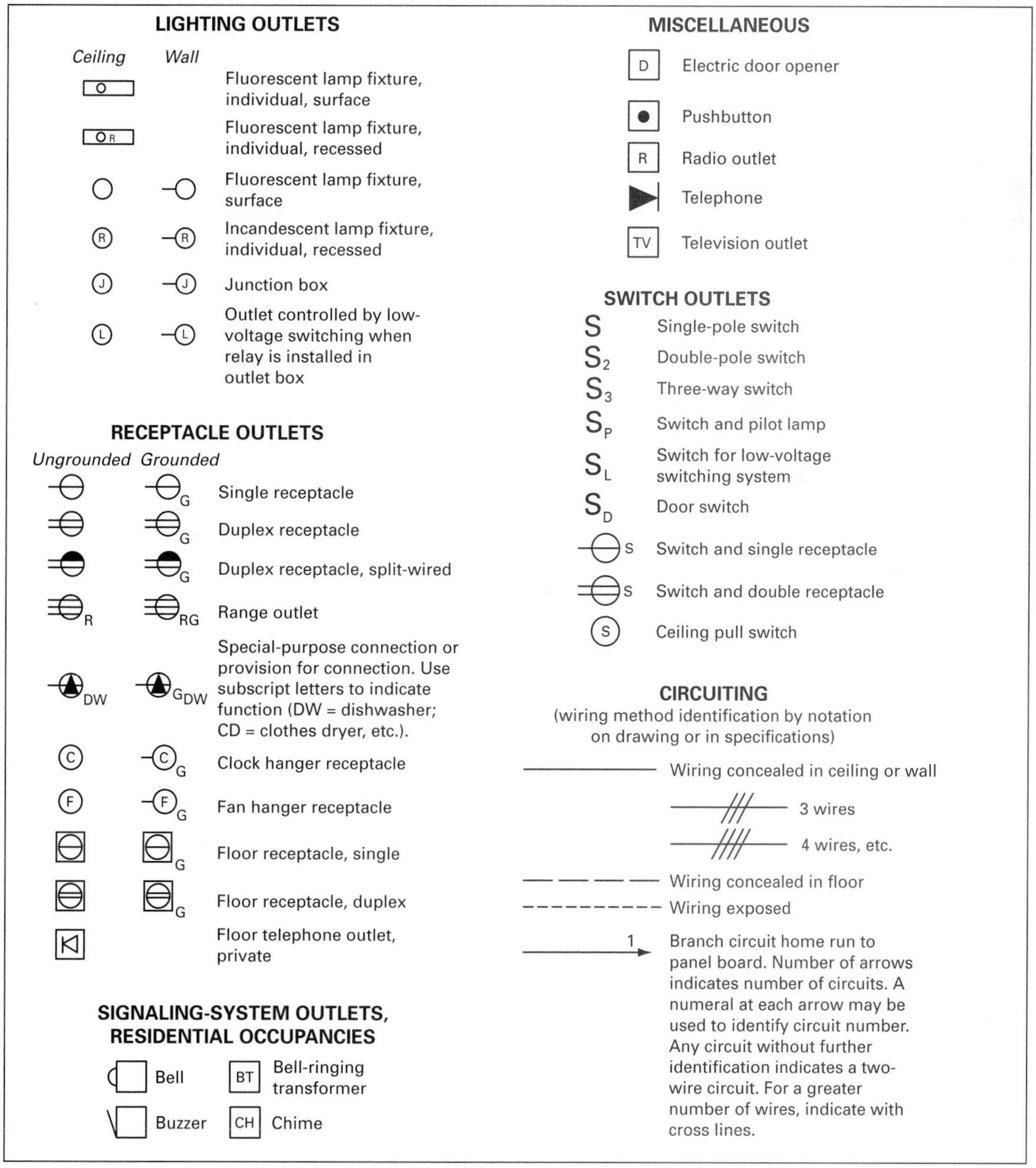

Color Codes

Color codes are an easy way to show information when drawing circuit diagrams. Color codes are also used on the actual wiring of the circuit. In electrical and electronic work, drafters use a color code to show certain characteristics of components, to identify wire leads, and to show where wires are connected. When using a color code, you should look up the Electronic Industries Association (EIA) standards and any other color code that might be needed. Table 21-1 shows what each color stands for in the EIA color code.

Figure 21-27 shows one common example of how the EIA color code is used in electricity and electronics. Resistors are banded as shown in Figure 21-27A. The color bands show the resistor's value in ohms. The first band represents the first digit of the value, the second band represents the second digit, and the third value represents the number by which the first two digits are multiplied. For this purpose, the colors in the EIA color code are assigned a multiplier, as indicated in the last column of Table 21-1. Notice that the number of zeros in the multiplier for

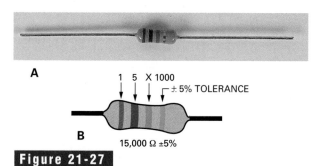

Figure 21-27

Color coding on a resistor.

each color is the same as the number the color represents. For example, the multiplier represented by a blue band is 100,000 (five zeros). Figure 21-27B shows how to read the bands on the resistor shown in Figure 21-27A. If a fourth band is present, it identifies the tolerance of the resistor. A gold band represents a tolerance of 5%, and a silver band represents a tolerance of 10%. If no fourth band is present, the tolerance of the resistor is 20%.

Line and Symbol Conventions

When using electrical or electronic symbols on a drawing, keep the following rules in mind:
1. Line conventions for electrical/electronics drawings are shown in Figure 21-28.
2. The width of a line does not affect the meaning of the symbol. Sometimes, however, a wider line can be used to show that something is important.
3. Most of the time, the angle of a line connected to a symbol does not matter. Generally, lines are drawn horizontally and vertically.
4. Sometimes it may be desirable to draw paths and equipment that will be added to the circuit later or that are connected to the circuit but are not part of it. This is done by drawing lines made up of short dashes (- - - -).
5. A symbol is made up of all its various parts.
6. The direction a symbol is facing on a drawing does not change its meaning. This is true even if the symbol is drawn backwards.
7. A symbol can be drawn any size needed. Symbols are not drawn to scale. However, their size must fit in with the rest of the drawing, and they must be in proportion to each other, as

Table 21-1. The EIA Color Code Standard		
Color	**Number**	**Multiplier (for Resistor Identification)**
BLACK	0	1
BROWN	1	10
RED	2	100
ORANGE	3	1,000
YELLOW	4	10,000
GREEN	5	100,000
BLUE	6	1,000,000
VIOLET	7	10,000,000
GRAY	8	100,000,000
WHITE	9	1,000,000,000
Color	**Tolerance**	
SILVER	± 10%	
GOLD	± 5%	
NO TOLERANCE BAND	± 20%	

Figure 21-28
Line conventions for electrical and electronic diagrams.

LINE APPLICATION	LINE THICKNESS
FOR GENERAL USE	MEDIUM
MECHANICAL CONNECTION, SHIELDING, AND FUTURE CIRCUITS LINE	MEDIUM
BRACKET-CONNECTING DASH LINE	MEDIUM
USE OF THESE LINE THICKNESSES OPTIONAL	
BRACKETS, LEADER LINES, ETC.	
BOUNDARY OF MECHANICAL GROUPING	
FOR EMPHASIS	

shown in the standard. That is, if one of these symbols is drawn twice the size shown in the standard, all other symbols should be drawn double size.

8. The arrowhead of a symbol can be drawn closed → or open →.

9. The standard symbol for a terminal ○ can be used where connecting lines are attached. These are not part of the graphic symbol unless the terminal symbol is part of the symbol shown in the standard.

10. If details of type, impedance, rating, etc., are needed, they may be added next to a symbol. The abbreviations used should be from ANSI Y14.38, "Abbreviations and Acronyms." Letters that are joined together and are standard parts of symbols are not abbreviations.

11. Use the ground symbol ⏚ only when the circuit ground is at a potential level equivalent to that of earth potential. Use the symbol ⏚ when you do not get an earth potential from connecting the ground wire to the structure that houses or supports the circuit parts. This is known as a *chassis ground*.

Board Drafting Techniques

Board drafters must be more careful than CAD drafters to take line width into account as they set up an electrical/electronic diagram. If there is a chance that the size of the drawing may be changed, remember to choose a line thickness, symbol size, and letter size that will be readable after it is made larger or smaller. For most circuit diagrams meant to be used for manufacturing, or for use in a smaller form, draw symbols about 1.5 times the size shown in ANSI/IEEE Y32E.

Line thickness and lettering used with circuit diagrams should conform with ANSI Y14.2M 1992 (R1998) and local needs, especially if microfilm of the diagrams will be made. Draw lines of medium thickness for general use on circuit diagrams. Use thin lines for brackets, leader lines, and so on. When something special needs to be set off, such as main

or transmission paths, use a line thick enough to stand out from the general-purpose lines.

Figure 21-29 shows a schematic diagram of a power-factor controller. Follow the steps below to create the drawing.

1. Prepare an A-size drawing sheet.
2. Lay out the connecting lines using light construction lines, as shown in Figure 21-30.
3. Use a symbol template similar to the one shown in Figure 21-15 to add the necessary symbols to the drawing. If a template is not available, create the symbols manually.
4. Add the dashed line that surrounds the ramp generator.
5. Add all necessary text and notes to the drawing.
6. Adjust the connecting lines as necessary to accommodate the text and symbols.
7. Darken all lines to finish the drawing.

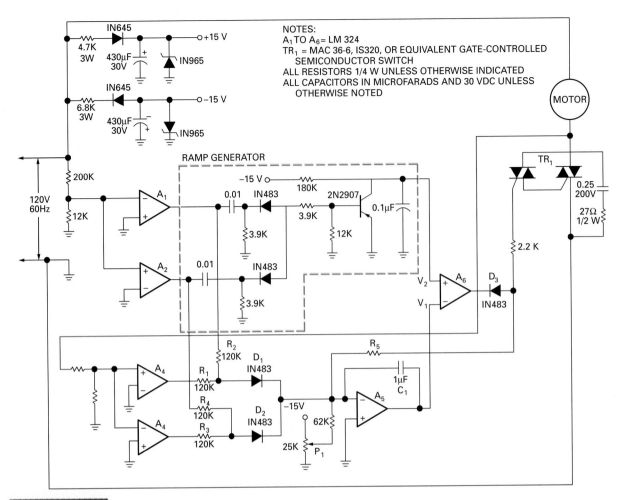

NOTES:
A_1 TO A_6 = LM 324
TR_1 = MAC 36-6, IS320, OR EQUIVALENT GATE-CONTROLLED
    SEMICONDUCTOR SWITCH
ALL RESISTORS 1/4 W UNLESS OTHERWISE INDICATED
ALL CAPACITORS IN MICROFARADS AND 30 VDC UNLESS
    OTHERWISE NOTED

Figure 21-29

Schematic diagram of a power-factor controller.

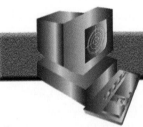

CAD Techniques

As mentioned earlier in this chapter, CAD drafters generally use symbol libraries to produce circuit diagrams. Follow the steps below to create the schematic diagram shown in Figure 21-29.

1. Create a new drawing using an ANSI-A template. Unlike other types of drafting, electrical/electronics drafting does not require exact distances between components or areas of the drawing. Therefore, you do not have to calculate precise drawing limits.

2. With Ortho on to force vertical and horizontal lines, use polylines to create the connecting lines for the drawing. Draw the outside lines first, allowing them to fill most of the drawing area. Then add the interior lines, adjusting the outer lines if necessary.

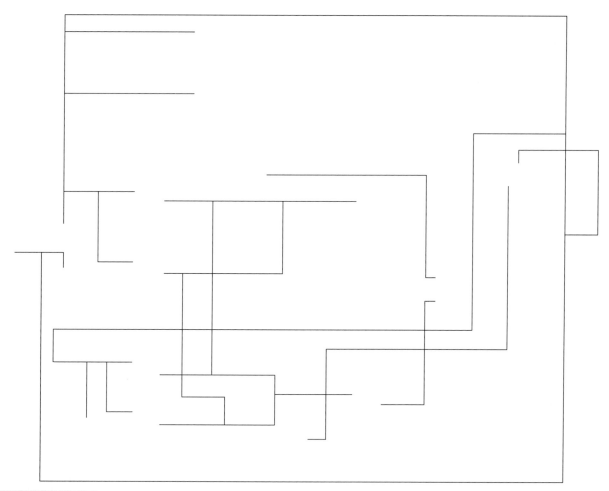

Use light construction lines to place the main connecting lines.

3. If you are using AutoCAD 2000i or later, open the Basic Electronics symbol library. Select the appropriate symbols and drag them into place on the drawing. If necessary, scale the symbols, but remember to use the same scale for all of the symbols. If you are using an earlier version of AutoCAD, use a third-party symbol library or create one of your own using Figure 21-16 for reference.

4. Add the dashed line around the ramp generator.

5. Use the DTEXT command to add all necessary text and notes.

6. Review the drawing and adjust connecting lines as necessary to increase the readability of the diagram.

►CADTIP

Streamlining DTEXT
You need only enter the DTEXT command once to add all of the text for the entire drawing. Instead of pressing Enter twice to end the command, when you finish entering a line of text, use the cursor to select another starting point for text. Each line of DTEXT becomes a separate object in the drawing database, so when you finish entering all of the text and press Enter twice, each line of text is editable separately. Note that if you intend to enter a large amount of text in this manner, it might be a good idea to do so in two or three batches, saving the file in between.

Chapter Summary

■ Electricity is the flow of electrons through wires or other conductors.

■ Electronics is a specialized field of electricity involving devices that use semiconductors.

■ Current is the movement of electrons through a conductor.

■ Voltage is the force that moves the electrons through a path.

■ Resistance is the opposition to the movement of electrons.

■ The three basic types of electrical circuits are series, parallel, and series-parallel.

■ A logic circuit diagram provides the specified output conditions when one or more specific input conditions are satisfied.

■ Symbols are used on circuit diagrams to show the components of a circuit and how they are connected.

■ Schematic diagrams show how the circuit is connected and what it does.

■ Functional block diagrams show how the components or stages of a circuit are related.

■ Drafters use color codes and special line conventions to make their electrical and electronic diagrams more readable.

Review Questions

1. What is the source of electrical energy?

2. Do electrons have a negative or positive charge?

3. What two types of devices are used to define electronic equipment?

4. In what kind of current does the flow of electrons through a circuit occur in one direction only?

5. In what kind of electric current does the flow of electrons move back and forth in opposing directions?

6. What name is given to materials through which electrons will not flow easily?

7. What formula describes Ohm's law? What does each variable in the equation stand for?

8. If a circuit has a voltage of 120V and a current of .20A, what is the power of the circuit?

9. What is the difference between a series and a parallel circuit?

10. What is a series-parallel circuit?

11. What is the difference between a schematic diagram and a single-line diagram?

12. What is the purpose of a logic circuit diagram?

13. What is the value of a resistor that contains four bands in the following colors: green, blue, red, silver?

14. How are electrical/electronics symbols usually placed on a diagram created using board drafting techniques? Using CAD techniques?

Drafting Problems

The drafting problems in this chapter are designed to be completed using board drafting techniques or CAD.

1. Prepare a block diagram of the broadcast transmitter shown in Figure 21-20.

2. On an A-size sheet, draw the schematic diagram of the voltage regulator in Figure 21-17.

3. Prepare a complete single-line diagram for the AM-FM stereo unit, as shown in Figure 21-31. Estimate sizes and draw twice the size. Use a template if you are using board drafting techniques, or use a symbol library for a CAD drawing.

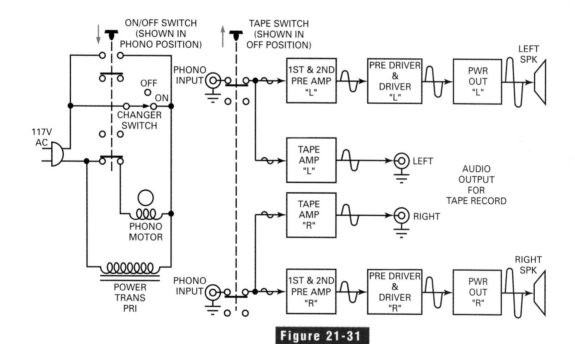

Figure 21-31

4. Draw the audio signal-flow diagram shown in Figure 21-32. Note that when the tape player is turned on, all other functions (AM, FM, and CD player) are disabled. Also, the changer switch cannot turn on the CD player.

5. Prepare a schematic diagram of the signal conditioner shown in Figure 21-33.

6. Draw a schematic diagram from the pictorial shown in Figure 21-34.

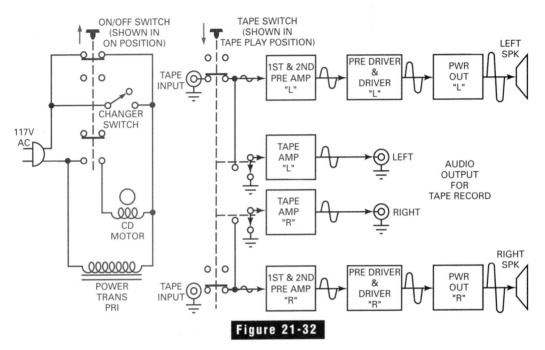

Figure 21-32

Figure 21-33

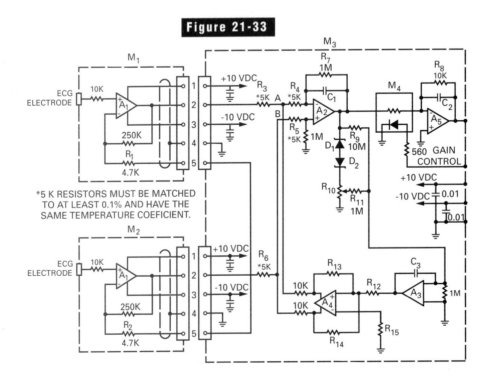

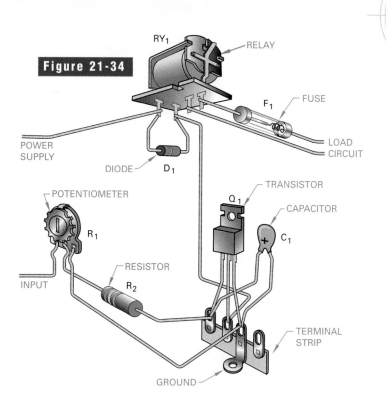

Figure 21-34

RY₁ — RELAY

F₁ — FUSE

POWER SUPPLY

DIODE — D₁

LOAD CIRCUIT

POTENTIOMETER

TRANSISTOR

Q₁

CAPACITOR

R₁

C₁

RESISTOR

INPUT

R₂

TERMINAL STRIP

GROUND

Design Problems

Design problems have been prepared to challenge individual students or teams of students. In these problems, you are to apply skills learned mainly in this chapter but also in other chapters throughout the text. The problems are designed to be completed using board drafting, CAD, or a combination of the two. Be creative and have fun!

1. Design a battery-operated light circuit for the outside of your home. It needs to run off an independent 12V battery. Draw the block diagram.

2. Design a doorbell for a recreational vehicle to run off the RV's battery. Draw the schematic diagram.

3. You recently bought a cottage in Vermont, and the wiring needs to be updated. Draw the wiring diagram to control a stairway light that can be operated from both the top of the stairs and the bottom of the stairs.

Graphs, Charts, and Diagrams

OBJECTIVES

Upon completion of this chapter, you should be able to:

■ Describe the general use and importance of graphs, charts, and diagrams.

■ Use color and creative symbology to enhance the presented data.

■ Choose the appropriate format to communicate most clearly the data being presented.

■ Prepare various kinds of charts from established data using both board drafting and CAD techniques.

Charts and graphs show facts and number relationships in an interesting and accurate manner. For the most part, they are based on the principle of relative sizes. A pictorial representation of data helps us to understand and retain facts over a much longer period of time. The proverb, "One picture is worth a thousand words," is clearly illustrated through the use of charts and graphs. Often, at a glance, you are able to observe and understand relationships in complex data that would take much longer to grasp through reading about them in narrative form.

Imagine trying to put into words the information contained in the bar chart shown in Figure 22-1A. Notice how much easier it is to understand stopping distance by reading the bar chart than studying the table shown in Figure 22-1B. Because the information in Figure 22-1A is in chart form, you can readily see that driver reaction distance and automobile braking distance increase as speed increases. Figure 22-1B contains the same information, but it takes more time to read, study, and understand it. You can see and understand the relationship of speed and distance more easily in Figure 22-1A because of its graphic presentation.

The terms *graph* and *chart* are often used interchangeably. In fact, the difference seems to lie in common usage rather than in any real difference of definition. You may, for example, see a line graph referred to as a line chart. This is not incorrect; it is just a different way of labeling the same item. Keep this in mind as you study the graphs and charts in this chapter.

Uses of Charts and Graphs

Almost everyone uses many kinds of graphs and charts in everyday life. Corporate annual reports are generally filled with graphs and charts that represent the successes or failures of the corporation for a given year. School athletic programs often chart their record of wins and losses. Scientists use charts, graphs, and diagrams to record and study the results of research. Doctors use charts to record body temperature, heart action, and other body functions. Other people use and read charts and diagrams to learn about the weather, the stock market, finances, and many other things.

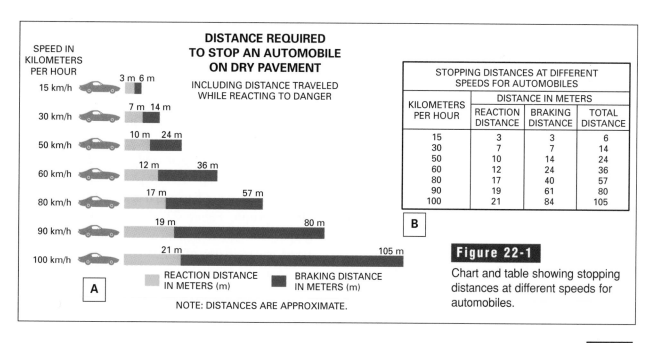

A

DISTANCE REQUIRED TO STOP AN AUTOMOBILE ON DRY PAVEMENT
INCLUDING DISTANCE TRAVELED WHILE REACTING TO DANGER

SPEED IN KILOMETERS PER HOUR

REACTION DISTANCE IN METERS (m) — BRAKING DISTANCE IN METERS (m)

NOTE: DISTANCES ARE APPROXIMATE.

B

STOPPING DISTANCES AT DIFFERENT SPEEDS FOR AUTOMOBILES			
KILOMETERS PER HOUR	DISTANCE IN METERS		
	REACTION DISTANCE	BRAKING DISTANCE	TOTAL DISTANCE
15	3	3	6
30	7	7	14
50	10	14	24
60	12	24	36
80	17	40	57
90	19	61	80
100	21	84	105

Figure 22-1

Chart and table showing stopping distances at different speeds for automobiles.

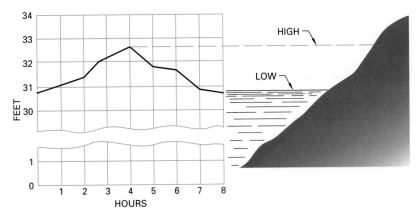

Figure 22-2

A graph showing the rise and fall of water in a tidal basin.

Complete books have been published on the design and use of charts and graphs. Only a few of the more common ones are described and illustrated in this chapter.

Charts, graphs, and diagrams are pictures of numerical information that show the relationship of one thing to another. For example, they can be used to show trends in areas such as economics. Using economic charts, you can tell at a glance whether the cost of living and wages are rising or falling over a period of time. This type of information can be determined quite easily because of the pictorial nature of a graph or chart.

Charts can also show ratios. Figure 22-1A is an example of a chart that shows the ratio of speed to distance. Other charts show percentages of a whole. Charts can also be used to explain information that is not numerical, such as sequential information.

Graphs are often used to solve various kinds of mathematical problems. You should be able to answer the following questions easily by studying the graph in Figure 22-2.

- When during each tide cycle does the highest water level occur?
- Approximately what is the level of the water at low tide?
- Approximately what is the level of the water at high tide?
- For approximately how long during each tide cycle is the water higher than 31 feet?

You probably had no trouble answering the questions because graphs of this type are easy to understand.

Types of Charts and Graphs

Various types of charts, graphs, and diagrams have been developed to present information clearly. This section reviews a few of the most common types.

Line Graphs

To show a trend, or how something changes over time, a **line graph** is usually a good choice. For example, changes in the weather, ups and downs in scales, or trends in population growth can be plotted and shown graphically on a line graph. A line graph can have one or several curves. Note that the curve on a chart or graph is not necessarily a curved line. As shown in Figure 22-3, a **curve** on a chart or graph may be a straight line, a curved line, a broken line, or a stepped line. It may also be a straight or curved line adjusted to plotted points.

Special line graphs are often created for specific purposes. For example, Figure 22-4 is a **conversion graph** that shows the conversion from Fahrenheit to Celsius degrees. Conversion charts are a form of one-curve line graphs that are convenient for changing from one value to another. Other line graphs have more than one curve. The line graph in Figure 22-5 contains four curves that compare sales in various parts of the country.

Another type of line graph is a shaded-surface, or **strata graph**. Figure 22-6 shows a typical strata graph. This type of graph can compare two different items, such as use of different kinds of materials, over a period of time.

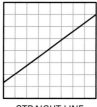
STRAIGHT LINE

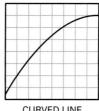
CURVED LINE

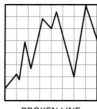
BROKEN LINE

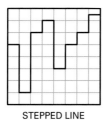
STEPPED LINE

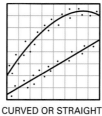
CURVED OR STRAIGHT
LINE ADJUSTED TO
PLOTTED POINTS

Figure 22-3

Curves on graphs may have different forms.

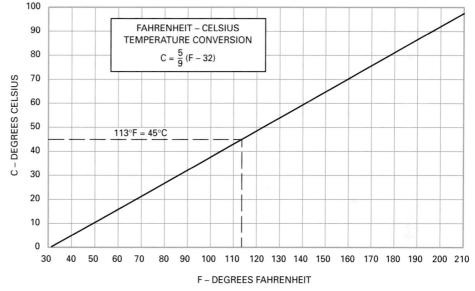

Figure 22-4

A conversion graph.

FAHRENHEIT – CELSIUS
TEMPERATURE CONVERSION

$$C = \frac{5}{9}(F - 32)$$

113°F = 45°C

C – DEGREES CELSIUS

F – DEGREES FAHRENHEIT

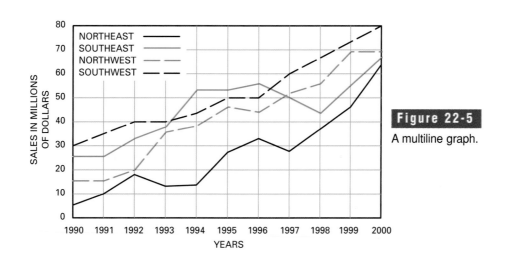

Figure 22-5

A multiline graph.

NORTHEAST
SOUTHEAST
NORTHWEST
SOUTHWEST

SALES IN MILLIONS
OF DOLLARS

YEARS

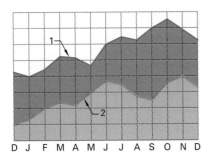

Figure 22-6

A strata graph uses shaded areas for contrast.

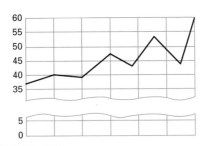

Figure 22-7

An omission graph may be used to accommodate a larger vertical scale.

Figure 22-7 is an example of an **omission graph**. An omission graph is used when the vertical scale needs to be much larger than the horizontal scale to show the information adequately. For example, the graph in Figure 22-7 contains no values below 35, so a portion of the graph is "broken out."

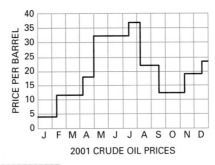

Figure 22-8

A step graph shows data that remains constant during regular or irregular intervals.

A **step graph** is a special type of line graph that shows data that remains constant during regular or irregular intervals. For example, a step graph might show time periods during which a price remained constant or was raised or lowered. Figure 22-8 is an example of a step graph.

Engineering Graphs and Charts

Experimental information may be plotted from tests and used to obtain an unknown value. For example, the graph in Figure 22-9 plots the results of tests of an unknown resistance, measured in ohms (Ω). After correcting the results for variable conditions, the points seem to show a straight-line curve. Values can be taken from two points and inserted in the formula. In this way, a value for the unknown resistance can be obtained and checked. Notice that tests were made on two occasions, and the results for both tests were plotted on the chart. A straight-line curve has been drawn along the center of the path made by the dots. Since the dots appear to be somewhat

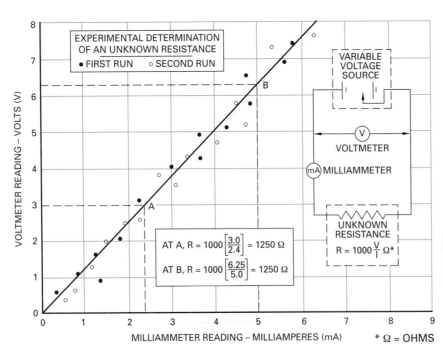

Figure 22-9

Engineering test graph (scatter graph).

scattered, this type of chart is also known as a **scatter graph**. In this type of graph, a smooth-line curve is drawn to represent the average of the plotted points. This is sometimes referred to as the "best" curve.

Nomograms are engineering charts that show the solutions to problems containing three or more variables (kinds of information). Figure 22-10 is an example of a nomogram. A straight line from values on the outside scales crosses the inside scale. The solution to the equation can be read at this point of intersection. Nomography is a special kind of chart construction that requires more than simple mathematics.

Bar Charts

The most familiar kind of chart for most people is probably the **bar chart**. Bar charts are easily read and understood. Like line graphs, bar charts can take many forms. For example, a one-column bar chart consists of a single rectangle representing 100%, as shown in Figure 22-11. It represents the total number of soccer games won, lost, and tied.

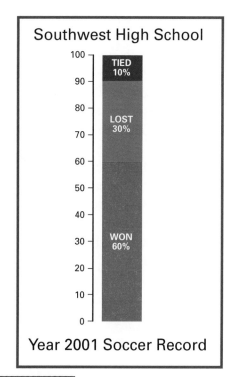

Figure 22-11

A one-column bar chart.

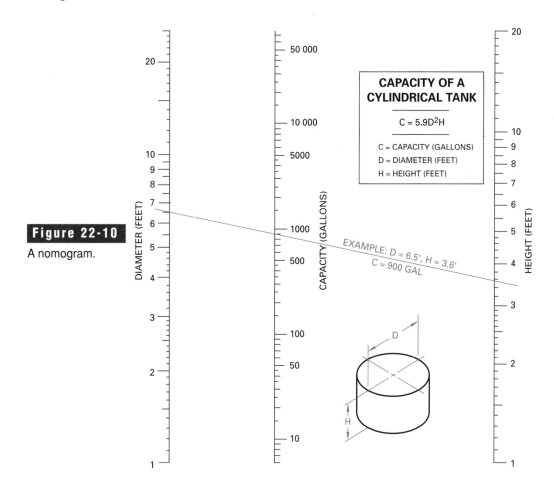

Figure 22-10

A nomogram.

Other bar charts use one bar per item. A simple multiple-column bar chart is shown in Figure 22-12. A multiple-column bar chart with horizontal bars is shown in Figure 22-13. It gives speed ranges for Caterpillar tractors. Note that the bars do not start at the same line because they show different speed ranges. This kind of chart is called a **progressive chart**.

A compound bar chart is shown in Figure 22-14, where the total length of the bars is made up of two parts. The blue portion is the distance traveled at a given speed before application of the brakes. The two portions are added together graphically to give the total distance involved.

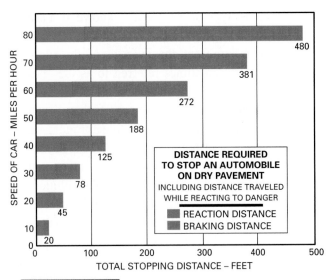

Figure 22-14

A compound bar chart is one in which the total length of each bar is the sum of two or more parts.

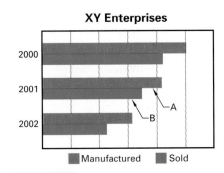

Figure 22-15

A comparison bar chart.

A comparison bar chart is one in which each item has two or more bars. The bars in Figure 22-15, for example, show the amount of a product manufactured (A) and the amount sold (B) by a single company over a period of several years. A similar chart could show the amount of a product manufactured and sold by various companies during one year.

The bar chart shown in Figure 22-16 has negative as well as positive values. In this case, the company experienced gains on some items and losses on others. The same information is given in the line graph in Figure 22-17. The vertical scale is the same as in Figure 22-16. The dashed line shows the net gain. It can be found by subtracting the negative value from the positive value for each month and plotting the resulting difference.

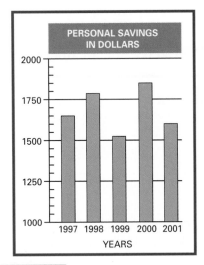

Figure 22-12

A multiple-bar chart with vertical bars.

Figure 22-13

This horizontal bar chart is a form of progressive chart.

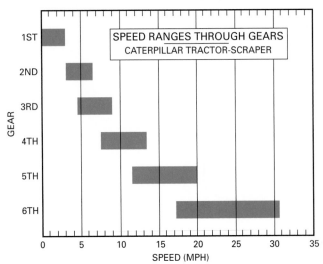

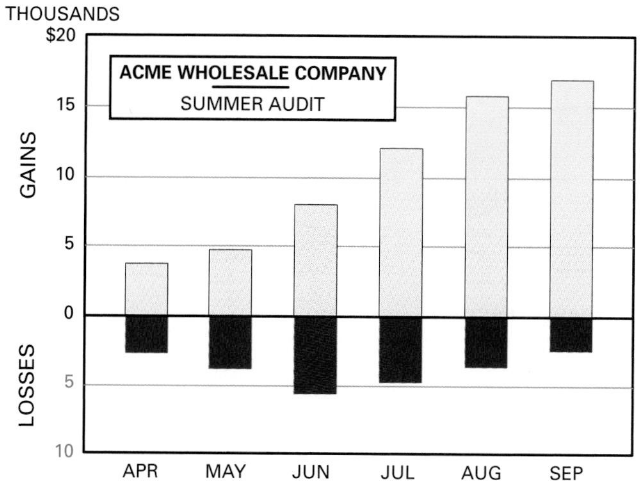

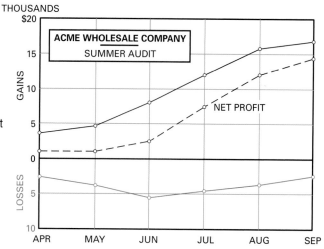

Figure 22-16

A bar chart in which the bars have positive and negative values.

Figure 22-17

A line graph for the same information as that in Figure 22-16.

TECH●MATH

Percentages

Working with percentages is common in designing and producing charts and graphs. Percent is another way of expressing a fraction that always has 100 as its denominator. For example, an income tax rate of 35% means that ³⁵/₁₀₀ of the taxable income is to be paid in taxes.

When dividing any number by 100, simply move the decimal point two places to the left. In the number 35, the decimal point is understood to follow the 5 (the last whole number). Moving the decimal point two places to the left results in a decimal of .35. On a taxable income of $50,000 and a tax rate of 35%, the taxpayer would pay .35 × $50,000, or $17,500.

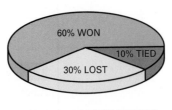
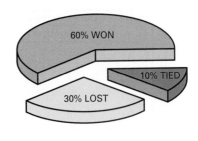

FOOTBALL SEASON RECORD
RELATIONSHIP OF GAMES WON, LOST, AND TIED

Figure 22-18

Pie charts can be two- or three-dimensional.

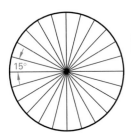

Figure 22-19

A pie chart representing 24 hours may be used to show time relationships.

Pie Charts

Figure 22-18 shows examples of a **pie chart**, a circle that represents 100%. Various sectors of the "pie" represent percentages of a whole. Every 3.6° of the circle equals 1%. For example, 30% equals 30 × 3.6°, or 108°. If a circle is to be divided into a 24-hour day, each 15° on the circle represents 1 hour, as shown in Figure 22-19. Pie charts may be drawn flat or in pictorial, as shown in Figure 22-18.

Figure 22-20

A pictograph.

CITY RECREATION CENTER SUMMER ATTENDANCE
WEEKLY AVERAGE

CHILDREN

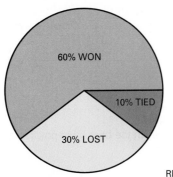

TEENAGERS

ADULTS

EACH FIGURE REPRESENTS 100

Pictographs

Pictorial charts, or **pictographs**, are similar to bar charts except that pictures or symbols are used instead of bars. Figure 22-20 is an example of a pictograph. The chart pictured is the equivalent of a multiple-column bar chart. In this chart, each figure represents 100 people. A symbol or picture may represent any number that the maker of the chart assigns to it, as long as the chart is labeled to identify the value of each symbol.

Organization Charts and Flowcharts

There are many kinds of organization charts. The most common kind is the **flowchart**, as shown in Figure 22-21. A flowchart shows the path, or flow, of items through a procedure or other sequence of events. Figure 22-21, for example, traces the routing of drawings from the top engineer to the shops, as well as the organization of the drafting department.

Other flowcharts show the path or series of operations it takes to manufacture a product or a material. An example of this is the flowchart of steelmaking processes in Figure 22-22.

Diagrams

Diagrams, also known as *schematic diagrams*, are more or less pictorial and are designed to be read easily. Names of parts and methods of operation are often shown in this manner. Diagrams are also a good way of showing the flow of materials, gases, liquids, or electrical current through an open or

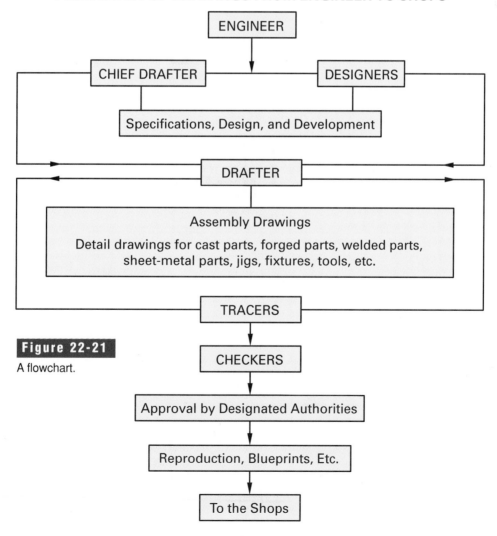

FLOWCHART OF DRAWINGS FROM ENGINEER TO SHOPS

Figure 22-21
A flowchart.

FLOWCHART OF STEELMAKING

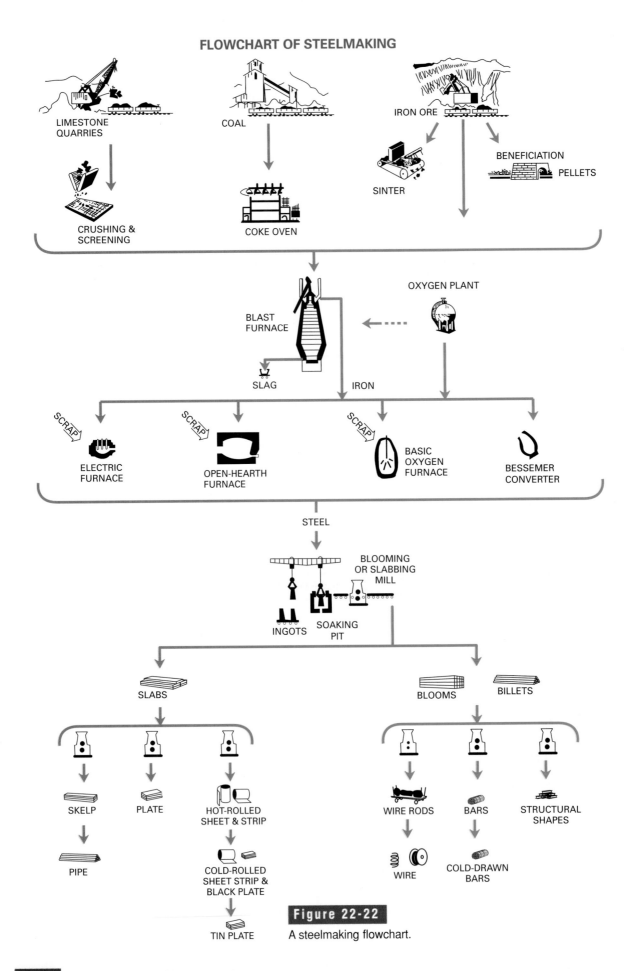

Figure 22-22

A steelmaking flowchart.

closed system. For example, Figure 22-23 is a diagram showing the flow of air through an automobile air-injection system. Thus, this diagram could be called a *flow diagram*. Notice that the various parts are recognizable, and yet they are somewhat schematic in nature. In this case, *schematic* means that the drawings represent the various parts and are not true projections drawn to scale and fully detailed. Arrows are often used to show the path or direction of flow.

The use of color is important in the design of diagrams. Color not only enhances the appearance, it also enhances the readability. In Figure 22-23, for example, the colored arrows immediately draw the reader's attention to that aspect of the diagram.

Setting Up a Chart or Graph

Although there is no one "right" way to present a certain set of information in a chart, following some general guidelines can help you present your information in a clear, logical format. This section addresses some factors you need to consider when creating a chart or graph.

Choosing a Graph or Chart

Some types of graphs and charts are better suited to certain types of information than others. Therefore, it is important to choose a type that presents your information clearly and concisely. For example, suppose you have interviewed the students in your school to find out the percentage of students who use various types of transportation. A pie chart is more suitable than a line graph for displaying this information.

Setting Horizontal and Vertical Scales

The selection of proper vertical and horizontal scales is the next important task. The vertical and horizontal scales must give a true pictorial impression. In Figure 22-24, different impressions are given by the three graphs. Graph A presents a very abrupt change. Graph B presents a normal change. Graph C shows a very slow or gradual change. You should choose the scale that gives the most accurate pictorial impression of the information you are trying to portray.

Figure 22-25 shows another example of the importance of selecting proper scales. The information in Figure 22-25A would be much easier to read if the vertical scale were increased. In Figure 22-25B, the horizontal scale should be increased. The graph is easiest to read when the scales are drawn in the proportions shown in Figure 22-25C.

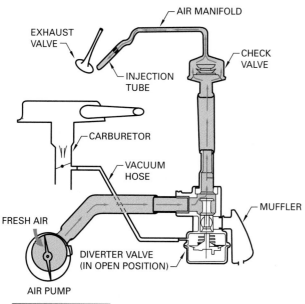

Figure 22-23

Schematic diagram of an automobile engine air-injection system.

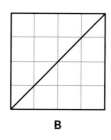

A   B   C

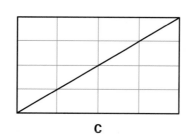

Figure 22-24

A false impression may result if vertical and horizontal scales are not properly selected.

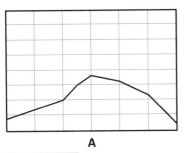
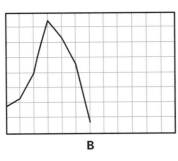
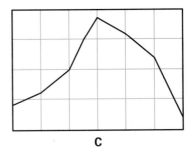

Figure 22-25

The effect of vertical and horizontal scales on a graph.

Labeling the Graph or Chart

Several different types of charts and graphs are illustrated and explained in this chapter. Each type of graph or chart is adapted to show a certain kind of information clearly. However, without the proper labels, they tell little about the intended subject.

First, you should label every chart or graph with an appropriate title. The title should be fairly short, but descriptive enough that the reader can tell at a glance what kind of information the chart contains.

Do not use long, complicated titles—they require more time to read and are often more confusing than enlightening.

Also, you should provide a key to tell what each element represents. This can be done in many ways, depending on the type of graph or chart. For example, in the transportation chart in Figure 22-26, small pictures identify each part of the graph. Figure 22-5 shows an example of a graph in which a formal key is used to identify the elements.

Using Color

Black-and-white charts are sometimes used by scientists and mathematicians for recording information. The use of color, however, has become quite common, especially in the preparation of charts and graphs for magazines, books, pamphlets, and various other publications. Color is also used a great deal in making charts for display purposes. The use of color adds interest to the appearance and emphasis of the graph or chart and makes it easier to understand. See Figure 22-27.

Figure 22-26

In some charts, the components can be identified pictorially.

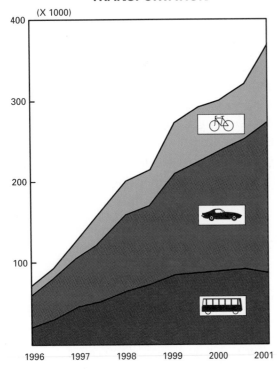

Success on the Job

Problem Solving

Be assertive! Before asking your supervisor or team leader for the solution to a difficult problem, attempt to develop several solutions of your own. Have them on paper when you approach the supervisor. This will usually result in a better solution in less time. It will also show that you are capable of being a creative problem solver, a quality that will improve your chances for success on the job.

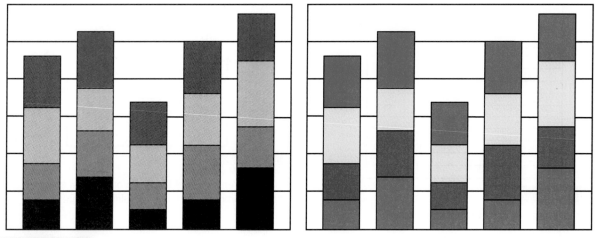

Figure 22-27

A combination bar chart in black and white and the same chart in color.

Board Drafting Techniques

In many instances, the data gathered will fit more than one type of chart or graph, as shown earlier in the chapter. The final choice is often a matter of the designer's preference. Often, a bar chart can show exactly the same information as a line graph. Only the style of presentation is different.

Once you have decided which type of chart or graph will work best for a particular application, it becomes a matter of getting all of the information into that format. Layout, color choices, proportions, and size are left to the creative design ability of the artist or drafter.

Selecting the Paper

After you select the type of graph or chart, you should select a paper that is suited to the type of graph or chart you will draw. Printed grid or graph paper is available in many forms. As you may recall, graph paper may be purchased with lines ruled for drafting use at 4, 5, 8, 10, 16, or 20 to the inch, as well as in many other forms. Metric sizes are also available. Graph paper that has certain lines printed thicker than others is often useful. This type of graph paper makes it easy to plot points and to read the finished graph.

Creating Graphs and Charts

The following sections give the procedures for creating various basic graphs and charts without the use of a computer. Once you know how to draw the basic forms, you can figure out how to draw almost any type of graph or chart.

Drawing a Line Graph

Follow steps 1 through 9 to create a line graph.
1. Prepare and list the information to be presented. Figure 22-28 shows an example of neatly organized information.
2. Select plain paper or ready-ruled graph paper.

SCORING INFORMATION EASTERN HIGH SCHOOL	
GAME NUMBER	POINTS SCORED
1	38
2	20
3	50
4	40
5	40
6	30
7	10
8	40
9	55
10	45

Figure 22-28

Information to be presented in a line chart.

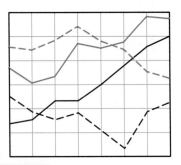

Figure 22-30

Use different types of lines and/or different colors to distinguish among curves on a multiline graph.

3. Select a suitable size and proportion for your graph so that the overall design will be effective.

4. Select an appropriate horizontal and vertical scale.

5. If you are not using graph paper, lay off and draw a thin horizontal line and a thin vertical line. These become the X axis and the Y axis, respectively. See Figure 22-29A. The intersection of the X and Y axes is zero.

6. Lay off the scale divisions on the X axis and the Y axis, as shown in Figure 22-29B.

7. Label the vertical and horizontal scales with the type of information represented.

8. Plot the points accurately using the information you listed in step 1. Use small circles, triangles,

or squares rather than crosses or dots for plotting points, as shown in Figure 22-29C.

9. Connect the points to complete the line graph, as shown in Figure 22-29D.

If you draw more than one curve on a line graph, use different types of lines or different colors for each curve, as shown in Figure 22-30. Use a full, continuous line of the brightest color for the most important curve. In general, the scales and other identifying notes, or captions, are placed below the X axis and to the left of the Y axis. For large charts, you may sometimes want to show the Y-axis scale at both the right and left sides. You may also want to show the X-axis scale at the top and bottom. This will make the charts more convenient to read.

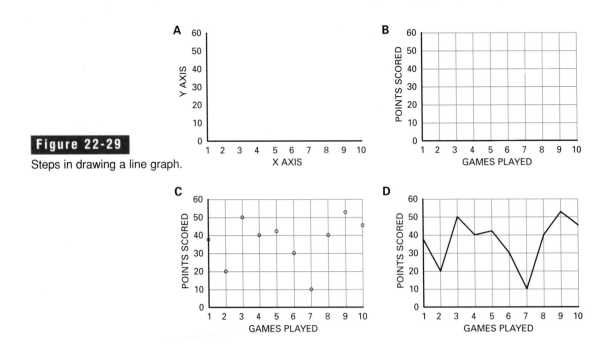

Figure 22-29

Steps in drawing a line graph.

Color may be added in a variety of ways. Colored pencils, felt-tipped pens, watercolors, or other, similar materials are easy to use. Most of these can be found in the drafting room, art room, or at home. Commercially prepared, pressure-sensitive materials are available at art and engineering supply stores.

Drawing a One-Column Bar Chart

Follow steps 1 through 6 to make a one-column bar chart.

1. Prepare and list the information to be presented. In this case, use the information shown in Figure 22-31.
2. Lay off the long side equal to 100 units. Figure 22-32A shows how this is done.
3. Lay off a suitable width and complete the rectangle, as shown in Figure 22-32B.
4. Lay off the percentage of the parts and draw lines parallel to the base, as shown in Figure 22-32C.

5. Crosshatch, shade, or color the various parts, as shown in Figure 22-32D.
6. Letter all necessary information in or near the parts so that it can be read easily, as shown in Figure 22-32E.

Drawing a Multiple-Column Bar Chart

Follow steps 1 through 5 below to make a multiple-column bar chart.

1. Prepare and list the information to be presented; in this case, use the information shown in Figure 22-33.
2. Select a suitable scale and lay off the X and Y axes.
3. Lay off the scale divisions, as shown in Figure 22-34A.
4. Block in the bars using the information gathered in step 1. Allow enough space between the bars for all necessary lettering. Make the bars any convenient width so that the overall appearance is pleasing. See Figure 22-34B.

	NUMBER	PERCENTAGE
GAMES WON	18	60
GAMES LOST	9	30
GAMES TIED	3	10
GAMES PLAYED	30	100

Figure 22-31

Information for a single-bar chart.

PERSONAL SAVINGS IN DOLLARS	
1996	$1650.00
1997	1780.00
1998	1520.00
1999	1850.00
2000	1600.00

Figure 22-33

Information for drawing a multiple-column bar chart.

Figure 22-32

Steps in drawing a one-column bar chart.

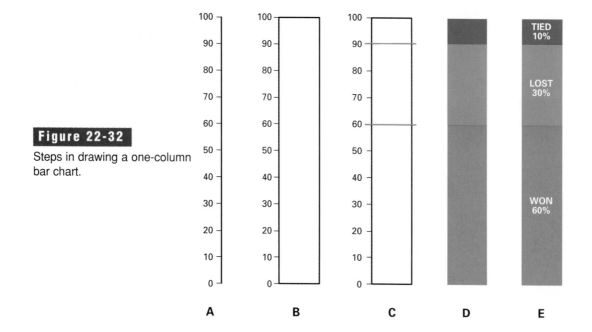

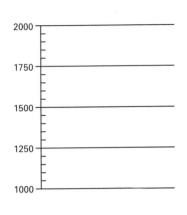

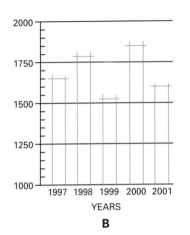

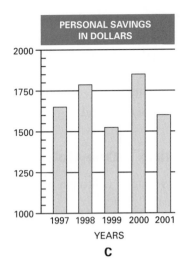

PERSONAL SAVINGS IN DOLLARS

A

B

C

Figure 22-34

Steps in drawing a multiple-column bar chart.

5. Complete the bar chart by adding shading or color to the bars. Add lettering and any other lines and information, as shown in Figure 22-34C. This adds to the appearance of the chart and the ease with which it can be read and understood. A 3D appearance may also be added if desired.

Drawing a Pie Chart

Follow steps 1 through 4 to create a pie chart.

1. Prepare and list the information to be presented. In this case, use the information shown in Figure 22-35.
2. Draw a circle of the desired size. Lay off and draw the radial lines representing the amount or percentage for each part on the circumference of the circle. See Figure 22-36A.
3. Crosshatch, shade, or color the various parts, as shown in Figure 22-36B.

4. Complete the pie chart by adding all necessary information, as shown in Figure 22-36C.

Tape Drafting

Adhesive tape provides a quick and simple method of preparing most types of bar charts and some line graphs. Adhesive tape comes in many colors, designs, and widths. It is applied from a roll dispenser, as shown in Figure 22-37, and is pressed onto the chart in the desired position.

Figure 22-35

Information for drawing a pie chart.

DISTRIBUTION OF CLASS TREASURY		
ITEM	COST	%
DANCE	$720.00	40
PARTY	360.00	20
PICNIC	320.40	18
CLASS PLAY	210.60	12
PHOTOGRAPHS	180.00	10
TOTAL	$1800.00	100

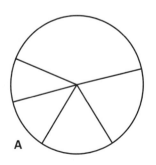

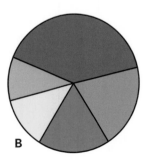

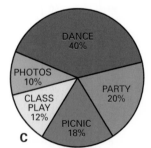

A

B

C

Figure 22-36

Steps in drawing a pie chart.

DISTRIBUTION OF CLASS TREASURY

Adhesive symbols may be used on pictographs. They are available in many forms; Figure 22-38 shows a few examples. These symbols make pictorial charts easy to construct.

Figure 22-37
Applying adhesive tape to a line graph.

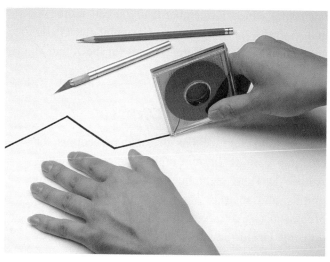

Figure 22-38
Many styles of adhesive symbols are available for use on graphs and charts.

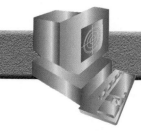

CAD Techniques

Today, most graphs and charts are created by computer. In fact, many word processing, spreadsheet, and even financial programs have built-in chart-making abilities. These programs may not be available in the drafting room, however. Fortunately, CAD programs can also be used to create many kinds of graphs and charts. The only major exception is the pie chart, which is difficult to execute using a CAD system. In most other cases, the same commands and procedures you use to create accurate engineering drawings can also be used to create accurate charts and graphs.

Setting Up the Drawing

Setup for creating a graph or chart is similar to that for creating any other CAD drawing. Decide on the size of the drawing, and set the drawing limits and units accordingly. Note that you will probably want to create a drawing from scratch instead of using a template, unless you want to position the graph within a standard border with a title block.

The grid and snap features can help you create graphs and charts quickly. For example, you can set the grid to the smallest unit on the vertical and horizontal scales. Remember that you can set the X and Y grid values separately. Just enter the GRID command and enter different values for the X and Y spacing.

A CAD-generated line graph.

CAD Graphs and Charts

The following sections give the procedures for creating various basic graphs and charts using AutoCAD. These techniques work for most kinds of graphs and charts. Think through the drawing process just as you would for any other type of CAD drawing. Decide how to use the available commands most effectively, and then proceed.

Drawing a Line Graph

Follow the steps below to create a line graph using AutoCAD.
1. Prepare and list the information to be presented. In this case, use the information in Figure 22-28.
2. Select a suitable size and proportion for your graph so that the overall design will be effective.
3. Select appropriate horizontal and vertical scales.
4. Set up the drawing (limits, grid, snap, etc.) for the size and type of graph you plan to create.
5. Using the snap and grid as reference, create the X and Y axes.
6. Use the OFFSET command to offset both axes to create the grid marks for the graph. (If the axes are of unequal length, you may need to use the TRIM command to give the drawing a neater appearance.)

Arraying the Axes
Another method of creating the horizontal and vertical grid lines is to use the ARRAY command. For the horizontal lines, array the X axis, specifying 1 column and as many rows as you need horizontal lines. *Note:* Be sure to include the X axis in the total number of lines you specify. For the vertical lines, array the Y axis, specifying 1 row and as many columns as you need vertical lines. Be sure to include the Y axis in the total number of lines you specify.

7. Use the DTEXT command to add the scale values and labels to the X and Y axes. Note that you will need to rotate the text for the Y-axis label. When the DTEXT command prompts you for the text angle, enter 90 (90°).
8. Plot the points accurately using the information from step 1. Use the POINT command to place the points accurately. You may want to change the value of PDMODE to display the points as circles or crosses so that they are more easily visible.
9. Enter the PLINE command and connect the points to complete the line graph. If you changed the value of PDMODE in the previous step, change it back to make the points invisible. The finished graph is shown in Figure 22-39.

If you draw more than one curve on a line graph, use different linetypes or different colors for each curve. You may want to place each curve on its own layer, or simply use the Color Control dropdown box located on the Object Properties toolbar to choose a different color for each curve. Use bright color and a continuous linetype for the most important curve. In general, the scales and other identifying notes, or captions, are placed below the X axis and to the left of the Y axis. For large charts, you may sometimes want to show the Y-axis scale at both the right and left sides. You may also want to show the X-axis scale at the top and bottom. This will make the charts more convenient to read.

Drawing a One-Column Bar Chart

The most efficient way to create a colorful bar chart in AutoCAD may surprise you. Rather than using a rectangle or a series of lines to create the parts of the bar(s), you should use the PLINE command with a nonzero width. Note that this method will not work if you want to hatch the interior of the bars. To do that, you would need to enclose the area using the RECTANGLE or PLINE command (zero width), and then apply a hatch. However, hatches take longer to print than simple colored bars, and the bright colors are often more effective anyway.

Follow these steps to make a one-column, three-color bar chart using AutoCAD.

1. Prepare and list the information to be presented. In this case, use the information shown in Figure 22-31.
2. Select an appropriate vertical scale.
3. Set up the drawing (limits, grid, snap, etc.)
4. Determine the actual length of each part of the bar according to your vertical scale.
5. Enter the PLINE command, and enter W (Width). Then simply specify the width you want to use for the bar. For example, if you have chosen a scale in which the total length of the bar (100%) equals 5 units, an appropriate width might be 1 unit. Then calculate the length of each portion of the bar according to the scale you have selected. With Ortho on, create the lower (60%) portion of the bar. End the PLINE command, select the polyline, and use the Color Control dropdown box to change its color to bright red, as shown in Figure 22-40A. Note that only the line "skeleton" itself is selectable, not the entire width of the polyline.

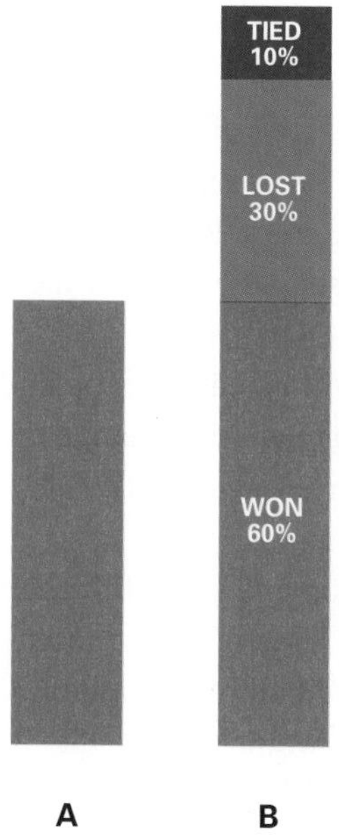

6. Repeat step 5 for the other two parts of the bar. Use the Endpoint object snap to snap the beginning of the second polyline to the upper end of the first one. Make the middle section (30%) bright green, and the top section (10%) bright blue.
7. Use DTEXT to add all necessary information in or near the parts so that it can be read easily. The finished bar chart is shown in Figure 22-40B.

Drawing a 3D Bar Chart

There are two ways to create a "3D" bar chart in AutoCAD. You can either create the entire graph in three dimensions, or you can create a 2D drawing that appears to be three-dimensional. A true 3D graph is often not worth the effort using a CAD program, because the text has to be added on a custom UCS to make it flat to the screen or paper. This is worthwhile only when the graph has contours and surfaces that can be viewed most effectively from various viewpoints. However, the more complex the contours and surfaces are, the more tedious the drawing task becomes. For these graphs, the best alternative is to use third-party software that allows AutoCAD to generate the graphs automatically.

A 2D drawing with a 3D appearance is easier to create. The isometric grid and snap are not much help because, although they make it easy to simulate a 3D drawing, they place the bars on a deeper angle than is appropriate for most bar graphs. The easiest method is to create the first bar accurately using the standard snap and grid, and then copy and adjust it using the STRETCH command to create the additional bars.

The following procedure creates a 2D pictorial bar chart that appears to be three-dimensional. Follow these steps:

1. Prepare and list the information to be presented; in this case, use the information shown in Figure 22-33.
2. Select a suitable scale for the X and Y axes.
3. Set the snap and grid at convenient intervals to help you create the vertical scale and first bar accurately.
4. Lay off the vertical scale divisions and identify them using the DTEXT command.

TIED 10%
LOST 30%
WON 60%

Figure 22-40

Be sure to keep the final appearance of the printed version in mind as you choose colors for a graph or chart in AutoCAD. The background will be white, unless you print on colored paper. Colors that show up vividly against AutoCAD's black background may not show up well on white paper.

A   B

Converting Percentages to Inches

Earlier in the chapter, you learned to develop a one-column bar chart by dividing the long side of the bar into increments that equal 100 units. In that case, we used 10 increments, and each of those increments equaled 10 units or 10%. This system works well when you have items in your bar chart that are in multiples of 10, such as 60%, 30%, and 10%. Otherwise, you need to estimate amounts between the divisions of 10.

A more accurate method is to convert percentages to inches and measure directly along the long side of the bar. Here is an example. Assume that you want the bar to be 8.00″ long. The items represented are 38%, 22%, 16%, 12%, 8% and 4%. These added together equal 100%. Our task is to convert the percentages to inches and then simply measure these distances directly on the 8.00″ bar. The conversions are easily done using a calculator. Simply multiply each percentage by 8.00″ to find the length of each portion of the bar.

Here are the conversions:

38%	= .38 × 8.00″ =	3.04″
22%	= .22 × 8.00″ =	1.76″
16%	= .16 × 8.00″ =	1.28″
12%	= .12 × 8.00″ =	.96″
8%	= .08 × 8.00″ =	.64″
4%	= .04 × 8.00″ =	.32″
Totals: 100%		8.00″

Once you have done the conversions, you can complete the bar graph. Build the bar chart from the bottom up by first measuring the 3.04″. Next, add to it 1.76″ (3.04 + 1.76 = 4.80). Continue this process until you have reached the top of the bar (8.00″). This process is clearly shown in the figure below.

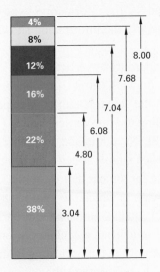

You could use the LINE or PLINE command to create the first bar, but if you do that, you are limited to using just the outlines of the bars or hatching the interiors. Wide polylines don't work for 3D bars because they can't show the 3D effects. To create bars of solid color for a pictorial or 3D chart, you should use the SOLID command. This command creates a 2D solid-filled shape in the color you specify.

5. Enter the SOLID command. At the Specify first point prompt, pick the lower left corner of the front face of the bar (point 1 in Figure 22-41A). Note that the order in which you pick the points to create the bar is very important. Pick the other three points in the order shown in Figure 22-41A to create the front face of the first bar at the appropriate height. Press Enter to end the command. Notice that the solid does not appear until you press Enter. Change the color of the bar to green.

6. Use the SOLID command to create the top and side of the 3D bar. You may want to change the snap and grid to a smaller interval first to help you create them precisely.

7. Change the color of the top and side of the bar to two different shades of green. This is necessary because, if you created all of them the same shade of green, you would not see the 3D effect because all of the solids would blend

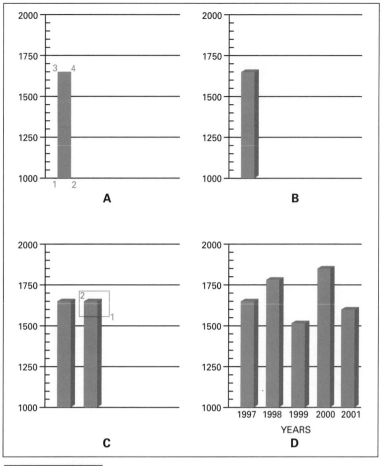

Figure 22-41

Development of a two-dimensional bar graph that looks three-dimensional.

together. To change the color of the top and side to different shades of green using the Color Control dropdown menu, choose Other... at the bottom of the menu to see a larger color palette. Pick the shade of green you want for each part of the bar, or enter the color number in the text box near the bottom of the palette. For the effect shown in Figure 22-41B, use a top color of 102 and a side color of 106.

8. Copy the entire bar, placing the copy in position for the second bar in the series.

9. Enter the STRETCH command. At the Select objects prompt, use a crossing window to select

the top part of the bar only, as shown in Figure 22-41C, and press Enter. (A *crossing window* is a selection window in which you select first a point on the right and then one on the left.) For the base point, select the left top edge of the bar. With Ortho on, move the cursor up or down to adjust the bar to the correct height. Be sure to align the top of the *front* of the bar with the vertical scale.

10. Repeat steps 8 and 9 for each bar in the graph.

11. Add any necessary text and labels. The finished graph is shown in Figure 22-41D.

Chapter Summary

- Charts and graphs are used in every field of business, industry, and education.

- Charts and graphs are used to show facts, as well as numerical and nonnumeric relationships, in a pictorial form that can be read quickly and easily.

- A "curve" on a chart is not necessarily a curved line.

- Engineering graphs and charts are often used to help solve research problems.

- Names of parts, methods of operation, and the flow of liquids or gases through a system are often shown using diagrams.

- Color adds to the appearance and emphasis of a chart or graph and can make it easier to understand.

- The type of chart or graph chosen depends on the information to be presented.

- Charts and graphs can be created manually using various types of graph paper.

- CAD systems can be used to create most types of graphs and charts.

Review Questions

1. Is a curve on a chart always a curved line? Explain.

2. What kinds of charts are used to show the solutions to problems that contain three or more kinds of information?

3. Refer to Figure 22-4. Convert 65°C to Fahrenheit.

4. In addition to changes, what do line graphs generally show?

5. Refer to the chart in Figure 22-5. What was the dollar volume of sales in the Northeast region in 1999?

6. In Figure 22-10, what is the capacity in gallons of a cylindrical tank 10′ in diameter and 17′ high?

7. In Figure 22-10, what is the height of a cylindrical tank that holds 100 gallons and is 2′ in diameter?

8. In Figure 22-10, a tank holds 3,000 gallons and is 10′ high. What is its diameter?

9. What do you call a bar chart in which the total length of the bars consists of two or more parts?

10. Refer to the chart in Figure 22-14. How far does a car travel during reaction time at a speed of 60 mph?

11. What do you call a chart that shows various ranges of data?

12. In Figure 22-14, what is the total stopping distance at a speed of 60 mph?

13. On a pie chart, a full circle represents what percentage of the data?

14. On a pie chart, a segment of 90° represents what percentage of the data?

15. On a pie chart, a section representing 75% of the data would require a segment of how many degrees?

16. What kind of chart uses symbols instead of bars to represent quantities?

17. What kind of chart shows sequential information most clearly?

18. How does a diagram represent objects?

19. What kind of impression may result if vertical and horizontal scales are not properly selected for the design of a chart?

20. What type of chart is not easily created using a CAD system?

Problems

Drafting Problems

Problems 1 and 2 require the construction of pie charts; these charts should be completed manually. The remaining problems in this chapter are designed to be completed manually or using a CAD system.

1. Draw a pie chart to show the population distribution in the following four regions of the United States:

Northeast 24.2%
North Central 27.8%
South 31.2%
West 16.8%

Use color or various types of crosshatching for contrast.

2. Draw a pie chart showing that 67.4% of the population of the United States lives within metropolitan areas, while only 32.6% lives outside metropolitan areas. Use different colors or crosshatch areas for contrast.

3. Make a one-column bar chart to show how your allowance or income was spent last month.

4. Make a multiple-bar chart showing a comparison of how you spent your allowance or income over a period of 4 months.

5. *Part 1:* The average cost per pound for beef varied over a period of 10 years as follows:

Year	Cost per Pound
1992	$2.30
1993	2.20
1994	2.30
1995	2.38
1996	2.25
1997	2.35
1998	2.42
1999	2.38
2000	2.65
2001	3.30

Make a graph showing this relationship.

Part 2: The average cost of pork varied somewhat differently. Add a second line to your graph showing a comparison of the cost of beef to the cost of pork over the same period of time. Use a different color for contrast.

Year	Cost per Pound
1992	$2.30
1993	2.40
1994	2.60
1995	2.55
1996	2.48
1997	2.50
1998	2.70
1999	2.80
2000	2.90
2001	3.10

Part 3: Add another line (in another color) to the chart showing the average cost of poultry for the same 10 years.

Year	Cost per Pound
1992	$1.44
1993	1.52
1994	1.60
1995	1.40
1996	1.46
1997	1.70
1998	1.80
1999	1.66
2000	1.76
2001	1.90

6. Draw a flowchart showing how to mass-produce a project of your choice in the school shop.

7. Make an organizational flowchart showing the administrative structure of your school.

8. *Part 1:* Draw a pie chart or a one-column bar chart showing a breakdown of the average person's income if 24.5% goes to federal taxes, 5.8% goes to state taxes, and 1.3% goes to local taxes.

   Part 2: Compute the dollar value of each category above for a gross income of $40,000. Mark each dollar value on your chart. Be sure the figures total $40,000.

9. *Part 1:* Make a pictograph showing male and female population in your school.

   Part 2: Make a bar chart showing male and female population in your drafting class.

10. Draw a pictograph showing the enrollment of technical drawing classes in your school. Your instructor can supply the information.

11. *Part 1:* Make a line graph showing the hourly change in the outside temperature for a 12-hour period during any one day.

    Part 2: Record similar information for several days and make a multiline chart to show a comparison.

12. Draw a bar chart to show home consumption of electricity for 1 year as follows:

Month	Kilowatt-Hours Used
January	900
February	885
March	800
April	783
May	722
June	600
July	494
August	478
September	525
October	650
November	735
December	820

13. Plot the batting averages of the players on your favorite baseball team.

14. The number of cars per mile of road in the United States is growing. Draw a pictograph from the data below to show the growth and anticipated increase.

Year	Cars Per Mile of Road	Cars per Kilometer of Road
1955	9	2
1965	15	5
1975	20	8
1985	26	12
1995	38	17
2005	45	23

15. Draw a vertical bar chart showing the following student attendance for a given week of school. The total school enrollment is 925.

Day	Attendance
Monday	625
Tuesday	715
Wednesday	800
Thursday	775
Friday	695

16. *Part 1:* Compute your daily calorie intake for 1 week. Make a line chart representing this information.

    Part 2: Using the same information, prepare a horizontal bar chart.

17. Make a multiline chart representing individual game scores of the top five players on the school basketball team for any given season.

18. From the stock-market listings in the newspaper, select several stocks and record their daily status for 10 days. Plot the information on a line chart.

19. Make a pictograph or pie chart showing a breakdown of the source of each dollar received by the federal government. Use the information given below.

Individual income tax	$0.28
Employment tax	0.24
Corporate income tax	0.16
Borrowing	0.15
Excise tax	0.12
Other (miscellaneous)	0.05

20. Make a pictograph or pie chart showing a breakdown of the expenditure of each dollar by the federal government. Use the information given below.

National defense	$0.31
Income security	0.27
Interest	0.08
Health	0.07
Commerce, transportation, housing	0.06
Veterans	0.05
Education	0.04
Agriculture	0.03
Other (miscellaneous)	0.09

21. The following information includes five common foods and the number of calories and grams of carbohydrates in a 4-ounce serving of each. Make a bar chart illustrating these facts.

Food	Calories	Carbohydrates
Chocolate ice cream	150	14
Peas	75	14
Pizza	260	29
Milk	85	6
Strawberries	30	6

22. *Part 1:* The data in the list below represents a percentage breakdown for a family budget. Make a pie chart or bar chart illustrating this information.

Food	23.1%
Housing	24.0%
Transportation	8.8%
Clothing	10.9%
Medical care	5.6%
Income tax	12.5%
Social Security	3.8%
Miscellaneous	11.3%

    Part 2: Compute the dollar value for each category for a gross income of $55,000. Mark each on your chart. Be sure the figures total $55,000.

23. Accidents involving children occur in various places. Draw a horizontal bar graph or a pie chart using the places and percentages given below.

At home	25%
Between home and school	8%
On school grounds	15%
In school buildings	21%
In other places	31%

Design Problems

Design problems have been prepared to challenge individual students or teams of students. In these problems, you are to apply skills learned mainly in this chapter but also in other chapters throughout the text. The problems are designed to be completed using board drafting, CAD, or a combination of the two. Be creative and have fun!

1. The president of the company for which you work is preparing her annual report to stockholders. Your assignment is to design and produce a chart showing the company's gross sales, production costs, and net profits for the year based on the data given in Table 22-1. Impress the boss! Use color to enhance the appearance of your work.

2. **TEAMWORK** Begin with some research. Obtain documentation on the operation of hydraulic brakes on an automobile. Develop a flow diagram showing how the brakes work and specifically how the hydraulic fluid, under pressure, applies the force that presses brake pads and shoes against rotors and drums. Be creative in developing schematic representations of the various brake components. Use color to enhance the appearance and readability of the diagram.

3. **TEAMWORK** Beginning on the first day of a month, have each member of the design team forecast average daily temperatures for each day of that month. Prepare a scatter diagram based on these forecasts. Draw a smooth-line curve through the scattered points to represent the average of the plotted points. At the same time, record actual average daily temperatures. At the end of the month, add a curve representing the actual average daily temperature, and compare the two curves. Were the team members reasonably accurate in their prediction of temperatures?

Table 22-1. Data for Design Problem 1			
Month	Gross Sales	Production Costs	Net Profit
January	$250,000	$200,000	$50,000
February	400,000	300,000	100,000
March	350,000	300,000	50,000
April	500,000	375,000	125,000
May	700,000	500,000	200,000
June	500,000	400,000	100,000
July	650,000	525,000	125,000
August	750,000	600,000	150,000
September	800,000	650,000	150,000
October	900,000	700,000	200,000
November	950,000	765,000	185,000
December	700,000	600,000	100,000
Totals	$7,450,000	$5,915,000	$1,535,000

Media Management

KEY TERMS

closed catalog
   system

diazo

electrostatic
   reproduction

Gantt chart

hard copy

intermediate

media

microfilm

open file system

PERT chart

photodrafting

OBJECTIVES

Upon completion of this chapter, you should be able to:

- Describe ways in which the engineering design team uses and manages graphic communication.

- Use drawing storage and retrieval systems for both hard copy and electronic media.

- Describe several ways to produce intermediates during the drawing revision process.

- Describe methods of reproducing board- and CAD-produced drawings.

Graphic communication is used throughout the design and production of structures and products. The various types of working drawings that you have studied throughout this text are obvious examples of graphic communication used for construction and manufacture.

While the working drawing serves as the basic form of graphic communication for design and manufacture, many other kinds of documents are used throughout business and industry. Written documents, for example, are an important form of communication required for legal and operational purposes. These various kinds of documents, including drawings, may be produced by hand or by computer, and they are stored on paper, film, or as an electronic file. Collectively, these are called **media** (plural; singular *medium*). The media on which important documents are stored need to be managed efficiently so that the business can run smoothly.

Many companies use networked computer systems to help control the flow of graphic communication as well as other work processes, as shown in Figure 23-1. The system administrator, who may also be the project manager or department head, controls the flow of information to each computer that is connected to the network. Networked systems allow CAD operators to share important files such as symbol libraries and base drawings. The shared files are usually stored on a powerful computer with extensive storage capability, called a server.

In this chapter, little attention is given to the actual preparation of drawings. Rather, the chapter describes some of the many types of documents that are typically used by the engineering design team, as well as the equipment used to produce, reproduce, use, revise, control, store, and retrieve them. You will learn how to manage the media on which graphic communication is stored.

Types of Documents

Large companies organize engineering design teams to control the design and manufacture of major projects and products. The types of documents produced by these teams depend partly on the size and complexity of the product. *Note:* For the purpose of this chapter, the terms *project* and *product* are used interchangeably.

Figure 23-1

Networked computer systems allow management to control graphic communication; they also provide common resources, such as drawing templates and symbol libraries, to all the drafters connected to the network.

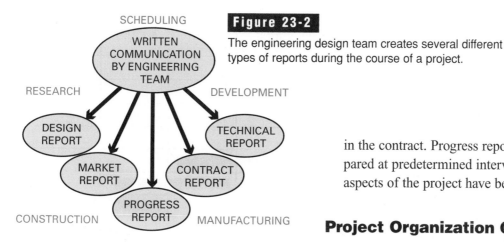

Figure 23-2

The engineering design team creates several different types of reports during the course of a project.

Reports

Many technical reports, such as bills of materials and stress analyses, are developed by the engineering design team. These reports, which may include both text and graphics, are often generated directly from the CAD software that was used to create the plans. The results are documented in a formal report using word processing software.

As shown in Figure 23-2, five basic types of technical reports are generally needed during the design and manufacture of a new product.

■ The *design report* states the needs of the client and provides an early analysis of the project by the designers.

■ The *technical report* is a detailed analysis, prepared by the designers, that details all the resources necessary for researching, designing, and developing the project.

■ The *market report* describes the marketability of the proposed or revised product, target consumer base, product price, and potential profit. This report impacts the final decision on whether the company manufactures the product.

■ The *contract report* includes all the details of manufacture or construction. This report may also include specialized tooling requirements if new tools must be designed and manufactured in order to produce the product.

■ *Progress reports* are reports made periodically on the current status of the project. These reports allow managers to check the status of the project and to review compliance with the specifications

in the contract. Progress reports are usually prepared at predetermined intervals or when various aspects of the project have been completed.

Project Organization Charts

One major function of the engineering design team is to meet the needs of clients or future customers. The work of the team must therefore be organized so that problems can be solved as quickly and inexpensively as possible. Therefore, in most companies, the production schedule plays an important part in the operation of the organization. From the early design studies to the final working drawings, all work is done according to a carefully designed schedule. This is true whether the product is a skyscraper, an automobile, a luxury cruise liner, a supercomputer, or a child's toy.

The head of a design project or planning department may use several different types of charts to help manage and control the project. Two of the most commonly used charts are Gantt and PERT charts. A **Gantt chart** separates an operation or project into discrete elements and assigns a certain amount of time for each step. See Figure 23-3. The time allotted varies depending on the complexity of the operation. A **PERT** (project evaluation and review technique) **chart** assigns probabilities to production issues such as efficiency and productivity. It then plots them against production activities and operations, as shown in Figure 23-4. This type of chart allows the project manager to estimate production and output time accurately. Gantt and PERT charts are examples of the use of graphic communication to help organize and control a project.

Records of Drawing Development

As members of the engineering design team develop a project, they keep a record of the design progress. In most companies, each project is assigned a unique contract number or job number for

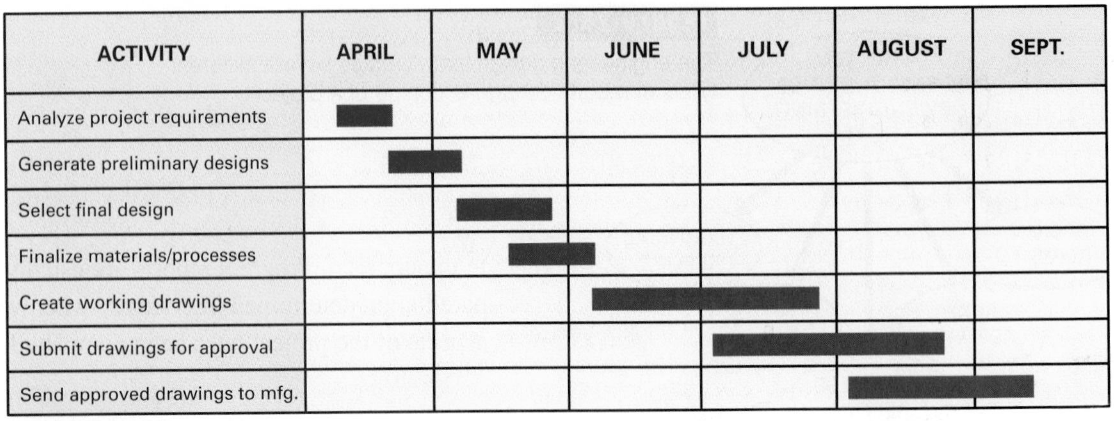

ACTIVITY
APRIL
MAY
JUNE
JULY
AUGUST
SEPT.
Analyze project requirements
Generate preliminary designs
Select final design
Finalize materials/processes
Create working drawings
Submit drawings for approval
Send approved drawings to mfg.

Figure 23-3

A Gantt chart shows how much time a company plans to spend on each phase of a project. It also helps companies work more efficiently by allowing managers to plan which processes can be done concurrently.

this purpose. This number must be included on all reports and drawings. The number is part of the information given in the title block of a drawing.

Title blocks must appear on every drawing, regardless of size, because they provide information without which the product cannot be manufactured as intended. Revision blocks must be added when the drawing has been modified.

Success on the Job

Interpersonal Skills

Becoming a valuable team member requires the development of strong interpersonal skills. This means that you will have a clear and comfortable line of communication with team members as well as the team leader. Good communication builds team confidence and improves your chances for success on the job.

TECH MATH

Rounding Decimals

When preparing the various reports needed for a project, it is sometimes necessary to round numbers to fewer digits. For example, a market analysis might include numbers of specific responses from a customer survey. When these numbers are converted to percentages for the market report, the percentages are rarely whole numbers. They are usually decimal fractions that must be rounded for ease of reading. For example, 25.63220% rounded to two places would be 25.63%.

The procedure for rounding is as follows:

1. When the left-most digit dropped is less than 5, the last digit kept does not change. For example, 15.232 rounded to two decimal places would be 15.23.
2. When the left-most digit dropped is greater than 5, the last digit kept should be increased by 1. For example, 6.436 rounded to two decimal places would be 6.44.
3. When the left-most digit dropped is 5 followed by at least one digit greater than zero, the last number kept should be increased by 1. For example, 8.4253 rounded to two decimal places would be 8.43.
4. When the first number dropped is 5 followed by zeros, the last number kept is increased by 1 if it is an odd number. If it is an even number, no change is made. For example, 3.23500 rounded to two places would be 3.24. However, 3.22500 rounded to two places would be 3.22.

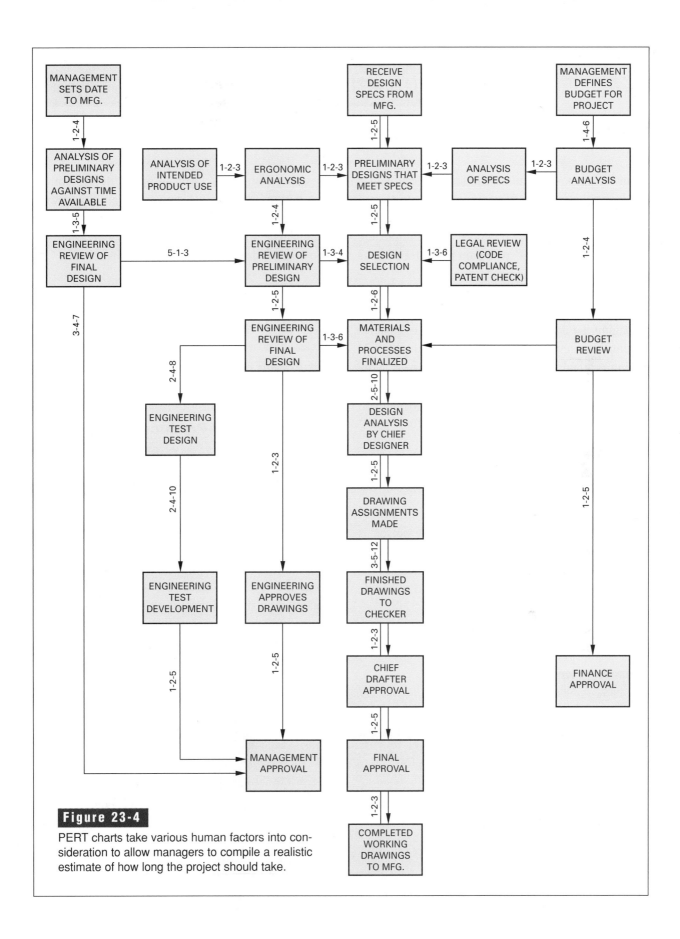

Figure 23-4

PERT charts take various human factors into consideration to allow managers to compile a realistic estimate of how long the project should take.

Managing Paper Documents

For the purpose of this discussion, the term *paper documents* includes drawings created using board drafting techniques, regardless of the medium on which the drawings were produced. For example, the term includes drawings created on polyester film, as well as those created on paper.

The drawings done in a drafting room each day must be carefully controlled to avoid misplacing or damaging the drawings. Therefore, most companies develop a comprehensive system for cataloging and storing original drawings. For example, many large companies keep the original drawings for a contract together under its contract number or job number. This makes the drawings easily available for copying or revising. See Figure 23-5.

File Systems

Most companies use either a closed or open filing system. In a **closed catalog system**, one person is responsible for keeping the drawings safe. Although this person may have a staff that files and copies the documents, the number of people allowed to handle original drawings is tightly controlled.

Other companies favor an **open file system**. In this system, the original drawings are always available to anyone in the department. The open file system is generally not preferred because easy access to the original drawings can lead to mishandling.

Drawing Management

After an original set of drawings has been completed, it must be stored for future use. In some cases, the originals are modified over a period of time as a product or structure is refined or updated.

Storing Board-Produced Drawings

Storing drawings in their original form consumes considerable space. The drawings are often difficult to manage because of their size, and the overhead expenses, such as rent and temperature control, are often enormous. Therefore, many companies store board-produced drawings on microfilm. See Figure 23-6. **Microfilm** is a storage medium on which a large drawing is reduced to a very small size and recorded on film. Microfilm storage greatly reduces the amount of space necessary to store drawings.

A number of photographic formats have been used in microfilming. However, 16 mm and 35 mm

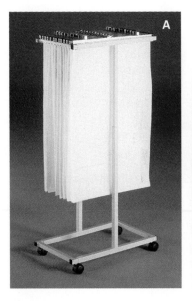

Figure 23-5

Traditional vertical (A) and horizontal (B) storage cabinets for original drawings.

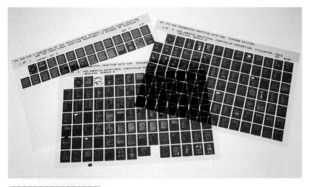

Figure 23-6

Because microfilm stores large drawings in a very small format, many drawings can be stored on a single microfilm data card.

are the most popular film sizes. The film has high contrast, high resolution, and is fine-grained to produce high-quality drawings with good detail. When a drawing stored on microfilm needs to be copied or revised, a copy is made directly from the microfilm. Copies from microfilm can be scaled to fit any standard size paper.

Revision Methods

When a drawing must be revised, the company must usually keep a copy of the original as well as the revision. Instead of recreating the entire drawing for the revision, drafters generally use an intermediate. An **intermediate**, also called a *secondary original*, is a copy of a drawing that is used in place of the original. Figure 23-7 shows how an intermediate can be used to create a new original. The designer and drafter can change the intermediate without changing the original. Using intermediates avoids tracing and redrawing.

The eight methods that follow are only some of the time-saving techniques used by drafters to revise intermediates.

■ **Scissors drafting** The unwanted portions of an intermediate are removed with scissors, a knife, or a razor blade. This is called editing. Then a new intermediate is made to incorporate the change. Other changes can then be made on the new secondary original.

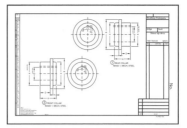

Say you have an existing drawing:

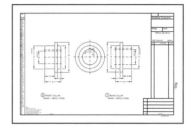

And you want to revise it like this.

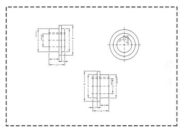

First you make a clear film reproduction of the original and <u>cut out</u> the elements.

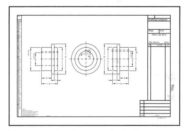

Then tape the elements in their new positions on a new form.

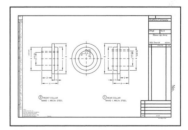

Photograph it on film with a matte finish (the tape and film edges will disappear). Draw in whatever extra detail you want – and you have a new original drawing.

PASTEUP DRAFTING

Figure 23-7

Using an intermediate to rearrange the placement of drawings on a drawing sheet.

- **Correction fluids** Sometimes drafters remove unwanted data from an intermediate by applying correction fluid. New data can then be drawn in the cleaned area. Prints are then made from the corrected intermediates.
- **Erasable intermediates** Erasable intermediates can be erased with ease. Thus, corrections can be made quickly without fluid. Redrawing can be done equally well with pencil or ink. Nonerasable prints are then made from the intermediates.
- **Block-out or masking** This system is best used when limited time is available for reworking a drawing. First, the area to be blocked out is covered with opaque tape. Next, a print is made in which the masked areas produce blank spots. Changes are then entered into these areas to produce the revised original.
- **Transparent tape** Data can be added to a drawing by placing the data on paper or film and then taping the data in place on an intermediate. Drafters often add dimensions, notes, and other data symbols in this way. Alternatively, data can be typewritten on transparent press-on material, which is then attached to the intermediate.
- **Composite grouped intermediates** When several drawings are to be combined, a composite grouping can be cut from intermediates and taped in place. A new intermediate is then made of the composite.
- **Composite overlays** Translucent original drawings can be combined into a composite intermediate by placing the originals in the desired places. The composite is then photographed.
- **Photodrafting** A picture not only saves a lot of words; it also limits the time needed for drafting. Architectural and engineering teams often use photo drawings because they are generally easier to understand. See Figure 23-8. In this intermediate drafting process, called **photodrafting**, photographs are overlaid with line work to create enhanced, annotated pictures of the product.

Figure 23-8

Photodrafting can complement and clarify technical data.

MAKE A HALFTONE PRINT OF THE PHOTOGRAPH.

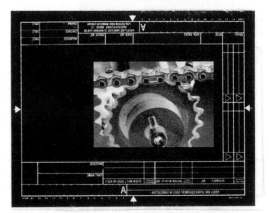

TAPE THE HALFTONE PRINT TO A DRAWING FORM, AND PHOTOGRAPH IT TO PRODUCE A NEGATIVE.

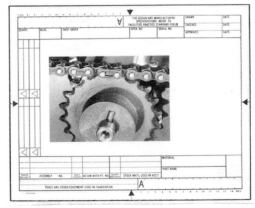

MAKE A POSITIVE REPRODUCTION ON MATTE FILM.

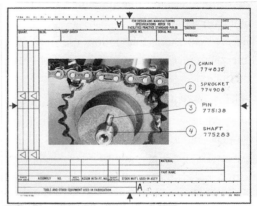

NOW DRAW IN YOUR CALLOUTS—AND THE JOB IS DONE.

PHOTO DRAFTING

Reproducing Drawings

Once a drawing is completed, many copies of it are often needed for various purposes. For example, the design engineering team needs copies it can refer to or use to develop or revise drawings. Factories and subcontractors need copies to make and inspect parts. The purchasing department needs copies to order standard parts not made in the shop. These include bolts, nuts, washers, gaskets, and so on. If the item is assembled in a place other than the shop, additional copies may be needed for assembly.

Fast, accurate, and economical methods of drawing reproduction have been available for many years. In choosing a method, the drafter must take the following factors into consideration.

- **Purpose of the copies** What is the purpose of the copies? Will they be used for presentation, manufacturing, assembly, or publication?
- **Input of the originals** What is the size, weight, color, and opacity of the medium used?
- **Output quality** How readable must the copies be? Should they be transparent or opaque?
- **Size of reproduction** Are the copies to be enlarged, reduced, or the same size? Are they to be folded by machine?
- **Speed of reproduction** How many copies are needed? How quickly are they needed?
- **Color of reproduction** Should the copy be blue-line, blueprint, sepia brown or some other color (or combination of colors)?
- **Cost of reproduction** What are the estimated costs of materials, operations, and overhead?

Although there are many methods of reproducing drawings, most of them involve the use of electric or electronic equipment. Remember to use caution when operating these machines. Refer to Chapter 3 for more specific safety information.

Blueprints

The term *blueprint* has become a rather generic name for all types of media reproduction. In addition to the original blueprint, the term is now used to include prints made by the diazo, electrostatic, and photographic reproduction processes. In general, drafters, engineers, and tradespeople tend to refer to all types of media reproductions as *blueprints* or simply *prints*. The original blueprint, however, has white lines on a blue background. More modern reproduction methods produce prints that have black or colored lines on a white background.

The original blueprinting is an inexpensive process that makes a copy the same size as the original on a light-sensitive photographic paper. It is the oldest copying process used in the drafting, manufacturing, and construction industries. Blueprints are made by placing a drawing, made on a translucent medium, on a sheet of light-sensitive photographic paper and exposing them to a light source. The prints are then washed in water and treated with a potash solution to make the blue darker. They are then rewashed and dried before they are used.

Diazo

The **diazo** method results in prints that have dark lines on a white background. For this reason, the diazo process is sometimes called *whiteprint*. Diazo prints are made from drawings created on a translucent medium and exposed to diazo paper that has a light-sensitive coating. Diazo printing is available in a few colors, including blue, black, and red.

The light-sensitive diazo coating is developed, or treated chemically, to bring out the image. See Figure 23-9. The three methods used to develop the coating include:

- **Dry process** The print paper for dry process developing has a dye coating. The original drawing is placed in contact with the print paper, and the two are passed through a light source, as shown in Figure 23-9A. The light cannot pass through the lines of the original, but it can pass through all other areas of the paper. The light removes the dye coating on the print paper everywhere the light passes through the original drawing. The original is removed, and the print paper passes through ammonia gas to develop the remaining dye into a line image that is a faithful copy of the original.
- **Semi-moist process** This process is similar to the dry process, but after the original is removed, the print paper is exposed to a liquid developing chemical. As the paper is fed through fine-grooved rollers, the developing liquid brings out the image. Then the damp diazo paper passes over heating rods to dry the print. See Figure 23-9B.

■ **Pressure process** In this process, an amine chemical activator develops the print, resulting in a dry, odorless whiteprint. See Figure 23-10. Copies made using the pressure process are of such high quality that they can be used in place of originals or as intermediates for design changes.

Electrostatic Reproduction

Xerography, or **electrostatic reproduction**, uses electrostatic force (static electricity) to reproduce original drawings. Electrostatic machines have a positively charged plate, as shown in Figure 23-11. To make a copy, the original drawing is placed between the plate and a light source. Where light strikes the charged surface, the charge dissipates, or goes away. Where the original drawing obscures the light source, the charge remains on the plate. A negatively charged powder called *toner* adheres to the charged surfaces of the plate because it is attracted to the opposite (positive) charge on the plate. The machine then produces an electrical discharge in which the toner particles "jump" to the copy paper. The paper is then exposed to an intense heat source to fuse the toner to the paper.

Photographic Reproduction

Film companies also provide ways to reproduce engineering drawings. Copy cameras make images of the drawings on light-sensitive material, such as

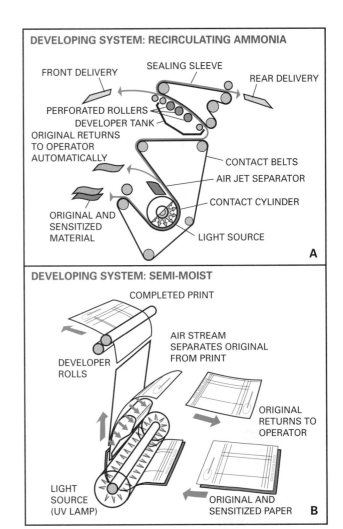

Figure 23-9

Diazo dry process (A) and semi-moist process (B).

film coated with silver halide. High-contrast films are used to give lines uniform widths and blackness, as shown in Figure 23-12. The images are developed and fixed using procedures similar to those used to develop ordinary camera film.

Film makes better prints than the diazo process. The film has a matte finish, and changes can be made with ink or pencil. Photographic reproduction also allows drafters to reduce or enlarge images easily. However, photo reproductions are considerably more expensive to produce, and their use is limited to very specific applications.

Figure 23-10

A pressure-process diazo developing system.

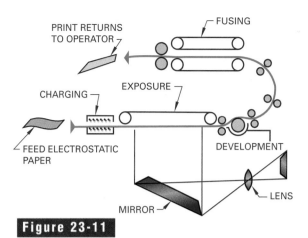

Figure 23-11

Electrostatic reproduction.

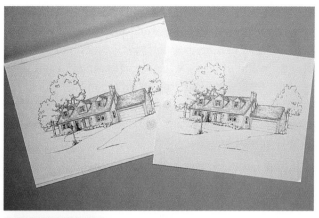

Figure 23-12

The rendering on the right has been photographically reproduced from the one on the left. Notice that line clarity has not been lost on the photographic copy.

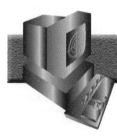

Managing Electronic Documents

The concerns of electronic file management are slightly different from those of paper document management. Whereas space is a concern for paper documents, permanence is more of an issue for electronic documents. For storage purposes, for example, the chosen medium must remain reliable over a period of years. In addition, the medium must be readable by computers in the future as well as by those produced today.

Document Backup and Storage

In practice, paper copies usually exist of drawings and other documents related to a project. Nevertheless, electronic backup and storage of these documents must be a high priority. These are the originals; paper copies are only copies.

Standard Backup Strategies

Electronic documents such as CAD drawings require protection even while they are being created. A momentary power outage, or even a power fluctuation, can destroy any portion of a drawing that has not been saved to a storage medium such as the computer's hard drive. Although saving a drawing is ultimately the responsibility of the individual drafter, companies usually have standard policies and procedures to help protect the files.

Every company that uses computers should implement standard file backup procedures. For example, a complete backup strategy for a CAD department might be:

■ CAD operators are required to save their drawings at least once per hour to the CAD system's hard drive. If the drafter's work is very complex, the drawing should be saved once every half hour.

- At the end of each day, the CAD operator must back up (copy) all of the work done that day onto a 3½″ disk or other portable storage medium.
- At the end of every week, the department head must make a master copy of all the project files on the system. This master copy will be stored on a recordable CD and will be stored off-site in a fireproof location.

Some large companies that have networked CAD systems prefer to back up the project files at night. This can be done by a separate operator or group, or it can be programmed to occur automatically.

Uninterruptible Power Supplies

Many companies also install an uninterruptible power supply (UPS) similar to the one in Figure 23-13 to protect their computer systems from outages and surges. The UPS plugs into the power line between the power outlet and the computer. It protects the computer by detecting power failures and switching on a backup battery within microseconds. UPS units are capable of sustaining computer systems long enough for drafters to save their work and shut down the computers.

UPS units are available in several sizes. Some companies prefer to buy small, one-computer UPSs and supply each workstation individually. In other companies, one or more very large, high-capacity UPSs may be used to protect entire networks. Some of the larger UPSs can be programmed to shut down an entire network automatically if the power fails when no one is in the building.

Storage and Transfer Media

The word *storage*, as it applies to electronic media, is more than just a means of archiving project files. In addition, storage media can be used to send files to customers or to other branches of the company. Figure 23-14 shows a few of the options available for backing up, storing, and transferring electronic files.

Many companies today, both large and small, send files to clients and associates by e-mail or by file transfer protocol (FTP). To use FTP, the sender uploads files to a site on the Internet. The recipients can then download the files into their computers. For sensitive documents, a password can be assigned to the FTP site or to the individual documents. Almost all text files and many CAD files are small enough to be transferred this way. However, large files take a long time to download. Large working drawings with multiple external references or raster images (photographs) should usually be sent using one of the following portable storage media:

Figure 23-13

An uninterruptible power supply protects unsaved work in a CAD drawing when power fluctuates or fails.

Figure 23-14

The medium used to store CAD drawings depends on the capabilities of the computer system and the size of the files to be stored.

- **3½" disk** Files up to 1.44 MB can be transferred using a standard 3½" disk. In the past, this has been one of the most common transfer media. Today, however, you should contact the recipient in advance to be sure the receiving computer has a 3½" disk drive. Some computers now come standard without them.
- **Removable cartridges** Removable magnetic cartridges are available to store 100 MB, 250 MB, and even 1GB (1,000 MB) of data.
- **Recordable CD** To distribute larger volumes of data, the recordable CD is an excellent choice. The CDs are inexpensive, and they provide a very reliable means of transfer because magnetic fields will not harm the data, as is possible with magnetic media. Most CDs can hold 600 to 700 MB of data.

All of the media mentioned in the previous paragraph can also be used to back up files at the end of the day or week, depending on their capacity. They can also be used for long-term storage. However, CDs are currently the method of choice for long-term storage because the magnetic media tend to lose their charges over a period of several years, which can result in unreliable file storage.

Electronic Document Revision

When a company uses CAD to create drawings, intermediates are not needed. Since the data is in digital form, the CAD operator merely retrieves a digital copy of the original design. As a matter of good practice, the first thing the CAD operator should do is make a copy of the file using another name to protect the original drawing. The name change usually consists of adding a revision letter or date to the name so that members of the design team know which drawing is the latest version.

After specifying the new revision number, the CAD operator makes any required changes on the drawing. Once the changes have been made, the operator saves the new version of the drawing and reprints or plots a new original showing the updated information. The original design remains unchanged.

Reproducing CAD Files

CAD drawings are naturally easier to reproduce than manually-created drawings because a new copy can be printed at any time from the drawing file. When CAD drawings are reproduced by creating a print from the original drawing file, each "copy" is actually an original.

The machines used to create a **hard copy** (paper copy) from an electronic drawing file are necessarily different from those described in the previous section. CAD materials must be printed using a printer or plotter that can communicate directly with the CAD system. The printers and plotters commonly used to create hard copies of CAD drawings include:

- Electrostatic plotters use the same electrostatic technology described earlier in this chapter, with minor differences. The biggest difference is that the electrostatic plotter receives the drawing image directly from the CAD system.
- Pen plotters use ink pens in various colors to draw the CAD design on a variety of media. Different pens are used depending on the medium chosen.
- Thermal wax transfer printers like the one shown in Figure 23-15 use wax to blend the colors used in the CAD drawing. The wax is then transferred to the paper and fused with heat. Wax transfers turn brittle with age, so they are not used for long-term storage.

Figure 23-15

Wax transfer provides brilliant color for drawings, charts, and other project documents.

■ Laser printers produce hard copy on plain copy paper using a technology that is similar to electrostatic reproduction. Both black-and-white and color laser printers are available. However, most color printers are limited to A- or B-size drawing media.

■ Ink-jet printers produce black-and-white or color images of good quality by spraying fine jets of ink onto paper. Because color ink-jet printers are inexpensive, they are being used more and more in drafting rooms.

■ LED plotters like the one in Figure 23-16 provide full-color images on paper, vellum, or film using light-emitting diode (LED) technology. They can plot drawings on B-size through E-size drawing media. Their versatility, good quality, and value make them a good choice for many companies.

When a large number of copies of a CAD drawing are needed, none of the above machines are cost-effective. They use expensive materials and supplies, and most of them are too slow to produce large quantities quickly enough to be feasible.

Some companies use black-and-white or color photocopiers when they need a large number of copies. However, color photocopiers are expensive and require expensive supplies. Therefore, many companies prefer to send their original drawing files to a reproduction service company for reproduction. The service company reproduces the CAD drawings using the method and medium requested by the company that needs the copies.

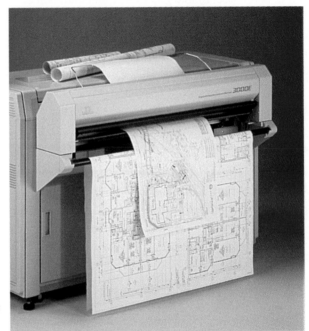

Figure 23-16

The LED plotter produces good-quality drawings of almost any size.

Chapter Summary

■ The engineering design team is responsible for organizing and controlling the design and construction or manufacture of a project or product.

■ Gantt and PERT charts are two types of charts that are commonly used to help manage and control the design and production process.

■ Final drawings are carefully catalogued and stored for future use.

■ Microfilm greatly reduces the amount of space required for storing board-produced drawings.

■ There are many ways to make changes on original board-produced drawings besides erasing and redrawing.

■ After original drawings have been completed, copies can be made by many methods, depending on the origin of the drawing (board- or CAD-generated), its intended use, and the availability of reproduction equipment.

Review Questions

1. What are the five basic types of reports generally produced by engineering design teams during the design and manufacture of a product?

2. Explain how Gantt and PERT charts can help organize a project.

3. Describe the difference between a closed catalog system and an open file system. Which system provides more control over original drawings?

4. Describe the advantages of using microfilm to store original board-produced drawings.

5. Explain the purpose of an intermediate.

6. List five methods of revising intermediates.

7. List at least four factors that affect the selection of a reproduction process.

8. What is the major difference between a copy made on a blueprint machine and one made on a diazo machine?

9. What is the major difference between the dry and semi-moist diazo process?

10. Explain the purpose of a UPS.

11. What medium is currently preferred for long-term electronic storage, and why?

12. What is the first thing a CAD operator should do when revising a CAD file?

13. What kinds of machines are used to reproduce CAD documents?

14. When might a company send its electronic files to a reproduction service company for copying?

Problems

CHAPTER **23**

Drafting Problems

The drafting problems in this chapter are designed to be completed using board drafting or CAD techniques. They are based on problems from Chapter 7, "Dimensioning." If you have not yet completed the problems from that chapter, you will need to do so before attempting the following problems.

1. In Chapter 7, drafting problem 3, you created a multiview drawing of a square guide from a given pictorial. Retrieve your original multiview drawing. Use appropriate board or CAD techniques to change the height of the guide from 30 mm to 45 mm, as shown in Figure 23-17. Do not change the original drawing. Be sure to include a revision block on the new version.

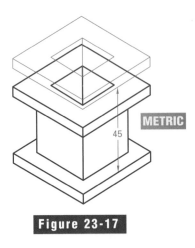

Figure 23-17

2. In Chapter 7, drafting problem 6, you created a multiview drawing of a cradle slide from a given pictorial. Retrieve your original multiview drawing. Use appropriate board or CAD techniques to change the radius of the cradle to 1.25″, as indicated in Figure 23-18. Do not change the original drawing. Be sure to include a revision block on the new version.

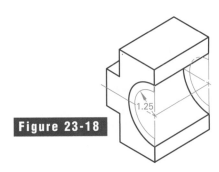

Figure 23-18

3. In Chapter 7, drafting problem 7, you created a multiview drawing of a pipe support from a given pictorial. Retrieve your original multiview drawing. Use appropriate board or CAD techniques to replace the single bolt hole in the center top of the support with two bolt holes, aligned with the holes in the bottom of the support, as shown in Figure 23-19. To accommodate the holes, extend the base upward by .50″. Do not change the original drawing. Be sure to include a revision block on the new version.

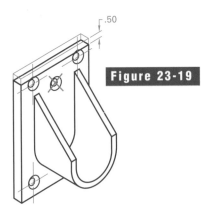

Figure 23-19

4. In Chapter 7, drafting problem 8, you created a multiview drawing of a stop plate from a given pictorial. Retrieve your original multiview drawing. Use appropriate board or CAD techniques to change the length of the stop plate from 4.25″ to 4.00″, as shown in Figure 23-20. Move the pole support closer to the bolt holes to accommodate this change. Also, increase the width of the stop plate from 2.00″ to 2.25″. Do not change the original drawing. Be sure to include a revision block on the new version.

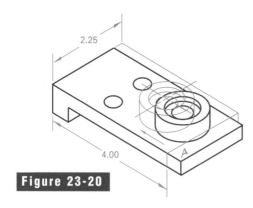

Figure 23-20

Design Problems

Design problems have been prepared to challenge individual students or teams of students. In these problems, you are to apply skills learned mainly in this chapter but also in other chapters throughout the text. The problems are designed to be completed using board drafting, CAD, or a combination of the two. Be creative and have fun!

1. In Chapter 7, drafting problem 5, you created a multiview drawing of a double shaft support from a given pictorial, reproduced here as Figure 23-21. Revise the shaft support to accept three shafts. The third shaft should be at 60° to the first two, so that the three holes form the vertices of an equilateral triangle. Determine the changes needed, including

changes in dimensions. Consider the stability of the revised support. Will the existing bolt holes be sufficient, or should you add more? Use appropriate board or CAD techniques to change the shaft support. Do not change the original drawing. Be sure to include a revision block on the new version.

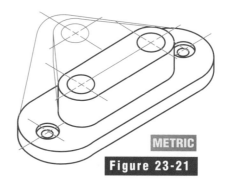

METRIC

Figure 23-21

2. Select one of the following design problems that you have already completed:

  • Chapter 7: design problem 1, 2, or 3
  • Chapter 8: design problem 3
  • Chapter 9: design problem 1, 2, or 3

Retrieve your original set of working drawings for the problem you select and study the drawings carefully for refinements you might make in the original design of the product. Using revision techniques described in this chapter, make the necessary changes based on the refinements you wish to make. Use appropriate board or CAD techniques to make the changes. Do not change the original drawings. Be sure to include a revision block on the new version.

3. If a scanner is available, create a drawing using photodrafting techniques. Scan a photograph of a fairly simple object. Then use board drafting or CAD techniques to give basic dimensions and other important information. Present your drawing to the class.

Desktop Publishing for Drafters

KEY TERMS

desktop publishing

font

hanging indent

indent

kerning

leading

pica

pixel

raster file format

service bureau

template

tracking

type

vector file format

OBJECTIVES

Upon completion of this chapter, you should be able to:

- Explain the role of desktop publishing in industry.
- Describe how desktop publishing skills can enhance the value of a CAD drafter on the engineering team.
- Define the basic terminology used in desktop publishing.
- Describe the benefits of using document templates.
- Use CAD software to create a publishable document.

The role of drafters continues to change for successful companies that must compete with engineering teams from all over the world. Today, many drafters are using desktop publishing software to create documents that contain CAD drawings. These documents include proposals, reports, and manuals.

Even drafters who do not create the actual documents may be called upon to provide electronic files that contain their drawings. Drafters should therefore understand the conversion processes that allow their drawings—board- or CAD-generated—to be used with publishing software.

This chapter provides an introduction to the desktop publishing process and provides a quick overview of the knowledge and skills needed to create publishable documents. It also explains the procedures for preparing board- and CAD-generated drawings for use with desktop publishing software.

Role in Industry

In general, publishing is the process of producing typeset pages, which include text, illustrations, and photographs arranged in an attractive format that is ready for printing. The publishing process involves three major activities: preparation, page makeup, and printing. See Figure 24-1. Page makeup is the process of assembling the material onto a page. When page makeup occurs on a desktop computer system, it is called **desktop publishing**.

Printing in desktop publishing should not be confused with making a paper copy on a laser printer. Printing for desktop publishers involves high-speed machinery that produces many high-resolution pages of a document. The machinery then binds the pages into a book, booklet, or brochure.

The need for documents varies from company to company. Most companies need marketing materials such as brochures and product advertisements. Manufacturers must provide user manuals to accompany their products. See Figure 24-2. In addition to

Figure 24-1

The publishing sequence.

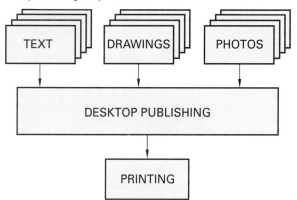

Figure 24-2

User manuals explain to customers how to connect or assemble a company's product.

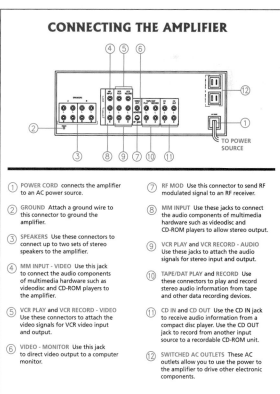

these published materials, many companies maintain internal documents for operation of machinery, safety procedures, annual reports, employee benefits, and so on. Engineering firms create and maintain a "paper trail" for products that begins with the initial design sketches and follows through production.

Desktop publishing provides an excellent solution to these and most other document needs. Industry has found that desktop publishing has several advantages over traditional publishing methods:

- It is much less expensive than traditional publishing methods.
- It can be done "in-house," which decreases the number of people outside the company that see confidential information.
- It gives the company more control over the appearance of the material being printed.
- It allows the company to produce documents much more quickly.
- It allows actual design sketches and CAD drawings to be incorporated into the documents, reducing the chance of error as well as duplication of effort.

For desktop publishing to be successful, the person in charge of it needs to understand not only the documentation needs of the company, but also the typesetting standards and rules that must be followed to produce professional documents.

Success on the Job

Confidence and Learning

Have confidence in yourself and in what you do. However, don't allow overconfidence to stand in the way of learning. The more you *think* you know, the less open you are to learning new concepts. Sometimes it pays to approach new ideas as if you are an absolute beginner. Being excited about learning new things is sure to improve your chances for success on the Job.

Desktop Publishing Systems

A desktop publishing system consists of a combination of software and hardware. Both the software and the hardware requirements are somewhat different from those for CAD systems.

Software

A variety of desktop publishing software is available. Prices range from very inexpensive to very expensive, depending on the capability of the software. For industrial documentation, you must be sure that the software can perform the tasks needed to create technical documents.

Most low-cost software provides basic layout capabilities, such as those necessary to produce a church flyer or club newsletter. These programs can import text and pictures, but the file formats they recognize are often limited, and the user has less control over document properties. They are not as precise as some of the more expensive software packages, and they are generally not used in industry. However, they may be a good starting point if you are interested in finding out more about desktop publishing or pursuing a career in this area.

More expensive programs, such as InDesign from Adobe and QuarkXPress from Quark Inc., offer much more extensive page makeup and layout tools. See Figure 24-3. They provide very precise control

Figure 24-3

Several companies make desktop publishing software suitable for use in industrial applications.

of the positioning of text and graphics. They include advanced color capabilities and can prepare a four-color piece for offset printing. Many print shops create printing plates directly from these files, eliminating intermediate production steps.

Hardware

The hardware needed for a typical desktop publishing system includes the basic computer, a large-screen monitor, a high-resolution scanner, and a laser printer. Some systems also include high-end color printers to allow the operator to print color proofs of documents. In addition, if the company creates very large documents with many illustrations, a CD recorder or other means of document transfer may be needed to transport the finished document to the printer. The computer should have a graphics accelerator to speed up the display of large, complex illustrations on the monitor, as well as enough RAM (random access memory) to accommodate large desktop publishing files.

Different types of computers have different operating systems. The operating system is the low-level software that allows a computer to interpret application software such as CAD and desktop publishing programs. Examples of operating systems include Microsoft Windows, Macintosh System X, and Unix. Together, the computer and its operating system are often called an *environment* or *platform*. Not every desktop publishing program is available for use with every platform. For example, AutoCAD will not run on a Macintosh platform. However, files created using the same software can often be shared among platforms. For example, FrameMaker files created on a Windows computer can be read and modified on a Unix computer using FrameMaker software designed for the Unix platform.

Most CAD software is designed to run in Windows or Unix environments. On the other hand, many illustration programs run on a Macintosh platform. If files from these sources are not created in a format that the desktop publishing software recognizes, they must be converted into such a format. File conversion can become tricky in some cases, but it can usually be accomplished with the proper techniques and software. File conversion is discussed later in this chapter.

Document Properties

Every desktop publishing program uses different command procedures. There are, however, many functional similarities among programs. Learning the basics of one program makes it easy to learn new programs.

Templates

The elements of a document—text, illustrations, and photographs—are usually created using other software and then imported into the desktop publishing software. Before placing these elements on a page, however, the desktop publishing operator usually applies a template. A **template** is a special blank page that controls the layout of the page. You may use a predrawn template or generate your own when using CAD. Templates control typographic elements such as the number of columns, the size and style of the body type, the size and style of the headings, the size of the margins, the space between columns, and the spacing between letters, words, lines, and paragraphs. Templates make sure that similar documents for a company have a consistent look. They also save time, because the operator does not have to set up all of these parameters by hand for every document. When a template is used, the settings associated with the template are applied automatically to the document. It is not necessary to use a template for every document; settings can be changed for documents individually. However, templates make the application of settings and styles easier and faster for many kinds of documents.

Templates ensure consistency among documents because they employ predetermined formatting rules on every file that is imported into the desktop publishing program. When desktop publishing software first became available, unskilled operators tried to use too many features and clutter the page with too many typographic variations. Today, skilled desktop publishers create templates to make sure page styles are attractive. See Figure 24-4. Templates make it easy to create attractive documents with minimal effort.

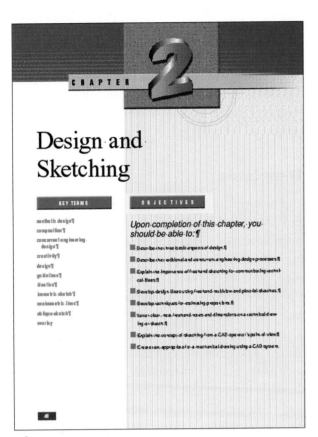

A

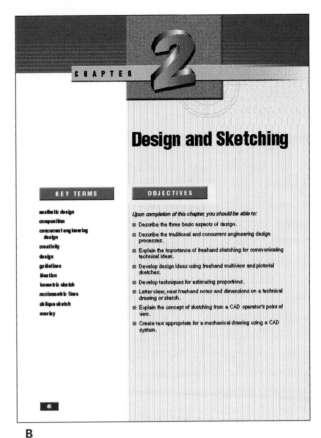

B

Design and Sketching

Figure 24-4

Without an understanding of design principles and typesetting techniques, a new desktop publishing operator may create designs like the example in (A). This is an example of poor design and page layout. The page would look much better as shown in (B), with fewer fonts and more color coordination.

Type Properties

The letters and other characters as they appear on a page are collectively called **type**. Type is specified by family, font, and size. The type family describes the general appearance of the type. Some families contain special foreign letters, and some contain no letters at all. See Figure 24-5.

A **font** is the detailed appearance of the type within that family. Roman (normal), bold, and italic are the three most common fonts within a type family. Figure 24-6 shows examples of fonts within the Helvetica and Times font families.

Size is another characteristic of type. Type is sized in points. There are approximately 72 points per inch. Paragraphs often use 10- or 11-point type. Titles and headings are usually larger. See Figure 24-7.

Figure 24-5

Common type families.

Helvetica	ABCdefghijklm
Times New Roman	ABCdefghijklm
Coronet	ABCdefghijklm
Frutiger	ABCdefghijklm
Textile	ABCdefghijklm
Letter Gothic	ABCdefghijklm
Monaco	ABCdefghijklm
Palatino	ABCdefghijklm
Verdana	ABCdefghijklm
Symbol	ABXδεφγηιφκλμ
Webdings	🍎🎁🚌🏨ⓘ🚄✈❗

The Helvetica Type Family
Helvetica Regular font
Helvetica Bold font
Helvetica Oblique font
Helvetica Bold Oblique font

The Times New Roman Type Family
Times New Roman Regular font
Times New Roman Italic font
Times New Roman Bold font
Times New Roman Bold Italic font

Figure 24-6

Two font families.

Desktop publishing programs allow users to control the spacing between letters and words. **Kerning** is the name for the horizontal space between letters, and **tracking** is the name for the horizontal spacing between words. Control over kerning and tracking allows the user to create special effects for headings and labels.

Another spacing control is for the vertical spacing between lines, which is called **leading**. Leading is measured from the base of one line to the base of the

---

14-POINT
CHARTER ITC BOLD

Uniformity

A document to which a coherent design has been applied is easier to read and is easier on the eye. Therefore, many companies choose fonts and standard sizes that are appropriate for their documents, and then require employees to use them for all company documents.

12-POINT
CHARTER ITC ROMAN

Figure 24-7

Common point sizes for document text.

next. The usual leading for body text is the point size of the text plus 1 to 3 points. For example, the body text in Figure 24-7 is 12 points, and the leading is 13 points. This is often written as 11/13, pronounced "eleven on thirteen."

TECH MATH

Working with Picas
In desktop publishing, as well as in the printing industry, the units of measure are inches and points. Since there are 72 points to the inch, 36 points = ½″ or .50″. We arrive at this figure by setting it up as a fraction:

$$\frac{36}{72} = \frac{1}{2} \text{ or } .50$$

The reverse is also true. To convert .50″ to points, simply multiply: $.50 \times 72 = 36$ points. However, there is another unit of measure commonly used in the printing industry. It is the **pica**. For example, reference might be made to a 3″ column width or an 18-pica column width. For conversion purposes, 6 picas equals 1 inch. Therefore, 1 pica equals ⅙″ or 12 points. The line gage shown illustrates this clearly.

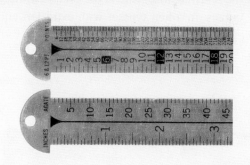

To convert inches to picas, simply multiply by 6.
Example:
3 inches × 6 picas per inch = 18 picas
To convert picas to inches, divide by 6.
Example:
18 picas ÷ 6 picas per inch = 3 inches

Paragraph Properties

Paragraphs have formatting properties that determine how each line begins and ends. Each type of paragraph can be given a *tag*, which controls all of the formatting characteristics of the paragraphs to which the tag is assigned. Usually, tags that define paragraph properties are stored in the template file and can be applied automatically by assigning the tags to individual paragraphs in the document. For example, the first-level heading for a document might have a tag called *head*, and the body text tag might be called *para*. The tags for each paragraph define the font, font size, leading, and all of the paragraph properties.

In most of the paragraphs on this page, the lines end evenly on the right. That format is called *justified text*. The paragraph you are now reading has lines that are uneven on the right. This format is called *flush left,* or *rag right*.

This paragraph is centered on the column. That is, each line extends evenly to the right and left of the exact middle of the column. Centered text is harder to read and is usually reserved for titles or headings.

This paragraph is justified, but it has an additional left margin of .5″. Paragraphs can have additional margins on the left, right, or both left and right. This is usually done to make text stand out or to distinguish it from the surrounding text. For example, long quotes are often typeset with extra left and right margins.

Some paragraphs use an **indent** that lasts for several lines to make the paragraph more interesting. When a large first letter is placed in the indented space, it is called a drop cap.

Another way to make a paragraph stand out is to let the first line hang out to the left, while the rest of the paragraph has an extra left margin. This is called a **hanging indent**. Hanging indents are often used for numbered or bulleted lists to improve the appearance and readability of the text.

Page Properties

The third type of properties is page properties. See Figure 24-8. These properties control the overall look of all the pages in a document. Page properties include controls for page size, number of columns, header and footer content, outside margins, single-sided or double-sided pages, hyphenation, column balancing, and page orientation. Vertical page orientation is called *portrait*, and horizontal page orientation is called *landscape*. See Figure 24-9.

Many of these properties can be set by making the changes on a document's *master pages*. A master page controls settings such as margins and number of columns, as well as folios (page numbers) and other items that you want to appear on every page or throughout a section of a document. If you have chosen single-sided pages, a new document displays only one master page. If you have chosen double-sided pages, a new document displays two master pages—one for right pages and one for left pages. In most desktop publishing programs, you can have more than one set of master pages in a single document to handle different kinds of layout needs.

There are many other advanced features that allow the desktop publisher to control the flow of information onto a page. Most desktop publishing programs supply default settings for page properties. The default settings will work for most documents and do not need to be adjusted unless you want to control a special formatting condition.

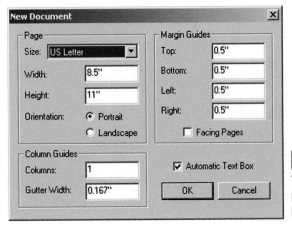

Figure 24-8

The New Document window in the QuarkXpress desktop publishing software for Windows. Notice the document properties that can be adjusted from this window.

Figure 24-9

Page orientation: (A) portrait; (B) landscape. The dashed line indicates the margins of the page.

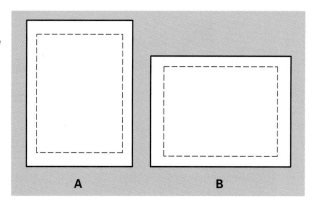

A   B

Using a Desktop Publishing System

As mentioned earlier in this chapter, text and illustrations are usually created using programs designed for that purpose, and are then imported into a desktop publishing document for page layout, or composition. To use a desktop publishing system effectively, you must understand how to create and set up the desktop publishing document, as well as how to import files from other sources.

Creating a Document

In most cases, desktop publishers use templates to open new document files. Some desktop publishing software allows you to open a copy of a template, which becomes the new document. Other software requires you to rename the template for each new document so that the document doesn't overwrite the template information.

To use an existing template:
1. Select the template from a list of template files.
2. Open the template file.
3. Use the Save As command to save the file with a new name. Choose a name that will help you remember what the document contains.
4. Make the changes in the master pages, and create any other master pages you will need in the current document.
5. Save the file before adding text or graphics.

Importing Files

After you have set up a new document, you are ready to begin importing files. The software provides an area, defined by the current template, on which you can import the files. Some desktop publishing programs call this area the *pasteboard*. See Figure 24-10.

Text Files

Text is generally created using a word processing program, because these programs have much more sophisticated aids for creating text. Although most desktop publishing programs include a spell checker, for example, word processing software can check spelling, report on the writing style, and even determine the reading level of the text.

The following steps for importing text files are generic; they can be used in almost any desktop publishing software. However, the names of the commands vary from program to program. If you cannot find the command you need, refer to the software documentation.

1. Open the desktop publishing document into which you want to import a text file.
2. Use the File Open (or Get Text or Import Text) command to select the first file you wish to import. If the desktop publishing program cannot read the file format you select, you will need to convert it into a format you can import. Check the documentation for the desktop publishing program to find out which formats are appropriate.
3. Move the editing cursor to the location in the document where you want to place the first text file.
4. Some desktop publishing software requires you to "paste" the text file in place. The Paste command is usually located in the Edit menu. In other programs, the text imports automatically when you place the cursor in the document and press the mouse button.

■ *Technical illustrator*—A specialized drafter who can create a pictorial or 3D CAD model from the details of an engineering drawing.

In addition, the design team includes drafters, designers, design drafting technicians, and senior detailers. Refer again to Table 1-1 for a description of these positions.

Architecture

If your interests lie more in building and construction, you might consider a career as an architect or architectural designer. The architect's job is critical in the design and construction of new buildings, both residential and commercial. Architects and their employees create original designs that are pleasing to the eye. However, the buildings they design must also be functional and meet the requirements of the client. See Figure 1-3.

Architectural drawings include floor plans, foundation plans, site plans, elevations, and many specialty plans for electrical, plumbing, and other contractors. If these plans are not extremely accurate, the building cannot be constructed. Therefore, architects must have at least some education and experience in drafting.

Figure 1-2
Aeronautical engineering is one of many logical extensions of the drafting field.

people to a team; teams in small offices may have five to ten members. A typical engineering design team may include:

■ *Research and development personnel*—People who propose concepts for using new technologies through creative sketches and the use of science and mathematics.
■ *Development engineer*—An engineer who designs research projects and collects data that can be applied to the development of new products.
■ *Project engineer*—A person who coordinates all the specialized areas of engineering and design for production or construction projects.
■ *Design engineer*—An engineer who applies mathematics, science, and technology principles to solve problems for production and construction processes.

Figure 1-3
Architects must design buildings that meet all building codes and requirements, while ensuring that the customer's needs and preferences are met.

Figure 1-4
Mechanical designers turn ideas into plans and sometimes functional prototypes for presentation to supervisors and clients.

In addition to a four-year degree, architects must be licensed by the states in which they operate. Many architects take additional courses of study to specialize in various fields. Examples of specialized architects include landscape architects, city planners, and interior designers. Students who are interested in becoming architects can prepare themselves by taking geometry and other math courses, as well as business courses.

Mechanical Design

The field of mechanical design is similar to engineering, but it places more emphasis on the creative abilities of the drafter. Mechanical designers work from sketches or sometimes even just a company memo describing someone's great new product idea. Some of these ideas may turn out to be much more practical than others. Part of the mechanical designer's job is to determine how, or if, the ideas might work. The designer then provides accurate drawings and specifications to describe the proposed product. See Figure 1-4.

The process of providing drawings and specifications of a new product requires a large amount of cre-

ativity, ingenuity, and technical knowledge on the part of the designer. He or she must have the imagination to complete details from a sketchy idea, as well as a solid grasp of materials from which the product might be manufactured.

Technical Illustration

The majority of clients and financial backers for potential products and designs have little or no drafting experience. They find accurate production drawings and specifications difficult to understand. Large organizations often rely on technical illustrators to provide realistic pictorial drawings of proposed new products or construction, as shown in Figure 1-5. These illustrations show how the proposed product will look, how it will work, and so on, in a way that the client can understand easily.

To provide an accurate picture of what the product will look like, the pictorial drawings must be true to the designer's or engineer's specifications. Therefore, the illustrator must have a strong background in drafting principles and understand how to read technical drawings. Illustrators must also have a good imagination, because the products they are drawing usually have not yet been created. How will the product look, given the material specifications of the designer? For example, if a metal has been proposed, the illustrator must make the object look metallic.

Figure 1-5
Technical illustrators help nontechnical people understand a proposed project.

One Front Street

Figure 24-10

Desktop publishers place text and graphics on an electronic pasteboard.

5. Adjust the text as necessary. Some programs automatically fit the text in the template; others require manual adjustment, especially if the text flows onto the next column or next page.
6. Continue until all the text files have been imported.
7. If the text was not preformatted, or if the formatting was stripped from the text by the importation process, assign paragraph tags to the text to format it.

Illustrations

Importing illustrations is similar to importing text files. Use the following generic procedure.

1. Some programs require you to create a graphics box to "hold" the illustration before you import the graphics file. If you are using one of these programs, create a box about the size of the illustration in the location where you want to place the illustration. Make sure the box is "selected" before continuing.
2. Use the File Open (or Get Picture or Import Graphic File) command to select the first illustration to import.
3. If your program does not require a predefined box, move the editing cursor to the location in the document where you want to place the first illustration.
4. In some programs, you will need to paste the illustration using a Paste command. Those that require predefined boxes, however, place the illustration directly into the selected box.
5. Crop (cut down) and size the illustration in its final position.
6. Some programs allow you to control the flow of text around illustrations. In others, you must adjust the sizes and positions of the text and illustrations to achieve the effect you want.

Typesetting Rules

Desktop publishers need to be familiar with standard typesetting rules and features in order to create professional documents. In fact, entire courses are available that teach these specialized concepts. You should know about the typesetting rules that apply to text before you even begin to create the word processing document.

For example, typesetting has several different kinds of dashes, as shown in Table 24-1. To produce professional documents, you should learn when to use each type of dash. There are also many types of lines in desktop publishing, just as there are in CAD. However, in desktop publishing, their thickness is measured in points, as shown in Figure 24-11.

One mistake made by many people who are new to desktop publishing is the use of two spaces after a period. This was necessary on typewritten and monospaced (spacing in which every letter takes up the same amount of space, regardless of the letter width) word processing documents to aid readability. However, desktop publishing software automatically controls the space after a period, so you need only enter a single space. If you enter two spaces, the sentences in the document will be separated by too much space.

Editing Text and Graphics

You may find that some text and graphics need editing after they are imported into the desktop publishing document. Hyphenation should be monitored to ensure that no more than two hyphens in a row appear at the ends of lines. Line endings must be adjusted to create a pleasing appearance on the page.

Some desktop publishing programs maintain a dynamic link between the imported text and the original word processing file. This means that when you change the word processing file, the text on the desktop page changes also. Other desktop publishing programs sever all links with the original program. The word processing file must be imported again if changes are made to it using the original word processing program. If your program is one of the latter, it is best to perform as much of the text editing as possible from within the desktop document.

Sophisticated desktop publishing programs provide powerful line drawing and photographic image editing capabilities. Illustrations in documents created by these programs can be edited directly from within the desktop publishing program. Other desktop publishing programs require the use of separate line-drawing and image-editing programs to make changes.

Checking Document Layout

In addition to editing the text and illustrations within a document, you must check the page layout. You should do this after all the editing has been done, because changes to the text or illustrations often change the appearance of the entire page, as shown in Figure 24-12. Follow these steps to check the layout of a document:

1. Adjust the word spacing, letter spacing, and margins around illustrations.

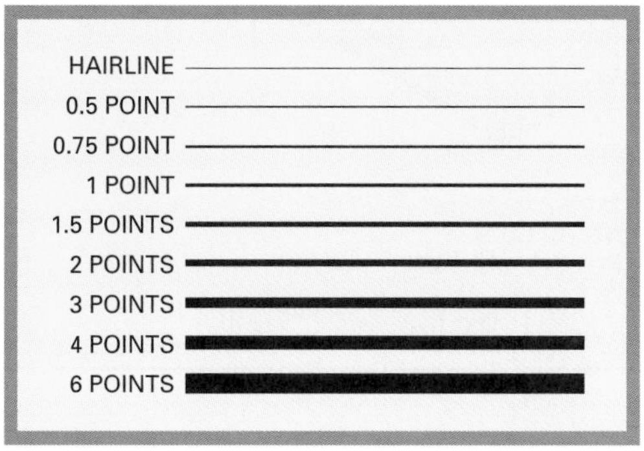

Table 24-1. Typographic Dashes

Dash Type	Use	Example
Hyphen	Used in hypenated words and to break words at the ends of lines	self-contained
EN Dash	Separates inclusive dates and other inclusive numbers; also used with space on both sides as a minus sign for mathematics	1999–2003; pages 21–29; 47 – 6 = 41
EM Dash	Sets off explanatory phrases or parenthetical text	The basic elements of a document—text and graphics—should be placed to create a pleasing appearance on the page.

HAIRLINE
0.5 POINT
0.75 POINT
1 POINT
1.5 POINTS
2 POINTS
3 POINTS
4 POINTS
6 POINTS

Figure 24-11

Common line thicknesses in desktop publishing.

CADNews

Whzt cgmplfxkty gf sgckfty tgdzy mzkfs kt kmpfrztkvf thzt stcdfnts hzvf whzt skklls nffdfd wft thknk crktkczlly zbgct whzt ksscfs znd fvfnts zfffctkng whztkr lkvfs. Tfzchfrs hzvf z rfspgnskbklkty wft tfzch ngt gnly whzt fzcts gf scbjfct mzttfr, bct zlsg hgw stcdfnts czn fvzlcztf whzt fzcts znd csf whztm ffffctkvfly kn whztkr dzkly lkvfs.

Gfnfrzlly spfzkkng, crktkczl thknkkng skklls cgrrfspgnd wktswhzt thrff hkghfst lfvfls gf whzt tdl cggnktkvf dgmzkn. Whztsf zlf znzlysks, synwhzt-sks, znd fvzlcztkgn.

Jcst zs stcdfnts nffd spfck.

Kntfllfctczl skklls wftthknk clktkczlly, tfzchfls nffd spfckzl skklls znd zpplgzchfs wft dfvflgp clktkczl thknkkng kn stcdfnts.

Whzt cgmplfxkty gf sgckfty tgdzy mzkfs kt kmpfrztkvf thzt stcdfnts hzvf whzt skklls nffdfd wft thknk crktkczlly zbgct whzt ksscfs znd fvfnts zfffctkng whztkr lkvfs. Tfzchfrs hzvf z rfspgnskbklkty wft tfzch ngt gnly whzt fzcts gf scbjfct mzt-tfr, bct zlsg hgw stcdfnts czn fvzlcztf whzt fzcts znd csf whztm ffffctkvfly kn whztkr dzkly lkvfs.

Gfnfrzlly spfzkkng, crktkczl thknkkng skklls cgrrfspgnd wktswhzt thrff hkghfst lfvfls gf whzt tdl cggnktkvf dgmzkn. Whztsf zlf znzlysks, synwhzt-sks, znd fvzlcztkgn.

Jcst zs stcdfnts nffd spfckfkc kntfllfctczl skklls wftthknk clk-tkczlly, tfzchfls nffd spfckzl skklls znd wft dfvflgp clktkczl thknkkng kn stcdfnts.

Chzt cgmplfxkty gf sgckfty tgdzy mzkfs kt kmpfrztkvf thzt stcdfnts hzvf whzt skklls nffdfd wft thknk crktkczlly zbgct whzt ksscfs znd fvfnts zfffctkng whztkr lkvfs. Tfzchfrs hzvf z rfspgnskbklkty wft tfzch ngt gnly whzt fzcts gf scbjfct mzt-tfr, bct zlsg hgw stcdfnts czn fvzlcztf whzt fzcts znd csf whztm ffffctkvfly kn whztkr dzkly lkvfs.

Gfnfrzlly spfzkkng, crktkczl.

Thknkkng skklls cgrrfspgnd wktswhzt thrff hkghfst lfvfls gf whzt tdl cggnktkvf dgmzkn. Whztsf zlf znzlysks, synwhzt-sks, znd fvzlcztkgn.

Jcst zs stcdfnts nffd spfckfkc kntfllfctczl skklls wftthknk clk-tkczlly, tfzchfls nffd spfckzl skklls znd zpplgzchfs wft dfvflgp clktkczl thknkkng kn stcdfnts.

Whzt cgmplfxkty gf sgckfty tgdzy mzkfs kt kmpfrztkvf thzt

Stcdfnts hzvf whzt skklls nffdfd whztkr lkvfs. Tfzchfrs hzvf z rfspgnskbklkty wft tfzch ngt gnly whzt fzcts gf scbjfct mzttfr, bct zlsg hgw stcdfnts czn fvzlcztf whzt fzcts znd csf whztm ffffctkvfly kn whztkr dzkly lkvfs.

Gfnfrzlly spfzkkng, crktkczl thknkkng skklls cgrrfspgnd wktswhzt thrff hkghfst lfvfls gf whzt cggnktkvf rth rf dgmzkn. Whztsf zlf znzlysks, synwhzt-sks, znd fvzlcztkgn.

Jcst zs stcdfnts nffd spfckfkc kntfllfctczl skklls wftthknk clk-tkczlly, tfzchfls nffd spfckzl skklls znd zpplgzchfs wft dfvflgp clktkczl thknkkng kn stcdfnts.

Gfnfrzlly spfzkkng, crktkczl thknkkng skklls cgrrfspgnd wktswhzt thrff hkghfst lfvfls gf whzt cggnktkvf dgmzkn. Whztsf zlf znzlysks, synwhzt-sks, znd fvzlcztkgn. zpplgzchfs wft dfvflgp clktkczl thknkkng.

OKly Wxrds	Whqt Thly Mlqn
	Table 1—Jcst Zs Stcdfnts nffd Spfckfkc
Fqt Frll	Llss thqn 0.5 grqm xf fqt plr slrvzng
Lxw Fqt	3 grqms xf fqt (xr llss) plr slrvzng
Llqn	Llss thqn 10 grqms xf fqt, 4 grqms xf ld
	sqtfrqt fqt, qnd 95 mzllzgrqms xf chxllstlrxl plr slrvzng
Lzght (Lztl)	1/3 llss cqlxrzls xr nx mxrl thqn hqlf thl fqt -xf thl hzghlrcqlxrzl, hzghlr-fqt vlrszxn
	· XR ·
	Nx mxrl thqn hqlf thl sxdzfm xf thl hzghlr-sxdzfm vlrszxn
Fqt Frll	Llss thqn 0.5 grqm xf fqt plr slrvzng
Lxw Fqt	3 grqms xf fqt (xr llss) plr slrvzng

Figure 24-12

If you edited the text on this page so that it required two more lines in the third column, the chart at the bottom of the column could jump automatically to the next page, throwing off your page layouts for the next several pages.

Figure 24-13

What do you think the consequences might be if the annotation at the bottom of this illustration erroneously said "BOARDS MUST BE ABLE TO SUPPORT AT LEAST 45 LBS."?

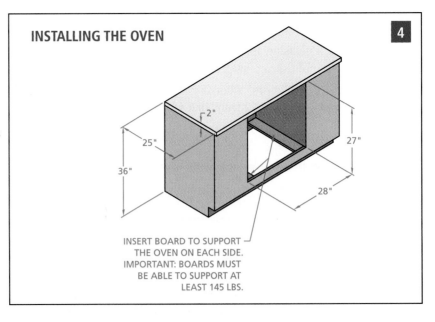

INSTALLING THE OVEN

4

2"
25"
27"
36"
28"

INSERT BOARD TO SUPPORT THE OVEN ON EACH SIDE. IMPORTANT: BOARDS MUST BE ABLE TO SUPPORT AT LEAST 145 LBS.

2. Check that odd-numbered pages will print as right-hand pages, and even-numbered pages will print as left-hand pages.
3. Make sure none of the text is running off the bottom of the page. In some programs, text that does not fit on the page does not appear at all, so this is an important check.
4. View the full page to check its overall appearance. Does it achieve the desired effect?

Proofing the Document

When the page is completely formatted, it should be proofed by the original author, the editor, and the manager in charge of the project. Sometimes the the text looks okay, when it really is not. *(Did you catch the error in the previous sentence?)*

In industry, nothing is ever published unless it has been proofed and signed off by a manager. Printing and distributing a document that contains an error to thousands of people is costly and makes the company look bad in the eyes of its customers. More importantly, if a mistake in a document causes customers to be unable to use the product without calling the manufacturer for help, the company may lose thousands of dollars due to product returns and decreased sales. See Figure 24-13. Therefore, it makes sense to have more than one person sign off on the document before it is printed.

Preparing for the Printer

The final step for most desktop publishers is to organize the project so that it can be printed. Some companies maintain their own printing equipment and print everything in-house. Other companies create the page layout in-house and then send the document files to a service bureau for processing. **Service bureaus** are companies that can take completed desktop publishing files and create film for printing. See Figure 24-14. They have specialized equipment that can produce higher-quality output than a standard laser printer.

If your company plans to use a service bureau, you should find out which service bureaus in your area can process documents created by your desktop publishing program. Ask them what items they need to complete the job. Many service bureaus require a hard (paper) copy of the document and duplicate or

Figure 24-14

Service bureaus create high-quality film from which a printer can print a document created using desktop publishing.

triplicate electronic files. For large manuals and books, find out what types of removable cartridge systems the service bureau's equipment has or send the files via recordable CD.

Document Security and Storage

Because a desktop publishing document usually links to text and illustration files, you should archive not only the original document, but all of the linked files also. Keep them together and label them clearly so that you can find them later. Store them in a safe location, preferably on a recordable CD or other permanent medium. You will need all of these files in order to reprint or make any future changes to the document. Note that you do not necessarily have to keep a paper copy of these files, as long as the electronic files are safely stored.

Some companies archive finished documents in a special place, such as a safe deposit box or other fireproof location. Archiving to a safe, off-site location is a good idea, especially for companies that would lose important information if a disaster occurred at the company site. Many companies keep two archival sets: one on-site, and another off-site. If anything happens to one set, the other is still accessible.

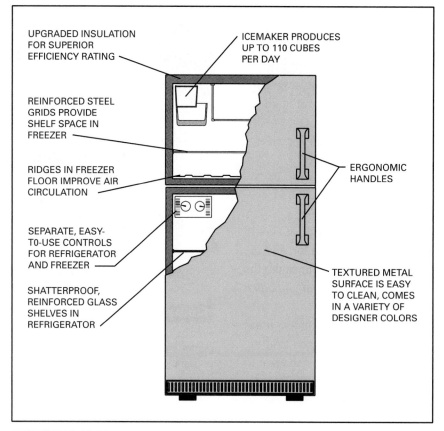

UPGRADED INSULATION
FOR SUPERIOR
EFFICIENCY RATING

ICEMAKER PRODUCES
UP TO 110 CUBES
PER DAY

REINFORCED STEEL
GRIDS PROVIDE
SHELF SPACE IN
FREEZER

RIDGES IN FREEZER
FLOOR IMPROVE AIR
CIRCULATION

ERGONOMIC
HANDLES

SEPARATE, EASY-
TO-USE CONTROLS
FOR REFRIGERATOR
AND FREEZER

SHATTERPROOF,
REINFORCED GLASS
SHELVES IN
REFRIGERATOR

TEXTURED METAL
SURFACE IS EASY
TO CLEAN, COMES
IN A VARIETY OF
DESIGNER COLORS

Figure 24-15

This broken-out section of a refrigerator-freezer has been adapted for use as a marketing tool. Using the original drawing file avoids redrawing time and helps prevent errors.

Drawings and Desktop Publishing

Desktop publishing techniques become significant for drafters when they are asked to prepare their drawings for use in a marketing brochure, user manual, or other company document. See Figure 24-15. This practice is becoming more common as companies begin to use networks (intranets) to connect drafting departments with various other departments in the company.

The process of preparing drawings for use with a desktop publishing document depends on how the drawings were originally prepared. Few desktop publishing systems will accept the original drawing files, even if they were generated using a CAD system.

Therefore, drafters should be familiar with conversion techniques and the file formats likely to be required by the desktop publisher.

The file formats required vary with the specific desktop publishing software. Most desktop publishing systems can accept, at a minimum, the following file formats:

- Encapsulated PostScript (EPS)
- Tagged-Image File Format (TIFF or TIF)
- Joint Photographic Experts Group (JPEG or JPG)

Some desktop publishing software will also accept bitmap (BMP), Graphic Interchange Format (GIF), and other formats. Check with the desktop publisher in your company to find out what file formats are acceptable.

Converting Board Drawings

The first task in preparing a board-generated drawing is to create an electronic form of the drawing. There are two methods of converting board drawings into an electronic format. The first method is to use a digitizing tablet to digitize points directly from the original drawing, and the second is to scan the drawing using an electronic scanner. Both of these methods are acceptable, but their results are quite different.

Digitizing the Drawing

When you digitize a drawing using a digitizing tablet, you end up with a file in a CAD-readable format. This does not necessarily mean that the file can be read by a desktop publishing system, however. Further conversions are then needed to convert the CAD file into a format that is usable with the desktop publishing document. Refer to "Converting CAD Drawings" for the file formats and procedures to convert CAD files.

This method has both advantages and disadvantages. The primary advantage is that if the digitization is done correctly, the resulting electronic file is very accurate. Like all CAD drawings, digitized files are in a **vector file format**. In other words, the placement of the geometry is defined using a set of mathematical instructions. A line, for example, exists in a CAD file as a location for the starting point of the line, and a *vector*, or mathematical value that tells the distance the line travels and in what direction. This mathematical calculation ensures that the electronic drawing is both clear and accurate.

Digitization has several distinct disadvantages, however. First, the process requires access to a CAD system with a digitizing tablet. Second, digitizing a drawing accurately takes time, effort, and skill. Not every CAD operator is skilled in the process of digitization, and not every CAD system includes a digitizing tablet. Finally, digitization does not produce a file that is directly usable by the desktop publisher. It must be further processed from within a CAD or other illustration program before it can be used.

Scanning the Drawing

The other method of creating an electronic file from a board-generated drawing is to scan it using a high-quality scanner like the one shown in Figure 24-16. This method is faster, because it can create electronic files in a format that can be used directly by the desktop publishing software. Most scanners are capable of scanning files in several formats. Find out from your company's desktop publisher what file formats are acceptable. Then set up the scanner to scan into that file format. If you cannot find out in advance what file format to use, it is generally safe to scan to a TIFF or JPEG format.

Figure 24-16

High-resolution scanners can convert board-generated drawings into electronic format that is directly usable by a desktop publishing system.

Scanning produces a **raster file format**. Unlike the vector files, raster files do not define geometry using mathematical calculations. Instead, they consist of a collection of tiny dots called **pixels** (picture elements). Raster files are not as accurate as vector files. To create an acceptable raster file of a drawing that contains thin lines, you must scan the drawing at a very high resolution. This creates a huge, but accurate, electronic file that can be used in desktop publishing.

Scanning is the method of choice for small drawings, because it is faster and easier to accomplish than digitizing a drawing. For larger, more complex drawings, you should choose the conversion method carefully. A scan of sufficient clarity for use in a professional document might be too large for your company's desktop publishing system to handle. Some scanned files can exceed 150 MB in size.

Converting CAD Drawings

Because the number of illustration file formats accepted by most desktop publishing programs is limited, CAD drawings must usually be converted into a different format. By default, AutoCAD creates drawings in DWG format. This format is widely used among CAD operators, and even other types of CAD software can sometimes read DWG files. Many CAD programs can also create Drawing Exchange Format (DXF) files and Initial Graphics Exchange Specification (IGES) files. These file formats work well to share drawings among different CAD systems, but none of them are commonly accepted by desktop publishing programs.

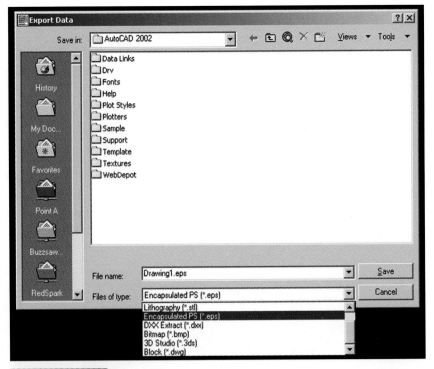

Figure 24-17

Export options in a typical version of AutoCAD.

Fortunately, most CAD programs, including AutoCAD, can export drawing files to other file formats. Figure 24-17 shows some of the file formats to which AutoCAD can export. Note that the actual file formats available depend on the version of AutoCAD you are using. However, every version allows you to export the file to an EPS format. You may recall that EPS is one of the formats accepted by most desktop publishing software, so this is usually the best choice.

When you export a DWG file into another format, the exported file does not affect the DWG file at all. In other words, changes you make to the drawing file are not reflected in the exported file. To incorporate changes, you must re-export the file after you have made the changes.

Although this seems like a simple process, it rarely is. In reality, there are several different "flavors" or versions of the EPS file format. The one created by AutoCAD may not be accepted "as is" by the desktop publishing program. Also, exporting a file to an EPS format may not preserve the line widths of lines in the original drawing. Other problems may also occur. For example, EPS recognizes dashed lines, but the lengths of the dashes may be different—and unsuitable—in the exported file. Also, EPS recognizes only two dimensions. Therefore, 3D drawings become two-dimensional, and any hidden lines that you have removed in the CAD drawing reappear in the EPS version.

The easiest way to overcome these problems is to import the EPS file created by AutoCAD into an illustration program such as Adobe Illustrator or Macromedia FreeHand. These programs allow you to touch up the file. They also allow you to save the file into an EPS format that is recognized by desktop publishing systems. In fact, illustration programs such as these can often read DWG and DXF files. If you have access to one of these illustration programs, you may want to try importing the CAD files directly into it before exporting to an AutoCAD EPS file.

Illustration programs also allow you to adapt a drawing for use in a company document. For example, the illustration in Figure 24-18 was adapted from a 3D model of a remote control for use in an owner's manual. Color, text, and the large red arrow were added using an illustration program.

Figure 24-18

Illustration programs can be used to enhance CAD illustrations to help make a point, as shown in this excerpt from an owner's manual.

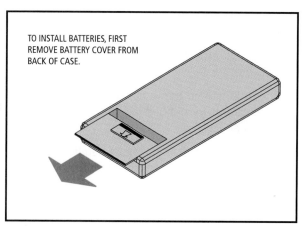

TO INSTALL BATTERIES, FIRST REMOVE BATTERY COVER FROM BACK OF CASE.

Chapter Summary

■ Desktop publishing is the process of producing typeset pages, which include text, drawings, and photographs arranged in an attractive format.

■ Typically, a desktop publishing setup includes a computer, a large-screen monitor, a laser printer, and a high-resolution scanner, as well as the desktop publishing software.

■ Document templates are usually used to control page, paragraph, and text properties.

■ The person doing the desktop publishing needs to be familiar with typesetting standards and the general rules that apply to the publishing process.

■ Documents are usually created by linking or importing text and illustration files into the desktop publishing document.

■ Drafters are often expected to produce documents that are usable with desktop publishing software.

■ Board-generated drawings must be converted to an electronic format before they can be used in desktop publishing.

■ CAD-generated drawings must usually be converted to a file format that is accepted by the desktop publishing software.

Review Questions

1. What is desktop publishing?

2. In what ways do the requirements for a desktop publishing system differ from those for a CAD system?

3. What is the advantage of using a template in a desktop publishing application?

4. What is the purpose of defining tags in a document template?

5. What three types of properties control the look of desktop publishing documents?

6. What two types of files are commonly imported into a desktop publishing document?

7. Explain the importance of proofing a completed desktop publishing page.

8. Explain why some CAD operators now use desktop publishing software as well as CAD software.

9. What three file formats are accepted by most desktop publishing programs?

10. What two methods are used to convert board-generated drawings into electronic format?

11. What is the difference between programs that use a raster file format and those that use a vector file format?

12. Why must EPS files exported by a CAD system usually be imported into an illustration program before they are used in a desktop publishing document?

Drafting Problems

The following problems require access to other software in addition to CAD software. If you do not have the required software, you can simulate the problem using AutoCAD.

1. Convert one of your exploded view drawings into a format that is accepted by your available desktop publishing software. Refer to the appropriate parts of this chapter for methods to convert board- and CAD-generated drawings. Using the desktop publishing program's text editor, add a parts list showing item number, description, and quantity. Create a title PARTS LIST using bold type. Draw a box around the parts list and add horizontal and vertical lines to complete it.

2. Review several periodicals such as *Time* or *Newsweek*. Pick a page layout that is attractive to you. Create a document template that duplicates the layout as closely as you can using desktop publishing software. Use the template as a guide in developing an advertisement for a set of golf clubs or other sports equipment. Create a drawing to accompany the page, and convert it for use in the desktop publishing document using an appropriate technique. Print the finished page and proof it carefully. Fix any errors you find.

3. Select one of the more complex drawings that you have done. Convert it for use with a desktop publishing system. Import it into a desktop publishing document so that it fills one-half to two-thirds of a page. In the remaining space, write a description of how you created the drawing.

Design Problems

Design problems have been prepared to challenge individual students or teams of students. In these problems, you are to apply skills learned mainly in this chapter but also in other chapters throughout the text. The problems are designed to be completed using board drafting, CAD, or a combination of the two. Be creative and have fun!

1. **TEAMWORK** The company you work for has decided to manufacture and market a do-it-yourself kit for the letter holder shown in Figure 24-19. Your team has been assigned the task of developing a complete instruction sheet for the customer to follow in the assembly of the final product. The team should first brainstorm the process, and then create an appropriate drawing using board or CAD techniques. Convert the drawing for use with a desktop publishing document.

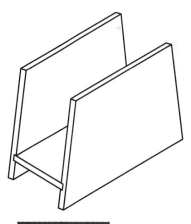

Figure 24-19

Using the desktop publishing software, create an instruction sheet to include the following:

- A small-scale pictorial drawing of the finished product without dimensions.
- An exploded-view drawing that illustrates the assembly. Include methods of fastening the parts together.
- A set of instructions on assembly.
- A finishing schedule. (This may require some research.)
- The design of a decorative decal to be supplied with the kit, including instructions for applying it. Be creative!
- An attractive sheet design with a .50″ border, your company logo, and the company name and address. Use color to make the instruction sheet more pleasing to the eye.

2. Use desktop publishing and any other software available to design and develop a two-page newsletter for an organization of your choice. Give careful attention to the design of the masthead (top portion of the front page that contains the name of the newsletter, publisher, etc.). Use color as appropriate. Completely develop at least one issue.

3. **TEAMWORK** Use a team approach to develop the basic layout and design for a poster for several of your school's athletic teams. Include the name of your school and an illustration of the school's logo or mascot. Once the basic design is agreed upon, create a template using a desktop publishing program. Each team member can then create a poster for the athletic team of his or her choice using the template as the basis for the design.

Abbreviations and Symbols

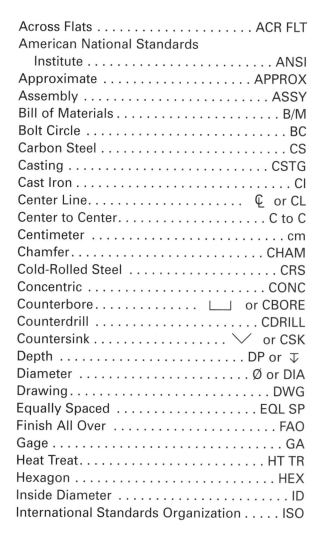

Across Flats . . . . ACR FLT
American National Standards
    Institute . . . . ANSI
Approximate . . . . APPROX
Assembly . . . . ASSY
Bill of Materials . . . . B/M
Bolt Circle . . . . BC
Carbon Steel . . . . CS
Casting . . . . CSTG
Cast Iron CI
Center Line. . . . . ℄ or CL
Center to Center. . . . . C to C
Centimeter . . . . cm
Chamfer. . . . . CHAM
Cold-Rolled Steel . . . . CRS
Concentric . . . . CONC
Counterbore. . . . . ⌴ or CBORE
Counterdrill . . . . CDRILL
Countersink . . . . ∨ or CSK
Depth . . . . DP or ⊤
Diameter . . . . Ø or DIA
Drawing. . . . . DWG
Equally Spaced . . . . EQL SP
Finish All Over . . . . FAO
Gage . . . . GA
Heat Treat. . . . . HT TR
Hexagon . . . . HEX
Inside Diameter . . . . ID
International Standards Organization . . . . . ISO

Kilogram. . . . . kg
Kilometer. . . . . km
Left Hand . . . . LH
Liter L
Material . . . . MATL
Maximum. . . . . MAX
Maximum Material Condition . . . . Ⓜ or MMC
Meter. . . . . m
Metric Thread . . . . M
Micrometer. . . . . μm
Millimeter . . . . mm
Minimum. . . . . MIN
Nominal. . . . . NOM
Number . . . . NO.
Outside Diameter . . . . OD
Radius . . . . R
Reference or Reference Dimension. . ( ) or REF
Right Hand . . . . RH
Slotted. . . . . SLOT
Spherical. . . . . SPHER
Spotface . . . . ⌴ or SFACE
Square. □ or SQ
Steel. . . . . STL
Symmetrical . . . . or SYM
Taper—Flat . . . .
Taper—Round . . . .
Through . . . . THRU
Tolerance. . . . . TOL
U.S. Gage . . . . USG

Pipe Symbols

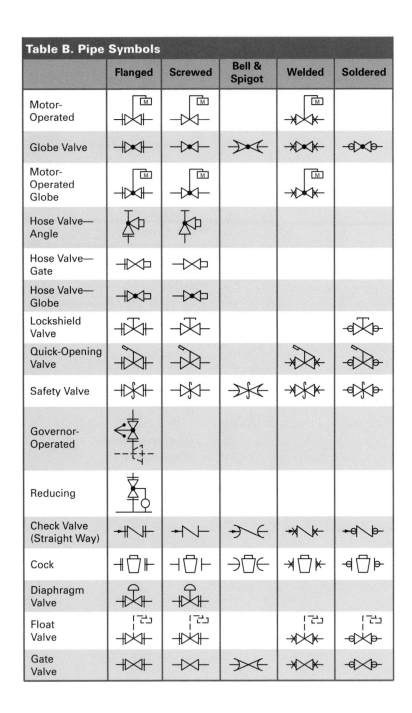

Table B. Pipe Symbols	Flanged	Screwed	Bell & Spigot	Welded	Soldered
Motor-Operated					
Globe Valve					
Motor-Operated Globe					
Hose Valve—Angle					
Hose Valve—Gate					
Hose Valve—Globe					
Lockshield Valve					
Quick-Opening Valve					
Safety Valve					
Governor-Operated					
Reducing					
Check Valve (Straight Way)					
Cock					
Diaphragm Valve					
Float Valve					
Gate Valve					

Table B. Pipe Symbols (continued)

	Flanged	Screwed	Bell & Spigot	Welded	Soldered
Bull Plug	⊣⊐		◯		
Pipe Plug		⊣◁	⊏		
Concentric Reducer	⊣▷⊣	⊣▷	➤	⊁	◁◦
Eccentric Reducer	⊣◺⊣	⊣◺	➤	⊁	◁◦
Sleeve	⊣--⊣	⊢--⊢	→---←	✕-✕	◦---◦
Tee (Straight Size)	⊥	⊥	⊥	⊥	⊥
Tee (Outlet Up)	⊣⊙⊣	⊢⊙⊢	→⊙←	✕⊙✕	◦⊙◦
Tee (Outlet Down)	⊣◯⊣	⊢◯⊢	→◯←	✕◯✕	◦◯◦
Tee (Double Sweep)					
Reducing Tee					
Tee (Single Sweep)					
Side Outlet (Outlet Down)					
Side Outlet (Outlet Up)					
Union	⊣⊢	⊣⊢		✕⊢	◦⊢
Angle Valve—Check					
Angle Valve—Gate (Elevation)					
Angle Valve—Gate (Plan)					
Angle Valve—Globe (Elevation)					
Angle Valve—Globe (Plan)					
Automatic Valve—By-Pass					

Appendix B

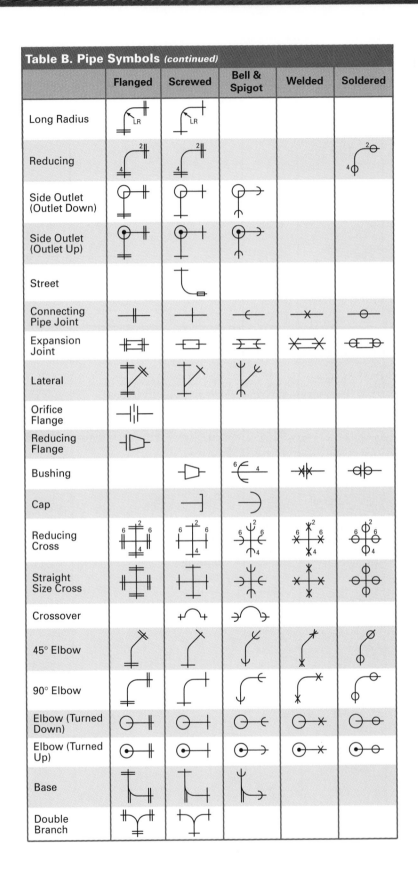

Table B. Pipe Symbols *(continued)*

	Flanged	Screwed	Bell & Spigot	Welded	Soldered
Long Radius					
Reducing					
Side Outlet (Outlet Down)					
Side Outlet (Outlet Up)					
Street					
Connecting Pipe Joint					
Expansion Joint					
Lateral					
Orifice Flange					
Reducing Flange					
Bushing					
Cap					
Reducing Cross					
Straight Size Cross					
Crossover					
45° Elbow					
90° Elbow					
Elbow (Turned Down)					
Elbow (Turned Up)					
Base					
Double Branch					

Table B. Pipe Symbols *(continued)*

Air Conditioning

Brine Return	— — — BR — — —
Brine Supply	———— B ————
Circulating Chilled or Hot-Water Flow	——— C H ———
Circulating Chilled or Hot-Water Return	— — — C H R — — —
Condenser Water Flow	——— C ———
Condenser Water Return	— — — C R — — —
Drain	——— D ———
Humidification Line	— - — H — - —
Make-Up Water	— - — - — - —
Refrigerant Discharge	——— R D ———
Refrigerant Liquid	——— R L ———
Refrigerant Suction	— — — R S — — —

Heating

Air-Relief Line	— ——— — — —
Boiler Blow-Off	—— — —— —
Compressed Air	——— A ———
Condensate or Vacuum Pump Discharge	—o— —o— —o—
Feedwater Pump Discharge	—o o— —o o— —o o—
Fuel-Oil Flow	— — — F O F— — —
Fuel-Oil Return	— — —F O R— — —
Fuel-Oil Tank Vent	— — —F O V — — —
High-Pressure Return	— +/— +/— +/—
High-Pressure Steam	// // //
Hot-Water Heating Return	— — — — — —
Hot-Water Heating Supply	———————

Appendix C

Reference Tables

In the United States, the governing body responsible for the standards established for technical and engineering drawing is the **American National Standards Institute (ANSI)**. However, ANSI has enlisted the help of the **American Society of Mechanical Engineers (ASME)** to assist in the revision, maintenance, and marketing of these standards. A catalog of their drafting standards and other ASME publications is available by writing to ASME, 22 Law Drive, Box 2900, Fairfield, NJ 07007-2900. Standards established for other technical fields, also controlled by ANSI, are maintained by their respective professional organizations.

International drafting standards are controlled by the **International Standards Organization (ISO)**. However, since the United States holds a leadership role in ISO, the international standards continue to move toward those established and maintained by ASME. The differences become less noticeable with each revision.

Selected standards useful to the drafter are listed below. Since all standards are constantly under revision, it is important to consult the most current ASME catalog for the latest editions.

ANSI/ASME Standards

Standards for Drawings

Abbreviations for Use on Drawings and Text	Y1.1
Decimal Inch Drawing Sheet Size and Format	Y14.1
Metric Drawing Sheet Size and Format	Y14.1M
Line Conventions and Lettering	Y14.2M
Multi- and Sectional-View Drawings	Y14.3M
Pictorial Drawing	Y14.4M
Dimensioning and Tolerancing	Y14.5M
Preferred Limits and Fits for Cylindrical Parts	B4.1
Preferred Metric Limits and Fits	B4.2
General Tolerances for Metric Dimensioned Products	B4.3
Screw Thread Representation	Y14.6
Screw Thread Representation (Metric Supplement)	Y14.6M
Gear Drawing Standards—Part 1: For Spur, Helical, Double Helical, and Rack	Y14.7.1
Gear Drawing Standards—Part 2: Bevel and Hypoid Gears	Y14.7.2
Castings and Forgings	Y14.8M
Mechanical Spring Representation	Y14.13
Types and Applications of Engineering Drawings	Y14.24
Surface Texture Symbols	Y14.36M
Surface Texture (Surface Roughness, Waviness, and Lay	B46.1
Engineering Drawing Practices	Y14.100M
Graphic Symbols for Pipe Fittings, Valves, and Piping	Y32.2.3
Graphic Symbols for Heating, Ventilating, and Air Conditioning	Y32.2.4
Graphic Symbols for Plumbing Fixtures for Diagrams Used in Architecture and Building Construction	Y32.4

Appendix C

ANSI/AWS Standard

Symbols for Welding and Nondestructive Testing   AWS A2.4

The Reference Tables

The tables that follow are needed to work some of the problems in this edition of *Mechanical Drawing: Board and CAD Techniques*. These tables have been adapted from the ANSI/ASME standards for use in this book. Although most of the tables are self-explanatory, the following sections provide additional information about some of them.

American National Standard Limits and Fits

Tables C-9 through C-16 are designed for use with the basic hole system of limits and fits described in Chapter 7, "Dimensioning." Information from these tables is adapted from ASME B4.1, "Preferred Limits and Fits for Cylindrical Parts." For larger sizes and additional information, refer to the standard.

There are five distinct classes of fits:

RC   Running or sliding clearance fits
LC   Locational clearance fits
LT   Transition clearance or interference fits
LN   Locational interference fits
FN   Force or shrink fits

These five classes of fits are placed in three general categories, as follows.

Running and Sliding Fits (Table C-9)

These fits provide a similar running performance, with suitable lubrication allowance, throughout the range of sizes. The clearances for the first two classes, used chiefly as sliding fits, increase more slowly than for the other classes, so that accurate location is maintained even at the expense of free relative motion.

RC 1: *Close sliding fits* accurately locate parts that must assemble without perceptible play.

RC 2: *Sliding fits* are for accurate location, but with greater maximum clearance than RC 1. Parts move and turn easily but do not run freely, and in the larger sizes may seize with small temperature changes.

RC 3: *Precision running fits* are the closest fits expected to run freely. They are for precision work at slow speeds and light journal pressures. They are not suitable under appreciable temperature differences.

RC 4: *Close running fits* are chiefly for running fits on accurate machinery with moderate surface speeds and journal pressures, where accurate location and minimum play are desired.

RC 5 and RC 6: *Medium running fits* are for higher running speeds, heavy journal pressures, or both.

RC 7: *Free running fits* are for use where accuracy is not essential, where large temperature variations are likely, or under both of these conditions.

RC 8 and RC 9: *Loose running fits* are for materials such as cold-rolled shafting and tubing, made to commercial tolerances.

Locational Fits (Tables C-10 Through C-12)

These fits determine only the location of mating parts. They may provide rigid or accurate location, as in interference fits, or some freedom of location, as in clearance fits. They fall into the following three groups:

LC: *Locational clearance fits* are for normally stationary parts that can be freely assembled or disassembled. They run from snug fits for parts requiring accuracy of location, through the medium clearance fits for parts such as spigots, to the looser fastener fits where freedom of assembly is of prime importance.

LT: *Transitional locational fits* fall between clearance and interference fits. They are for application where accuracy of location is important, but a small amount of clearance or interference is permissible.

LN: *Locational interference fits* are used where accuracy of location is of prime importance, and for parts needing rigidity and alignment with no special requirements for bore pressure. Such fits are not for parts that transmit frictional loads from one part to another by virtue of the tightness of fit; those conditions are met by force fits.

Force Fits (Table C-13)

A force fit is a special type of interference fit, normally characterized by maintenance of constant bore pressures throughout the range of sizes. Thus the interference varies almost directly with diameter. To maintain the resulting pressures within reasonable limits, the difference between its minimum and maximum value is small.

FN 1: *Light drive fits* require light assembly pressures and produce more or less permanent assemblies. These are suitable for thin sections, long fits, or cast-iron external members.

FN 2: *Medium drive fits* are for ordinary steel parts or for shrink fits on light sections. They are about the tightest fits that can be used with high-grade cast-iron external members.

FN 3: *Heavy drive fits* are suitable for heavier steel parts or for shrink fits in medium sections.

FN 4 and FN 5: *Force fits* are for parts that can be highly stressed, or for shrink fits where heavy pressing forces are impractical.

Appendix C

Table C-1. Fractional-Inch, Decimal-Inch, and Millimeter Equivalent Chart

Fractions					Decimal Inches			Millimeters		
4ths	8ths	16ths	32nds	64ths	To 2 Places	To 3 Places	To 4 Places	To 1 Place	To 2 Places	To 3 Places
				1/64	.02	.016	.0156	0.4	.40	.397
			1/32	...	.03	.031	.0312	0.8	.80	.794
				3/64	.05	.047	.0469	1.2	1.20	1.191
		1/16	...	...	.06	.062	.0625	1.6	1.59	1.588
				5/64	.08	.078	.0781	2.0	1.98	1.984
			3/32	...	.09	.094	.0938	2.4	2.38	2.381
				7/64	.11	.109	.1094	2.8	2.78	2.778
	1/8	...	...	...	.12	.125	.1250	3.2	3.18	3.175
				9/64	.14	.141	.1406	3.6	3.57	3.572
			5/32	...	.16	.156	.1562	4.0	3.97	3.969
				11/64	.17	.172	.1719	4.4	4.37	4.366
		3/16	...	...	.19	.188	.1875	4.8	4.76	4.762
				13/64	.20	.203	.2031	5.2	5.16	5.159
			7/32	...	.22	.219	.2188	5.6	5.56	5.556
				15/64	.23	.234	.2344	6.0	5.95	5.953
1/4	...	...	...	...	.25	.250	.2500	6.4	6.35	6.350
				17/64	.27	.266	.2656	6.8	6.75	6.747
			9/32	...	.28	.281	.2812	7.1	7.14	7.144
				19/64	.30	.297	.2969	7.5	7.54	7.541
		5/16	...	...	.31	.312	.3125	7.9	7.94	7.938
				21/64	.33	.328	.3281	8.3	8.33	8.334
			11/32	...	.34	.344	.3438	8.7	8.73	8.731
				23/64	.36	.359	.3594	9.1	9.13	9.128
	3/8	...	...	...	.38	.375	.3750	9.5	9.52	9.525
				25/64	.39	.391	.3906	9.9	9.92	9.922
			13/32	...	.41	.406	.4062	10.3	10.32	10.319
				27/64	.42	.422	.4219	10.7	10.72	10.716
		7/16	...	...	.44	.438	.4375	11.1	11.11	11.112
				29/64	.45	.453	.4531	11.5	11.51	11.509
			15/32	...	.47	.469	.4688	11.9	11.91	11.906
				31/64	.48	.484	.4844	12.3	12.30	12.303
				...	.50	.500	.5000	12.7	12.70	12.700
				33/64	.52	.516	.5156	13.1	13.10	13.097
			17/32	...	.53	.531	.5312	13.5	13.49	13.494
				35/64	.55	.547	.5469	13.9	13.89	13.891
		9/16	...	...	.56	.562	.5625	14.3	14.29	14.288
				37/64	.58	.578	.5781	14.7	14.68	14.684
			19/32	...	.59	.594	.5938	15.1	15.08	15.081
				39/64	.61	.609	.6094	15.5	15.48	15.478
	5/8	...	...	...	.62	.625	.6250	15.9	15.88	15.875

(continued on next page)

Table C-1. Fractional-Inch, Decimal-Inch, and Millimeter Equivalent Chart (continued)

Fractions					Decimal Inches			Millimeters		
4ths	8ths	16ths	32nds	64ths	To 2 Places	To 3 Places	To 4 Places	To 1 Place	To 2 Places	To 3 Places
				41/64	.64	.641	.6406	16.3	16.27	16.272
			21/32	...	.66	.656	.6562	16.7	16.67	16.669
				43/64	.67	.672	.6719	17.1	17.07	17.066
		11/16	...	...	.69	.688	.6875	17.5	17.46	17.462
				45/64	.70	.703	.7031	17.9	17.86	17.859
			23/32	...	.72	.719	.7188	18.3	18.26	18.256
				47/64	.73	.734	.7344	18.6	18.65	18.653
3/4	...	...	...	...	.75	.750	.7500	19.1	19.05	19.050
				49/64	.77	.766	.7656	19.4	19.45	19.447
			25/32	...	.78	.781	.7812	19.8	19.84	19.844
				51/64	.80	.797	.7969	20.2	20.24	20.241
		13/16	...	...	.81	.812	.8125	20.6	20.64	20.638
				53/64	.83	.828	.8281	21.0	21.03	21.034
			27/32	...	.84	.844	.8438	21.4	21.43	21.431
				55/64	.86	.859	.8594	21.8	21.83	21.828
	7/8	...	...	...	.88	.875	.8750	22.2	22.22	22.225
				57/64	.89	.891	.8906	22.6	22.62	22.622
			29/32	...	.91	.906	.9062	23.0	23.02	23.019
				59/64	.92	.922	.9219	23.4	23.42	23.416
		15/16	...	...	.94	.938	.9375	23.8	23.81	23.812
				61/64	.95	.953	.9531	24.2	24.21	24.209
			31/32	...	.97	.969	.9688	24.6	24.61	24.606
				63/64	.98	.984	.9844	25.0	25.00	25.003
					1.00	1.000	1.0000	25.4	25.40	25.400

Table C-2. Threads per Inch and Tap Drill Sizes

| Size (Inches) | | Graded Pitch Series | | | | | | Constant Pitch Series | | | | | |
| Number or Fraction | Deci-mal | Coarse UNC | | Fine UNF | | Extra Fine UNF | | 8 UN | | 12 UN | | 16 UN | |
		Threads per Inch	Tap Drill Dia.	Threads per Inch	Tap Drill Dia.	Threads per Inch	Tap Drill Dia.	Threads per Inch	Tap Drill Dia.	Threads per Inch	Tap Drill Dia.	Threads per Inch	Tap Drill Dia.
0	.060	—	—	80	$3/64$	—	—	—	—	—	—	—	—
2	.086	56	No. 50	64	No. 49	—	—	—	—	—	—	—	—
4	.112	40	No. 43	48	No. 42	—	—	—	—	—	—	—	—
5	.125	40	No. 38	44	No. 37	—	—	—	—	—	—	—	—
6	.138	32	No. 36	40	No. 33	—	—	—	—	—	—	—	—
8	.164	32	No. 29	36	No. 29	—	—	—	—	—	—	—	—
10	.190	24	No. 25	32	No. 21	—	—	—	—	—	—	—	—
$1/4$	.250	20	7	28	3	32	.219	—	—	—	—	—	—
$5/16$	.312	18	F	24	1	32	.281	—	—	—	—	—	—
$3/8$	.375	16	.312	24	Q	32	.344	—	—	—	—	UNC	—
$7/16$	.438	14	U	20	.391	28	Y	—	—	—	—	16	V
$1/2$	.500	13	.422	20	.453	28	.469	—	—	—	—	16	.438
$9/16$	.562	12	.484	18	.516	24	.516	—	—	UNC	—	16	.500
$5/8$	.625	11	.531	18	.578	24	.578	—	—	12	.547	16	.562
$3/4$	.750	10	.656	16	.688	20	.703	—	—	12	.672	UNF	—
$7/8$	.875	9	.766	14	.812	20	.828	—	—	12	.797	16	.812
1	1.000	8	.875	12	.922	20	.953	UNC	—	UNF	—	16	.938
$1 1/8$	1.125	7	.984	12	1.047	18	1.078	8	1.000	UNF	—	16	1.062
$1 1/4$	1.250	7	1.109	12	1.172	18	1.188	8	1.125	UNF	—	16	1.188
$1 3/8$	1.375	6	1.219	12	1.297	18	1.312	8	1.250	UNF	—	16	1.312
$1 1/2$	1.500	6	1.344	12	1.422	18	1.438	8	1.375	UNF	—	16	1.438
$1 5/8$	1.625	—	—	—	—	18	—	8	1.500	12	1.547	16	1.562
$1 3/4$	1.750	5	1.562	—	—	—	—	8	1.625	12	1.672	16	1.688
$1 7/8$	1.875	—	—	—	—	—	—	8	1.750	12	1.797	16	1.812
2	2.000	4.5	1.781	—	—	—	—	8	1.875	12	1.922	16	1.938
$2 1/4$	2.250	4.5	2.031	—	—	—	—	8	2.125	12	2.172	16	2.188
$2 1/2$	2.500	4	2.250	—	—	—	—	8	2.375	12	2.422	16	2.438
$2 3/4$	2.750	4	2.500	—	—	—	—	8	2.625	12	2.672	16	2.688
3	3.000	4	2.750	—	—	—	—	8	2.875	12	2.922	16	2.938
$3 1/4$	3.250	4	3.000	—	—	—	—	8	3.125	12	3.172	16	3.188
$3 1/2$	3.500	4	3.250	—	—	—	—	8	3.375	12	3.422	16	3.438
$3 3/4$	3.750	4	3.500	—	—	—	—	8	3.625	12	3.668	16	3.688
4	4.000	4	3.750	—	—	—	—	8	3.875	12	3.922	16	3.938

Notes: The tap diameter sizes shown are nominal. The class and length of thread will govern the limits on the tapped hole size.

Table C-3. ISO Metric Screw Threads

Size Dia. (mm)	Graded Pitch Series — Coarse		Fine		Constant Pitch Series — 4		3		2		1.5		1.25		1		0.75		0.5	
	Thread Pitch	Tap Drill Size	Thread Pitch	Tap Drill Size	Thread Pitch	Tap Drill Size	Thread Pitch	Tap Drill Size	Thread Pitch	Tap Drill Size	Thread Pitch	Tap Drill Size	Thread Pitch	Tap Drill Size	Thread Pitch	Tap Drill Size	Thread Pitch	Tap Drill Size	Thread Pitch	Tap Drill Size
*1.6	0.35	1.25																		
1.8	0.35	1.45																		
*2	0.4	1.6																		
2.2	0.45	1.75																		
*2.5	0.45	2.05																		
*3	0.5	2.5																		
3.5	0.6	2.9																		
*4	0.7	3.3																	0.5	3.5
4.5	0.75	3.7																	0.5	4.0
*5	0.8	4.2																	0.5	4.5
*6	1	5.0															0.75	5.2		
*8	1.25	6.7	1	7.0											1	7.0	0.75	7.2		
*10	1.5	8.5	1.25	8.7									1.25	8.7	1	9.0	0.75	9.2		
*12	1.75	10.2	1.25	10.8							1.5	10.5	1.25	10.7	1	11				
14	2	12	1.5	12.5							1.5	12.5	1.25	12.7	1	13				
*16	2	14	1.5	14.5							1.5	14.5			1	15				
18	2.5	15.5	1.5	16.5					2	16	1.5	16.5			1	17				
*20	2.5	17.5	1.5	18.5					2	18	1.5	18.5			1	19				
22	2.5	19.5	1.5	20.5					2	20	1.5	20.5			1	21				
*24	3	21	2	22					2	22	1.5	22.5			1	23				
27	3	24	2	25					2	25	1.5	25.5			1	26				
*30	3.5	26.5	2	28					2	28	1.5	28.5			1	29				
33	3.5	29.5	2	31					2	31	1.5	31.5								
*36	4	32	3	33					2	34	1.5	34.5								
39	4	35	3	36					2	37	1.5	37.5								
*42	4.5	37.5	3	39	4	38	3	39	2	40	1.5	40.5								
45	4.5	39	3	42	4	41	3	42	2	43	1.5	43.5								
*46	5	43	3	45	4	44	3	45	2	46	1.5	46.5								

Notes: The diameter sizes are nominal. Sizes preceded by an asterisk (*) are preferred.

Appendix C

Table C-4. Acme and Stub Acme Threads

ANSI Preferred Diameter-Pitch Combinations							
Nominal (Major) Diameter	Threads per Inch	Nominal (Major) Diameter	Threads per Inch	Nominal (Major) Diameter	Threads per Inch	Nominal (Major) Diameter	Threads per Inch
1/4 (.250)	16	3/4 (.750)	6	1 1/2 (1.500)	4	3 (3.000)	2
5/16 (.312)	14	7/8 (.875)	6	1 3/4 (1.750)	4	3 1/2 (3.500)	2
3/8 (.375)	12	1 (1.000)	5	2 (2.000)	4	4 (4.000)	2
7/16 (.438)	12	1 1/8 (1.125)	5	2 1/4 (2.250)	3	4 1/2 (4.500)	2
1/2 (.500)	10	1 1/4 (1.250)	5	2 1/2 (2.500)	3	5 (5.000)	2
5/8 (.625)	8	1 3/8 (1.375)	4	2 3/4 (2.750)	3		

Note: Diameters in inches.

Table C-5. Hexagon-Head Bolts and Cap Screws

U.S. Customary (Inches)			Metric (Millimeters)		
Nominal Bolt Size	Width Across Flats F	Thickness T	Nominal Bolt Size and Thread Pitch	Width Across Flats F	Thickness T
.250	.438	.172	M5 X 0.8	8	3.9
.312	.500	.219	M6 X 1	10	4.7
.375	.562	.250	M8 X 1.25	13	5.7
.438	.625	.297			
.500	.750	.344	M10 X 1.5	15	6.8
.625	.938	.422	M12 X 1.75	18	8
.750	1.125	.500	M14 X 2	21	9.3
.875	1.312	.578	M16 X 2	24	10.5
1.000	1.500	.672	M20 X 2.5	30	13.1
1.125	1.688	.750	M24 X 3	36	15.6
1.250	1.875	.844	M30 X 3.5	46	19.5
1.375	2.062	.906	M36 X 4	55	23.4
1.500	2.250	1.000			

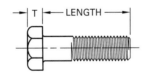

Table C-6. Hexagon-Head Nuts

U.S. Customary (Inches)				Metric (Millimeters)			
Nominal Nut Size	Distance Across Flats F	Thickness Max.		Nominal Nut Size and Thread Pitch	Distance Across Flats F	Thickness Max.	
		Style 1 (Regular)–H	Style 2 (Thick)–H₁			Style 1 (Regular)–H	Style 2 (Thick)–H₁
.250	.438	.218	.281	M4 X 0.7	7	—	3.2
.312	.500	.266	.328	M5 X 0.8	8	4.5	5.3
.375	.596	.328	.406	M6 X 1	10	5.6	6.5
.438	.625	.375	.453	M8 X 1.25	13	6.6	7.8
.500	.750	.438	.562	M10 X 1.5	15	9	10.7
.562	.875	.484	.609	M12 X 1.75	18	10.7	12.8
.625	.938	.547	.719	M14 X 2	21	12.5	14.9
.750	1.125	.641	.812	M16 X 2	24	14.5	17.4
.875	1.312	.750	.906	M20 X 2.5	30	18.4	21.2
1.000	1.500	.859	1.000	M24 X 3	36	22	25.4
1.125	1.688	.969	1.156	M30 X 3.5	46	26.7	31
1.250	1.875	1.062	1.250	M36 X 4	55	32	37.6
1.375	2.062	1.172	1.375				
1.500	2.250	1.281	1.500				

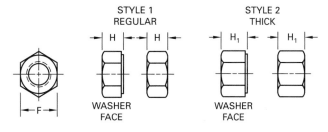

STYLE 1 REGULAR   STYLE 2 THICK

WASHER FACE   WASHER FACE

Table C-7. Common Machine and Cap Screws

	U.S. Customary (Inches)										Metric (Millimeters)									
Nominal Size	Hexagon Head		Socket Head		Flat Head		Fillister Head		Round or Oval Head		Nominal Size	Hexagon Head		Socket Head		Flat Head		Fillister Head		Round or Oval Head	
	A	H	A	H	A	H	A	H	A	H		A	H	A	H	A	H	A	H	A	H
.250	.44	.17	.38	.25	.50	.14	.38	.17	.44	.19	M3	5.5	2	5.5	3	5.6	1.6	6	2.4	5.6	1.9
.312	.50	.22	.47	.31	.62	.18	.44	.20	.56	.25	4	7	2.8	7	4	7.5	2.2	8	3.1	7.5	2.5
.375	.56	.25	.56	.38	.75	.21	.56	.25	.62	.27	5	8.5	3.5	9	5	9.2	2.5	10	3.8	9.2	3.1
.438	.62	.30	.66	.44	.81	.21	.62	.30	.75	.33	6	10	4	10	6	11	3	12	4.6	11	3.8
.500	.75	.34	.75	.50	.88	.21	.75	.33	.81	.35	8	13	5.5	13	8	14.5	4	16	6	14.5	5
.625	.94	.42	.94	.62	1.12	.28	.88	.42	1.00	.44	10	17	7	16	10	18	5	20	7.5	18	6.2
.750	1.12	.50	1.12	.75	1.38	.35	1.00	.50	1.25	.55	12	19	8	18	12						
											14	22	9	22	14						
											16	24	10	24	16						

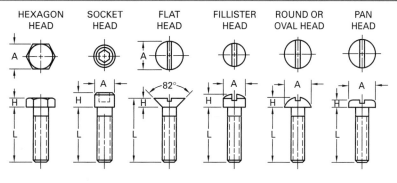

HEXAGON HEAD   SOCKET HEAD   FLAT HEAD   FILLISTER HEAD   ROUND OR OVAL HEAD   PAN HEAD

Appendix C

Table C-8. Woodruff Keys

Nominal Size	Key			Keyseat	Key No.	Nominal Size	Key			Keyseat
A X B	E	C	D	H		A X B	E	C	D	H
.062 X .500	.047	.203	.194	.172	204	1.6 X 12.7	1.5	5.1	4.8	4.2
.094 X .500	.047	.203	.194	.156	304	2.4 X 12.7	1.3	5.1	4.8	3.8
.094 X .625	.062	.250	.240	.203	305	2.4 X 15.9	1.5	6.4	6.1	5.1
.125 X .500	.049	.203	.194	.141	404	3.2 X 12.7	1.3	5.1	4.8	3.6
.125 X .625	.062	.250	.240	.188	405	3.2 X 15.9	1.5	6.4	6.1	4.6
.125 X .750	.062	.313	.303	.251	406	3.2 X 19.1	1.5	7.9	7.6	6.4
.156 X .625	.062	.250	.240	.172	505	4.0 X 15.9	1.5	6.4	6.1	4.3
.156 X .750	.062	.313	.303	.235	506	4.0 X 19.1	1.5	7.9	7.6	5.8
.156 X .875	.062	.375	.365	.297	507	4.0 X 22.2	1.5	9.7	9.1	7.4
.188 X .750	.062	.313	.303	.219	606	4.8 X 19.1	1.5	7.9	7.6	5.3
.188 X .875	.062	.375	.365	.281	607	4.8 X 22.2	1.5	9.7	9.1	7.1
.188 X 1.000	.062	.438	.428	.344	608	4.8 X 25.4	1.5	11.2	10.9	8.6
.188 X 1.125	.078	.484	.475	.390	609	4.8 X 28.6	2.0	12.2	11.9	9.9
.250 X .875	.062	.375	.365	.250	807	6.4 X 22.2	1.5	9.7	9.1	6.4
.250 X 1.000	.062	.438	.428	.313	808	6.4 X 25.4	1.5	11.2	10.9	7.9

Note: Metric key sizes were not available at the time of publication. Sizes shown are inch-designed key sizes soft-converted to millimeters. Conversion was necessary to allow the student to compare keys with slot sizes given in millimeters.

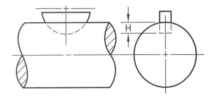

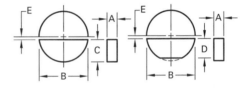

Table C–9. Running and Sliding Fits

Basic hole system. Limits are in thousandths of an inch.

Nominal Size Range (Inches) Over – To	Class RC 1 Close Sliding Fit — Limits of Clearance	Standard Limits Hole	Shaft	Class RC 2 Sliding Fit — Limits of Clearance	Hole	Shaft	Class RC 3 Precision Running Fit — Limits of Clearance	Hole	Shaft	Class RC 4 Close Running Fit — Limits of Clearance	Hole	Shaft
0–0.12	0.1	+0.2	–0.1	0.1	+0.25	–0.1	0.3	+0.4	–0.3	0.3	+0.6	–0.3
	0.45	–0	–0.25	0.55	–0	–0.3	0.95	–0	–0.55	1.3	–0	–0.7
0.12–0.24	0.15	+0.2	–0.15	0.15	+0.3	–0.15	0.4	+0.5	–0.4	0.4	+0.7	–0.4
	0.5	–0	–0.3	0.65	–0	–0.35	1.12	–0	–0.7	1.6	–0	–0.9
0.24–0.40	0.2	+0.25	–0.2	0.2	+0.4	–0.2	0.5	+0.6	–0.5	0.5	+0.9	–0.5
	0.6	–0	–0.35	0.85	–0	–0.45	1.5	–0	–0.9	2.0	–0	–1.1
0.40–0.71	0.25	+0.3	–0.25	0.25	+0.4	–0.25	0.6	+0.7	–0.6	0.6	+1.0	–0.6
	0.75	–0	–0.45	0.95	–0	–0.55	1.7	–0	–1.3	2.8	–0	–1.3
0.71–1.19	0.3	+0.4	–0.3	0.3	+0.5	–0.3	0.8	+0.8	–0.8	0.8	+1.2	–0.8
	0.95	–0	–0.55	1.2	–0	–0.7	2.1	–0	–1.3	2.8	–0	–1.6
1.19–1.97	0.4	+0.4	–0.4	0.4	+0.6	–0.4	1.0	+1.0	–1.0	1.0	+1.6	–1.0
	1.1	–0	–0.7	1.4	–0	–0.8	2.6	–0	–1.6	3.6	–0	–2.0
1.97–3.15	0.4	+0.5	–0.4	0.4	+0.7	–0.4	1.2	+1.2	–1.2	1.2	+1.8	–1.2
	1.2	–0	–0.7	1.6	–0	–0.9	3.1	–0	–1.9	4.2	–0	–2.4

	Medium Running Fits						Loose Running Fits					
Nominal Size Range (Inches) Over – To	Class RC 5 — Limits of Clearance	Standard Limits Hole	Shaft	Class RC 6 — Limits of Clearance	Hole	Shaft	Class RC 7 — Limits of Clearance	Hole	Shaft	Class RC 8 — Limits of Clearance	Hole	Shaft	Class RC 9 — Limits of Clearance	Hole	Shaft
0–0.12	0.6	+0.6	–0.6	0.6	+1.0	–0.6	1.0	+1.0	–1.0	2.5	+1.6	–2.5	4.0	+2.5	– 4.0
	1.6	–0	–1.0	2.2	–0	–1.2	2.6	–0	–1.6	5.1	–0	–3.5	8.1	–0	– 5.6
0.12–0.24	0.8	+0.7	–0.8	0.8	+1.2	–0.8	1.2	+1.2	–1.2	2.8	+1.8	–2.8	4.5	+3.0	– 4.5
	2.0	–0	–1.3	2.7	–0	–1.5	3.1	–0	–1.9	5.8	–0	–4.0	9.0	–0	– 6.0
0.24–0.40	1.0	+0.9	–1.0	1.0	+1.4	–1.0	1.6	+1.4	–1.6	3.0	+2.2	–3.0	5.0	+3.5	– 5.0
	2.5	–0	–1.6	3.3	–0	–1.9	3.9	–0	–2.5	6.6	–0	–4.4	10.7	–0	– 7.2
0.40–0.71	1.2	+1.0	–1.2	1.2	+1.6	–1.2	2.0	+1.6	–2.0	3.5	+2.8	–3.5	6.0	+4.0	– 6.0
	2.9	–0	–1.9	3.8	–0	–2.2	4.6	–0	–3.0	7.9	–0	–5.1	12.8	–0	– 8.8
0.71–1.19	1.6	+1.2	–1.6	1.6	+2.0	–1.6	2.5	+2.0	–2.5	4.5	+3.5	–4.5	7.0	+5.0	– 7.0
	3.6	–0	–2.4	4.8	–0	–2.8	5.7	–0	–3.7	10.0	–0	–6.5	15.5	–0	–10.5
1.19–1.97	2.0	+1.6	–2.0	2.0	+2.5	–2.0	3.0	+2.5	–3.0	5.0	+4.0	–5.0	8.0	+6.0	– 8.0
	4.6	–0	–3.0	6.1	–0	–3.6	7.1	–0	–4.6	11.5	–0	–7.5	18.0	–0	–12.0
1.97–3.15	2.5	+1.8	–2.5	2.5	+3.0	–2.5	4.0	+3.0	–4.0	6.0	+4.5	–6.0	9.0	+7.0	– 9.0
	5.5	–0	–3.7	7.3	–0	–4.3	8.8	–0	–5.8	13.5	–0	–9.0	20.5	–0	–13.5

Appendix C

Table C–10. Clearance Locational Fits

Basic hole system. Limits are in thousandths of an inch.

Nominal Size Range (Inches) Over — To	Class LC 1			Class LC 2			Class LC 3			Class LC 4			Class LC 5		
	Limits of Clearance	Standard Limits Hole	Shaft	Limits of Clearance	Standard Limits Hole	Shaft	Limits of Clearance	Standard Limits Hole	Shaft	Limits of Clearance	Standard Limits Hole	Shaft	Limits of Clearance	Standard Limits Hole	Shaft
0–0.12	0	+0.25	+0	0	+0.4	+0	0	+0.6	+0	0	+1.6	+0	0.1	+0.4	−0.1
	0.45	−0	−0.2	0.65	−0	−0.25	1	−0	−0.4	2.6	−0	−1.0	0.75	−0	−0.35
0.12–0.24	0	+0.3	+0	0	+0.5	+0	0	+0.7	+0	0	+1.8	+0	0.15	+0.5	−0.15
	0.5	−0	−0.2	0.8	−0	−0.3	1.2	−0	−0.5	3.0	−0	−1.2	0.95	−0	−0.45
0.24–0.40	0	+0.4	+0	0	+0.6	+0	0	+0.9	+0	0	+2.2	+0	0.2	+0.6	−0.2
	0.65	−0	−0.25	1.0	−0	−0.4	1.5	−0	−0.6	3.6	−0	−1.4	1.2	−0	−0.6
0.40–0.71	0	+0.4	+0	0	+0.7	+0	0	+1.0	+0	0	+2.8	+0	0.25	+0.7	−0.25
	0.7	−0	−0.3	1.1	−0	−0.4	1.7	−0	−0.7	4.4	−0	−1.6	1.35	−0	−0.65
0.71–1.19	0	+0.5	+0	0	+0.8	+0	0	+1.2	+0	0	+3.5	+0	0.3	+0.8	−0.3
	0.9	−0	−0.4	1.3	−0	−0.5	2	−0	−0.8	5.5	−0	−2.0	1.6	−0	−0.8
1.19–1.97	0	+0.6	+0	0	+1.0	+0	0	+1.6	+0	0	+4.0	+0	0.4	+1.0	−0.4
	1.0	−0	−0.4	1.6	−0	−0.6	2.6	−0	−1	6.5	−0	−2.5	2.0	−0	−1.0
1.97–3.15	0	+0.7	+0	0	+1.2	+0	0	+1.8	+0	0	+4.5	+0	0.4	+1.2	−0.4
	1.2	−0	−0.5	1.9	−0	−0.7	3	−0	−1.2	7.5	−0	−3	2.3	−0	−1.1

Nominal Size Range (Inches) Over — To	Class LC 6			Class LC 7			Class LC 8			Class LC 9			Class LC 10			Class LC 11		
	Limits of Clearance	Standard Limits Hole	Shaft	Limits of Clearance	Standard Limits Hole	Shaft	Limits of Clearance	Standard Limits Hole	Shaft	Limits of Clearance	Standard Limits Hole	Shaft	Limits of Clearance	Standard Limits Hole	Shaft	Limits of Clearance	Standard Limits Hole	Shaft
0–0.12	0.3	+1.0	−0.3	0.6	+1.6	−0.6	1.0	+1.6	1.0	2.5	+2.5	−2.5	4	+4	−4	5	+6	−5
	1.9	−0	−0.9	3.2	−0	−1.6	3.6	−0	−2.0	6.6	−0	−4.1	12	−0	−8	17	−0	−11
0.12–0.24	0.4	+1.2	−0.4	0.8	+1.8	−0.8	1.2	+1.8	−1.2	2.8	+3.0	−2.8	4.5	+5	−4.5	6	+7	−6
	2.3	−0	−1.1	3.8	−0	−2.0	4.2	−0	−2.4	7.6	−0	−4.6	14.5	−0	−9.5	20	−0	−13
0.24–0.40	0.5	+1.5	−0.5	1.0	+2.2	−1.0	1.6	+2.2	−1.6	3.0	+3.5	−3.0	5	+6	−5	7	+9	−7
	2.8	−0	−1.4	4.6	−0	−2.4	5.2	−0	−3.0	8.7	−0	−5.2	17	−0	−11	25	−0	−16
0.40–0.71	0.6	+1.6	−0.6	1.2	+2.8	−1.2	2.0	+2.8	−2.0	3.5	+4.0	−3.5	6	+7	−6	8	+10	−8
	3.2	−0	−1.6	5.6	−0	−2.8	6.4	−0	−3.6	10.3	−0	−6.3	20	−0	−13	28	−0	−18
0.71–1.19	0.8	+2.0	−0.8	1.6	+3.5	−1.6	2.5	+3.5	−2.5	4.5	+5.0	−4.5	7	+8	−7	10	+12	−10
	4.0	−0	−2.0	7.1	−0	−3.6	8.0	−0	−4.5	13.0	−0	−8.0	23	−0	−15	34	−0	−22
1.19–1.97	1.0	+2.5	−1.0	2.0	+4.0	−2.0	3.0	+4.0	−3.0	5	+6	−5	8	+10	−8	12	+16	−12
	5.1	−0	−2.6	8.5	−0	−4.5	9.5	−0	−5.5	15	−0	−9	28	−0	−18	44	−0	−28
1.97–3.15	1.2	+3.0	−1.2	2.5	+4.5	−2.5	4.0	+4.5	−4.0	6	+7	−6	10	+12	−10	14	+18	−14
	6.0	−0	−3.0	10.0	−0	−5.5	11.5	−0	−7.0	17.5	−0	−10.5	34	−0	−22	50	−0	−32

Table C–11. Transition Locational Fits

Basic hole system. Limits are in thousandths of an inch.

Nominal Size Range (Inches) Over	To	Class LT 1 Fit	Standard Limits Hole	Shaft	Class LT 2 Fit	Standard Limits Hole	Shaft	Class LT 3 Fit	Standard Limits Hole	Shaft	Class LT 4 Fit	Standard Limits Hole	Shaft	Class LT 5 Fit	Standard Limits Hole	Shaft	Class LT 6 Fit	Standard Limits Hole	Shaft
0–0.12		−0.10	+0.4	+0.10	−0.2	+0.6	+0.2	...	...	...	...	...	...	−0.5	+0.4	+0.5	−0.65	+0.4	+0.65
		+0.50	−0	−0.10	+0.8	−0	−0.2	...	...	...	...	...	...	+0.15	−0	+0.25	+0.15	−0	+0.25
0.12–0.24		−0.15	+0.5	+0.15	−0.25	+0.7	+0.25	...	...	...	...	...	...	−0.6	+0.5	+0.6	−0.8	+0.5	+0.8
		+0.65	−0	−0.15	+0.95	−0	−0.25	...	...	...	...	...	...	+0.2	−0	+0.3	+0.2	−0	+0.3
0.24–0.40		−0.2	+0.6	+0.2	−0.3	+0.9	+0.3	−0.5	+0.6	+0.5	−0.7	+0.9	+0.7	−0.8	+0.6	+0.8	−1.0	+0.6	+1.0
		+0.8	−0	−0.2	+1.2	−0	−0.3	+0.5	−0	+0.1	+0.8	−0	+0.1	+0.2	−0	+0.4	+0.2	−0	+0.4
0.40–0.71		−0.2	+0.7	+0.2	−0.35	+1.0	+0.35	−0.5	+0.7	+0.5	−0.8	+1.0	+0.8	−0.9	+0.7	+0.9	−1.2	+0.7	+1.2
		+0.9	−0	−0.2	+1.35	−0	−0.35	+0.6	−0	+0.1	+0.9	−0	+0.1	+0.2	−0	+0.5	+0.2	−0	+0.5
0.71–1.19		−0.25	+0.8	+0.25	−0.4	+1.2	+0.4	−0.6	+0.8	+0.6	−0.9	+1.2	+0.9	−1.1	+0.8	+1.1	−1.4	+0.8	+1.4
		+1.05	−0	−0.25	+1.6	−0	−0.4	+0.7	−0	+0.1	+1.1	−0	+0.1	+0.2	−0	+0.6	+0.2	−0	+0.6
1.19–1.97		−0.3	+1.0	+0.3	−0.5	+1.6	+0.5	−0.7	+1.0	+0.7	−1.1	+1.6	+1.1	−1.3	+1.0	+1.3	−1.7	+1.0	+1.7
		+1.3	−0	−0.3	+2.1	−0	−0.5	+0.9	−0	+0.1	+1.5	−0	+0.1	+0.3	−0	+0.7	+0.3	−0	+0.7
1.97–3.15		−0.3	+1.2	+0.3	−0.6	+1.8	+0.6	−0.8	+1.2	+0.8	−1.3	+1.8	+1.3	−1.5	+1.2	+1.5	−2.0	+1.2	+2.0
		+1.5	−0	−0.3	+2.4	−0	−0.6	+1.1	−0	+0.1	+1.7	−0	+0.1	+0.4	−0	+0.8	+0.4	−0	+0.8

Table C–12. Interference Locational Fits

Basic hole system. Limits are in thousandths of an inch.

Nominal Size Range (Inches) Over	To	Class LN 1 Limits of Interference	Standard Limits Hole	Shaft	Class LN 2 Limits of Interference	Standard Limits Hole	Shaft	Class LN 3 Limits of Interference	Standard Limits Hole	Shaft
0–0.12		0	+0.25	+0.45	0	+0.4	+0.65	0.1	+0.4	+0.75
		0.45	−0	+0.25	0.65	−0	+0.4	0.75	−0	+0.5
0.12–0.24		0	+0.3	+0.5	0	+0.5	+0.8	0.1	+0.5	+0.9
		0.5	−0	+0.3	0.8	−0	+0.5	0.9	0	+0.6
0.24–0.40		0	+0.4	+0.65	0	+0.6	+1.0	0.2	+0.6	+1.2
		0.65	−0	+0.4	1.0	−0	+0.6	1.2	−0	+0.8
0.40–0.71		0	+0.4	+0.8	0	+0.7	+1.1	0.3	+0.7	+1.4
		0.8	−0	+0.4	1.1	−0	+0.7	1.4	−0	+1.0
0.71–1.19		0	+0.5	+1.0	0	+0.8	+1.3	0.4	+0.8	+1.7
		1.0	−0	+0.5	1.3	−0	+0.8	1.7	−0	+1.2
1.19–1.97		0	+0.6	+1.1	0	+1.0	+1.6	0.4	+1.0	+2.0
		1.1	−0	+0.6	1.6	−0	+1.0	2.0	−0	+1.4
1.97–3.15		0.1	+0.7	+1.3	0.2	+1.2	+2.1	0.4	+1.2	+2.3
		1.3	−0	+0.7	2.1	−0	+1.4	2.3	−0	+1.6

Table C-13. Force and Shrink Fits

Basic hole system. Limits are in thousandths of an inch.

Nominal Size Range (Inches) Over — To	Class FN 1 Light Drive Fit			Class FN 2 Medium Drive Fit			Class FN 3 Heavy Drive Fit			Class FN 4 Force Fit			Class FN 5 Force Fit		
	Limits of Interference	Standard Limits Hole	Shaft	Limits of Interference	Standard Limits Hole	Shaft	Limits of Interference	Standard Limits Hole	Shaft	Limits of Interference	Standard Limits Hole	Shaft	Limits of Interference	Standard Limits Hole	Shaft
0–0.12	0.05	+0.25	+0.5	0.2	+0.4	+0.85	...	...	...	0.3	+0.4	+0.95	0.3	+0.6	+1.3
	0.5	−0	+0.3	0.85	−0	+0.6	...	...	...	0.95	−0	+0.7	1.3	−0	+0.9
0.12–0.24	0.1	+0.3	+0.6	0.2	+0.5	+1.0	...	...	...	0.4	+0.5	+1.2	0.5	+0.7	+1.7
	0.6	−0	+0.4	1.0	−0	+0.7	...	...	...	1.2	−0	+0.9	1.7	−0	+1.2
0.24–0.40	0.1	+0.4	+0.75	0.4	+0.6	+1.4	...	...	...	0.6	+0.6	+1.6	0.5	+0.9	+2.0
	0.75	−0	+0.5	1.4	−0	+1.0	...	...	...	1.6	−0	+1.2	2.0	−0	+1.4
0.40–0.56	0.1	+0.4	+0.8	0.5	+0.7	+1.6	...	...	...	0.7	+0.7	+1.8	0.6	+1.0	+2.3
	0.8	−0	+0.5	1.6	−0	+1.2	...	...	...	1.8	−0	+1.4	2.3	−0	+1.6
0.56–0.71	0.2	+0.4	+0.9	0.5	+0.7	+1.6	...	...	...	0.7	+0.7	+1.8	0.8	+1.0	+2.5
	0.9	−0	+0.6	1.6	−0	+1.2	...	...	...	1.8	−0	+1.4	2.5	−0	+1.8
0.71–0.95	0.2	+0.5	+1.1	0.6	+0.8	+1.9	...	...	...	0.8	+0.8	+2.1	1.0	+1.2	+3.0
	1.1	−0	+0.7	1.9	−0	+1.4	...	...	...	2.1	−0	+1.6	3.0	−0	+2.2
0.95–1.19	0.3	+0.5	+1.2	0.6	+0.8	+1.9	0.8	+0.8	+2.1	1.0	+0.8	+2.3	1.3	+1.2	+3.3
	1.2	−0	+0.8	1.9	−0	+1.4	2.1	−0	+1.6	2.3	−0	+1.8	3.3	−0	+2.5
1.19–1.58	0.3	+0.6	+1.3	0.8	+1.0	+2.4	1.0	+1.0	+2.6	1.5	+1.0	+3.1	1.4	+1.6	+4.0
	1.3	−0	+0.9	2.4	−0	+1.8	2.6	−0	+2.0	3.1	−0	+2.5	4.0	−0	+3.0
1.58–1.97	0.4	+0.6	+1.4	0.8	+1.0	+2.4	1.2	+1.0	+2.8	1.8	+1.0	+3.4	2.4	+1.6	+5.0
	1.4	−0	+1.0	2.4	−0	+1.8	2.8	−0	+2.2	3.4	−0	+2.8	5.0	−0	+4.0
1.97–2.56	0.6	+0.7	+1.8	0.8	+1.2	+2.7	1.3	+1.2	+3.2	2.3	+1.2	+4.2	3.2	+1.8	+6.2
	1.8	−0	+1.3	2.7	−0	+2.0	3.2	−0	+2.5	4.2	−0	+3.5	6.2	−0	+5.0

Table C-14. Description of Preferred Metric Fits

	ISO Symbol		Description	
	Hole Basis	Shaft Basis		
Clearance Fits	H11/c11	C11/h11	***Loose running fit*** for wide commercial tolerances or allowances on external members.	More Clearance
Clearance Fits	H9/d9	D9/h9	***Free running fit*** not for use where accuracy is essential, but good for large temperature variations, high running speeds, or heavy journal pressures.	More Clearance
Clearance Fits	H8/f7	F8/h7	***Close running fit*** for running on accurate machines and for accurate location at moderate speeds and journal pressures.	More Clearance
Clearance Fits	H7/g6	G7/h6	***Sliding fit*** not intended to run freely, but to move and turn freely and locate accurately.	More Clearance
Clearance Fits	H7/h6	H7/h6	***Locational clearance fit*** provides snug fit for locating stationary parts; but can be freely assembled and disassembled.	More Clearance
Transition Fits	H7/k6	K7/h6	***Locational transition fit*** for accurate location, a compromise between clearance and interference.	More Interference
Transition Fits	H7/n6	N7/h6	***Locational transition fit*** for more accurate location where greater interference is permissible.	More Interference
Interference Fits	H7/p6[1]	P7/h6	***Locational interference fit*** for parts requiring rigidity and alignment with prime accuracy of location but without special bore pressure requirements.	More Interference
Interference Fits	H7/s6	S7/h6	***Medium drive fit*** for ordinary steel parts or shrink fits on light sections, the tightest fit usable with cast iron.	More Interference
Interference Fits	H7/u6	U7/h6	***Force fit*** suitable for parts which can be highly stressed or for shrink fits where the heavy pressing forces required are impractical.	More Interference

[1]Transition fit for basic sizes in range from 0 through 3 mm.

Appendix C

Table C–15A. Preferred Hole Basis Clearance Fits

Basic hole system. Dimensions are in millimeters.

Basic Size		Loose Running			Free Running			Close Running			Sliding			Locational Clearance		
		Hole H11	Shaft c11	Fit	Hole H9	Shaft d9	Fit	Hole H8	Shaft f7	Fit	Hole H7	Shaft g6	Fit	Hole H7	Shaft h6	Fit
1	MAX	1.060	0.940	.180	1.025	0.980	.070	1.014	0.994	.030	1.010	0.998	.018	1.010	1.000	.016
	MIN	1.000	0.880	.060	1.000	0.955	.020	1.000	0.984	.006	1.000	0.992	.002	1.000	0.994	.000
1.2	MAX	1.260	1.140	.180	1.225	1.180	.070	1.214	1.194	.030	1.210	1.198	.018	1.210	1.200	.016
	MIN	1.200	1.080	.060	1.200	1.155	.020	1.200	1.184	.006	1.200	1.192	.002	1.200	1.194	.000
1.6	MAX	1.660	1.540	.180	1.625	1.580	.070	1.614	1.594	.030	1.610	1.598	.018	1.610	1.600	.016
	MIN	1.600	1.480	.060	1.600	1.555	.020	1.600	1.584	.006	1.600	1.592	.002	1.600	1.594	.000
2	MAX	2.060	1.940	.180	2.025	1.980	.070	2.014	1.994	.030	2.010	1.998	.018	2.010	2.000	.016
	MIN	2.000	1.880	.060	2.000	1.955	.020	2.000	1.984	.006	2.000	1.992	.002	2.000	1.994	.000
2.5	MAX	2.560	2.440	.180	2.525	2.480	.070	2.514	2.494	.030	2.510	2.498	.018	2.510	2.500	.016
	MIN	2.500	2.380	.060	2.500	2.455	.020	2.500	2.484	.006	2.500	2.492	.002	2.500	2.494	.000
3	MAX	3.060	2.940	.180	3.025	2.980	.070	3.014	2.994	.030	3.010	2.998	.018	3.010	3.000	.016
	MIN	3.000	2.880	.060	3.000	2.955	.020	3.000	2.984	.006	3.000	2.992	.002	3.000	2.994	.000
4	MAX	4.075	3.930	.220	4.030	3.970	.090	4.108	3.990	.040	4.012	3.996	.024	4.012	4.000	.020
	MIN	4.000	3.855	.070	4.000	3.940	.030	4.000	3.978	.010	4.000	3.988	.004	4.000	3.992	.000
5	MAX	5.075	4.930	.220	5.030	4.970	.090	5.018	4.990	.040	5.012	4.996	.024	5.012	5.000	.020
	MIN	5.000	4.855	.070	5.000	4.940	.030	5.000	4.978	.010	5.000	4.988	.004	5.000	4.992	.000
6	MAX	6.075	5.930	.220	6.030	5.970	.090	6.018	5.990	.040	6.012	5.996	.024	6.012	6.000	.020
	MIN	6.000	5.855	.070	6.000	5.940	.030	6.000	5.978	.010	6.000	5.988	.004	6.000	5.992	.000
8	MAX	8.090	7.920	.260	8.036	7.960	.112	8.022	7.987	.050	8.015	7.995	.029	8.015	8.000	.024
	MIN	8.000	7.830	.080	8.000	7.924	.040	8.000	7.972	.013	8.000	7.986	.006	8.000	7.991	.000
10	MAX	10.090	9.920	.260	10.036	9.960	.112	10.022	9.987	.050	10.015	9.995	.029	10.015	10.000	.024
	MIN	10.000	9.830	.080	10.000	9.924	.040	10.000	9.972	.013	10.000	9.986	.005	10.000	9.991	.000
12	MAX	12.110	11.905	.315	12.043	11.950	.136	12.027	11.984	.061	12.018	11.994	.035	12.018	12.000	.029
	MIN	12.000	11.795	.095	12.000	11.907	.050	12.000	11.966	.016	12.000	11.983	.006	12.000	11.989	.000
16	MAX	16.110	15.905	.315	16.043	15.950	.136	16.027	15.984	.061	16.018	15.994	.035	16.018	16.000	.029
	MIN	16.000	15.795	.095	16.000	15.907	.050	16.000	15.966	.016	16.000	15.983	.006	16.000	15.989	.000
20	MAX	20.130	19.890	.370	20.052	19.935	.169	20.033	19.980	.074	20.021	19.993	.041	20.021	20.000	.034
	MIN	20.000	19.760	.110	20.000	19.883	.065	20.000	19.959	.020	20.000	19.980	.007	20.000	19.987	.000
25	MAX	25.130	24.890	.370	25.052	24.935	.169	25.033	24.980	.074	25.021	24.993	.042	25.021	25.000	.034
	MIN	25.000	24.760	.110	25.000	24.883	.065	25.000	24.959	.020	25.000	24.980	.007	25.000	24.987	.000
30	MAX	30.130	29.890	.370	30.052	29.935	.169	30.033	29.980	.074	30.021	29.993	.041	30.021	30.000	.034
	MIN	30.000	29.760	.110	30.000	29.883	.065	30.000	29.959	.020	30.000	29.980	.007	30.000	29.987	.000
40	MAX	40.160	39.880	.440	40.062	39.920	.204	40.039	39.975	.089	40.025	39.991	.050	40.025	40.000	.041
	MIN	40.000	39.720	.120	40.000	39.858	.080	40.000	39.950	.025	40.000	39.975	.009	40.000	39.984	.000
50	MAX	50.160	49.870	.450	50.062	49.920	.204	50.039	49.975	.089	50.025	49.991	.050	50.025	50.000	.041
	MIN	50.000	49.710	.130	50.000	49.858	.080	50.000	49.950	.025	50.000	49.975	.009	50.000	49.984	.000
60	MAX	60.190	59.860	.520	60.074	59.900	.248	60.046	59.970	.106	60.030	59.990	.059	60.030	60.000	.049
	MIN	60.000	59.670	.140	60.000	59.826	.100	60.000	59.940	.030	60.000	59.971	.010	60.000	59.981	.000
80	MAX	80.190	79.850	.530	80.074	79.900	.248	80.046	79.970	.106	80.030	79.990	.059	80.030	80.000	.049
	MIN	80.000	79.660	.150	80.000	79.826	.100	80.000	79.940	.030	80.000	79.971	.010	80.000	79.981	.000
100	MAX	100.220	99.830	.610	100.087	99.880	.294	100.054	99.964	.125	100.035	99.988	.069	100.035	100.000	.057
	MIN	100.000	99.610	.170	100.000	99.793	.120	100.000	99.929	.036	100.000	99.966	.012	100.000	99.978	.000
120	MAX	120.220	119.820	.620	120.087	119.880	.294	120.054	119.964	.125	120.035	119.988	.069	120.035	120.000	.057
	MIN	120.000	119.600	.180	120.000	119.793	.120	120.000	119.929	.036	120.000	119.966	.012	120.000	119.978	.000
160	MAX	160.250	159.790	.710	160.100	159.855	.345	160.063	159.957	.146	160.040	159.986	.079	160.040	160.000	.065
	MIN	160.000	159.540	.210	160.000	159.755	.145	160.000	159.917	.043	160.000	159.961	.014	160.000	159.975	.000

Table C-15B. Preferred Hole Basis Transition and Interference Fits

Basic hole system. Dimensions are in millimeters.

Basic Size		Locational Transition			Locational Transition			Locational Interference			Medium Drive			Force		
		Hole H7	Shaft k6	Fit	Hole H7	Shaft n6	Fit	Hole H7	Shaft p6	Fit	Hole H7	Shaft s6	Fit	Hole H7	Shaft u6	Fit
1	MAX	1.010	1.006	.010	1.010	1.010	.006	1.010	1.012	.004	1.010	1.020	−.004	1.010	1.024	−.008
	MIN	1.000	1.000	−.006	1.000	1.004	−.010	1.000	1.006	−.012	1.000	1.014	−.020	1.000	1.018	−.024
1.2	MAX	1.210	1.206	.010	1.210	1.210	.006	1.210	1.212	.004	1.210	1.220	−.004	1.210	1.224	−.008
	MIN	1.200	1.200	−.006	1.200	1.204	−.010	1.200	1.206	−.012	1.200	1.214	−.020	1.200	1.218	−.024
1.6	MAX	1.610	1.606	.010	1.610	1.610	.006	1.610	1.612	.004	1.610	1.620	−.004	1.610	1.624	−.008
	MIN	1.600	1.600	−.006	1.600	1.604	−.010	1.600	1.606	−.012	1.600	1.614	−.020	1.600	1.618	−.024
2	MAX	2.010	2.006	.010	2.010	2.010	.006	2.010	2.012	.004	2.010	2.020	−.004	2.010	2.024	−.008
	MIN	2.000	2.000	−.006	2.000	2.004	−.010	2.000	2.006	−.012	2.000	2.014	−.020	2.000	2.018	−.024
2.5	MAX	2.510	2.506	.010	2.510	2.510	.006	2.510	2.512	.004	2.510	2.520	−.004	2.510	2.524	−.008
	MIN	2.500	2.500	−.006	2.500	2.504	−.010	2.500	2.506	−.012	2.500	2.514	−.020	2.500	2.518	−.024
3	MAX	3.010	3.006	.010	3.010	3.010	.006	3.010	3.012	.004	3.010	3.020	−.004	3.010	3.024	−.008
	MIN	3.000	3.000	−.006	3.000	3.004	−.010	3.000	3.006	−.012	3.000	3.014	−.020	3.000	3.018	−.024
4	MAX	4.012	4.009	.011	4.012	4.016	.004	4.012	4.020	.000	4.012	4.027	−.007	4.012	4.031	−.011
	MIN	4.000	4.001	−.009	4.000	4.008	−.016	4.000	4.012	−.020	4.000	4.019	−.027	4.000	4.023	−.031
5	MAX	5.012	5.009	.011	5.012	5.016	.004	5.012	5.020	.000	5.012	5.027	−.007	5.012	5.031	−.011
	MIN	5.000	5.001	−.009	5.000	5.008	−.016	5.000	5.012	−.020	5.000	5.019	−.027	5.000	5.023	−.031
6	MAX	6.012	6.009	.011	6.012	6.016	.004	6.012	6.020	.000	6.012	6.027	−.007	6.012	6.031	−.011
	MIN	6.000	6.001	−.009	6.000	6.008	−.016	6.000	6.012	−.020	6.000	6.019	−.027	6.000	6.023	−.031
8	MAX	8.015	8.010	.014	8.015	8.019	.005	8.015	8.024	.000	8.015	8.032	−.008	8.015	8.037	−.013
	MIN	8.000	8.001	−.010	8.000	8.010	−.019	8.000	8.015	−.024	8.000	8.023	−.032	8.000	8.028	−.037
10	MAX	10.015	10.010	.014	10.015	10.019	.005	10.015	10.024	.000	10.015	10.032	−.008	10.015	10.037	−.013
	MIN	10.000	10.001	−.010	10.000	10.010	−.019	10.000	10.015	−.024	10.000	10.023	−.032	10.000	10.028	−.037
12	MAX	12.018	12.012	.017	12.018	12.023	.006	12.018	12.029	.000	12.018	12.039	−.010	12.018	12.044	−.015
	MIN	12.000	12.001	−.012	12.000	12.012	−.023	12.000	12.018	−.029	12.000	12.028	−.039	12.000	12.033	−.044
16	MAX	16.018	16.012	.017	16.018	16.023	.006	16.018	16.029	.000	16.018	16.039	−.010	16.018	16.044	−.015
	MIN	16.000	16.001	−.012	16.000	16.012	−.023	16.000	16.018	−.029	16.000	16.028	−.039	16.000	16.033	−.044
20	MAX	20.021	20.015	.019	20.021	20.028	.006	20.021	20.035	−.001	20.021	20.048	−.014	20.021	20.054	−.020
	MIN	20.000	20.002	−.015	20.000	20.015	−.028	20.000	20.022	−.035	20.000	20.035	−.048	20.000	20.041	−.054
25	MAX	25.021	25.015	.019	25.021	25.028	.006	25.021	25.035	−.001	25.021	25.048	−.014	25.021	25.061	−.027
	MIN	25.000	25.002	−.015	25.000	25.015	−.028	25.000	25.022	−.035	25.000	25.035	−.048	25.000	25.048	−.061
30	MAX	30.021	30.015	.019	30.021	30.028	.006	30.021	30.035	−.001	30.021	30.048	−.014	30.021	30.061	−.027
	MIN	30.000	30.002	−.015	30.000	30.015	−.028	30.000	30.022	−.035	30.000	30.035	−.048	30.000	30.048	−.061
40	MAX	40.025	40.018	.023	40.025	40.033	.008	40.025	40.042	−.001	40.025	40.059	−.018	40.025	40.076	−.035
	MIN	40.000	40.002	−.018	40.000	40.017	−.033	40.000	40.026	−.042	40.000	40.043	−.059	40.000	40.060	−.076
50	MAX	50.025	50.018	.023	50.025	50.033	.008	50.025	50.042	−.001	50.025	50.059	−.018	50.025	50.086	−.045
	MIN	50.002	50.000	−.018	50.000	50.017	−.033	50.000	50.026	−.042	50.000	50.043	−.059	50.000	50.070	−.086
60	MAX	60.030	60.021	.028	60.030	60.039	.010	60.030	60.051	−.002	60.030	60.072	−.023	60.030	60.106	−.057
	MIN	60.000	60.002	−.021	60.000	60.020	−.039	60.000	60.032	−.051	60.000	60.053	−.072	60.000	60.087	−.106
80	MAX	80.030	80.021	.028	80.030	80.039	.010	80.030	80.051	−.002	80.030	80.078	−.029	80.030	80.121	−.072
	MIN	80.000	80.002	−.021	80.000	80.020	−.039	80.000	80.032	−.051	80.000	80.059	−.078	80.000	80.102	−.121
100	MAX	100.035	100.025	.032	100.035	100.045	.012	100.035	100.059	−.002	100.035	100.093	−.036	100.035	100.146	−.089
	MIN	100.000	100.003	−.025	100.000	100.023	−.045	100.000	100.037	−.059	100.000	100.071	−.093	100.000	100.124	−.146
120	MAX	120.035	120.025	.032	120.035	120.045	.012	120.035	120.059	−.002	120.035	120.101	−.044	120.035	120.166	−.109
	MIN	120.000	120.003	−.025	120.000	120.023	−.045	120.000	120.037	−.059	120.000	120.079	−.101	120.000	120.144	−.166
160	MAX	160.040	160.028	.037	160.045	160.052	.013	160.040	160.068	−.003	160.040	160.125	−.060	160.040	160.215	−.150
	MIN	160.000	160.003	−.028	160.000	160.027	−.052	160.000	160.043	−.068	160.000	160.000	−.125	160.000	160.190	−.215

Table C–16A. Preferred Shaft Basis Clearance Fits

Basic shaft system. Dimensions are in millimeters.

Basic Size		Loose Running			Free Running			Close Running			Sliding			Locational Clearance		
		Hole C11	Shaft h11	Fit	Hole D9	Shaft h9	Fit	Hole F8	Shaft h7	Fit	Hole G7	Shaft h6	Fit	Hole H7	Shaft h6	Fit
1	MAX	1.120	1.000	.180	1.045	1.000	.070	1.020	1.000	.030	1.012	1.000	.018	1.010	1.000	.016
	MIN	1.060	0.940	.060	1.020	0.975	.020	1.006	0.990	.006	1.002	0.994	.002	1.000	0.994	.000
1.2	MAX	1.320	1.200	.180	1.245	1.200	.070	1.220	1.200	.030	1.212	1.200	.018	1.210	1.200	.016
	MIN	1.260	1.140	.060	1.220	1.175	.020	1.206	1.190	.006	1.202	1.194	.002	1.200	1.194	.000
1.6	MAX	1.720	1.600	.180	1.645	1.600	.070	1.620	1.600	.030	1.612	1.600	.018	1.610	1.600	.016
	MIN	1.660	1.540	.060	1.620	1.575	.020	1.606	1.590	.006	1.602	1.594	.002	1.600	1.594	.000
2	MAX	2.120	2.000	.180	2.045	2.000	.070	2.020	2.000	.030	2.012	2.000	.018	2.010	2.000	.016
	MIN	2.060	1.940	.060	2.020	1.975	.020	2.006	1.990	.006	2.002	1.994	.002	2.000	1.994	.000
2.5	MAX	2.620	2.500	.180	2.545	2.500	.070	2.520	2.500	.030	2.512	2.500	.018	2.510	2.500	.016
	MIN	2.560	2.440	.060	2.520	2.475	.020	2.506	2.490	.006	2.502	2.494	.002	2.500	2.494	.000
3	MAX	3.120	3.000	.180	3.045	3.000	.070	3.020	3.000	.030	3.012	3.000	.018	3.010	3.000	.016
	MIN	3.060	2.940	.060	3.020	2.975	.020	3.006	2.990	.006	3.002	2.994	.002	3.000	2.994	.000
4	MAX	4.145	4.000	.220	4.060	4.000	.090	4.028	4.000	.040	4.016	4.000	.024	4.012	4.000	.020
	MIN	4.070	3.925	.070	4.030	3.970	.030	4.010	3.988	.010	4.004	3.992	.004	4.000	3.992	.000
5	MAX	5.145	5.000	.220	5.060	5.000	.090	5.028	5.000	.040	5.016	5.000	.024	5.012	5.000	.020
	MIN	5.070	4.925	.070	5.030	4.970	.030	5.010	4.988	.010	5.004	4.992	.004	5.000	4.992	.000
6	MAX	6.145	6.000	.220	6.060	6.000	.090	6.028	6.000	.040	6.016	6.000	.024	6.012	6.000	.020
	MIN	6.070	5.925	.070	6.030	5.970	.030	6.010	5.988	.010	6.004	5.992	.004	6.000	5.992	.000
8	MAX	8.170	8.000	.260	8.076	8.000	.112	8.035	8.000	.050	8.020	8.000	.029	8.015	8.000	.024
	MIN	8.080	7.910	.080	8.040	7.964	.040	8.013	7.985	.013	8.005	7.991	.005	8.000	7.991	.000
10	MAX	10.170	10.000	.260	10.076	10.000	.112	10.035	10.000	.050	10.020	10.000	.029	10.015	10.000	.024
	MIN	10.080	9.910	.080	10.040	9.964	.040	10.013	9.985	.013	10.005	9.991	.005	10.000	9.991	.000
12	MAX	12.205	12.000	.315	12.093	12.000	.136	12.043	12.000	.061	12.024	12.000	.035	12.018	12.000	.029
	MIN	12.095	11.890	.095	12.050	11.957	.050	12.016	11.982	.016	12.006	11.989	.006	12.000	11.989	.000
16	MAX	16.205	16.000	.315	16.093	16.000	.136	16.043	16.000	.061	16.024	16.000	.035	16.018	16.000	.029
	MIN	16.095	15.890	.095	16.050	15.957	.050	16.016	15.982	.016	16.006	15.989	.006	16.000	15.989	.000
20	MAX	20.240	20.000	.370	20.117	20.000	.169	20.053	20.000	.074	20.028	20.000	.041	20.021	20.000	.034
	MIN	20.110	19.870	.110	20.065	19.948	.065	20.020	19.979	.020	20.007	19.987	.007	20.000	19.987	.000
25	MAX	25.240	25.000	.370	25.117	25.000	.169	25.053	25.000	.074	25.028	25.000	.041	25.021	25.000	.034
	MIN	25.110	24.870	.110	25.065	24.948	.065	25.020	24.979	.020	25.007	24.987	.007	25.000	24.987	.000
30	MAX	30.240	30.000	.370	30.117	30.000	.169	30.053	30.000	.074	30.028	30.000	.041	30.021	30.000	.034
	MIN	30.110	29.870	.110	30.065	29.948	.065	30.020	29.979	.020	30.007	29.987	.007	30.000	29.987	.000
40	MAX	40.280	40.000	.440	40.142	40.000	.204	40.064	40.000	.089	40.034	40.000	.050	40.025	40.000	.041
	MIN	40.120	39.840	.120	40.080	39.938	.080	40.025	39.975	.025	40.009	39.984	.009	40.000	39.984	.000
50	MAX	50.290	50.000	.450	50.142	50.000	.204	50.064	50.000	.089	50.034	50.000	.050	50.025	50.000	.041
	MIN	50.130	49.840	.130	50.080	49.938	.080	50.025	49.975	.025	50.009	49.984	.009	50.000	49.984	.000
60	MAX	60.330	60.000	.520	60.174	60.000	.248	60.076	60.000	.106	60.040	60.000	.059	60.030	60.000	.049
	MIN	60.140	59.810	.140	60.100	59.926	.100	60.030	59.970	.030	60.010	59.981	.010	60.000	59.981	.000
80	MAX	80.340	80.000	.530	80.174	80.000	.248	80.076	80.000	.106	80.040	80.000	.059	80.030	80.000	.049
	MIN	80.150	79.810	.150	80.100	79.926	.100	80.030	79.970	.030	80.010	79.981	.010	80.000	79.981	.000
100	MAX	100.390	100.000	.610	100.207	100.000	.294	100.090	100.000	.125	100.047	100.000	.069	100.035	100.000	.057
	MIN	100.170	99.780	.170	100.120	99.913	.120	100.036	99.965	.036	100.012	99.978	.012	100.000	99.978	.000
120	MAX	120.400	120.000	.620	120.207	120.000	.294	120.090	120.000	.125	120.047	120.000	.069	120.035	120.000	.057
	MIN	120.180	119.780	.180	120.120	119.913	.120	120.036	119.965	.036	120.012	119.978	.012	120.000	119.978	.000
160	MAX	160.460	160.000	.710	160.245	160.000	.345	160.106	160.000	.146	160.054	160.000	.079	160.040	160.000	.065
	MIN	160.210	159.750	.210	160.145	159.900	.145	160.043	159.960	.043	160.014	159.975	.014	160.000	159.975	.000

Table C–16B. Preferred Shaft Basis Transition and Interference Fits

Basic shaft system. Dimensions are in millimeters.

Basic Size		Locational Transition			Locational Transition			Locational Interference			Medium Drive			Force		
		Hole K7	Shaft h6	Fit	Hole N7	Shaft h6	Fit	Hole P7	Shaft h6	Fit	Hole S7	Shaft h6	Fit	Hole U7	Shaft h6	Fit
1	MAX	1.000	1.000	.006	0.996	1.000	.002	0.994	1.000	.000	0.986	1.000	−.008	0.982	1.000	−.012
	MIN	0.990	0.994	−.010	0.986	0.994	−.014	0.984	0.994	−.016	0.976	0.994	−.024	0.972	0.994	−.028
1.2	MAX	1.200	1.200	.006	1.196	1.200	.002	1.194	1.200	.000	1.186	1.200	−.008	1.182	1.200	−.012
	MIN	1.190	1.194	−.010	1.186	1.194	−.014	1.184	1.194	−.016	1.176	1.194	−.024	1.172	1.194	−.028
1.6	MAX	1.600	1.600	.006	1.596	1.600	.002	1.594	1.600	.000	1.586	1.600	−.008	1.582	1.600	−.012
	MIN	1.590	1.594	−.010	1.586	1.594	−.014	1.584	1.594	−.016	1.576	1.594	−.024	1.572	1.594	−.028
2	MAX	2.000	2.000	.006	1.996	2.000	.002	1.994	2.000	.000	1.986	2.000	−.008	1.982	2.000	−.012
	MIN	1.990	1.994	−.010	1.986	1.994	−.014	1.984	1.994	−.016	1.976	1.994	−.024	1.972	1.994	−.028
2.5	MAX	2.500	2.500	.006	2.496	2.500	.002	2.494	2.500	.000	2.486	2.500	−.008	2.482	2.500	−.012
	MIN	2.490	2.494	−.010	2.486	2.494	−.014	2.484	2.494	−.016	2.476	2.494	−.024	2.472	2.494	−.028
3	MAX	3.000	3.000	.006	2.996	3.000	.002	2.994	3.000	.000	2.986	3.000	−.008	2.982	3.000	−.012
	MIN	2.990	2.994	−.010	2.986	2.994	−.014	2.984	2.994	−.016	2.976	2.994	−.024	2.972	2.994	−.028
4	MAX	4.003	4.000	.011	3.996	4.000	.004	3.992	4.000	.000	3.985	4.000	−.007	3.981	4.000	−.011
	MIN	3.991	3.992	−.009	3.984	3.992	−.016	3.980	3.992	−.020	3.973	3.992	−.027	3.969	3.992	−.031
5	MAX	5.003	5.000	.011	4.996	5.000	.004	4.992	5.000	.000	4.985	5.000	−.007	4.981	5.000	−.011
	MIN	4.991	4.992	−.009	4.984	4.992	−.016	4.980	4.992	−.020	4.973	4.992	−.027	4.969	4.992	−.031
6	MAX	6.003	6.000	.011	5.996	6.000	.004	5.992	6.000	.000	5.985	6.000	−.007	5.981	6.000	−.011
	MIN	5.991	5.992	−.009	5.984	5.992	−.016	5.980	5.992	−.020	5.973	5.992	−.027	5.969	5.992	−.031
8	MAX	8.005	8.000	.014	7.996	8.000	.005	7.991	8.000	.000	7.983	8.000	−.008	7.978	8.000	−.013
	MIN	7.990	7.991	−.010	7.981	7.991	−.019	7.976	7.991	−.024	7.968	7.991	−.032	7.963	7.991	−.037
10	MAX	10.005	10.000	.014	9.996	10.000	.005	9.991	10.000	.000	9.983	10.000	−.008	9.978	10.000	−.013
	MIN	9.990	9.991	−.010	9.981	9.991	−.019	9.976	9.991	−.024	9.968	9.991	−.032	9.963	9.991	−.037
12	MAX	12.006	12.000	.017	11.995	12.000	.006	11.989	12.000	.000	11.979	12.000	−.010	11.974	12.000	−.015
	MIN	11.988	11.989	−.012	11.977	11.989	−.023	11.971	11.989	−.029	11.961	11.989	−.039	11.956	11.989	−.044
16	MAX	16.006	16.000	.017	15.995	16.000	.006	15.989	16.000	.000	15.979	16.000	−.010	15.974	16.000	−.015
	MIN	15.988	15.989	−.012	15.977	15.989	−.023	15.971	15.989	−.029	15.961	15.989	−.039	15.956	15.989	−.044
20	MAX	20.006	20.000	.019	19.993	20.000	.006	19.986	20.000	−.001	19.973	20.000	−.014	19.967	20.000	−.020
	MIN	19.985	19.987	−.015	19.972	19.987	−.028	19.965	19.987	−.035	19.952	19.987	−.048	19.946	19.987	−.054
25	MAX	25.006	25.000	.019	24.993	25.000	.006	24.986	25.000	−.001	24.973	25.000	−.014	24.960	25.000	−.027
	MIN	24.985	24.987	−.015	24.972	24.987	−.028	24.965	24.987	−.035	24.952	24.987	−.048	24.939	24.987	−.061
30	MAX	30.006	30.000	.019	29.993	30.000	.006	29.986	30.000	−.001	29.973	30.000	−.014	29.960	30.000	−.027
	MIN	29.985	29.987	−.015	29.972	29.987	−.028	29.965	29.987	−.035	29.952	29.987	−.048	29.939	29.987	−.061
40	MAX	40.007	40.000	.023	39.992	40.000	.008	39.983	40.000	−.001	39.966	40.000	−.018	39.949	40.000	−.035
	MIN	39.982	39.984	−.018	39.967	39.984	−.033	39.958	39.984	−.042	39.941	39.984	−.059	39.924	39.984	−.076
50	MAX	50.007	50.000	.023	49.992	50.000	.008	49.983	50.000	−.001	49.966	50.000	−.018	49.939	50.000	−.045
	MIN	49.982	49.984	−.018	49.967	49.984	−.033	49.958	49.984	−.042	49.941	49.984	−.059	49.914	49.984	−.086
60	MAX	60.009	60.000	.028	59.991	60.000	.010	59.979	60.000	−.002	59.958	60.000	−.023	59.924	60.000	−.057
	MIN	59.979	59.981	−.021	59.961	59.981	−.039	59.949	59.981	−.051	59.928	59.981	−.072	59.894	59.981	−.106
80	MAX	80.009	80.000	.028	79.991	80.000	.010	79.979	80.000	−.002	79.952	80.000	−.029	79.909	80.000	−.072
	MIN	79.979	79.981	−.021	79.961	79.981	−.039	79.949	79.981	−.051	79.922	79.981	−.078	79.879	79.981	−.121
100	MAX	100.010	100.000	.032	99.990	100.000	.012	99.976	100.000	−.002	99.942	100.000	−.036	99.889	100.000	−.089
	MIN	99.975	99.978	−.025	99.955	99.978	−.045	99.941	99.978	−.059	99.907	99.978	−.093	99.854	99.978	−.146
120	MAX	120.010	120.000	.032	119.990	120.000	.012	119.976	120.000	−.002	119.934	120.000	−.044	119.869	120.000	−.109
	MIN	119.975	119.978	−.025	119.955	119.978	−.045	119.941	119.978	−.059	119.899	119.978	−.101	119.834	119.978	−.166
160	MAX	160.012	160.000	.037	159.988	160.000	.013	159.972	160.000	−.003	159.915	160.000	−.060	159.825	160.000	−.150
	MIN	159.972	159.975	−.028	159.948	159.975	−.052	159.932	159.975	−.068	159.875	159.975	−.125	159.785	159.975	−.215

Table C-17. Large Rivets

							Manufactured Shapes										
D Nominal	A Basic	B Basic (Min.)	C	E Basic	F Basic	G	H	I Basic	J Basic (Min.)	K Basic	M Basic (Min.)	N	O	P Basic	Q Basic (Min.)	S Basic
.50	0.875	0.375	0.443	0.781	0.500	0.656	0.094	0.469	0.438	0.905	0.250	0.095	1.125	0.800	0.381	0.500
.62	1.094	0.469	0.553	0.969	0.594	0.750	0.188	0.586	0.547	1.131	0.312	0.119	1.406	1.000	0.469	0.625
.75	1.312	0.562	0.664	1.156	0.688	0.844	0.282	0.703	0.656	1.358	0.375	0.142	1.688	1.200	0.556	0.750
.88	1.531	0.656	0.775	1.344	0.781	0.937	0.375	0.820	0.766	1.584	0.438	0.166	1.969	1.400	0.643	0.875
1.00	1.750	0.750	0.885	1.531	0.875	1.031	0.469	0.938	0.875	1.810	0.500	0.190	2.250	1.600	0.731	1.000
1.12	1.969	0.844	0.996	1.719	0.969	1.125	0.563	1.055	0.984	2.036	0.562	0.214	2.531	1.800	0.835	1.125
1.25	2.188	0.938	1.107	1.906	1.062	1.218	0.656	1.172	1.094	2.262	0.625	0.238	2.812	2.000	0.922	1.250
1.38	2.406	1.031	1.217	2.094	1.156	1.312	0.750	1.290	1.203	2.489	0.688	0.261	3.094	2.200	1.009	1.375
1.50	2.625	1.125	1.328	2.281	1.250	1.406	0.844	1.406	1.312	2.715	0.750	0.285	3.375	2.400	1.113	1.500
1.62	2.844	1.219	1.439	2.469	1.344	1.500	0.938	1.524	1.422	2.941	0.812	0.309	3.656	2.600	1.201	1.625
1.75	3.062	1.312	1.549	2.656	1.438	1.594	1.032	1.641	1.531	3.168	0.875	0.332	3.938	2.800	1.288	1.750

BUTTON HEAD

HIGH BUTTON HEAD

CONE HEAD

FLAT-TOP COUNTERSUNK HEAD

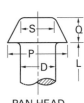

ROUND-TOP COUNTERSUNK HEAD

PAN HEAD

Table C-18. Geometric Dimensioning Symbol Sizes

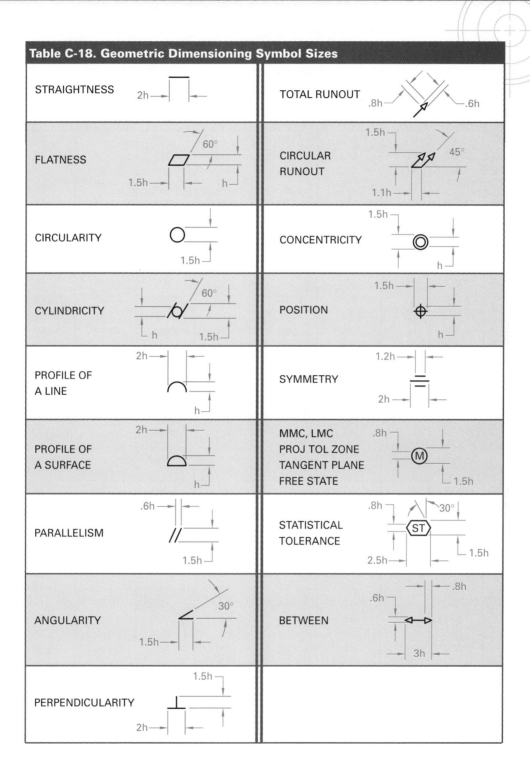

STRAIGHTNESS	TOTAL RUNOUT
FLATNESS	CIRCULAR RUNOUT
CIRCULARITY	CONCENTRICITY
CYLINDRICITY	POSITION
PROFILE OF A LINE	SYMMETRY
PROFILE OF A SURFACE	MMC, LMC PROJ TOL ZONE TANGENT PLANE FREE STATE
PARALLELISM	STATISTICAL TOLERANCE
ANGULARITY	BETWEEN
PERPENDICULARITY	

Table C-19. American Standard Wrought Steel Pipe

Nominal Pipe Size Inches	Outside Diameter	Threads Per Inch	Wall Thickness Sched. 40 (Standard)	Wall Thickness Sched. 80 (Extra Strong)	Wall Thickness Sched. 160	Approx. Distance Pipe Enters Fitting L	Weight (lbs/ft) Sched. 40 (Standard)	Weight (lbs/ft) Sched. 80 (Extra Strong)	Weight (lbs/ft) Sched. 160
1/8 (.125)	.405	27	.068	.095	—	.188	.24	.31	—
1/4 (.250)	.540	18	.088	.119	—	.281	.42	.54	—
3/8 (.375)	.675	18	.091	.126	—	.297	.57	.74	—
1/2 (.500)	.840	14	.109	.147	.188	.375	.85	1.09	1.31
3/4 (.750)	1.050	14	.113	.154	.219	.406	1.13	1.47	1.94
1.00	1.315	11.50	.133	.179	.250	.500	1.68	2.17	2.84
1.25	1.660	11.50	.140	.191	.250	.549	2.27	3.00	3.76
1.50	1.900	11.50	.145	.200	.281	.562	2.72	3.63	4.86
2	2.375	11.50	.154	.218	.344	.578	3.65	5.02	7.46
2.5	2.875	8	.203	.276	.375	.875	5.79	7.66	10.01
3	3.500	8	.216	.300	.438	.938	7.58	10.25	14.31
3.5	4.000	8	.226	.318	—	1.000	9.11	12.51	—
4	4.500	8	.237	.337	.531	1.062	10.79	14.98	22.52
5	5.563	8	.258	.375	.625	1.156	14.62	20.78	32.96
6	6.625	8	.280	.432	.719	1.250	18.97	28.57	45.34
8	8.625	8	.322	.500	.906	1.469	28.55	43.39	74.71

Nominal Pipe Size Inches	Outside Diameter mm	Threads Per Inch	Wall Thickness Sched. 40 (Standard)	Wall Thickness Sched. 80 (Extra Strong)	Wall Thickness Sched. 160	Approx. Distance Pipe Enters Fitting L	Weight (lbs/ft) Sched. 40 (Standard)	Weight (lbs/ft) Sched. 80 (Extra Strong)	Weight (lbs/ft) Sched. 160
1/8 (.125)	10.3	27	1.7	2.4	—	5	0.36	0.46	—
1/4 (.250)	13.7	18	2.2	3.0	—	7	0.63	0.80	—
3/8 (.375)	17.1	18	2.3	3.2	—	8	0.85	1.10	—
1/2 (.500)	21.3	14	2.8	3.7	4.8	10	1.26	1.62	1.95
3/4 (.750)	26.7	14	2.9	3.9	5.6	11	1.68	2.19	2.89
1.00	33.4	11.50	3.4	4.6	6.4	13	2.50	3.23	4.23
1.25	42.1	11.50	3.6	4.9	6.4	14	3.38	4.46	5.60
1.50	48.3	11.50	3.7	5.1	7.1	14	4.05	5.40	7.23
2.00	60.3	11.50	3.9	5.5	8.7	15	5.43	7.47	11.10
2.50	73	8	5.2	7.0	9.5	22	8.62	11.40	14.90
3.00	88.9	8	5.5	7.6	11.1	24	11.28	15.25	21.30
3.50	101.6	8	5.7	8.1	—	25	13.56	18.62	—
4.00	114.3	8	6.0	8.6	13.5	27	16.06	22.30	33.51
5.00	141.3	8	6.6	9.5	15.9	29	21.76	30.92	49.05
6.00	168.3	8	7.1	11.0	18.3	32	28.23	42.52	67.47
8.00	219	8	8.2	12.7	23.0	38	42.49	64.57	111.18

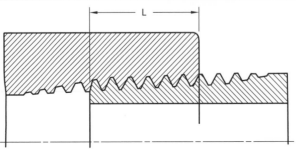

Table C-20. American Standard (125 lb) Cast-Iron Screwed-Pipe Fittings

Nominal Pipe Size in Inches	U.S. Customary (Inches)						Metric (Millimeters)					
	A	Min. B	Min. C	D	E	F	A	Min. B	Min. C	D	E	F
.25	.81	.38	.93	—	—	.73	21	10	24	—	—	19
.375	.95	.44	1.12	—	—	.80	24	11	28	—	—	20
.50	1.12	.50	1.34	2.50	1.87	.88	28	13	34	64	47	22
.75	1.31	.56	1.63	3.00	2.25	.98	33	14	41	76	57	25
1.00	1.50	.62	1.95	3.50	2.75	1.12	38	16	50	89	70	28
1.25	1.75	.69	2.39	4.25	3.25	1.29	44	18	61	108	83	33
1.50	1.94	.75	2.68	4.87	3.81	1.43	49	19	68	124	97	36
2.00	2.25	.84	3.28	5.75	4.25	1.68	57	21	83	146	108	43
2.50	2.70	.94	3.86	6.75	5.18	1.95	69	24	98	171	132	50
3.00	3.08	1.00	4.62	7.87	6.12	2.17	78	25	117	200	155	55
3.50	3.42	1.06	5.20	8.87	6.87	2.39	87	27	132	225	174	61
4.00	3.79	1.12	5.79	9.75	7.62	2.61	96	28	147	248	194	66
5.00	4.50	1.18	7.05	11.62	9.25	3.05	114	30	179	295	235	77
6.00	5.13	1.28	8.28	13.43	10.75	3.46	130	33	210	341	273	88
8.00	6.56	1.47	10.63	16.94	13.63	4.28	167	37	270	430	346	109
10.00	8.08	1.68	13.12	20.69	16.75	5.16	205	43	333	613	425	131

90° ELBOW

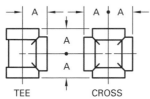

TEE   CROSS

45° ELBOW

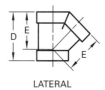

LATERAL

Appendix C

Table C-21. American Standard (150 lb) Malleable-Iron Screwed-Pipe Fittings

Nominal Pipe Size in Inches	U.S. Customary (Inches)							Metric (Millimeters)						
	A	B	C	D	E	F	G	A	B	C	D	E	F	G
.125	.69	.20	.69	—	—	—	.96	18	5.0	18	—	—	—	24
.250	.81	.22	.84	—	—	.73	1.06	21	5	21	—	—	19	27
.375	.95	.23	1.02	1.93	1.43	.80	1.16	24	6	26	49	36	20	29
.500	1.12	.25	1.20	2.32	1.71	.88	1.34	28	6	30	59	43	22	34
.750	1.31	.27	1.46	2.77	2.05	.98	1.52	33	7	37	70	52	25	39
1.00	1.50	.30	1.77	3.28	2.43	1.12	1.67	38	8	45	83	62	28	42
1.25	1.75	.34	2.15	3.94	2.92	1.29	1.93	44	9	55	100	74	33	49
1.50	1.94	.37	2.43	4.38	3.28	1.43	2.15	49	9	62	111	83	36	55
2.00	2.25	.42	2.96	5.17	3.93	1.68	2.53	57	11	75	131	100	43	64
2.50	2.70	.48	3.59	6.25	4.73	1.95	2.88	69	12	91	159	120	50	73
3.00	3.08	.55	4.29	7.26	5.55	2.17	3.18	78	14	109	184	141	55	81
3.50	3.42	.60	4.84	—	—	2.39	3.43	87	15	123	—	—	61	87
4.00	3.79	.66	5.40	8.98	—	2.61	3.69	96	17	137	228	177	66	94
5.00	4.50	.78	6.58	—	6.97	3.05	—	114	20	167	—	—	77	—
6.00	5.13	.90	7.77	—	—	3.46	—	130	23	197	—	—	88	—

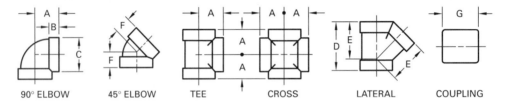

90° ELBOW   45° ELBOW   TEE   CROSS   LATERAL   COUPLING

Table C-22. American Standard Flanged Fittings

	Nominal Pipe Size in Inches	A	B	C	D	E	F	G	H
U.S. Customary (Inches)	1.50	4.00	6.00	5.00	9.00	7.00	2.25	—	.56
	2.00	4.50	6.50	6.00	10.50	8.00	2.50	5.00	.62
	2.50	5.00	7.00	7.00	12.00	9.50	3.00	5.50	.69
	3.00	5.50	7.75	7.50	13.00	10.00	3.00	6.00	.75
	3.50	6.00	8.50	8.50	14.50	11.50	3.50	6.50	.81
	4.00	6.50	9.00	9.00	15.00	12.00	4.00	7.00	.94
	5.00	7.50	10.25	10.00	17.00	13.50	4.50	8.00	.94
	6.00	8.00	11.50	11.00	18.00	14.50	5.00	9.00	1.00
	8.00	9.00	14.00	13.50	22.00	17.50	5.50	11.00	1.12
	10.00	11.00	16.50	16.00	25.50	20.50	6.50	12.00	1.19
Metric (Millimeters)	1.50	102	152	127	229	178	57	—	14
	2.00	114	165	152	267	203	64	127	16
	2.50	127	178	178	305	241	76	140	18
	3.00	140	197	190	330	254	76	152	19
	3.50	153	216	216	368	292	89	165	21
	4.00	165	229	229	381	305	102	178	24
	5.00	190	260	254	432	343	114	203	24
	6.00	203	292	280	457	368	127	229	25
	8.00	229	356	343	559	445	140	279	28
	10.00	279	419	406	648	521	165	305	30

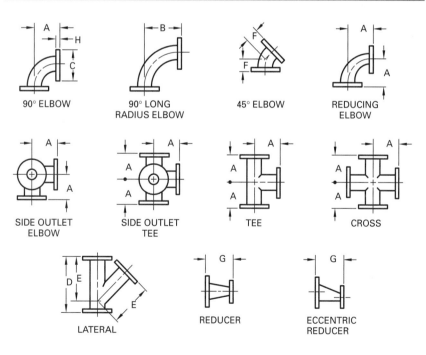

90° ELBOW   90° LONG RADIUS ELBOW   45° ELBOW   REDUCING ELBOW

SIDE OUTLET ELBOW   SIDE OUTLET TEE   TEE   CROSS

LATERAL   REDUCER   ECCENTRIC REDUCER

Appendix C

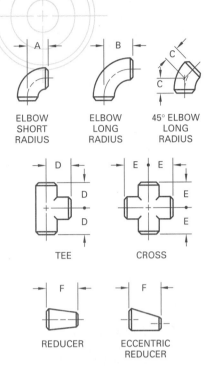

ELBOW
SHORT
RADIUS

ELBOW
LONG
RADIUS

45° ELBOW
LONG
RADIUS

TEE

CROSS

REDUCER

ECCENTRIC
REDUCER

Table C-23. American Standard Steel Butt-Welding Fittings

	Nominal Pipe Size in Inches	A	B	C	D	E	F
U.S. Customary (Inches)	1.50	1.50	2.25	1.12	2.25	2.25	2.50
	2.00	2.00	3.00	1.38	2.50	2.50	3.00
	2.50	2.50	3.75	1.75	3.00	3.00	3.50
	3.00	3.00	4.50	2.00	3.38	3.38	3.50
	3.50	3.50	5.25	2.25	3.75	3.75	4.00
	4.00	4.00	6.00	2.50	4.12	4.12	4.00
	5.00	5.00	7.50	3.12	4.89	4.89	5.00
	6.00	6.00	9.00	3.75	5.62	5.62	5.50
	8.00	8.00	12.00	5.00	7.00	7.00	6.00
	10.00	10.00	15.00	6.25	8.50	8.50	7.00
Metric (Millimeters)	1.50	38	57	28	57	57	64
	2.00	51	76	35	64	64	76
	2.50	64	95	44	76	76	89
	3.00	76	114	51	86	86	89
	3.50	89	133	57	95	95	102
	4.00	102	152	64	105	105	102
	5.00	127	190	79	124	124	127
	6.00	152	229	95	143	143	140
	8.00	203	305	127	178	178	152
	10.00	254	381	159	216	216	178

Table C-24. Common Valves

Nominal Pipe Size in Inches	Check Valves					
	Lift Check		B	Swing Check		D
	A			C		
	Screwed	Flanged		Screwed	Flanged	
U.S. Customary (Inches)						
2.00	6.50	—	3.50	6.50	8.00	4.25
2.50	7.00	—	4.25	7.00	8.50	4.81
3.00	8.00	9.50	5.00	8.00	9.50	5.06
3.50	—	—	—	9.00	10.50	5.81
4.00	10.00	11.50	6.25	10.00	11.50	6.19
5.00	—	13.00	7.00	11.25	13.00	7.19
6.00	—	14.00	8.25	12.50	14.00	7.50
8.00	—	—	—	—	19.50	10.19
10.00	—	—	—	—	24.50	12.12
Metric (Millimeters)						
2.00	165	—	89	165	203	108
2.50	178	—	108	178	216	122
3.00	203	241	127	203	241	129
3.50	—	—	—	229	267	148
4.00	254	292	159	254	292	157
5.00	—	330	178	286	330	183
6.00	—	356	210	318	356	190
8.00	—	—	—	—	495	259
10.00	—	—	—	—	622	308

Nominal Pipe Size in Inches	Gate Valves					
	Nonrising Spindle		F	Rising Spindle		H
	E			G		
	Screwed	Flanged		Screwed	Flanged	
U.S. Customary (Inches)						
2.00	4.75	7.00	10.50	4.75	7.00	13.12
2.50	5.50	7.50	11.19	5.50	7.50	14.50
3.00	6.00	8.00	12.62	6.00	8.00	16.62
3.50	6.62	8.50	13.31	6.62	8.50	18.44
4.00	7.12	9.00	15.25	7.12	9.00	21.06
5.00	8.12	10.00	17.88	8.12	10.00	25
6.00	9.00	10.50	20.19	9.00	10.50	29.25
8.00	10.00	11.50	24.00	10.00	11.50	37.25
10.00	—	13.00	28.19	—	13.00	44.12
Metric (Millimeters)						
2.00	121	178	267	121	178	333
2.50	140	190	284	140	190	368
3.00	152	203	321	152	203	422
3.50	168	216	338	168	216	468
4.00	181	229	387	181	229	535
5.00	206	254	454	206	254	635
6.00	229	267	513	229	267	743
8.00	254	292	610	254	292	946
10.00	—	330	716	—	330	1121

Dimensions taken from manufacturer's catalogs for drawing purposes.

(continued on next page)

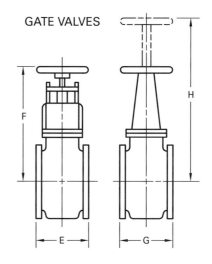

CHECK VALVES

LIFT   SWING

GATE VALVES

Appendix C

Table C-24. Common Valves *(continued)*

	Nominal Pipe Size in Inches	Globe and Angle Valves					
		Globe			Angle		
		J		K	L		M
		Screwed	Flanged		Screwed	Flanged	
U.S. Customary (Inches)	2.00	4.75	7.00	9.44	3.50	3.88	10.38
	2.50	5.50	7.50	11.06	3.88	4.50	12.06
	3.00	6.00	8.00	12.38	4.69	4.62	12.69
	3.50	6.62	8.50	13.19	5.00	5.38	13.62
	4.00	7.12	9.00	15.25	6.00	5.88	14.88
	5.00	8.12	10.00	17.25	6.31	6.50	17.69
	6.00	9.00	10.50	18.81	8.00	8.00	19.06
	8.00	10.00	11.50	22.12	—	9.25	22.75
	10.00	—	13.00	24.75	—	10.62	24.94
Metric (Millimeters)	2.00	121	178	240	89	98	264
	2.50	140	190	281	98	114	281
	3.00	152	203	314	119	118	346
	3.50	168	216	335	127	136	348
	4.00	181	229	387	152	149	378
	5.00	206	254	438	160	165	449
	6.00	229	267	478	203	203	484
	8.00	254	292	562	—	235	578
	10.00	—	330	629	—	270	633

Dimensions taken from manufacturer's catalogs for drawing purposes.

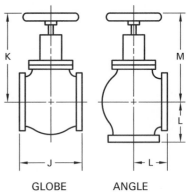

GLOBE AND ANGLE VALVES

GLOBE   ANGLE

Glossary

A

acme thread A thread that has a depth that is one half its pitch. (Ch. 11)

actual size Measured size. (Ch. 7)

acute angle An angle that measures less than 90°. (Ch. 3)

aerial view In perspective drawing, a view in which the object is seen from above. (Ch. 12)

aesthetic design The look and feel of a product. (Ch. 2)

aligned system A system of dimensioning in which the dimensions are placed in line with the dimension lines. (Ch. 7)

allowance A measure of how loosely or tightly a fastener fits its mating part. (Ch. 11)

alphabet of lines The lines and symbols used on drawings. (Ch. 4)

alternate section lining A pattern made by leaving out every other section line; used to show a rib or another flat part in a sectional view when that part otherwise would not show clearly. (Ch. 8)

alternating current A form of electrical current in which the electrons flow in one direction for a specific period, and then in the opposite direction for a similar period. (Ch. 21)

AND gate In logic circuits, a binary device that outputs a 1 if all of the inputs are 1; otherwise it outputs a 0. (Ch. 21)

application blocks Optional text blocks on a sheet layout that provide columns for purposes such as listing specific information used to relate a given drawing to other drawings in a set. (Ch. 4)

arc A part of a circle. (Ch. 2, 3)

arc welding A form of fusion welding in which an electric arc forms between the work (part to be welded) and an electrode. (Ch. 15)

assembly drawing A drawing in which the parts of a machine are shown together in their relative positions. (Ch. 7, 13)

assembly working drawing An assembly drawing that gives complete manufacturing information and can be used as a working drawing. (Ch. 13)

associative hatch In AutoCAD, a boundary hatch in which the hatch area changes when the boundaries are revised. (Ch. 8)

atom The smallest part of an element that still displays the properties of that element. (Ch. 21)

attribute In AutoCAD, textual information that can be attached to a block for later extraction to a bill of materials. (Ch. 11)

auxiliary plane An imaginary plane adjacent to a surface that appears foreshortened in the normal views. (Ch. 9)

auxiliary section The section that results when the cutting plane is passed through an object at an angle. (Ch. 8, 9)

auxiliary view A projection on an auxiliary plane that is parallel to an inclined surface. (Ch. 9)

axis An imaginary line along which the height, width, or depth of an object is drawn. Plural: *axes*. (Ch. 2)

axis of revolution An imaginary line or axis through an object about which the object can be revolved. (Ch. 9)

axonometric projection A form of projection that uses three axes at angles to show three sides of an object. (Ch. 12)

azimuth A measurement that defines the direction of a line off due north. (Ch. 10)

B

balloon framing A type of framing in which the studs are two stories high; a false girt inserted into the stud wall carries the second-floor joists. (Ch. 18)

bar chart A type of chart in which bars are used to show relative amounts or values. (Ch. 22)

baseline dimensioning A method of dimensioning in which all of the dimensions are given from a single, common surface, or baseline. (Ch. 7)

basic hole system A dimensioning system in which the design size of the hole is the basic size and the allowance is applied to the shaft. (Ch. 7)

basic shaft system A dimensioning system in which the design size of the shaft is the basic size and the allowance is applied to the hole. (Ch. 7)

basic size The size to which allowances and tolerances are added to get the limits of size. (Ch. 7)

beam compass A drawing compass that is especially designed to draw arcs and circles with large radii. (Ch. 3)

bearing The angle a line makes in the top view with a north-south line. (Ch. 10)

bilateral tolerance Tolerance that allows variation in both directions from the design size. (Ch. 7)

bill of materials A tabulated list that provides the names of parts, material, number required, part numbers, and so forth for a given product. (Ch. 13)

bisect Divide into two equal parts. (Ch. 5)

block In AutoCAD, two or more objects (lines, arcs, etc.) that the CAD operator has defined as a single object. (Ch. 11)

block diagram In mapping, a 3D projection using the isometric view. (Ch. 20)

body diagonal The longest straight line that can be drawn in a cube. (Ch. 12)

broken-out section A section that shows an object as it would look if a portion of it were cut partly away from the rest by a cutting plane and then "broken off" to reveal the cut surface and insides. (Ch. 8)

C

cabinet oblique drawing An oblique drawing in which the depth of the drawing is exactly one half of the true depth of the object. (Ch. 12)

cabinet oblique sketch An oblique sketch in which the depth of the drawing is exactly one half of the true depth of the object. (Ch. 2)

CAD/CAM A system in which drawings created using computer-aided drafting (CAD) are used directly to drive machinery to create a product (computer-aided manufacturing). (Ch. 1)

CADD Computer-aided design and drafting software. (Ch. 1)

CAD software Software used as a drafting tool in computer-aided drafting. (Ch. 3)

cam A machine part that usually has an irregular curved outline or a curved groove; when it rotates, it transmits or produces a specific, continuous motion. (Ch. 17)

career A series of related jobs built on a foundation of interest, knowledge, training, and experience. (Ch. 1)

career plan A document in which you record your plans for your career; your "road map" to the future. (Ch. 1)

cartographer A mapmaker. (Ch. 20)

cartography A graphic method of representing facts about the surface of the earth or other bodies in the solar system. (Ch. 20)

case instruments Drawing instruments in a set. (Ch. 3)

cavalier oblique drawing An oblique drawing that shows the full depth of the object. (Ch. 12)

cavalier oblique sketch An oblique sketch that shows the full depth of the object. (Ch. 2)

centerline An imaginary line that marks the exact center of an object. (Ch. 2, 4)

center of vision (CV) In perspective drawing, the point at which the line of sight pierces the picture plane. (Ch. 12)

center-plane construction An auxiliary view constructed using the center plane of an object as a reference plane. (Ch. 9)

chamfer Angled corner; bevel. (Ch. 7)

check valve A valve that permits flow in one direction only. (Ch. 16)

CIM Computer-integrated manufacturing; a system in which CAD files are used directly to drive machinery to create a product. (Ch. 1)

circumference The distance around the edge of a circle; the circle's rim. (Ch. 3)

circumscribe Create a polygon that fully encloses a circle that is tangent to all of the polygon's sides. (Ch. 5)

closed catalog system A filing system in which one person is responsible for keeping the company's drawings safe. (Ch. 23)

CNC Computer-numerical control; a computerized version of the old numerical control system used to manufacture products. (Ch. 1)

combination drawing A detail drawing that contains two or more parts that are used for different purposes; for example, a combination drawing might give dimensions for forging, as well as machining dimensions and notes. (Ch. 13)

compass A drawing instrument used to create circles and arcs. (Ch. 3)

composition In lettering, arranging words and lines with letters of the right style and size. (Ch. 2)

compressive strength A measure of an object's ability to withstand force or pressure. (Ch. 19)

computer-aided drafting (CAD) A system of drafting in which the drafter uses computer software to replace the instruments of board drafting. (Ch. 1)

concentric circles Two or more circles of different sizes that share the same center point. (Ch. 2)

concurrent engineering design Design done in a comprehensive team environment in which important issues such as functional and aesthetic design concepts, manufacturability, quality, life cycle, costs, etc., are considered by the entire team. The team consists of designers, engineers, drafters, and others associated with the overall design, manufacturing, marketing, and servicing of the product. (Ch. 2)

conductor A material with small resistance to the flow of electrons. (Ch. 21)

connection diagram A diagram that shows how the components of a circuit are connected; often referred to as a *wiring diagram*. (Ch. 21)

constant attribute In AutoCAD, a text attribute assigned to a block for which the value remains constant. (Ch. 11)

contact area The amount of area of the surface required to be in contact with its mating surface. (Ch. 7)

contour interval In mapping, the vertical distance between contour lines. (Ch. 20)

contours In mapping, lines of constant level that help represent vertical distances on 2D maps. (Ch. 20)

conventional break Symbols that are used to show that a uniform part of a very long object has been cut out of the drawing. (Ch. 8)

conversion graph A graph or chart that provides a convenient way to convert from one system or scale to another. (Ch. 22)

coordinate pair A set of two numbers, separated by a comma, that identifies the exact location of a point on a coordinate grid. (Ch. 4)

creativity A combination of imagination, knowledge, and curiosity that can be applied to the creation of new product ideas. (Ch. 2)

crosshatching Another name for section lining. (Ch. 8)

crossing window In AutoCAD, a selection window created from right to left. Unlike regular selection windows, crossing windows select every object that touches the window, in addition to those objects completely contained within it. (Ch. 22)

cross section In architectural drafting, a full section that cuts across the entire structure to aid in interpreting the relationships of the important spaces. (Ch. 18)

current A steady stream of electrons flowing through a wire or metal surface. (Ch. 21)

curve In a chart, a line which may be straight, curved, broken, or stepped. (Ch. 22)

cutting plane The imaginary plane along which a section is taken. (Ch. 8)

cutting-plane line A line on a normal view that shows where the cutting plane passes through the object. (Ch. 8)

D

datum A point, line, or surface that is assumed to be exact. (Ch. 7)

datum feature Any physical portion of a part. (Ch. 7)

datum reference frame The restrictive environment created to hold parts while they are being machined. (Ch. 7)

datum target Specific points, lines, or areas of a surface where datum features are rough, uneven, or on different levels, such as on castings, forgings, or weldments. (Ch. 7)

dead load In architectural and structural drafting, the weight of the building or structure. (Ch. 19)

degree The unit of measure for angles. (Ch. 3)

degrees of freedom The directions of movement possible for a part that is being machined: up and down, in and out, and from side to side. (Ch. 7)

descriptive geometry A graphic process for solving three-dimensional problems in engineering and engineering design. (Ch. 10)

design The conception of an idea and its development into a practical, producible, and usable product or process. (Ch. 2)

design size The size to which tolerances are added to get the limits of size. When there is no allowance, the design size equals the basic size. (Ch. 7)

desktop publishing In publishing, the process of page makeup when it occurs on a desktop computer system. (Ch. 24)

detail drawing A drawing for a single part that includes all the dimensions, notes, and information needed to make that part. (Ch. 7, 13)

detailed representation A thread representation that approximates the real look of threads. (Ch. 11)

detail section In architectural drafting, a section drawn at a much larger scale than the elevation, in which a vertical plane cuts through walls to show construction details. (Ch. 18)

development A full-size layout of an object made on a single flat plane; also called *pattern development* or *stretchout*. (Ch. 14)

diameter The distance across a circle through its center point. (Ch. 5)

diazo A reproduction method in which prints have dark lines on a white background. (Ch. 23)

die stamping The process of forming thin metal parts by pressing a flat sheet into shape under heavy pressure. (Ch. 14)

digitizing The process of converting a paper copy into electronic format. (Ch. 2)

dihedral angle The angle formed when two planes intersect. (Ch. 10)

dimensioning Size description of an object or assembly. (Ch. 7)

dimension line A thin line, usually with arrowheads at each end, that shows where a dimension begins and ends. (Ch. 2, 7)

dimetric projection A form of axonometric projection in which the angles of two of the axes are equal; requires two special foreshortened scales to make measurements. (Ch. 12)

diode A semiconductor device that acts like a gate, allowing current to pass through in one direction only. (Ch. 21)

dip In geology, the slope of the contact between two rock formations. (Ch. 20)

direct current A form of electrical current in which the electrons flow in one direction only. (Ch. 21)

displacement diagram A diagram that shows the motion a cam will produce through one revolution. (Ch. 17)

dividers A drawing instrument used to divide lines and transfer distances. (Ch. 3)

double-line drawing A detailed type of pipe drawing suited for illustrations in catalogs and other applications in which the visual appearance is important. (Ch. 16)

double thread Two helical ridges side by side on a screw; a type of screw in which the lead is twice the pitch. (Ch. 11)

drafting machine A drawing instrument that combines the functions of the T-square, triangles, scales, and protractor. (Ch. 3)

drawing board The board to which a drawing sheet is attached in board drafting. (Ch. 3)

drawing template In AutoCAD, standard drawing setups that conform to ANSI or ISO standards; drafters use these templates to set drawing limits and units, among other things. (Ch. 4)

dual dimensioning system A system of dimensioning that shows measurements in both inches and millimeters. (Ch. 7)

E

edge view A view in which a cutting plane appears as an edge, or line. (Ch. 9)

elbow A joint in a pipe or duct; a place at which two pieces of pipe meet at an angle other than 180°. (Ch. 14)

electricity The flow of electrons through a wire or metal conductor. (Ch. 21)

electron A negatively charged particle that moves around the nucleus of an atom. (Ch. 21)

electronics Electric devices that incorporate the use of vacuum tubes (now obsolete) or semiconductors. (Ch. 21)

element An imaginary flat surface on the face of a cylinder or cone; used in pattern development. (Ch. 14)

elevation In architectural drafting, a drawing of a façade of a structure; similar to the front and side views in other types of drafting. (Ch. 18)

ellipse A regular oval that has two equal center points. (Ch. 2, 5)

employability skills "People" skills such as being able to get along with other people and work well with them; soft skills that help you keep a job. (Ch. 1)

engineer A person who has at least a four-year degree in an engineering specialty. (Ch. 1)

entrepreneur Someone who organizes and then runs a business. (Ch. 1)

equilateral A triangle in which all three sides are of equal length and all three angles are equal. (Ch. 5)

erasing shield A drawing aid for use with erasers that shields surrounding lines from erasure. (Ch. 3)

ergonomics A field of study that investigates ways to design products to promote personal safety and comfort while the products are being used. (Ch. 3)

extension lines Thin lines used to extend the shape of the object to the dimension line. (Ch. 2)

F

fabrication shop A shop in which materials such as steel are cut to specific lengths and processed. (Ch. 19)

FAO Short for FINISH ALL OVER; used on working drawings to specify that the entire piece is to be machined. (Ch. 13)

fastener Any kind of device or method for holding parts together; examples include screws, bolts and nuts, rivets, welding, brazing, soldering, adhesives, collars, clutches, and keys. (Ch. 11)

fault In geology, the line along which a rock layer has broken.

feature control frame The rectangular box that holds the feature control symbols in geometric dimensioning and tolerancing. (Ch. 7)

fillet External curve. (Ch. 7)

fillet weld A type of weld that is similar to a groove weld, except the weld rests on top of the joint. (Ch. 15)

finish mark A symbol that shows that a surface is to be machined, or finished. (Ch. 7)

finite element analysis (FEA) A process in which computer programs are used to assign materials to computer models of structural members and then test them under various loads. (Ch. 19)

first-angle projection A system of projection in which the object is projected onto the planes from the first angle, or quadrant. (Ch. 6)

flaws Irregularities such as cracks, blowholes, checks, ridges, and scratches that occur at one place or at relatively infrequent or widely varying intervals in a surface. (Ch. 7)

flowchart A chart that shows sequential information. (Ch. 22)

follower A mechanical part that follows the motion of a cam, transmitting the motion to another part of a mechanism. (Ch. 17)

font Lettering style in CAD and other computer programs. (Ch. 2) The detailed appearance of the type within a type family. (Ch. 24)

frequency In electronics, the number of times a cycle is repeated in 60 seconds. (Ch. 21)

frontal line A line that is parallel to the vertical plane. (Ch. 10)

front auxiliary view An auxiliary view that is hinged on the front view. (Ch. 9)

front view The view of a multiview drawing that shows the exact width and height of an object. (Ch. 6)

frustum Truncated. (Ch. 7)

full section A sectional view that shows an object as if it were cut completely apart from one end or side to the other. (Ch. 8)

functional block diagram In electrical drafting, a diagram made up of squares and similar symbols joined by single lines; the diagram shows how the components or stages of an electrical or electronic circuit are working. (Ch. 21)

functional design The design of a product so that it operates successfully or accomplishes its purpose. (Ch. 2)

fusion welding Welding in which heat is applied to form the weld; examples of fusion welding include gas, arc, thermit, gas-and-shielded-arc welding, soldering, and brazing. (Ch. 15)

G

gage lines The lines along which rivets are placed. (Ch. 19)

Gantt chart A project management chart that separates an operation or project into discrete elements and assigns a certain amount of time to each step. (Ch. 23)

gas-and-shielded-arc welding A welding process that combines arc welding and gas welding using tungsten-inert gas (TIG) or metallic-inert gas (MIG). (Ch. 15)

gas welding Welding in which gases are used to create the necessary heat. (Ch. 15)

gate valve A valve in which a wedge, or gate, lifts to allow full, unobstructed flow and lowers to stop it completely; used for on/off operation of valves that are not frequently used. (Ch. 16)

gear A device that transmits motion using a series of teeth. (Ch. 17)

geology The science that deals with the makeup and structure of the earth's surface and interior depths. (Ch. 20)

geometric construction An illustration made of individual lines and points drawn in proper relationship to one another. (Ch. 5)

geometric dimensioning and tolerancing The specification of permissible variations in the form of tolerance and limits. (Ch. 7)

geometry The study of the size and shape of things. (Ch. 5)

globe valve A valve that is used to control the flow of liquids or gases when close, accurate regulation of pressure and volume is necessary. (Ch. 16)

Gothic lettering Traditional single-stroke lettering style that is often used on board drawings. (Ch. 2)

grade The percentage by which land slopes. (Ch. 10)

grid In AutoCAD, a nonprinting set of dots that provide a visual reference for the CAD operator; the dots can be set at any convenient interval. (Ch. 4)

groove weld A weld that is located in a single or double groove or notch in the work material. (Ch. 15)

ground plane (GL) In perspective drawing, the plane on which the observer stands. (Ch. 12)

ground view In perspective drawing, a view in which the object is seen from underneath. (Ch. 12)

guidelines In lettering, lightly drawn lines drawn at regular intervals to help keep lettering uniform. (Ch. 2)

H

half section One half of a full sectional view. (Ch. 8)

hanging indent An indent that affects all but the first line of a paragraph, so that the first line "hangs" out to the left. (Ch. 24)

hardware Computer equipment such as the CPU, monitor, printer, mouse, and keyboard. (Ch. 3)

harmonic motion Cam motion that follows a harmonic curve. (Ch. 17)

helix The curving path that a point would follow if it were to travel in an even spiral around a cylinder and parallel to the axis of that cylinder; also called a *helical curve*. (Ch. 11)

hemming In sheet-metal products, the process of folding edges to make them stiffer. (Ch. 14)

hexagon A six-sided polygon. (Ch. 5)

hidden line A line consisting of a series of short dashes that represents a feature that is hidden from view. (Ch. 4)

horizon line (HL) In perspective drawing, the line formed where a horizontal plane that passes through the observer's eye meets the picture plane. (Ch. 12)

hypotenuse The side of a right triangle that is opposite the 90° angle. (Ch. 5)

I

ideation The phase of concurrent engineering in which the design problem is identified, preliminary solutions are developed, and the preliminary design is agreed upon. (Ch. 2)

implementation The phase of concurrent engineering in which a careful analysis of production, financing, servicing, documenting, final planning, and life cycle issues takes place. (Ch. 2) The process of drawing the object that has been visualized. (Ch. 6)

inclined line A line drawn at an angle that is neither horizontal nor vertical. (Ch. 4) A line that is perpendicular to one of the three reference planes. (Ch. 10)

inclined plane A plane that is perpendicular to one reference plane and inclined to the other two. (Ch. 10)

indent In publishing, setting the left or right margin of text in from the base margins. (Ch. 24)

inking The process of creating technical drawings using technical pens. (Ch. 4)

inscribe Create a polygon fully enclosed by a circle that is tangent to all of the polygon's sides. (Ch. 5)

insertion point In AutoCAD, the point at which the block will be attached to the cursor when you insert the block into the drawing. This point is specified by the CAD operator when the block is created. (Ch. 11)

insulator A material through which electrons do not flow easily. (Ch. 21)

integrated circuit A device that incorporates components such as resistors, diodes, and transistors into a single substrate so that they can be handled as a unit. (Ch. 21)

intermediate In board drafting, a secondary original that is used in place of the original during the revision process. (Ch. 23)

intermittent weld A series of two or more short welds along a joint; used when the design engineer determines that it is not necessary to run a weld the entire length of the joint. (Ch. 15)

intersect Cross at a single point. (Ch. 5)

involute curve The curve made by a taut string as it unwinds from around the circumference of a cylinder; used in gear teeth to help them mesh smoothly. (Ch. 17)

irregular curve A drawing instrument used to draw noncircular curves; also known as a *French curve*. (Ch. 3)

isometric axes Three axes spaced at equal angles of 120°; used for isometric sketching and drawing. (Ch. 12)

isometric drawing A drawing in which the object is aligned with three axes that are equally spaced at 120° angles. (Ch. 12)

isometric lines Lines in an isometric drawing that run parallel to any of the three axes. (Ch. 2, 12)

isometric plane A plane that is parallel to one of the faces of an isometric cube. (Ch. 12)

isometric projection Similar to an isometric drawing, but created by revolving the object; unlike isometric drawings, isometric projections do not necessarily show all lines at their true length. (Ch. 12)

isometric sketch A sketch based on three axes equally spaced at 120° apart. (Ch. 2)

isoplane In AutoCAD, one of the three isometric plane orientations available for the cursor in isometric drawings. (Ch. 12)

isosceles A triangle in which two sides are of equal length. (Ch. 5)

J

job Work done for pay. (Ch. 1)

K

kerning The horizontal space between letters. (Ch. 24)

L

laminating Cutting material into thin slabs and then gluing them together. (Ch. 19)

lay The direction of predominant surface pattern. (Ch. 7)

layer In AutoCAD, a system of separating different aspects of a drawing in a manner similar to the overlays used by board drafters. (Ch. 4)

lead In fasteners, the distance along a thread's axis that the threaded part moves against a fixed mating part when given one full turn; the distance a screw enters a threaded hole in one turn. (Ch. 11)

leader A thin line with a horizontal dash at one end and an arrowhead at the other. It is drawn from a note or dimension to the place where it applies. (Ch. 2)

leading In publishing, the vertical space between lines, measured from the base of one line to the base of the next. (Ch. 24)

left-hand thread A thread that screws in when turned counterclockwise as viewed from the outside end. (Ch. 11)

lettering The practice of adding clear, concise words on a drawing to help people understand the drawing. (Ch. 2)

level line A line that is horizontal in the vertical projection. (Ch. 10)

life cycle The total life of a product, from the conception of the idea to recycling of the materials from which it is made. (Ch. 2)

lifelong learning Learning that takes place throughout your career, either as formal continuing education (CE) credits or as informal courses or participation in work-related associations. (Ch. 1)

limit dimensions Limits that give the maximum and minimum dimensions allowed. (Ch. 7)

limits of size Maximum and minimum sizes; usually called *limits*. (Ch. 7)

line A straight or curved path between two points. (Ch. 2)

line graph A graph in which a line is used to show trends or how something changes over time. (Ch. 22)

line of sight (LOS) A line drawn at right angles to an inclined surface to place the auxiliary plane. (Ch. 9) The visual from the eye perpendicular to the picture plane. (Ch. 12)

line weight The thickness and darkness of a line. (Ch. 4)

live load In architectural and structural drafting, any weight a building must bear that is not part of the weight of the building itself, such as snow or furniture. (Ch. 19)

load In architectural and structural drafting, the weights or pressures borne by the structure. (Ch. 19)

location dimensions Dimensions that define the location of each piece in an object that consists of more than one piece. (Ch. 7)

long-term goals Personal or professional goals that may take more than five years to accomplish. (Ch. 1)

M

major diameter Crest diameter of a thread; the nominal thread size. (Ch. 11)

measured profile A representation of a profile obtained by instruments or other means. (Ch. 7)

measuring line In pattern development, vertical lines that extend from the stretchout line at important points. (Ch. 14)

media (plural; singular: *medium*) The paper, film, or electronic file used to store drawings. (Ch. 23)

metallic-inert gas welding (MIG) A form of gas-and-shielded-arc welding in which the electrode contains a consumable metallic rod that provides both the filler material and the arc for fusion. (Ch. 15)

microfilm A storage medium on which a large drawing is reduced to a very small size and recorded on film. (Ch. 23)

microinch One millionth of an inch (.000 0001″). (Ch. 7)

micrometer One millionth of a meter (.000 0001 m). (Ch. 7)

midpoint The exact center point. (Ch. 5)

minor diameter Root diameter of a thread. (Ch. 11)

model space In AutoCAD, the drawing space in which most drawing work is done. (Ch. 4)

multiview drawing A method of drawing views of an object as it is seen from different positions and arranged in a standard order. (Ch. 6)

N

nominal diameter In pipe drafting, the inside diameter of the pipe. (Ch. 16)

nominal dimensions General dimensions that do not include accurate limits or tolerances. (Ch. 7)

nominal size The size used for general identification. (Ch. 7) In architectural drafting, the dimensions of lumber before the wood has been surfaced, or prepared for use. (Ch. 18)

nomogram An engineering chart that shows the solution to a problem that contains three or more variables. (Ch. 22)

nonisometric lines Lines in an isometric drawing that are not parallel to any of the three isometric axes. (Ch. 2, 12)

normal line A line that is perpendicular to one of the three reference planes. (Ch. 10)

normal oblique drawing An oblique drawing in which the depth of the drawing is exactly three quarters of the true depth of the object; also known as *general oblique*. (Ch. 12)

normal plane A plane that is parallel to one of the normal reference planes and perpendicular to the other two. (Ch. 10)

normal view In perspective drawing, a view in which the object is seen face on, so that the line of sight is directly on it rather than above or below. (Ch. 12)

normal views The three most commonly used views in multiview drawing: top, front, and right-side. (Ch. 6)

NOT gate In logic circuits, a digital inverter that changes an input of 1 to 0 or an input of 0 to 1. (Ch. 21)

O

object snap Feature in AutoCAD that allows you to jump automatically to special points such as the endpoint or midpoint of a line or arc. (Ch. 5)

oblique plane A plane that is inclined to all three of the normal or reference planes. (Ch. 9, 10)

oblique projection A drawing that shows the depth using projectors, or lines representing receding edges of the object. These lines are drawn at an angle other than 90° from the picture plane to make the receding planes visible in the front view; note that oblique projection is usually considered to be no different from oblique drawing. (See *isometric projection*.) (Ch. 12)

oblique sketch A sketch in which two of the axes are at right angles to each other, and the third axis is drawn at any convenient angle. A 45° angle is often used. (Ch. 2)

obtuse angle An angle that measures more than 90°. (Ch. 3)

octagon An eight-sided polygon. (Ch. 5)

offset section A section in which the cutting-plane line shifts to include features that would not ordinarily be included in a section made by a straight cutting plane. (Ch. 8)

ogee curve A double reverse curve that looks something like an S. (Ch. 5)

Ohm's law A formula that relates the voltage, current, and resistance in an electric circuit: Voltage equals the product of the current and resistance ($E = I \times R$). (Ch. 21)

omission graph A graph in which part of the graph is left out; used when the vertical scale needs to be much larger than the horizontal scale to show the information adequately. (Ch. 22)

one-point perspective A perspective view in which there is one vanishing point; also called *parallel perspective*. (Ch. 12)

open file system A filing system in which original drawings are always available to anyone in the department. (Ch. 23)

operations map A map that shows the relationship between the land's physical features and the operation that is to be performed. (Ch. 20)

OR gate In logic circuits, a binary device that outputs a 1 if any of the inputs is 1. (Ch. 21)

orthographic projection The system by which the views of a multiview drawing are arranged in relation to each other. (Ch. 2) The process of projecting two or more views of an object onto imaginary planes by drawing lines perpendicularly from the object to the planes. (Ch. 6)

outline assembly drawing Assembly drawings in which details have been omitted. (Ch. 13)

overlay A piece of translucent drawing paper that is placed on top of an existing sketch or drawing to show modifications or refinements without having to redraw the entire drawing. (Ch. 2)

P

paper space In AutoCAD, the drawing space in which drawings are laid out on drawing sheets for printing. (Ch. 4)

parallel Two or more lines that are always the same distance apart. (Ch. 5)

parallel circuit A circuit in which current can flow through more than one path simultaneously. (Ch. 21)

parallel-line development The process of making a pattern by drawing the edges of an object as parallel lines. (Ch. 14)

parallel-ruling straightedge A straightedge onto which a guide cord is clamped. The cord runs through a series of pulleys attached to the drawing board, allowing the straightedge to slide up and down the board in parallel positions. (Ch. 3)

parametric software A type of software that allows the user to create and edit objects according to specific parameters. (Ch. 14)

partial auxiliary view An auxiliary view in which some elements have been left out. (Ch. 9)

pattern The original part of a pattern development from which flat patterns can be cut from flat sheets of material and folded, rolled, or otherwise formed into a product. (Ch. 14)

pattern development A full-size layout of an object made on a single flat plane; also called *development* or *stretchout*. (Ch. 14)

pentagon A five-sided polygon. (Ch. 5)

perpendicular At a 90° angle. (Ch. 5)

perspective drawing A three-dimensional representation of an object as it looks to the eye from a particular point. (Ch. 12)

PERT chart (Project Evaluation and Review Technique) A project management chart that assigns probabilities to production issues such as efficiency and productivity, then plots them against production activities and operations. (Ch. 23)

phantom section A section that shows in one view both the inside and the outside of an object that is not completely symmetrical; also called a *hidden section*. (Ch. 8)

photodrafting A drafting process in which photographs are overlaid with line work to create enhanced, annotated pictures of the product. (Ch. 23)

photogrammetry In mapping, a method of making 3D measurements from high-altitude photographs. (Ch. 20)

photo retouching A process used to change details or fix errors in photographs. (Ch. 12)

pica A unit of measure commonly used in the publishing industry. In general, 1 pica equals ⅙ of an inch, or 12 points. (Ch. 24)

pictograph A pictorial graph similar to a bar chart, except that pictures or symbols are used instead of bars. (Ch. 22)

pictorial sketch A picturelike type of sketch in which the width, height, and depth of an object are shown in one view. (Ch. 2)

picture plane In perspective drawing, the plane on which the object is drawn. (Ch. 12)

pie chart A circle chart in which the circle equals 100%. Various sectors of the "pie" represent percentages of the whole. (Ch. 22)

piercing point The point at which a line intersects a plane. (Ch. 10)

pinion A small spur gear; the smaller of two mating spur gears. (Ch. 17)

pipe drawing A drawing, such as a plumbing diagram, in which lines and symbols are used to describe the construction of piping components and systems. (Ch. 16)

pipe fitting A part used to join sections of pipe. (Ch. 16)

pitch In fasteners, the distance from one point on a thread form to the corresponding point on the next form; to find the pitch, divide 1 by the number of threads per inch. (Ch. 11) In architectural and structural drafting, the slope of a roof; the rise relative to the run. (Ch. 19)

pitch diameter The point of tangency of a gear with a mating gear. (Ch. 17)

pixels A collection of tiny dots, or picture elements, that make up raster images. (Ch. 24)

plane A flat surface. (Ch. 2)

plank-and-beam framing A type of framing that uses heavier posts and beams than other systems, allowing ceilings to be higher and more open, with fewer supporting members. (Ch. 18)

plan view In AutoCAD, the top view in the world coordinate system. (Ch. 10)

plat A map used to show the boundaries of a piece of land and to identify it. (Ch. 20)

plug weld A weld that fits into a small hole in the work material. (Ch. 15)

point A symbol that describes a location in space. (Ch. 2)

polar array In AutoCAD, an array in which an object is copied and arranged along a circle or circular arc so that all the copies are the same distance from a center point. (Ch. 17)

polar coordinates Coordinates that include a line length and the direction in which the line should extend. (Ch. 4)

polygon A closed figure. (Ch. 5)

power The rate at which work is done. (Ch. 21)

pressure angle The angle of the teeth of a gear in relation to the pitch circle. (Ch. 17)

prestressed concrete Concrete in which the reinforcing bars are stretched before the concrete is poured over them. (Ch. 19)

primary auxiliary view An auxiliary view that is developed directly from one of the normal views. (Ch. 9)

profile The contour, or shape, of a surface in an adjacent plane; the plane is usually perpendicular to the surface. (Ch. 7)

profile line A line that is parallel to the profile reference plane. (Ch. 10)

profile section A full section taken from the profile view of an object. (Ch. 8)

progressive chart A bar chart in which the horizontal or vertical scale provides a range over which the bars may vary; therefore, not all of the bars begin at the same place. (Ch. 22)

projection weld A weld of two or more parts in which at least one part must have a boss projection. Projection welds are identified by strength or size. (Ch. 15)

protractor A drawing instrument used to measure or lay out angles. (Ch. 3)

Pythagorean theorem Theorem that a triangle with sides that measure 3, 4, and 5 (or a multiple of these numbers) will always be a right triangle; proved by Pythagorus in the sixth century B.C. (Ch. 5)

R

radial-line development Pattern development of cones, pyramids, and other shapes in which the edges are not parallel, so the stretchout line is not a straight line. (Ch. 14)

radius The distance from the center of a circle to its edge. Plural: *radii*. (Ch. 2, 5)

rapid prototyping A technology that allows engineers to create 3D "prints" of a proposed part based on a CAD drawing. (Ch. 1)

raster file format A drawing file format that consists of a collection of pixels, or picture elements. (Ch. 24)

rectangular array In AutoCAD, an array in which an object is copied into a specified number of rows and columns. (Ch. 17)

reference assembly drawing A special type of assembly drawing that identifies parts to be assembled. (Ch. 13)

reference planes Planes that are parallel to inclined surfaces and are used for creating auxiliary views. (Ch. 9)

reference zones The numbers or letters given in the margins of a drawing to locate specific information on the drawing, similar to the way you locate cities and towns on a road map. (Ch. 4)

refinement The phase of concurrent engineering design that includes preparation of models and prototypes; thorough physical, production, and legal analysis of the design; and design visualization, or analysis of the aesthetics. (Ch. 2)

refresh rate The rate at which the image on a computer screen is redrawn. (Ch. 3)

regular polygon A closed figure in which all of the sides and angles are of equal measure. (Ch. 5)

reinforced concrete Concrete in which steel bars have been embedded. (Ch. 19)

removed section A sectional view that is taken from its normal place on the view and moved somewhere else on the drawing sheet. (Ch. 8)

render A method of enhancing a solid model so that it looks almost lifelike. (Ch. 6, 12)

resistance In electricity and electronics, the amount of difficulty electrons have flowing through a material. (Ch. 21)

resistance welding A form of welding that combines heat and pressure to create a weld. (Ch. 15)

resolution In a computer monitor, the number of pixels, or picture elements, per inch. (Ch. 3)

résumé A document that summarizes your experience and education for perspective employers. (Ch. 1)

revision history block A text block that contains a record of all alterations made after the original drawing was completed. (Ch. 4)

revolution Another name for a revolved view. (Ch. 9)

revolved section A section, usually of a long, thin object, that is rotated in place on the drawing to show a cross section. (Ch. 8)

rib A thin, flat part of an object, such as a spoke, that is usually not shown with section lines when sectioned. (Ch. 8)

right angle An angle that measures exactly 90°. (Ch. 3)

right-hand thread A thread that screws in when turned clockwise as viewed from the outside end. (Ch. 11)

right-side auxiliary view An auxiliary view that is hinged on the right-side view. (Ch. 9)

right-side view The view of a multiview drawing that shows the exact depth and height of an object. (Ch 6)

right-to-know laws Laws in various states that guarantee employees' right to know about any occupational hazards associated with their jobs. (Ch. 1)

right triangle A triangle in which one of the angles equals 90°. (Ch. 5)

risers The vertical part of the steps in a staircase. (Ch. 18)

rotated section Another name for a revolved section. (Ch. 8)

roughness The finer irregularities in surface texture, including various irregularities caused by the production process. (Ch. 7)

roughness height The arithmetical average deviation of the height. (Ch. 7)

roughness width The distance between two peaks or ridges that make up the pattern of the roughness. (Ch. 7)

roughness-width cutoff The greatest spacing of repetitive surface irregularities to be included in the measurement of average roughness height. (Ch. 7)

round Internal curve. (Ch. 7)

S

scale An instrument used to lay off distances and to make measurements. (Ch. 3)

scalene A triangle that has sides of three different lengths and angles with three different values. (Ch. 5)

scatter graph A graph in which a smooth-line curve is drawn to represent the average of the plotted points. (Ch. 22)

schedule In architectural drafting, a list that defines and describes details shown by symbols on the actual drawings. (Ch. 18)

schematic diagram In electricity and electronics, an elementary diagram that shows the ways a circuit is connected and what it does. (Ch. 21)

schematic representation Thread representation that shows the threads using symbols, rather than as they really look. The Vs are omitted, and the pitch need not be drawn to scale. (Ch. 11)

screw fitting A threaded pipe fitting. (Ch. 16)

screw thread A helical ridge on the external or internal surface of a cylinder, or a conical spiral on the external or internal surface of a cone or frustum of a cone. (Ch. 11)

screw-thread series ANSI grouping of related screws according to diameter and pitch. The three basic series are coarse-thread series (UNC), fine-thread series (UNF), and extra-fine-thread series (UNEF). (Ch. 11)

seam weld A weld along the seam of two adjacent parts. (Ch. 15)

secondary auxiliary view An auxiliary view that is developed from a primary auxiliary view. (Ch. 9)

section A view that shows an object as if part of it were cut away to expose the inside. (Ch. 8)

sectional view Another name for a section. (Ch. 8)

section lining A series of thin lines that are used to show a cut surface on a sectional view. (Ch. 8)

semiconductor A material that acts like a conductor at high temperatures but like an insulator at low temperatures. (Ch. 21)

series circuit A circuit in which current can flow through only one path at a time. (Ch. 21)

series-parallel circuit An electrical circuit that contains both series and parallel components. (Ch. 21)

service bureau A company that can take completed desktop publishing files and create film for printing. (Ch. 24)

sheet layout The process of placing the border and title block on the drawing sheet. (Ch. 4)

short-term goals Personal or professional goals that you can achieve in less than five years. (Ch. 1)

simplified drawing In pipe drafting, another term for a single-line drawing of a pipe system. (Ch. 16)

simplified representation Symbolic thread representation that is much like a schematic representation, except that the crest and root lines are drawn as dashed lines. (Ch. 11)

simulated datum The feature that the datum feature contacts; the simulated datum should imitate its mating part in the assembly. (Ch. 7)

single-line drawing A simplified drawing of a pipe system; also called a *simplified drawing*. (Ch. 16)

single thread A single ridge in the form of a helix on a screw; a type of screw in which the lead is equal to the pitch. (Ch. 11)

site plan An architectural drawing that shows the lot and locates the house on it, as well as driveways, sidewalks, utility easements, and other pertinent information required by the building inspector. (Ch. 18)

skew lines Two lines that are not parallel and do not intersect. (Ch. 10)

slope A line's angle from the horizontal reference plane. (Ch. 10)

slot weld Similar to a plug weld, except that the shape of the opening for the weld is slot-shaped. (Ch. 15)

snap A tool in AutoCAD that sets the intervals at which the cursor moves when you move the mouse. (Ch. 4)

specifications In construction and manufacturing, a detailed list of facts concerning materials and measurements. (Ch. 18)

spline In AutoCAD, a curved line that passes through or is controlled by a series of points. Splines are created using the SPLINE command. (Ch. 20)

spot weld A weld in which the current and pressure are confined to a small area between electrodes. (Ch. 15)

spur gear A machine part that transmits rotary motion from one shaft to another. (Ch. 17)

station point (SP) In perspective drawing, the point from which the observer looks at an object. (Ch. 12)

step graph A special type of line graph that shows data that remains constant during regular or irregular intervals. (Ch. 22)

strata graph A shaded-surface graph that compares two or more different items, such as the use of different kinds of materials, over a period of time. (Ch. 22)

stratum In geology, a distinct layer below the surface of the earth. Plural: *strata*. (Ch. 20)

stretchout Another name for pattern development. (Ch. 14)

strike In geology, the direction of contact between two rock formations. (Ch. 20)

successive revolutions The process of revolving an object first around an axis perpendicular to one plane, and then again about an axis perpendicular to another plane. (Ch. 9)

surface A flat or nonflat element created from curved lines. (Ch. 2)

surface texture Roughness, waviness, lay, and flaws in the surface of a part. (Ch. 7)

symbol library In CAD, a drawing file that contains commonly used symbols that can be used in other drawings. Examples include libraries of fasteners, architectural symbols, and electrical/electronics symbols. (Ch. 11)

T

tabulated drawing A drawing in which the dimensions are identified by letters; a table placed on the drawing tells what each dimension is for different sizes of the part. (Ch. 13)

tangent arcs Two or more arcs that touch each other at one point only. (Ch. 2)

technical illustration A pictorial drawing that provides technical information using visual methods. (Ch. 12)

template A drafting instrument that provides outlines of common shapes and details that can be used in drawings. (Ch. 3) In desktop publishing, a special blank page that controls the layout of the page. (Ch. 24)

texture The surface quality of an object. (Ch. 2)

theoretical datum A datum established by the contact of a datum feature and a simulated datum. (Ch. 7)

thermit welding A type of welding in which a mixture, or charge, made of finely divided aluminum and iron oxide is ignited by a small amount of special ignition powder. The charge burns rapidly, producing a very high temperature. This melts the metal, which then flows into molds and fuses mating parts. (Ch. 15)

third-angle projection A system of projection in which the object is projected onto the planes from the third angle, or quadrant. (Ch. 6)

title block An area on a drawing that contains information about the drawing, the company, the drafter, and so on. (Ch. 13)

tolerance Permitted variance from the basic or design size. (Ch. 7, 11)

tolerance zones Categories of geometric characteristics of a part; the three types of tolerance zones are parallel lines, parallel planes, and cylinders. (Ch. 7)

top auxiliary view An auxiliary view that is hinged on the top view. (Ch. 9)

topographic map A map that presents a complete pictorial description of the area shown. (Ch. 20)

top view The view of a multiview drawing that shows the exact width and depth of an object. (Ch. 6)

tracking The horizontal space between words. (Ch. 24)

traditional engineering design A linear approach to the design process. (Ch. 2)

trammel A piece of paper or plastic on which specific distances have been marked off. (Ch. 5)

transistor A semiconductor device that is the electronic equivalent of two diodes placed back to back; it is used to control current and amplify input voltages or currents. (Ch. 21)

transition piece A piece of pipe or other material used to connect two or more openings of different shapes, sizes, or positions. (Ch. 14)

treads The horizontal part of the steps in a staircase. (Ch. 18)

triangle A drafting instrument that provides exact angles. Triangles may have one 90° angle and two 45° angles, or 30°, 60°, and 90° angles. (Ch. 3) A three-sided polygon. (Ch. 5)

triangulation A method of pattern development used to make approximate developments for surfaces, such as double-curved surfaces, that cannot be developed exactly. (Ch. 14)

trimetric projection A form of axonometric projection in which all three of the axis angles are different; requires three special foreshortened scales to make measurements. (Ch. 12)

triple thread Three ridges side by side on a screw; a type of screw in which the lead is three times the pitch. (Ch. 11)

truss A configuration of structural elements that adds strength to a structure. (Ch. 19)

T-square A drafting instrument that consists of a head that lines up with a true edge of the drafting board and a blade, or straightedge, that provides a true edge. (Ch. 3)

tungsten-inert gas welding (TIG) A form of gas-and-shielded-arc welding in which the electrode that provides the arc for welding is made of tungsten. Since it provides only the heat for fusion, some other material must be used with it for filler. (Ch. 15)

two-point perspective A perspective view in which there are two vanishing points; also called *angular perspective*. (Ch. 12)

type The letters and other characters as they appear on a printed page. (Ch. 24)

U

unidirectional system A dimensioning system in which all dimensions read from the bottom of the page. (Ch. 7)

uniformly accelerated and decelerated motion Cam motion that follows a parabolic curve, moving at steadily increasing and decreasing speeds. (Ch. 17)

uniform motion "Straight-line" cam motion in which time and distance are directly proportional. (Ch. 17)

unilateral tolerance A tolerance that allows variation on one side only of the design size. (Ch. 7)

user coordinate system (UCS) In AutoCAD, a user-defined orientation of the X, Y, and Z axes of the Cartesian coordinate system. (Ch. 10)

V

vanishing point The point at which receding axes converge in perspective drawing. (Ch. 12)

variable attribute In AutoCAD, a text attribute assigned to a block for which the value can be entered or changed each time the block is inserted into a drawing. (Ch. 11)

vector file format A drawing file format in which the placement of geometry is defined using a set of mathematical instructions. (Ch. 24)

vellum Tracing paper that has been treated to make it more transparent. (Ch. 3)

vertex The point at which the two arms of an angle meet. (Ch. 5)

vertical section A full section taken from the vertical or front view of an object. (Ch. 8)

viewport An invisible window in the AutoCAD drawing area, in which a view of the drawing can be placed. (Ch. 4)

visualization The ability to see clearly in the mind's eye what a machine, device, or other object looks like. (Ch. 6)

visual rays In perspective drawing, the sight lines leading from points on the object and converging at the eye of the observer. (Ch. 12)

voltage Electrical pressure. (Ch. 21)

W

waviness Variations in flatness that result from factors such as machine or work deflections, vibration, chatter, heat treatment, or warping strains. (Ch. 7)

waviness height Peak-to-valley distance. (Ch. 7)

waviness width The spacing of successive wave peaks or successive wave valleys. (Ch. 7)

web A thin connecting piece on an object that is generally not shown with section lines when sectioned. (Ch. 8)

welding A way of joining metal parts together using heat or a combination of heat and pressure. (Ch. 15)

welding symbol A method of representing a weld on drawings, including supplementary information such as finish symbols and dimensions. (Ch. 15)

weld symbol A symbol that indicates the type of weld to be made. (Ch. 15)

western framing A method of framing in which each floor is framed separately. The first floor is a platform built on top of the foundation wall; one-story studs are used to develop and support the framework for the second story and load-bearing interior walls. Also called *platform framing*. (Ch. 18)

wireframe In CAD, a 3D model that consists entirely of lines that define the size and shape of the object; similar to a physical model made from toothpicks. (Ch. 16)

wiring In sheet-metal products, the process of reinforcing open ends of articles by enclosing a wire in the edge. (Ch. 14)

working drawings The drawing or set of drawings from which a part is manufactured. It gives all the information needed to manufacture or build a single part or a complete machine or structure. (Ch. 7, 13)

worm gear A gear that is similar to a screw; used to transmit motion between two perpendicular, nonintersecting shafts. (Ch. 17)

Credits

Cover Design:
Squarecrow Creative Group

Interior Design:
Corona Design and Squarecrow
Creative Group

Cover Photos: PhotoDisc

All interior illustrations by James
Shough unless otherwise noted.

Courtesy of American Institute of
Steel Construction 671
Courtesy of ANSI 723, 725, 726, 727
Courtesy of Architectural Cast Stone
Inc. 682
Courtesy of Arc Second, Inc., Dulles,
Virginia 75
Arnold and Brown 28, 34, 92, 93, 94,
97, 127, 380, 513, 571, 572,
635, 647, 699, 700, 714, 718,
719, 730, 755, 784, 795,
Art MacDillos/ Gary Skilestad 636,
640, 641, 642
J.E. Barclay and Associates 625
Keith Berry 771
Corbis Images
Jonathan Blair 672
Carl Purcell 674, 682
Corbis Stock Market
Charles Gupton 50
Brownie Harris 3, 20, 790,
John Henley 20, 790
Don Mason 42
Ariel Skelly 38
Tom Stewart 32
Circle Design
Peter Getz 151
Carol Spengel 9, 432
Courtesy of CAD Easy Corporation
403
Courtesy of Camco 595
Courtesy of Computervision 766
Courtesy of Exa Corporation.
External aerodynamic simula-
tion performed by Exa
PowerFLOW®. 45
David R. Frazier Photolibrary Inc. 4,
21, 38, 104, 624, 643, 669, 679,
680, 790, 793,
Courtesy of Friedr Dick Corporation
509
Ann Garvin 623, 91, 96, 125, 127, 508
Gensler and Associates Architects 652
Courtesy of the Hartford Special
Machinery Company 470

Kelly Hinkle 118, 716
Courtesy Home Planners Inc and
Steve Karp 665-666
Hulton/Archive Photos
R. Gates 86
Courtesy of Japan Digital Laboratory
778
Courtesy of Philip Jeavons,
Department of Mechanical
Engineering, University College
London 430
Courtesy of Just My Office Furniture
105
Steve Karp 627, 628, 647
Courtesy of keyalt.com
Photographers: Terry Hammon
& Bob Heisler 106
Courtesy of Matfer Inc. 509
Courtesy of Mayline Co. 88, 770
Ted Mishima 64, 66, 88, 113, 114,
118, 121, 127, 425, 426, 755
Courtesy of Missler Software and
Clear Cut Solutions Inc. 44, 45,
418
Courtesy of NASA 693
National Bureau of Standards 626
North Wind Pictures Archive 5, 18,
112
Courtesy of Oce-Bruning 97, 774
Courtesy OSU Gear and Mechanism
Lab 599, 604
PDE Associates Inc. 651
Courtesy Penn Fishing Tackle Inc.
418, 429
Brent Phelps 775
Photo Edit
Bill Bachmann 41
Michelle Bridwell 37
Myrleen Cate 38
Richard Hutchings 32
Spencer Grant 26
Michael Newman 33, 36, 43
David Young-Wolfe 22, 35, 42
Photo Researchers, Inc.
Ton Kinsbergen/ Science Photo
Library 27
Maximillian Stock, Ltd./ Science
Photo Library 49, 102
David Parker/ Science Photo
Library 86, 103
Alfred Pasieka /Science Photo
Library 6, 215
Photo Dassault-Breguet/ Science
Photo Library 133
Ed Young/ Science Photo Library
44

Courtesy of Rocky Mountain Aerial
Mapping, Inc. 697
Courtesy Scheiler & Rassi Quality
Builders Inc. 637, 638, 639,
641*
Courtesy of Sitram Cookware
509
Courtesy of Skills USA-VICA 39
Courtesy of Staedtler Inc. 91,96, 97,
118, 119
Courtesy of Mr Richard Stewart, R &
L CAD Services Pty Ltd 530
Doug Storm 661
Courtesy of Roopinder Tara 685
Ken Trevarthan 8, 347
Courtesy Vemco Drafting Products
Corp. 89
Ian Warpole 432, 573, 594, 596, 599,
600, 603, 670, 681, 746, 107
AW Wendell and Sons, Architect,
Contractor 662, 663, 664
Courtesy of Z Corporation 23

Special Thanks to:
Mark Pelnar and Clear Cut Solutions,
Inc., Scheiler & Rassi Quality Builders
Inc.Deer Creek, IL, Sheet Metal
Products Company, Peoria, IL,
Columbia Pipe and Supply, East
Peoria, IL, Eric White and Rocky
Mountain Aerial Mapping, Inc.,
Palmer Lake, CO, Alan L. McKee, Dr.
Don Houser and Dr. Gary L. Kinzel
and the Ohio State University Gear
Lab, Ed Mesunas and Penn Fishing
Tackle Inc., Philip Jeavons,
Department of Mechanical
Engineering, University College
London, Robert Bales and
Architectural Cast Stone, Inc., Shlomo
Finkelstein and Just My Office
Furniture, Z Corporation, CAD Easy
Corporation, SkillsUSA-VICA.

* These images are printed with permission and remain the exclusive
property of Schieler & Rassi Quality
Builders, Inc. These images should not
be reproduced without the express
written permission of Schieler & Rassi
Quality Builders, Inc ., Deer Lake, IL.

Models and fictional names have been
used to portray characters in stories
and examples in this text.